ISBN 978-0-331-10979-5
PIBN 11015121

TWENTY-EIGHTH ANNUAL REPORT

OF

THE LOCAL GOVERNMENT BOARD,

1898–99.

SUPPLEMENT

CONTAINING THE

REPORT OF THE MEDICAL OFFICER

For 1898–99.

Presented to both Houses of Parliament by Command of Her Majesty.

LONDON:
PRINTED FOR HER MAJESTY'S STATIONERY OFFICE,
BY DARLING & SON, LTD., 1–3, GREAT ST. THOMAS APOSTLE, E.C.

And to be purchased, either directly or through any Bookseller, from
EYRE & SPOTTISWOODE, EAST HARDING STREET, FLEET STREET, E.C., and
32, ABINGDON STREET, WESTMINSTER, S.W.; or
JOHN MENZIES & CO., 12, HANOVER STREET, EDINBURGH, and
90, WEST NILE STREET, GLASGOW; or
HODGES, FIGGIS, & CO., Limited, 104, GRAFTON STREET, DUBLIN.

1899.

[C.—9445.] *Price 7s. 4d.*

PUBLIC HEALTH.

ANNUAL REPORT

OF THE

MEDICAL OFFICER

OF

THE LOCAL GOVERNMENT BOARD

FOR THE YEAR

1898-99.

2362—1250—8/99 Wt 29216 D & S a 2

CONTENTS.

REPORT.

MEDICAL OFFICER'S REPORT.

TO THE RIGHT HONOURABLE HENRY CHAPLIN, M.P., PRESIDENT OF THE LOCAL GOVERNMENT BOARD.

SIR,

I HAVE the honour herewith to submit to you, for presentation to Parliament, my report on the proceedings of the Medical Department for the year 1898–99.

In view of the ever-increasing amount of work devolving on the Medical Department, two temporary medical inspectors were appointed during the course of the year; but before its termination it was found necessary to make the appointments permanent. The two officers in question are Dr. L. W. Darra Mair, D.P.H., and Dr. Ralph W. Johnstone, D.P.H. Certain other additional appointments were made in connexion with the supply of vaccine lymph, to which I propose to refer later on.

I.—Administrative Relations of the Medical Department.

1. VACCINATION AND PUBLIC VACCINATION.

VACCINATION OFFICERS' RETURNS.

The digest of the Vaccination Officers' Returns contained in Appendix A, No. 1, relates to children whose births were registered during the calendar year 1896, the most recent period for which final information was procurable at the close of the official year 1898–99. It is compiled from the twenty-fifth annual return under the Vaccination Act, 1871, and it is concerned with 914,205 births. Of the children in question 602,922, or 66 per cent., were returned as successfully vaccinated; and after deducting those registered either as having died before vaccination, or as being "insusceptible" of vaccination, or as having had small pox before vaccination, there still remained 209,007 to be accounted for. The proportion of children born in England and Wales not finally accounted for as regards vaccination was therefore 22·9 per cent.; or 26·4 per cent. in the metropolis and 22·3 per cent. in the rest of England and Wales. These are the highest percentages of default under the Vaccination Acts that are on record.

INSUSCEPTIBILITY TO VACCINATION.

Of 605,807 children on whom vaccination had been performed, as many as 2,885 were certified to be insusceptible of vaccination, that is to say, they had been vaccinated unsuccessfully at least three times. No such case was met with during 1898–99 amongst the 3,548 primary vaccinations performed by the Board's operators, and this number being added to the 107,180 similar cases referred to in my last report, *gives a total of* 110,728 consecutive primary [illegible] without the [illegible]

[illegible] including [illegible] districts [illegible]

[illegible] from the [illegible]

[illegible] the Board's [illegible] the operation was [illegible] cases [illegible] The number of [illegible] cases [illegible] Appendix A. [illegible]

[illegible] the Medical Department [illegible]

Subsequently to the date of the latter report the Vaccination Act, 1898, was passed [illegible] that Act it is provided that [illegible] the Public Vaccinators [illegible] glycerinated calf lymph or such other lymph as may be issued by the Local Government Board." Instructions to Public Vaccinators issued under the General Order relating to the Vaccination Acts, 1867 to 1898, of the 18th of October, 1898, also require that :—"If the parent or other person having the custody of a child requires that it shall be vaccinated with lymph issued by the Local Government Board, the vaccination must be performed with such lymph." Having regard to these requirements and to the circumstances that the Vaccination Act 1898, came into operation on January 1st, 1899, it became necessary to provide from that date for a supply of glycerinated calf lymph prepared by the Board's officers for the public vaccinators of England and Wales.

A brief account of some of the arrangements made in order to carry into effect this new system of lymph supply, including an announcement of the appointment of Dr. Frank R. Blaxall as Bacteriologist to the Board for this purpose, has already been given in my last report, but I have thought it desirable on this occasion to deal with the subject more in detail, and for this purpose I append a report by Dr. Blaxall on the work which has been carried on at the Glycerinated Calf Lymph Establishment during so much of the year 1898–99 as it was in operation. (Appendix A., No. 5.)

The new scheme has involved a "Quarantine Station" for the detention of calves prior to their vaccination; stabling in which the calves are stalled during the process of vaccination; operating rooms, &c. This provision was already in existence in connexion with the Board's Animal Vaccine Establishment; and although the arrangements were not such as would have been provided in a set of new buildings, yet it was determined after mature consideration to modify the then existing stabling, &c., somewhat to increase the staff at the station, and provisionally, at least, to carry on all the work involved in the supervision and vaccination of the calves and in the collection of their lymph at the existing quarantine establishment, and at the vaccination station in Lamb's Conduit Street. Laboratories and other necessary accommodation were taken on lease at the premises of the British (now Jenner) Institute of Preventive Medicine on the Chelsea Embankment, where Dr. Blaxall was placed in charge. Mr. Fremlin, L.R.C.P., was appointed Assistant Bacteriologist, Mr. W. F. Mulcahy, who had long been connected with the Medical Department, took over the superintendence of the clerical and allied work, and a small staff of assistants was appointed. The organisation of the National Vaccine Establishment was, after re-inforcement, retained for the distribution of the lymph supplied from the laboratories.

Having regard to the fact that the number of public vaccinators in England and Wales at the end of 1898 did not much exceed 3,000; that about half of the children vaccinated in England and Wales had hitherto been vaccinated by medical practitioners other than Public Vaccinators; and that, apart from the requirements above quoted, the glycerinated calf lymph which public vaccinators must use need not necessarily be that prepared and issued by the Board, it was considered that if arrangements were made for the issue of such lymph at the rate of 1,000 tubes a day —each tube to serve for the vaccination of one child—all the requirements of public vaccinators during non-epidemic periods, would, in the absence of any prevalence of small-pox, be sufficiently met. But it soon became evident that the demand which the Board were called upon to meet was considerably in excess of this estimate. As will be seen from Dr. Blaxall's report, the number of tubes of glycerinated calf lymph issued from the laboratories in the first three months of the present year reached a daily average of 1,616; the total for the three months being 126,038 tubes. This demand necessitated a further increase in laboratory accommodation and staff, and a further addition to the staff of the National Vaccine Establishment, from which the *lymph is issued to the* individual applicants.

DICAL
ICER'S
PORT.

The operations connected with the preparation of glycerinated calf lymph are set out in some detail in Dr. Blaxall's report, and it will suffice for me to summarise the more prominent of the measures adopted. These are :—1°, selection of healthy calves; 2°, supervision of calves as to their healthiness at a quarantine station; 3°, vaccination of calves and subsequent collection of calf lymph at the Animal Vaccine Establishment; 4°, preparation at the laboratories of an emulsion of glycerinated calf lymph; 5°, testing of the emulsion by culture and otherwise so as to ensure that it contains no harmful organisms; 6°, transference of the emulsion to capillary tubes for issue by the National Vaccine Establishment; and 7°, compilation of the records received from public vaccinators as to the use of each series of lymph issued. The laboratory work is throughout carried out under strict aseptic precautions, and the preparation of the lymph is accompanied by experimental work with a view of enhancing the "keeping" properties of the glycerinated lymph, and of preventing, as far as practicable, its undergoing any deterioration, as, for example, in consequence of its necessary exposure, after being tubed and issued, to a degree of warmth considerably exceeding the low temperature at which it is stored in the laboratory.

It should be added that after the collection of lymph each calf is slaughtered under the personal superintendence of a qualified veterinary surgeon, Mr. Wm. F. Shaw, F.R.C.V.S., who is appointed by the Board for this purpose, and that no lymph is sent out from the laboratory until a certificate has been received, as regards the calf from which it was derived, to the effect that the animal, including its internal organs, glands, &c., was healthy. In the period ending March 31st the lymph of one calf was rejected because some tubercular lesions were discovered after slaughter; and it is interesting to note that the most careful investigation of this lymph failed to reveal in it any evidence whatever of the tubercle bacillus.

INATION
T, 1898.

The principal changes introduced into the English system of public vaccination under the Vaccination Act, 1898, are as follows:—Firstly, instead of parents being compelled to take their children to a vaccination station where they have heretofore been habitually vaccinated with humanised lymph from arm to arm, the public vaccinator is now required to visit the homes of the children and to offer to vaccinate them with glycerinated calf lymph; stations must, however, be provided, when the Local Government Board deem it expedient by Order so to direct. Secondly, the statutory period after the expiration of which a parent having an unvaccinated child falls into default has been extended from three to six months. Thirdly, no parent is liable to a penalty under the compulsory clauses of the Vaccination Acts who affords proof that he has, within four months from the birth of a child, satisfied a stipendiary magistrate or two justices in petty sessions that he conscientiously believes that vaccination would be prejudicial to the health of the child. And, fourthly, in no case can proceedings be taken more than twice against a defaulting parent, namely, once under Section 29 of the Act of 1867, and once under Section 31 provided that the child has reached the age of four years.

Dealing, as this report does, with a period which only includes the first three months of the operation of this Act, it would be quite premature to express any definite opinion as to its working. But it may be stated that such evidence as has been obtained goes to indicate that the number of children vaccinated under the new Act during that period is distinctly in excess of the number vaccinated during the corresponding period in recent years under the former Acts. There is also evidence that this increase in the number of children vaccinated was being maintained beyond the close of the official year 1898–99.

VACCINATION ORDER, 1898.

With a view of giving effect to the new state of the law the Board, on October 18th, 1898, issued a General Order embodying Amended Regulations under the Vaccination Acts, 1867–98. This lays down amongst other things the duties of public vaccinators and of vaccination officers, and it contains in a schedule a set of instructions to vaccinators under contract. These instructions differ from those which were formerly in force; modifications having become necessary in view of the new law and the abandonment of the use of humanised lymph. They also contain a definite requirement which has been introduced with a view of ensuring that, as in all surgical operations, whether major or minor, aseptic precautions shall be observed at all stages of the operation of vaccination. The Order will be found at Appendix A., No. 6.

2. Other Administrative Business of the Medical Department.

The Medical Department was during 1898, as in previous years, concerned with a variety of matters connected with public health administration, respecting which no public reports are issued. Not only has this been so as regards medical inspectors, but the officers of the department stationed at Whitehall have spent much time in conferences with representatives of local sanitary authorities on such subjects as proposed hospital buildings, the adaptation of byelaws to the needs of particular sanitary areas, &c. Indeed, such conferences, which are becoming increasingly numerous, result in the saving of much correspondence, and generally in a satisfactory solution of difficulties which have arisen or would be likely to arise in the absence of some such means of bringing local proposals into consonance with the principles which the experience of the Board has led them to adopt.

As regards conferences and inquiries of the above sort undertaken by the Medical Inspectorate elsewhere than at the office, I may state that these had reference last year to cases where mention was made of vaccination in certificates of death received by local registrars; to conferences with local medical officers of health in regard to the performance of their duties; to local investigation on the subject of byelaw proposals; to the provision of isolation hospitals for infectious diseases and means of sewerage and sewage disposal; and to matters such as port sanitary administration, public water supplies, and public means of disinfection.

With reference to hospital accommodation for the isolation of cases of infectious disease, inquiries, having for the most part concern with the suitability of proposed sites and the erection of hospital buildings, were instituted in the following localities:—Aston Manor, Auckland, Shildon and Willington (Joint), Barnoldswick, Barry, Bromley and Beckenham (Joint), Canterbury, Cardiff, Castleford, Chippenham (Joint County Council proposal), Cromer, Croydon, Dewsbury (Joint), Hanwell, Hebburn, High Peak (Joint County Council proposal), Keswick, Kingston (Joint districts of Ham, The Maldens and Coombe, Surbiton, The Dittons, and Esher), Leicester, Liverpool, Lowestoft, Maldon Rural, Manchester, Mutford and Lothingland Rural, Newport, Mon., North Bierley, Penmaenmawr, Reigate, Richmond, Heston and Isleworth (Joint), Romford Urban and Rural (Joint), Royston Urban and Ashwell and Melbourn Rural (Joint), Spennymoor, Stratford-on-Avon, Ulverston, Dalton and Grange (Joint), Walthamstow, Wath-upon-Dearne and North Rotherham (Joint), Weston-super-Mare, Widnes, Wirral (Joint), Withington, and York.

As regards questions of sewage disposal the following districts were visited by Medical Inspectors, some of them in association with an Engineering Inspector:—The Borough of St. Helens (Lancashire), where Dr. Buchanan held inquiry as to the manner in which the Corporation had fulfilled the obligations imposed upon them by section 52 of the St. Helens Corporation Act, 1893, by which section it became necessary for them to carry out a scheme for the treatment in subsidence tanks of the sewage of the borough by "the best known available chemical process as shall be approved by the Local Government Board;" the Hemsworth Rural District where inquiry was held by Dr. Bulstrode, as to a proposal of the Rural District Council to raise loan in respect of the disposal of sewage on the "septic tank" system in the parishes of South Elmsall and South Kirby; the Borough of Guildford, where Dr. Sweeting held inquiry as to a proposal for a loan in respect of sewage disposal by the "bacterial filter" process; the Borough of Portsmouth, where Dr. Bulstrode made inquiry in relation to a complaint by the Portsmouth Waterworks Company under section 299 of the Public Health Act, 1875, that the water supplied by the Company was endangered by default of the Havant Urban and Rural District Councils in the matter of sewage disposal; and the town of Keswick and the Cockermouth Rural District, where Dr. Fletcher investigated a complaint by the Corporation of Maryport that pollution of the River Derwent was caused by mismanagement of the sewage farms of the Councils of those two districts.

A proposal by the Chorlton and Manchester Joint Asylum Committee to establish an asylum for imbeciles on a site which formed part of the gathering ground of the water supply of the Liverpool Corporation was made the subject of inquiry by Dr. Buchanan in association with an engineering inspector, and later, in view of important considerations involved, renewed inquiry was held by Dr. Thomson, again in association with an engineering inspector, with the result that the Board confirmed the conclusions of the earlier inquirers, namely, that with due precautions the proposed asylum could be erected on the site without danger to the Liverpool water supply; but they deferred for a time their final decision on the application of the Joint Committee with a view of giving the Town Council of Liverpool a further opportunity of entering into an agreement for the purchase of the land in question.

In the circumstances of the excessive drought of last year and of the restricted supply of water to districts served by the East London Water Company, Dr. Bruce Low made inquiry into certain allegations as to the effect on health of the diminished and intermittent supply furnished by that Company. At Croydon, Dr. Wheaton was associated with an engineering inspector in an inquiry as to a proposal to supplement the Corporation water supply by the sinking of a well in the Caterham Valley. The same inspector made detailed investigation and report, which was subsequently published, as to pollution of the River Glaslyn above the intake of the Carnarvon Corporation waterworks and within the Rural Districts of Glaslyn and Gwyrfai. Local complaints received from Aldeburgh, in Suffolk, as to the quality of the water furnished by the private company that serves that town led to inspection of and report on the gathering grounds on behalf of the Board by Dr. Sweeting. On request by the River Blyth Port Sanitary Authority, Dr. Fletcher held a local inquiry on the subject of the constitution of the authority and the apportionment of expenses.

In regard of local prevalences of infectious disease and of questions concerning local sanitary administration, many inquiries were held during the past year, as the "Abstract of Medical Inspections" in Appendix A., No. 7, will show. Thus, fatal outbreaks of enteric fever formed the subject of investigation by Dr. Bruce Low at Camborne and in the adjoining Rural Districts of Redruth and Helston; by Dr. Thomson at Chichester and at Swinton and Pendlebury; and by Dr. Mair at Gainsborough.

Outbreaks of diphtheria were investigated by Dr. Wheaton in the Claypole Rural District, by Dr. Johnstone in the Eton Rural District, by Dr. Fletcher at Longton and Fenton, and by Dr. Buchanan at Tunbridge Wells.

Dr. Thomson made inquiry into an extensive and fatal outbreak of measles at Burton-on-Trent, and reported fully on the

circumstances under which the disease had prevailed, and on the measures of control that had been adopted.

The fatal occurrence of smallpox in and around Middlesbrough during the early part of last year led you to place at the disposal of the local authorities the services of a medical inspector with a view of advising them as to the best measures to adopt under the then existing emergency, and Dr. Reece was instructed to proceed to Middlesbrough and to several neighbouring towns for this purpose.

It was deemed advisable during the year to make inquiry in several districts with regard to their sanitary condition and administration, these districts comprising Christchurch (South Hants), where inquiry was made by Dr. Mivart; Glyncorwg, Ormskirk, and Bettws-y-coed Rural District, enquiries as to which were entrusted severally to Mr. Royle, Dr. Copeman, and Dr. Wheaton; and in the following districts, with a view of determining how far the powers of section 42 of the Public Health Act, 1875, should be enforced, namely :—Blackrod and Tow Law, where formal inquiry was made respectively by Dr. Sweeting and Dr. Mivart; and Altofts, Ardsley, Bolsover, and Gorton, where preliminary local inspections were instituted by Drs. Reece, Fletcher, and Thomson on the same subject. Dr. Buchanan was also instructed to re-inspect the borough of West Bromwich with a view of ascertaining what action of a remedial character had been taken by the Town Council as the result of his report to the Board on the sanitary administration of that place in 1895.

Inquiries having to do with the sanitary condition and administration of districts where a plurality of health officers have been appointed were made in the cases of the Rural Districts of Dore, by Dr. Fletcher, Wortley, by Dr. Reece, and Stow-on-the-Wold, by Dr. Sweeting; whilst in the case of the Lunesdale Rural District, inquiry was instituted by Dr. Fletcher as to the circumstances in which the Inspector of Nuisances holds multiple appointments.

The County Council for Northumberland having made representation to the Board as to default of the Alnwick Urban District Council in the matter of the housing of the working classes in that town, Dr. Buchanan made inquiry and report on the subject. Dr. Bulstrode made inquiry at the Bromley and Beckenham Joint Hospital in reference to certain "return cases" of scarlet fever. He was also instructed to proceed to the River Colne for the purpose of representing the Board at some float experiments made to determine the flow of currents in relation with certain oyster layings in the river.

Dr. Bulstrode also enquired into the circumstances attending the arrival at Plymouth and in the Thames respectively, of the plague-infected steamships "Caledonia" and "Golconda," whilst Dr. Thomson made enquiry in the case of the plague-infected steamship "Carthage" on her arrival in the port of London.

Dr. Copeman was engaged during a considerable part of the year in connexion with matters relating to our establishment for the manufacture and distribution of glycerinated calf-lymph. He also made arrangements at several of our ports for the taking and transmission to us of samples of imported milk and cream, and as a result some few samples have been submitted to expert chemical and bacteriological examination.

Dr. Parsons, your second Assistant Medical Officer, was appointed a member of the Departmental Committee which has made inquiry into the manufacture and use of water gas and other gases containing a large proportion of carbonic oxide, and the report of that Committee has been presented to both Houses of Parliament.

I may state that I was myself during the past year engaged on two Royal Commissions having to do respectively with the subjects of Tuberculosis and Sewage Disposal. As regards the former, the report of the Commission was presented to Parliament in the early part of the year, and has since received consideration at your hands, with result that the Board have issued (1) an Order making disease of the udder of a cow certified by a veterinary surgeon to be tubercular, to be a "disease" within the terms of Article 15 of the Dairies, Cowsheds, and Milkshops Order of 1885; and (2) Model Regulations under those Orders for the guidance of local authorities. In transmitting a copy of the Order and Regulations to urban and rural district councils, the Board in a circular letter addressed to the Councils, drew their special attention to those parts of the Report of the Royal Commission on Tuberculosis which relate to milk and tuberculous meat, and to the qualifications which it was considered desirable that Meat Inspectors should possess. (I reproduce in App. A., No. 8, the Circular Letter of the Board to Urban Councils, the Order, and the Model Regulations.)

I proceed now to comment at greater length on certain of the inquiries to which the above general reference relates.

ON THE HOUSING OF THE WORKING CLASSES IN ALNWICK.

I have from time to time drawn attention in my annual report to inspections undertaken by Inspectors of the Medical Department, owing to the receipt by the Board of representations made by County Councils under Section 19 (2) of the Local Government Act, 1888, to the effect that the Public Health Act had not been properly put in force in the districts in question.

One inspection of this class which has been carried out during the year now reported on is of interest, inasmuch as it relates primarily to the question of the housing of the working classes. The report referred to is one by Dr. Buchanan; it deals with the general sanitary condition and administration of Alnwick, and will be found in Appendix A., No. 9.

With the exception of one principal roadway, the streets of this ancient town are narrow, a large portion of the dwellings of the lower classes are huddled together on areas lying at the backs of houses, and which once formed long strips of gardens. The resulting courts are severally approached by an "entry," driven through the ground floor of houses fronting the street, and generally ending in a *cul-de-sac*. Houses in these courts, as a rule, only have doors and windows facing a passage some four to six feet wide which is common to the row, the other side of the passage being bounded by a high wall which forms the back of similar property erected behind an adjoining house. The windows, too, are commonly small, and are only arranged to open to a trifling extent, the resulting absence of due light and ventilation leading in turn to produce another evil, namely, a state of extreme dirtiness within the dwellings. Other houses have little or no back space or other open space belonging to them, and there is a general prevalence of close aggregation of dwellings on area.

Notwithstanding this, however, animals are allowed to be kept where there is insufficient room for human beings; thus, the report states :—" Where horses and cows are kept in such courts as these, nuisance inevitably arises from accumulations of dung and stable refuse." A large number of the houses stand on a wet soil; though constructed of porous stone they have not been provided with a damp-proof course, and at times their wetness is further ensured by the fact of their being built against the hill side.

It has now long been known that certain conditions of housing are intimately associated with a large mortality from phthisis; these being dampness of site, want of light and ventilation, and overcrowding of dwellings on area and of persons in dwellings. These conditions are evidently prominent features of Alnwick. The borough may be generally described as lying on impermeable soil in a hollow situated at an elevation above the river Aln; the higher land is rich in springs which drain towards the town, and wetness of site is hence a natural characteristic. This, in itself, should have ensured the adoption of precautions against such dampness of dwellings as can be avoided by reasonable building regulations; but until quite recently no regulations as to this were imposed. Aggregation of houses on area also prevails in its worst form, for whilst "back-to-back" houses, in which phthisis is so wont to be excessive, do not actually exist, all the worst features of this class of house seem to have been contrived. Inside the houses, too, the evils of excessive aggregation are accentuated. According to data obtained a few years ago, and as to which no material alteration could be heard of, some 800 to 900 persons in Alnwick lived in 300 single-room tenements; and in numerous instances from four to eight persons lived and slept in a single room. And further, 1,395 person were accommodated in 333 two-roomed tenements, the second room being occasionally little better than a wash-house. So also, light and free movement of air, which acting together are so destructive of the infection of tubercle and so inimical to dirt, are markedly absent in a large proportion of the houses.

As a natural result it transpired that this little borough of some 6,700 inhabitants was found to have a mortality from phthisis reaching 28·6 per 10,000 living, or more than 12 per 10,000 in excess of that for England and Wales as a whole, whilst the rate of death from all causes also greatly exceeded that for the country generally.

Some of the conditions leading to this waste of human life in Alnwick have been unavoidable. Amongst these must be included the enclosure of the town in former centuries by walls for the purposes of defence, and the difficulty of obtaining building land outside the walled area. But others are obviously the result of neglect which has been maintained in the face of numerous warnings. In 1866, the late Sir George Buchanan, formerly Medical Officer of the Board, pointed out that death from phthisis was increasing in Alnwick, notwithstanding the provision of works of sewerage and of water supply, and he attributed this increase essentially to dampness of soil and of dwellings. In 1885, Mr. Spear, one of the Board's Medical Inspectors, found that the conditions for the housing of the poorer classes had not only undergone no improvement in recent years, but that overcrowding was worse than ever. "Wherever individual interests have to be opposed," Mr. Spear observed, "or seemingly opposed, sanitary administration has been paralyzed." Since then some of the inhabitants appear to have become increasingly impressed with the prevailing state of affairs, but, says Dr. Buchanan, "little or nothing . . . . has come of advice, representations and complaints."

Perhaps the most important administrative action that has been taken in connexion with the evils here referred to was the adoption in 1889 by the then Local Board of a code of byelaws as to new buildings, and if these are strictly and impartially enforced they must in time effect good results. Apart from local administration it should also be recorded that for some time past excellent dwellings have been in course of erection at the instance of the Duke of Northumberland and on his estate. These are let on conditions under which the tenant can eventually acquire the freehold, and when Dr. Buchanan was in Alnwick every such dwelling had been taken in advance. But these measures, beneficial as they are in themselves, cannot be expected to undo the maintained mischief of the past. For this, the only remedy is comprehensive action by which unhealthy areas shall be dealt with under the Housing of the Working Classes Act, 1890.

Dr. Buchanan's report which was forwarded to the Alnwick Urban Council in September, 1898, indicated three "unhealthy areas" which called for action of this sort. In October of that year the District Council forwarded to the Board a report of a committee of their body in which it was proposed to deal with a portion of one of these three areas. The locality in question is admittedly one of the worst in Alnwick, and it would appear to present peculiar facilities for the action proposed, for the committee in reporting on it says:—"There are no buildings upon it of special value, and taking it would not in any way interfere with the trade of the town." But, up to the present date, no further information on this proposal has been forthcoming.

MEDICAL OFFICER'S REPORT.

SOURCES OF WATER SUPPLY.

In several of my recent reports I have drawn attention to the dangers to which the public are exposed by reason of the grave sources of pollution to which water supplies, over which they have no control, are subjected. This is especially the case where statutory powers are acquired, outside the Board's jurisdiction, by companies and other bodies to collect and impound waters derived from gathering grounds the proper control of which, in so far as the wholesomeness of the water is concerned, has not been acquired by purchase or otherwise. In such instances both the purveyors and the consumers of the water are often alike helpless, even though the most superficial examination of the catchment area may give obvious indication of multiple sources of pollution of the gravest sort. As further examples of the mischief which is year by year revealed as the outcome of the method of procedure which has often governed the acquirement of these powers, I would refer to three reports prepared in the Medical Department.

ENTERIC FEVER AT CAMBORNE AND NEIGHBOURHOOD.

In the early part of 1898 Dr. Bruce Low submitted a report on an outbreak of enteric fever which affected three adjoining sanitary districts in Cornwall, namely, the Camborne Urban District; the parish of Illogan, in the Redruth Rural District; and the parish of Crowan, in the Helston Rural District (App. A., No. 10). The outbreak, as such, commenced in December, 1897, and continued into January, 1898, but neither sanitary district had been free from occasional attacks of the disease for some time before the epidemic prevalence set in.

From the first there was suspicion that the epidemic had been brought about by the water supplied by the Camborne Water Company; and by a process of elimination it soon became possible, as the result of detailed enquiry, to show that no other circumstance, such as is at one and another time identified with the causation of enteric fever, could have been concerned with the production of the disease. Of 3,127 houses in Camborne 2,365 were provided with the Company's water, and 106 of these, yielding 116 cases, were invaded. The remaining 762 houses were supplied from other sources, and in six of these six single attacks occurred. But of these six persons three had habitually drunk the Camborne Company's water; two others worked in Camborne and had daily opportunities of using the water, and the sixth, a visitor, whilst denying that she had drunk the water, admitted that she had on numerous occasions partaken of meals in the town. We thus have 121 attacks in 2,365 houses, all of which attacks are identified either with the habitual use of, or with daily opportunities for using the Company's water; whereas only one case remains of which it can be said that there is no sufficient reason to associate the attack from fever with the use of the water supply in question. The parish of Illogan is the only one in the Redruth Rural District to which the Company's water is laid on. It contains 2,107 houses, 810 of which receive this water and 1,297 do not. The 810 houses using the public water service yielded 31 attacks of enteric fever; the 1,297 having a different supply yielded only two attacks, occurring in two separate houses. Of the two latter patients one was in the habit of drinking the Company's water freely, the other, whilst admitting visits to Illogan, denied the use of that water In the Helston Rural District the Camborne water is only supplied to two villages in Crowan parish, namely, Praze, with about 119 houses, and Churchtown, with 39 houses, which are supplied by means of a public tap. Irrespective of three cases in Crowan parish which occurred in two houses quite apart from the invaded area and altogether unconnected with the outbreak in question, six cases of fever occurred in Praze, all those attacked being accustomed to use the incriminated water. In addition to the attacks referred to, four persons who had visited the infected area, but who lived elsewhere, sickened with fever in their respective homes. In all, 165 cases and 12 deaths occurred; and of the 165 cases all but two were associated with habitual use of the Camborne Company's water.

The water supplied by that Company is partly from springs, partly surface water, and it is derived from two nearly adjacent watersheds, known as Cargenwyn and Boswyn respectively. On Cargenwyn watershed 72 persons reside in 19 houses, none of which are provided with means of drainage, and only six of which have any closet accommodation. Human and animal excreta were found to be spread on the meadows, and by this and other means a byewash leading to a storage reservoir, which is stated to have been in use up to November 12th, had ample means of becoming contaminated. On the Boswyn watershed a dam had been so arranged across a brook as to raise the water to such a level that it should flow by means of a connecting pipe to the Company's service tank, and thus augment the supply. Nine houses, occupied by 42 persons, are comprised within the catchment area of this brook. None of the nine houses have any proper drainage arrangements, and six of them have no closet accommodation; slops and liquid filth soak away as they can. Rain fell on November 12th and 13th, and again on November 24th and 25th, to such an extent as to ensure the conveyance of surface pollutions to the brook.

It thus happened that the water derived from both these watersheds had opportunity for becoming fouled with human and other filth at a period, antecedent to the commencement of the epidemic which corresponds with the incubation period of enteric fever. But this is not the whole story as regards the Boswyn watershed, for in one of the houses standing on it, which had no closet and which was situated within 700 yards of the service tank referred to, four cases of enteric fever occurred during October and November. A pond near by served as the receptacle for all the fluid dejecta of the sick and for the washings of linen, &c., whilst the ground around was found by Dr. Low to be profusely littered with solid human excreta. Under ordinary circumstances this pond has no outlet; but on heavy rainfall, such as that which occurred in mid-November, it overflows and runs down a steep hill side to the brook which augments the

Company's supply. Here then we have the story of a watershed for a public water service so neglected and mal-administered as always to have involved risk of water-pollution; and so circumstanced that it only needed the accidental concurrence of one or more cases of specific disease on the gathering ground with a heavy rainfall in order that the water delivered to the public should serve as a vehicle for the distribution of a dangerous specific infection to the water consumers.

The second of these two reports is concerned with the nature and circumstances of the water supply of Aldeburgh-on-Sea, investigation as to which was undertaken by Dr. Sweeting, in consequence of complaints made to the Board by residents in this seaside resort. The report will be found in Appendix A., No. 11. The water is supplied by a local company, formed pursuant to the Companies Acts, 1862-67, who obtained their powers under a Provisional Order in 1871. After severe condemnation of one part of their supply in 1896, alterations were made in the works, and the water now in question is impounded sub-soil or ground water, locally deemed to be "spring water," derived from the Coralline Crag, which is described as a shelly sand, with thin bands of crystalline limestone, overlaid by soft yellow rock. This formation, which rests upon the London Clay, comes to the surface where the waterworks and a collecting reservoir are situated. It is altogether of a porous nature; and amongst the means of fouling to which it is subjected within a distance that may be termed the neighbourhood of the water sources, are a considerable number of leaky cesspools, privy pits, drainage from pigstyes and foldyards, percolation from manure heaps, and soakage from garden ground and fields manured with human and other excreta. Overlying a portion of the Coralline Crag are certain narrow bands of porous Red Crag. One of these lies in the direction of the town of Aldeburgh, and on this formation are certain houses situated within 300 yards of the waterworks. These houses are provided with cesspits, sunk in some instances to a depth of 18 and 26 feet, almost uniformly pervious, and generally having their bottoms at a higher level than the bottom of the neighbouring collecting reservoir. Coming still nearer to the town is a narrow overlying band of Chillesford Clay, and then comes an expanse of glacial drift which is occupied by the town proper. The Coralline Crag extends underneath the whole of the shingle on which the main part of the town is immediately built, and it forms the bed of the foreshore. The majority of the town cesspools, 375 in number, are sunk into the shingle and are the reverse of impervious. The Coralline Crag beneath the shingle is porous. Its opportunities for contamination with cesspool filth are increased by tidal action; and its subsoil "water," which is dammed back at high tides, can hardly fail to be drawn towards, if not into, the collecting reservoir on the occasion of repeated pumping operations which are at times resorted to. "It may indeed be suspected," writes Dr. Sweeting, "that the Aldeburgh inhabitants are, in a manner and on occasion, forced into drinking their own diluted sewage." Be this as it may, the surroundings of this water service are open to the gravest suspicion; complaints have been made to the effect that when heated the water has given forth a nauseating odour; and analysis of it has been such that in commenting on it in July, 1898, one expert chemist found himself compelled to explain that whilst he was not prepared to assert that the results "afford serious ground for alarm," yet "searching investigation by adequate inspection was called for." That investigation has been made, and its outcome is recorded by Dr. Sweeting. It is true that, as the result of a previous examination of a particular sample taken from the collecting reservoir on a particular day by another expert, no results definitely condemnatory of the water appear to have been arrived at; but this, it is almost needless to say, is a common experience as regards waters which are unquestionably exposed to risk of pollution, and it fails to explain either the evidence of obvious risk afforded by the topographical considerations referred to, or the other considerations set forth in the report, and which justify grave suspicion that this water is at recurring intervals exposed to the chances of serious contamination.

THE WATER SUPPLY OF CARNARVON.

The water supply for the Borough of Carnarvon is another case in point. In 1865 the Town Council obtained power to take water from the Gwyrfai; but it being found later on that the water was polluted by the drainage of the village of Waenfawr, the intake of the water was removed, in 1879, to a point some distance higher up the stream. In 1893 the Board's attention was drawn to the fact that the Carnarvon water supply was subject to pollution by a report of the local medical officer of health; and in 1895 Dr. Bruce Low, one of the Board's officers, in submitting a general report on sanitary progress in Carnarvonshire, gave an account of the risk of pollution to which this water was subjected, by reason especially of the liquid refuse from the village of Rhyd-ddu, the refuse from privies, and the soakage or overflow from cesspools. This was later on supplemented by private complaints. In the meantime correspondence ensued between the Board and the Rural District Councils of Gwyrfai and Glaslyn, within whose jurisdictions the village of Rhyd-ddu is situated; and these bodies were urged to take such measures for dealing with the sewage of the village as to prevent the pollution of the River Gwyrfai. But, apart from the substitution of pail closets for a number of privies which had been so contrived as to discharge direct into the stream, nothing substantial was found to have been done as the result of Dr. Low's visit in 1895. Early in 1898 the Board, after having received a communication from the Town Council of Carnarvon stating that they were not aware how they could themselves do anything more in the matter, instructed Dr. Wheaton to report in detail on the circumstances.

From Dr. Wheaton's report, which will be found in Appendix A., No. 12, it appears that the Carnarvon water supply is taken from the River-Gwyrfai about half a mile below the point where it issues from Lake Quellyn, the lower of two lakes lying in the Gwyrfai Valley, at the head of which are the steep slopes of Snowdon. The report sets out the numerous sources of pollution to which river and lake are alike subjected; and the details become the more revolting because, from their abundance, they have to be repeated again and again in Dr. Wheaton's account of one and another point where the fouling of the water supply takes place. Briefly summarised, the many sources of pollution are as follows:—Human excreta washed, at times in abundance, from the surrounding land by reason of the lack or complete absence of closets, or deliberately thrown into the river from pail closets; privy and cesspool contents, washed down ditches or small affluents leading to the main stream or to the lake, and derived from separate houses, from a hotel by the lake side, from a woollen mill, and from a quarry where 250 men are employed, of whom 60 live there in barracks; and lastly, the liquid and other refuse from the altogether unsewered village of Rhyd-ddu. The population of this village is about 150 persons, living partly on the north and partly on the south side of the river which here divides the Rural Districts of Gwyrfai and Glaslyn; and referring to it Dr. Wheaton writes:—"There can be no doubt that the sanitary circumstances of this village are such as to form a serious danger to the consumers of the water from the Gwyrfai River, and there would be grave risks to such consumers should enteric fever or cholera at any time break out in this village."

Such are the circumstances affecting the sources of the water supply for some 10,000 people who normally inhabit Carnarvon, and for the visitors who frequent the borough in the holiday season. And yet on the one hand we have the Town Council of the Borough who, having acquired no powers over the water above the intake in the river, declare themselves helpless; and on the other hand two rural district councils who evidently do not see their way, on the ground of cost, to provide a remedy for these evils merely because a borough some six or seven miles away has obtained Parliamentary powers to take a water supply from sources arising within their jurisdiction.

The circumstances of this case have been cited by way of general illustration of a by no means rare conditions of things.

A further reason for giving prominence to these recent experiences is that it affords an opportunity for placing on record the fact that the services of the Medical Department are now in frequent requisition in regard to powers as to new schemes of water supply or as to extension of existing supplies which it is sought to obtain through Parliament by means of Bills for Local Acts. Such Bills being now habitually examined by the Board, it is their practice to include in the report that is made on each of them any information which is in the possession of the Medical Department and which has any bearing upon the question of the wholesomeness or otherwise of the water in question.

MEDICAL OFFICER'S REPORT.

THE SANITARY CIRCUMSTANCES OF STOW-ON-THE-WOLD.

Another report which I reproduce (Appendix A, No. 18) is one prepared by Dr. Sweeting on the sanitary circumstances and administration of the rural district of Stow-on-the-Wold. It affords an example of a class of work in which the Medical Department is constantly engaged, and it gives prominence to some of the least satisfactory features which still attach to English sanitary work and administration. Sanitary progress, which has made immense strides in this country during the last generation, is, speaking generally, much less advanced in some divisions of England and Wales than in others, and is markedly behind the times in many of our scattered agricultural districts. In such districts, lessons taught by former epidemics seem to be more easily set aside than is the case in districts of an urban and manufacturing character. The causes of this are doubtless many, and they often have to do with financial considerations; but the fact is unquestionable.

The village of Bourton-on-the-Water in the Stow-on-the-Wold rural district was reported on in 1874 by the late Dr. Ballard, F.R.S., on account of an epidemic of enteric fever. The disease had, in the first instance, been imported, but it was associated with and "fostered by defective drainage, excremental nuisances and contaminated water-supply;" and Dr. Sweeting, reporting after an interval of twenty-four years, found that in many essential respects the same condition of things remained. It is true that certain steps had been taken as regards the sewerage of the village of Bourton-on-the-Water, which, after a lapse of twenty years, had led in 1894 to a scheme being formulated and carried out; and also that proposals for a water supply, frustrated first on one ground and then on another, including a declaration as late as 1897 by the Parish Council to the effect that there was no need for a fresh supply, had been made from time to time; but taking the rural area as a whole its water supply and its means of drainage were found still to consist largely of the primitive two holes sunk in a pervious soil, into one of which, called a cesspool, filth is poured, and from the other of which, called a well, water is drawn.

Another point, one to which the state of affairs in Stow-on-the-Wold must be largely attributed, also deserves consideration; it is the system of appointing multiple medical officers of health to a single district. When medical officers of health were first appointed under statute in 1872, it was deemed desirable, in many parts of England and Wales, to elect poor law medical officers to be *ex officio* medical officers of health, on the ground that since they had regularly to visit their districts in one capacity they would best perform the duties of another office. This practice was largely responsible for the division of rural sanitary districts into multiple health areas; but after long experience the system has been found in almost all cases to be a thoroughly faulty one. Although, therefore, there are certain exceptions, due in the main to topographical circumstances, in which it is advisable to divide a rural district for the purposes of the appointment of medical

MEDICAL OFFICER'S REPORT.

officers of health, yet it has now been the policy of the Board for many years past to discourage the appointment by one authority of more than one adviser in matters relating to public health.

The rural district of Stow-on-the-Wold is a case in which the serious drawbacks to the system of multiple health officers are well shown. Thus, Dr. Sweeting points out that the two officers here in question very rarely attend the meetings of the District Council; that, being engaged in many other duties in their respective areas, they never meet officially and "no joint or concerted action is ever taken by them." One of them, it is true, is stated for a while to have braved unpopularity by his severe strictures and conscientious attitude on the water supply question, and to have given excellent advice in his reports to his district council; but his colleague, who is stated to have paid attention to little else than questions of water-supply, "chronicled the state of things" in this respect "in more favourable terms than the actual circumstances appeared to warrant"; and his journal seldom contained any record of visits paid by him or of action taken. Such conflicting attitudes could, apparently, not be long maintained, and, in the end, the more energetic of the two was constrained to make irrelevant reports to the effect that, as regards his sanitary area also, the health of the district was "good." Here, as elsewhere, where this system is still maintained, efficient sanitary administration is practically unknown. There is no system of public scavenging, no means for the isolation of first cases of infectious diseases, no proper means of disinfection, no bye-laws of any kind under the Public Health Act of 1875, and, as I have already stated, the area as a whole lacks, among other things, proper means of drainage and of water supply. Happily the Board have done much to discountenance this system of multiple medical officers of health, and they have succeeded in a large number of instances in putting an end to it.

ENTERIC FEVER AND SOIL POLLUTION.

In my report for 1896–97 I submitted certain statistical data and charts which went to show that when comparison was made between the death-rate from "fever" in England and Wales during the two decennial periods 1871–80 and 1881–90, it appeared that the rate of death registered under this heading during the second period was only about half that which had taken place during the first. The disease registered as "fever" is mainly enteric or typhoid fever, and this being typical of the class of diseases which can be so largely prevented by the adoption of sanitary measures, the saving of life indicated in these statistics afforded opportunity for considerable gratification. But whilst this was the case, study of the available statistical evidence went to show that in certain localities in England and Wales diminution in the rate of death from this preventable disease had in more recent years practically come to a stand-still, and that this state of affairs was due, not so much to the occasional occurrence of local epidemics brought about by some casual circumstance such as a sudden and accidental pollution of the water-supply, but, to either a maintained persistence of the disease or frequent periodic recurrences of it, which could in no sense be explained by a newly imported infection or by an occasional contamination of water, sewers, &c. And, I ventured to observe that the failure to make further substantial reduction in the amount of disease and death from this cause was likely to be owing to the maintenance, in the localities in question, of conditions which involved organic pollution of the soil on which the communities in question lived, and this especially where the means of excrement and refuse disposal known as the old-fashioned midden-privy system was maintained.

AT SWINTON AND PENDLEBURY.

A report by Dr. Theodore Thomson on the persistence of enteric fever in the Swinton and Pendlebury Urban District, which I reproduce in Appendix A., No. 14, affords one of the many contributions which the Medical Department could supply in illustration of the point referred to. Here we have a small and prosperous town of only some 22,000 inhabitants lying about four miles to the north-west of Manchester; and although its general death-rate from all causes is not so low as it ought to be, yet it is just about on a level with that for England and Wales as a whole, and, as might have been expected, it is distinctly lower than that for the Registrar General's selected 33 Great Towns. But when we compare the deaths which in the respective areas most nearly represent enteric fever, we find that the rate in Swinton and Pendlebury is two-and-a-half times greater than that which prevails in England and Wales, and that, in recent times and over a period of years, it is more than double that for the Great Towns. Dr. Thomson also gives information as to the annual rates of attack from enteric fever in recent years as regards fifty towns concerning which such information was available. These fifty towns comprise none having so small a population as Swinton and Pendlebury, but on the other hand they include a considerable number, the population of which ranges from 50,000 upwards until that of Liverpool with over 600,000 is reached. Taking these fifty towns as a group, the annual attack rate from enteric fever has been 1·03 per thousand living, whereas that for Swinton and Pendlebury reached 2.13.

In discussing the causes of this exceptional amount of disease and death in Swinton and Pendlebury, Dr. Thomson finds nothing which enables him to refer the persistence of enteric fever either to the water supply or to the sewerage of the district. But, on the other hand, he found, on the occasion of his inspection, that the place abounded in nuisances, mainly arising from the offensive midden-privies which form the chief method of excrement and refuse disposal in the district, and which lead to persistent fouling of the soil in the neighbourhood of dwellings. I must refer to the report itself for those details which show how, in Swinton and Pendlebury, well nigh every possible source of nuisance with which this midden-privy system can be associated has been allowed to grow up; but I would observe that in the face of the knowledge of which this country has been possessed during the past generation, such a system must be regarded as nothing short of a relic of ignorance and barbarism which ought not any longer to be imposed on or permitted within civilised communities.

OF AT N-

An epidemic of measles in the Borough of Burton-on-Trent was inquired into by Dr. Theodore Thomson, whose report upon the occurrence, reproduced in Appendix A., No. 15, well exemplifies the difficulties with which sanitary authorities are confronted in coping with this disease. During the quinquennium, 1887–91, at the close of which this borough had a population of 46,047, the average annual death-rate from measles was 0·29 per thousand living; but in the following quinquennium the rate had risen to 0·74, and in the year 1896 it reached 1·7 per thousand. In the month of January 1898, when the population may be taken to have been some 51,600, a fresh epidemic commenced, which increased with occasional exacerbations until it led to 758 attacks in the four weeks ending June 11th. Between January 2nd and August 13th, 2,015 cases and 29 deaths had occurred, the chief incidence of attack being on the 2nd–5th years of life, and the chief incidence of death being on the 1st–3rd years; the fatality being greatest during the 2nd year of life.

It will be recollected that in my Report for the year 1894–95,* I called attention to the results of a special inquiry which had been made by Dr. Thomson in different parts of the country into the general question of the control of measles, the outcome of which was to show that in the case of this disease no really efficient control could be expected from the adoption of one or more of the necessary means of prevention so long as one or more of the remainder were neglected; or, to recall the words I then used that "if any approach to complete success be aimed at, each one of the several measures indicated must be regarded as necessary and supplementary to the others." The chief interest of Dr. Thomson's report on the outbreak in Burton-on-Trent lies in the fact that the occurrence well exemplifies the failure experienced as the result of the non-observance of this maxim, and this although the measures of repression which were adopted were carried out with exemplary persistence.

Measles was made a notifiable disease in Burton as far back as 1893, and there is reason to believe that during 1898 there were but few cases which did not receive medical attention, and but a small number can have escaped notification. On the day of the receipt of the notification certificate the household was visited by an officer of the authority, who gave verbal instructions as to the best available isolation of the patient in its own home, and handbills were circulated setting out the symptoms by which the disease might be recognised in its early stages, and calling the attention of householders to their several duties in relation to infectious disease. Information as to the existence of measles in a household was furnished daily by the local authority to the Clerk to the School Board, and this led to the exclusion from

* Supplement by the Medical Officer to the 24th Annual Report of the Local Government Board (C.—7906), 1896.

school of members of families living in the particular house invaded; households and invaded neighbourhoods received additional visits from sanitary officials whose object was to see that the instructions laid down were being observed; and school closure was resorted to not only as regards the public elementary schools, but also in respect of certain private schools and Sunday schools. Measures of disinfection and cleansing were also resorted to. But, as the disease progressed, some of these measures, such as the repetition of visits by the sanitary officers, could no longer be carried out by the staff which was available, and this practice fell into abeyance. Above all, no action was taken at any stage to secure the removal of those first affected in households to the isolation hospital; and, although the local authority supplied full information as to households invaded to the local School Board, they did not take the same action as regards the private and Sunday schools, neither was there anything approaching to reciprocal action on the part of the officials connected with the Board Schools.

On reviewing the circumstances of this outbreak from the administrative point of view, it becomes clear that, in order successfully to control measles in a borough of this description, where children, suffering from the early stages of and incubating that disease, are constantly liable to come into contact, the measures that are most likely to avail are isolation in hospital of the children first attacked, and the strict exclusion from school of all children living in houses where the disease prevails, to be followed, if need be, by actual closure of the schools. To attain the desired end with the least interference with the system of elementary education, it is as necessary for the school officials daily to seek the required information from their scholars and to transmit it to the sanitary officers, as it is for the sanitary officers to transmit to the school officials information gained from the notification returns. And, above all, the time when these measures have most chance of success is at the very onset of the outbreak. Indeed, not only can no sanitary authority be expected to maintain in readiness means of isolation for more than a limited number of cases of measles, but the value of hospital isolation for preventing the extension of this disease is generally limited to quite the initial period of an outbreak.

I am aware that special difficulties, such as the extreme youth of some of the patients, are involved in the control of measles, especially in the early stages of an epidemic, before parents have become alarmed by the general prevalence and fatality of the disease; but when it is remembered that close upon 13,000 lives are annually sacrificed to this one disease in England and Wales, and that, co-incidently with the increasing aggregation of children in elementary schools, the tendency of this mortality is to increase rather than to decrease, the question of its control must be regarded as one calling for a mutual energetic and sustained action on the part of the authorities who are concerned with the maintenance of public health, and of those who are charged with the education *of young children.*

AL
ER'S
RT.

II.—Plague Precautions at Home.

TE
IONS
ISH
S.

Since bubonic plague has been prevalent in certain parts of our Indian Empire it has been necessary to maintain careful supervision in our home ports over arrivals from Indian ports. In my report for 1896–97 I gave an account of the measures which were adopted on the occurrence of three fatal cases of this disease which occurred amongst natives serving on board two vessels recently arrived in the ports of London; and I now submit in Appendix A, Nos. 16 and 17, two reports by Dr. Bulstrode, which will serve to illustrate the measures of precaution adopted in our port sanitary districts as regards all vessels that are deemed to be either "suspected" or "infected" in relation to plague. In the case of one vessel, the suspected patient, and other members of the crew who had been in intimate contact with him, had been landed at Moses' Wells by the authorities at Suez, and the vessel had only to be dealt with in this country as one concerning which we had information that a case suspected to have been plague had occurred on board some time before her arrival. In the second instance, a patient believed to have been at the time in the convalescing stage of plague in a mild form—pestis ambulans—was actually removed from a vessel at Plymouth. Briefly stated, the system adopted in this country in such cases is the medical examination of all persons on board; the removal to hospital of any person either suffering from plague or suspected to be so suffering; the disinfection of articles believed to have had opportunity of becoming infected, and of those portions of the vessel occupied by the sick; the registering on board of the names and addresses of all the remaining persons, including the crew, such persons being then free to leave the ship and to go to the addresses given; and lastly, the transmission to the sanitary authorities of the names and places of residence of persons leaving the vessel for their respective districts with a view to such persons being maintained under supervision of the medical officer of health during the ten days which have been determined on as representing, in so far as administrative purposes are concerned, the period of incubation of plague. The system embodied in these measures is that which England has now for a long period adopted with regard to exotic diseases; it aims at arresting at our ports actual cases of foreign disease, plague, cholera, and yellow fever, and of securing the disinfection of articles which may reasonably be held to have incurred risk of infection. For the rest it imposes no restrictions on either individuals or articles imported; but it relies on the internal sanitary administration of the country to control at the onset any infection which may perchance evade the precautions adopted at our ports. Thus far it has been singularly successful, as regards all the three diseases to which it has been applied, and the responsibility which devolves under it on local sanitary authorities so to organize their local public health departments that they shall be always prepared effectually to deal with any chance infection which may reach them from abroad has had the inestimable advantage of helping to secure at the same time a standard of health at home which has resulted in an immense saving of life.

III.—Prevalence of Bubonic Plague.

In continuation of the study which for some time has been made in the Medical Department as to the diffusion of foreign epidemics, especially of such exotic diseases as are apt to be imported into Europe and into this country, I submit in Appendix A, No. 18, a report compiled by Dr. Bruce Low on the Diffusion of Bubonic Plague. The more recent contributions of the Medical Department to the history of this class of diseases have had almost sole concern with cholera; it is as far back as 1881, when the last report was made on the subject of diffusion of plague. In that year a special supplement was issued containing a report and papers by the late Mr. Netten Radcliffe on the then Recent Progress of Levantine Plague, in which the available history of that disease was recorded down to the end of 1879.* Dr. Low takes up the story from that date, supplying, as far as he is able, the links necessary to a consecutive history of this disease in different parts of the world between 1879 and 1894, when its epidemic incidence on the population of southern China, and notably of Hong Kong, constituted the first stage of a renewed prevalence of the malady in the East which involved danger to Europe, and which has been maintained up to the present date. It had long been known that there were endemic plague centres in certain parts of Asia, notably in Arabia, Mesopotamia, and China; but as new countries have been opened up and as the means of acquiring knowledge as to the conditions obtaining in the interior of semi-civilised and other states have increased, it has now come to be known that plague has long been persistent, at times in a quiescent form, amongst many more populations than had heretofore been believed. And thus it happens that at the present date numerous centres of plague have to be recorded from portions of Central and Eastern Africa, from Assyr in Arabia, Mesopotamia, Persia, Upper India, Central Asia, and eastward as far as Southern China.

PLAGUE CENTRES.

With increasing facilities for the movement of peoples, a number of these centres have gradually become sources of danger, not only to the communities living near them, but also by reason of traffic, by land and by sea, to more distant communities. A considerable number of outbreaks of plague of greater or less intensity, and which have thus involved risk of further spread, are recorded by Dr. Low, but amongst them all the greatest importance, in so far as the British Empire is concerned, attaches to the epidemic in Hong Kong.

It is known that for some twenty years plague has prevailed off and on in Southern China, and that in 1890 it broke out in certain

* Supplement to the Medical Officer's Report for 1879 on the Recent Progress of Levantine Plague and on Quarantine in the Red Sea. (C.—2905) 1881.

MEDICAL OFFICER'S REPORT.

towns on the coast line between Pakhoi and Canton, many thousands of lives being sacrificed to it. In February, 1894, the disease began to devastate Canton, and by May it had been imported into Hong Kong, between which place and the mainland an almost unceasing traffic is maintained. The story of the recurrences of the disease in this British possession and in its vicinity in 1895, 1896, and, to some extent, down to the present year, is recorded in detail by Dr. Low ; and some account is given of extensions of the disease to places which were in commercial relations with Hong Kong, including Formosa and Japan.

PLAGUE IN BOMBAY.

On September 23rd, 1896, the City of Bombay was declared to be infected with plague ; the disease recurred in epidemic form in the winter of 1897–98 ; and a third epidemic prevalence continued into the spring of 1899. The heavy incidence of the disease on the miserable inhabitants of the huge tenement buildings, known as "chawls," in the native part of the city is notorious, as is also the terribly insanitary condition of many of these dwellings, the construction and surroundings of which are such that they might have been actually contrived so as to ensure to the utmost absence of light and air as well as filthiness. Equally notorious is the almost complete immunity from the disease of persons living close at hand, under the opposite sanitary conditions, in the European part of the same city.

For reasons to be referred to, trustworthy statistics are not available as to the number of attacks or of deaths from plague in this city or elsewhere in India ; but, in so far as deaths in Bombay are concerned, Dr. Low records 11,577 fatal attacks in the epidemic of 1896–97 and 16,606 in that of 1897–98. But the disease was in no way limited to Bombay city ; it spread to many parts of the Presidency, including the important port of Karachi, where 3,398 plague deaths were reported between December, 1896, and July, 1897, and the City of Poona, which, together with its suburbs, &c., yielded 6,278 deaths between January and December, 1897. Indeed, a report quoted by Dr. Low states that the number of deaths in the Bombay Presidency, excluding Bombay city, between September, 1896, and March, 1899, reached 191,000.

PESTIS MINOR.

Dr. Low records in some detail the general distribution of plague since 1896 in other parts of India, such as the Punjaub, the North-West Provinces and Oudh, Central India, and Bengal. In the latter Presidency the incidence of the disease on Calcutta is of the greater interest, inasmuch as it affords some explanation of one of the main difficulties in controlling the spread of plague amongst many of the communities comprising our Indian Empire, and it incidentally affords proof of the imperative necessity for the strictest control, by way of medical inspection, to prevent the exportation of the infection from infected ports. I refer to the obscurity which attaches to plague in certain forms of its manifestation, such as are known by the names Pestis minor or Pestis ambulans.

MEDICAL OFFICER'S REPORT.

It was stated, and it was likewise denied, that cases of plague had occurred in Calcutta towards the end of 1896 and the beginning of 1897; it seems, however, that it was not until the spring of 1898 that the prevalence of the disease in that city "could no longer be denied." But, writes Dr. Low, "cases of fever with glandular enlargements are stated to have been not uncommon in the city for some few months, if not longer, previous to plague being declared." Now this is precisely the same experience as had taken place on a smaller scale at the end of 1896, when "a few cases of glandular swellings with fever" were reported to the Calcutta municipal authorities, and when, notwithstanding the alleged discovery of the plague bacillus in five out of eleven cases, "the diagnosis of plague was not upheld."

In the same way "fever with glandular swellings" prevailed in Bombay before it was recognised that plague had reached that city; and it is impossible to read the medical history of this disease in almost every part of the world without being impressed with the frequency with which recognised plague has been preceded by ailments of such slight severity, involving some bubonic enlargement of glands and some rise in body temperature, as to mask the real nature of the malady. In this respect plague would appear at times, if not commonly, to resemble other specific epidemic diseases in their pre-epidemic behaviour; and just as seemingly simple sore throats or slight diarrhœal attacks, both of them so trivial in their nature as to allow the subjects of them to move about freely amongst their fellow subjects, turn out ultimately to have been the means of sowing broadcast the infections of diphtheria and typhoid fever respectively, so does the ambulatory plague patient, who feels somewhat out of sorts and has some commencing enlargement in his groin or armpit, often constitute a grave danger to public health. Indeed, if plague differs from the diseases referred to in this sense, it would be by the frequency with which indolent buboes form a solitary outward indication of mischief, and by the facility with which these otherwise ominous symptoms evade detection. The striking story recorded by Dr. Low, at the end of this report, of the state of uncertainty which existed when fatal plague occurred in Vienna, as to the clinical and bacteriological interpretation to be assigned to the indications then being witnessed by experts who had made special study of the disease in India, is peculiarly significant in this sense. It also serves as a warning to those having administrative responsibility for the prevention of plague to regard symptoms which serve in common as signs of plague and of other maladies as being at the present juncture indications of that disease until the contrary can with confidence be asserted.

PREVENTIVE MEASURES (a) REASONS FOR FAILURE

These considerations go to emphasize the immense difficulties with which the Indian Government have had to contend in preventing the diffusion of the infection of plague from one part of the Empire to another. Even amongst communities who are willing, in the interests of the public health, to submit themselves

MEDICAL OFFICER'S REPORT.

to a careful medical inspection, cases of the class described must often escape detection or present great administrative difficulties by reason of doubtful diagnosis. But when added to this, racial and religious obstacles present themselves involving the deepest resentment at any medical examination of either the living or the dead, and when in order to avoid this or any other form of preventive intervention hundreds of thousands of persons take flight from their homes; when plague cases are concealed and after death the bodies are buried inside the actual dwellings of the survivors; and when in carrying out one of the highest duties which one man can perform towards his fellow subjects and to those under whom he serves he is met by fanatical outbreaks, riots and wrecking of property, and now by assassination, then the difficulties are almost overwhelming. And yet these are precisely the circumstances which Dr. Low has again and again to record in connexion with the measures adopted to control the spread of plague in India and elsewhere in the East.

(b) SUCCESSFUL RESULTS.

Fortunately there are, side by side with these records, others which tell of marked and immediate success as resulting from the evacuation of infected villages and other localities, from the formation of segregation camps, from the adoption of measures of disinfection, and from a medical examination of the suspected, both sick and dead. These successes would, happily, appear to be on the increase, and this increase would seem to have relation to a more intelligent apprehension by the natives of the desire on the part of the governing authorities to avoid all unnecessary interference with traditions of race or of creed. In this connexion the employment by the Government, in certain cases, of qualified native and of female medical practitioners may be noted.

MEDICAL INSPECTION.

Amongst the measures of prevention which have been resorted to, the medical inspection of those travelling by land from one part of India to another, and of persons leaving India by sea, whether as ordinary passengers or in connexion with the Mecca pilgrimage, deserves more than a passing notice. Thus the removal by night of plague-stricken persons from Poona led to the organisation of a system of picquets and of inspections at railway stations, with the result that the picquets found 103 plague patients and 23 plague corpses on the road, whilst no less than 799 persons attacked with the disease were detected at the railway depôt. But the immense task involved in the adoption of such measures in our Indian Empire may be best estimated by the fact that when danger of the importation of the disease into Calcutta was realised, no less than 1,800,000 travellers were inspected, of whom 40,000 were detained as suspects until it was ascertained that all but six of them were free from the disease.

In a similar way, medical inspection is carried out by male and female medical practitioners at Indian ports of departure, whether these ports are in themselves infected or whether they are liable to serve as points of departure for persons coming from infected areas. In this way all persons going on board ship, whether passengers, pilgrims, or crew, are examined with a view of preventing the transmission of plague to other ports or countries; and as regards certain groups of persons, such as coolies, special camps are provided for their detention under observation before the last stage of their journey to the port of departure. As an example, it may be noted that during 1897 no less than 1,313,117 persons entering or leaving Bombay harbour were medically inspected, and that out of 42,552 of these who were rejected or segregated for the purposes of observation, 169 were ultimately found to be suffering from plague. Since that date the system of medical inspection has been further matured: medical inspections which take place before going on board are verified by a second inspection before the departure of the vessel, and any glandular enlargements with febrile symptoms are deemed to be suspicious of plague unless they clearly admit of other clinical explanation.

In all essential respects these medical inspections, disinfections, &c., at ports of departure are carried out in accordance with the provisions of the International Sanitary Conference of Venice, 1897, at which gathering I had the honour to act as one of the delegates for Great Britain, and to serve as Her Majesty's Plenipotentiary to sign the resulting Convention on behalf of this country. As, however, the ratification of the Convention by the signatory powers has been temporarily deferred, I reserve further reference to the subject to a future occasion.

PROHIBITIO[N] OF THE MEC[CA] PILGRIMAG[E]

One other measure of prevention calls for notice here. In view of the exceptional circumstances which prevailed early in 1897, the Government of India, for the first time in the history of our Indian Empire, directed on February 16th that no person resident permanently or temporarily in the Bombay Presidency or Sind, and no person who though not so resident, had entered the Bombay Presidency or Sind with the object of proceeding on a pilgrimage to the Hedjaz, should be permitted to embark on any ship at any port in British India with the object of making the pilgrimage. This prohibition, which put an end to the departure of all pilgrim vessels, took immediate effect, and resulted in a stoppage of the pilgrimage.

INFECTED CLOTHING, &

The mild and ambulatory attacks of plague to which Dr. Low refers, and which obtain so commonly in connexion with epidemics of this disease, have a further interest. As will be seen from Dr. Low's report, the recognised prevalence of plague in one and another place or country is often set down to infection conveyed by means of clothing, and of other articles that are imported from an infected district. It must be admitted that certain articles, notably clothing, bedding, &c., which have been in contact with the sick, may unquestionably serve as vehicles for the communication of the infection of plague, and it is of the utmost importance that such articles should always be either destroyed or efficiently disinfected. But, as the result of increasing knowledge as to the methods by which the infection of plague is communicated from one place to another, I have *no hesitation* in venturing to assert that the individual who suffers from Pestis minor or Pestis ambulans, and whose movements no one has found cause to control, is an infinitely greater danger as a vehicle of infection than even infected clothing; and this, especially when it is a question of conveying the disease to a hitherto healthy country. Here, again, we have the analogy of our current infectious fevers, such as scarlatina, the communication of which was formerly so often set down to the retention of infection for almost untold periods in odd pieces of rag or flannel which had at some time or another been seen in the hands of a scarlatina patient. Now, however, we have come to learn that one of the commonest methods of infection of this disease is, not a piece of clothing or a few scales of effete epithelium, but a recrudescence, during convalescence, of the disease itself in the form of a recurring throat malady which is often so trivial as to escape detection. In brief, even more importance attaches to an investigation into the clinical features of comparatively recent, though perhaps obscure, attacks of such diseases, as a cause of further mischief, than to the discovery of some chance of a remote infection through some indirect channel.

PLAGUE IN RELATION TO MERCHANDISE.

Some nations and persons have, in sheer panic, attempted to impose almost universal prohibition on the importation of all articles from countries where plague prevails; and in their most pronounced form these attempts have, in so far as the interests of this country are concerned, culminated in occasional suggestions to prohibit the export of grain from plague-infected ports, or to destroy cargoes of grain, such as corn and rice, when carried by a vessel on which one or more cases of plague have occurred. With regard to such proposals, I have taken the responsibility of advising that no prohibition either as to exportation or importation, under the circumstances referred to, was called for; and further, that no futile attempts at the disinfection of cargoes such as those of grain were justifiable, except in so far as the local medical officer of health had evidence that any individual sack or sacks of grain had had opportunity of becoming infected, or afforded indication of having been soiled by or of having been eaten into by rats whose dead bodies gave evidence of the occurrence of the disease in them. Thus far, my action, which has often been followed by the use for food of the cargoes concerned, has been fully justified by the results.

In taking this attitude as regards the importation and exportation of articles of commerce in general, I have been influenced by the following considerations. Firstly, as regards grain cargoes there is an absence of anything that can be regarded as evidence to show that plague is communicable to man by means of an article of diet; and there is, further, evidence which goes to show that even when grain is purposely infected with the plague bacillus that organism dies in a few days when the grain is dry, as when packed for transit. Secondly, there is overwhelming evidence to the effect that cargoes from infected vessels have failed to communicate the infection of plague, even under circumstances when

nearly every condition present seemed to favour such communication.*

In this connexion I would recall a report of a Special Committee appointed in 1846 by the Academy of Medicine of France, to inquire into the question of quarantine restrictions in relation to plague. The report states as follows:—"In 1835 plague prevailed in epidemic form amongst the employés of every grade who were engaged in the warehouses of the Egyptian Government. Notwithstanding this, bales of cotton handled by labourers were transmitted from January to June, that is to say, throughout the whole course of the epidemic, to all the principal ports of Europe, without a single case of plague resulting. There were thus forwarded, in 1835, 31,700 bales to England, 33,812 to Marseilles, 424 to Leghorn, 150 to the Netherlands, 32,263 to Trieste, and 32 to sundry other places. These bales of cotton, we repeat, did not convey plague to a single person, and this although no precautions whatever were taken by way of disinfection. After being subjected to pressure by machinery before being placed on board, the bales were packed in the smallest possible space, the hatchways were fastened down, and the vessel took her departure. Out of sixteen English vessels, laden with cotton, which left Alexandria between the commencement of January and the end of June, eight suffered from plague on board; but, notwithstanding this, the cotton carried by these vessels was no more dangerous than that carried by non-infected vessels. We conclude our statement as to the transmissibility of plague by means of merchandise by drawing your attention to a fact of the highest importance, and which is based on positive and official authority. Since 1720 not a single porter employed in the lazaret at Marseilles in the unlading and handling of merchandise has contracted plague."

Having regard to our present knowledge concerning the etiology of bubonic plague, I feel justified in holding the view that, in so far as danger of the introduction of that disease into this country is concerned, the first and most important point to be held in view in our port sanitary administration is such strict medical inspection of persons arriving from infected countries as shall go to secure

* Since the above was written a report from the Governor of the Straits Settlements to the Secretary of State for the Colonies, dated 4th August, 1899, has been communicated to this Board. That report refers to the measures adopted in the Straits Settlements since May, 1894, on account of the prevalence and repeated recurrences of plague in Hong Kong, and it proceeds as follows:—

"One remarkable result of the experience of the past five years has been this, that though science teaches us that certain cargo (called susceptible) is capable of carrying plague germs and of communicating the disease, yet practically the risk arising from such cargo must be a negligible quantity as, owing to the vast amount of commercial business done at the large entrepôts of Hong Kong and Singapore, no disinfection of such cargo is possible or has been contemplated.

"Nevertheless no single case of plague has occurred here from the free admittance of susceptible cargo indiscriminately with non-susceptible cargo, or from its being unpacked, *packed or handled.*"

MEDICAL OFFICER'S REPORT.

the detection and immediate isolation of those mild attacks which are mainly identified with slight or indolent bubonic enlargements and indications of general mal-aise. Next to this, and always to be observed, is the disinfection of all articles of clothing, &c., which have had opportunity of becoming infected either before they were packed for the journey or since. Thirdly, comes the disinfection of those portions of the vessel which may have been used by patients suffering from or suspected to have had plague. And, lastly, measures should be adopted to prevent the conveyance of the disease to the shore, from an infected ship, by means of rats which are peculiarly susceptible to the plague infection.

But no system of inspection or other restriction is perfect: and hence for the rest we must trust to such local sanitary administration of our towns, villages, and hamlets, as will deprive a disease such as plague of the means of diffusing itself should a chance case make its way inland. The experience derived from India is abundantly encouraging in this respect; for the extraordinary immunity of Europeans living under more or less modern sanitary conditions, on the one hand, and the terrible incidence of the disease upon natives in the same cities and places, but whose surroundings are typical of that pollution of soil and air which always favours the diffusion of filth diseases, on the other hand, forms a contrast which not only emboldens us to hope that the sanitary administration of this country may continue to show itself antagonistic to the diffusion of imported plague, but should act as a fresh incentive to the removal of the conditions favourable to infection of this type which still remain in our midst.

NOTIFICATION AND DEATH RETURNS.

I include in Appendix A., No. 19, a compilation, quarter by quarter for the year 1898, of the notified attacks and the certified deaths from certain infectious diseases in 93 urban sanitary districts of England and Wales for which the necessary data were available. In the case of the sanitary districts of the Metropolis, the same information is set out also week by week.

INTERNATIONAL SANITARY CONFERENCE OF PARIS.

In the spring of 1894 I had the honour, in accordance with the instructions of the Earl of Rosebery, then Secretary of State for Foreign Affairs, of representing Her Majesty's Government at the International Sanitary Conference of Paris, which had for its object the compilation of regulations to prevent the diffusion of cholera through the agency of the Mecca Pilgrimage, and also by reason of the navigation of the Persian Gulf. My colleague for Great Britain was Mr. E. Constantine Phipps, C.B., Minister Plenipotentiary, and at that date Secretary to Her Majesty's Embassy in Paris, who acted as diplomatic delegate. British India was on the same occasion represented by Surgeon-General Cuningham, C.S.I. The Conference lasted from the 7th of February until April 3rd, and on its termination Her Majesty was pleased to appoint me one of Her Plenipotentiaries to sign the Convention which had been agreed to.

The subject of this Conference has not been referred to in my previous reports because, owing both to questions which arose after its conclusion, and to the desire to secure, if possible, the co-operation of Turkey in carrying the provisions of the Convention into effect in Turkish territory, the ratification of the instrument was delayed until June 20th, 1898. Thus far, the countries that have ratified the Convention are—Great Britain, including British India and the Straits Settlements, Germany, Austria-Hungary, Belgium, Denmark, Spain, France, Greece, Italy, the Netherlands, Persia, Portugal, and Russia. Her Majesty's Government, however, in becoming a party to the Convention, did so subject to three reservations

The first of these related to a requirement that all pilgrims shall, before leaving a port of departure, give evidence that they are in possession of the necessary pecuniary means to accomplish the journey both to and from the Hedjaz. But after it had been announced to the Conference, on the authority of the Turkish delegation, and according to a decision communicated from Constantinople, that no one could under the Mussulman law be prohibited from undertaking the pilgrimage, it was held by Her Majesty's Government that no such general condition could be imposed on Her Majesty's Mussulman subjects without a serious interference with the religious liberty they were entitled to enjoy. In view of this, it was decided that the condition in question was not to be held to apply to Indian pilgrims except when the local circumstances made its imposition permissible.

The second reservation had concern with the requirement that provision should be made between-decks on all pilgrim vessels at the rate of 2 square metres (some 6 feet 7 inches by 3 feet 3½ inches) per head. Such a condition could not have been objected to if sanitary considerations had alone been involved; but, having regard to a number of other considerations, such as the certainty that the increase of passage money which such a requirement carried with it would be a serious restriction on the poorer Mahommedans desirous of fulfilling a religious rite, coupled with the fact that many of the pilgrims from British India habitually sleep on the upper deck, which is required to be kept clear for them, Her Majesty's Government reserved to themselves the right to substitute a superficial area of 16 square feet for the two metres referred to.

The third reservation took account of a code of regulations which had been drawn up for the control of the navigation of the Persian Gulf. In the opinion of the British and Indian delegation these regulations were not only in excess of the requirements of their professed object, but they were otherwise objectionable. So, also, the cost of carrying them out, which would fall almost exclusively on shipping carrying the British flag, was deemed to be excessive. These views were endorsed by Her Majesty's Government, and the ratification of the Convention was made subject to the condition that the stipulations relating to the Persian Gulf should not be applicable to the Governments of

Medical Officer's Report.

Great Britain and of British India, nor to British and Indian shipping.

The provisions of the Paris Convention regulate in much detail the matters of which it takes cognizance, but, speaking generally, its principal provisions are as follows. Dealing in the first place with the journey of pilgrims to the Hedjaz, it provides for an adequate medical examination of all intending pilgrims before they go on board ship at a port of departure; it lays down in precise terms the conditions of sanitation and administration which shall govern the pilgrim ships and which shall be applied to the pilgrims carried by them both on the outward and the return voyage; it determines the conditions under which pilgrims may be detained under supervision on the island of Camaran in the Red Sea before they are landed at Jeddah, and it limits the period of detention at Camaran, in the case of healthy vessels to forty-eight hours, instead of the term of ten days which heretofore had been applied to all vessels alike; and it specifies the sanitary improvements which are called for both at Camaran and at the other sanitary stations in the Red Sea which serve pilgrims of different nationalities. The control of the navigation of the Persian Gulf is dealt with in another portion of the Convention; but, as already stated, Her Majesty's Government are not a party to the regulations made under this heading. And, lastly, the measures to be adopted for the superintendence and execution of the terms of the Convention are laid down.

With regard to the value of this Convention in preventing the diffusion of cholera through the agency of the Mecca pilgrimage, and in securing the adoption of more reasonable measures of restriction at sanitary stations in the Red Sea, it will be evident that much must depend upon the co-operation of the Turkish Government, whose territories in the Red Sea and in Arabia are so intimately involved; and hence it is much to be regretted that Turkey has not yet given in her adhesion to it. This has, however, not stood in the way of the adoption of those measures of precaution, set out in the Convention, which have to do with the voyage to and from the holy places of Mussulman subjects of Her Majesty; and the measures which have been taken in this direction go to increase the security against the infection of cholera which was the primary aim of the Convention, and, at the same time, to add to the general healthiness and well-being of the pilgrims.

IV.—Auxiliary Scientific Investigations.

Bacillus enteritidis sporogenes.

The anaërobic bacillus enteritidis, an account of which, as regards its morphology and biology, and its observed relations with infantile diarrhœa and with English cholera, was given in my last Report, has in 1898–99 been made the subject of further investigation. Dr. Klein has (Appendix B, No. 1) sought to ascertain what may be the concern of this bacillus in other forms of serious and dangerous diarrhœa, as for instance, the diarrhœa which accompanies enteric fever. Above forty cases of the latter disease have been closely scrutinised by him at various stages of the malady and whether diarrhœa was or was not at the moment a prominent symptom, with a view to detecting the presence or absence, the abundance or the scarcity, of bacillus enteritidis in the bowel evacuations of the patients. As a result, in the phases of enteric fever in which the stools were "typical" of the disease, Dr. Klein almost invariably detected in abundance in such stools spores of this bacillus; whereas in phases of this fever wherein the evacuations of the patient were not characteristic of the disease he failed to demonstrate these spores in the stools or demonstrated them only in small numbers and with difficulty.

The observations recorded by Dr. Klein in this connexion, though not inconsistent with a relation of effect and cause between the "pea-soup" diarrhœa of enteric fever and the abundant presence for the time being of bacillus enteritidis in the bowel evacuations of sufferers by the complication, comprise, as Dr. Klein was not slow to observe, facts of much wider bearing. His difficulty, whatever the stage of the fever, in detecting spores of bacillus enteritidis in the evacuations when these were for the occasion solid or semi-solid, involved test in numerous instances of large amount of fœcal matter before a positive result was arrived at; and accordingly Dr. Klein asked himself whether this microbe might not after all be found to inhabit, though sparsely, the human intestine *in health* as well as in disease, if only it were, in like manner and equally diligently, sought for. This indeed he ascertained on further investigation to be the case. As a consequence the latter part of his report is given to a discussion from this new point of view of the specificity so to speak of bacillus enteritidis sporogenes, and to a record of some further observations he has been making of the biology of this microbe and of its interaction with the typhoid bacillus.

BACTERIA AND THE BURIED ANIMAL BODY.

The extent to which graveyards and cemeteries may, through the medium of the underground water yielded by wells and springs in their neighbourhood, contribute dangerous contamination to water-supplies, has been an ever recurring subject of dispute. On the one hand chemistry has found that the water draining below ground from graveyards is apt to contain abundantly in solution dead organic matter derived from the corpses undergoing dissolution in the superincumbent soil; and as a consequence the chemist has been prone to condemn such water as a source of domestic supply. On the other hand bacteriology, while showing abundance of bacteria in drainage water from graveyards, has failed to afford evidence that the micro-organisms derived from the graveyard are at all commonly identifiable with microbes capable of inducing in the human subject specific disease. In the circumstances it has been judged profitable to subject to bacterial test actually buried bodies; in

MEDICAL OFFICER'S REPORT.

order in the first place to ascertain the nature (whether or not harmful) of the bacteria proper to graveyards, those for instance which prey so to speak on corpses; and further to ascertain the fate in graveyards of pathogenic microbes which, having multiplied in the animal body during life, are, after death of that body, buried along with it in soils of different sorts.

Dr. Klein gives, in Appendix B, No. 2, an account in regard of rodents buried by him in diverse ways in different soils, of the bacteria, adventitious or other, which are chiefly concerned in the disintegration and dissolution of the dead animal body. Also he records, in reference to similar experimentally buried amimals, the ultimate fate of certain pathogenic bacteria which previous to death had been made to multiply in the animal bodies in question.

Dr. Klein's observations in both subject matters are thus far reassuring. He does not find that disintegration and dissolution of the animal body after burial is brought about by microbes harmful to man, by aërobic microbes for instance such as bacteria of the coli and proteus groups, which are on occasion pathogenic to the human subject and which heretofore have been regarded as the main agents in the putrefaction of dead animal matter. As result of his experiments, dissolution of corpses would seem to be almost wholly an affair of anaërobiosis, and indeed practically the work of a particular anaërobic bacterium, that is in his view inherent in all buried bodies and which Dr. Klein proposes to term *bacillus cadaveris sporogenes*. Similarly Dr. Klein finds that under his conditions of burial, and in the presence within the buried bodies of the rapidly multiplying bacillus cadaveris, no single one of a great variety of pathogenic microbes which during life had been made to proliferate in experimental guineapigs, is capable of maintaining its vitality in the decomposing bodies of these animals for a period measured by more than a very few weeks.

SOIL INVESTIGATION.

Drs. Martin and Houston in continuation of their researches with soil as a multiplying medium for pathogenic microbes, have, in 1898 99, been dealing with the more difficult problem of the viability of one and another morbific micro-organism when placed in competition, under soil-conditions, with the bacteria proper to soil.

(a) TYPHOID BACILLUS.

The typhoid bacillus has been further tested by Dr. Martin (Appendix B, No. 3), under laboratory conditions and under the conditions of an outdoor shed, as to its ability to maintain itself in *unsterilised* soils of different sorts:—namely, I. and II. soils from Chichester, III. soil from the grounds in connexion with University College, London, and IV. soil from a nursery garden. In several of these soils, *when they had been sterilised*, the typhoid bacillus had previously been found to persist for many months. But now that under the new conditions of experiment this bacillus was brought, in the very same soils, into competition with bacteria natural to soil, in no single instance did it survive beyond twenty-four hours; seemingly it at once died out.

This diverse behaviour of the typhoid bacillus according as one and the same soil was sterile or non-sterile, afforded presumption that the bacteria, or some of them, proper to the unsterilised soils had been the agencies inhibiting in these experiments the life-processes of the pathogenic microbe in question. Dr. Martin therefore proceeded to sort from the Chichester soils and from the University College soil their several aërobic bacteria, and then went on to put each soil-bacterium thus isolated separately in competition with the typhoid bacillus. This he effected by means, in each instance, of two sorts of culture-media: the one a nutritive liquid, the other the soil *after* sterilization whence the saprophytic bacterium of experiment had been derived. As a result he found that while certain of these bacteria proper to one and another soil quickly suppressed the typhoid bacillus, and others of them flourished so to speak side by side with the microbe, others again not only could not compete with, but were inhibited by, the micro-organism of enteric fever.

(*b*) KOCH'S VIBRIO.

Dr. Houston (Appendix B, No. 4) after some preliminary laboratory testings of the viability under these artificial conditions, of bacillus prodigiosus, bacillus diphtheria, and Koch's vibrio in unsterilised soils from diverse sources, instituted by well contrived experiment sustained observation of Koch's vibrio and bacillus prodigiosus in circumstances of exceptional interest for the epidemiologist. These microbes he implanted side by side and in great amount in each instance so as to cover a considerable area, in the natural soil of the open country, with a view to subjecting them to conditions, seasonal, meteorological and other, such as would be likely to be encountered by Koch's vibrio were it to become in the course of a cholera-outbreak self sown as it were in the soil, garden or other, of town or country. His reason for choosing the non-pathogenic prodigiosus bacillus for the control experiment in this connection was that this micro-organism, while easily identifiable in a mixture of bacteria of various sort, is, like Koch's vibrio, a non-sporing microbe, and one therefore to be expected to react to physical agencies to an extent comparable with Koch's bacterium.

Tested "in rure" in the above sense, under conditions which, though they varied greatly from time to time, always remained parallel, the two microbes of experiment behaved in very different fashion. Bacillus prodigiosus, the indigenous micro-organism, proved, as a result of a single sowing, able to maintain its existence, though in diminishing number, in the soil during a period covered by months; whereas the exotic Koch's vibrio, though repeatedly sown in large numbers over several square feet of surface, could with difficulty be recovered from the soil after lapse of a few days. There could be no doubt therefore that the sum of the conditions in the above series of experiments had proved hostile to Koch's vibrio to a far greater degree than to bacillus prodigiosus. Wherefore, lest the different behaviour of the two microbes might have been due solely to scarcity in the soil of nutrient material proper for Koch's vibrio, Dr. Houston instituted a further experiment.

Once again he re-sowed this vibrio, in vast amount and broad cast as before: but on this occasion, antecedent and subsequent to the sowing, he supplied to the soil additional nutriment in the form of dilute liquid horse manure. The result, however, did not differ conspicuously from that obtained in the previous experiments. All that can with certainty be said is that Koch's vibrio did not die out quite so quickly in the soil; there was no indication whatever that it multiplied therein. Nevertheless, there was suggestion in these experiments, as in former experiments by Dr. Klein, of a tendency in Koch's vibrio under adverse cultural conditions to undergo morphological modification to an extent rendering it extremely difficult of identification.

With reference to the negative results of his experiments, Dr. Houston deprecates any inference from the "particular" to the "general." As he observes, the provisional interpretation of his facts concerns the particular soil in question, it does not necessarily apply to other soils, other seasons of the year, other climates, and other conditions of experiments.

SOIL WASHINGS.

Commenting in my last year's report on Dr. Houston's investigations of the chemistry and bacteriology of different soils, I noted not only the value of these researches in helping toward determination of the degree of excremental pollution of a given soil, but also their promise of especial utility to the bacteriologist who, dealing with a water service to which flood-water has access, seeks to ascertain whether during time of storm the washings of one or another doubtfully pure surface soil have been contaminating the supply. In Appendix B, No. 5, Dr. Houston gives account of experiments undertaken by him during 1898–99 in this direction.

Flood-water, for Dr Houston's purposes, was represented in each instance by 1,000 c.c. of distilled water to which had been added, after settling, 10 c.c. of fluid from the surface of an intimate mixture of 10 grammes of soil in 100 c.c. of water. Ten soils of diverse origin were thus dealt with, and samples of these "soil washings," as Dr. Houston terms them, were tested chemically and bacterially as in his previous investigation.

To *chemical* tests, Dr. Houston's artificial flood-waters yielded results which, if they were obtained in the case of drinking water, would be regarded as affording in some samples a slight, and in others a stronger, presumption of contamination with organic matter of vegetable or animal origin. But—and this was to be anticipated from Dr. Houston's previous researches—as between one soil washing and another the degree of response to chemical tests in no way corresponded to the relative amount of contamination of their respective soils with animal organic matter, and these tests altogether failed in discriminating samples derived from soils the contamination of which had been recent.

Bacterially, and first seeking among conditions common to all the samples an index of their derivation from surface soil, Dr. Houston found in each sample bacteria which he had already shown to be common in soils of diverse origins, but which cannot be included among the numerous microbes whose normal habitat is water. Bacillus mycoides, for example, was present, though in greatly varying amount, in each of the ten soil washings, and cladothrix and bacillus fluorescens liquefaciens in a considerable proportion of the samples. Another feature common to all ten, and in Dr. Houston's view characteristic of their soil-origin, was the large proportion of bacteria present in spore form.

Next, as to bacterial conditions helpful in differentiating the experimental flood-waters according as the contamination of their corresponding soils by animal organic matter had been great or little, recent or remote, Dr. Houston in each instance made special search for bacillus coli, for bacillus enteritidis sporogenes, and for varieties of streptococcus. In none of the ten samples did he find a bacillus which answered in subculture to every one of the tests to which "normal" bacillus coli is in the laboratory called upon to respond. Nevertheless, in three of the samples, bacteria were found responding in subculture to the majority of these tests, and whose kinship therefore to bacillus coli may be thought of as close. These samples were derived from the three experimental soils which appeared to have been subject to most recent and most considerable contamination, namely, a field irrigated with sewage, Thames mud at Blackfriars, and mud from the bank of the River Brent below the outfall of an effluent from sewage works. Bacillus enteritidis sporogenes was found (as spores) in seven of the ten samples: in the three already mentioned, and also in the washings of each of four other soils which had a history of recent or remote exposure to contamination with animal organic matter. In one sample of manured meadow soil spores of this bacillus were not detected, and they appeared also to be absent from an unmanured pasture soil, and from a moorland peat soil. Finally, from the three samples which yielded bacteria most nearly responding culturally to bacillus coli, and from no other of the ten soils under observation, Dr. Houston obtained streptococcus of one or another variety. This circumstance, suggesting as it does a possible relationship between the presence of streptococcus and recent contamination, has led Dr. Houston to some further tentative bacterial investigation of water supplies, an account of which is given separately by him in Appendix B., No. 6.

SCARLATINA "RETURN CASES."

It is, as I pointed out in my last year's report, in all probability *solely* those persons who retain, or reproduce, in their throats streptococcus scarlatinæ that are able on their return home from hospital to infect with scarlatina other persons brought into close relation with them.

This subject, "Return Cases," has in 1898-99 been further studied by Dr. Mervyn Gordon (Appendix B., No. 7). First he set himself to scrutinise anew the streptococcus groups of micro-organisms, with a view to the better differentiation, for his own purposes, of streptococcus scarlatinæ from allied bacteria. Having achieved this differentiation, which he duly sets forth in detail for the guidance of other observers, he proceeded to exact and sustained observation of the streptococci to be obtained at various intervals after scarlatina attack from the aural and nasal discharges, and from the throat-secretions, of persons the subjects of that malady.

The results obtained by Dr. Gordon go far to confirm the inferences that had been provisionally arrived at. In no single instance, out of twelve cases examined, did he identify the streptococcus scarlatinæ in the *ear-discharge* of scarlatina convalescents, such ear-discharge being tested at periods ranging between one day and eight weeks from its commencement. So too, as regards *nasal discharge*, tested (again in twelve cases) at periods varying from two days to eleven weeks from its first appearance, the results were all but negative; in only two samples of this material, both obtained in the second week of the discharge, did Dr. Gordon identify the streptococcus scarlatinæ, and in each instance the micro-organism was but sparsely present. But as regards the throat-secretions of scarlatina convalescents the evidence is altogether the other way. Eighteen cases were dealt with by Dr. Gordon in this connexion, and 27 examinations of their throats undertaken. Of the 27 tests thus applied no less than 20 gave positive result as regards presence of streptococcus scarlatinæ; moreover 13 of these positive results were obtained later than the beginning of the fifth week since attack—indeed, in two instances the patient was on the point of discharge from the hospital.

The above experience goes to suggest that, as we had begun to suspect, streptococcus scarlatinæ is present, though not abundantly perhaps in many instances, in the throats of a large proportion of scarlatina patients up to a late stage of their convalescence, and even after seeming complete recovery. Whether, however, this streptococcus always maintains in such circumstances its full virulence is open to doubt, ; indeed certain physiological tests by Dr. Gordon of this micro-organism when taken from the throats of scarlatina convalescents tend to reassurance in this particular.

BACTERIOSCOPIC EXAMINATION OF WATER.

In my report for last year I pointed out that in bacteriology we have a ready means of detecting contaminating matters in waters which, from the chemist's point of view, would be classed as "of high degree of purity"; that, for instance, taking the presence of bacillus coli and bacillus enteritidis in a water as evidence of its previous sewage contamination, Drs. Klein and Houston had been able to detect these microbes in a liquid to which raw sewage had been added in the small proportion of only 1 part in 20,000 of water. During 1898-99 Drs. Klein and Houston have (Appendix B., No. 8) made further test of their process, applying it systematically to several sorts of sewage, and in dilutions ranging from 1 : 1000 to 1 : 100,000. As before, side by side with their bacterial examinations of these dilute sewage samples, they have employed the usual chemical tests as a check to their other procedure. But their main object has been to learn whether under their bacterial testings the ratio of bacillus coli to bacillus enteritidis, and the ratios of both these bacilli to the total aërobic microbes of all sorts in the raw sewage, remained constant, or nearly so, for each sample in each of its several dilutions. Incidentally only have they sought to ascertain how far the biological qualities of a given sewage afford index of the general character of such sewage; whether, for instance, sewage comprising much manufacturing refuse can be readily differentiated biologically, again in each of several dilutions, from every day domestic sewage.

So far as their main object is concerned, Drs. Klein and Houston would seem to have attained a considerable measure of success. From Table II. of their report it appears that for a given sample of sewage their bacterioscopic processes, whatever the dilution (and provided always a sufficient amount of sample be dealt with), yield, as regards relative amount of bacillus coli and bacillus enteritidis, data fairly in accord with the already observed facts respecting the raw sewage of experiment. At the same time the facts of the table give small promise that as between one class of sewage and another class of sewage their bacterial method can afford any sufficient basis of differentiation. Not only is the proportion of bacillus coli and of bacillus enteritidis to total aërobic microbes, as also the ratio of these two bacilli to one another, found to vary within very wide limits in sewage of different origin, but each separate sewage would seem to possess in these respects its own flora; so that any inference as to the general character of a polluting sewage, from the greater or less abundance of bacillus coli and of bacillus enteritidis in a water contaminated thereby, would seem to be altogether precluded.

In Appendix C. I submit a Joint Report on the Histology of Vaccinia, by Dr. S. Monckton Copeman and Dr. Gustav Mann.

The investigations on which the report is based were not made for the Board in connexion with the usual Treasury grant for scientific purposes. It was freely placed at the disposal of the Board by the authors: and having regard to the great interest and importance which attach to the subject to which it affords so interesting a contribution, it was at once decided to accept the report, and to reproduce it in the form of an appendix to their Annual volume, in the belief that it will aid in the elucidation of some of the problems relating to the histology of vaccinia which have hitherto remained unsolved.

MEDI OFFIC REPC —.

HISTOL VACCIN.

MEDICAL OFFICER'S REPORT.

LOCAL INSPECTIONS OF MEDICAL INSPECTORS.

In Appendix D. will be found an Index to the Local Inspections of Medical Inspectors since the date of the formation of the Board. It has been compiled in the Medical Department, and it is certain to be of value for administrative purposes within, if not also beyond, the offices of the Board.

I have the honour to be,
Sir,
Your obedient Servant,
RICHD THORNE THORNE.

December, 1899.

APPENDICES.

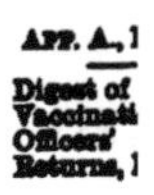

APPENDIX A.

No. 1.

Digest of the Vaccination Officers' Returns with regard to Children whose Births were registered in the Year 1896.

The following is a summary of the twenty-fifth annual return under the Vaccination Act, 1871. Of 914,205 births returned to the Board by the several vaccination officers in England and Wales as registered during the year 1896, the number which, at the time the return was made, had been registered as successfully vaccinated was 602,922 (being 66·0 per cent. of the whole), and the number registered as having died unvaccinated was 99,386 (or 10·9 per cent. of the whole). Of the remaining 211,897 children, 2,885 (or 0·3 per cent of the whole) had been registered as insusceptible of vaccination; 5 (or 0·0005 per cent.) as having contracted small-pox before they had been vaccinated; 14,682 (or 1·6 per cent.) as having their vaccination postponed by medical certificate; leaving 194,325 (or 21·3 per cent.) as "removed," "not to be traced," or otherwise unaccounted for. If from the 914,205 births returned by these officers deduction be first made of the deaths that took place before vaccination, it it appears that of the surviving 814,819 children, there were registered at the time of the return 74·0 per cent. as successfully vaccinated; 0·4 per cent. as either insusceptible of vaccination, or as having had small-pox; and, including 1·8 per cent. as under medical certificate of postponement, 25·6 per cent. as at that time not finally accounted for as regards vaccination.

The proportion of cases not finally accounted for in England and Wales for 1896 is 22·9 per cent; in the metropolitan returns,

A., No. 1. — of nation rs' 1896.

26·4 per cent.; in the provincial returns, 22·3. Of the registered births of the 25 years, 1872–1896, the proportion not finally accounted for in regard to vaccination in each year respectively has been as follows :—

—	Metropolis.	Rest of England.	—	Metropolis.	Rest of England.
1872	8·8	4·5	1885	7·0	5·5
1873	8·7	4·2	1886	7·8	6·1
1874	8·8	4·1	1887	9·0	6·7
1875	9·3	3·8	1888	10·3	8·2
1876	6·5	4·0	1889	11·6	9·6
1877	7·1	4·1	1890	13·9	10·9
1878	7·1	4·3	1891	16·4	12·9
1879	7·8	4·5	1892	18·4	14·3
1880	7·0	4·5	1893	18·2	15·7
1881	5·7	4·3	1894	20·6	19·0
1882	6·6	4·5	1895	24·9	19·8
1883	6·5	4·9	1896	26·4	22·3
1884	6·8	5·3			

In 1896 the proportion of cases unaccounted for, after deduction of the postponed cases, in the Metropolis and in the rest of England, was 24·9 and 20·6 per cent. respectively.

App. A., Digest of Vaccination Officers' Returns.

—	Returns, 1896.							
	Births.	Successfully Vaccinated.	Insusceptible of Vaccination.	Had Small-pox.	Died unvaccinated.	Vaccination postponed.	Remaining.	Children not finally accounted for (including cases postponed), per cent. of births.
ENGLAND & WALES ..	914,205	602,922	2,885	5	99,386	14,682	194,325	22·9
Ditto (excluding Metropolitan Unions).	780,372	518,629	2,358	4	85,716	12,638	161,027	22·3
METROPOLITAN UNIONS ..	133,833	84,293	527	1	13,670	2,044	33,298	26·4
COUNTIES.								
ENGLAND:								
Bedford	4,460	407	1	..	531	3	3,518	78·9
Berks	7,207	5,554	24	..	623	148	858	14·0
Bucks	4,523	2,780	13	..	381	40	1,309	29·8
Cambridge	5,278	4,216	9	..	459	53	541	11·2
Chester	21,644	17,851	104	..	2,331	361	997	6·3
Cornwall	8,123	5,112	5	..	789	91	2,126	27·3
Cumberland	7,717	5,298	5	..	806	182	1,426	20·8
Derby	14,646	6,939	19	..	1,721	133	5,834	40·7
Devon	16,580	13,505	60	..	1,512	437	1,066	9·1
Dorset	5,097	3,821	11	..	393	57	815	17·1
Durham	38,925	26,103	117	..	4,640	692	7,373	20·7
Essex	27,072	18,680	93	..	2,539	711	5,049	21·3
Gloucester	15,025	8,868	64	3	1,528	555	4,007	30·4
Hereford	3,056	2,577	8	..	245	62	164	7·4
Hertford	7,058	4,914	18	..	543	74	1,509	22·4
Huntingdon	1,262	1,068	3	..	99		92	7·4
Kent (extra-metropolitan)	22,662	17,293	93	..	2,026	486	2,764	14·3
Lancaster	132,496	83,640	412	..	16,492	1,875	30,077	24·1
Leicester	12,503	765	4	..	1,857	14	9,863	79·0
Lincoln	13,169	7,671	39	..	1,577	354	3,528	29·5
Middlesex (ex.-metropolitan)	15,151	10,867	71	..	1,333	577	2,303	19·0
Monmouth	10,332	7,398	26	..	1,151	350	1,407	17·0
Norfolk	12,873	7,274	17	..	1,405	102	4,075	32·4
Northampton	9,503	2,180	3	..	1,077	38	6,205	65·7
Northumberland	17,637	12,366	48	..	2,227	245	2,751	17·0

—	Returns, 1896.							
	Births.	Successfully Vaccinated.	Insusceptible of Vaccination.	Had Small-pox.	Died unvaccinated.	Vaccination postponed.	Remaining.	Children not finally accounted for (including cases, postponed), per cent. of births.
COUNTIES—*cont.*								
ENGLAND—*cont.*								
Nottingham	17,807	8,103	29	..	2,250	265	6,960	41·0
Oxford	4,916	3,459	14	..	473	45	925	19·7
Rutland	491	419	..	..	38	9	25	6·9
Salop	6,885	5,473	11	..	586	107	708	11·8
Somerset	14,159	9,335	37	..	1,272	296	3,219	24·8
Southampton	19,132	15,604	89	..	1,653	343	1,443	9·3
Stafford	34,804	20,103	82	..	4,653	472	9,494	28·6
Suffolk	9,983	7,566	21	..	888	117	1,391	15·1
Surrey (extra-metropolitan)	15,606	10,866	63	..	1,247	392	3,038	22·0
Sussex	13,982	9,791	37	..	1,136	296	2,722	21·6
Warwick	25,948	17,088	106	..	3,491	317	4,946	20·3
Westmorland	1,648	1,442	1	..	125	28	52	4·9
Wilts	6,819	3,660	12	..	483	139	2,535	39·1
Worcester	19,008	15,075	77	..	1,951	186	1,719	10·0
York, East Riding	12,929	9,627	59	..	1,455	157	1,631	13·8
York, North Riding	10,669	8,024	26	..	1,124	185	1,310	14·0
York, West Riding	79,541	52,764	349	..	8,979	754	16,695	21·9
WALES:								
[illegible]	836	730	..	..	74	8	24	3·8
[illegible]	1,521	1,212	..	..	170	36	103	9·1
[illegible]	1,504	1,248	1	..	146	13	96	7·2
[illegible]	4,232	3,678	..	..	382	50	122	4·1
[illegible]	3,321	2,800	3	..	331	54	133	5·6
[illegible]	2,962	2,471	2	..	322	84	103	6·3
[illegible]	2,393	1,995	3	..	217	42	136	7·4
[illegible]	29,282	23,961	63	1	3,347	456	1,454	6·5
[illegible]	1,755	1,456	1	..	220	50	28	4·4
[illegible]	1,633	1,395	3	..	147	29	59	5·4
[illegible] ..	2,208	1,830	1	..	212	61	104	7·5
[illegible] ..	[illegible]	[illegible]	1	..	59	5	205	36·3

METROPOLITAN UNIONS.	Returns, 1896.							
	Births.	Successfully Vaccinated.	Insusceptible of Vaccination.	Had Small-pox.	Died unvaccinated.	Vaccination postponed.	Remaining.	Children not finally accounted for (including cases postponed), per cent. of births.
Bethnal Green	4,776	485	1	..	632	31	3,687	76·8
Camberwell	7,571	3,792	29	..	669	197	2,884	40·7
Chelsea	2,604	2,031	10	..	255	27	281	11·8
Fulham	6,987	5,595	47	..	715	68	562	9·0
George, St., Hanover Square	2,579	2,185	12	..	240	20	122	5·5
George, St., in the East ..	2,060	1,316	3	..	170	66	505	27·7
Giles, St., and St. George ..	1,040	704	2	..	125	6	203	20·1
Greenwich	5,648	4,608	17	..	578	58	387	7·9
Hackney	7,163	1,979	6	..	619	25	4,534	63·6
Hampstead	1,452	1,185	24	..	120	7	116	8·5
Holborn	5,000	3,401	15	..	572	78	934	20·2
Islington	9,752	6,851	47	..	998	195	1,661	19·0
Kensington	3,677	2,842	25	..	357	56	397	12·3
Lambeth	9,439	5,854	34	1	1,041	209	2,300	26·6
Lewisham	2,567	1,638	13	..	248	3	655	25·7
London, City of	534	405	2	..	51	12	64	14·2
Marylebone	4,356	3,417	12	..	381	..	546	12·5
Mile End Old Town	4,237	1,112	3	..	469	..	2,653	62·6
Olave, St.	4,896	3,262	11	..	535	84	1,004	22·2
Paddington	3,042	2,443	28	..	286	11	274	9·4
Pancras, St.	6,932	5,023	32	..	620	262	995	18·1
Poplar	5,926	3,284	19	..	672	63	1,888	32·9
Saviour, St.	7,160	4,814	21	..	855	33	1,437	20·5
Shoreditch	4,322	1,379	9	..	536	3	2,395	55·5
Stepney	2,012	1,063	8	..	222	60	659	35·7
Strand	496	355	..	..	55	7	79	17·3
Wandsworth and Clapham ..	10,059	6,824	72	..	991	447	1,725	21·6
Westminster	767	589	..	..	64	2	112	14·9
Whitechapel	3,201	2,761	11	..	256	9	164	5·4
Woolwich	3,588	3,096	14	..	348	5	125	3·6
	133,838	84,293	527	1	13,670	2,044	33,298	26·4

App. A., No. 1.

Digest of Vaccination Officers' Returns, 1896.

—	Returns, 1896.							
	Births.	Successfully [illegible]d.	In [illegible] of Vaccination.	Had Small-pox.	Died [illegible]ated.	Vaccination [illegible]d.	[illegible]aining.	Children at 1 [illegible]lly accounted for (including cases postponed), per cent. of births.
BEDFORD.								
Ampthill	354	108	..	..	36	..	210	59·3
Bedford	1,320	145	1	..	144	2	1,028	78·0
Biggleswade	772	51	..	..	85	..	636	82·4
Leighton Buzzard	462	19	..	..	34	..	409	88·5
Luton	1,327	23	..	..	214	..	1,090	82·1
Woburn	225	61	..	..	18	1	145	64·4
	4,460	407	1	..	531	3	3,518	78·9
BERKS.								
Abingdon	469	418	..	..	36	4	11	3·2
Bradfield	442	395	1	..	29	4	13	3·8
Easthampstead	365	310	2	..	20	6	27	9·0
Faringdon..	359	279	1	..	24	23	32	15·3
Hungerford and Ramsbury..	439	393	..	..	26	12	8	4·6
Maidenhead	608	431	..	..	56	8	113	19·9
Newbury	531	431	1	..	48	25	26	9·6
Reading	1,826	1,136	8	..	223	48	411	25·1
Wallingford	336	289	..	..	30	3	14	5·1
Wantage	472	390	1	..	45	4	32	7·6
Windsor	911	741	7	..	60	11	92	11·3
Wokingham	449	341	3	..	26	..	79	17·6
	7,207	5,554	24	..	623	148	858	14·0
BUCKINGHAM.								
Amersham	564	217	1	..	52	1	283	51·3
Aylesbury..	647	577	4	..	43	13	10	3·6
Buckingham	301	244	1	..	19	3	34	12·3
Eton..	786	666	4	..	63	7	46	6·7
Newport Pagnell	744	520	2	..	64	9	149	21·2
Winslow	209	26	..	..	21	..	162	77·5
Wycombe	1,282	530	1	..	119	7	625	49·3
	4,523	2,780	13	..	381	40	1,309	29·8
CAMBRIDGE.								
Cambridge	930	647	4	..	74	12	193	22·0
Caxton and Arrington ..	245	228	..	..	11	..	6	2·4
Chesterton	790	597	..	..	66	6	121	16·1
Ely..	534	374	2	..	44	5	109	21·3
Linton	321	274	..	..	29	2	16	5·6
Newmarket	926	808	..	..	76	12	30	4·5
North Witchford	472	413	..	..	47	2	10	2·5
Whittlesey	218	195	..	..	20	1	2	1·4
Wisbech	842	680	3	..	92	13	54	6·0
	5,278	4,216	9	..	459	53	541	11·3

—	Returns, 1896.							
	Births.	Successfully Vaccinated.	Insusceptible of Vaccination.	Had Small-pox.	Died unvaccinated.	Vaccination postponed.	Remaining.	Children not finally accounted for (including cases postponed), per cent. of births.
CHESTER.								
Birkenhead	4,619	3,998	18	..	511	21	71	2·0
Bucklow	1,817	1,604	8	..	145	28	32	3·3
Chester	1,489	1,271	10	..	145	18	45	4·2
Congleton	977	812	5	..	89	44	27	7·3
Macclesfield	1,530	1,283	2	..	176	36	34	4·6
Nantwich	2,083	1,654	13	..	205	62	149	10·1
Northwich	2,069	1,870	3	..	176	13	38	2·0
Runcorn	1,318	1,105	7	..	126	40	41	6·1
Stockport	4,200	2,954	34	..	618	82	512	14·1
Tarvin	409	367	3	..	35	2	2	1·0
Wirral	1,113	934	2	..	106	15	56	6·4
	21,644	17,851	104	..	2,331	361	997	6·3
CORNWALL.								
Austell, St.	943	679	2	..	82	3	177	19·1
Bodmin	466	391	..	..	36	9	30	8·4
Camelford	164	137	..	..	11	..	16	9·8
Columb Major, St.	409	188	..	..	28	8	185	47·2
Falmouth	581	38	1	..	55	..	487	83·8
Germans, St.	480	415	..	..	35	12	18	6·2
Helston	566	432	..	..	58	29	47	13·4
Launceston	375	319	..	..	28	9	19	7·5
Liskeard	638	445	1	..	51	2	139	22·1
Penzance	1,276	917	..	..	132	7	220	17·8
Redruth	1,260	638	1	..	174	3	444	35·5
Stratton	186	151	..	..	11	2	22	12·9
Truro	779	362	..	..	88	7	322	42·2
	8,123	5,112	5	..	789	91	2,126	27·3
CUMBERLAND.								
Alston-with-Garrigill	62	55	..	..	2	3	2	8·1
Bootle	435	352	..	..	39	25	19	10·1
Brampton	254	188	..	..	27	14	25	15·4
Carlisle	1,664	1,391	2	..	151	77	43	7·2
Cockermouth	2,110	571	..	..	296	10	1,233	58·9
Longtown	183	156	1	..	13	1	12	7·1
Penrith	574	495	1	..	52	13	13	4·5
Whitehaven	1,761	1,488	1	..	176	33	63	5·5
Wigton	674	602	..	..	50	6	16	3·3
	7,717	5,298	5	..	806	182	1,426	20·8
DERBY.								
Ashbourne	576	445	..	..	45	8	78	14·9
Bakewell	863	649	1	..	82	29	102	15·2
Belper	2,231	1,451	1	..	221	10	548	25·0
Chapel-en-le-Frith	661	523	1	..	60	19	58	11·6
Chesterfield	4,556	2,498	9	..	579	38	1,432	32·3
Derby	2,876	310	3	..	406	..	2,257	78·5
Glossop	699	84	..	..	121	1	493	70·7
Hayfield	343	260	1	..	24	1	57	16·9
Shardlow	1,841	819	3	..	183	27	809	45·4
	14,846	6,939	19	..	1,721	133	5,834	40·7

pp. A., No. 1.

igest of
accination
fficers'
eturns, 1896.

	Returns, 1896.							
	Births.	Successfully Vaccinated.	Insusceptible of Vaccination.	Had Small-pox.	Died unvaccinated.	Vaccination postponed.	Remaining.	Children not finally accounted for (including cases postponed), per cent. of births.
DEVON.								
Axminster	372	325	..	..	27	7	13	5·4
Barnstaple	1,066	886	3	..	100	41	36	7·2
Bideford	569	487	2	..	52	16	12	4·9
Crediton	428	386	..	..	27	5	10	3·5
East Stonehouse	491	344	2	..	62	14	69	16·9
Exeter	975	746	9	..	130	30	60	9·2
Holsworthy	251	233	..	..	17	1	..	0·0
Honiton	494	447	2	..	27	5	13	3·6
Kingsbridge	431	378	..	..	40	4	9	3·0
Newton Abbot	1,782	1,585	6	..	129	25	37	3·5
Okehampton	404	342	..	..	32	8	22	7·4
Plymouth	2,610	1,805	14	..	331	51	409	17·6
Plympton St. Mary	710	545	..	..	63	16	86	14·4
South Molton	367	322	1	..	31	7	6	3·5
Stoke Damerel	1,574	1,159	12	..	145	91	167	16·4
Tavistock	636	558	4	..	51	4	19	3·6
Thomas, St.	1,306	1,055	1	..	111	96	43	10·6
Tiverton	728	633	2	..	56	10	27	5·1
Torrington	326	296	..	..	14	2	12	4·3
Totnes	1,060	971	2	..	67	4	16	1·9
	16,580	13,505	60	..	1,512	437	1,066	9·1
DORSET.								
Beaminster	258	227	..	..	19	4	8	4·7
Blandford	298	208	..	..	37	11	42	17·8
Bridport	317	292	3	..	16	3	3	1·9
Cerne	140	128	..	..	9	2	1	2·1
Dorchester	501	424	..	..	35	6	36	8·4
Poole	943	605	4	..	89	7	238	26·0
Shaftesbury	301	246	1	..	23	9	22	10·3
Sherborne	260	217	..	..	20	..	23	8·8
Sturminster	218	186	..	..	26	1	5	2·8
Wareham and Purbeck ..	414	377	..	..	19	5	13	4·3
Weymouth	977	560	..	..	72	..	345	35·3
Wimborne and Cranborne ..	470	351	3	..	28	9	79	18·7
	5,097	3,821	11	..	393	57	815	17·1
DURHAM.								
Auckland	3,181	1,921	4	..	391	59	806	27·2
Chester-le-Street	2,174	1,711	10	..	279	49	125	8·0
Darlington	1,429	322	..	..	187	..	920	64·4
Durham	2,528	1,946	9	..	266	45	262	12·1
Easington	2,073	1,759	4	..	233	28	49	3·7
Gateshead	5,566	2,371	6	..	772	..	2,417	43·4
Hartlepool	2,565	2,045	19	..	233	42	226	10·4
Houghton-le-Spring	1,577	1,319	8	..	191	21	38	3·7
Lanchester	2,612	1,898	11	..	321	34	348	14·6
Sedgefield	794	672	1	..	86	5	30	4·4
South Shields	5,334	3,600	17	..	624	165	928	20·5
Stockton	2,017	1,568	6	..	237	46	160	10·2
Sunderland	6,015	4,307	21	..	699	180	808	16·4
Teesdale	605	291	..	..	80	8	226	38·7
Weardale	455	373	1	..	41	10	30	8·8
	38,925	26,103	117	..	4,640	692	7,373	20·7

—	Returns, 1896.							
	Births.	Successfully Vaccinated.	Insusceptible of Vaccination.	Had Small-pox.	Died unvaccinated.	Vaccination postponed.	Remaining.	Children not finally accounted for (including cases postponed), per cent. of births.
ESSEX.								
Billericay	471	394	4	..	29	14	30	9·3
Braintree	635	529	..	..	63	31	12	6·8
Chelmsford	872	743	2	..	57	17	53	8·0
Colchester	1,003	669	2	..	97	13	222	23·4
Dunmow	359	319	..	..	19	2	19	5·8
Epping	723	619	3	..	55	13	33	6·4
Halstead	395	333	..	..	34	9	19	7·1
Lexden and Winstree ..	593	536	..	..	34	10	13	3·9
Waldon	632	552	1	..	38	5	36	6·5
Ongar	269	223	..	..	24	9	13	8·2
Orsett	906	305	..	..	105	28	468	54·7
Rochford	1,103	652	4	..	107	30	310	30·[illegible]
Romford	2,070	1,386	5	..	212	45	422	22·6
Saffron Walden	409	355	2	..	24	2	26	6·8
Tendring	1,221	1,018	2	..	106	20	75	7·8
West Ham	15,411	10,047	68	..	1,535	463	3,298	24·4
	27,072	18,680	93	..	2,539	711	5,049	21·3
GLOUCESTER.								
Barton Regis	5,872	4,000	17	..	601	475	779	21·4
Bristol	1,450	1,021	1	..	207	2	219	15·2
Cheltenham	1,228	553	5	..	146	..	524	42·7
Chipping Sodbury	468	343	2	..	30	7	86	19·9
Cirencester	509	413	3	..	35	..	58	11·4
Dursley	302	128	..	..	16	5	153	52·3
Gloucester	1,500	695	29	3	167	18	597	40·3
Newent	243	190	1	..	24	7	21	11·5
Northleach	222	185	..	..	19	..	18	8·1
Stow-on-the-Wold	204	152	2	..	16	2	32	16·7
Stroud	958	219	2	..	88	8	641	67·7
Tetbury	186	98	..	..	21	1	66	36·0
Tewkesbury	355	133	2	..	39	2	179	51·0
Thornbury	434	365	..	..	32	7	30	8·5
Westbury-on-Severn ..	693	75	..	..	52	..	566	81·7
Wheatenhurst	153	101	..	..	17	11	24	22·9
Winchcomb	239	197	..	..	18	10	14	10·0
	15,025	8,868	64	3	1,528	555	4,007	30·4
HEREFORD.								
Bromyard	316	250	1	..	22	14	29	13·6
Dore	205	158	1	..	23	9	14	11·2
Hereford	874	757	1	..	70	11	35	5·3
Kington	288	237	2	..	26	7	16	8·0
Ledbury	374	324	1	..	26	4	19	6·1
Leominster	320	276	1	..	24	6	13	5·9
Ross	491	415	1	..	37	10	28	7·7
Weobley	188	160	..	..	17	1	10	5·9
	3,056	2,577	8	..	245	62	164	7·4

APP. A., No. 1.

Digest of Vaccination Officers' Returns, 1896

—	Returns, 1896.							
	Births.	Successfully Vaccinated.	Insusceptible of Vaccination.	Had Small-pox.	Died unvaccinated.	Vaccination postponed.	Remaining.	Children not finally accounted for (including cases postponed), per cent. of births.
HERTS.								
Albans, St.	775	147	..	..	70	4	554	72·0
Barnet	1,282	872	5	..	101	17	287	23·7
Berkhampstead	430	334	..	..	33	7	56	14·7
Bishop Stortford	531	451	..	..	44	5	31	6·8
Buntingford	134	100	..	..	16	..	8	6·5
Hatfield	180	156	..	..	12	2	10	6·7
Hemel Hempstead	361	230	1	..	31	8	91	27·4
Hertford	420	374	3	..	29	3	11	3·3
Hitchin	692	609	4	..	50	1	28	4·2
Royston	439	387	3	..	30	2	17	4·3
Ware	561	481	..	..	36	2	42	7·8
Watford	1,218	733	2	..	88	21	374	32·4
Welwyn	45	40	..	..	3	2	..	4·4
	7,058	4,914	18	..	543	74	1,509	22·4
HUNTINGDON.								
Huntingdon	514	442	2	..	42	1	27	5·4
Ives, St.	396	347	1	..	37	1	10	2·8
Neots, St.	352	277	..	..	20	..	55	15·6
	1,262	1,066	3	..	99	2	92	7·4
KENT (EXTRA METROPOLITAN).								
Ashford, East	347	307	..	..	27	2	11	3·7
Ashford, West	463	346	2	..	39	29	47	16·4
Blean	538	441	1	..	54	5	37	7·8
Bridge	302	260	1	..	17	6	18	7·9
Bromley	1,849	1,238	13	..	164	55	379	23·5
Canterbury	493	378	1	..	42	5	67	14·6
Cranbrook	307	259	1	..	13	2	32	11·1
Dartford	2,317	1,904	13	..	203	51	146	8·5
Dover	1,114	696	..	..	107	38	273	27·9
Eastry	754	671	5	..	52	4	22	3·4
Elham	1,156	942	2	..	102	50	60	9·5
Faversham	796	710	2	..	65	1	18	2·4
Gravesend and Milton	643	462	4	..	61	4	112	18·0
Hollingbourn	321	272	..	..	26	1	22	7·2
Hoo	102	85	2	..	12	..	3	2·9
Maidstone	1,254	859	6	..	139	49	201	19·9
Malling	747	649	1	..	60	6	31	5·0
Medway	2,419	1,895	11	..	197	39	277	13·1
Milton	818	618	2	..	66	29	103	16·1
Romney Marsh	189	170	1	..	13	5	..	2·6
Sevenoaks	739	615	2	..	63	12	47	8·0
Sheppey	488	373	5	..	64	37	9	9·4
Strood	1,293	1,060	4	..	143	3	83	6·7
Tenterden	243	205	1	..	18	5	14	7·8
Thanet, Isle of	1,419	1,136	7	..	132	22	122	10·1
Tonbridge	1,551	742	6	..	147	26	630	42·3
	22,662	17,298	93	..	2,026	486	2,764	14·3

—	Returns, 1896.							
	Births.	Successfully Vaccinated.	Insusceptible of Vaccination.	Had Small-pox.	Died unvaccinated.	Vaccination postponed.	Remaining.	Children not finally accounted for (including cases postponed), per cent. of births.
LANCASTER.								
Ashton-under-Lyne	4,947	886	5	..	743	43	3,271	67·0
Barrow-in-Furness	1,552	1,338	..	..	143	19	52	4·6
Barton-upon-Irwell	2,812	1,828	17	..	311	91	565	23·3
Blackburn	6,214	4,432	34	..	714	138	896	16·6
Bolton	7,679	6,157	15	..	852	34	621	8·5
Burnley	5,990	998	20	..	751	68	4,153	70·5
Bury	3,641	792	9	..	522	6	2,312	63·7
Chorley	1,848	1,349	4	..	254	29	212	13·0
Chorlton	9,679	6,396	15	..	1,276	105	1,867	20·6
Clitheroe	583	462	1	..	42	19	59	13·4
Fylde, The	2,042	1,370	16	..	225	65	366	21·1
Garstang	262	234	..	..	21	3	4	2·7
Haslingden	3,015	1,612	3	..	461	38	901	31·1
Lancaster	1,556	1,270	5	..	172	38	71	7·0
Leigh	2,886	2,167	5	..	333	11	370	13·2
Liverpool	5,244	4,412	21	..	609	17	185	3·9
Lunesdale	167	99	..	..	11	..	57	34·1
Manchester	5,170	4,285	2	..	647	49	187	4·6
Oldham	6,010	212	..	..	1,025	..	4,773	79·4
Ormskirk	2,782	2,341	17	..	242	48	134	6·5
Prescot	5,334	4,372	16	..	587	51	308	6·7
Preston	4,681	2,935	15	..	784	104	843	20·2
Prestwich	5,398	3,823	15	..	524	103	933	19·2
Rochdale	2,847	192	3	..	401	..	2,251	79·1
Salford	7,626	4,791	43	..	1,159	265	1,368	21·4
Todmorden	909	82	4	..	97	..	726	79·9
Toxteth Park	4,184	3,161	33	..	468	217	305	12·5
Ulverston	1,063	949	1	..	82	12	19	2·9
Warrington	3,278	2,592	21	..	395	23	247	8·2
West Derby	16,152	13,168	48	..	1,765	216	955	7·2
Wigan	6,945	4,936	24	..	876	63	1,046	16·0
	132,496	83,640	412	..	16,492	1,875	30,077	24·1
LEICESTER.								
Ashby-de-la-Zouch	1,439	222	..	..	190	6	1,021	71·4
Barrow-on-Soar	793	12	..	..	89	1	691	87·3
Billesdon	127	84	..	..	10	1	32	26·0
Blaby	786	23	..	..	107	..	656	83·5
Hinckley	792	27	..	..	112	..	653	82·4
Leicester	6,105	52	1	..	1,041	..	5,011	82·1
Loughborough	1,022	48	1	..	144	..	829	81·1
Lutterworth	297	103	..	..	24	..	170	57·2
Market Bosworth	534	152	..	..	48	6	328	62·5
Market Harborough*	..	..	..	..	..	..	..	..
Melton Mowbray	608	42	2	..	92	..	472	77·6
	12,503	765	4	..	1,857	14	9,863	79·0
LINCOLN.								
Boston	1,074	515	2	..	134	25	398	39·4
Bourne	467	320	..	..	42	4	101	22·5
Caistor	399	310	..	..	43	13	33	11·5
Gainsborough	973	152	1	..	124	4	692	71·5
Glanford Brigg	1,330	532	2	..	131	1	664	50·0
Grantham	840	663	5	..	86	9	77	10·2

* *No return relating to children born in the Market Harborough* Union during the Year *1896 has been received. The number of births registered was 455.*

App. A., No. 1.

Digest of Vaccination Officers' Returns, 1896.

	Returns, 1896.							
	Births.	Successfully Vaccinated.	Insusceptible of Vaccination.	Had Small-pox.	Died unvaccinated.	Vaccination postponed.	Remaining.	Children not finally accounted for (including cases postponed), per cent. of births.
LINCOLN—*cont.*								
Grimsby	2,367	1,422	12	..	361	195	377	24·2
Holbeach	465	110	..	..	71	..	284	61·1
Horncastle	474	305	2	..	40	5	122	26·8
Lincoln	1,905	1,470	4	..	253	65	113	9·3
Louth	733	553	1	..	68	22	89	15·1
Sleaford	641	386	2	..	71	4	178	28·4
Spalding	554	122	..	..	66	3	363	66·1
Spilsby	561	483	3	..	48	4	23	4·8
Stamford	386	328	5	..	39	..	14	3·6
	13,169	7,671	39	..	1,577	354	3,528	29·5
MIDDLESEX (Extra Metropolitan).								
Brentford	4,102	3,328	39	..	379	19	337	8·7
Edmonton	8,097	5,144	16	..	733	518	1,686	27·2
Hendon	1,185	961	16	..	102	23	83	8·9
Staines	866	659	..	..	56	11	140	17·4
Uxbridge	901	775	..	..	63	6	57	7·0
	15,151	10,867	71	..	1,333	577	2,303	19·0
MONMOUTH.								
Abergavenny	779	610	1	..	91	9	68	9·9
Bedwellty	2,987	1,674	..	..	387	204	722	31·0
Chepstow	564	410	4	..	51	10	89	17·6
Monmouth	873	566	2	..	85	15	205	25·2
Newport	3,707	3,061	16	..	406	57	167	6·0
Pontypool	1,422	1,077	3	..	131	55	156	14·8
	10,332	7,398	26	..	1,151	350	1,407	17·0
NORFOLK.								
Aylsham	486	410	3	..	52	..	21	4·3
Blofield	299	225	..	..	38	6	30	12·0
Depwade	551	473	..	..	32	6	40	8·3
Docking	415	362	..	..	34	2	17	4·6
Downham	476	417	1	..	34	3	21	5·0
Erpingham	597	456	2	..	44	12	83	15·9
Faith, St.	332	172	..	..	41	4	115	35·8
Flegg, East and West	294	241	2	..	36	5	10	5·1
Forehoe	317	238	..	..	23	3	53	17·7
Freebridge Lynn	329	271	..	..	34	4	20	7·3
Guiltcross	251	218	..	..	18	1	14	6·0
Henstead	280	223	..	..	17	..	40	14·3
King's Lynn	570	27	..	..	55	..	488	85·6
Loddon and Clavering ..	341	280	..	..	32	4	25	8·5
Mitford and Launditch ..	612	491	..	..	41	6	74	13·1
Norwich	3,353	233	1	..	528	..	2,591	77·3
Smallburgh	460	341	..	..	40	5	74	17·2
Swaffham	302	225	..	..	23	5	49	17·9
Thetford	415	350	1	..	39	2	23	6·0
Walsingham	511	411	1	..	47	11	41	10·2
Wayland	259	217	..	..	28	4	10	5·4
Yarmouth, Great	1,423	993	6	..	169	19	236	17·9
	12,873	7,274	17	..	1,405	102	4,075	32·4

App. A., No. —

Digest of Vaccination Officers' Returns.

—	Returns, 1896.							
	Births	Successfully Vaccinated.	Insusceptible of Vaccination.	Had Small-pox.	Died unvaccinated.	Vaccination postponed.	Remaining.	Children not finally accounted for (including cases postponed), per cent. of births.
NORTHAMPTON.								
Brackley	299	167	..	..	45	2	85	29·1
Brixworth	283	99	1	..	33	..	150	53·0
Daventry	470	188	..	..	46	4	232	50·2
Hardingstone	316	55	..	..	28	..	233	73·7
Kettering	1,384	17	..	..	198	..	1,169	84·5
Northampton	2,566	90	..	..	339	4	2,133	83·3
Oundle	281	226	..	..	22	4	29	11·7
Peterborough	1,262	1,049	2	..	100	18	93	8·8
Potterspury	325	174	..	..	18	6	127	40·9
Thrapston..	389	35	..	..	37	..	317	81·5
Towcester..	320	64	..	..	34	..	222	69·4
Wellingborough..	1,608	16	..	..	177	..	1,415	88·0
	9,503	2,180	3	..	1,077	38	6,205	65·7
NORTHUMBERLAND.								
Alnwick	651	512	..	..	62	1	76	11·8
Belford	122	109	..	..	12	..	1	0·8
Bellingham	127	110	1	..	6	..	10	7·9
Berwick-upon Tweed ..	555	467	1	..	66	11	10	3·8
Castle Ward	733	574	2	..	57	9	91	13·6
Glendale	214	190	..	..	17	1	6	3·3
Haltwhistle	230	78	..	..	20	..	132	57·4
Hexham	896	729	1	..	75	13	78	10·2
Morpeth	1,810	839	..	..	406	8	557	31·2
Newcastle-on-Tyne	7,064	5,297	27	..	836	50	854	12·8
Rothbury	147	133	..	..	11	2	1	2·0
Tynemouth	5,088	3,328	16	..	659	150	935	21·3
	17,637	12,366	48	..	2,227	245	2,751	17·0
NOTTINGHAM.								
Basford	6,005	2,534	9	..	848	60	2,554	43·5
Bingham	350	276	1	..	30	21	22	12·3
East Retford	691	569	1	..	54	10	57	9·7
Mansfield	2,710	1,248	4	..	322	17	1,119	41·9
Newark	777	630	3	..	66	11	67	10·0
Nottingham	5,359	1,465	6	..	776	99	3,013	58·1
Southwell	505	383	4	..	38	31	49	15·8
Worksop	1,210	998	1	..	116	16	79	7·9
	17,607	8,103	29	..	2,250	265	6,960	41·0
OXFORD.								
Banbury	736	146	1	..	89	2	498	67·9
Bicester	347	275	1	..	27	4	40	12·7
Chipping Norton	453	199	1	..	54	6	193	43·9
Headington	920	802	4	..	81	7	26	3·6
Henley	611	493	3	..	54	7	54	10·0
Oxford	631	519	3	..	65	5	39	7·0
Thame	327	266	1	..	29	3	28	9·5
Witney	528	447	..	..	50	9	22	5·9
Woodstock	363	312	..	..	24	2	25	7·4
	4,916	3,459	14	..	473	45	925	19·7

A., No. 1.
of
1898.

	Returns, 1896.							
	Births.	Successfully Vaccinated.	Insusceptible of Vaccination	Had Small-pox.	Died unvaccinated.	Vaccination postponed.	Remaining.	Children not finally accounted for (including cases postponed), per cent. of births.
RUTLAND.								
Oakham	262	222	..	..	24	8	8	6·1
Uppingham	229	197	..	..	14	1	17	7·9
	491	419	..	..	38	9	25	6·9
SALOP.								
Atcham	1,314	1,058	1	..	106	40	129	12·9
Bridgnorth	438	361	..	..	39	2	24	6·1
Church Stretton	128	118	..	..	10	..	..	0·9
Cleobury Mortimer	238	194	..	..	21	6	7	5·7
Clun*	141	109	1	..	11	5	15	14·2
Drayton	389	349	..	..	35	2	4	1·5
Ellesmere	389	349	1	..	25	1	11	3·1
Ludlow	471	405	..	..	25	5	33	8·1
Madeley	662	567	..	..	40	18	27	6·9
Newport	348	295	..	..	33	7	13	5·7
Oswestry	787	694	5	..	69	11	18	3·7
Shifnal	298	234	1	..	23	8	32	13·4
Wellington	733	249	1	..	106	1	376	51·4
Wem	249	247	1	..	8	1	2	1·2
Whitchurch	320	276	..	..	27	..	17	5·3
	6,086	5,473	11	..	586	107	708	11·8
SOMERSET.								
Axbridge	1,040	338	4	..	98	15	585	57·7
Bath	1,817	1,302	1	..	191	117	206	17·8
Bedminster	2,667	1,398	12	..	373	32	851	37·0
Bridgwater	951	573	1	..	94	35	248	29·8
Chard	694	626	1	..	41	10	16	3·7
Clutton	750	540	..	..	47	15	148	21·7
Dulverton	139	118	1	..	8	1	1	1·6
Frome	547	404	2	..	39	1	101	18·6
Keynsham	1,070	520	..	..	108	20	422	41·3
Langport	363	276	..	..	39	3	43	13·2
Shepton Mallet	399	259	1	..	30	..	109	27·3
Taunton	1,130	955	6	..	107	15	48	5·6
Wellington	469	382	3	..	40	10	34	9·4
Wells	610	460	1	..	43	4	102	17·4
Williton	383	334	..	..	27	10	12	5·7
Wincanton	416	342	4	..	31	4	34	9·4
Yeovil	745	517	1	..	66	12	149	21·6
	14,159	9,336	37	..	1,372	396	3,219	24·8
SOUTHAMPTON.								
Alresford	153	139	..	..	5	1	7	5·3
Alton	394	321	..	..	33	3	37	7·6
Alverstoke	832	737	2	..	65	3	15	2·2
Andover	416	325	2	..	31	15	43	13·9
Basingstoke	511	327	2	..	32	5	145	29·4
Catherington	76	55	..	..	15	..	6	7·9
Christchurch	1,361	815	9	..	137	48	352	29·6
Droxford	268	241	..	..	20	1	6	2·6
Fareham	491	446	3	..	31	2	9	2·2
Fordingbridge	171	151	..	..	10	..	10	5·8

* No return relating to children born in the Clun District of the Clun Union during the Year 1896 has been received. The number of births registered was 106.

APP. A., No. 1.

Digest of Vaccination Officers' Returns, 1896.

—	Returns, 1896.							
	Births.	Successfully Vaccinated.	Insusceptible of Vaccination.	Had Small-Pox.	Died unvaccinated.	Vaccination postponed.	Remaining.	Children not finally accounted for (including cases postponed), per cent. of births.
SOUTHAMPTON—*cont.*								
Hartley Wintney	713	547	2	..	57	18	89	15·0
Havant	261	232	2	..	17	4	6	3·8
Hursley	99	91	..	..	4	2	2	4·0
Kingsclere..	243	223	3	..	9	1	7	3·3
Lymington	344	286	..	..	21	3	34	10·8
New Forest	382	320	1	..	19	8	34	11·0
Petersfield..	276	236	..	..	18	3	19	8·0
Portsea Island	4,920	4,329	25	..	476	31	59	1·8
Ringwood..	164	147	..	..	13	..	4	2·4
Romsey	255	227	..	..	14	3	11	5·5
Southampton	1,912	1,558	18	..	230	16	90	5·5
South Stoneham	1,895	1,437	9	..	184	56	209	14·0
Stockbridge	156	132	..	..	13	4	7	7·1
Whitchurch	169	140	1	..	11	8	9	10·1
Wight, Isle of	1,949	1,525	9	..	132	83	200	14·5
Winchester, New	752	627	1	..	66	15	43	7·7
	19,132	15,614	89	..	1,653	333	1,443	9·3
STAFFORD.								
Burton-on-Trent	2,795	1,162	..	..	350	3	1,280	45·9
Cannock	1,458	1,035	2	..	159	7	255	18·0
Cheadle	753	604	..	..	79	5	65	9·3
Leek	1,239	955	..	..	130	85	69	12·4
Lichfield	1,250	1,058	12	..	121	20	39	4·7
Newcastle-under-Lyme ..	1,355	1,219	2	..	118	1	15	1·2
Seisdon	471	348	3	..	36	3	81	17·8
Stafford	887	745	2	..	73	19	48	7·6
Stoke-upon-Trent	5,115	3,884	9	..	750	63	409	9·2
Stone	657	551	1	..	66	12	27	5·9
Tamworth	701	451	2	..	61	15	172	26·7
Uttoxeter	385	292	..	..	37	7	49	14·5
Walsall	3,954	1,826	19	..	649	33	1,427	36·9
West Bromwich	5,157	1,353	16	..	725	86	2,977	59·4
Wolstanton and Burslem ..	3,540	2,726	..	..	507	26	281	8·7
Wolverhampton	5,067	1,894	14	..	792	67	2,300	46·9
	34,804	20,103	82	..	4,653	472	9,494	28·6
SUFFOLK.								
Blything	700	642	2	..	41	3	12	2·1
Bosmere and Claydon ..	397	339	..	..	27	6	25	7·8
Bury St. Edmunds	419	347	..	..	39	19	14	7·9
Cosford	433	381	..	..	39	5	8	3·0
Hartismere	365	305	1	..	29	2	28	8·2
Hoxne	322	297	1	..	17	..	7	2·2
Ipswich	1,794	552	7	..	242	6	987	55·4
Mildenhall	224	210	..	..	10	..	4	1·8
Mitford and Lothingland ..	1,245	1,001	4	..	100	26	114	11·2
Plomesgate	543	453	..	..	49	6	35	7·6
Risbridge	455	382	1	..	49	6	17	5·1
Samford	323	279	..	..	26	3	14	5·3
Stow	549	489	..	..	48	3	9	2·2
Sudbury	744	614	..	..	66	18	46	8·6
Thingoe	413	386	..	..	19	1	7	1·9
Wangford	409	359	..	..	41	1	8	2·2
Woodbridge	649	530	5	..	46	12	56	10·5
	9,963	7,568	21	..	888	117	1,391	15·1

APP. A., No. 1.

Digest of Vaccination Officers' Returns, 1896.

—	Births.	Successfully Vaccinated.	Insusceptible of Vaccination.	Had Small-Pox.	Died unvaccinated.	Vaccination postponed.	Remaining.	Children not finally accounted for (including cases postponed), per cent. of births.
	RETURNS, 1896.							
SURREY (EXTRA METROPOLITAN).								
Chertsey	836	538	1	..	53	6	238	29·2
Croydon	4,146	2,178	13	..	368	189	1,398	38·3
Dorking	384	280	..	..	19	2	83	22·1
Epsom	1,153	826	7	..	102	23	195	18·9
Farnham	1,576	1,253	3	..	125	6	189	12·4
Godstone	469	347	1	..	27	7	87	20·0
Guildford	1,473	1,028	5	..	128	43	269	21·2
Hambledon	527	471	..	..	30	3	23	4·9
Kingston	3,103	2,441	28	..	244	72	318	12·6
Reigate	868	570	2	..	65	33	198	26·6
Richmond	1,071	934	3	..	86	8	40	4·5
	15,606	10,866	63	..	1,247	392	3,038	22·0
SUSSEX.								
Battle	495	288	..	..	48	8	151	32·1
Brighton	2,622	2,109	4	..	251	111	147	9·8
Chailey	224	116	..	..	19	10	79	39·7
Chichester	254	213	1	..	22	8	10	7·1
Cuckfield	541	458	6	..	29	10	38	8·9
Eastbourne	1,109	208	..	..	108	4	789	71·5
East Grinstead	525	189	..	..	35	2	299	57·3
East Preston	854	709	2	..	65	33	45	9·1
Hailsham	411	302	..	..	29	19	61	19·5
Hastings	1,428	1,000	7	..	129	3	278	19·7
Horsham	611	508	4	..	42	11	51	10·1
Lewes	258	52	..	..	31	..	175	67·8
Midhurst	336	299	1	..	22	7	7	4·2
Newhaven	284	222	..	..	22	1	39	14·1
Petworth	244	220	..	..	18	1	5	2·5
Rye	287	215	1	..	31	1	39	13·9
Steyning	1,559	1,344	9	..	100	27	79	6·8
Thakeham	185	167	1	..	11	3	3	3·2
Ticehurst	402	320	1	..	27	5	49	13·4
Uckfield	605	232	..	..	46	8	319	54·0
Westbourne	179	161	..	..	11	..	7	3·9
West Firle	60	32	..	..	8	4	16	33·3
Westhampnett	511	423	..	..	32	20	36	11·0
	13,962	9,791	37	..	1,136	296	2,722	21·6
WARWICK.								
Alcester	551	457	1	..	53	9	31	7·3
Aston	9,438	6,636	55	..	1,387	169	1,191	14·4
Atherstone	576	291	..	..	85	..	200	34·7
Birmingham	8,160	6,319	34	..	1,175	55	577	7·7
Coventry	1,659	337	7	..	242	..	1,073	64·7
Foleshill	744	309	..	..	77	..	358	48·1
Meriden	284	220	..	..	30	5	29	12·0
Nuneaton	821	65	..	..	112	..	644	78·4
Rugby	804	151	1	..	85	8	559	70·5
Solihull	893	677	3	..	75	29	109	15·5
Southam	282	220	..	..	18	7	37	15·6
Stratford-on-Avon	522	402	..	..	48	1	71	13·8
Warwick	1,214	1,004	5	..	104	34	67	8·3
	25,948	17,068	106	..	3,491	317	4,946	20·3

—	Returns, 1896.							
	Births.	Successfully Vaccinated.	Insusceptible of Vaccination.	Had Small-Pox.	Died unvaccinated.	Vaccination postponed.	Remaining.	Children not finally accounted for (including cases postponed), per cent. of births.
WESTMORLAND.								
East Ward	358	287	..	..	41	9	21	8·4
Kendal	1,086	981	1	..	65	19	20	3·6
West Ward	204	174	..	..	19	..	11	5·4
	1,648	1,442	1	..	125	28	52	4·9
WILTS.								
Amesbury	165	139	..	..	16	..	10	6·1
Bradford	262	176	1	..	14	2	69	27·1
Calne	186	83	..	..	24	..	80	43·0
Chippenham	584	352	..	..	37	15	180	33·4
Cricklade & Wootton Bassett	343	77	..	..	29	2	235	69·1
Devizes	476	252	..	..	32	20	172	40·3
Highworth and Swindon ..	1,716	335	..	..	102	63	1,216	74·5
Malmesbury	320	198	..	..	29	6	87	29·1
Marlborough	199	182	..	..	13	..	4	2·0
Melksham	429	98	..	..	43	..	288	67·1
Mere	176	155	1	..	8	4	8	6·8
Pewsey	342	306	3	..	20	1	12	3·8
Salisbury	691	589	4	..	47	12	39	7·4
Tisbury	182	150	2	..	15	6	9	8·2
Warminster	285	218	..	..	22	..	45	15·8
Westbury and Whorwellsdown.	224	133	1	..	26	..	64	28·6
Wilton	239	218	..	..	6	8	7	6·3
	6,819	3,660	12	..	483	139	2,525	39·1
WORCESTER.								
Bromsgrove	902	790	5	..	96	6	5	1·2
Droitwich	516	440	2	..	47	3	24	5·2
Dudley	5,568	4,438	16	..	617	107	390	8·9
Evesham	494	388	..	..	22	..	84	17·0
Kidderminster	1,042	898	10	..	99	6	29	3·4
King's Norton	4,284	2,978	26	..	447	14	799	19·1
Martley	452	399	..	..	27	..	26	5·8
Pershore	323	293	..	..	22	1	6	2·2
Shipston-on-Stour	406	325	3	..	36	1	41	10·3
Stourbridge	2,965	2,581	10	..	287	30	57	2·9
Tenbury	196	172	1	..	15	2	6	4·1
Upton-on-Severn	532	448	2	..	54	7	21	5·3
Worcester	1,349	925	2	..	182	9	231	17·8
	19,008	15,075	77	..	1,961	186	1,719	10·0
YORK, East Riding.								
Beverley	766	615	2	..	86	8	55	8·2
Bridlington	496	352	2	..	61	1	80	16·3
Driffield	521	451	3	..	47	2	18	3·8
Howden	345	288	..	..	36	..	21	6·1
Kingston-upon-Hull	2,533	1,697	19	..	310	26	481	20·0
Patrington	216	178	..	..	15	3	20	10·6
Pocklington	388	340	..	..	37	4	7	2·8
Sculcoates	4,906	3,484	13	..	565	55	789	17·2
Skirlaugh	253	218	..	..	20	5	10	5·9
York	2,506	2,004	20	..	278	53	150	8·1
	12,929	9,627	59	..	1,455	157	1,631	13·8

A., No. 1.

of

1896.

—	Returns, 1896.							
	Births.	Successfully Vaccinated.	Insusceptible of Vaccination.	Had Small-Pox.	Died unvaccinated.	Vaccination postponed.	Remaining.	Children not finally accounted for (including cases postponed), per cent. of births.
YORK, North Riding.								
Aysgarth	113	98	..	..	15	..	..	0·0
Bedale	246	210	..	..	29	..	7	2·8
Easingwold	267	242	..	..	14	2	9	4·1
Guisborough	1,281	1,066	4	..	123	4	87	7·1
Helmsley	115	92	2	..	8	..	13	11·3
Kirkby Moorside	130	100	1	..	13	1	15	12·3
Leyburn	167	154	..	..	10	..	3	1·8
Malton	574	460	..	..	59	13	42	9·6
Middlesbrough	4,360	3,487	12	..	533	101	227	7·5
Northallerton	288	261	..	..	19	4	4	2·8
Pickering	301	234	..	..	29	19	19	12·6
Reeth	61	44	..	..	5	..	12	19·7
Richmond	320	240	1	..	34	1	44	14·1
Scarborough	1,260	316	4	..	133	4	803	64·0
Stokesley	278	245	..	..	25	5	3	2·9
Thirsk	330	293	1	..	27	..	9	2·7
Whitby	578	485	1	..	48	31	13	7·6
	10,869	8,034	26	..	1,124	185	1,310	14·0
YORK, West Riding.								
Barnsley	4,300	3,338	42	..	471	26	423	10·4
Bierley, North	3,669	2,223	12	..	403	62	963	28·0
Bradford	5,520	3,049	23	..	663	59	1,726	32·3
Bramley	4,300	3,338	42	..	471	26	423	21·9
Dewsbury	4,574	1,162	2	..	787	74	2,549	57·3
Doncaster	2,512	1,809	11	..	296	37	360	15·8
Ecclesall Bierlow	4,761	4,040	37	..	476	32	176	4·4
Goole	923	744	3	..	96	5	75	8·7
Halifax	4,615	158	3	..	542	..	3,912	84·8
Hemsworth	719	571	2	..	82	8	56	8·9
Holbeck	959	773	5	..	110	33	38	7·4
Huddersfield	3,806	3,316	20	..	357	13	100	3·0
Hunslet	2,548	2,095	11	..	288	54	100	6·0
Keighley	2,194	44	..	..	160	..	1,990	90·7
Knaresborough	751	373	..	..	82	21	275	39·4
Leeds	7,271	6,017	39	..	756	93	366	6·3
Ouseburn, Great	262	236	1	..	18	2	5	2·7
Pateley Bridge	215	164	..	..	14	..	37	17·2
Penistone	465	397	..	..	46	5	17	4·7
Pontefract	2,468	1,979	14	..	313	5	157	6·6
Ripon	388	312	..	..	33	5	38	11·1
Rotherham	3,715	3,064	11	..	370	24	246	7·3
Saddleworth	440	58	..	..	54	1	327	74·5
Sedbergh	97	86	..	..	4	3	4	7·2
Selby	443	385	1	..	37	3	16	4·3
Settle	341	259	..	..	31	3	48	15·0
Sheffield	7,823	6,115	63	..	973	55	629	8·7
Skipton	1,168	189	1	..	140	2	836	71·7
Tadcaster	916	770	..	..	89	7	50	6·2
Thorne	431	290	..	..	54	15	73	20·2
Wakefield	3,638	2,879	13	..	448	37	262	8·2
Wetherby	373	326	..	..	27	6	14	5·4
Wharfedale	1,417	979	3	..	120	18	288	21·6
Wortley	1,527	1,236	2	..	162	20	117	9·0
	79,541	52,764	349	..	8,979	754	16,695	21·9

App. A., No.
Digest of Vaccination Officers' Returns, 1896.

—	Returns, 1896. Births.	Successfully Vaccinated.	Insusceptible of Vaccination.	Had Small-Pox.	Died unvaccinated.	Vaccination postponed.	Remaining.	Children not finally accounted for (including cases postponed), per cent. of births.
ANGLESEY.								
Anglesey	321	284	..	..	31	1	5	1·9
Holyhead	515	446	..	..	43	7	19	5·0
	836	730	..	..	74	8	24	3·8
BRECKNOCK.								
Brecknock	402	350	..	..	36	5	11	4·0
Builth	221	193	..	..	20	3	5	3·6
Crickhowell	645	494	..	..	82	28	41	10·7
Hay	253	175	..	..	32	..	46	18·2
	1,521	1,212	..	..	170	36	103	9·1
CARDIGAN.								
Aberayron	269	240	..	..	22	..	7	2·6
Aberystwith	491	411	1	..	54	11	14	5·1
Cardigan	341	256	..	..	20	1	64	19·1
Lampeter	217	182	..	..	31	1	3	1·8
Tregaron	186	159	..	..	19	..	8	4·3
	1,504	1,248	1	..	146	13	96	7·2
CARMARTHEN.								
Carmarthen	929	837	..	..	71	6	15	2·3
Llandilo Fawr	689	607	..	..	67	4	11	2·2
Llandovery	275	244	..	..	21	3	7	3·6
Llanelly	1,902	1,610	..	..	185	28	79	5·6
Newcastle-in-Emlyn	437	380	..	..	38	9	10	4·3
	4,232	3,678	..	..	382	50	122	4·1
CARNARVON.								
Bangor and Beaumaris ..	1,005	874	2	..	98	13	18	3·1
Carnarvon	1,083	879	..	..	120	27	57	7·8
Conway	743	625	..	..	63	5	50	7·4
Pwllheli	490	422	1	..	50	9	8	3·5
	3,321	2,800	3	..	331	54	133	5·6
DENBIGH.								
Llanrwst	327	263	..	..	36	23	5	8·6
Ruthin	318	286	..	..	26	4	2	1·9
Wrexham	2,337	1,922	2	..	260	57	96	6·5
	2,982	2,471	2	..	322	84	103	6·3

. A., No. 1.

—

...st of ...ination ...ers' ...rns, 1896.

—	Returns, 1896.							
	Births.	Successfully Vaccinated.	Insusceptible of Vaccination.	Had Small-Pox.	Died unvaccinated.	Vaccination postponed.	Remaining.	Children not finally accounted for (including cases postponed), per cent. of births.
FLINT.								
Asaph, St.	709	556	2	..	74	5	72	10·9
Hawarden	488	400	1	..	42	10	35	9·2
Holywell	1,196	1,039	..	..	101	27	29	4·7
	2,393	1,995	3	..	217	42	136	7·4
GLAMORGAN.								
Bridgend and Cowbridge ..	2,061	1,735	3	..	239	28	76	5·0
Cardiff	7,586	6,026	37	..	861	36	626	8·7
Gower	296	240	3	..	17	..	36	12·2
Merthyr Tydfil	4,710	3,802	3	..	567	274	64	7·2
Neath	2,351	2,073	2	..	208	21	47	2·9
Pontardawe	869	760	..	..	94	9	6	1·7
Pontypridd	7,556	6,037	9	..	1,003	88	419	6·7
Swansea	3,833	3,288	6	1	358	..	180	4·7
	29,282	23,961	63	1	3,347	456	1,454	6·5
MERIONETH.								
Bala	136	109	..	..	19	4	4	5·9
Corwen	473	398	..	..	50	16	9	5·3
Dolgelley	357	306	1	..	39	4	5	2·5
Festiniog	789	641	..	..	112	26	10	4·6
	1,755	1,456	1	..	220	50	28	4·4
MONTGOMERY.								
Forden	394	339	1	..	35	10	9	4·8
Llanfyllin	440	386	..	..	42	7	5	2·7
Machynlleth	260	234	..	..	17	3	6	3·5
Newtown and Llanidloes ..	539	436	2	..	53	9	39	8·9
	1,633	1,395	3	..	147	29	59	5·4
PEMBROKE.								
Haverfordwest	919	707	1	..	105	30	76	11·5
Narberth	450	403	..	..	32	3	12	3·3
Pembroke	839	720	..	..	75	28	16	5·2
	2,208	1,830	1	..	212	61	104	7·5
RADNOR.								
Knighton	332	124	1	..	38	5	164	50·9
Rhayader	247	185	..	..	21	..	41	16·6
	579	309	1	..	59	5	205	36·3

No. 2.

Inspection of Public Vaccination.

List (alphabetically arranged) of 286 Unions inspected during the Year 1898, with reference to the Proceedings under the Vaccination Acts, 1867 and 1871, and an Account of the Awards certified by the Board as payable to the respective Public Vaccinators out of County Funds.

UNION.	No. of Vaccination Districts in the Union.	No. of Public Vaccinators recommended for Award.	Range of Awards in each Union.		Total Sum awarded in each Union.	Medical Inspector.
			Minimum.	Maximum.		
			£ s. d.	£ s. d.	£ s. d.	
Aberaeron	2	2	6 9 0	9 14 0	16 3 0	Dr. Mair.
Abergavenny	3	3	4 13 0	47 16 0	69 2 0	„ Mivart.
Aberystwith	4	3	8 14 0	16 8 0	34 9 0	„ Mair.
Abingdon	6	4	2 17 0	9 4 0	23 0 0	„ Bulstrode.
Albans, St.	5	4	0 4 0	3 10 0	6 6 0	„ Thomson.
Alcester	5	4	6 19 0	11 0 0	36 16 0	„ Johnstone.
Alresford	2	2	5 19 0	7 2 0	13 1 0	Mr. Royle.
Alston-with-Garrigill	2	2	2 1 0	2 1 0	4 2 0	Dr. Fletcher.
Alton	4	4	4 13 0	13 0 0	32 16 0	Mr. Royle.
Andover	5	4	1 17 0	14 11 0	27 5 0	Do.
Anglesey	3	2	6 4 0	8 5 0	14 9 0	Dr. Wheaton.
Aston	7	5	7 12 0	100 13 0	208 2 0	„ Johnstone.
Atherstone	2	2	7 18 0	20 19 0	28 17 0	Do.
Auckland	9	6	8 16 0	21 12 0	84 12 0	Dr. Mivart.
Axminster	11	5	1 9 0	3 14 0	14 11 0	Mr. Royle.
Aylsham	7	7	3 11 0	6 14 0	32 9 0	Dr. Copeman.
Banbury	6	2	2 5 0	2 14 0	4 19 0	Dr. Bulstode.
Barnet	5	5	3 6 0	12 16 0	41 11 0	„ Thomson.
Barnsley	6	3	9 1 0	28 10 0	52 0 0	„ Reece.
Barnstaple	10	6	2 3 0	14 0 0	42 9 0	Mr. Royle.
Basingstoke	7	6	0 11 0	7 4 0	22 17 0	Do.
Battle	7	5	0 17 0	7 9	18 2 0	Dr. Bruce Low.
Bedwellty	7	4	16 12 0	63 3 0	123 9	„ Mivart.
Berkhampstead ..	3	3	4 8 0	10 9 0	20 13 0	„ Thomson.
Bethnal Green ..	2	1	..	..	6 11 9	„ Sweeting.

UNION.	No. of Vaccination Districts in the Union.	No. of Public Vaccinators recommended for Award.	Range of Awards in each Union.		Total Sum awarded in each Union.	Medical Inspector
			Minimum.	Maximum.		
			£ s. d.	£ s. d.	£ s. d.	
Bicester	7	5	0 16	10 0 0	22 5 0	Dr. Bulstrode.
Bideford	6	6	0 12	15 14 0	30 19 0	Mr. Boyle.
Billericay	4	4	1 16	14 16 0	27 1 0	Dr. Mivart.
Bishops Stortford ..	7	6	1 12	7 3 0	21 19 0	" Thomson.
Blofield	3	3	1 2	9 19 0	16 14 0	" Copeman.
Bolton	9	9	7 14	45 8 0	166 19 0	" Fletcher.
Bootle	3	3	2 16	6 17 0	13 0 0	Do.
Boston	8	5	3 16	12 0 0	31 14 0	" Mair.
Bourne	7	4	3 15	7 18 0	21 9 0	Do.
Brackley	4	4	0 19	3 2 0	8 6 0	" Buchanan.
Bradfield	5	5	2 19	7 11 0	27 14 0	" Bulstrode.
Braintree	7	5	5 5 0/6	14 1 0	46 10 0	" Mivart.
Brampton	1	1	..	..	8 16 0	" Fletcher.
Brentford	8	7	6 1	26 0 0	94 1 0	" Sweeting.
Brixworth	5	3	0 18	2 7 0	4 3 0	" Buchanan.
Bromsgrove	5	3	0 18	17 13 0	21 9 0	" Wheaton.
Bucklow	6	5	2 7	25 14 0	50 7 0	Do.
Buntingford	2	2	4 1 0/6	4 13 0	8 14 0	" Thomson.
Burnley	7	..	..	..	..	" Fletcher.
Calne	1	1	..	..	7 15 0	Dr. Johnstone.
Camberwell	5	4	22 16	44 7 0	146 11 0	" Sweeting.
Cardiff	9	8	2 6	155 16 0	458 18 0	" Bruce Low.
Carlisle	5	5	1 18	41 6 0	56 18 0	" Fletcher.
Chelmsford	9	8	2 14	10 4 0	38 5 0	" Mivart.
Chepstow	5	4	4 3	10 0 0	27 15 0	Do.
Chertsey	6	6	0 15	9 0 0	35 11 0	" Sweeting.
Chester	3	3	0 17	36 18 0	40 8 0	" Wheaton.
Chesterfield	15	11	4 1	24 15 0	153 0 0	" Buchanan.
Chester-le-Street ..	4	3	14 15 0/6	38 10 0	74 11 0	" Mivart.
Chichester	1	1	..	..	28 6 0	" Reece.
Chippenham	6	3	2 12	11 12 0	19 19 0	" Johnstone.
Chipping Norton ..	4	3	2 6	8 18 0	14 16 0	" Bulstrode.
Chorley	5	4	7 11	20 0 0	47 2 0	" Fletcher.
Clitheroe	5	5	1 4	8 1 0	20 14 0	Do.
Cockermouth	5	4	0 18 0/0	14 8 0	27 8 0	Do.

	No. of Vaccin. Districts in the Union.	No. of Public Vaccinators recommended for Award.	Range of Awards in each Union. Minimum.	Maximum.	Total Sum awarded in each Union.	Medical Inspector.
			£ s. d.	£ s. d.	£ s. d.	
Colchester	1	1	..	..	42 1 0	Dr. Mivart.
Congleton	3	2	4 11 0	22 10 0	27 1 0	„ Wheaton.
Coventry	1	1	..	..	1 0 0	„ Johnstone.
Crediton	11	8	1 6 0	10 9 0	37 5 0	Mr. Royle.
Cricklade and Wootton Bassett.	3	..	..	..	..	Dr. Johnstone.
Croydon..	6	3	1 9 0	25 19 0	42 16 0	„ Sweeting.
Cuckfield	6	5	1 19 0	16 0 0	39 0 0	„ Reece.
Darlington	5	5	1 15 0	10 9 0	19 5 0	Dr. Mivart.
Daventry	6	4	0 5 0	4 12 0	7 10 0	„ Buchanan.
Devizes	7	1	..	..	4 10 0	„ Johnstone.
Dewsbury	13	9	0 4 0	11 15 0	42 17 0	„ Reece.
Docking..	4	3	4 0 0	11 6 0	19 19 0	„ Copeman.
Doncaster	7	5	2 2 0	21 3 0	35 11 0	„ Reece.
Dorking..	4	3	0 14 0	5 15 0	12 0 0	„ Sweeting.
Dudley	8	9	2 15 0	44 14 0	163 5 0	„ Wheaton.
Dunmow	6	3	2 6 0	7 8 0	15 4 0	„ Mivart.
[illegible]am	5	4	15 16 0	36 19 0	102 10 0	Do.
[illegible]asington	6	1	..	..	42 2 0	Dr. Mivart.
[illegible]asthampstead ..	4	3	6 16 0	9 1 0	22 18 0	„ Bulstrode.
[illegible]ast Stonehouse ..	1	1	..	..	8 4 0	Mr. Royle.
East Ward	6	5	2 0 0	3 13 0	15 14 0	Dr. Fletcher.
Ecclesall Bierlow ..	3	3	2 18 0	40 13 0	65 19 0	„ Reece.
Edmonton	16	16	0 4 0	30 14 0	156 14 0	„ Sweeting.
Epsom	9	8	0 3 0	10 4 0	38 14 0	Do.
Erpingham	3	3	4 14 0	8 9 0	21 0 0	„ Copeman.
Evesham	5	5	0 15 0	13 2 0	24 8 0	„ Wheaton.
Faith, St.	5	3	2 5 0	5 16 0	12 2 0	Dr. Copeman.
Farnham	7	6	1 5 0	17 16 0	46 5 0	„ Sweeting.
Faringdon	4	4	1 13 0	7 3 0	18 17 0	„ Bulstrode
Flegg, East and West	4	3	0 14 0	9 2 0	13 4 0	„ Copeman
Foleshill	4	3	1 1 0	14 14 0	21 12	Mr. Royle.

. A. No. 2.
of
ac-
and
to
ac-
1896.

UNION.	No. of Vaccination Districts in the Union.	No. of Public Vaccinators recommended for Award.	Range of Awards in each Union. Minimum.	Range of Awards in each Union. Maximum.	Total Sum awarded in each Union.	Medical Inspector.
			£ s. d.	£ s. d.	£ s. d.	
Forden	4	1	..	..	8 5 0	Dr. Wheaton.
Fulham	3	3	94 4 0	100 4 0	293 13 0	„ Sweeting.
Fylde, The	5	4	2 2 0	14 3 0	28 9 0	„ Fletcher.
Garstang	3	3	3 16 0	7 16 0	17 18 0	Dr. Fletcher.
Gateshead	5	4	9 1 0	29 18 0	67 13 0	„ Mivart.
George's, St.	3	3	2 17 0	48 12 0	81 0 0	„ Sweeting.
Giles, St., and St. George, Bloomsbury.	1	1	..	..	9 10 0	Do.
Godstone	5	4	2 14 0	4 8 0	14 0 0	Do.
Goole	4	2	1 4 0	13 17 0	15 1 0	„ Reece.
Grantham	7	7	2 3 0	11 14 0	41 11 0	„ Mair.
Greenwich	2	2	37 9 0	73 3 0	110 12 0	„ Sweeting.
Guildford	8	8	2 7 0	13 6 0	57 6 0	Do.
Hackney	4	4	7 16 0	15 0 0	43 5 0	Dr. Sweeting.
Halstead	5	5	1 11 0	9 6 0	20 14 0	„ Mivart.
Hambledon	5	5	3 16 0	9 8 0	29 5 0	„ Sweeting.
Hampstead	1	1	..	..	16 1 0	Do.
Hardingstone	3	2	0 14 0	3 5 0	3 19 0	„ Buchanan.
Hartlepool	3	3	6 18 0	46 17 0	87 12 0	„ Mivart.
Hartley Wintney ..	7	7	1 11 0	8 14 0	37 8 0	Mr. Royle.
Haslingden	5	3	7 7 0	22 15 0	39 11 0	Dr. Fletcher.
Hastings	2	2	4 0 0	38 0 0	42 0 0	„ Bruce Low.
Hatfield	3	2	1 14 0	6 14 0	8 0	„ Thomson.
Headington	2	2	10 12 0	16 11 0	27 3 0	„ Bulstrode.
Hemel Hempstead ..	4	3	2 15 0	5 6 0	13 4 0	„ Thomson.
Hemsworth	4	3	1 6 0	18 13 0	29 5 0	„ Reece.
Hendon	5	4	1 7 0	20 3 0	31 1 0	„ Sweeting.
Henley	6	5	2 11 0	12 2 0	30 17 0	„ Bulstrode.
Henstead	5	4	1 2 0	3 12 0	8 13 0	„ Copeman.
Hertford	5	2	5 11 0	12 16 0	18 7 0	„ Thomson.
Highworth and Swindon.	4	1	..	..	5 9 0	„ Johnstone.
Hitchin	5	4	5 2 0	18 1 0	37 1 0	„ Thomson.
Holbeach	6	4	0 4 0	1 13 0	4 9 0	„ Mair.
Holborn	5	5	0 8 0	26 15 0	91 2 0	„ Sweeting.

UNION.	No. of Vaccination Districts in the Union.	No. of Public Vaccinators recommended for Award.	Range of Awards in each Union.		Total Sum awarded in each Union.	Medical Inspector.
			Minimum.	Maximum.		
			£ s. d.	£ s. d.	£ s. d.	
Holsworthy	5	4	2 0 0	10 15 0	18 0	Mr. Royle.
Holyhead	3	3	7 14 0	10 6 0	20 2 0	Dr. Wheaton.
Honiton	13	6	1 6 0	16 6 0	[illegible] 16 0	Mr. Royle.
Horsham	7	6	2 1 0	9 5 0	31 3 0	Dr. Reece.
Houghton-le-Spring ..	3	3	10 5 0	14 9 0	35 9 0	" Mivart.
Huddersfield	16	9	2 10 0	21 6 0	89 17 0	" Reece.
Hungerford and Ramsbury.	5	2	2 10 0	5 5 0	7 15 0	" Bulstrode.
Huntingdon	5	5	4 5 0	9 4 0	31 4 0	" Thomson.
Hursley	1	1	..	..	6 2 0	Mr. Royle.
Ives, St.	5	4	3 0 0	8 11 0	23 1 0	Dr. Thomson.
Kendal	9	8	1 19 0	12 7 0	41 5 0	Dr. Fletcher.
Kensington	3	4	22 10 0	37 4 0	112 14 0	" Sweeting.
Kettering	5	..	..	..	..	" Buchanan.
Kingsbridge	7	6	1 12 0	6 8 0	24 17 0	Mr. Royle.
Kingsclere	3	3	4 19 0	12 15 0	22 16 0	Do.
Kings Norton	7	7	0 18 0	44 9 0	112 4 0	Dr. Wheaton.
Kingston-on-Thames	11	11	1 13 0	30 9 0	104 10 0	" Sweeting.
Knighton	5	4	1 4 0	1 13 0	5 7 0	" Mair.
Lambeth	5	5	10 14 0	59 17 0	214 15 0	Dr. Sweeting.
Lampeter	2	2	8 9 0	9 17 0	18 6 0	" Mair.
Lanchester	4	2	14 19 0	41 2 0	56 1 0	" Mivart.
Leeds	7	6	1 6 0	64 2 0	171 5 0	" Bruce Low.
Leigh	3	3	12 4 0	38 7 0	76 13 0	" Fletcher.
Lexden and Winstree	10	9	1 19 0	11 19 0	49 18 0	" Mivart.
Liverpool	3	3	40 10 0	96 3 0	194 4 0	" Fletcher.
Llanfyllin	5	4	3 17 0	8 17 0	24 18 0	" Wheaton.
Llanrwst	3	3	5 4 0	10 2 0	20 11 0	Do.
Longtown	2	1	..	..	5 0 0	" Fletcher.
Macclesfield	6	4	1 19 0	37 18 0	48 3 0	Dr. Wheaton.
Machynlleth	4	3	4 15 0	6 3 0	16 8 0	Do.
Maidenhead	3	3	1 14 0	14 14 0	29 0 0	" Bulstrode.

A. No. 2.

...ction of ...c Vac-...ion, and ...rds to ...c Vac-...ors, 1898.

UNION.	No. of Vaccination Districts in the Union.	No. of Public Vaccinators recommended for Award.	Range of Awards in each Union.		Total Sum awarded in each Union.	Medical Inspector.
			Minimum.	Maximum.		
			£ s. d.	£ s. d.	£ s. d.	
Maldon	8	8	1 14 0	12 16 0	50 11 0	Dr. Mivart.
Malmesbury	4	1	..	..	5 3 0	" Johnstone.
Manchester	3	3	25 9 0	62 5 0	132 0 0	" Fletcher.
Melksham and Trowbridge.	2	1	..	..	4 4 0	" Johnstone.
Mere	2	2	4 12 0	9 16 0	14 8 0	Do.
Meriden..	3	3	2 8 0	7 13 0	10 1 0	Mr. Royle.
Midhurst	4	4	5 0 0	11 7 0	30 11 0	Dr. Reece.
Monmouth	6	6	2 8 0	26 2 0	59 9 0	" Mivart.
Nantwich	7	6	1 19 0	30 5 0	51 8 0	Dr. Wheaton.
Neot's, St.	6	5	0 14 0	10 3 0	23 13 0	" Thomson.
Newbury	3	3	6 4 0	11 1 0	24 19 0	" Bulstrode.
Newport, Mon. ..	12	5	0 6 0	57 17 0	147 17 0	" Mivart.
Newton Abbot ..	13	9	1 14 0	19 3 0	69 3 0	Mr. Royle.
Newtown and Llanidloes.	4	4	4 1 0	17 12 0	38 15 0	Dr. Wheaton.
Northampton	4	..	..	..	..	" Buchanan.
Northwich	5	5	11 5 0	35 10 0	114 11 0	" Wheaton.
Norwich	2	1	..	..	4 1 0	" Copeman.
Nuneaton	4	2	0 19 0	1 10 0	2 9 0	" Johnstone.
Okehampton	6	5	1 15 0	8 17 0	26 7 0	Mr. Royle.
Olave, St.	3	3	15 17 0	47 11 0	84 16 0	Dr. Sweeting.
Oundle	4	3	0 17 0	7 14 0	12 19 0	" Buchanan.
Oxford	1	1	..	..	51 11 0	" Bulstrode.
Paddington	1	1	..	..	108 14 0	Dr. Sweeting.
Pancras, St.	1	1	..	..	142 18 0	Do.
Penistone	3	3	2 8 0	10 16 0	19 8 0	" Reece.
Penrith	6	6	1 15 0	6 9 0	22 3 0	" Fletcher.
Pershore	5	1	..	..	7 5 0	" Wheaton.
Peterborough	7	6	1 19 0	33 2 0	61 9 0	" Buchanan.
Petworth	4	4	1 17 0	8 0 0	21 8 0	" Reece.
Pewsey	5	3	1 5 0	7 16 0	11 8 0	" Johnstone.
Plymouth	1	1	..	..	29 4 0	Mr. Royle.
Plympton, St. Mary ..	5	3	4 4 0	5 15 0	14 7 0	Do.

UNION.	No. of Vaccination Districts in the Union.	No. of Public Vaccinators recommended for Award.	Range of Awards in each Union.		Total Sum awarded in each Union.	Medical Inspector.
			Minimum.	Maximum.		
			£ s. d.	£ s. d.	£ s. d.	
Pontefract	7	7	2 3 0	35 3 0	107 13 0	Dr. Reece.
Pontypool	4	2	20 5 0	29 8 0	49 13 0	" Mivart.
Poplar	4	3	32 1 0	67 9 0	135 4 0	" Sweeting.
Potterspury	4	3	1 8 0	10 8 0	13 5 0	" Buchanan.
Preston	4	3	8 12 0	49 18 0	90 18 0	" Fletcher.
Reading	1	1	..	..	27 3 0	Dr. Bulstrode.
Reigate	4	4	3 12 0	12 5 0	22 19 0	" Sweeting.
Rhayader	2	1	..	..	10 11 0	" Mair.
Richmond, Surrey ..	3	2	9 2 0	9 16 0	18 18 0	" Sweeting.
Rochford	6	5	0 17 0	15 0 0	32 10 0	" Copeman.
Romsey	5	2	1 5 0	5 4 0	6 9 0	Mr. Royle.
Rotherham	9	8	3 14 0	30 15 0	83 18 0	Dr. Reece.
Royston	5	5	4 10 0	8 16 0	30 19 0	" Thomson.
Rugby	8	5	0 5 0	2 3 0	6 0 0	" Johnstone.
Runcorn	4	4	5 0 0	27 10 0	53 11 0	" Wheaton.
Ruthin	3	2	1 18 0	6 17 0	8 15 0	Do.
Rye	3	2	5 10 0	14 3 0	19 13 0	" Bruce Low.
Saddleworth	1	1	..	..	1 19 0	Dr. Reece.
Saffron Walden ..	7	2	1 9 0	5 16 0	7 5 0	" Mivart.
Salford	5	5	8 10 0	38 6 0	91 15 0	" Bruce Low.
Salisbury	5	2	9 5 0	33 2 0	42 7 0	" Johnstone.
Saviour, St.	5	4	27 6 0	50 5 0	160 3 0	" Sweeting.
Sedgefield	4	3	12 11 0	20 3 0	50 7 0	" Mivart.
Selby	4	3	0 18 0	14 7 0	20 18 0	" Reece.
Sheffield..	6	5	19 7 0	48 7 0	172 8 0	" Bruce Low.
Shipston-on-Stour ..	6	4	2 0 0	3 16 0	12 2 0	" Wheaton.
Shoreditch	3	3	4 15 0	11 0 0	24 1 0	" Sweeting.
Sleaford	7	6	1 4 0	9 14 0	31 0 0	" Mair.
Smallburgh	4	4	3 5 0	5 16 0	17 18 0	" Copeman.
Solihull	4	4	5 7 0	18 17 0	40 19 0	" Johnstone.
Southam	4	2	1 5 0	9 14 0	10 19 0	" Buchanan.
South Molton	8	4	2 8 0	12 5 0	26 5 0	Mr. Royle.

No. 2 of ac- and to ac- 1898.

UNION.	No. of Vaccination Districts in the Union.	No. of Public Vaccinators recommended for Award.	Range of Awards in each Union.		Total Sum awarded in each Union.	Medical Inspector.
			Minimum.	Maximum.		
			£ s. d.	£ s. d.	£ s. d.	
South Shields ..	5	3	25 6 0	54 9 0	118 3 0	Dr. Mivart.
Spalding	7	3	1 9 0	4 2 0	8 11 0	„ Mair.
Staines	6	6	1 18 0	9 14 0	36 1 0	„ Sweeting.
Stamford	7	4	0 7 0	10 8 0	16 9 0	„ Thomson.
Stepney	1	1	..	..	43 16 0	„ Sweeting.
Steyning	6	4	5 9 0	40 10 0	62 0 0	„ Reece.
Stockbridge	2	2	8 0 0	8 11 0	16 11 0	Mr. Royle.
Stockport	6	6	2 19 0	23 1 0	64 18 0	Dr. Wheaton.
Stockton	3	3	11 19 0	22 8 0	50 15 0	„ Mivart.
Stoke Damerel ..	2	2	9 12 0	12 19 0	22 11 0	Mr. Royle.
Stourbridge	7	5	5 4 0	29 4 0	91 16 0	Dr. Wheaton.
Strand	1	1	..	..	9 9 0	„ Sweeting.
Stratford-on-Avon ..	7	7	2 1 0	14 3 0	37 13 0	„ Buchanan.
Sunderland	6	6	10 10 0	22 4 0	89 17 0	„ Mivart.
Tadcaster	5	4	0 14 0	16 6 0	40 13 0	Dr. Reece.
Tarvin	4	4	0 5 0	12 8 0	36 13 0	„ Wheaton.
Tavistock	8	6	2 15 0	12 7 0	33 1	Mr. Royle.
Teesdale..	5	4	2 0	9 9 0	16 13 0	Dr. Mivart.
Tenbury	2	1	..	..	6 17 0	„ Wheaton.
Tendring	8	3	7 4 0	20 3 0	40 5 0	„ Copeman.
Thakeham	2	2	9 3 0	9 12 0	18 15 0	„ Reece.
Thame	6	5	0 12 0	10 15 0	30 1 0	„ Bulstrode.
Thomas, St.	15	14	0 14 0	24 16 0	84 19 0	Mr. Royle.
Thorne	6	6	1 12 0	5 17 0	19 9 0	Dr. Reece.
Thrapston	5	2	0 10 0	0 19 0	1 9 0	„ Buchanan.
Tisbury	3	3	4 6 0	6 13	17 2 0	„ Johnstone.
Tiverton	13	9	0 6 0	12 7 0	40 12 0	Mr. Royle.
Torrington	5	5	3 1 0	10 5 0	26 9 0	Do.
Totnes	13	11	0 14 0	19 19	68 1 0	Do.
Towcester	4	4	0 9 0	1 13 0	5 0	Dr. Buchanan.
Toxteth Park	2	2	24 5 0	25 9	49 14 0	„ Bruce Low.
Tregaron	2	1	..	..	12 9 0	„ Mair.
Uxbridge		6	1	8 16	9	Dr. Sweeting.

UNION.	No. of Vaccination Districts in the Union.	No. of Public Vaccinators recommended for Award.	Range of Awards in each Union.		Total Sum awarded in each Union.	Medical Inspector.
			Minimum.	Maximum.		
			£ s. d.	£ s. d.	£ s. d.	
Wakefield	6	6	7 11 0	52 17 0	147 9 0	Dr. Reece.
Wallingford	4	2	7 16 0	8 9 0	16 5 0	" Bulstrode.
Walsingham	5	2	4 13 0	5 12 0	10 5 0	" Copeman.
Wantage	6	5	2 0 0	6 2 0	21 5 0	" Bulstrode.
Ware	8	7	0 8 0	9 7 0	24 6 0	" Thomson.
Warminster	5	3	1 8 0	9 8 0	13 17 0	" Johnstone.
Warrington	3	3	5 1 0	100 14 0	137 8 0	" Fletcher.
Warwick	3	2	16 17 0	17 18 0	34 15 0	" Buchanan.
Watford..	4	4	3 2 0	14 18 0	30 18 0	" Thomson.
Wayland	2	1	..	..	9 4 0	" Copeman.
Weardale	5	5	1 2 0	4 2 0	12 2 0	" Mivart.
Wellingborough ..	5	..	..	..	..	" Buchanan.
Welwyn..	1	1	..	..	2 19 0	" Thomson.
Westbourne	3	3	2 11 0	7 3 0	14 18 0	" Reece.
Westbury and Whorwelsdown	4	1	..	..	1 7 0	" Johnstone.
West Derby	7	7	22 2 0	154 6 0	444 2 0	" Fletcher.
West Hampnett ..	5	5	6 0 0	15 15 0	50 5 0	" Reece.
Westminster	1	1	..	..	35 13 0	" Sweeting.
West Ward	4	3	1 18 0	3 1 0	6 18 0	" Fletcher.
Whitchurch (Hants)	3	3	3 3 0	7 16 0	15 11 0	Mr. Royle.
Whitechapel	1	1	..	..	168 13 0	Dr. Sweeting.
Whitehaven	6	5	2 4 0	30 14 0	74 17 0	" Fletcher.
Wigan	10	6	3 19 0	49 5 0	160 3 0	Do.
Wigton	7	7	2 12 0	11 14 0	36 14 0	Do.
Willesden	2	1	..	..	14 17 0	Dr. Sweeting.
Wilton	4	3	4 9 0	8 3 0	17 3 0	" Johnstone.
Winchester, New ..	4	4	4 9 0	30 6 0	53 17 0	Mr. Royle.
Windsor..	3	3	3 10 0	13 15 0	27 7 0	Dr. Bulstrode.
Witney	4	2	13 6 0	14 10 0	27 16 0	Do.
Wokingham	5	3	2 15 0	12 17 0	21 18 0	Do.
Woodstock	4	2	4 15 0	9 7 0	14 2 0	Do.
Wortley..	5	2	2 3 0	33 7 0	35 10 0	Dr. Reece.
Wrexham	5	6	0 17 0	54 7 0	106 11 0	" Wheaton.
Yarmouth, Great ..	2	2	14 17 0	23 15 0	38 12 0	Dr. Copeman.
Totals.. ..	1,481	1,099	..	..	12,846 11 0	

No. 3.

PP. A.—No. 3. Estab. 1898.

STATISTICS OF THE NATIONAL VACCINE ESTABLISHMENT AND EDUCATIONAL VACCINATION STATIONS.

I.—Staff at end of 1898.

N.B.—The Stations named in *italics* are Educational Vaccination Stations authorized by the Local Government Board.

Description of Vaccinator.	Name of Vaccinator.	Vaccination Stations.	Days and Hours of Attendance.
Vaccinators supplying lymph for the public service and salaried from the Parliamentary grant.	1. Dr. R. Cory... ...	*Surrey Chapel..*	Tues., Thurs.; 2.
	2. Mr. Joseph Loane...	*Tottenham Court Chapel.*	Mon., Wed.; 1.
Parochial and other Vaccinators not salaried from the Parliamentary grant, but contributing lymph at a fixed rate of payment.	1. Dr. A. C. Clarke ...	*Salford* ...	Thursday; 2.
	2. Dr. Edmund Robinson	*Birmingham...*	Tuesday; 11.
	3. Mr. N. E. Roberts...	*Liverpool* ...	Tuesday; 1.
	4. Dr. A. E. Cope ...	*Westminster* ...	Monday; 10.
	5-6. Mr. E. C. Greenwood	*Marylebone*	Tuesday; 2. Wednesday; 2.
	7. Mr. J. Loane ...	*Whitechapel* ...	Wednesday; 11.
	8. Mr. Frederick Holmes	*Leeds*	Tuesday; 2.30.
	9. Dr. Edward Lynes...	Coventry ...	Tuesday; 12.
	10-11. Mr. R. H. Henderson.	*Glasgow* ...	Monday; 12. (Women). Thursday; 12. (Men).
	12. Mr. J. F. Staines ...	*Endell Street...*	Tuesday; 10.
	13. Mr. W. Skinner ...	*Sheffield* ...	Tuesday; 3.
	14. Dr. G. A. Miskin ...	Waterloo ...	Tuesday; 2.
	15. Dr. J. B. Buist ...	*Edinburgh* ...	Thursday; 3.
	16. Mr. F. Cadell ...	*Edinburgh* ...	Tuesday; 12.
	17. Dr. R. Cory... ...	*St. Thomas's Hospital.*	Wednes.; 11.30.
	18. Mr. J. W. Nicol ...	*Glasgow, West.*	Monday; 12.
	19. Dr. William A. Budd	*Exeter* ...	Thursday; 3.
	20. Mr. R. G. McKerron	*Aberdeen* ...	Wednes.; 2.30.
	21. Mr. J. Ll. Treharne	*Cardiff* ...	Tuesday; 11.
Teachers of Vaccination not supplying lymph.	Dr. W. Husband ...	*Edinburgh* ...	Wed., Sat.: 12.
	Mr. G. S. Page ...	*Bristol* ...	Wednesday; 10.
	Mr. V. A. Jaynes ...	*Horsleydown...*	Wednesday; 3.
	Dr. A. N. Montgomery	*Dublin* ...	Tues., Fri.; 10.
	Dr. R. Cory... ...	*Cambridge* ...	Friday; 11.
	Mr. Frank Hawthorn	*Newcastle-on-Tyne.*	Wednesday; 3.

II.—Animal Vaccine Station.

The Animal Vaccine Station is at 95, Lamb's Conduit Street, where Dr. R. Cory and Mr. T. S. Stott attend for the Vaccination of Children on Tuesdays and Thursdays, at 10.30 A.M.

III.—Sources and Amount of Lymph Supply in 1898.

N.B.—The Stations named in *italics* are Educational Vaccination Stations authorised by the Local Government Board.

Description of Vaccinator.	Vaccination Stations.	Number of Vaccinations performed at the Stations respectively.		Number of Charges of Lymph supplied from the Stations respectively.	
		Primary.	Re-vaccinations.	Charged Ivory Points.	Charged Tubes, each estimated as equal to 10 Ivory Points.
Vaccinators salaried from the Parliamentary grant.	1. *Surrey Chapel* ...	107	20	—	—
	2. *Tottenham Court Chapel.*	181	4	—	20
		288	24	—	20
Parochial and other Vaccinators not salaried from the Parliamentary grant, but contributing lymph at a fixed rate of payment.	1. *Salford*	84	—	—	—
	2. *Birmingham* ...	2,654	184	—	269
	3. *Liverpool* ...	408	20	—	—
	4. *Westminster* ...	400	10	—	—
	5–6. *Marylebone* ...	708	22	—	—
	7. *Whitechapel* ...	1,509	—	—	578
	8. *Leeds*	535	2	—	—
	9. Coventry ...	45	1	—	—
	10–11. *Glasgow* ...	515	4	—	—
	12. *Endell Street* ...	72	3	—	—
	13. *Sheffield* ...	331	—	—	—
	14. Waterloo ...	756	3	—	1,293
	15. *Edinburgh* ...	274	—	—	—
	16. *Edinburgh* ...	518	—	—	163
	17. *St. Thomas's Hospital.*	119	—	—	—
	18. *Glasgow, West*...	412	3	—	—
	19. *Exeter*	381	13	—	—
	20. *Aberdeen* ...	185	—	—	—
	21. *Cardiff*	884	—	—	—
	Total	10,790	265	—	2,303
	Grand Total ...	11,078	289	—	2,323

No. 3.

III.—Sources and Amount of Lymph—*continued.*

Estab-, 1898.

During the year additional supplies (to the extent of 1,618 charged tubes) were obtained from the following gentlemen :—

Mr. J. Bark, Kirkdale.
" A. Meeson, Toxteth Park.
Dr. G. A. Miskin, Kennington.

IV.—Distribution of Human Lymph, 1898.

Number of applications :—

From Medical Practitioners in England and Wales ...	1,903
" " " Scotland	86
" the Navy	31
" India and the Colonies	30
" Diplomatic and other Foreign Services	6

Supplies sent out :—

Charged capillary tubes	3,759

V.—Distribution of Calf Lymph, 1898.

Number of applications	2,47[illegible]

Supplies received :—

Charged ivory points	19,910
" capillary tubes	105

Supplies sent out :—

Charged ivory points	20,312
" capillary tubes	105

No. 4.

Report on the Operations of the Animal Vaccine Station at Lamb's Conduit Street, during the Year 1898–99, by Dr. Cory.

App. A. No. 4. On the Operations of the Animal Vaccine Establishment, 1898–99; by Dr. Cory.

During the year, 1st April, 1898, to 31st March, 1899, 232 calves (males 222, females 10) were vaccinated for the purposes of the station at Lamb's Conduit Street.

The aggregate weight on reception at Lamb's Conduit Street of the 232 calves was—males, 69,448 lbs.; females, 3,153 lbs. On dismissal from the station their aggregate weights were respectively 74,120 lbs. and 3,366 lbs.; so that during retention for vaccination purposes calves of both sexes gained considerably in weight; males by an average of 21·05 lbs., females by an average of 21·3 lbs. each.

Of the above calves, 229 were vaccinated with lymph directly derived from other calves, and three were vaccinated with calf lymph which had been stored. As usual, vaccinations performed with fresh lymph proved much more successful than the others. Insertions to the number of 14,970 in 229 calf-to-calf operations produced 14,582 vesicles; whereas in the three cases vaccinated with stored lymph, 128 insertions only produced 107 vesicles; giving rates of insertion success respectively of 97·4 and 83·6 per cent.

No material difference in the results of calf-to-calf vaccinations was observed whether the lymph used was from calves vaccinated 96 hours or from calves vaccinated 120 hours previously; in both cases the rate of insertion success was practically 97·4 per cent.

Primary Vaccinations.—A.—Cases Vaccinated Calf-to-arm.—During the year 1898–99 the primary vaccinations performed at the station with lymph direct from calf-to-arm numbered 2,853, five separate insertions of lymph being made in each instance. Of the persons thus vaccinated, 1,432 were males, and 1,421 were females. All but 21 of the 2,853 primary vaccinations succeeded at the first attempt, and in no case was a third attempt at vaccination requisite. The amount of insertion success obtained by each of four several operators was as follows:—Of 463 persons primarily vaccinated with calf-to-arm lymph by Mr. Stott, two failed to return for inspection. Of the 461 remaining, 374 were on examination found to have taken in five places, 33 in four, 19 in three, 11 in two, 5 in one place, and 19 failed on the first attempt. Mr. Stott's insertion success rate, therefore, was 90·5 per cent.—Doctors Savory and Leslie C. Thorne-Thorne acted for Mr. Stott or myself in the absence of one or other of us. Dr. Savory vaccinated 152 cases, one of which failed to return for inspection. Of the 151 remaining, 126 were found to have taken in five places, 18 in four, 4 in three, 1 in two, and 2 in one place. Dr. Savory's insertion success rate, therefore, was 95·1 per cent. Dr. Thorne primarily vaccinated 48 cases with calf-to-arm lymph, 39 of which took in five places, 7 in four, and 2 in three places. Dr. Thorne's insertion success rate was, therefore, 95·4 per cent. Students vaccinated 4 cases, all of which were successful in five places. Of 1,491 persons primarily vaccinated by myself, 33 failed to return for inspection, and other three, who were operated on for the cure of naevi, are not taken into account in these statistics. Of the 1,455 remaining, 1,301 were found to have taken in five places, 87 in four, 38 in three, 16 in two, and 11 in one place, while in two cases vaccination proved unsuccessful on first trial, giving an insertion success rate of 96·36 per cent.

B.—Cases Vaccinated with Glycerinated Calf Lymph.—During the same period, 695 primary vaccinations were performed with glycerinated calf lymph. In a number of instances the lymph used was prepared before the Board's scheme for the issue of glycerinated calf lymph had been matured, and in others samples of stored lymph from Continental sources were employed. In these circumstances the experience derived from the use of glycerinated calf lymph during the year in question is not comparable with the ordinary practice of the Animal Vaccine Station.

Re-vaccinations.—These numbered 801, and all were performed calf-to-arm. Mr. Stott performed 55; Dr. Savory 265; Dr. Thorne 44; students 10; and I performed 427. Of Mr. Stott's 55 cases, 49 were successful in five places, 5 in four, and 1 in three places; giving an insertion success rate of 97·45 per cent. Of Dr. Savory's 265 cases, 240 took in five places, 15 in four, 6 in three, 1 in two, and 3 in one place; yielding an insertion success rate of 96·83 per cent. Of Dr. Thorne's 44 cases, 28 took in five places, 7 in four, 5 in three, 2 in two, 1 in one place, and 1 failed to return, giving an insertion success rate of 87·44 per cent. Students vaccinated 10 cases, of which 8 were successful in five places, 1 in four, and 1 in two places, giving an insertion success rate of 92 per cent. Of the 427 cases I myself re-vaccinated, 337 were successful in five places, 43 in four, 23 in three, 12 in two, 5 in one place, 5 failed, and two did not return for inspection; yielding an insertion success rate of 92 per cent. for re-vaccination.

There were 25 cases brought back after "inspection" on account of some abnormal course of their vaccination. In the great majority of cases, namely, in 19 instances, the abnormality consisted of sore arm. In one case the child was brought back in the fourth week of vaccination on account of an axillary abscess; four children were brought back, in the second, third, and fifth weeks, and the third month of vaccination respectively, on account of transient eruptions; and in the case of one child erysipelas commenced on the twenty-first day of vaccination.

No. 5.

Report on the Operations of the Glycerinated Calf Lymph Establishment, 1898–99; by Dr. F. R. Blaxall.

App. A. No. 5.

On the Operations of the Glycerinated Calf Lymph Establishment, 1898–99; by Dr. Blaxall.

In April, 1898, arrangements were made by the Local Government Board with the authorities of the British (now Jenner) Institute of Preventive Medicine at Chelsea, for the rental of certain of their rooms, to serve as laboratories for the preparation of glycerinated calf lymph, which should be available for distribution to public vaccinators in England and Wales on January 2nd, 1899, the date when the Vaccination Act of 12th August, 1898, so far as this Department was concerned, came into operation.

At the same time I was appointed Bacteriologist to the Department, and I at once proceeded to visit some of the principal establishments for the preparation of glycerinated calf lymph on the Continent. On my return I made arrangements to ensure that the structural fittings of these laboratories, including, amongst other things, a supply of water power and electric motor power, should be designed in the best possible manner to carry out the work in question.

The rooms at the Institute were ready for occupation as regards the structural fittings in June of last year, and measures were at once taken to equip them suitably with laboratory apparatus, &c. The rooms in question at the time comprised two laboratories, one office, and one sterilising room; the laboratories and office being situated on the second floor, the sterilising room in the basement.

At the Board's Animal Vaccine Establishment, in Lamb's Conduit Street, alterations were carried out as regards the stabling accommodation for the calves, the work being in the direction of obtaining the utmost procurable cleanliness, thorough ventilation, and the maintenance of a suitable temperature.

For the initial work of the Laboratories a small staff was appointed under the Bacteriologist, the members being trained with reference to the special nature of the work. The cultivation of lymph was carried on from calf to calf at Lamb's Conduit Street, and, after preparations and testing as to the presence of harmful organisms, the resulting glycerinated lymph was stored to meet the demands anticipated in the first weeks of January, 1899. Before the end of that month, however, it was found that the demand exceeded the estimate made, and additions to the laboratory accommodation and to the staff employed became necessary. The latter now consists of one bacteriologist, one assistant bacteriologist, four laboratory assistants, one laboratory boy, and two clerks.

The methods employed in the production, preparation, and storage, of vaccine lymph, and the work performed by the staff are, in brief, as follows:—

1. *Vaccination of the Calf.*—Calves of suitable age (three to six months), breed, and condition, are placed in a quarantine stable for a week, and, their health being ascertained to be satisfactory, they

are transferred to the Animal Vaccine Establishment. Each calf, on admission is examined as to its general health, is weighed, and its temperature taken; a record as to these points being kept. When required for vaccinating purposes, the calf is strapped to a large tilting table, and the lower part of the abdomen, extending as far forward as the umbilicus and backwards into the flanks, is carefully shaved. This shaved area is first washed with a five per cent. solution of carbolic acid or lysol, then well syringed with tap water, and finally cleansed with sterilised water. The moisture from such washing is removed from this shaved area, and from the adjacent skin, by means of sterilised gauze sponges. By these means, I have found that this area of skin can be freed from organisms, as evidenced by absence of growth on surface agar-agar or gelatine culture media, inoculated with scrapings.

The calf is then vaccinated with glycerinated calf lymph, introduced into the skin in numerous parallel linear incisions by a sharp scalpel, previously sterilised, which is dipped from time to time in the vaccinating fluid. The incisions are designed to penetrate the epidermis and to open up the rete malpighii, if possible, without drawing blood; and as they are made, additional glycerinated lymph is run in along the whole length by the aid of a sterilised blunt instrument, such as an ivory or bone spatula. The inoculation of the incisions is effected immediately they are made, otherwise the lips of the wound are apt to swell and close the opening. After vaccination, the calf is removed from the table and is then so stalled in a stable as to prevent any injury to the vaccinated surface. The temperature of this stable is not allowed to fall below 60° Fahr.

2. *Collection of the Vaccine Material.*—After five days (120 hours) the calf is again placed on the table, and the vaccinated surface is thoroughly washed with soap and warm water, gently rubbed over it by the clean hands of the operator. It is again washed with tap water and finally cleansed with sterilised water. Next, any crusts that may have formed upon the vesicular lines, and any epidermal débris are removed by the careful use of a sterilised India rubber pad. Superfluous moisture is absorbed by sterilised gauze sponges. At this stage the site of each incision should present a line of continuous vesiculation.

The skin having been put firmly on the stretch, the vesicles and their contents are collected with a sterilised Volkmann's spoon, each line being treated in turn and scraped once only, care being taken that the edge of the spoon does not touch the neighbouring lines of vesicles. In this way the vesicular pulp is removed without admixture of blood. The pulp obtained by the above procedure is received into a previously sterilised stoppered bottle of known weight.

The abraded surface of the calf is gently washed with warm water, and dusted over with starch powder or boracic acid powder. The animal is then removed from the table and is weighed. Nearly all calves show a considerable gain in weight during their stay at the station and during the vaccination process. Each calf is then transferred to the slaughter-houses attached to the Islington Cattle Market and is there slaughtered. A complete examination is made of the carcase and all the viscera, on behalf of the Board, by a veterinary surgeon especially appointed for the purpose. A report of this examination is received at the laboratories next morning. No lymph is used, for any vaccination of the human subject, until the animal in question is certified to have been healthy.

3. *Glycerination of the Vaccine Material.*—The bottle containing the lymph pulp from each calf is taken to the laboratories where the exact weight of the material is ascertained. The pulp is next transferred to a triturating machine; that employed being either one invented by Dr. Chalybäus, of Dresden, or a modified form of it. All the parts of the machine which come in contact with the lymph pulp are previously sterilised by prolonged steaming. The vaccine material, just as it is derived from the calf, is then passed through the machine, which is worked by an electric motor. When the pulp has been triturated in this way, the amount of subdivision it has undergone can be ascertained by suspending a loopful of the ground-up material in a watch glass containing distilled water. If the trituration has been effectual, such suspension should show only the minutest particles of pulp; causing the water to appear merely cloudy. The pulp is then passed through the machine a second time together with six times its weight of a sterilised mixture of 50 per cent. pure glycerine in distilled water. The resulting mixture is then once more passed through the machine; thus producing a fine and intimate emulsion. At this stage a loopful of the emulsion is withdrawn with a sterilised platinum needle, and agar-agar plates are established, in order to estimate both the number and the quality of organisms present in the lymph.

4. *Storage of Emulsion.*—The emulsion is next received into conical glass receptacles, previously sterilised. By means of a stopcock at the point of the cone, the glycerinated lymph is run into small sterilised test tubes capable of holding 4 to 10 c.c. Each tube is filled as completely as possible, so that very little air remains in contact with the emulsion. It is plugged with a sterilised cork, is sealed with melted paraffin, which has been rendered aseptic with carbolic acid, and is then placed in a dark cool cupboard or ice-chest. Week by week, agar-agar plates are established from the emulsion with the result that the number of colonies is shown to diminish successively in the several plate cultures. At the end of a month the plates rarely show colonies of any sort.

5. *Use of the Lymph at the Animal Vaccine Establishment prior to distribution.*—When the stage is reached at which agar plates show no growth after inoculation with the emulsion, samples of the lymph are drawn up into capillary tubes and despatched to the Animal Vaccine Establishment for the vaccination of children. The results of these vaccinations are recorded a week later, and from the number and size of the vesic es obtained, an estimate is made as to the potency of the lymph.

6. *Transference of the Glycerinated Lymph to Capillary Tubes for distribution.*—When the lymph of a given calf has been shown

. A. No. 5. he ions inated ph ment, ; by . Blaxall.

to be satisfactory, the bulk of it is transferred to sterilised capillary tubes by means of special tube filling machines worked by water power. These tubes are next sealed in a small gas flame, every care being taken to prevent overheating of the lymph during the process. These sealed tubes are then stored in an ice-chest in boxes in such numbers that any quantity demanded up to some 2,000 tubes per day can be at once despatched to the National Vaccine Establishment at Whitehall, from whence the lymph is distributed to public vaccinators.

7. *Recording of the Results of the Vaccination by the Public Vaccinators.*—Each public vaccinator receives, in response to application made to the National Vaccine Establishment (*see* appended form A), a consignment of lymph together with a schedule in which to record the results of its use (*see* appended form B), and these schedules, after having been examined in the National Vaccine Establishment, are sent to the Laboratories. The Schedules indicate the series number of the lymph, the date of its despatch from the National Vaccine Establishment, the name of the public vaccinator to whom it was supplied, the number of tubes sent, the dates when the several tubes were used, the number of persons vaccinated, the number of scarifications made, and the number of vesicles obtained. All these details are recorded at the Laboratories, and from the last two items, information as to the success which has resulted, both as regards individuals vaccinated, and insertions of lymph made, is obtained, and set forth both in full, and in the form of a percentage.

In addition to these records, a register is kept stating the particulars of the calves employed, the details of the lymph obtained from each, including the results of the bacteriological examinations, the results of the use of the lymph at the Animal Vaccine Establishment, and also the number of tubes of each series despatched to the National Vaccine Establishment.

Certain experimental work has also been commenced on the bacteriological flora of vaccine lymph; on the influence of external conditions acting on or likely to act on glycerinated lymph; and on other matters bearing upon the methods best calculated to produce efficient vaccine lymph. Efforts are also being made to devise a means by which to standardise the lymph. This work is still in progress.

Up to and including March 31st, 1899, lymph from 103 calves has been sent out from the Department. All these calves were, after slaughter, examined by the veterinary surgeon and certified to be healthy. One other calf, which appeared to be in health during the course of its vaccination, was found at the autopsy by the veterinary surgeon to exhibit tubercular lesions in certain glands and other organs. The lymph obtained from this calf was at once destroyed.

The number of capillary tubes of lymph sent to the National Vaccine Establishment during the three months in question amounts to 126,038. Each tube is charged with glycerinated calf lymph amply sufficient for the vaccination of one person. The average number of tubes thus despatched between Monday, January 2nd and Saturday, April 1st amounted to 9,695 per week, or 1,616 per day. The following table gives the numerical details as to the despatch of the lymph to the National Vaccine Establishment :—

App. A.

On the Operatio of the Glycerin Calf Lyn Establisl 1898-99; Dr.

Week ending.	No. of Tubes sent to N.V.E.	—	Week ending.	No. of Tubes sent to N.V.E.	—
January 7	10,711		March 4	9,844	
" 14	12,538		" 11	10,200	
" 21	7,569		" 18	9,800	
" 28	9,190		" 25	9,900	
February 4	9,738		April 1	6,100	
" 11	9,500				
" 18	10,253				
" 25	10,700		Total ...	126,038	

The results of 100,000 vaccinations, performed by 9,880 Public Vaccinators had been tabulated at the date of the compilation of this report. These are made up of 99,390 primary vaccinations, and 610 re-vaccinations; the case success-rate and the insertion success-rate being, per cent., respectively as under :—

	Primary Vaccinations.	Re-vaccinations.
Case Success	96·4	91·8
Insertion Success	89·1	82·0

[Addendum A.

ADDENDUM A.

Form of Application for Glycerinated Lymph.

COUNTERFOIL.

Date of Application ____________

To

The National Vaccine Establishment.

Glycerinated Calf Lymph for ____________ *Children.*

Please send sufficient Glycerinated Calf Lymph for the Vaccination of ______ *Children.*

[N.B.—It is undesirable to store the lymph longer than necessary after it has been placed in capillary tubes. Hence such number of tubes only as are immediately required should be applied for; and the lymph should be kept in the dark, in a cool place.]

Signed ____________
Public Vaccinator.

Date ____________

No.

Address ____________

____

Date of transmitting to the Local Government Board, statement as to results obtained.

ADDENDUM B.

App. A. No. 5.

On the Operations of the Glycerinated Calf Lymph Establishment 1898-99; by Dr. Blaxall.

SCHEDULE OF RESULTS OF USE OF GLYCERINATED LYMPH ISSUED BY THE NATIONAL VACCINE ESTABLISHMENT.

NATIONAL VACCINE ESTABLISHMENT,
ST. STEPHEN'S HOUSE,
CANNON ROW,
LONDON, S.W.

*Series*____________

*No. of Tubes sent*__________ ____________189 .

*To*________________

It is requested that the Inspector of the National Vaccine Establishment may be furnished with the information required in the schedule below concerning the vaccinations performed with the accompanying calf lymph. The form should be returned to him as soon as convenient after the results of the vaccinations have been ascertained.

N.B.—All future applications for glycerinated calf lymph should specify the number of persons for whose vaccination lymph is immediately required.

No.	Date of using Lymph.	Whether for a primary or revaccination.	Age of person vaccinated.	No. of scarifications to which lymph was applied.	No. of vesicles obtained.	Remarks.
1						
2						
3						
4						
5						
6*						

N.V.E. 12.

*Signed*________________

*Date*________________

* Some schedule forms have 12, and others, again, 20 spaces provided; and such number of forms is sent out on each separate occasion as will suffice for record of the results of use of each tube issued in response to the application of a Public Vaccinator.

No. 6.

'P. A. No. 6.
Vaccination Order, 1898.

GENERAL ORDER of the LOCAL GOVERNMENT BOARD, under the VACCINATION ACTS, 1867 to 1898, embodying the AMENDED REGULATIONS of the BOARD. (18th October, 1898). To the Board of Guardians of every Poor Law Union in England and Wales; to the Public Vaccinators of the several Vaccination Districts in England and Wales; and to all others whom it may concern.

Whereas by Section 4 of the Vaccination Act, 1867 (hereinafter referred to as "the Act of 1867"), it was enacted that no person should be appointed a Public Vaccinator, or act as deputy for a Public Vaccinator, who should not possess the qualification theretofore prescribed by the Lords of Her Majesty's Council, or such as should be from time to time prescribed by them, except when such Lords should, upon sufficient cause, sanction any departure from their directions; and that all such regulations as the said Lords had theretofore made, or should thereafter make, which they were thereby authorised to make, to secure the efficient performance of Vaccination, should be duly observed by the several persons to whom they applied;

And whereas by Section 8 of the Act of 1867 the Lords of Her Majesty's Council were authorised to issue regulations in respect of the re-vaccination of persons who might apply to be re-vaccinated;

And whereas by Section 5 of the Vaccination Act, 1871 (hereinafter referred to as "the Act of 1871"), it was enacted that, subject to the provisions of that Act, the Poor Law Board should have the same powers with respect to Guardians and Vaccination Officers in matters relating to Vaccination as they had with respect to Guardians and Officers of Guardians in matters relating to the relief of the Poor, and might make rules, orders, and regulations accordingly, and that all enactments relating to such powers, and to such orders, rules, and regulations, should apply, mutatis mutandis, and that the Poor Law Board should also frame appropriate Books and Forms for the use of Vaccination Officers, Public Vaccinators, and Medical Practitioners under the Act of 1867 and the Act of 1871;

And whereas by Section 15 of the Act of 1871 it was enacted that the Poor Law Board might, by Order, from time to time repeal, alter, and add to, the Forms contained in the Schedule to the Act of 1867;

And whereas by virtue of the Local Government Board Act, 1871, and of Section 16 of the Act of 1871, all the powers and duties vested in or imposed on Her Majesty's Most Honourable Privy Council by the Act of 1867, and any Acts amending the said Act, and conferring powers on the said Privy Council, and all powers and duties vested in or imposed on the Poor Law Board by the several Acts of Parliament relating to the relief of the Poor, and any other Acts, are now vested in and imposed on Us, the Local Government Board;

And whereas by Section 1 of the Vaccination Act, 1874 (herein-after referred to as "the Act of 1874"), it was enacted that the powers conferred by the above-recited Section 5 of the Act of 1871 should be deemed to extend to and include the making of rules, orders, and regulations prescribing the duties of Guardians and their Officers in relation to the institution and conduct of the proceedings to be taken for enforcing the provisions of the Act of 1867 and the Act of 1871, and the payment of the costs and expenses relating thereto; and that rules, orders, and regulations under the Act of 1874 should be deemed to be made under the said Section 5 of the Act of 1871;

And whereas by divers general and special orders the Lords of Her Majesty's Privy Council, the Poor Law Board, and We Ourselves have made regulations under, or which had effect under, the herein-before recited enactments; and the said Orders or some parts thereof are still in force;

And whereas by Section 6 of the Vaccination Act, 1898 (herein-after referred to as "the Act of 1898"), We are empowered to make rules and regulations with respect to the duties and remuneration of Public Vaccinators, whether under contracts made before or after the passing of that Act;

And whereas by Section 7 of the Act of 1898 We are further empowered by Order, if in our opinion it is expedient by reason of serious risk of outbreak of small-pox or of other exceptional circumstances, to require the Guardians of any Poor Law Union to provide vaccination stations for the vaccination of children with glycerinated calf lymph or such other lymph as may be issued by Us, and to modify as respects the area to which the Order applies and during the period for which it is in force, the provisions of that Act requiring the Public Vaccinator to visit the home of the child otherwise than on request of the parent;

Now therefore, We, the Local Government Board, in pursuance of the powers given to Us by the Statutes in that behalf, do hereby Order that from and after the Thirty-first day of December, One thousand eight hundred and ninety-eight (herein-after referred to as "the commencement of this Order"), the following provisions shall, unless We otherwise direct, have effect; viz.,—

Article 1.—All the Orders of Council and General and other Orders made by the Lords of Her Majesty's Privy Council, the Poor Law Board, and by Us, under, or which have effect under, any of the herein-before recited enactments, shall be rescinded:

Provided that nothing in this Article shall affect—

1. The validity of any contract for public vaccination made under any Order hereby rescinded; but the Public Vaccinator thereunder shall, as from the commencement of this Order, perform the duties, and be paid the remuneration substituted by this Order for the duties and remuneration fixed by such contract; or
2. The appointment or tenure of office of any Vaccination Officer appointed under any of those Orders; but he shall, as from the commencement of this Order, perform the duties prescribed by this Order, and be remunerated in manner provided by this Order.

Contracts with Public Vaccinators.

ARTICLE 2.—The Guardians of any Poor Law Union shall not enter into a contract for public vaccination with any registered medical practitioner, or approve of any such practitioner as deputy for a Public Vaccinator, unless he shall produce a certificate of proficiency in vaccination given, under such conditions as We from time to time fix, by some person whom We shall have authorised to act for the purpose and by whom he shall have been duly instructed and examined in the practice of vaccination; but it shall not be necessary to produce the certificate to the Guardians, if such certificate was required as a condition of obtaining any diploma, licence, or degree which the contractor possesses.

ARTICLE 3.—(1.) Every contract for public vaccination, other than a contract with the Medical Officer of a Workhouse for the vaccination of the persons resident therein, shall be made in the Form set out in the First Schedule hereto, with such modifications, if any, as the Guardians and the contractor shall, with Our approval, adopt; and shall provide for payments to be made to the Public Vaccinator as follows; that is to say,—

(*a*) a payment of not less than one shilling in respect of every child whose birth shall have been registered in his district after the Thirty-first day of August One thousand eight hundred and ninety-eight, or who shall be resident in his district and whose birth shall have been registered in some other district after that date, or shall not have been registered at all, except children who shall have died or been removed from the district before attaining the age of four months, or who shall have been duly certified to have been successfully vaccinated otherwise than by the Public Vaccinator, or to be insusceptible of vaccination, or to have had small-pox, before reaching that age, or with regard to whom a certificate under Section 2 of the Act of 1898 is in force;

(*b*) a payment of not less than five shillings in respect of every successful primary vaccination or re-vaccination performed by him at the home of the person vaccinated; and

(*c*) a payment of not less than two shillings and sixpence in respect of every successful primary vaccination of any person other than a child, or of any successful re-vaccination, such primary vaccination or re-vaccination having been performed by him at his surgery or elsewhere than at the home of the person vaccinated:

Provided that no payment shall be made in respect of any vaccination unless the same shall have been performed in accordance with the conditions herein prescribed, nor unless the provisions of the Vaccination Acts, 1867 to 1898, and of this Order as to certificates and otherwise have been observed with reference thereto:

Provided also that no payment shall be made in respect of the re-vaccination of any person who shall be less than ten years old, or who shall have been previously re-vaccinated within a period of ten years.

(2.) The number of children in respect of whom payments are to be made under paragraph (*a*) of this Article shall be the number of children in the Lists to be sent by the Vaccination Officer to the Public Vaccinator as provided by paragraph 6 (*a*) of the "Instructions to Vaccination Officers" in the Fourth Schedule hereto, together with the number of children not included in such Lists but vaccinated by the Public Vaccinator himself.

(3.) Each of the said payments to be made in accordance with this Article shall, subject to the above provisions as to the minimum, be of such amount, and shall be made at such times, and subject to such conditions as may be fixed in the contracts approved by Us.

(4.) The payments made to the Public Vaccinator in accordance with this Article shall be deemed to include any expense in respect of postage incurred by the Public Vaccinator, unless otherwise agreed between him and the Guardians.

ARTICLE 4.—Where a Workhouse is a Vaccination District, every vaccination contract made after the commencement of this Order with the Medical Officer of the Workhouse for the vaccination of persons resident therein shall be made in the Form set out in the Second Schedule hereto, with such modifications, if any, as the Guardians and the contractor shall, with Our approval, adopt; and shall provide for a payment of not less than two shillings and sixpence in respect of each successful primary vaccination or re-vaccination performed by him under his contract.

ARTICLE 5.—(1.) Any contract for public vaccination, other than a contract made with the Medical Officer of a Workhouse for the vaccination of the persons resident therein, which shall be in force at the commencement of this Order shall continue in force until the same shall have been determined by the death of the contractor or by notice as therein provided, or by Us, or until a new contract shall be entered into with the contractor and shall have been approved by Us in place thereof; and such contract shall, as from the commencement of this Order, be deemed, as regards the duties of the Public Vaccinator, in lieu of the provisions in that behalf therein contained, to require the Public Vaccinator to perform the duties prescribed by this Order, or specified in the form of contract in the First Schedule hereto, and as regards the remuneration of the Public Vaccinator, in lieu of the payments in such contract mentioned, to provide for the payment by the Guardians to the Public Vaccinator of such payments as are prescribed by paragraphs (*a*), (*b*), and (*c*) of Article 3 of this Order, the amount of each of such payments being such as may have been agreed upon by the Guardians and Public Vaccinator before the First day of January, One thousand eight hundred and ninety-nine, and may be approved by Us, or if the amount of each of such payments shall not be so settled then, as may be determined by Us.

(2.) Any contract made under the Vaccination Act of 1867 with the Medical Officer of a Workhouse for the vaccination of the persons resident therein which shall be in force at the commencement of this Order shall continue in force until the same shall

. A. No 6.
on
. 1898.

have been determined by the death of the contractor or by notice or otherwise as therein provided, or by Us, or until a new contract shall be entered into with the contractor, and shall have been approved by Us, and such contract shall be deemed, as regards the duties of the Public Vaccinator, in lieu of the provisions in that behalf therein contained, to require the Public Vaccinator to perform the duties prescribed by this Order or specified in the form of contract in the Second Schedule hereto, and, as regards the remuneration of the Medical Officer as Public Vaccinator, in lieu of the payments in such contract specified, to provide for the payment by the Guardians of such payment as is prescribed by Article 4 of this Order, the amount of such payment being such as may be agreed upon by the Guardians and the Medical Officer before the First day of January, One thousand eight hundred and ninety-nine, and may be approved by Us, or if the amount of such payments shall not be so settled then, as may be determined by Us.

Duties of Public Vaccinator.

ARTICLE 6.—Every Public Vaccinator shall diligently perform the duties imposed on him by his contract or by this Order; and shall perform such duties in person, except when, on account of reasonable absence from the District, or on some other sufficient ground, he shall be obliged to leave any of them to be performed by a duly qualified deputy approved by the Guardians.

ARTICLE 7.—(1.) The visit of the Public Vaccinator to the home of a child, whether on request of the parent or other person having the custody of the child, or after notice from the Vaccination Officer, shall be made not earlier than 9 o'clock in the morning, nor later than 4 o'clock in the afternoon, unless some other time shall have been arranged between the Public Vaccinator and the parent or such other person. In either case at least twenty-four hours' notice shall be given by the Public Vaccinator of his intention to visit the home of the child, unless the parent or other person having the custody of the child otherwise agrees in any case where the visit is made at his request. In the case of a visit after notice from the Vaccination Officer, the notice of the intention to visit shall be in the Form I. set out in the Fifth Schedule hereto, or to the like effect.

(2.) The visit of the Public Vaccinator to the home of a child, whether on request of the parent or other person having the custody of the child, or after notice from the Vaccination Officer, shall, in the absence of any sufficient reason for delay, be made within two weeks after receipt of the request or notice, as the case may be.

(3.) The Public Vaccinator shall enter in the proper columns of the list of children sent to him in the Form H. set out in the Fifth Schedule hereto by the Vaccination Officer in respect of whom the necessary certificates have not been received by such Officer the several particulars as to each visit made by him; and shall, within one month from the receipt of such notice, return the same to the Vaccination Officer, with such particulars duly inserted.

(4.) This Article shall not apply to Public Vaccinators under contracts for the vaccination of persons resident in a Workhouse.

ARTICLE 8.—In the performance and inspection of all vaccinations under contract and otherwise in relation thereto. every Public Vaccinator shall observe the "Instructions to Vaccinators under Contract" in the Third Schedule hereto.

ARTICLE 9.—Every Public Vaccinator shall duly register all vaccinations performed by him in a register in the Form O. set out in the Fifth Schedule hereto, and in manner directed in the "Instructions to Vaccinators under Contract" in the Third Schedule hereto.

Appointment of Vaccination Officers.

ARTICLE 10.—(1.) Where the number of Vaccination Officers already appointed or hereafter appointed in any Poor Law Union shall at any time, in the opinion of the Guardians or in Our opinion, be insufficient for the purpose of securing the due execution of the Vaccination Acts, 1867 to 1898, in such Union, the Guardians shall, with Our approval, or on Our requisition, appoint a sufficient number of such Officers.

(2.) Whenever, in consequence of an outbreak of small-pox, or for other cause, it may appear to the Guardians or to Us to be requisite that temporary assistance should be provided for any Vaccination Officer in the discharge of his duties, the Guardians may and, if so directed by Us, shall appoint an Assistant or Assistants to the Vaccination Officer for such time as the Guardians may deem necessary or We may direct.

ARTICLE 11.—(1.) Every appointment of a Vaccination Officer made after the commencement of this Order shall be subject to Our approval.

(2.) Notice of a proposal to appoint a Vaccination Officer shall be given at one of the two ordinary meetings next preceding the meeting at which the appointment is to be made, such notice being duly entered on the minutes, or else an advertisement specifying the District for which, and the date of the meeting at which, such appointment is proposed to be made, together with the rate of remuneration to be paid, shall be published in some newspaper circulating in the Union at least seven days before the day fixed for the appointment.

ARTICLE 12.—The Guardians shall furnish the Vaccination Officer with a copy of the resolution appointing him signed by the Chairman of the meeting at which the appointment was made, or of the ensuing meeting; or, in the case of any Vaccination Officer appointed before the commencement of this Order, with a copy of such resolution under the seal of the Guardians.

ARTICLE 13.—Every appointment of a Vaccination Officer shall, within seven days after it is made, be reported to Us by the Clerk to the Guardians, who shall furnish such particulars relating thereto as We may require.

No. 6.

ARTICLE 14.—If any Vaccination Officer is at any time prevented by sickness or accident or other sufficient reason from performing his duties, the Guardians shall appoint a competent person to act as his temporary substitute, and such person shall be deemed to be the Vaccination Officer. It shall not be necessary in any such case that the foregoing Articles as to appointment, except Article 12, should be complied with, nor shall Our approval be required to any such temporary appointment.

ARTICLE 15.—(1.) In the event of a vacancy in the office of Vaccination Officer at or after the commencement of this Order, the Clerk to the Guardians shall report it to Us, and the Guardians shall make a fresh appointment without delay, unless We shall otherwise direct.

(2.) If the Guardians are unable to fill up the vacancy forthwith, they shall appoint a person to act temporarily, subject to Our approval.

Tenure of Office of Vaccination Officers.

ARTICLE 16.—Every Vaccination Officer appointed under this Order shall continue to hold the office until he shall die, or resign, or be removed by the Guardians with Our consent, or by Us, or shall be proved to be insane by evidence which We shall deem sufficient.

ARTICLE 17.—Where a Vaccination Officer is appointed for a particular District, and any change in the extent of the District may be deemed necessary, and he shall decline to acquiesce therein, the Guardians may after six months' notice in writing, signed by their Clerk, and given to such Vaccination Officer, determine his office subject to Our consent.

ARTICLE 18.—No person shall be appointed a Vaccination Officer who does not agree to give one month's notice previous to resigning the office, or to forfeit such sum as may be agreed upon as liquidated damages.

ARTICLE 19.—If any such Officer gives notice of an intended resignation to take effect on a future day, the Guardians may appoint a successor at any time subsequent to such notice.

Remuneration of Vaccination Officers.

ARTICLE 20.—(1.) The remuneration of every Vaccination Officer, whether appointed before or after the commencement of this Order, shall consist of—

(*a*) a payment of not less than threepence in respect of each child entered on the Birth Lists sent to him after the Thirty-first day of December, One thousand eight hundred and ninety-eight, by the Registrar of Births and Deaths;

(*b*) a payment of not less than ninepence in respect of the registration by him after the same date in his Vaccination Register of the successful vaccination of any child born in his district; and

(c) a payment of not less than ninepence in respect of the transmission by him after the same date to the Vaccination Officer of the district where the birth was registered of a copy, certified by him, of the certificate of successful vaccination in his district of any child not born in the district, a note of which he shall have entered in Column 17 of his Report Book.

(2.) Subject to the above provisions as to the minimum, the amount of such payments shall be such as We may approve or direct; and shall be increased or reduced, and such additional payment shall be made for extraordinary services of the Vaccination Officer, or under other unforeseen circumstances, as We shall from time to time approve or direct.

(3.) The remuneration of the Vaccination Officer shall be deemed to include any expense in respect of postage incurred by him, unless otherwise agreed between him and the Guardians.

ARTICLE 21.—The remuneration of every Vaccination Officer shall be payable up to the day on which he ceases to hold the office, and no longer, subject to any deduction which the Guardians may be entitled to make under Article 18.

ARTICLE 22.—Subject to the provisions of Article 23, the remuneration assigned to every Vaccination Officer shall be payable quarterly, namely, at Lady Day, Midsummer Day, Michaelmas Day, and Christmas Day; but the Guardians may pay to him at the expiration of every calendar month such proportion as they may think fit on account of the remuneration to which he may become entitled at the termination of the quarter.

ARTICLE 23.—Every Vaccination Officer shall make out an Account at the end of each quarter, and submit it to the Guardians, together with the Books which he may be required to keep, and the Certificates in his possession; and until such Account, Books, and Certificates have been so submitted, the Guardians may postpone the payment of the remuneration which may then remain due.

ARTICLE 24.—(1.) The Guardians may pay a reasonable remuneration to any temporary assistant to the Vaccination Officer, or to any temporary substitute for the Vaccination Officer, whom they appoint, and shall pay such remuneration as We may direct to any such assistant whose appointment has been made in pursuance of Our directions.

(2.) No remuneration to any temporary assistant or substitute shall be paid for a longer period than six weeks, unless Our consent is obtained thereto.

Duties of Vaccination Officers.

ARTICLE 25.—Every Vaccination Officer shall, subject to the provisions of Article 14, perform the duties of his office in person, unless, with Our permission given on the application of the Guardians, he shall be allowed to entrust the performance of all or any of them to some deputy approved by such Guardians.

Article 26.—Every Vaccination Officer shall duly observe the "Instructions to Vaccination Officers" contained in the Fourth Schedule hereto.

Article 27.—(1.) Every Vaccination officer shall, when required by the Guardians, produce to them his books and the certificates in his possession, and the lists sent by him to the Public Vaccinator of children in respect of whom the necessary certificates have not been received by him, when such notices have been returned to him filled up by the Public Vaccinator; and shall within seven days after the expiration of each month submit to the Guardians a copy of the Summary of his Proceedings under the Vaccination Acts, 1867 to 1898, which he is required to keep by paragraph 12 of the "Instructions to Vaccination Officers" in the Fourth Schedule hereto, so far as it relates to that month.

(2.) He shall, when required by the Guardians, give them full information as to any legal proceedings taken by him as Vaccination Officer; and, subject to the provisions of the Vaccination Acts, 1867 to 1898, and of this Order, shall obey all lawful orders of the Guardians which are applicable to his office.

Article 28.—The Guardians shall, from time to time, ascertain whether the Vaccination Officer is performing the duties imposed on him by the Vaccination Acts, 1867 to 1898, of enforcing the provisions of those Acts, and the duties imposed on him by this Order, and shall require the due performance by him of such duties; and, in case of any continued neglect on his part, shall report the same to Us.

Costs and Expenses of Vaccination Officer.

Article 29.—(1.) The Guardians shall pay the reasonable costs and expenses incurred by the Vaccination Officer in any proceedings taken by him for enforcing the provisions of the Vaccination Acts, 1867 to 1898, including the reasonable costs of obtaining any necessary legal assistance in connection with the institution and conduct of any such proceedings; and the Vaccination Officer shall within seven days after the receipt of any sum of money recovered or received by him from any defendant in respect of such costs or expenses, or in respect of any penalty under the said Acts, pay the same to the Treasurer of the Guardians to their credit.

(2.) The Guardians shall also pay the costs incurred by the Vaccination Officer in binding the Birth List sheets as provided in paragraph 4 of the "Instructions to Vaccination Officers" contained in the Fourth Schedule hereto.

(3.) If the Guardians think fit to direct that the name and address of the Vaccination Officer, or of the Public Vaccinator, as the case may be, shall be printed on any of the Forms and Notices prescribed by this Order, the Vaccination Officer shall cause the same to be so printed, and the Guardians shall pay the cost of such printing.

Forms.

Article 30.—The Forms to be used for the purposes of the Vaccination Acts, 1867 to 1898, and Orders thereunder, shall be those set out in the Fifth Schedule hereto, and the same shall be used as follows :—

App. A. No.
Vaccination Order, 1898.

(1.) The notice to be given by the Registrar of Births and Deaths under section 15 of the Act of 1867 shall be in Form A. or to the like effect. Copies of Forms B., C., D., and E. shall be attached thereto, and the name and address of the Vaccination Officer shall be written or printed on the back thereof.

(2.) The Certificate of Postponement of Vaccination to be given under Section 18 of the Act of 1867, shall be in Form B. or to the like effect; provided that if such certificate is given by a Public Vaccinator in pursuance of sub-section (4) of Section 1 of the Act of 1898, the same shall be in Form C. or to the like effect.

(3.) The Certificate of Insusceptibility of Vaccination to be given under Section 20 of the Act of 1867 shall be in Form D. or to the like effect.

(4.) The Certificate of successful Vaccination to be given under Section 21 or Section 23 of the Act of 1867, or Section 7 of the Act of 1871, shall be in Form E. or to the like effect.

(5.) The Certificate to be given by the Public Vaccinator in the cases provided for in Section 12 of the Act of 1871 shall be in Form F. or to the like effect.

(6.) The Request by a parent, or other person having the custody of a child, under sub-section (2) of Section 1 of the Act of 1898, may be in Form G. or to the like effect.

(7.) The List to be sent by the Vaccination Officer to the Public Vaccinator as required by paragraph 6 (*a*) of the "Instructions to Vaccination Officers" shall be in Form H.

(8.) The Notice to be given by the Public Vaccinator of his visit to the home of a child under sub-section (3) of Section I of the Act of 1898 shall be in Form I. or to the like effect.

(9.) The Notice of Default to be given by the Vaccination Officer to the parent, or other person having the custody of a child, as provided in paragraph 6 (*d*) of the "Instructions to Vaccination Officers" shall be in Form K. or to the like effect.

(10.) The Vaccination Register to be kept by the Vaccination Officer as provided in the "Instructions to Vaccination Officers" shall be in Form L., and the columns for the same shall be printed on the same sheet as the Birth List sheets to be sent to the Vaccination Officer by the Registrar of Births and Deaths.

(11.) The Report Book to be kept by the Vaccination Officer as provided by paragraph 8 of the "Instructions to Vaccination Officers" shall be in Form M.

(12.) The Summary of Proceedings under the Vaccination Acts, 1867 to 1898, required to be kept by the Vaccination Officer as provided in paragraph 12 of the "Instructions to Vaccination Officers" shall be in Form N.

(13.) The Vaccinator's Register to be kept by the Public Vaccinator as provided in Article 9 of this Order shall be in Form O.

(14.) The Notice to be given by the Public Vaccinator to the Medical Officer of Health under sub-section (4) of Section 1 of the Act of 1898 shall be in Form P. or to the like effect:

Provided that the Forms B., C., D., and E. in the Schedule to the Order issued by Us on the Thirtieth day of November, One thousand eight hundred and seventy-one, may continue to be used, in place of the Forms B., D., E., and F. in the Schedule to this Order, respectively, until We otherwise direct.

Definitions.

ARTICLE 31.—In this Order—

The term "Workhouse" means any Workhouse, School, or Infirmary which is under the management of a Board of Guardians.

The term "child" means any person not more than fourteen years of age.

Short Title.

ARTICLE 32.—This Order may be cited as the Vaccination Order, 1898.

THE SCHEDULES ABOVE REFERRED TO.

FIRST SCHEDULE.

Form of Vaccination Contract.

ARTICLES OF AGREEMENT entered into this day of , One thousand eight hundred and between of the one part, and the Guardians of the Poor of the Union, in the County of , of the other part.

Whereas the said Guardians have, in pursuance of the several Statutes in that behalf, with the approval of the Local Government Board, divided the Union aforesaid into Districts for the purpose of Vaccination, one of which Districts comprises the Parishes and Places following; that is to say,

and the said Guardians have agreed with the said to enter into a Contract for the performance of Vaccination in the said District:

Now, therefore, the said doth hereby covenant and agree with the said Guardians and their successors, that, from and after the day of , he will (subject to any Order as to Vaccination made by the Local Government Board under Section 7 of the Vaccination Act, 1898) by himself, or (when he shall be unable to perform his duties in person) by the deputy herein-after mentioned, or who may hereafter be approved by the Guardians, and whose name may be endorsed hereon, duly and according to the requirements of the Acts and Orders relating to Vaccination perform the following duties:—

(1.) In the case of every child resident in the District, on the request of the parent or other person having the custody of the child, he will, within two weeks after the receipt of such request, visit the home of the child for the purpose of vaccinating the child.

(2.) In the case of every child resident in the District who has reached the age of four months, and as to whom he has received the requisite notice from the Vaccination Officer, he will visit the home of the child within two weeks after receipt of the notice, and offer to vaccinate the child with glycerinated calf lymph, or such other lymph as may be issued by the Local Government Board, and if his offer is accepted will thereupon (or after such postponement, if any, as may in his opinion be necessary) so vaccinate the child.

(3.) In the case of any person not being a child, and applying to him for primary vaccination, or of any person applying to him for re-vaccination, who shall be not less than ten years old, and shall not have been previously re-vaccinated within a period of ten years, he will, if so requested, visit the home of such person for the purpose of vaccinating or re-vaccinating him, or will, if not so requested, perform the operation at his surgery, or at such other place as may be arranged by him with the person so applying.

(4.) In every such case he will do and perform all such acts and things as, to the best of his judgment, and in accordance with the requirements of the Orders in force as to Vaccination, shall seem to him necessary for the purpose of causing the vaccination to be successfully terminated.

(5.) In every case in which he has performed vaccination he will, not less than six days nor more than fourteen days after the operation, attend at the place where the vaccination was performed, or, in the case of a re-vaccination, at such other place as may be arranged, and inspect the result of such vaccination; and will thereafter do such acts, and give such directions, and otherwise treat the case as may be necessary.

(6.) If any child vaccinated by him shall, in his opinion, require medical treatment in consequence of the vaccination, he will, if the parent or other person having the custody of the child consent, attend the child and prescribe such treatment as may be required.

(7.) He will keep a book, to be termed "The Vaccinator's Register," according to the form prescribed by the Local Government Board, to be provided for him by the said Guardians, and will, on the same day on which he shall have vaccinated any person to whom this Contract shall apply, and on the same day on which he shall have inspected the results of the vaccination of such person, make the entries respectively applicable to the vaccination and the inspection of the results of the vaccination, and will on the day next before the first ordinary meeting of the said Guardians in every quarter of the year, and also at such other times as may be required by the Guardians or for purposes of audit, deliver, or cause to be delivered to their Clerk, the book in which he shall have made such entries during the interval preceding such meeting or audit.

(8.) He will make out an account at the end of every quarter of the sums payable to him under this Contract, and will cause the same to be delivered to the Clerk to the Guardians as soon as practicable after the end of the quarter.

(9.) He will perform any other duties in regard to vaccination which may be imposed on him by the Vaccination Acts, 1867 to 1898, or by any Order of the Local Government Board under those Acts.

And the said Guardians do, for themselves and their successors, covenant and agree with the said as follows:—

That is to say—to pay him, his executors or administrators, within one calendar month after Lady Day, Midsummer Day, Michaelmas Day, and Christmas

App. A. No. 4.

Vaccination Order, 1898.

Pay respectively during the continuance of this Contract, and within one month after its termination, the following sums:—

(1.) A sum of   in respect of every child whose birth shall have been registered in his District after the 31st day of August, 1898, or who shall be resident in his District and whose birth shall have been registered in some other District after that date, or shall not have been registered at all, except children who shall have died or been removed from the District before attaining the age of four months, or shall have been duly certified to have been successfully vaccinated otherwise than by the Public Vaccinator, or to be insusceptible of vaccination, or to have had small-pox, before reaching that age, or with regard to whom a certificate under section 2 of the Vaccination Act, 1898, is in force: the number of children in respect of whom payments are to be made being the number in the Lists to be sent by the Vaccination Officer to the Public Vaccinator as provided by paragraph 6 (a) of the "Instructions to Vaccination Officers" in the Fourth Schedule to the Vaccination Order, 1898, together with the number of children not included in such Lists, but vaccinated by the Public Vaccinator himself.

(2.) A sum of   for every successful primary vaccination or re-vaccination performed by him under this Contract at the home of the person vaccinated.

(3.) A sum of   for every successful primary vaccination of any person other than a child and every successful re-vaccination, such primary vaccination or re-vaccination having been performed by him under this Contract at the Vaccinator's Surgery or elsewhere than at the home of the person vaccinated:

Provided that no payment shall be made in respect of any primary vaccination or re-vaccination unless the same shall have been performed in accordance with the conditions prescribed by the Vaccination Order, 1898, nor unless the provisions of the Vaccination Acts, 1867 to 1898, and of that Order in regard to certificates and their transmission, and otherwise shall have been observed in relation thereto; nor shall any payment be made in respect of any vaccination or re-vaccination, the particulars of which shall not have been duly entered in the Vaccinator's Register, except in the case of any omission which shall be explained to the satisfaction of the said Guardians.

The said Guardians approve of   as the occasional deputy of the said   for the purposes of this Contract.

And it is mutually agreed that this Contract may be put an end to by either of the parties thereto, by giving twenty-eight days' notice to the other party of the intention to put an end to the same.

In witness whereof the said   hath hereunto set his hand and seal, and the said Guardians their Common Seal, the day and year first above written.

Signed, sealed, and delivered by the above-named   in the presence of   (L.S.)

(Guardians' Seal.)

The Common Seal of the Guardians of the above-named Union was hereto affixed at a Meeting of the Board of Guardians, held on the day of the date hereof by   Chairman of the Board at the said Meeting, in the presence of

---

Clerk to the Guardians of the said Union.

SECOND SCHEDULE.

Form of Vaccination Contract with the Medical Officer of a Workhouse.

Articles of Agreement entered into this day of , One thousand eight hundred and , between of the one part, and the Guardians of the Poor of the Union, in the County of , of the other part.

Whereas the said Guardians have, in pursuance of the several Statutes in that behalf, with the approval of the Local Government Board, divided the Union aforesaid into Districts for the purpose of Vaccination, one of which Districts consists of the Workhouse of the said Union; and the said Guardians have agreed with the said to enter into a Contract for the performance of Vaccination at the said Workhouse:

Now, therefore, the said doth hereby covenant and agree with the said Guardians and their successors that, from and after the day of , he will by himself, or (when he shall be unable to perform his duties in person) by the deputy herein-after mentioned, or who may hereafter be approved by the Guardians, and whose name may be endorsed hereon, duly and according to the requirements of the Acts and Orders relating to Vaccination perform the following duties:—

(1.) In the case of every child resident in the Workhouse, on the request of the parent or other person having the custody of the child, he will, as soon as practicable, after such request, attend at the Workhouse for the purpose of vaccinating the child.

(2.) In the case of every child resident in the Workhouse who has reached the age of two months without having been vaccinated he will offer to vaccinate the child with glycerinated calf lymph, or such other lymph as may be issued by the Local Government Board, and if his offer is accepted will thereupon (or after such postponement, if any, as may in his opinion be necessary) so vaccinate the child.

(3.) He will vaccinate any other person resident in the Workhouse applying to him for primary vaccination or re-vaccination, provided that in the case of re-vaccination such person shall not be less than ten years old, and shall not have been previously re-vaccinated within a period of ten years.

(4.) In every such case he will do and perform all such acts and things as, to the best of his judgment, and in accordance with the requirements of the Orders in force as to Vaccination, shall seem to him necessary for the purpose of causing the vaccination to be successfully terminated.

(5.) In every case in which he has performed vaccination or re-vaccination he will, if the person vaccinated is still in the Workhouse, not earlier than the fifth day, nor later than the tenth day, after the operation, inspect the result; and will thereafter do such acts, and give such directions, and otherwise treat the case as may be necessary.

(6.) He will keep a book, to be termed "The Vaccinator's Register," according to the form prescribed by the Local Government Board, to be provided for him by the said Guardians, and will, on the same day on which he shall have vaccinated any person to whom this Contract shall apply, and on the same day on which he shall have inspected the results of the vaccination of such person, make the entries respectively applicable to the vaccination and the inspection of the results of the vaccination, and will on the day next before the first ordinary meeting of the said Guardians in every quarter of the year, and also at such other times as may be required by the Guardians, or for purposes of audit, deliver, or cause to be delivered to their Clerk, the book in which he shall have made such entries during the interval preceding such meeting or audit.

(7.) He will perform any other duties in respect of vaccination which may be imposed on him by the Vaccination Acts, 1867 to 1898, or by any Order of the Local Government Board under those Acts.

And the said Guardians do, for themselves and their successors, covenant and agree with the said as follows :—

That is to say—to pay him, his executors or administrators, within one calendar month after Lady Day, Midsummer Day, Michaelmas Day, and Christmas Day respectively during the subsistence of this Contract, and within one month after its termination, the sum of for every successful primary vaccination or re-vaccination :

Provided that no payment shall be made in respect of any primary vaccination or re-vaccination unless the same shall have been performed in accordance with the conditions prescribed by the Vaccination Order, 1898, nor unless the provisions of the Vaccination Acts, 1867 to 1898, and of that Order in regard to certificates and their transmission, and otherwise shall have been observed in relation thereto, nor shall any payment be made in respect of any vaccination or re-vaccination, the particulars of which shall not be duly entered in the Vaccinator's Register, except in the case of any omission which shall be explained to the satisfaction of the said Guardians.

The said Guardians approve of as the occasional deputy of the said for the purposes of this Contract.

And it is mutually agreed that this Contract may be put an end to by either of the parties thereto, by giving twenty-eight days' notice to the other party of the intention to put an end to the same.

In witness whereof the said hath hereunto set his hand and seal, and the said Guardians their Common Seal, the day and year first above written.

Signed, sealed, and delivered by the above-named in the presence of } (L.S.)

(Guardians' Seal.)

The Common Seal of the Guardians of the above-named Union was hereto affixed at a Meeting of the Board of Guardians, held on the day of the date hereof by Chairman of the Board at the said Meeting, in the presence of

Clerk to the Guardians of the said Union.

THIRD SCHEDULE.

Instructions to Vaccinators under Contract.

(1.) Except so far as any immediate danger of small-pox may require, the Public Vaccinator must vaccinate only subjects who are in good health. As regards infants, he must ascertain that there is not any febrile state, nor any irritation of the bowels, nor any unhealthy state of the skin, especially no chafing or eczema behind the ears, or in the groin, or elsewhere in folds of skin. He must not, except of necessity, vaccinate in cases where there has been recent exposure to the infection of diseases such as measles, scarlatina, or diphtheria, nor where erysipelas is prevailing in or about the place of residence.

(2.) A certificate of postponement must be given by the Public Vaccinator in the form prescribed by the Local Government Board or to the like effect—

(*a*.) If in his opinion the child is not in a fit and proper state to be vaccinated : or

(*b*.) If in his opinion the child cannot be safely vaccinated on account of the condition of the house in which it resides or because there is or has been a recent prevalence of infectious disease in the District; and in any such case the Public Vaccinator is required to forthwith give notice of such certificate to the Medical Officer of Health for the District in the Form P. set out in the Fifth Schedule to this Order or to the like effect.

(3.) All public vaccinations are to be performed with glycerinated calf lymph, or with such other lymph as may be issued by the Local Government Board. If the parent or other person having the custody of a child requires that it shall be vaccinated with lymph issued by the Local Government Board, the vaccination must be performed with such lymph.

(4.) The Public Vaccinator must keep such record of the lymph he uses for vaccinating as will enable him always to identify the origin of the lymph used in each operation. He must not employ lymph supplied by any person who does not keep an exact record of its source.

(5.) The Public Vaccinator must keep in good condition the lancets or other instruments which he uses for vaccinating, and he must not use them for any other purpose whatever. When he vaccinates he must cleanse and sterilise his instrument after one operation before proceeding to another, and must always, when vaccinating, have with him the means of doing this. When once he has unsealed a tube of lymph he must never attempt to keep any part of its contents for the purposes of vaccination on a future occasion. Under no circumstances should the mouth be applied directly to the tube in which the lymph is contained for the purpose of expelling the lymph. In the case of ordinary capillary tubes an artificial blower may properly be used for this purpose.

(6.) Vaccination should at every stage be carried out with aseptic precautions. These should include: 1st, the cleansing of the surface of the skin before vaccination; 2nd, the use of sterilised instruments; and 3rd, the protection of the vaccinated surface against extraneous infection both on the performance of the operation and on inspection of the results.

Advice as to the precautions to be taken in this respect until the scabs have fallen and the arm has healed should always be given to the person having the custody of the child.

(7.) In all ordinary cases of primary vaccination the Public Vaccinator must aim at producing four separate good sized vesicles or groups of vesicles, not less than half an inch from one another. The total area of vesiculation resulting from the vaccination should not be less than half a square inch.

(8.) The Public Vaccinator must enter all cases in his Register on the day when he vaccinates them, together with all particulars required in the Register up to and including the column headed "Initials of person performing the vaccination." The results of the vaccination, which must be attested by the initials of the person who inspects the case, are to be entered upon the day of inspection. In cases of successful primary vaccination the Public Vaccinator must record the number of separate scarified areas, punctures, or groups of punctures made and the number of separate normal vaccine vesicles or groups of vesicles which have been produced. In cases of re-vaccination he must register as "successful" only those cases in which either vesicles, normal or modified, or papules surrounded by areolæ have resulted. When any operation (whether vaccination or re-vaccination) has to be repeated owing to want of success in the first instance it should be entered as a fresh case in the Register.

FOURTH SCHEDULE.

Instructions to Vaccination Officers.

1. The duties of the Vaccination Officer will be to act as Registrar of Vaccination for the district to which he is appointed; to see that all children resident therein are duly accounted for as regards Vaccination; and generally to carry into effect all such provisions of the Vaccination Acts, 1867 to 1898, and the orders made thereunder as are applicable to his office.
2. The Vaccination Officer shall receive from the Registrars of Births and Deaths the "Monthly Lists" of births and deaths which will be sent to him under the provisions of the Vaccination Act, 1871, and he shall be responsible for the safe custody of such lists, and of any lists which were sent to any former Vaccination Officer and have been transferred to him. In the columns which are provided for the purpose, in the part of the "Monthly Birth-List" sheets headed "Vaccination Register," he shall

[illegible]

[illegible]

[illegible]

If any list of births or deaths is not received from a Registrar within one week from the time [illegible], the Vaccination Officer shall report this in writing to the Guardians at their next meeting, with a view to the Registrar being immediately called upon for an explanation. The Vaccination Officer shall [illegible] forward to the Local Government Board a copy of such report. If the Vaccination Officer shall lose any of the lists of births or deaths, he shall obtain another from the Registrar at his own cost.

4. The Vaccination Officer shall see that all children entered in the birth-lists of his district are either duly vaccinated, or are otherwise properly accounted for in accordance with the law relating to vaccination.

He shall keep the birth-lists examined from week to week, and—

(a) on the expiration of seven days after any child entered therein shall have attained the age of four calendar months, if none of the certificates mentioned in paragraph 2 of these Instructions shall have been received by the Vaccination Officer and he shall not have in his possession a valid certificate of postponement in respect of such child, he shall proceed to make personal inquiries with a view to obtain the requisite certificate. If on making these inquiries he is unable to ascertain that the child has been vaccinated, but the child is still resident in the district, the Vaccination Officer shall include the name and home of the child in a List which he shall send to the Public Vaccinator in the Form H. set out in the Fifth Schedule to this Order. The Lists shall be sent once in each week, and the name of each child shall be included in such list that it may reach the Public Vaccinator within three weeks of the child having attained the age of four months. The date of sending the notice to the Public Vaccinator must be entered by the Vaccination Officer in his Report Book.

(b) Where the Vaccination Officer shall have excluded from the List sent by him to the Public Vaccinator the name and home of any child owing to his having in his possession a valid certificate of postponement in respect of such child, he shall on the expiration of such certificate forthwith deal with the case in the manner prescribed by subdivision (*a*) of this paragraph, unless such certificates shall be immediately renewed, or he shall have received in respect of the child one of the certificates mentioned in paragraph 2 of these Instructions.

(*c*.) If the Vaccination Officer shall receive any certificate of postponement relating to any child not resident in his district he shall, so far as possible, transmit such certificate to the Vaccination Officer of the district where the child is resident.

(*d*.) If the Vaccination Officer has not received in respect of any child a certificate under Section 2 of the Vaccination Act, 1898, within the time limited by that section, and at the end of seven days after the expiration of six calendar months from the birth of the child, has not received any other of the certificates mentioned in subdivision (*a*) of this paragraph, the Vaccination Officer shall forthwith give a notice in the Form K. set out in the Fifth Schedule to this Order or to the like effect to the parent or other person having the custody of the child by delivering the same by post or otherwise at the last known residence of such parent or person. If that notice is not duly complied with within the time specified therein, it will become the duty of the Vaccination Officer under the Vaccination Act, 1871, to take proceedings for the enforcement of the law.

7. The Vaccination Officer shall at all times use his best endeavours to ascertain whether children resident in his district, but not born in it, or, if so born, not having had their births registered in it, are unvaccinated, and it will be his duty in such cases, subject to the provisions of the Vaccination Acts, 1867 to 1898, to take the requisite steps for procuring their vaccination. Paragraph 6 of these Instructions shall apply to such cases with the necessary modifications.

8. The Vaccination Officer shall keep a book, to be called "The Vaccination Officer's Report Book," in the Form M. set out in the Fifth Schedule to this Order, in which he shall forthwith enter the particulars required with regard to children as to whom personal inquiries may have been made, with the dates of such inquiries, and also all certificates of postponement with the date of the certificate, the cause for which it was given, the name of the practitioner who signed it, and the period for which it was given, with a view to any inquiries which may be necessary at the expiration of that period. When certificates of postponement are delivered to him on the form attached to the "Notice of the Requirement of Vaccination," he shall see that the parent or other person having the custody of the child is supplied with a new form of this Notice, with the required particulars duly filled in. Copies of the form of "Notice of the Requirement of Vaccination" can be obtained by the Vaccination Officer on his applying to the Registrar. He shall note in the Report Book any further action taken in any case, and make any remarks which the case calls for. He shall take care to make the necessary reference in Column VI. of the "Vaccination Register" to each case entered in the Report Book.

9. When the Vaccination Officer finds that a child has been successfully vaccinated, but that the vaccination has not been duly certified, or that any certificate of postponement, of insusceptibility, or of the child having had small-pox, has been given but has not been transmitted, he shall ascertain with whom the default rests, having regard to the requirements of Sections 21, 23, and 30 of the Vaccination Act, 1867, and Section 7 of the Vaccination Act, 1871, and shall forthwith take the necessary steps for obtaining the certificate required.

10. The Vaccination Officer shall carefully examine every certificate received by him, and shall not accept any certificate not signed by a registered medical practitioner, or, in the case of a certificate under Section 2 of the Vaccination Act, 1898, not signed as required by that section.

11. When the Vaccination Officer shall find that the parent or other person having the custody of any child, respecting whom he has not received a certificate of the kind referred to in paragraph 2 of these Instructions, has removed from the district, he shall take pains to ascertain the Vaccination

App. A. No. 6.

Vaccination Order, 1898.

Officer's district to which such removal has taken place, and shall give notice to the Vaccination Officer of that district, with a view to the vaccination of the child, and the due transmission to him of a copy of the necessary certificate. And whenever a certificate respecting a child whose birth was registered in the district of some other Vaccination Officer is sent to him, he shall take pains to ascertain the district in which the birth took place, and when he has ascertained it, he shall forward to the Vaccination Officer of that district a copy of the certificate.

12. The Vaccination Officer shall keep a book in the Form N. set out in the Fifth Schedule to this Order, in which he shall enter a Summary of his Proceedings under the Vaccination Acts, 1867 to 1898, in each month.

13. The Vaccination Officer shall prepare at the end of every half-year a summary of the vaccinations in his district, and at the commencement of each year a supplemental return of vaccination in his district, in the forms prescribed and issued by the Local Government Board, and shall submit one copy of each of such summaries to the Guardians, and shall transmit another copy to the Local Government Board, and shall himself preserve another copy for reference. He shall also furnish such other returns to the Guardians and Local Government Board as the latter may direct.

14. The Vaccination Officer shall, on any outbreak of small-pox, make such house to house visitations as the Local Government Board or the Guardians may direct in reference to vaccination, and carry out any special instructions which the Board or the Guardians may issue on the subject.

15. The Vaccination Officer shall see that the Registrars of Births and Deaths in his district are kept informed of his own name and place of abode or office, in order that the address on the notices of the requirement of vaccination delivered by the Registrar to parents may be correct.

16. The Vaccination Officer shall undertake the distribution of the certificates, books, and other forms issued by the Local Government Board to the Public Vaccinators and Medical Practitioners in his district, and shall, on request, furnish any parent or other person having the custody of a child with a copy of the Form G. set out in the Fifth Schedule to this Order duly filled in, and with the name and address of the Public Vaccinator written on the back thereof.

17. The Vaccination Officer shall be responsible for the safe custody of the Vaccination Registers, except any bound Registers which do not contain any entry of a birth registered within the last preceding seven years, and which he may, with the Guardians' consent, have deposited in the Union Offices.

18. The Vaccination Officer shall preserve every certificate received by him, and the lists, in the Form H. set out in the Fifth Schedule to this Order, which shall have been returned to him by the Public Vaccinator, until after the inspection by an Inspector of the Local Government Board of public vaccination in the Union next following the date of the certificate, and shall not, in any case, destroy any of them until two years have elapsed from its date.

FIFTH SCHEDULE.

FORM A.

THE VACCINATION ACTS, 1867 TO 1898.

Notice of the Requirement of Vaccination.

Copy hereunder the No. of the Entry of the Child's Birth from the Register Book.
Entry No. }

To the Father, or Mother, or other Person having the Custody of the Child herein named.

1. I, the undersigned, hereby give you Notice to have the Child named [1] , whose birth is now registered, vaccinated by a Public Vaccinator or some other Medical Practitioner, pursuant to the provisions of the Vaccination Acts, 1867 to 1898.

2. These Acts require every child to be vaccinated before it is six months old. The Vaccination may, however, be postponed by Medical Certificate, if the child is not in a fit state to be vaccinated, or if, in the opinion of the Public Vaccinator, the condition of the house in which the child resides is such, or there is or has been such a recent prevalence of infectious disease in the district, that the child cannot be safely vaccinated.

3. If you desire the child to be vaccinated by the Public Vaccinator before it is four months old, you should give notice to him in the following form, or to the like effect:—

To ,
Public Vaccinator of the District of
the Union.

In accordance with Section 1 (2) of the Vaccination Act, 1898, I hereby request that you will visit [1]
for the purpose of vaccinating [2] who is now residing at that address.

Dated this day of 18 .

(*Signed*) ,
Parent or other Person having the custody of the said Child.

If you desire it, you can obtain from the Vaccination Officer a copy of this form, with the name and address of the Public Vaccinator. The Public Vaccinator will visit the child's home for the purpose of vaccinating the child not earlier than 9 o'clock in the morning nor later than 4 o'clock in the afternoon, unless some other time shall have been arranged between him and you.

4. If within a week after the child has attained the age of four months, the Vaccination Officer has not received a certificate of its successful vaccination, or of its insusceptibility to vaccination, or of its having had small-pox, and has not in his possession a valid certificate of postponement of the vaccination of the child, and has not received such a certificate as is herein-after mentioned in paragraph 6, the Vaccination Officer will give notice to the Public Vaccinator, and the Public Vaccinator will call at the home of the child before the child attains the age of six months and will offer to vaccinate the child with glycerinated calf lymph, or such other lymph as may be issued by the Local Government Board.

5. The Public Vaccinator will give you at least 24 hours' notice of his intention to visit the home of the child as mentioned above in paragraphs 3 and 4; and the visit will, in the absence of any sufficient reason for delay, be made within two weeks after receipt of the notice from you or from the Vaccination Officer, as the case may be. If, when the Public Vaccinator visits the home of the child for the purpose of vaccinating it, or of offering to vaccinate it, you request that the vaccination should be performed with lymph issued by the Local Government Board. the Public Vaccinator will use such lymph.

6. You will be exempt from any penalty under Section 29 or Section 31 of the Vaccination Act, 1867, for not having the child vaccinated, if within four months from the birth of the child you satisfy two justices, or a stipendiary or Metropolitan Police Magistrate, in petty sessions, that you conscientiously believe that vaccination would be prejudicial to the health of the child, and within seven days thereafter deliver to the Vaccination Officer for the district a certificate by such justices or magistrate that they are satisfied accordingly.

7. After the vaccination has been performed the child must be inspected by the Vaccinator, in order that, if the operation has been successful, he may fill up and sign the requisite certificate.

8. The Vaccinator will give his certificate in one of the annexed forms, and for this purpose *this paper* should be produced to him. If he is a Public Vaccinator it will be *his* duty to forward the paper to the Vaccination Officer; but if he is not a Public Vaccinator it will be *your* duty, after the Certificate has been duly filled up and signed, to forward this paper to the Vaccination Officer, whose name and address are on the back.

Dated this day of , 18 .

(*Signature of Registrar*)

Registrar of Births and Deaths for the Sub-District of
in the Superintendent Registrar's District of

APP. A. N[o]
Vaccinatio[n] Order, 1898.

1 Child's na[me] and surnam[e]

1. Address [of] the child.
2. Child's na[me] and surnam[e]

App. A. No 8.

Vaccination Order, 1898.

FORM B.

THE VACCINATION ACTS, 1867 TO 1898.

Medical Certificate of Postponement of Vaccination owing to the state of the Child's Health.

Directions for filling up this Certificate.

[1] Child's name and surname.
[2] Father's or (if the child is illegitimate) Mother's name and surname.
[3] Child's age.
[4] No. or name of the House, and name of the Street or Road, and Parish, and County or Borough.
[5] This must not exceed two calendar months from the date of the Certificate.
[6] If the person signing is not a Public Vaccinator, strike out this line.

I, the undersigned, hereby certify that I have this day examined [1] the child of [2] aged born at [4] in the Parish (Township) of [4] in the County (Borough) of [4] and residing at [4] in the Parish (Township) of [4] in the County (Borough) of [4] and am of opinion that the said child is in the following state of health, namely

and is therefore not in a fit and proper state to be successfully vaccinated. I do hereby postpone the Vaccination until the [5] day of .

Dated this day of , 18 .

(*Signed*) ,

[6][Public Vaccinator of the Union of .]

Medical Practitioner duly registered.

FORM C.

THE VACCINATION ACTS, 1867 to 1898.

Medical Certificate of Postponement of Vaccination owing to the condition of the House, or the recent prevalence of Infectious Disease in the District.

Directions for filling up this Certificate.

[1] Child's name and surname.
[2] Father's or (if the child is illegitimate) Mother's name and surname.
[3] Child's age.
[4] No. or name of the House, and name of the Street or Road, and Parish, and County or Borough.
[5] Strike out the words which do not apply to the case.
[6] This must not exceed two calendar months from the date of the Certificate.

I, the undersigned, hereby certify, that [1] the child of [2] aged [3] born at [4] in the Parish (Township) of [4] in the County (Borough) of [4] and residing at [4] in the Parish (Township) of [4] in the County (Borough) of [4] cannot be safely vaccinated because [5] of the condition of the house in which the child resides [or because of the recent prevalence of infectious disease in the District].

For the above reason I do hereby postpone the vaccination of the child until the [6] day of .

Dated this day of , 18 .

(*Signed*) .

Public Vaccinator of the Union of

FORM D.

THE VACCINATION ACTS, 1867 TO 1898.

Medical Certificate of Insusceptibility of successful Vaccination, or of Child having had Small Pox.

I, the undersigned, hereby certify that [1]
the child of [2] aged [3]
born at [4] in the Parish
(Township) of [4] in the County (Borough)
of [4] and residing at [4]
in the Parish (Township) of [4] in the County (Borough)
of [4] [5] [has been [6] times unsuccessfully vaccinated by
me, and is, in my opinion, insusceptible of successful Vaccination] *or* [5] [has already
had Small Pox].

Dated this day of 18 .

(Signed)

[7] [Public Vaccinator of the Union of]
Medical Practitioner duly registered.

Directions for [filling] up this Certifi[cate].

[1] Child's nam[e and] surname.
[2] Father's or [if] child is illegit[imate] Mother's name [and] surname.
[3] Child's age.
[4] No. or name [of] House, and na[me of] the Street or Ro[ad,] Parish, and Cou[nty or] Borough.
[5] Strike out [the] words which [do not] apply to the cas[e.]
[6] This numbe[r must] not be less than [...]
[7] If the perso[n sign]ing is not a [Public] Vaccinator stri[ke out] this line.

FORM E.

THE VACCINATION ACTS, 1867 TO 1898.

Medical Certificate of successful Vaccination.

The Registrar to insert the No. of the Entry of the Child's birth in the Register Book.
Entry No. }

I, the undersigned, hereby certify, that [1]
the child of [2] aged [3]
born at [4] in the Parish
(Township) of [4] in the County
(Borough) of [4] and residing
at [4] in the Parish (Township) of [4]
in the County (Borough) of [4] has been successfully
vaccinated by me.

Dated this day of 18 .

(Signed)

[5] [Public Vaccinator of the Union of]
Medical Practitioner duly registered.

Directions for [filling] up this Certifi[cate].

[1] Child's nam[e and] surname.
[2] Father's or [if] child is illegit[imate] Mother's name [and] surname.
[3] Child's age.
[4] No. or name [of] House, and na[me of] the Street or Ro[ad,] Parish, and Cou[nty or] Borough.

[5] If the perso[n sign]ing is not a [Public] Vaccinator, stri[ke out] this line.

APP. A. No. 6.

Vaccination Order, 1898.

FORM F.

THE VACCINATION ACTS, 1867 TO 1898.

Medical Certificate under Section 12 of the Vaccination Act, 1871, of successful Vaccination.

Directions for filling up this Certificate.

[1] Child's name and surname.
[2] Father's or (if the child is illegitimate) Mother's name and surname.
[3] Child's age.
[4] No. or name of the House, and name of the Street or Road, and Parish, and County or Borough.
[5] Child's present residence.

I, the undersigned, being a Public Vaccinator of the Union of hereby certify, that [1] the Child of [2] aged [3] born at [4] in the Parish (Township) of [4] in the County (Borough) of [4] and residing at [5] has been examined by me, and that I find the said child to have been successfully vaccinated.

Dated this day of 18 .

(*Signed*)

Public Vaccinator of the Union of

FORM G.

THE VACCINATION ACTS, 1867 TO 1898.

Request for the Attendance of the Public Vaccinator to vaccinate a Child.

[1] Address of the child.
[2] Child's name and surname.

To

Public Vaccinator of the District of the Union.

In accordance with Section 1 (2) of the Vaccination Act, 1898, I hereby request that you will visit [1] for the purpose of vaccinating [2] who is now residing at that address.

Dated this day of 18 .

(*Signed*) ,

Parent or other Person having the custody of the said Child.

FORM H.

APP. A. No.

Vaccination Order, 1898.

THE VACCINATION ACTS, 1867 TO 1898.

List of Children in respect of whom the necessary Certificates have not been received by the Vaccination Officer.

To

Public Vaccinator of the District of the Union.

In accordance with paragraph 6 (*a*) of the Instructions to Vaccination Officers in the Fourth Schedule to the Vaccination Order, 1898, I hereby give you notice that the children whose names and addresses are stated below, and with respect to whom I have not received the necessary certificates under the Vaccination Acts, 1867 to 1898, attained the age of four months on the dates respectively specified in Column 3.

Dated this day of , 18 .

(*Signed*)

Vaccination Officer for the .

Child's Name. (1.)	Child's Address. (2.)	Date on which the Child attained the Age of Four Calendar Months. (3.)	Dates of Notice by Public Vaccinator to Parent. (4.)	Date of Visit. (5.)	Result of Visit. (6.)	Remarks. (7.)

N. B. — Columns (4) to (6) are to be filled up by the Public Vaccinator.

NOTE.—It is the duty of the Public Vaccinator to visit the homes of these children within two weeks after receipt of this notice, and to offer to vaccinate them in manner provided by the Vaccination Act, 1898, and the Order of the Local Government Board made thereunder.

APP. A. No. 6.

Vaccination Order, 1898.

FORM I.

THE VACCINATION ACTS, 1867 TO 1898.

Notice from Public Vaccinator to Parent or other Person having Custody of Child of intended Visit.

Directions for filling up this Notice.

[1] Name and address of parent or other person having custody of the child.
[2] Name of child.
[3] Date of intended visit.

To [1]

In accordance with Section 1 (3) of the Vaccination Act, 1898, I hereby giv you notice that I shall visit the home of the child [2] on [3] , and shall offer to vaccinate it wit glycerinated calf lymph, or such other lymph as may be issued by the Loc Government Board.

Dated this day of , 18 .

(*Signed*) ,

Public Vaccinator of the District

of the Union.

Address of Public Vaccinator

FORM K.

THE VACCINATION ACTS, 1867 TO 1898.

Notice of Default.

Directions for filling up this Notice.

[1] Child's name.

[2] Strike out the words which do not apply to the case.

To .

WHEREAS you are in default under the above Acts, respecting the Chil [1]

I hereby require you [to have the said Child vaccinated within fourteen days fro the date hereof, and do all other things the law requires touching the sa Vaccination [2]] *or* [to transmit to me within seven days from the date hereof t requisite Certificate concerning the Vaccination of the said Child [2]], failing whi it will be my duty to take the proper steps for securing the enforcement the law.

Dated this day of , 18 .

(*Signed*)

Vaccination Officer for

Address of Vaccination Officer

FORM L.

Vaccination Register.

Union.

District.

Vaccination Officer.

Register of Certificates.			Date of Certificate under Section 2 of the Vaccination Act, 1898.	Date of Death in case of Child having died before Vaccination.	Reference to consecutive number in the Officer's "Report Book," in cases transferred thereto.
Date of Medical Certificate of *Successful* Vaccination.	Date of Certificate of *Insusceptibility* or of having had *Small Pox.* (Enter "Ins." or "S.P.," as the case may be.)	Name of the Medical Man by whom the Certificate is signed.			
I.	II.	III.	IV.	V.	VI.

. A. No. 6.

ination
r, 1898.

FORM B

Vaccination Officer's Report Book

UNION. *Vaccination Officer.*

Consecutive Number in this Book.	Birth Registration District.	No. on Birth Register.	Name of Child.	Date of Birth.	Address of Parent.	Date or Dates of Personal Inquiries.	Vaccination Postponed by Medical Certificate.				Case not Found; or Parent Removed out of District, and where.	Date of Notice to Public Vaccinator to visit Home of Child.	Case duly accounted for, and entered in "Vaccination Register."	Date of Notice sent in case of Default.	Note of any Proceedings taken.	REMARKS.
							Date of Certificate.	By whom given.	Cause for which it was Postponed.	Date to which Postponed.						
1.	2.	3.	4.	5.	6.	7.	8.	9.	10.	11.	12.	13.	14.	15.	16.	17.
64																
65																
66																
67																
68																
69																
70																
71																
72																
73																
74																
75																
76																
77																
78																
79																
80																

FORM N.

Summary of Proceedings under the Vaccination Acts, 1867 *to* 1898.

UNION.

Month of , 18 .

Vaccination Officer.

No. of Cases in Birth Lists received during Month.	No. of Certificates of Vaccination received.	No. of Certificates of Postponement owing to			No. of Certificates under Section 2 of Vaccination Act, 1898.	No. of Certificates of Insusceptibility or of having had Small Pox.	No. of Cases.		No. of Entries in Lists sent to Public Vaccinator.	Proceedings taken, showing in each Class of Cases the stage at the end of the Month.			Costs incurred in Proceedings.	Costs received.	Dates of several Payments to Treasurer.	REMARKS.
		Health of Child.	Condition of House.	Prevalence of Infectious Disease.			Parents removed out of District.	Otherwise not found.		Under Sec. 29 of Vaccination Act, 1867, or Sec. 7 of Vaccination Act, 1871.	Under Section 31 of Vaccination Act, 1867. Applications for Justices' Orders.	Under Section 31 of Vaccination Act, 1867. Proceedings for Penalties on Default under Orders.				
1.	2.	3.	4.	5.	6.	7.	8.	9.	10.	11.	12.	13.	14.	15.	16.	17.
										No. of Cases.	No. of Cases.	No. of Cases.				
										Summonses taken out.	Orders applied for.	Summonses taken out.				
										Convictions.	Orders granted.	Penalties imposed.				
										Cases dismissed.	Orders refused.	Cases dismissed.				
										Cases adjourned.	Cases adjourned.	Cases adjourned.				

FORM O.

Vaccinator's Register.

VACCINATOR'S REGISTER of the DISTRICT of the *Public Vaccinator.* UNION.

No. of Case (consecutive to 500, and then to be repeated).	2. Date of Vaccination.		5. Age.		6. Residence.	7. Where vaccinated: state Address. — If a primary Vaccination or Re-vaccination at Surgery, mark S	8. Record by which the source of Lymph may be traced. Thus, if it is obtained from the National Vaccine Establishment insert the letters N.V.E., together with the Official Reference Number.	9. No. of separate Scarified areas, Punctures, or Groups of Punctures made.	10 Initials of Person performing the Vaccination.	11. Date and Place of Inspection.	12. Result.		13. Initials of the Person inspecting.	14 Date of sending Certificate to the Vaccination Officer.	15. Fee due in respect of each successful case of primary Vaccination or Re-vaccination.	REMARKS.
			Years.	Months.							Successful Number of separate Vesicles or Groups of Vesicles produced.					

FORM P

APP. A. No. 6.

Vaccination Order, 1898.

THE VACCINATION ACTS, 1867 TO 1898.

Notice from Public Vaccinator to Medical Officer of Health.

To

Medical Officer of Health of the District.

In accordance with Section 1 (4) of the Vaccination Act, 1898, I hereby give you notice that I have this day postponed the vaccination of [1] the child of [2] who resides at [3] on account of
[4] the condition of the house in which the child resides.
[4] the recent prevalence of infectious disease in the district.

Dated this day of , 18 .

(*Signed*) ,

Public Vaccinator of the District.
Union.

[1] Child's name.
[2] Father's or (if the child is illegitimate) Mother's name and surname.
[3] Child's residence.
[4] Strike out that cause which does not apply.

Given under the Seal of Office of the Local Government Board, this Eighteenth day of October, in the year One thousand eight hundred and ninety-eight.

(L.S.) HENRY CHAPLIN,
President.

J G H OWEN,
Secretary.

No. 7.

App. A. No. 7.

Abstract of Medical Inspections.

ABSTRACT of MEDICAL INSPECTIONS made in 1898 with regard to the INCIDENCE of DISEASE on particular places, and to questions concerning LOCAL SANITARY ADMINISTRATION.

1. ALDEBURGH-ON-SEA (SUFFOLK); population (1891), 2,159; Dr. Deane Sweeting.

Authority concerned: Town Council of Aldeburgh.

Ground of Inquiry: Local complaints as to the quality of the water supplied to the town by the Aldeburgh Water Company.

Chief Facts reported by Inspector: Old source of supply, condemned by Sir E. Frankland in 1896, obtained from a shallow well, in relation with an elongated ditch impounding water from a marsh. New source of supply, condemned by the same chemist, and by Mr. Cassal in 1898, but reported upon not unfavourably by Dr. Thresh in the same year. New supply merely soil water from a gathering ground co-terminous with the "main mass" of the Suffolk Coralline Crag. This area very porous, and subject to pollution on its surface and within the soil, by various kinds of filth. Excessive pumping operations at works not unlikely to facilitate entrance to the supply of cesspool impurities from foreshore. Filtration altogether inefficient. Supply inadequate for town during visitors' season.

2. ALNWICK URBAN DISTRICT (NORTHUMBERLAND); population (1891), 6,746; Dr. Buchanan.

Authority concerned: Alnwick Urban District Council.

Ground of Inquiry: Complaint by Northumberland County Council of neglect of District Council to deal with unhealthy dwellings of poorer classes in Alnwick.

Chief Facts reported by Inspector: An old town, formerly walled, the centre of a large agricultural district. Large proportion of population of poorer class, living in unhealthy, damp, and overcrowded dwellings packed away in narrow courts behind main thoroughfares. For decades, excess in mortality from all causes, and notably in phthisis mortality. Overcrowding of persons somewhat diminished, but still very common. Houses in the courts subdivided into one or two-roomed tenements; no through ventilation; much dampness owing to thick porous sandstone walls; dwellings crowded on area; frequently dilapidated. Courts badly paved; nuisances from refuse of stables and cowsheds therein. Some houses for working classes of satisfactory sort, recently built at instance of Earl Percy, much in demand. Local authority inactive as regards steps that can be considered at all adequate to remedy unhealthiness of courts. Unhealthy areas which should be dealt with by schemes under Housing of Working Classes Act defined, and statistics given.

District Council and preceding Local Board to be credited with useful work in other directions. Sewerage on the whole satisfactory; water-closets general, all with flushing apparatus; scavenging fairly efficient; improvement in water supply, although certain contributory sources of supply hardly satisfactory. Sanitary staff capable. Need for attention to cowsheds and common lodging houses.

3. **Bettws-y-coed Rural District (Carnarvon)**; population (1891), 5,598; Dr. Wheaton.

Authority concerned: Bettws - y - coed Rural District Council.

Ground of Inquiry: Repeated outbreaks of enteric fever and diphtheria, and local complaints as to persistent dangerous nuisances.

Chief Facts reported by Inspector: Dwellings mostly well-built, but their surroundings unsatisfactory. Water supply of Bettws, Trefriw, and Cwm Penmachno, unsatisfactory. No sewers at Capel Curig or Cwm Penmachno. All liquid refuse from dwellings discharged into streams without purification. Excrement disposal chiefly by privies, which are often built over watercourses. No public scavenging; accumulations of filth in privy receptacles a marked feature. Condition of cowsheds very unsatisfactory. The Infectious Disease (Notification) Act, 1889, not adopted. No hospital provision, no disinfecting apparatus.

Enteric fever occurred at Bettws-y-coed and at Trefriw in 1897. At Bettws the majority of the attacks occurred in a locality which was unsewered, and in which house drainage and excrement disposal were very defective. An outbreak of diphtheria commenced in the summer of 1897 at Penmachno and at Cwm Penmachno, still in progress at date of visit; 116 attacks had occurred in a population of 1,550 persons; the majority were at Cwm Penmachno. Almost complete absence of all sanitary requirements at the latter village.

4. **Burton-on-Trent (Stafford)**; population (1891), 46,047; Dr. Theodore Thomson.

Authority concerned: Burton-on-Trent Town Council.

Ground of Inquiry: Outbreak of measles; Report of Medical Officer of Health.

No. 7. *Chief Facts reported by Inspector:* The outbreak commenced in January 1898, and persisted in the town until the month of August: its spread from one to another part of the district, not being rapid. Parts of the town escaped the disease, save for a few scattered cases: nevertheless, some 2,000 cases are known to have occurred. The source of origin of the earliest cases not ascertained. Subsequent spread of the disease was materially contributed to by schools: there was also, however, insufficient isolation of the sick from the healthy at their homes. Information as to occurrence of measles in families of which members attend school, was not systematically furnished to the Local Authority by the School Authorities. The sanitary staff proved insufficient to carry out the measures called for in attempt to control the epidemic.

5. **Camborne**; population (1891), 14,700; the **Parish** of **Illogan**; population, 2,107; and the **Village** of **Praze**; population, 400 (**Cornwall**); Dr. Bruce Low.

Authorities concerned: The Urban District Council of Camborne; the Rural District Council of Redruth; and the Rural District Council of Helston.

Ground of Inquiry: Sudden outbreak of enteric fever Reports from the local Medical Officers of Health.

Chief Facts reported by Inspector: The town of Camborne, the parish of Illogan, and the village of Praze were invaded by enteric fever in December, 1897; the outbreak lasted about six weeks, 165 cases and 12 deaths being recorded. The only condition common to these three localities was their water supply, which was from the service of the Camborne Water Company. One of the two sources of this supply was from some springs at Boswyn, to augment which section of the supply a pipe had been laid from a brook to the service tank. The brook was fed partly by surface water from a hill side on which, in a cottage, four cases of enteric fever occurred in October and November, 1897. The bowel discharges of these cases reached a pond close to the house which had no privy. In wet weather the pond overflowed, and the water from it ran down a channel to the brook, entering it about 100 yards above the intake pipe of the Camborne Company. In Mid-November the rainfall was excessive, and the flood-water washed out the pond and carried the pollutions to the brook, and thence to the service reservoir. After removal of the pipe connecting the brook with the reservoir, and after cleansing the service tank and flushing the mains, the epidemic abruptly ceased. The sanitary condition of Camborne, Illogan, and Praze far from satisfactory.

. **Carnarvon Borough**; population (1891), 9,804; Dr. Wheaton.

Authorities concerned: Carnarvon Town Council; Gwyrfai and Glaslyn Rural District Councils.

Ground of Inquiry: Complaint from Carnarvon Town Council as to pollution of the water supply of the Borough by liquid refuse from the village of Rhyd-ddu, situate in the Gwyrfai and Glaslyn Rural Districts.

Chief Facts reported by Inspector: Water supply of the Borough derived from River Gwyrfai, the gathering-ground consisting of the whole of the valley of the Gwyrfai above the intake. No filtration of the water. In the course of the river above the intake are two lakes. Situate above the lower of the two lakes is the village of Rhyd-ddu. This village not sewered, and all liquid refuse from it discharged into the river, together with a certain amount of solid refuse from dwellings and of contents of pail privies. Other sources of possible pollution of the water of the Gwyrfai, viz., the liquid and solid refuse from farm buildings, of which there are a number situate on the gathering ground, and in some instances the contents of privies pertaining to farm-houses. A large quarry, at which 250 men are employed, forms an additional source of pollution, since all water draining from it falls into the lower of the two lakes. The surface water from main roads which falls into this lake, or into the river above the intake, forms an additional source of possible pollution of the water.

A certain amount of work, carried out of late, in the way of diversion of liquid refuse from the streams falling into the Gwyrfai, and also of providing dwellings on the gathering ground with moveable privy receptacles. Nothing hitherto done by the Gwyrfai or Glaslyn Rural District Councils to prevent the discharge of unpurified liquid refuse from the village of Rhyd-ddu into the river.

Chichester (Sussex); population (1891), 7,830; Dr. Theodore Thomson.

Authority concerned: Chichester Town Council.

Ground of Inquiry: Enteric Fever: Memorial from rate-payers in Chichester.

Chief Facts reported by Inspector: With only occasional intermissions, enteric fever has prevailed in Chichester in serious amount for many years, as indicated by statistical information on this subject during the period 1870–98, and by a report to the Privy Council anterior to 1870. In recent years the incidence of the disease has been characterised by marked preference for particular neighbourhoods in Chichester; a preference, however, not observable in certain earlier years. Adequate explanation of this fever prevalence not to be found in

:. A. No. 7.
ract of
ical
ections.

circumstances connected with milk, shellfish, sewerage, water supply, rainfall, or subsoil water. Soil conditions favourable to the viability and growth of the infective material of enteric fever perhaps capable of affording such explanation. But, at present, the knowledge of conditions of this sort in general, and of the degree to which they prevail at Chichester in particular, insufficient to warrant definite conclusion as to the concern of soil circumstances with enteric fever in Chichester. Further investigation of this question at Chichester and elsewhere desirable.

8. CHRISTCHURCH (SOUTH HANTS); population (estimated), 4,432; Dr. Mivart.

Authority concerned: Christchurch Town Council.

Ground of Inquiry: General sanitary circumstances and administration of the Borough: Reports of the Medical Officer of Health; and local complaints.

Chief Facts reported by Inspector: Dwellings of the poorer class very defective. Water supply of a dangerous character. Out of 900 houses in Borough, 575 said to be supplied from public and private wells. Wells shallow; sunk in gravel and alluvial soil. Water often admittedly unwholesome. Some houses without means of water supply. Sewers badly constructed, gradients flat. Principal sewers still draining to river Stour, in contravention of Local Government Board's Order of 4th June, 1897. House drainage often imperfect. Excrement disposal and privy arrangements generally neglected and not seldom of a dangerous kind. Public scavenging very unsatisfactory. No isolation hospital. No disinfector. Urgent need of a new sewerage scheme. Administration by the Town Council generally dilatory as regards important sanitary questions.

9. DORE RURAL DISTRICT (HEREFORDSHIRE); population (1891), 7,112; Dr. Fletcher.

Authority concerned: Dore Rural District Council.

Ground of Inquiry: Plurality of Medical Officers of Health, and need for information as to local sanitary administration.

Chief Facts reported by Inspector: Indifference and inaction on the part of the District Council, and inefficient supervision on the part of the Medical Officers of Health, together with neglect of duties by, and incompetence of, Inspectors of Nuisances. Want of satisfactory water supplies, drainage for house slops, and proper closet accommodation. Existence of many unhealthy dwellings. Unsatisfactory isolation hospital accommodation in proximity to the workhouse. Absence of action under the Dairies, Cowsheds, and Milkshops Order of 1885.

10. Eton Rural District (Buckinghamshire); population (1891), 21,505; Dr. Johnstone.

Authority concerned: Eton Rural District Council.

Ground of Inquiry: Prevalence of diphtheria.—Local complaints.

Chief Facts reported by Inspector: Undue prevalence of diphtheria since 1890 as compared with Rural England. Increase in 1898, when epidemic specially affected three parishes: Langley Marish, Stoke Poges, and Denham. Type of epidemic severe under age of 15 years. There were 127 cases notified in 1898, with 34 deaths; 105 of these cases under age of 15 years, with 33 deaths. Disease spread by personal contact. School teachers not warned of disease in scholars' families.

Accumulations of London house refuse locally thought of as a cause of the diphtheria. No diphtheria amongst hands (some young) working at the refuse. No cause apparent for any special infectivity of refuse in 1898. No sufficiently marked incidence of diphtheria on dwellings in neighbourhood of refuse accumulations, nor other evidence to show that the refuse in question conduced to the prevalence of diphtheria in 1898 in the district.

Absence of efficient scavenging. No systematic cleansing of cesspools. Need, in places, of sewers. Damp sites. No available hospital provision.

11. Gainsborough (Lincs); population (estimated in 1897), 18,000; Dr. Darra Mair.

Authority concerned: Gainsborough Urban District Council.

Ground of Inquiry: Insanitary condition of district and prevalence of "fever."

Chief Facts reported by Inspector: Tendency of "fever" to appear all over town and throughout the year, with considerable increase in summer and autumn months; death rate from "fever" for past decade half as much again as in England and Wales, or in the great towns. Fever prevalence probably due to extensive pollution of soil of district, and possibly, occasionally, to use of imperfectly filtered river water. Overcrowding of dwellings on area in centre of town; numerous confined courts and "yards," for the most part very imperfectly paved and drained; excrement and refuse disposed of mainly in privies (a few pails) and in enormous middens; scavenging very inefficiently done under contract; sewers antiquated, of inadequate size, and liable to dangerous heading back of sewage; discharge of sewage directly into River Trent; water supply partly from a well in Sandstone 1,350 feet deep, and partly from River

Trent; river water, filtered, occasionally the only water supplied; under these circumstances filter beds inadequate; new well 1,450 feet deep being sunk in Sandstone.

Notification system not adopted; salary of Medical Officer of Health and Inspector of Nuisances inadequate; the latter superintends scavenging work, his ordinary routine duties being much neglected; small isolation hospital exists, but has not been used since 1888, in consequence of the high tariff of charges for treatment; a hot air disinfector, seldom used.

12. GLYNCORWG (GLAMORGANSHIRE); population (estimated in 1898), 6,000; Mr. Arnold Royle.

Authority concerned: Glyncorwg Urban District Council.

Ground of Inquiry: Insanitary condition of district and need for detailed information on the part of the Board.

Chief Facts reported by Inspector: No proper system of drainage; pollution of river; pollution of soil around houses; great difficulty in disposing of sewage owing to peculiar conformation of the valleys in which the inhabitants of the district reside; competent engineering advice needed; no isolation hospital; no proper disinfecting apparatus.

13. LONGTON AND FENTON (STAFFORDSHIRE); Borough of Longton: population (1891), 34,327; Fenton Urban District; population (1891), 16,998; Dr. Fletcher.

Authorities concerned: Longton Town Council and Fenton Urban District Council.

Ground of Inquiry: Long sustained and highly fatal prevalence of diphtheria in both districts; Registrar-General's returns.

Chief Facts reported by Inspector: Longton: Area, 1,948 acres. Water supply from the North Staffordshire Water Company. Prior to 1896, no special incidence of diphtheria; subsequently, attack-rates per 1,000 inhabitants equalled, in 1896, 2·41; in 1897, 15·29; and in first half of 1898, 9·25 (= 18·5 annual rate), with corresponding death rates of 0·596; 2·673; and 2·953. Death rates of "67 other large towns" for like periods were 0·25, 0·24, 0·28. Spread and maintenance of infection due to personal intercommunication; school influences being very clearly indicated, as also, but much less markedly, home influences. Absence of isolation hospital accommodation, and consequent neglect of isolation of patients. Insufficient steps taken, and inefficient methods adopted, to cope with the disease. Neglect of "throat sickness" not definitely notified as diphtheria. Regulation of attendance at school unsatisfactory. School closure not resorted to until the epidemic had got beyond control, when all twelve schools were closed indiscriminately. House-to-house inquiries in invaded neighbourhoods neglected. Dairies, Cowsheds, and Milkshops Order of 1885 not enforced. Bye-laws in need of revision.

Fenton: Area, 1,748 acres. Water supply from the North Staffordshire Water Company. Prior to 1896, no special incidence of diphtheria; subsequently, attack-rates per 1,000 inhabitants equalled, in 1896, 2·006; in 1897, 5·585; and in the first half of 1898, 5·8265 (= 11·653 annual rate), with corresponding annual death rates of 0·527, 1·913, and 2·431. Spread and maintenance of infection due to personal intercommunication, but home influences apparently more marked, and school influences less marked, than in the case of Longton. Attack-rate greatest on children under three years of age, and not on children of school age, as in Longton. Case mortality on females aged 3 to 15 years greatly in excess of that on males of corresponding age-period, and total case mortality much greater in Fenton than in Longton. Isolation hospital provided, but removals of patients relatively few. Insufficient steps taken, and inefficient methods adopted, to cope with the disease. Neglect of "throat sickness," as in Longton. School attendance regulated better than in Longton, but no school closure on account of diphtheria till the summer of 1898, and then only one school closed for a few days. House-to-house inquiries in invaded districts neglected.

14. LUNESDALE RURAL DISTRICT (LANCASHIRE); population (1891), 7,347; Dr. Fletcher.

Authority concerned: Lunesdale Rural District Council.

Ground of Inquiry: Joint tenure, by one person, of the Offices of Surveyor of Highways and Inspector of Nuisances for the whole District.

Chief Facts reported by Inspector: Area of the District, 75,734 acres. Length of District Roads, 201½ miles. (Seventeen county bridges, also, under the care of the Local Surveyor of Highways.) Water supplies for the larger villages may, in most instances, be looked on as satisfactory. For smaller villages, and for isolated houses, the water supplies are frequently of doubtful character. Sewerage is provided for the larger villages, but the disposal of sewage is, in some instances, unsatisfactory. Drainage of dwellings defective in places, and in some cases absent. Excrement disposal and removal unsatisfactory. Many privy nuisances. Deposits of refuse on banks of rivers. Generally,

A. No. 7.

house accommodation fair, but insufficient provisio of dwellings in Claughton. Cowsheds observe badly constructed, and overcrowded. Medical Offic of Health receives a salary of only £20 a yea The Inspector of Nuisances receives £50 as sala for that office, and £110 as salary for the office Surveyor of Highways. He also acts as Distri Surveyor, and as Canal Boat Inspector, witho remuneration. A capable officer, but not supported b his District Council. Much room for improvement i the District.

15. MARSTON, HOUGHAM, AND LONG BENNINGTON (VILLAGES) (LINCOLNSHIRE); population (1891), 302, 287, an 804 respectively; Dr. Wheaton.

Authority concerned: Claypole Rural District Council.

Ground of Inquiry: Prevalence of diphtheria; a vis having been made to Fulbeck, another village in t district, in 1896, on account of a similar prevalence diphtheria.

Chief Facts reported by Inspector: Forty-three attacks diphtheria occurred in the three villages betwe April 19th, 1897, and February 2nd, 1898, with sev deaths. The disease began in Marston and Hougha which are neighbouring villages, early in the year, b did not appear at Long Bennington until Novemb A number of instances of multiple attacks in househol Grossly unwholesome conditions found in connexi with the majority of invaded dwellings. Local suspici of sewage farm belonging to Grantham, and situate parish of Marston, as concerned with the diphtheria; evidence to confirm this. Extensive prevalence of so throat at Marston and Hougham previous to outbreak diphtheria; similar prevalence found among infants Long Bennington at time of inquiry.

Dwellings well constructed, but their surroundings ve unsatisfactory. Drainage very defective, several wat courses in the villages containing stagnant sewage; ma roughly constructed catchpits in connexion with hou drains. Water supply from wells and watercours Excrement disposal by objectionable vault privies; isolation hospital provision; no disinfecting apparatus.

16. ORMSKIRK (LANCASHIRE); population (1898), 6,797; I Copeman.

Authority concerned: Ormskirk Urban District Council.

Ground of Inquiry: Complaint from Justices of Pet Sessional Division of Ormskirk as to local insanita conditions.

Chief Facts reported by Inspector: Of the total population, no less than 1,298 persons, mostly of Irish nationality, find accommodation in various courts and alleys. These constitute the worst feature of the town, from a sanitary point of view. No cottages on neighbouring farmsteads, so that farm labourers employed within a radius of about seven miles from Ormskirk occupy dwellings in the town. Houses in courts closely packed together, and often built back-to-back; many without direct water supply. Present arrangements for scavenging insufficient.

17. STOW-ON-THE-WOLD RURAL DISTRICT (GLOUCESTERSHIRE AND WORCESTERSHIRE); population (1891), 7,800; Dr. Deane Sweeting.

Authority concerned: Stow-on-the-Wold Rural District Council.

Ground of Inquiry: General sanitary circumstances and adminstration; plurality of Medical Officers of Health.

Chief Facts reported by Inspector: Water supplies exposed to pollution; lack of or inefficient main drainage; faulty methods of sewage disposal, including pollution of rivers and streams; imperfect modes of excrement and refuse disposal; neglect of scavenging. No hospital provided; no proper disinfection; entire absence of bye-laws. Adoptive Acts not in force. Two medical officers of health, who rarely attend the Council meetings, and whose work is generally unsatisfactory. No concerted action taken or representation made by them. Work of Inspector of Nuisances satisfactory on whole.

18. SWINTON AND PENDLEBURY (LANCASHIRE); population (1891), 21,637; Dr. Theodore Thomson.

Authority concerned: Swinton and Pendlebury Urban District Council.

Ground of Inquiry: Marked prevalence of enteric fever in the district in recent years.

Chief Facts reported by Inspector: The death rate from enteric and continued fever in the district during the 10 years 1888–97 more than double that in the 33 great towns, and two-and-a-half times that of England and Wales during the same period. No evidence tending to incriminate water supply or sewerage conditions. Some localisation of the disease, suggesting its association with filthy conditions obtaining in certain localities in this district. These conditions mainly are, unpaved and badly paved yards, defective house and yard drains, and leaky privy middens: all being conditions conducive to serious fouling of the soil.

19. Tunbridge Wells (Kent); population (1891), 27,895; Dr. Buchanan.

Authority concerned: Tunbridge Wells Town Council.

Ground of Inquiry: Sustained high mortality from diphtheria during 1898.

Chief Facts reported by Inspector: Thirty deaths from diphtheria between January 1 and end of September, 1898, and 251 notified diphtheria cases during same period. Incidence particularly on poorer quarter of the town, in East Ward. Borough isolation hospital not used for diphtheria as it should have been. Action in repression of the outbreak in certain other respects open to criticism.

Health resort: large proportion of better class houses. Dwelling accommodation for persons of working class inadequate; overcrowding of persons in the smaller dwellings of the place. House drainage in many cases faulty; water-closets frequently "hand-flushed." Most sewers laid 40 to 50 years ago, often leaky, their ventilation sometimes unsatisfactory. Water supply from springs and deep wells near Pembury. Sanitary administration by Local Board up to 1889, lax. Some improvement since that date, but numerous sanitary shortcomings call for further action.

20. West Bromwich County Borough (Stafford); population (1891), 59,474; Dr. Buchanan.

Authority concerned: West Bromwich Town Council.

Ground of Inquiry: Re-inspection to ascertain progress made in remedying numerous insanitary conditions reported in 1895 by Dr. Buchanan, after local inquiry into enteric fever prevalence in West Bromwich.

Chief Facts reported by Inspector: During three years under review, some satisfactory progress in dealing with the numerous insanitary premises in the borough; action taken in this respect more systematic than formerly. Progress made in dealing with houses unfit for habitation, but as to this much remains to be done, and certain unhealthy areas should be dealt with as such. Paving and drainage of yards attended to, but nature of work effected at instance of Town Council often of a make-shift character. Opposition to substitution of water-closets for midden privies largely successful; policy of Town Council is now to offer "improvement" of privy midden as alternative to water-closet—an alternative accepted in most cases. Scavenging more efficient, but necessarily large accumulations of refuse on premises under the midden system. No improvement in cowsheds and milkshops, which, in several instances, contravene borough regulations. Improvement in common lodging-houses. Inspectorial staff increased, but increase hardly sufficient. Delay in adopting new byelaws. Byelaws for new streets and buildings badly needed. Hospital accommodation enlarged, and staff re-organised, but still difficulty in accommodating enteric fever patients.

21. WORTLEY RURAL DISTRICT (YORKS, WEST RIDING); population (estimated), 35,789; Dr. Reece.

Authority concerned: Wortley Rural District Council.

Ground of Inquiry: Plurality of Medical Officers of Health and need for information as to sanitary administration.

Chief Facts reported by Inspector: It would be feasible for one Medical Officer of Health, who was not employed in private practice, or who could give a large section of his time to the duties of that office, to act for the whole district. The Rural District Council had not appointed such an officer on the score of expense.

No. 8.

App. A., No. 8.

As to Dairies, Cowsheds and Milkshops Order, and Report of Royal Commission on Tuberculosis.

A.—Circular to Councils of Boroughs and Urban Districts as to the Dairies, Cowsheds, and Milkshops Order and Report of Royal Commission on Tuberculosis.

Local Government Board,

Whitehall, S.W.,

11th March, 1899.

Sir,

I am directed by the Local Government Board to advert to (1) the Dairies, Cowsheds and Milkshops Order of 1885, and (2) the Report made last year by the Royal Commission on Tuberculosis, and especially to the paragraphs in that Report which relate (*a*) to Milk, (*b*) to the qualifications of Meat Inspectors, and (*c*) to the principles which should be observed as regards the seizure of tuberculous meat intended for the food of man.

Dairies, Cowsheds and Milkshops Order.

The Council are empowered by Article 13 of the Dairies, Cowsheds and Milkshops Order of 1885 to make regulations for the following purposes, or any of them :—

(*a*.) For the inspection of cattle in dairies.

(*b*.) For prescribing and regulating the lighting, ventilation, cleansing, drainage and water supply of dairies and cowsheds in the occupation of persons following the trade of cowkeepers or dairymen.

(*c*.) For securing the cleanliness of milk-stores, milk-shops, and of milk-vessels used for containing milk for sale by such persons.

(*d*.) For prescribing precautions to be taken by purveyors of milk and persons selling milk by retail against infection or contamination.

Representations have been made to the Board to the effect that it would be desirable that they should issue Model Regulations for the guidance of Councils in making regulations under this Article, but they have deferred doing so pending the Report of the Royal Commission on Tuberculosis.

That Report having been made, the Board have caused some Model Regulations to be prepared, and two copies of them are enclosed.

It will be observed that No. 8 of the Regulations, which deals with the question of air space in cowsheds, does not apply to cowsheds the cows from which are habitually grazed on grass land during the greater part of the year, and, when not so grazed, are habitually turned out during a portion of each day, and it is obvious that a regulation on this subject which might be adapted to cowsheds in towns, where the cows are kept and fed within the building, might be unsuitable for cowsheds in the country, where the cows are regularly grazed on grass land during the greater part of the year, and are during the rest of the year usually turned out for a portion of each day.

The Royal Commission, in their recommendations, drew a distinction between the rules which should be observed on this subject as regards cowsheds situate in populous and those situate in non-populous places, but no indication was given as to the means by which this distinction was to be made. It is clear that it could not be accomplished by any test of population or by adopting the geographical limits of urban and rural districts without creating anomalies which would be indefensible.

Neither is it easy to see how the distinction can be carried out except upon the plan suggested by the Board, which seeks to give effect to the chief difference between cowsheds in towns and cowsheds in the country, or, in other words, between the case of cows which are kept entirely or as a rule indoors, and that of cows which are usually turned out to graze.

It will be noticed that No. 4 of the Model Regulations, which provides that every cowkeeper must cause every cowshed in his occupation to be sufficiently ventilated, and for this purpose to be provided with a sufficient number of openings into the external air to keep the air in the cowshed in a wholesome condition, applies in both classes of cases.

If the Council have not already made regulations under the Order of 1885, the Board think that they should do so, and that any such regulations would with advantage be based on the Model Clauses. If the Council have already made regulations under the Order, the Model Clauses may usefully be considered in connection with any fresh regulations or amendment of the existing code which the Council may propose to make.

The Board's confirmation of any regulations which may be made by the Council will not be required, but if at any time the Board are satisfied on enquiry with respect to any regulation that the same is of too restrictive a character, or otherwise objectionable, they may direct its revocation; and the Board suggest that the draft of any regulations which the Council may propose to make should be sent to them for consideration before the regulations are formally adopted.

Report of the Tuberculosis Commission.—Milk.

Article 15 of the Dairies, Cowsheds and Milkshops Order of 1885 provides that if at any time disease exists among the cattle in a dairy or cowshed, or other building or place, the milk of a diseased cow therein (*a*) shall not be mixed with other milk; and (*b*) shall not be sold or used for human food.

?. A. No. 3. — ? Dairies, sheds and shops ?r, and ?rt of ?l Com- ?ion on ?rculosis.

The term "disease" in the Order is limited to those diseases which were included under the Contagious Diseases (Animals) Act, 1878, of which tuberculosis is not one, and the Royal Commission on Tuberculosis state in paragraph 39 of their Report that "the evidence abundantly shows how this fact has precluded local "authorities from any attempt to deal with tuberculosis in milch "cows, although they may have shown themselves alive to the "danger and anxious to provide a remedy," and they express the opinion that "it is desirable that the Order should be made "applicable to all diseases of the udder in cows of which the milk "is offered for sale."

The Board have issued an Order to amend Article 15 of the Order of 1885 by providing that, for the purposes of paragraphs (*a*) and (*b*) of the Article, reference to disease shall include, in the case of a cow, such disease of the udder as shall be certified by a veterinary surgeon to be tubercular. The Board think that it will be competent for the Council to employ and pay a veterinary surgeon with a view of obtaining a certificate under the Article, as amended, or to appoint him as an Officer for this purpose, if they think fit to do so. Two copies of the amending Order are enclosed.

Qualification of Meat Inspectors.

Under Section 116 of the Public Health Act, 1875 (38 & 39 Vict. c. 55), any medical officer of health or inspector of nuisances may at all reasonable times inspect and examine any animal, carcase or meat exposed for sale, or deposited in any place for the purpose of sale, or of preparation for sale, and intended for the food of man. If the animal, carcase or meat appears to the medical officer of health or inspector of nuisances to be diseased or unsound or unwholesome, or unfit for the food of man, he may seize and carry it away in order to have it dealt with by a Justice. Under Section 131 of the Towns Improvement Clauses Act, 1847 (10 and 11 Vict. c. 34), which is incorporated with the Public Health Act, 1875, the inspector of nuisances, the officer of health or any other officer appointed by the Council for the purpose, may at all reasonable times enter and inspect any building or place within the district kept or used for the sale of butchers' meat or for slaughtering cattle, and examine whether any cattle or the carcase of any cattle is deposited there. If the officer finds any cattle or the carcase or part of the carcase of any beast which appears unfit for the food of man, he may seize and carry the same before a Justice so that it may be dealt with.

Moreover, where the Council are in a position to establish or regulate markets under Section 167 of the Public Health Act, any inspector of provisions appointed by them may, under Section 15 of the Markets and Fairs Clauses Act, 1847 (10 and 11 Vict. c. 14), which is incorporated with the Public Health Act, seize any unwholesome meat sold or exposed for sale in the market and carry the same before a Justice to be dealt with.

The Royal Commission on Tuberculosis considered that meat inspectors should possess certain qualifications. Their recommendation on the subject will be found on page 21 of their Report, and is as follows:—

APP. A., No. 8

As to Dairies, Cowsheds and Milkshops Order, and Report of Royal Commission on Tuberculosis.

"We recommend that in future no person be permitted to act "as a meat inspector until he has passed a qualifying examination, "before such authority as may be prescribed by the Local Government Board (or Board of Agriculture), on the following subjects :—

"(a.) The law of meat inspection, and such bye-laws, regulations, &c., as may be in force at the time he presents "himself for examination.

"(b.) The names and situations of the organs of the body.

"(c.) Signs of health and disease in animals destined for food, "both when alive and after slaughter.

"(d.) The appearance and character of fresh meat, organs, fat, "and blood, and the conditions rendering them, or "preparations from them, fit or unfit for human food."

At present a person cannot be required to pass a qualifying examination of the kind referred to before he acts as a meat inspector; but it appears to the Board that, in the case of a borough or urban district, where the work connected with the proper discharge of the duty of meat inspection is sufficient to justify the appointment of a separate officer for the purpose, it is very desirable that such an appointment should be made, and that the Council should satisfy themselves that the person appointed possesses adequate knowledge of the subjects mentioned in the recommendation of the Royal Commission.

In the smaller districts, where the work of meat inspection is not sufficient to render necessary the appointment of a separate officer, the Board consider that regard should be had to these qualifications in making future appointments to the office of Inspector of Nuisances.

Instructions to Meat Inspectors with regard to Tuberculosis in Animals intended for Food.

The Royal Commission recommended that the Board should "be empowered to issue instructions from time to time for the "guidance of meat inspectors, prescribing the degree of tubercular "disease which, in the opinion of the Board, should cause a "carcase, or part thereof, to be seized.

"Pending the issue of such instructions we are of opinion that "the following principles should be observed in the inspection of "tuberculous carcases of cattle :

"(a.) When there is miliary tuberculosis of both lungs ... ...

"(b.) When tuberculous leisons are "present on the pleura and "peritoneum .. ..

"(c.) When tuberculous lesions are "present in the muscular system "or in the lymphatic glands embedded in or between the "muscles ... ...

"(d.) When tuberculous lesions exist "in any part of an emaciated "carcase ... ...

} The entire carcase and all the organs may be seized.

A., No 8.

Dairies, [Cow-]sheds and [Milk-]shops, and [Memorandum] of Com[mittee] on [Tuber]culosis.

"(a.) When the lesions are confined "to the lungs and the thoracic "lymphatic glands ... ...
"(b.) When the lesions are confined "to the liver ... ...
"(c.) When the lesions are confined "to the pharyngeal lymphatic "glands ... ...
"(d.) When the lesions are confined "to any combination of the fore-"going, but are collectively small "in extent .. ..

} The carcase, if otherwise healthy, shall not be condemned, but every part of it containing tuberculous lesions shall be seized.

"In view of the greater tendency to generalisation of tuber-"culosis in the pig, we consider that the presence of tubercula[r] "deposit in any degree should involve seizure of the whole "carcase and of the organs."

"In respect of foreign dead meat, seizure shall ensure in every "case where the pleura have been 'stripped.'"

The Board do not consider it necessary at present, that anything should be added to these Instructions, or that they should be modified, and the Board think that the Council should direct those of their Officers who are employed as Meat Inspectors to act in accordance with the principles thus laid down.

The Board may at the same time draw attention to Article 19 (7) of their General Order of the 23rd March, 1891, with respect to the duties of an Inspector of Nuisances in relation to the inspection and seizure of meat. They may point out that where an Inspector of Nuisances is appointed under that Order, or under any Order superseded by that Order, he is required by the Article, in any case of doubt arising under it, to report the matter to the Medical Officer of Health with the view of obtaining his advice thereon. The Board think it desirable that any such Inspector of Nuisances should be reminded of this provision.

I am, Sir,
Your obedient Servant,
S. B. PROVIS,
Secretary.

The Town Clerk *or* The Clerk
to the Urban District Council.

B.—The Dairies, Cow-sheds and Milk-shops Order of 1899.

To the London County Council;—
To the Mayor and Commonalty and Citizens of the City of London, acting by the Mayor, Aldermen, and Commons of that City in Common Council assembled;—
To the Mayor, Aldermen, and Burgesses of the several County Boroughs for the time being in England and Wales;—
To the several Urban and Rural District Councils for the time being in England and Wales;—
And to all others whom it may concern.

Whereas on the 15th day of June, 1885, Her Majesty's Most Honourable Privy Council (herein-after referred to as "the Privy

Council"), in pursuance of the statutory provisions in that behalf, made an Order (herein-after referred to as "the Order") which is known as "The Dairies, Cow-sheds and Milk-shops Order of 1885";

And whereas certain powers of the Privy Council, including the power of altering or revoking the Order, have been transferred to Us, the Local Government Board; and, in pursuance of such powers, the Order has been altered by an Order (herein-after referred to as "the Amending Order"), which was made by Us on the 1st day of November, 1886, and is known as "The Dairies, Cow-sheds and Milk-shops Amending Order of 1886";

And whereas it is expedient that the Order as altered by the Amending Order should be further altered:

Now therefore, in pursuance of the powers vested in Us in that behalf, We hereby Order as follows:—

Article I.—This Order may be cited as "The Dairies, Cow-sheds and Milk-shops Order of 1899."

Article II.—Article 15 of the Order shall be altered so that, for the purposes of the provisions of paragraphs (a) and (b) thereof the expressions in the said Article which refer to disease shall include, in the case of a cow, such disease of the udder as shall be certified by a veterinary surgeon to be tubercular; and the Order and the Amending Order shall apply and be construed with the modifications necessary to give effect to this Article.

Given under the Seal of Office of the Local Government Board, this Seventh day of February, in the year One thousand eight hundred and ninety-nine.

(L.S.) HENRY CHAPLIN, *President.*

S. B. PROVIS, *Secretary.*

DRAFT FORM of MODEL REGULATIONS: DAIRIES, COWSHEDS AND MILKSHOPS.

REGULATIONS made by the[1]

WITH RESPECT TO DAIRIES, COWSHEDS AND MILK-SHOPS IN THE[2]

INTERPRETATION.

1. Throughout these regulations the expression "The Council" means the[1]

the expression "the District" means the[2]

the expression "Cowshed" includes any dairy in which milking cows may be kept, and the expression "Cowkeeper" means any person following the trade of a cowkeeper or dairyman who is, or is required to be, registered under the Dairies, Cowsheds and Milkshops Order of 1885.

[1] "Mayor, Aldermen, and Burgesses of the Borough of , acting by the Council" *or* "Urban (*or* Rural) District Council of ," *as the case may be.*

[2] "Borough" *or* "Urban (*or* Rural) District of ," *as the case may be.*

App. A. No. 2

[illegible] to Dairies, Cowsheds and Milkshops Order, and Report of Royal Commission on Tuberculosis.

For the Inspection of Cattle in Dairies.

2. Every occupier of a dairy wherein any cattle may be kept, and which the Medical Officer of Health, or the Inspector of Nuisances, or any other officer of the Council specially authorised by them in that behalf, may visit for the purpose of inspecting cattle, and every person for the time being having the care or control of any such dairy, or of any cattle therein, shall afford such Medical Officer of Health, Inspector of Nuisances, or officer, all reasonable assistance that may, for the purpose of the inspection, be required by him.

For Prescribing and Regulating the Lighting, Ventilation, Cleansing, Drainage and Water Supply of Cowsheds and Dairies in the Occupation of Persons following the Trade of Cowkeepers or Dairymen.

Part I.

The Regulations in this Part shall apply to Cowsheds the cows from which are habitually grazed on grass land during the greater part of the year, and, when not so grazed, are habitually turned out during a portion of each day.

Lighting.

3. Every cowkeeper shall provide that every cowshed in his occupation shall be sufficiently lighted with windows, whether in the sides or roof thereof.

Ventilation.

4. Every cowkeeper shall cause every cowshed in his occupation to be sufficiently ventilated, and for this purpose to be provided with a sufficient number of openings into the external air to keep the air in the cowshed in a wholesome condition.

Cleansing.

5. (1.) Every cowkeeper shall cause every part of the interior of every cowshed in his occupation to be thoroughly cleansed from time to time as often as may be necessary to secure that such cowshed shall be at all times reasonably clean and sweet.

(2.) Such person shall cause the ceiling or interior of the roof, and the walls of every cowshed in his occupation to be properly limewashed *twice* at least in each year, that is to say, once during the month of May and once during the month of October, and at such other times as may be necessary.

Provided that this requirement shall not apply to any part of such ceiling, roof or walls, that may be properly painted, or varnished, or constructed of or covered with any material such as to render the lime-washing unsuitable or inexpedient, and that may be otherwise properly cleansed.

(3.) He shall cause the floor of every such cowshed to be thoroughly swept, and all dung and other offensive matter to be removed from such cowshed as often as may be necessary, and not less than *once* in every day.

Drainage.

6. (1.) Every cowkeeper shall cause the drainage of every cowshed in his occupation to be so arranged that all liquid matter which may fall or be cast upon the floor may be conveyed by a suitable open channel to a drain inlet situate in the open air at a proper distance from any door or window of such cowshed, or to some other suitable place of disposal which is so situate.

(2.) He shall not cause or suffer any inlet to any drain of such cowshed to be within such cowshed.

Water Supply.

7. (1.) Every cowkeeper shall keep in, or in connection with, every cowshed in his occupation a supply of water suitable and sufficient for all such purposes as may from time to time be reasonably necessary.

(2.) He shall cause any receptacle which may be provided for such water to be emptied and thoroughly cleansed from time to time as often as may be necessary to prevent the pollution of any water that may be stored therein, and where such receptacle is used for the storage only of water he shall cause it to be properly covered and ventilated, and so placed as to be at all times readily accessible.

PART II.

The regulations in Part I., and also the following regulation, shall apply to all Cowsheds other than those the cows from which are habitually grazed on grass land during the greater part of the year, and, when not so grazed, are habitually turned out during a portion of each day.

8. A cowkeeper shall not cause or allow any cowshed in his occupation to be occupied by a larger number of cows than will leave not less than *eight hundred feet* of air space for each cow.

Provided as follows :—

(*a*.) In calculating the air space for the purposes of this regulation, no space shall be reckoned which is more than *sixteen feet* above the floor ; but if the roof or ceiling is inclined, then the mean height of the same above the floor may be taken as the height thereof for the purposes of this regulation.

(*b*.) This regulation shall not apply to any cowshed constructed and used before the date of these regulations coming into effect, until two years after that date.

PART III.

9. In this Part, the expression " Dairy " means a dairy in which cattle are not kept.

Lighting.

10. Every cowkeeper shall provide that every dairy in his occupation shall be sufficiently lighted with windows, whether in the sides or roof thereof.

Ventilation.

11. Every cowkeeper shall cause every dairy in his occupation to be sufficiently ventilated, and for this purpose to be provided

. A., No. 8. Dairies, [Sh]eds and [Sh]ops and of [C]om-[m]on [L]odgs.

with a sufficient number of openings into the external air to keep the air in the dairy in a wholesome condition.

Cleansing.

12. (1.) Every cowkeeper shall cause every part of the interior of every dairy in his occupation to be thoroughly cleansed from time to time as often as may be necessary to secure that such dairy shall be at all times reasonably clean and sweet.

(2.) He shall cause the floor of every such dairy to be thoroughly cleansed with water at least *once* in every day.

Drainage.

13. (1.) Every cowkeeper shall cause the drainage of every dairy in his occupation to be so arranged that all liquid matter which may fall or be cast upon the floor may be conveyed by a suitable open channel to the outside of such dairy, and may there be received in a suitable gulley communicating with a proper and sufficient drain.

(2.) He shall not cause or suffer any inlet to any drain of such dairy to be within such dairy.

Water Supply.

14. (1.) Every cowkeeper shall cause every dairy in his occupation to be provided with an adequate supply of good and wholesome water for the cleansing of such dairy and of any vessels that may be used therein for containing milk, and for all other reasonable and necessary purposes in connection with the use thereof.

(2.) He shall cause every cistern or other receptacle in which any such water may be stored to be properly covered and ventilated, and so placed as to be at all times readily accessible.

(3.) He shall cause every such cistern or receptacle to be emptied and thoroughly cleansed from time to time as often as may be necessary to prevent the pollution of any water that may be stored therein.

FOR SECURING THE CLEANLINESS OF MILK-STORES, MILK-SHOPS, AND OF MILK-VESSELS USED FOR CONTAINING MILK FOR SALE BY PERSONS FOLLOWING THE TRADE OF COW-KEEPERS OR DAIRYMEN.

Cleanliness of Milk-Stores and Milk-Shops.

15. Every cowkeeper who is the occupier of a milk-store or milk-shop shall cause every part of the interior of such milk-store or milk-shop to be thoroughly cleansed from time to time as often as may be necessary to maintain such milk-store or milk-shop in a thorough state of cleanliness.

Cleanliness of Milk-Vessels.

16. (1.) Every cowkeeper shall from time to time as often as may be necessary cause every milk-vessel that may be used by him for containing milk for sale to be thoroughly cleansed with

steam or clean boiling water, and shall otherwise take all proper precautions for the maintenance of such milk-vessel in a constant state of cleanliness.

(2.) He shall, on every occasion when any such vessel shall have been used to contain milk, or shall have been returned to him after having been out of his possession, cause such vessel to be forthwith so cleansed.

For prescribing Precautions to be taken by Purveyors of Milk and Persons Selling Milk by Retail against Infection or Contamination.

17. (1.) Every purveyor of milk or person selling milk by retail shall take all reasonable and proper precautions, in and in connection with the storage and distribution of the milk, and otherwise, to prevent the exposure of the milk to any infection or contamination.

(2.) He shall not deposit or keep any milk intended for sale—

(a.) in any room or place where it would be liable to become infected or contaminated by impure air, or by any offensive, noxious, or deleterious gas or substance, or by any noxious or injurious emanation, exhalation, or effluvium ; or

(b.) in any room used as a kitchen or as a living room ; or

(c.) in any room or building, or part of a building communicating directly by door, window, or otherwise with any room used as a sleeping room, or in which there may be any person suffering from any infectious or contagious disease, or which may have been used by any person suffering from any such disease and may not have been properly disinfected ; or

(d.) in any room or building or part of a building in which there may be any direct inlet to any drain.

(3.) He shall not keep milk for sale, or cause or suffer any such milk to be placed, in any vessel, receptacle or utensil which is not thoroughly clean.

(4.) He shall cause every vessel, receptacle or utensil used by him for containing milk for sale to be thoroughly cleansed with steam or clean boiling water after it shall have been used, and to be maintained in a constant state of cleanliness.

(5.) He shall not cause or suffer any cow belonging to him or under his care or control to be milked for the purpose of obtaining milk for sale—

(a.) Unless, at the time of milking, the udder and teats of such cow are thoroughly clean ; and

(b.) Unless the hands of the person milking such cow, also, are thoroughly clean and free from all infection and contamination.

Penalties.

18. Every person who shall offend against any of the foregoing regulations shall be liable for every such offence to a penalty of *five pounds*, and in the case of a continuing offence to a further penalty of *forty shillings* for each day after written notice of the offence from the Council.

. A., No. 8.
to Dairies,
wsheds and
kshops
er, and
rt of
al Com-
on
berculosis.

Provided, nevertheless, that the justices or court before whom any complaint may be made or any proceedings may be taken in respect of any such offence may, if they think fit, adjudge the payment as a penalty of any sum less than the full amount of the penalty imposed by this regulation.

Commencement of the Regulations.

19. These regulations shall come into force on and after t— day of 18 .

Revocation of Regulations.[1]

20. From and after the date on which these regulations s come into force, all regulations heretofore made under, or havi effect in pursuance of the Dairies, Cowsheds and Milkshops O of 1885, shall, so far as the same are now in force in the distri be revoked.

[1] If this clause is not included in the series submitted to the Local Gove Board for approval, it should be stated whether or not there are any regulati in force upon the subject.

No. 9.

REPORT upon the SANITARY CONDITION of the URBAN DISTRICT of ALNWICK, and upon HOUSING of the WORKING CLASSES therein; by DR. G. S. BUCHANAN.

App. A.
On the ...tary ... of th... Distr... wick Hou... W... Clas... in: Buc...

In December, 1897, the Board received representations from the County Council of Northumberland concerning the sanitary condition of Alnwick, an urban district with a population estimated at 6,700, situated within that county. The County Council Set forth that in February, 1897, Dr. Hembrough, the County Medical Officer of Health, had reported to them certain unwholesome conditions which he had met with in course of inspections at Alnwick: that in particular, he had drawn attention to the existence within that town of clusters of dwellings inhabited by the poorer class of the population, which owing to bad arrangement and faulty construction could not fail to be unhealthy; that objection to this report by Dr. Hembrough had, however, been taken by the Alnwick Urban District Council, who instructed their Medical Officer of Health, Dr. Easton, to review its statements in detail; and that in May, 1897, Dr. Easton, who, in consultation with the chairman and other officers of the District Council, had made the investigations desired of him, presented a report in which he confirmed in material particulars the account already given by Dr. Hembrough. Nevertheless, the County Council informed the Board, no action in remedy of the insanitary conditions complained of had since been inaugurated by the Alnwick District Council.

On receipt of this information, and having in view that for many years past annual reports of successive Medical Officers of Health of Alnwick had drawn attention to unhealthiness of habitations of the poorer classes there, but had failed to show that effective action in the matter was being taken by the Sanitary Authority, the Board directed local inquiry. In accordance with instructions I visited Alnwick at the end of March.

I. SANITARY HISTORY.

Brief reference may be made to the sanitary history of Alnwick. In this urban area, then designated "Alnwick and Canongate," a variety of unwholesome conditions were reported in 1849 by Mr. (Sir Robert) Rawlinson to the General Board of Health: unhealthy, damp, and overcrowded dwellings, packed away in narrow courts

APP. A. No. 8.

On the sanitary condition of the Urban District of Alnwick, and upon Housing of the Working Classes therein; by Dr. Buchanan.

and alleys behind the main thoroughfares; abundance of privy and midden nuisances: unpaved or badly paved yards: no public water supply and no system of sewerage.

Between 1852 and 1854 a large improvement scheme was carried out by the newly appointed Alnwick and Canongate Local Board. The town was sewered, under Mr. Rawlinson's direction: privies were abolished and water closets were provided: a public water supply was obtained. In 1866, my father, then Inspector of the Medical Department of the Privy Council, reported upon the sundry improvements which the Local Board had effected, but drew attention to the disappointing fact that completion of the works of improvement had not been followed by diminution in the general mortality of the place. An important amount of reduction in mortality had been experienced in the case of "fever" only, while on the other hand death from phthisis and lung diseases had if anything increased. In relation to his observations, Dr. (Sir George) Buchanan pointed out that the construction of the courts and dwellings occupied by the poorer classes in Alnwick remained as unwholesome as before, and that the provision of sewers had effected no drying of the subsoil.

Later administration by the Alnwick and Canongate Local Board appears to have been conspicuously lax. In 1885, Mr. Spear reported to the Local Government Board that certain useful public works such as the erection of public slaughter-houses and laying additional sewers had indeed been executed, and that in certain few matters, such as scavenging, administration by this Authority might be considered efficient; but that no attempt had been made to remedy a multiplicity of serious sanitary defects pointed out by successive Medical Officers of Health. Above all, the courts and dwellings occupied by the poorer classes had not been improved, and overcrowding of persons therein had been allowed to become even more common than formerly. "Wherever individual interests have to be opposed," Mr. Spear observed, "or seemingly opposed, sanitary administration has been paralysed."

In 1894 the district administered by the Alnwick and Canongate Local Board became, without alteration of area, the present Alnwick Urban District. Administration by Local Board and by District Council since the date of Mr. Spear's report will best be considered in connection with the description of the present sanitary condition of the town given below.

It will, however, be convenient here to set out certain statistics which I have obtained of mortality in Alnwick during the 12 years, 1886 to 1897. It would seem from these figures that in general the experience of previous years has been repeated.* Contrasted with the corresponding rate for the whole of England and Wales, the mortality from all causes in Alnwick has remained unduly high. And among particular causes of death the mortality from phthisis has in this, as in previous periods, been conspicuously excessive.

---

* *Conf.* tables at page 3 of Mr. Spear's Report.

	Alnwick Urban District.													England and Wales.
—	Total Deaths* during												Annual Mortality per 10,000 living during 12 years 1886-97 (on a population of 6,700.)	Annual Mortality per 10,000 living during 10 years, 1885-94.
	1886.	1887.	1888.	1889.	1890.	1891.	1892.	1893.	1894.	1895.	1896.	1897.		
Measles	—	36	—	—	—	—	2	—	—	14	—	—	6.5	4.54
Scarlatina	—	1	—	—	1	8	3	—	—	4	4	—	2.6	2.2
Diphtheria and Croup	2	—	3	1	2	—	2	—	—	1	—	—	1.4	2.02
Whooping Cough	12	4	—	4	—	1	3	—	—	15	—	—	4.8	4.37
Enteric Fever	2	2	2	1	—	—	—	1	2	—	—	—	1.2	1.76
Diarrhœa	7	2	8	9	8	8	5	3	—	7	—	—	7.1	6.11
Phthisis	19	22	23	30	20	16	16	18	13	23	17	13	28.6	15.87
Other lung diseases	28	26	20	26	31	28	22	24	7	17	19	21	33.5	36.8
From all causes...	186	185	155	172	163	165	149	135	108	195	128	138	233.7	189.0
Deaths under five years of age	54	69	49	44	35	44	36	38	26	60	35	38	65.7	(For 1881-90.) 56.82
Infantile mortality per 1,000 registered births	?	158	150	157	133	121	102	156	130	154	152	127	Average of rates 1887-97. 141	(For 1881-90.) 142

* Including deaths in Alnwick Workhouse and in the small "general hospital" of the town. Certain deaths in these institutions of persons who were not inhabitants of the district are here included. They are probably more than balanced by deaths of Alnwick inhabitants occurring outside the district.

.A. No.]

... sani- condition Urban of Aln- ... upon of the there- Dr.

II. General Description and Present Sanitary Condition.

The town of Alnwick is situate on the south bank of the river Aln, about six miles from its mouth. It lies on the hillside, and is separated from the river by the grounds of Alnwick Castle, the principal seat of the Duke of Northumberland. The physical and geological characters of the district were thus described by Mr. Spear :—"The geological formation is that of the 'northern drift,' consisting of beds of sand, gravel, and clay. These are irregularly interstratified, often apparently in such a manner (as when gravel or sand rests in an impermeable basin) as to produce the highest degree of dampness. The position of the town likewise contributes to this condition : for, although well elevated above the bed of the neighbouring river, in relation to widely extending tracts of land on the south and west, it lies in a hollow The higher land, moreover, is rich in springs, and the natural drainage is directed upon the town. Wetness of site is accordingly one of the characteristics of the district."

The area of Alnwick Urban District is 4,777 acres, of which the town itself occupies some 150 acres. The population of the district, which in the census returns for 1871 and 1881 was 6,218 and 6,693 respectively, in 1891 was 6,746. It is believed that there has been no material increase during the past six years. Some 300 inhabitants of the district dwell in open country outside the town, principally on Alnwick Moor.

Alnwick is the market of a large agricultural area, and contains the shops and places of business usual in a considerable country town. On quite a small scale a few manufactures—fishing tackle, mineral waters, and cabinet-making—are carried on. In Alnwick the proportion of persons belonging to the poorer class for a long while has been, and still is, a large one. Many of these (both men and women) find work in agriculture ; others, classed as "general labourers," obtain uncertain employment in the town and neighbourhood, principally with masons and builders. Others, again, have no apparent means of livelihood. A not inconsiderable number, frequently persons no longer able-bodied, appear to subsist by taking lodgers. Although the floating population of labourers and navvies to which Mr. Spear referred in 1885 decreased after the completion of the railway then being constructed in the district, the town is still frequented by a somewhat excessive number of hawkers and tramps.

The total rateable value of the district for the current year is £26,868 ; that assessable for the general district rate is £23,440. The district rate (including water rate) has averaged 2*s*. 5*d*. during the past seven years. Loan for sewage works executed in 1897 will entail, I am informed, an addition of 6*d*. in the £ to the rates of the current year.

Streets, Dwellings, and House Accommodation.—There are five miles of main road and 13 miles of dedicated highways in the district. In the town itself the principal road, from Bondgate on the east to Clayport on the west, is broad and macadamised. Other streets are as a rule narrow, and for the most part are paved unevenly with stone setts. In certain streets new stone paving is now being provided by the District Council.

Private dwellings and shops facing the main thouroughfares have usually been substantially buit of local sandstone ; a few are of brick. Save in the case of a small number of houses on the outskirts of the town, mostly of recent construction, these dwellings abut on the road in front, and usually have at the back little or no open space which belongs exclusively to the dwelling.

APP. A. N

On the sanitary condition of the Urban District of Alnwick, and Housing of Working Classes therein ; by Dr. Buchanan

A large proportion of the dwellings of the town, principally those occcupied by the poorer classes, are to be found huddled on small areas at the back of the main thouroughfares. This close aggregation of dwellings, which has long been a characteristic feature of Alnwick, must in part be attributed to the enclosure of the town in former centuries by walls ; to a greater extent, however, it is the outcome of later conditions.

Until comparatively recently few dwellings have been erected on the open country which immediately surrounds Alnwick town, and which for the most part is comprised in the Northumberland Estate, and accomodation for the labouring population seems to have been provided almost entirely by numerous owners of small freeholds within the town. These small "properties" as they are locally termed, have in many instances been in the possession of the same family for several generations. Almost always it has happened at one or another time in the history of a property that whatever open space originally belonged to it has been utilized for building. The commonest example is where a property at first consisted of a single dwelling in a row abutting on the main street, having at the back a long strip of garden or yard, the width of which is no greater than the breadth of the house to which it belongs. Buildings have subsequently been packed on this strip of back yard, and access to the "court" so formed is had by a narrow "entry" driven through the ground floor of the house in the main street. Going through an entry into a court of this kind one finds a passage, some four to six feet wide, extending the length of the property, sometimes terminating blindly, sometimes leading, by a second "entry" at its far end, into a neighbouring street or court. Along one side of the passage is a high wall, which forms the back of structures in the adjoining "property ;" along the other side is a row of buildings, most of them two-storey dwelling houses, outhouses, stables, or cowsheds. Here and there an irregularly shaped common yard is met with, surrounded by buildings of a similar sort. Dwellings in these courts have thick walls rudely constructed out of the local stone. Some habitations appear to have originally been stables or outhouses. As a rule, the only windows and doors are those facing the passage or common yard. These dwellings are thus without through ventilation, and commonly they receive insufficient light. Windows are frequently small, and often can be opened only to a trifling extent. Standing on a wet soil, constructed of porous stone, and unprovided with damp-proof course, these houses are conspicuously damp. In places the wall of the dwelling has been built against the hill-side, and here dampness is increased. In a few cases an additional source of wetness arises where eaves-gutters and down-spouting are wanted or are deficient. Dwellings in these courts vary considerably in their state of repair and cleanliness. In certain courts, where the inhabitants are workmen of better class or are persons employed at offices or shops in the

APP. A. No. 9.
On the sanitary condition of the Urban District of Alnwick, and upon Housing of the Working Classes therein; by Dr. Buchanan.

town, the interior of the dwelling is found to be in good repair and tidily kept. In other courts, the floors, roofs, stairs, and windows are frequently dilapidated, and inside the houses are extremely dirty. Although the main streets of the town are lighted at night by gas lamps, in hardly any instance has provision been made for the public lighting of the courts. Passages and common yards of these courts generally have an irregular paving of stone setts, pebbles, or flagstones. Here and there they have no paving whatever. Each court is drained to a sewer in the main street by means of a gully or gullies placed in the yard or passage. Open gutters designed to convey to these gullies rain-water and slop-water from the dwellings have often been constructed in rudimentary fashion. Liquid refuse thrown into the gutters or on to the surface of the yard is thus apt to accumulate in pools in front of the houses. Where horses or cows are kept in such courts as these, nuisance inevitably arises from accumulations of dung and stable refuse. These accumulations are from time to time carried out into the main street in barrows, and in the course of their removal objectionable matters frequently become scattered about the yard, passage, and "entry."

On both sides of Clayport Street, at the south-western end of Alnwick, every few paces brings one to a door or narrow archway leading to a court formed by a collection of dwellings such as I have described. Such courts, however, are to be found in every quarter of the town.

Details of unwholesome conditions found in certain of these courts are set out in the reports by Dr. Hembrough and by Dr. Easton, to which I have already referred. I had frequent occasion to consult the descriptions given by these Medical Officers of Health, almost always to find them in accord with the conditions obtaining at the time of my visit.

To the unhealthiness produced by crowding together of insanitary dwellings upon area are added the evils attending overcrowding of persons. Dwellings in these courts are in almost all cases subdivided into tenements. According to inquiries made for the District Council by Mr. Waters, the Inspector of Nuisances, in 1895 and 1896, there were then in Alnwick some 300 tenements, each consisting of only a single room, and these between them accommodated from 800 to 900 persons. There were also 333 tenements which consisted of two rooms each, and were occupied by a total of 1,395 persons. Instances were numerous in which from four to eight persons lived and slept in a single room; of the two-roomed tenements nearly one half were occupied by five persons or more. In connection with these figures it should be borne in mind that the cubic space of these tenement rooms is often scanty, and that in the case of "two-roomed" tenements the second room is occasionally little more than a wash-house. Moreover, a statement of the amount of cubic space and of the total number of occupants does not always tell the whole tale of overcrowding. Thus, in a two-roomed dwelling the larger room may be made over to one or two lodgers, while the smaller one constitutes the living and sleeping room of a whole family.

During the past two years, I was informed, overcrowding of persons has here and there been diminished, occasionally at the instance of the District Council. But I could not learn that the diminution thus effected had been sufficient to make any material alteration in the figures above given. In the course of my inspection I met with many instances of overcrowding. Extreme cases were a dwelling off Pottergate New Road, where I found a single room of barely 1,000 cubic feet capacity inhabited by a family of eight; and a house in New Street, containing six rooms and sublet in tenements, which was said to be occupied by as many as 31 persons.

For single-room tenements in Alnwick rents appear for the most part to be from 1*s.* 3*d.* to 2*s.* a week. Rents of tenements consisting of more than one room vary considerably, but in view of the accommodation provided they must, on the whole, be considered high.

Of recent years a considerable number of new dwelling houses have been erected by private enterprise within the district. About 50 of them have been built on Alnwick Moor, at some distance from the town. Elsewhere most of the new buildings are to be found at the south-eastern end of Alnwick, near to the railway station. Here there are some short streets containing between them about 70 dwellings, many of which are occupied by more well-to-do workmen and artisans.* On the whole, the newer dwellings of the place are a conspicuous improvement upon older houses in the town. Several have been built since 1889, and appear to conform to byelaws for New Streets and Buildings, based upon the Board's model code, which were adopted by the District Council in that year. In Wagon Way is a row of buildings, each divided into two tenements, one comprising the upper, the other the lower storey. These tenements are let at rentals of £7 10*s.* and £8 respectively. They appear well suited for their purpose, except that sculleries in the upper tenements are small and awkwardly placed, while all water used upstairs has to be brought by hand from a tap in the yard below.

With a view to obtaining better house accommodation for persons of the working classes, a co-operative building society recently formed at the instance of Earl Percy is now causing 31 two-storey houses to be built along Wagon Way, on land granted to them on favourable terms from the Northumberland Estate. Provisionally it is arranged that these dwellings, each of which consists of four rooms, a scullery, and out-buildings, shall be let at annual rentals of £10 10*s.* By payment for 30 years at the increased rate of £12 per annum the tenant can eventually acquire the freehold. All these new buildings, I learnt, have been taken in advance. I was informed that the owners of tenements which will soon be vacated by persons going to these new dwellings have found no difficulty in securing new tenants ready to come in at the earliest opportunity—a good illustration of the demand in Alnwick for dwelling accommodation.

* It may be supposed that as the population of Alnwick has for many years remained nearly stationary, overcrowding in the town must have been materially diminished by the construction of new dwellings. This is, however, only partly the case. In the course of the past 20 years a considerable number of unhealthy tenement houses in the part of Canongate which belongs to the Northumberland Estate have been demolished. It is estimated that 30 "properties" and about 100 families have been displaced from Canongate during this period, and that nearly all these persons have found accommodation in other parts of the town.

P. A. No. 9.

the sani- condition the Urban ct of Aln- and upon ing of the ing there- ; by Dr. uchanan.

Sewerage.—Natural drainage from upland slopes above Alnwick, as well as surface water from the principal roadways, finds its way to the Aln independently of sewers, partly by open gutters, partly by rude covered channels laid just beneath the surface.

The greater part of the town was sewered, as has been said, by 1854. The main outfall sewer until recently passed along the bank of the Aln and discharged its contents, unpurified, at a point about a mile below the Castle. By additional works constructed in 1897, however, a new main outfall sewer has been laid at a higher level. This sewer serves all except a small section of the town, and delivers its contents to sewage disposal works above Denwick Mill. At these works sewage is precipitated by an "alumino-ferric" preparation, after which it passes to 5 acres of land laid out in irrigation plots. The effluent passes into the Aln, about half a mile below the former outfall. At the date of my visit the treatment of sewage on these works appeared to be carried out in a satisfactory manner.

Sewers serving a population of about 350, dwelling in houses situate on the low ground of Canongate, to the west of the Castle, are connected to a separate outfall sewer. At points along the line of this Canongate sewer are boxes, from which a precipitant (sulphate of iron) is automatically added to the sewage. Matters thus precipitated in the sewer are received in small covered catch pits, the overflow from which passes without further treatment into the Aln. I am informed that the District Council have undertaken to secure adequate purification of this Canongate sewage at an early date. Levels do not permit it to be taken by gravitation to the new sewage works; but a lower plot of land between these works and the river is believed to be suitable and available for the purpose.

No sewer has yet been provided to serve certain private dwellings of good class along Alnmouth Road. Cesspools which receive slop water and water-closet sewage from these dwellings are here situated close to the high road. They are said frequently to overflow, and so to give rise to nuisance. The cleansing of these cesspools is not undertaken by the District Council.

The main sewers and their principal branches are laid at such a gradient that artificial flushing is in most instances rarely necessary. Of recent years ventilating shafts have been put up at certain important terminals, but the older sewers are still somewhat sparsely provided with ventilating openings.

Sink pipes are as a rule disconnected from the drain; they discharge their contents outside the dwelling over a gully or open gutter. In the courts of Alnwick the dwellings usually have no inside drains. Gullies serving these courts are in some cases untrapped. Many trapped gullies have, however, been recently provided at the instance of the District Council.

Excrement disposal is in almost all instances effected by water-closets, which are commonly placed outside the dwelling. Mr. Spear reported in 1885 that the majority of water-closets were supplied with water direct from the main. This matter has been attended to by the Sanitary Authority, and flushing boxes have now been provided in every case. In several of the courts the number of water-closets, although perhaps the maximum which can be constructed owing to the limited space available, is nevertheless insufficient for the number of inhabitants.

APP. A.

On the sanitary condition of the Urban District of Alnwick, and Housing of Working Classes therein; by Dr. Buchanan

House refuse is deposited in movable receptacles, provided by the occupier. In tenement houses boxes or pails for refuse are frequently stowed away beneath the common stair. On specified mornings these receptacles are carried out to the public street, and their contents are removed by the scavengers of the District Council, who make a complete round of the town twice a week. The receptacles provided are often old pails, broken boxes, or other makeshifts, and usually they have no cover. In consequence, before the arrival of the scavengers' cart much of the refuse may become blown about the public streets.

Refuse collected by the scavengers is "tipped" on high ground half-a mile out of the town.

The water supply of the district is obtained from land at Rugeley, about three miles south-west of Alnwick. On this land there are numerous springs which rise at the outcrop of a series of sandstones, limestones, and shales belonging to the Coal Measures. The supply of Alnwick is obtained principally from these springs, but partly also from the natural drainage of high ground in their neighbourhood.

At the date of Mr. Spear's visit much of the water reaching the waterworks was liable to contamination from drainage of public highways and by farm-house and cottage sewage. At that date also small open streams contributing to the supply appeared exposed to pollution from excrement deposited in their neighbourhood. Within recent years considerable improvements have been effected. At each of the springs a chamber has been constructed with a view to excluding surface drainage, and water is carried thence to the mains in iron pipes. Additional springs have been brought within the collecting system. Such water as is not derived from springs, however, is principally contributed, as before, by drainage of agricultural land. This source of supply cannot be considered satisfactory, not only because much of the land furnishing the water is from time to time manured, but also because it is apt to fail altogether in dry seasons. Water from the various "spring chambers" and "collecting chambers" at Rugeley is conveyed in a main by gravitation to a service reservoir on high ground immediately above Alnwick. This reservoir, which has a capacity of 210,000 gallons, is capable of storing little more than 24 hours' supply. By the side of the service reservoir is a filter bed of sand, the area of which, however, is too small to be of much value for purposes of purification. The water obtained from Rugeley is estimated to be seldom less than 30 gallons per head per day, although in certain dry seasons the daily yield has been smaller. Save on one or two occasions during the past few years, a constant supply has been maintained. Mr. Wilson, the District Surveyor, is now preparing a scheme whereby water from certain additional springs at Rugeley may be utilised.

There are ten registered *cowsheds* within the town, each nominally subject to regulations made by the Sanitary Authority in 1886, under the Dairies, Cowsheds, and Milkshops Order, 1885. Notwithstanding that their attention has been called to these cowsheds by the reports of the Inspector of Nuisances, the District

. A. No. 9

he sani-
condition
e Urban
rict of Aln-
, and upon
ing of the
king
es there-
by Dr.
anan.

Council appears to have neglected to enforce their regulations. I found that in every instance but one no heed had been paid to providing anything approaching the 800 cubic feet per cow which the regulations prescribe. Most of these cowbyres, indeed, seemed to contain as many cows as could be crowded within their four walls. Some of these byres have no ventilating openings whatever, others are provided with a few air-shafts insufficient for the purpose. Often byres are badly paved and drained, and have no convenient supply of water. Almost all are very imperfectly lighted. The superstition that a cow kept in foul air and in the dark is a "better milker" is here still prevalent: the risk run by persons who consume milk from cows so kept is unheeded. Locally neglect to enforce regulations as to cowsheds is condoned for the insufficient reason that cows are kept in their sheds during certain winter months only.

The 41 *milkshops* and 9 *bakehouses* of the district receive periodical inspection by officers of the District Council. A row of six small *public slaughter-houses* was erected by the Local Board in 1877; and other two, somewhat larger and better lighted, were provided by the District Council in 1896. These slaughter-houses are of satisfactory construction, and appear tidily kept. There is only one private slaughter-house in the town.

There is one registered *common lodging-house* in Alnwick. This is a house situate at the end of a narrow court called Turk's Head Yard. By reason of dilapidations it appears unfit for human habitation. Although the total cubic contents of all its rooms does not exceed 3,700 cubic feet, it has been registered, under bye-laws adopted in 1881, for as many as 16 lodgers.

There are several houses in the town, not registered, which are practically used as common lodging-houses. I visited certain of these houses late one evening. In one instance I found six men sleeping in three beds in a single room, with cubic capacity of 1,676 feet. Overcrowding of this sort in these unregistered lodging-houses is said to be common.

Shortly before my visit a registered common lodging-house in Angel Inn Yard had been closed by order of the District Council on account of dilapidations and other defects. The Council consider themselves unable to deal with the unregistered houses, or to close the remaining registered house until they have provided satisfactory common lodging-house accommodation in the town. In 1897, I learnt, they had applied to the Local Government Board for sanction to borrow money for the purpose. Their proposal, however, was to reconstruct an old building not well adapted for the purpose, and which had insufficient open space belonging to it, and, in consequence, sanction to loan was refused. At the date of my visit no further steps in the matter had been taken.

A small *isolation hospital*, about a quarter of a mile from the town, was erected in 1887 under loan sanctioned by the Board. This hospital comprises two wards of three beds each, a caretaker's house, and a block consisting of laundry, mortuary, and house for the ambulance carriage.

Their is no efficient *disinfecting apparatus* for the district. A small chamber for sulphur fumigation has, however, been provided at the hospital.

III. Sanitary Administration.

The Urban District Council consist of 18 members and meet twice a month. They have committees for various purposes. In particular, the town has been divided into six sub-areas, for each of which a separate visiting committee has been appointed.

The *Medical Officer of Health*, Dr. G. F. Easton, M.D., who has held the office since 1886, receives an annual salary of £50, half of which is repaid from County funds. Dr. Easton has kept himself acquainted with the sanitary circumstances of the district, and in annual and special reports to the Sanitary Authority he has been at pains to bring to their knowledge conditions adverse to the public health of the place, and to encourage them in taking remedial action.

The office of *Surveyor*, at a salary of £150, is held by Mr. G. F. Wilson, who appears to perform with thoroughness a multiplicity of duties attaching to the office. The *Inspector of Nuisances*, Mr. D. Waters, who receives a salary of £80, was appointed in 1893. He appears a willing and capable officer. The Council also have in their employ a waterworks inspector, a man in charge of sewage works, five scavengers, and men employed at road making and sewer construction.

The Infectious Disease (Notification) Act and Parts II. and III. of the Public Health Acts Amendment Act (1890) were adopted in this district in 1891. No bye-laws have been framed under the latter Act. No part of the Infectious Disease (Prevention) Act (1890) has been adopted.

As has been indicated, the Authority have series of bye-laws in respect of Slaughter Houses (adopted in 1854), Common Lodging-Houses and Nuisances (1884), and New Streets and Buildings (1889).

Comparing the present condition of the district with Mr. Spear's description of it in 1885, it is apparent that in some respects considerable improvements have been effected. The Sanitary Authority must, for instance, be given credit for having made satisfactory provision for the disposal of the greater part of the sewage of the district, for improvement in the water supply, for providing for isolation of cases of infectious disease, and for having adopted bye-laws for the adequate regulation of new streets and buildings. It is satisfactory to note that several of these improvements have been effected since the creation of the District Council in 1894, and that in respect of sundry day-to-day details of administrations this Council appear to have shown greater readiness to take efficient action than was the case when similar matters came before the Local Board.

But in the matter of dealing, during the thirteen years in question, with some of the gravest insanitary conditions of Alnwick, and above all with its notoriously unhealthy courts and dwellings and with the overcrowding of persons therein, Local Board and District Council alike must be held to have seriously neglected their duty.

The importance to the public health of the district that housing of the poorer classes should be dealt with in thorough fashion was represented, in no uncertain terms, in Mr. Spear's report, and in later years has been pointed out by the present Medical Officer of Health as well as by his predecessors. In 1889, the Local Board themselves appointed a Committee to investigate the high death rate of the district. This Committee, with the aid of Dr. Easton, obtained statistics of mortality from one and another cause during a series of years in various sub-divisions of the town. From these figures they arrived at a conclusion which might have been anticipated, namely, that the excess in general mortality, and in particular the excess in phthisis mortality, was sustained by the inhabitants of the unhealthy and overcrowded dwellings in the various courts of the place. As a result of their inquiries, the Committee informed the Local Board that they had nothing to add to the recommendations already made by successive Medical Officers of Health, and urged the Sanitary Authority to take action to remedy the unhealthy condition of courts and dwellings, and to abate overcrowding of persons. In the eight years which have since elapsed an important section of the inhabitants of the town have, I learnt, become increasingly impressed by the evils which attend the present housing of the poorer classes, and their dissatisfaction has found expression on many occasions, both within and without the Council. Little or nothing, however, has come of advice, representations, and complaints. Proposals for improvement have again and again fallen through on account of the reluctance of the Sanitary Authority to make demands upon the property owners, or to make any increase in the district rate. Particular properties, it is true, have here and there been singled out, and the owner has been required to provide eave spouting to the dwelling or to pave some part of a back yard or passage. Now and again, but very seldom, the use as a dwelling of some dilapidated building or stable has been forbidden; and occasionally an outrageous case of overcrowding of persons has been dealt with. But the effect of these few makeshift performances upon the general unwholesomeness of poorer class dwellings has been insignificant.

It is, however, just to observe that since the publication of the reports of Dr. Hembrough and Dr. Easton, the Alnwick District Council have shown desire to consider the whole question anew. In January of the present year a "Town Improvement Committee" was appointed to report upon means for securing a general improvement of the insanitary areas of the town. It was satisfactory to find, on conferring with this Committee, that they appeared desirous to formulate practical recommendations for dealing with the question in a comprehensive fashion.

IV. CONCLUSION.

The several sanitary shortcomings of Alnwick which I have above indicated manifestly call for action on the part of the District Council more vigorous and more sustained than heretofore. Above all, it would seem essential that they should no longer delay to apply adequate remedy to the principal evil, namely, the unwholesome housing of the working class population.

Question arises in respect of many unhealthy dwellings, courts, and groups of courts in the town, whether any remedy, short of actual demolition, can be applied with reasonable prospect of satisfactory result. Having regard to the diversity of the conditions which go to make these places unhealthy, the answer can in most cases be only negative. They are beyond remedy. Thus, even if it were practicable to put these various habitations into good internal repair, and to provide them with windows which gave adequate light and through ventilation, no amendment would be made in the dampness due to their porous sandstone walls. Moreover, whatever structural alterations were effected, the evils resulting from the lack of open space belonging to the dwelling, and from the want of air and light about the court in which it stands would remain unabated, and the problem of overcrowding of persons would still be untouched.

A proposal that certain of the most objectionable groups of courts should be "improved" by here and there removing a particular building which limits ventilation and light rather more than its neighbours, and also by constructing passages to connect one court with another, had, I learnt, been under consideration by the Town Improvement Committee to which I have referred. But I found that they had already come to be doubtful of the value of such a proposal. Not only was it questionable whether the results of such measures would justify the expenditure, but also there was danger, and a real one, that by expending public money in patchwork of these courts, the District Council would conduce to their permanence, and so to perpetuation of some of the worst sanitary evils of the place.

The only alternative appears to be that already indicated by Dr. Hembrough in his Report to the Northumberland County Council, namely, demolition of the whole of the dwellings upon, particular, larger or smaller, areas; while making provision, in each instance, for the housing of a population equivalent to that displaced. The action needed, in fact, is that which legislature has made available to sanitary authorities by enabling unhealthy areas to be dealt with by means of a scheme or schemes under Parts I. and II. of the Housing of the Working Classes Act, 1890. And in this connection it should be observed that within the town there are certain blocks of courts and dwellings which must needs be considered to form "unhealthy areas." I indicate below three such areas. Each of these comprises a congeries of narrow courts, closely packed one against the other. Within each area almost every court and dwelling presents numerous sanitary defects such as I have already described; many dwellings are dilapidated; while in addition a large proportion are sub-divided into one-room or two-room tenements, some of which are further sub-let to lodgers, and thus overcrowding of persons is common.

Area I. An area, roughly a triangle, and comprising about 2¼ acres, bounded by (*a*) Dispensary Street from the corner opposite the brewery to Clayport Street; (*b*) the north side of Clayport Street from Dispensary Street to Fenkel Street; (*c*) a straight line drawn from the Fenkel Street end of Robin Hood Yard to the corner of Dispensary Street opposite the brewery.

APP. A. No. 9.

On the sanitary condition of the Urban District of Alnwick, and upon Housing of the Working Classes therein; by Dr. Buchanan.

This area includes, among other unhealthy properties, the courts known as Spittle's, Spour's, and Trobe's Yard; Angel Inn Yard, Robin Hood Yard, and Union Court. It also includes certain dwellings which form the east side of the narrow Dispensary Street, and certain other dwellings facing the main thoroughfare of Clayport Street. Although the latter dwellings are of somewhat larger size than most houses on this area, and are not usually sub-divided into tenements, several must be considered to be, by reason of dilapidation and damp, scarcely better fitted for habitation than dwellings in the courts behind them.

Area II., of about half an acre, comprises four adjoining courts which open into the south side of Clayport Street, namely, Blythe's, Hood's, and Shield's Yards, and Tailor's Arms Yard.

Area III., of about three-quarters of an acre, comprises a series of adjoining courts which open into the part of the north side of Clayport Street, which lies to the west of Dispensary Street. These courts are Mills', Trotter's, Stamp's, Elliot's, and Lockey's Yards.

As in the case of Area I., and for similar reasons, it would seem that certain dwellings fronting Clayport Street should be included in each of the last two areas.

Certain statistics of population and of the number of inhabited dwellings and tenements in these areas have been supplied to me by Mr. E. G. Gibson, Assistant Overseer of the district, who has been good enough to make special house-to house inquiry for the purpose. I have summarised Mr. Gibson's statistics as follows :—

—	Number of Inhabited Dwellings.	Total Inhabitants.	Number of Tenements.*	Number of Tenements consisting of					
				One Room.	Inhabitants.	Two Rooms.	Inhabitants.	More than two rooms of the whole dwelling.	Inhabitants.
AREA I. Inclusive of 13 dwellings in Clayport Street, with 52 inhabitants.†	78	462	109	37	111	37	168	35	183
AREA II. Inclusive of five dwellings in Clayport Street, with 30 inhabitants.†	28	225	49	23	79	15	91	11	55
AREA III. Inclusive of one house in Clayport Street, with six inhabitants.†	18	188	40	21	75	12	72	7	41
Total of areas I., II., III.	124	875	198	81	265	64	331	53	279

* Each dwelling which is occupied by only one family is here returned as constituting a single tenement.

† As well as certain other dwellings abutting on Clayport Street, but reckoned as comprised in the courts which lie behind them.

Of the total number of unhealthy courts and buildings in Alnwick, and particularly of those in which insanitary conditions are most conspicuous, a large proportion are comprised within the areas above indicated. Apart from these areas, however, all of which are in the south-western portion of the town, known as Clayport, there are in other quarters of Alnwick several insanitary courts and "properties" which would require to be individually included in an additional scheme or schemes.

In connection with any scheme for dealing with unhealthy areas such as I have indicated, it is of course essential that adequate accommodation should be provided for persons displaced. In view of the overcrowding which at present exists, and in view also of the necessity that new streets and buildings should comply with the byelaws of the district, it would seem probable that upon the areas dealt with new house accommodation could be provided for only a small proportion of these persons. It would seem indeed to be an essential part of any such scheme that the District Council should acquire land on which the erection of suitable dwellings for the working classes could be undertaken. I am informed that little difficulty is anticipated in obtaining land for this purpose on the outskirts of the town.

Having regard to the unsatisfactory character of common lodging house accommodation in their district, the District Council might find advantage in including the construction of a satisfactory common lodging house in a scheme for the erection of working class dwellings.

Before taking leave of this subject it may be well briefly to consider such objections as I learnt were likely to be made locally to proposals for dealing with unhealthy areas of the town by a scheme or schemes under the Housing of the Working Classes Act. An objection raised that new and healthy dwellings for persons of the working class would be highly rented and so would not be in request can, I think, hardly be sustained in view of the local demand for dwellings such as those now being erected by the co-operative building association to which I have referred. The rent of these buildings is about four shillings a week, but healthy habitations at cheaper rates could probably be secured by the erection of suitably contrived buildings, each accommodating two or more families. Moreover, even if new buildings are resorted to rather by the more well-to-do members of the working class than by the poorer inhabitants displaced from the unhealthy areas, it may be anticipated that the latter will find sufficient and considerably improved accommodation in dwellings and tenements vacated by the former. Other objectors, I learnt, maintain that the erection of new dwellings by private enterprise will in course of time be of itself sufficient to solve the problem. Ten or twenty years hence, it is argued, plenty of good dwellings will be available and there will be little demand for the unhealthy tenements in the courts. Why, therefore, should the District Council now go to the trouble and expense of demolishing these courts? The answer is, that so long as these unhealthy dwellings are allowed to remain standing, so long will they almost certainly remain occu ed. Low rents, in addition to certain other

APP. A. No. 9.

On the sanitary condition of the Urban District of Alnwick, and upon Housing of the Working Classes therein; by Dr. Buchanan.

incidental advantages which Alnwick seems already to offer to the thriftless class, will be certain to attract to the town an undesirable class of immigrants.

Lastly, objection is made on the ground of expense involved in carrying out a scheme such as is in question. In this connection it should be remembered that the terms upon which unhealthy areas can be acquired under the provisions of the Housing of the Working Classes Act appear to offer exceptional advantages to the District Council, and that the expense of a satisfactory scheme under this Act can be distributed over a term of years by means of loan. Nor does it seem that objection need be taken to a scheme whereby in the first instance certain of the especially insanitary areas of the town alone are dealt with, provided that in such case due regard is had to future requirements in respect of the remaining areas of similar class.

No. 10.

REPORT on an OUTBREAK OF ENTERIC FEVER in the URBAN DISTRICT OF CAMBORNE, and in Portions of the adjoining RURAL DISTRICTS OF REDRUTH AND HELSTON; by DR. R. BRUCE LOW.

On December 15th last the Board received notice from Dr. Telfer Thomas, Medical Officer of Health for the Town of Camborne, that an outbreak of enteric fever had occurred in his district; and a few days later a similar intimation was received from Dr. E. Permewan, Medical Officer of Health for the Redruth Rural District, respecting the parish of Illogan, which lies close to Camborne. Subsequently, Dr. Haswell, Medical Officer of Health for the Helston Rural District, informed the Board that cases of enteric fever had occurred in a portion of the parish of Crowan, which also lies near Camborne. Later reports by Drs. Thomas and Permewan attributed the outbreak to the public water supply, which is common to the three places invaded by the fever. Seeing that three separate sanitary districts were involved, the Board decided to have the outbreak investigated by one of their medical staff, and I was instructed to undertake the inquiry.

I visited the locality on January 11th and found that the epidemic, which was by that time rapidly diminishing, was limited to (1) the Urban District of Camborne, (2) a portion of the parish of Illogan, and (3) the village of Praze, in the parish of Crowan. The parishes of Camborne, Illogan, and Crowan are contiguous, Camborne being situated between the other two.

The following is a brief description of each of these places.

1.—CAMBORNE URBAN DISTRICT.

The Camborne Urban District has an area extending over 6,931 acres. At the last census it had 3,264 inhabited houses and a population of 14,700. The town is situated in the centre of the Cornish mining district, and lies 12 miles west of Truro. Some portions of the urban district are rural in character; and there are several villages, *e.g.*, Troon (or Trewin), Perponds, and Barreppa, lying away from the town, though included within its boundaries. The subsoil is clay. A considerable proportion of the inhabitants are of the mining class, whose wages at the present time are said to average not more than 20*s.* a week. Owing to the extreme depression from which the Cornish mining industry has been suffering of late years, many miners have left the locality and gone to work in Indian, South African, or American mines, leaving their families behind them, and maintaining them by the money earned abroad. There is, therefore, just now a considerable preponderance of females among the population. It is estimated that the population at present is about 14,000; that, in fact, the population has diminished through the emigration of the miners during the last few years.

Sewerage.—The town, or central Camborne, is sewered to two outfalls, where the effluent is dealt with by irrigation on land; means for ventilation and flushing of the sewers are provided. One of the larger detached villages, Troon, population about 900, has a separate sewerage system of a similar kind; but the remaining outlying portions of the district are unsewered.

The water supply of the town is chiefly from the mains of the Camborne Water Company; but the villages of Troon, Perponds, and Barreppa, as well as the rural outlying portions of the urban district, have each a separate supply.

Excrement disposal is mainly by means of privies, with cesspits or middens attached. Some houses in the outlying portions of the district have no closets at all. Many of the privies are of rude construction and some are built of wood. The pits and middens are sunk below the level of the ground; many of them receive surface water as well as the drip from the privy roof. The majority of them cause nuisance. In some 50 instances the privy contents have, in process of removal, to be carried through the house from the back yard to the front street. Privies are, as a rule, emptied at the expense of occupiers at irregular intervals; the contents are mixed with ashes and carted off by farmers. Scavenging of privies as at present practised is altogether unsatisfactory. Some 65 houses are provided with box closets, *i.e.*, privies with movable wooden receptacles, emptied weekly by the Urban Council's servants. But the joints of the boxes that I saw were leaking, and fluid filth was wetting the surface of the ground, and at times dribbling out upon the yard. There are 477 houses provided with water closets, some 20 of them flushed by hand; the rest, with about seven exceptions, are supplied with proper flushing apparatus. The exceptions just mentioned are would-be slop water closets of an objectionable type and which do not act satisfactorily.

Removal of refuse.—House refuse is stored in middens or in ashpits, mostly uncovered, in which it is retained for long periods before any serious attempt is made at its removal. When farmers cannot be induced to cart it away, it is frequently thrown out in heaps in back lanes or on waste ground. As a result some lanes are almost blocked by accumulations of house refuse and rubbish of all sorts.

Many of the artisan dwellings are damp from want of eaves-spouting and some are dilapidated. A number have little open space at their backs. A good deal of house property is held on life-leases, and difficulty is experienced in getting any improvements executed on such property.

2.—The Parish of Illogan.

The parish of Illogan, in the Redruth Rural District, adjoins the town of Camborne. It has an area of 8,493 acres, and at the last census had 2,107 inhabited houses, with a population of 9,312. There are several large aggregations of population in this extensive parish, the chief of them with which this report is concerned comprise part of Tuckingmill,* Pool, and Illogan Highway; in the latter is situated the Redruth Union Workhouse which has an average population of 250. For the most part the people are engaged in the mining industry. *Sewerage.*—Only parts of this parish are sewered: at Pool there are sewers serving only a portion of the houses; some rows of houses have rubble drains behind them, with imperfect gulleys in the yards near the back doors. These rubble drains are often choked and cause nuisance. Some slop drains deliver into the gutter in the highway. The County Council, I am informed, have decided to pay no further contribution towards upholding the highways thus polluted until the Redruth Rural Council take steps to prevent discharge of sewage into the road gutters. *The Water Supply* for less than a third of the parish is from the mains of the Camborne Water Company. The rest of the parish has local supplies. *Excrement disposal* is by means of cesspit or midden privies, many of them dirty and dilapidated, and emptied at long irregular intervals by the occupiers. Many of them create grave nuisance. Some are built of wood, resembling sentry boxes, and soon get rotten or in bad repair. Rough householders occasionally chop up the privy door for firewood or knock out a board or two from the sides of the privy for the same purpose. Some houses have no privy at all. *House Refuse* is stored in ash-pits or middens, some of them close to back doors, and as they are not emptied for long intervals the contents overflow on to the surface of the yard and are trampled underfoot or washed into the yard gulleys, causing blocking of the drains. Occupiers are responsible for emptying these ashpits and middens, and the result is seen in accumulations of refuse and rubbish in back lanes and waste land on which the refuse has been surreptitiously tipped.

Many houses in Illogan, as at Camborne, have no eaves-spouting, and are consquently damp. Some have downfall pipes delivering into and wetting their foundations. There are some back-to-back houses. The soil here also is clay.

3.—The Parish of Crowan.

The parish of Crowan, in the Helston Rural District, adjoins Camborne Urban District on the south-west. The area of Crowan parish is 7,496 acres. At the last census it had 585 inhabited houses and a population of 2,468. There are several small villages or hamlets scattered over this wide parish, but this report has only concern with one of them, namely, Praze, which, I am informed, has a population of about 400. It lies about three miles south of Camborne, and consists of a single street with houses on either side. It has no proper sewer. Its *water supply* is from the service of the Camborne Water Company. The water is piped to only 38 of the houses, but the company have set up a tap in the centre of the village, whence all of the inhabitants may draw water and carry it to their homes. *Excrement disposal* is chiefly by means of cesspit or midden privies of the same objectionable type as already mentioned in connection with Camborne and Illogan. *House refuse* is stored in the middens or in ashpits, and disposed of along with privy contents upon garden land.

* Part of the Ecclesiastical Parish of Tuckingmill is included in the Urban District of Camborne, and part is in the Parish of Illogan.

A considerable part of the gathering grounds of the Camborne Water Company is situated, as will be seen further on, in the parish of Crowan.

THE OUTBREAK OF ENTERIC FEVER.

During the greater part of 1897 there appeared to be no marked prevalence of enteric fever in any of the three localities which subsequently suffered in December from an epidemic of this disease. The subjoined Table I. gives month by month for 1897 the number of cases of enteric fever reported in each of the three localities implicated in the outbreak, and shows the sudden development of the epidemic.

TABLE I.

Showing, month by month, the Number of cases of Enteric and Continued Fever notified in each of the three Parishes concerned in the Epidemic, from January, 1897, to January, 1898, inclusive :—

—	Population, Census of 1891.	1897.												1898
		January.	February.	March.	April.	May.	June.	July.	August.	September.	October.	November.	December.	January.
Camborne Urban District	14,700	2	1	3	2	—	—	1	2	10	2	2	85	37
Parish of Illogan, in the Redruth Rural District	9,312	—	1	2	1	2	—	—	1	2	—	2	21	12
Parish of Crowan in the Helston Rural District	2,468	—	—	—	—	—	—	—	1	-	2*	2*	4	2
Total	26,480	2	2	5	3	2	—	1	4	12	4	6	110	51

* All of these (4) cases occurred in the same house; one mild case, though ill at the same time as the others in October, was not notified. These four cases will be referred to again further on in the report.

The enteric fever epidemic began during the first week in December, and the largest proportion of the sufferers by it appear to have been attacked about the middle of the month. In addition to the cases given above, it was known that at least four persons, visitors or workers within the invaded area but having their homes outside it, contracted the fever. So that the known and reported cases comprised in the epidemic amounted to 165. But, in addition to these notified cases, there occurred just before the outbreak, and at times during its continuance, a large number of instances of transient illness; some of them marked by acute diarrhoea and vomiting, and others by feverish symptoms, with headache and prostration lasting some time. Such cases were locally regarded by some of the medical men as minor manifestations of the epidemic disease, or as abortive cases. One informant spoke of "scores" of such attacks having occurred within his own knowledge. The notified cases were, as a rule, of a mild type, and it is believed that a number of attacks that were truly enteric fever were never notified at all. Some patients, indeed, only came under notice some weeks after their attacks by the disease.

owing to a relapse through want of care. Two deaths from relapse were mentioned to me where, in one instance, purgative pills, and in another, improper food, had been given to the fever patient, with fatal effect. On the other hand, it was alleged by some that temporary illness of various kinds was notified as enteric fever during the course of the epidemic, and that some of these cases had no connexion with enteric fever at all.

Table II. gives the age and sex of each of the 165 notified cases.

TABLE II.

Showing the ascertained Age and Sex of 165 consecutive cases of Enteric Fever occurring in the Camborne Epidemic; and also the Age and Sex of twelve fatal cases during December, 1897, and January, 1898 :—

Age Periods.	Cases.			Deaths.		
	Males.	Females.	Total.	Males.	Females.	Total.
Under 1 year	—	—	—	—	—	—
1 to 5 years	11	8	19	—	1	1
5 „ 15 „	36	27	63	—	4	4
15 „ 25 „	28	23	51	1	2	3
25 „ 40 „	11	12	23	—	1	1
40 „ 65 „	4	5	9	1	2	3
65 and upwards... ...	—	—	—	—	—	—
	90	75	165	2	10	12

The case mortality among females was 13 per cent., and was high compared with that of males, 2 per cent. The case mortality, taking both sexes together, was a little over 7 per cent.

Altogether 149 houses were invaded. In the next table (III.) will be found particulars as to the number of invaded houses, notified cases, and deaths in each of the invaded localities during the outbreak.

TABLE III.

Showing the Number of Houses Invaded, the Number of Notified cases of Enteric Fever, and Deaths therefrom, in each of the three Districts affected by the Epidemic of December, 1897, and January, 1898 :—

Areas Invaded.	Houses Invaded by Enteric Fever.	Cases Notified as Enteric Fever.	Cases Fatal from Enteric Fever.
Camborne Urban District	112	122	7*
Illogan Parish (part of)	28	33	5
Crowan Parish (part of)	5	6	—
Places outside the above localities where occurred fever attacks contracted within the area	4	4	—
	149	165	12

* Including a case certified "acute tuberculosis," but which the certifying Practitioner now believes to have been enteric fever.

App. A. No. 10.

On an outbreak of Enteric Fever in the Urban District of Camborne and the Rural Districts of Redruth and Helston; by Dr. Bruce Low.

The progress of the epidemic, week by week, from start to finish is shown in Table IV. Some of the attacks credited to January commenced in December, but were not notified till the disease had far advanced, or till a relapse developed more marked symptoms. Of the later January cases some were undoubtedly secondary to attacks that had occurred in the dwelling during December. It may be added that multiple cases were met with in eleven houses, viz., in one house, five cases; in each of two houses, three cases; in each of eight houses, two cases.

Table IV.

Showing, week by week, the Number of Enteric Fever Cases notified in the Invaded Localities during the three months, November, 1897, to January, 1898, inclusive :—

The Epidemic Area.	Weeks Ending.												
	November 6.	November 13.	November 20.	November 27.	December 4.	December 11.	December 18.	December 25.	January 1.	January 8.	January 15.	January 22.	January 29.
Camborne Urban District.. ..	1	—	1	—	1	16	32	19	20	21	7	4	2
Illogan Parish (part of)	—	—	1	—	—	1	8	6	6	3	6	3	—
Crowan Parish (part of)	—	—	—	—	—	1	3	—	—	1*	1*	—	—
Cases in persons residing outside these places, but who contracted the disease in one of them.	—	—	—	—	—	—	1	—	—	2	1	—	—
Total	1	—	2	—	1	18	44	25	26	27	15	7	2

* Both of these cases became ill during the middle of December.

The Cause of the Outbreak.

Before my arrival careful inquiry had been made into the probable cause of the outbreak by Drs. Thomas and Permewan, the Medical Officers of Health for the two districts that suffered most. These gentlemen separately had come to the conclusion that the public water supply was the cause of the mischief. Locally, however, other opinions were offered. The opinion receiving most support was that the accumulations of filth and refuse, the result of neglected scavenging, were the direct occasion of the epidemic, at least in Camborne.* But, while the methods of excrement disposal in all the places invaded leave much to be desired, and though the scavenging of Camborne and Illogan has

* One enthusiastic amateur epidemiologist wrote to the local papers announcing that: as a result of long study he had definitely proved to his own satisfaction that the emanations from sewers and from filth heaps giving out a "deadly gas," called sulphuretted hydrogen, were the source of all the trouble. This self-constituted adviser of the public was so rash as to assert that boiling of water or milk was unnecessary as a preventive measure.

been much neglected by the District Councils concerned, yet the history of the outbreak and the behaviour of the fever differed totally from what might have been expected had the epidemic been produced by excremental contamination of the soil. In Camborne, for example, better class houses as well as those of a poorer sort were invaded. Houses having water closets, as well as houses served by box closets or by midden privies, suffered from fever. So too houses with well-cared for yards, as well as those with neglected back premises. In a word, the cases occurred practically all over the town. Moreover, the season when outbreaks of enteric fever, due to filth pollution of the soil, chiefly occur had gone by; such outbreaks generally follow the hot weather, and are usually not of the sudden character that marks epidemics due to infected milk or contaminated water.

With regard to milk, careful inquiry at Camborne, Illogan, and Crowan showed that the supplies of the invaded houses were many and various. In Camborne, for instance, there were some 40 milk supplies in question, with no perceptible excess of fever cases upon any one of them. A number of persons attacked denied that they ever took milk at all. Others had taken it only in small quantities in tea. Milk, then, afforded no explanation of the outbreak. But it was otherwise when the facts relating to drinking water came to be examined.

The water supply of a large portion of the three areas invaded by the epidemic is from the mains of the Camborne Water Company. This Company was established in 1867 by "An Act of Parliament for the better supplying with water of the parishes of Camborne, Crowan, and Illogan," the precise area, it may be remarked, affected by the recent epidemic.

In Camborne, out of a total of 3,127 houses occupied at present in the town there are 2,365 connected to the Company's mains. Among the 2,365 houses in Camborne using the Company's service 106 were invaded, yielding 116 cases; while of the 762 houses supplied from other sources there were only six invaded, yielding six cases. But when these six cases, occurring in dwellings not using the public water service, came to be investigated, it was found that most of them in all probability had been using the Camborne supply. One person had been attacked by enteric fever the day after his arrival at a relative's house from his home at Prazo, where the incriminated water was used. Two cases were of children coming daily from the outskirts to attend a school in Central Camborne, where the Company's service was laid on, and of which the children drank daily. Two other cases were of young persons coming from outlying parts of Camborne to the town every day to work, and where they had opportunities for drinking from the Camborne public supply. The sixth and last case was that of a frequent visitor in Central Camborne, who denied having drunk any Camborne water, though she admitted having on numerous occasions partaken of meals at the Camborne houses which she visited.

In Illogan parish, at the last census, there were 2,107 inhabited houses; this number, I am informed, represents very nearly those

A considerable part of the gathering grounds of the Camborne Water Company is situated, as will be seen further on, in the parish of Crowan.

THE OUTBREAK OF ENTERIC FEVER.

During the greater part of 1897 there appeared to be no marked prevalence of enteric fever in any of the three localities which subsequently suffered in December from an epidemic of this disease. The subjoined Table I. gives month by month for 1897 the number of cases of enteric fever reported in each of the three localities implicated in the outbreak, and shows the sudden development of the epidemic.

Table I.

Showing, month by month, the Number of cases of Enteric and Continued Fever notified in each of the three Parishes concerned in the Epidemic, from January, 1897, to January, 1898, inclusive :—

—	Population, Census of 1891.	1897.												1898
		January.	February.	March	April.	May.	June.	July.	August	September.	October.	November.	December.	January
Camborne Urban District	14,700	2	1	3	2	—	—	1	2	10	2	2	85	37
Parish of Illogan, in the Redruth Rural District	9,312	—	1	2	1	2	—	—	1	2	—	2	21	12
Parish of Crowan in the Helston Rural District	2,468	—	—	—	—	—	—	—	1	-	2*	2*	4	2
Total	26,480	2	2	5	3	2	—	1	4	12	4	6	110	51

* All of these (4) cases occurred in the same house ; one mild case, though ill at the same time as the others in October, was not notified. These four cases will be referred to again further on in the report.

The enteric fever epidemic began during the first week in December, and the largest proportion of the sufferers by it appear to have been attacked about the middle of the month. In addition to the cases given above, it was known that at least four persons, visitors or workers within the invaded area but having their homes outside it, contracted the fever. So that the known and reported cases comprised in the epidemic amounted to 165. But, in addition to these notified cases, there occurred just before the outbreak, and at times during its continuance, a large number of instances of transient illness ; some of them marked by acute diarrhœa and vomiting, and others by feverish symptoms, with headache and prostration lasting some time. Such cases were locally regarded by some of the medical men as minor manifestations of the epidemic disease, or as abortive cases. One informant spoke of "scores" of such attacks having occurred within his own knowledge. The notified cases were, as a rule, of a mild type, and it is believed that a number of attacks that were truly enteric fever were never notified at all. Some patients, indeed, only came under notice some weeks after their attacks by the disease, owing to a relapse through want of care. Two deaths from relapse were mentioned to me where, in one instance, purgative pills, and in another, improper food, had been given to the fever patient, with fatal effect. On the other hand, it was alleged by some that temporary illness of various kinds was notified as enteric fever during the course of the epidemic, and that some of these cases had no connexion with enteric fever at all.

Table II. gives the age and sex of each of the 165 notified cases.

TABLE II.

Showing the ascertained Age and Sex of 165 consecutive cases of Enteric Fever occurring in the Camborne Epidemic; and also the Age and Sex of twelve fatal cases during December, 1897, and January, 1898 :—

Age Periods.	Cases.			Deaths.		
	Males.	Females.	Total.	Males.	Females.	Total.
Under 1 year	—	—	—	—	—	—
1 to 5 years	11	8	19	—	1	1
5 " 15 "	36	27	63	—	4	4
15 " 25 "	28	23	51	1	2	3
25 " 40 "	11	12	23	—	1	1
40 " 65 "	4	5	9	1	2	3
65 and upwards... ...	—	—	—	—	—	—
	90	75	165	2	10	12

The case mortality among females was 13 per cent., and was high compared with that of males, 2 per cent. The case mortality, taking both sexes together, was a little over 7 per cent.

Altogether 149 houses were invaded. In the next table (III.) will be found particulars as to the number of invaded houses, notified cases, and deaths in each of the invaded localities during the outbreak.

TABLE III.

Showing the Number of Houses Invaded, the Number of Notified cases of Enteric Fever, and Deaths therefrom, in each of the three Districts affected by the Epidemic of December, 1897, and January, 1898 :—

Areas Invaded.	Houses Invaded by Enteric Fever.	Cases Notified as Enteric Fever.	Cases Fatal from Enteric Fever.
Camborne Urban District	112	122	7*
Illogan Parish (part of)	28	33	5
Crowan Parish (part of)	5	6	—
Places outside the above localities where occurred fever attacks contracted within the area	4	4	—
	149	165	12

* Including a case certified "acute tuberculosis," but which the certifying practitioner now believes to have been enteric fever.

App. A. No. 18. On an outbreak of Enteric Fever in the Urban District of Camborne and the Rural Districts of Redruth and Helston: by Dr. Bruce Low.

The progress of the epidemic, week by week, from start to finish is shown in Table IV. Some of the attacks credited to January commenced in December, but were not notified till the disease had far advanced, or till a relapse developed more marked symptoms. Of the later January cases some were undoubtedly secondary to attacks that had occurred in the dwelling during December. It may be added that multiple cases were met with in eleven houses, viz., in one house, five cases; in each of two houses, three cases; in each of eight houses, two cases.

TABLE IV.

Showing, week by week, the Number of Enteric Fever Cases notified in the Invaded Localities during the three months, November, 1897, to January, 1898, inclusive:—

The Epidemic Area.	Weeks Ending. November 6.	November 13.	November 20.	November 27.	December 4.	December 11.	December 18.	December 25.	January 1.	January 8.	January 15.	January 22.	January 29.
Camborne Urban District	[illegible]	—	[illegible]	—	[illegible]	[illegible]	[illegible]	[illegible]	[illegible]	[illegible]	[illegible]	4	2
Illogan Parish, part of	—	—	[illegible]	—	—	[illegible]	4	4	4	1	4	3	—
Crowan Parish, part of	—	—	—	—	—	[illegible]	[illegible]	—	—	[illegible]*	[illegible]*	—	—
Cases in persons residing outside these places, but who contracted the disease in one of them.	—	—	—	—	—	—	[illegible]	—	—	[illegible]	[illegible]	—	—
Total	[illegible]	—	[illegible]	—	[illegible]	[illegible]	[illegible]	[illegible]	[illegible]	[illegible]	[illegible]	[illegible]	2

* Both of these cases became ill during the middle of December.

THE CAUSE OF THE OUTBREAK.

Before my arrival careful inquiry had been made into the probable cause of the outbreak by Drs. Thomas and Permewan, the Medical Officers of Health for the two districts that suffered most. These gentlemen separately had come to the conclusion that the public water supply was the cause of the mischief. Locally, however, other opinions were offered. The opinion receiving most support was that the accumulations of filth and refuse, the result of neglected scavenging, were the direct occasion of the epidemic, at least in Camborne.* But, while the methods of excrement disposal in all the places invaded leave much to be desired, and though the scavenging of Camborne and Illogan has

* One enthusiastic amateur epidemiologist wrote to the local papers announcing that as a result of long study he had definitely proved to his own satisfaction that the emanations from sewers and from filth heaps giving out a "deadly gas," called sulphuretted hydrogen, were the source of all the trouble. This self-constituted adviser of the public went so far as to assert that boiling of water or milk was unnecessary as a preventive measure.

been much neglected by the District Councils concerned, yet the history of the outbreak and the behaviour of the fever differed totally from what might have been expected had the epidemic been produced by excremental contamination of the soil. In Camborne, for example, better class houses as well as those of a poorer sort were invaded. Houses having water closets, as well as houses served by box closets or by midden privies, suffered from fever. So too houses with well-cared for yards, as well as those with neglected back premises. In a word, the cases occurred practically all over the town. Moreover, the season when outbreaks of enteric fever, due to filth pollution of the soil, chiefly occur had gone by; such outbreaks generally follow the hot weather, and are usually not of the sudden character that marks epidemics due to infected milk or contaminated water.

With regard to milk, careful inquiry at Camborne, Illogan, and Crowan showed that the supplies of the invaded houses were many and various. In Camborne, for instance, there were some 40 milk supplies in question, with no perceptible excess of fever cases upon any one of them. A number of persons attacked denied that they ever took milk at all. Others had taken it only in small quantities in tea. Milk, then, afforded no explanation of the outbreak. But it was otherwise when the facts relating to drinking water came to be examined.

The water supply of a large portion of the three areas invaded by the epidemic is from the mains of the Camborne Water Company. This Company was established in 1867 by "An Act of Parliament for the better supplying with water of the parishes of Camborne, Crowan, and Illogan," the precise area, it may be remarked, affected by the recent epidemic.

In Camborne, out of a total of 3,127 houses occupied at present in the town there are 2,365 connected to the Company's mains. Among the 2,365 houses in Camborne using the Company's service 106 were invaded, yielding 116 cases; while of the 762 houses supplied from other sources there were only six invaded, yielding six cases. But when these six cases, occurring in dwellings not using the public water service, came to be investigated, it was found that most of them in all probability had been using the Camborne supply. One person had been attacked by enteric fever the day after his arrival at a relative's house from his home at Praze, where the incriminated water was used. Two cases were of children coming daily from the outskirts to attend a school in Central Camborne, where the Company's service was laid on, and of which the children drank daily. Two other cases were of young persons coming from outlying parts of Camborne to the town every day to work, and where they had opportunities for drinking from the Camborne public supply. The sixth and last case was that of a frequent visitor in Central Camborne, who denied having drunk any Camborne water, though she admitted having on numerous occasions partaken of meals at the Camborne houses which she visited.

In Illogan parish, at the last census, there were 2,107 inhabited houses; this number, I am informed, represents very nearly those

App. A. No. 16.

On an outbreak of Enteric Fever in the Urban District of Camborne and the Rural Districts of Redruth and Helston; by Dr Bruce Low.

that are in occupation at the present time. Of these 2,107 houses, 810 derive their water supply from the Camborne Company's service: and of the 810, 26 were invaded by enteric fever in December, 1897, and January, 1898, yielding 31 cases. On the other hand, during the same period, of the 1,297 houses having water supply other than that of the Camborne Company, only two were invaded. In one of these, the patient, a man, worked previous to his attack at a place where the Company's water was laid on, and he had been in the habit of drinking this water freely. The second of these houses was invaded in the person of a servant girl who was brought home from an adjoining district suffering from illness, notified as enteric fever, on January 15th. She had been visiting Illogan on each Sunday previous to her illness, but she denied drinking the Camborne Company's water there. This was the only Illogan case during the epidemic period that had not, so far as the evidence showed, consumed water from the public service.

In the parish of Crowan only two villages are supplied with the Camborne Company's water, viz., Praze, with about 119 houses, and Churchtown, with 39 houses. At the latter place no house has the water laid on; a public tap supplies the houses in the village. At Praze only 38 houses are connected to the Company's mains, but the rest ot the village obtain water from a public tap supplied by the Company in the middle of the village. At Praze six fever cases were notified in the two months December and January, all of the persons attacked having had access to the inculpated water service. To these six cases must be added that of the youth taken ill at Camborne the day after his arrival there from Praze, where he too had been accustomed to use the incriminated water.

In the remaining portions of Crowan parish, with some 19[illegible] houses, there were, during the epidemic period, only two dwellings invaded, yielding three fever cases. These two houses lay far apart from Praze, and, in the judgment of Dr. Haswell, the attacks in them were unconnected with the Camborne outbreak, having been infected from other sources, and they are not taken account of in connexion with the outbreak.

It has been already mentioned (page) that four cases occurred in persons visiting the infected area, but whose homes were in other districts not affected by the fever. Two of these cases had been temporarily residing in Camborne and using its public water service, but developed enteric fever shortly after their return home. The remaining two cases worked daily at Illogan, returning at night to their homes in Redruth Urban District. Both had drunk the Camborne Company's water, which was laid on at their places of work.

[illegible] within the three districts involved the incidence of the epidemic on houses supplied, and not supplied, with the Water Company's water, it was found that the rate of house invasion of houses supplied by the Company was more than ten tim[illegible] greater than on houses not so supplied.

This is shown in tabular form below:—

TABLE V.

Showing the Number of Houses supplied, and not supplied, by the Camborne Water Company's Service in each of the three localities involved in the Outbreak; with the Number of Houses Invaded in each class during December, 1897, and January, 1898; along with the respective Rates of Invasion.

Localities invaded by Enteric Fever.	Houses supplied by the Camborne Company.			Houses NOT supplied by the Camborne Company		
	Number of houses supplied.	Invaded by Fever	Per cent. invaded by fever.	Number of houses not supplied.	Invaded by fever.	Per cent. invaded by fever.
Camborne Urban District	2,365	106	4·5	762	5*	0·7
Illogan Parish ...	810	26	3·2	1,297	2	0·2
Crowan Parish ...	158†	7	4·4	427	2	0·5
	3,333	139	4·2	2,486	9	0·4

* I have deducted the case already mentioned that fell ill the day after arrival in the Camborne district from Praze, and have transferred it to that village where the disease was contracted.

† In this number are included all the houses taking water from the public taps at Praze and Churchtown.

Contrast in the above sense would be even greater if those persons, including at least four at Camborne and two at Illogan, who were known to have contracted their illness by drinking the incriminated water supply while at work, had not in the above table been reckoned as inmates of houses not supplied by Camborne water. In each of these instances the patient, though he had another supply at home, used the Camborne public service when at work.

The evidence against this water supply is strong; for out of a total of 165 cases that were reported during the outbreak, there were practically only two attacks in which there was doubt as to the persons having had opportunities of drinking the incriminated water supply, and whose illness, therefore, was not due directly or indirectly, so far as could be learned, to the water supplied by the Camborne Company.

The Sources of the Camborne Public Water Supply.

The Camborne Water Company's supply comes from two separate sources, Cargenwyn and Boswyn. The watersheds of the whole of the former source and half of the latter are in the parish of Crowan. These two gathering grounds, which are situated not far distant one from another, are shown on the annexed map. In both instances the supply is partly derived from springs and partly from surface water. There is no filtration of the water before delivery; the service is constant.

1. *Cargenwyn Source.*—The main source of this supply is from a spring which issues from a hollow in a field, the water passing

thence in an open channel for about 200 yards to a reservoir, named No. 2* (see map). The overflow from this No. 2 reservoir flows into an adjoining one, named No. 3, constructed parallel to, but on a slightly lower level than, the other. When these two reservoirs are full, water overflows from the lower of them into an open brook or so-called "byewash," which can, by means of an arrangement of wooden troughs, be connected to a still lower reservoir, named No. 1—see map; or, if the water is not needed, it can be diverted round this reservoir to discharge into a stream lower down. Surface water gains admission to the stream or byewash flowing to No. 1 reservoir from the surrounding land, which slopes to it. On this watershed are situated 19 houses, with a total population of 72 persons. No less than 13 out of the 19 houses have no closet or privy accommodation. The inmates in such instances have to defæcate in the fields, or they use an utensil at home, the contents of which are commonly thrown upon the dunghill. There is at least one slaughter-house on this area. A group of six houses (called High Cargenwyn), five of them unprovided with privies, lies on the slope parallel with the two higher reservoirs, and distant less than 100 yards from the stream or byewash, which ultimately enters the lower reservoir, No. 1—see map. These houses have no means of drainage. Two of the dwellings are occupied by farmers, who between them have 46 head of cattle and nearly 20 pigs on the farms. The dung from these animals is spread, along with human excrement, on the meadows on the watershed.

The Cargenwyn water, the original supply of the Camborne Company, is not now used in winter. It is collected during the winter months, and impounded for use during the summer, when the Boswyn springs, the other source to be presently described, run low and become insufficient for the demands of the district. The total storage capacity of the three Cargenwyn reservoirs is 34 million gallons. This supply, I am informed, ceased to be used in 1897, on November 12th.

The obvious sources of pollutions of the byewash and of the other open channels which convey to the reservoir water from the Cargenwyn spring, have since the outbreak attracted the attention of the Water Company. The water from the spring is now being piped direct to No. 2 reservoir, and the overflows from Nos. 2 and 3 reservoirs are to be piped to No. 1. By excluding the byewash in this way it is believed that pollution will be prevented. A sample of water taken from the byewash was sent for bacteriological analysis during December, 1897, and was reported on unfavourably—see Appendix I. The absence of privy accommodation for so many of the houses on this watershed is a source of danger. I learn from Dr. Haswell that as far back as 1883 he reported this defect to his Sanitary Authority, and orders were issued to erect privies. But nothing was done to enforce the Rural Sanitary Authority's orders. Some years later these orders were repeated, but were not enforced, it is said, owing to the apathy of the late Inspector of Nuisances. Dr. Haswell informs me that the matter is once again in hand, and that the new Inspector, Mr. J. J. Jenkin, will this time enforce the orders.

---

* The reservoirs have locally been numbered according to the order of their construction.

Copper Hill

No cases of enteric fever were known to have occurred on this watershed during the summer or autumn of 1897. It is evident that if such a case or cases had occurred in any of the houses, more particularly in any of those draining to the byewash, this source of supply might have been seriously endangered. This the Directors of the Water Company now recognise.

2. *The Boswyn Source* is derived mainly from a series of springs in the Boswyn valley. The water from these springs is piped to a service tank holding when full 60,000 gallons. In order to augment the supply, a brook running parallel to the tank and only a few yards distant from it, has been connected to it by means of a 6-inch earthenware pipe; a small dam across the brook serving to raise the water high enough to flow into the pipe leading to the service tank. This brook in dry weather receives little pollution, but in rainy weather surface pollutions are carried into it by the storm water from the area included in the watershed above the tank. Upon this area are situated nine inhabited houses, with a population of 42 persons. Six of the houses are unprovided with closet accommodation, and none of them have any suitable drainage arrangements. Slops and liquid filth are allowed to soak away upon the clayey soil as best they can. Upon this watershed, in a house situated about 700 yards from, and above, the service tank, there occurred during October and November, 1897, four cases of enteric fever; the first, of a girl, being notified on October 27th, and that of the mother of the family, the second case, on November 1st. The third case, of a child, had a mild attack before or about the same time as the mother, but her case was not notified at all; the fourth case, that of another girl, was notified on November 24th. This family was living in great poverty, and when the illness began the father of the family was in prison, so that the mother was left to support herself and the children by going out to work, or by begging from charitably disposed persons in the neighbourhood. The woman had some employment early in October in Troon (in the Urban District of Camborne), where, in September, at least two cases of enteric fever had occurred. It has been suggested that this family living on the Boswyn gathering ground was infected from Troon, in the Camborne Urban District; but on this point there is no sufficient evidence. It is possible, however, that in begging from door to door at Troon the children may have had given to them food from one or other of the two infected houses in the village. But for the purpose of accounting for this epidemic it is not material to trace whence the infection of this family was derived. It is, however, of the first importance to know that a house on one of the gathering grounds of the Camborne Water Company was invaded by enteric fever towards the end of October, and that this house remained infected by enteric fever during the whole of November, and possibly later.

The house occupied by the infected family in question is an isolated one, and stands alone towards the higher part of the Boswyn Watershed, about 700 yards from the service tank (see map). Close to the house is a pond nearly 40 yards in circumference and about 2 feet in depth. This pond receives the surface water from grass fields and moorland draining towards it. It also

P. A. No. 10.

—

[...]an outbreak of [...]teric Fever [...]he Urban [...]trict of [...]nborne and Rural [...]tricts of [...]ruth and [...]ston; by Bruce Low.

has hitherto served as a receptacle for all slops and liquid refuse from the cottage for which no other means of disposal have been provided. The house has no privy accommodation whatever, and at the time of my two visits to it the surface of the ground, round about the house and near the pond, was profusely littered with human excrement. When the members of this family were first taken ill the mother says they waited on one another as best they could. Ultimately, when all of them became ill, a charitable lady hearing of the condition of the family sent a woman from the neighbourhood to attend to the fever-smitten folk. Previous to the arrival of this untrained nurse, it is certain that little or no care was taken in disposing of the dejections of the patients. All the fluid excreta, if not the solid dejecta as well, went direct into the pond. If any solid fæces were thrown out near the house they must have been washed into the pond as soon as rain fell. The water in which the soiled linen was washed would have been thrown, too, into the pond. It is evident, then, that the water in the pond was, early in November, largely polluted by the specific poison of enteric fever, and that in this way specific pollution of the pond water was renewed from time to time throughout the months of November and December.

The pond in fine weather has no outlet, but, after heavy rainfall, the water accumulates till it overflows into an open ditch or channel which leads down the steep hillside direct to the brook. The point at which the overflow from the pond gains access to the brook is about 100 yards above the dam in the brook where is situated the pipe which connects it to the Camborne Water Company's service tank. It needed then but some heavy rainfall to carry specifically polluted water from the pond to the service tank. It may be mentioned that in its course from the pond the channel, about half-way down the hill, passes through a smaller pond or pool about 8 yards long, 3 yards wide, and about 2½ feet deep. When, in ordinary weather, the channel or water course is dry, the second pool like the pond still retains water; so that if both were infected, a considerable stock of fever-contaminated water might be stored up in them ready to be carried down the hill to the brook as soon as rain caused the upper pond to again overflow.

The local rainfall records show that rain fell on only one day from November 1st to 11th inclusive; but that on the 12th and 13th there fell respectively 1·11 and 0·97 inch, or *a total for the two consecutive days of 2·08 inches*. During the next five days small amounts of rain were recorded on each day, the total, however, being only 0·30 inch, making for the week a total rainfall of 2·38 inches, an excessive amount. Allowing an incubation period of fourteen days to enteric fever, and further allowing a week or ten days, or perhaps longer, during which the disease has to make sufficient progress to be with any certainty diagnosed for notification purposes, it is found that the outburst commenced at the time when it might have been expected to occur if specifically contaminated water had gained access to the public supply on the 12th of November and succeeding days. The family at the cottage tell me that the pond was very full of water during November. Reference to Table IV. (*see* page), shows that in the fortnight ending December 18th no less than sixty-two enteric fever cases

On an ou break of Enteric I in the Ur District c Camborn the Rura Districts Redruth Helston; Dr. Bruc

were notified in Camborne, Illogan, and Crowan, as compared with only one in the preceding fortnight. I have no doubt that this specific pollution of the public water service did, as a matter of fact, occur through overflow of the water of the above-named pond in the week ending November 18th, and that in this way the epidemic was set going. And there is reason to believe that the specific pollution in a similar fashion recurred on one or more occasions following the above date. On the two days, November 24th and 25th, a total rainfall of 1·08 inches was measured. Again, for the seven days ending December 7th, a fall of 1·49 inches was recorded; and last of all for the week ending December 14th no less than 3·40 inches of rainfall were measured. At some at least of these dates there is probability that the pond and the pool, already highly charged with fever organisms from the cottage, sent down a further supply of specifically infective material to the brook and thence to the tank, thus dangerously polluting the supply, so long as there was connexion between the brook and the service tank.

Drs. Thomas and Permewan, as has been stated, were convinced from an early period of their inquiry that the public water service was concerned in the outbreak. The first-named officer made an inspection of the Company's waterworks on December 13th and again on December 17th. On December 21st the two Medical Officers of Health together inspected the sources of the water supply, making particular examination of the Boswyn source, since they had learned from Dr. Haswell of the occurrence of the fever cases in the isolated house mentioned above. At this date the Company's caretaker at Boswyn Waterworks denied positively to Dr. Thomas that there was any connexion between the brook and the service tank. It was not till December 27th that Dr. Thomas obtained information direct from the Chairman of the Water Company that a connexion pipe did exist. On December 29th, Dr. Thomas, accompanied by the Secretary of the Water Company, again visited Boswyn, and the connecting pipe was then plugged with clay. There is however some conflict of evidence as to the precise date when the brook water was really excluded from the service tank. In the local newspapers a statement was published in January, 1898, from the Water Company, to the effect that water had not been admitted to the service tank through this connecting pipe "since the floods in the early part of the month" (December). The wet weather in December began on the 5th and continued up to the 16th, so that the flood-time may have comprised some 10 or 11 days. Another communication from the Secretary of the Company gives the date of closing of the pipe as December 1st. With this diversity of statement it is not easy to get at the exact truth. The Secretary of the Company resides at Camborne, and his information of actual procedures at the Boswyn Waterworks must be derived mainly from the resident caretaker. The results of the bacteriological and chemical analyses of the Boswyn water taken from town taps on the 15th and 17th of December do not, it is true, go to indicate that at this date polluted brook water was entering the service tank (*see* Appendix I.). But it is of course possible that owing to imperfect plugging, storm water was, even in December, now and again finding access along the connecting pipe to the service tank; in which case the continuation of the outbreak up to the middle of January would be fully accounted for.

Measures taken to check the Epidemic.

As soon as suspicion fell upon the Camborne public water service the Medical Officers of Health urged strongly upon the inhabitants of the infected localities the necessity of boiling all water and milk before use. Posters were issued in Camborne and Illogan directing the public attention to this precaution. As has been said, the connexion between the service tank and the Boswyn brook was severed some time in December. But owing to the existence of the pipe having been concealed from Dr. Thomas, valuable time was lost, and it is probable that infected water continued to gain admission to the service tank till the pipe was finally disconnected. On December 30th, the service tank was emptied, and carefully scrubbed out and cleansed under Dr. Thomas' directions. The water mains were afterwards freely flushed. On January 3rd, the Boswyn service was resumed, the Cargenwyn supply having been used from December 30th to January 3rd.

Disinfectants were supplied freely to invaded houses by the Nuisance Inspectors of Camborne Urban, and Redruth and Helston Rural Districts. Advice was given as to the proper dealing with the infected bowel discharges from patients.

I am informed, since my visit, that a large amount of scavenging in the invaded localities has been done by the Camborne Urban and the Redruth Rural District Councils, and that efforts are being made to disinfect, by means of quicklime and perchloride of mercury solution, the cesspits and middens which have been contaminated by the bowel discharges. It is to be hoped that the cleansing of privies and middens in Camborne and Illogan will be permanently undertaken by the respective District Councils, and not left, as heretofore, to the occupiers of houses, many of whom have no means for disposing of the contents of their privies and ashpits.

The Camborne District Council possess an isolation hospital—a private dwelling house, which is leased for a term of years. Diseases of only one sort can be treated in the building at one and the same time. The rooms are not adapted for fever wards, and the accommodation must be pronounced as inadequate and unsatisfactory for a district of the size of Camborne. It can only be regarded at best as a temporary makeshift.

During the epidemic, only nine fever cases were treated in this hospital, and even this small number taxed to the full the resources of the establishment. Four trained nurses were engaged to wait upon the cases in the hospital; while other two trained district nurses were employed in visiting daily some of the poorer families invaded by the fever. I heard of six trained nurses being employed by private persons of the better class to nurse sick members of their families.

The Redruth Rural Council also possess a makeshift hospital, which was last used for the isolation of small-pox cases about two years ago. It is altogether insufficient for the needs of the district.

None of the fever patients in the present epidemic at Illogan were removed to it for treatment. The cases were treated at their own homes, often amid difficulty and danger alike to the sick persons themselves and to those who were trying to nurse them.

Before leaving the locality I met in conference a committee appointed by the Camborne Urban Council to deal with the epidemic. At this conference I explained fully the circumstances under which the fever outbreak had occurred, and pointed out the dangers that might arise in the future if the specific pollutions of cesspits, middens, and yards were not fully dealt with at the present time. I discussed at considerable length the sanitary condition of the town and the means to be used for its amelioration. In this connection I reminded the Committee that my colleague, Dr. Reece, inspected Camborne under the Cholera Survey, in the summer of 1894, and afterwards presented to the late local board a series of recommendations, which I repeated to the Committee—enlarging upon each item, and explaining its purpose. I append a copy of Dr. Reece's recommendations (*see* Appendix II.). Apparently the late local board did not act upon the advice given to them by my colleague, for the conditions mentioned by him in 1894 remain practically unaltered in 1898. The Committee promised to bring before the Urban Council the facts I had submitted to them. I learned from a local newspaper that at the Council's next meeting the final discussion upon the recommendations was adjourned till this report on the fever outbreak could be placed in the hands of the members.

As has been stated, the Water Company have piped the Cargenwyn spring to No. 2 reservoir. The byewash water, too, has now been excluded from No. 1 reservoir. At Boswyn, the connecting pipe between the brook and the service tank has been taken up, and no storm or surface water can now find access to the supply.

I conferred also with the Clerk to the Redruth Rural District Council, and with the Medical Officer of Health of the District as to the condition of Illogan. To these gentlemen I submitted a number of recommendations dealing with (1) the present unsatisfactory methods of the excrement disposal and refuse removal in vogue in Illogan ; (2) the need for providing sewerage for those portions of the parish yet unsewered ; (3) the necessity for providing a satisfactory isolation hospital with the proper equipment for dealing with infectious diseases.

It may be remarked here that Drs. Thomas and Permewan have urged all or most of the above recommendations at one or another time on their respective Councils. It is a matter for regret that the sound advice of these efficient officers does not in either case receive at all times the attention that it deserves. The policy of both Councils seems in the past to have been more in favour of economy than efficiency.

With regard to the Helston Rural District I conferred with Dr. Haswell, the Medical Officer of Health, and Mr. J. J. Jenkin, the Inspector of Nuisances. I found that active measures for dealing with infected privies and middens had been taken at Praze. With regard to the absence of closet accommodation in the outlying portions of Crowan, Dr. Haswell promised to press this matter on the attention of his District Council.

Summary.

1. In December, 1897, a sudden outbreak of enteric fever appeared simultaneously in the three adjacent parishes of Camborne, Illogan, and Crowan, situate in three separate sanitary districts. Altogether, 165 attacks with 12 deaths were recorded in about six weeks.

2. The only condition common to the three localities was that of water supply derived from the same source, the Camborne Water Company's waterworks.

3. This Company's supply, immediately previous to the outbreak, came from Boswyn, where some springs are piped to a service tank, and where, to augment the supply, water from a brook was piped to the same tank.

4. Within the catchment area of this brook there occurred in October and November, 1897, in a single cottage, four cases of enteric fever, the bowel discharges from which fouled a pond close to the dwelling.

5. In wet weather the pond in question overflows, and the water therefrom runs down the hill in a channel to the above mentioned brook, entering it 100 yards above the pipe leading from the brook to the service tank.

6. In mid-November the rainfall was excessive, and by the flood thus caused the fever poison emanating from the infected cottage was no doubt swept into the Boswyn supply.

7. After the brook water was shut off from the service tank, and after the service tank had been scrubbed out and cleansed and the water mains well flushed, the outbreak came suddenly to an end.

8. The sanitary condition of Camborne and Illogan, as well as of parts of Crowan is far from being satisfactory.

APPENDIX I.

Sample I.—Camborne Tap Water (Boswyn Water).

"Analysis of, and report on, a sample of water received on December 15th, 1897 from Dr. J. Telfer Thomas, Camborne.

	Grains per gallon.
Total solid matter dried at 100° C.	12·0
Chlorine as Chlorides	1·8
Saline Ammonia	0·00224
Albuminoid Ammonia	0·00504
Nitrites	Absent.
Nitrogen as Nitrates	Merest trace.
Physical Examination	Bright; some deposit; free from odour.

Remarks.

"'The above figures show this water to be a water of very good quality, and in our opinion well suited for drinking and domestic purposes.

"'(Signed) Southall Bros. & Barclay.

"'Tower Priory, Birmingham,
"'December 18th, 1897.'"

SAMPLE II.

Laboratory Report on a sample of water taken from a tap at Camborne on December 17th.

"The specimen of water marked Boswyn, received here on 18th December, has been duly examined, and I have been instructed to forward the following report thereon :—

"'This sample of water does not contain many organisms—probably less than 100 per cubic centimetre as it is drawn. No bacilli typhi abdominalis could be found, and only a very few bacilli coli communis. There is not one of the latter in 30 cubic centimetres of the water, but there is one in 150 cubic centimetres. We should look upon this water as fit for drinking purposes.

"'(Signed) C. H. WELLS,

"'1, Southwark Street, "'Secretary of the Clinical Research Association.
"'London Bridge, S.E.,
"'31st December, 1897.'"

SAMPLE III.

Laboratory report on a sample of water taken from the Cargenwyn byewash and received by the Clinical Research Association on December 23rd.

"This specimen of water, marked Cargenwyn Reservoir Byewash, received here on December 23rd, has been duly examined, and I have been instructed to forward the following report thereon :—

"'The sample contains a fairly large number of organisms, between 800 and 900 per cubic centimetre. We have not been able to isolate the bacillus typhi abdominalis. The bacillus coli communis was not found in 1 cubic centimetre nor in 2 cubic centimetres, but was found in 5 cubic centimetres of the water. The large number of organisms in conjunction with the presence of the bacillus coli communis in every 5 cubic centimetres of the water is, we think, sufficient to render the water suspicious and undesirable for drinking purposes.

"'(Signed) C. H. WELLS,
"'Secretary of the Clinical Research Association.

"'1, Southwark Street, London Bridge, S.E.,
"'December 30th, 1897.'"

APPENDIX II.

Copy of Recommendations presented by Dr. Reece to the late Local Board of Camborne on November 2nd, 1894 :—

A. GENERAL DUTIES UNDER THE PUBLIC HEALTH ACTS.

1. *Water Supply.*—The Sanitary Authority should themselves provide, or cause to be provided for each household, an adequate supply of wholesome water. They should see that existing supplies are protected from becoming fouled, and that polluted wells are closed under section 70 of the Public Health Act, 1875.

2. *Isolation Provision.*—The Sanitary Authority are urged to provide sufficient and proper hospital accommodation for infectious diseases in place of the inadequate provision they now possess. It is not necessary that the accommodation provided in the first instance should be on a large or costly scale, but it is essential that it should be ready beforehand in order that the first persons attacked may be promptly isolated. Such hospital provision should include—

(*a.*) A properly equipped laundry.
(*b.*) A mortuary.
(*c.*) A proper disinfecting apparatus for the efficient disinfection of infected articles.
(*d.*) A suitable and sufficient ambulance.

3. *Excrement and Refuse Disposal.*—The Authority should take into immediate consideration what method for the disposal of excrement and refuse will be best adapted to the circumstances of their district, and their attention is directed to the uncemented privies, which, as at present constructed, are a source of nuisance of the gravest kind, allowing soakage of their contents and consequent contamination of the surrounding soil, which cannot fail to be injurious to health.

Where fixed receptacles for excrement are retained, they should be reduced to the smallest practicable dimensions, and so constructed as to keep out all unnecessary moisture, and to facilitate the mingling of ashes with excrement. All privies that cause nuisance should be efficiently dealt with as such.

4. *Insanitary Dwellings.*—The Sanitary Authority should cause all dwellings which from dampness, want of ventilation, dilapidation and other structural defects are unfit for habitation, to be placed in proper repair or permanently closed.

The attention of the Authority should specially be given to the danger to health arising from the dampness of foundations and walls, due to the absence of proper spouting for the conveyance of rain water from the roofs of houses.

Yards and open spaces about houses should be properly levelled and paved, or laid with suitable materials, so as to secure efficient drainage and cleanliness.

5. *The Housing of the Working Classes Act, 1890.*—The Sanitary Authority should seriously consider the advisability of dealing in a comprehensive manner with insanitary property in their district under the Housing of the Working Classes Act of 1890.

6. *Overcrowding of Persons in Houses.*—The Sanitary Authority should, without further delay, exercise their Statutory powers to put a stop to the present overcrowding of persons in houses; such overcrowding as exists cannot fail to be prejudicial to Public Health.

7. *Common Lodging Houses, Slaughter-houses, and Dairies, Cowsheds, and Milkshops.*—The attention of the Sanitary Authority is called to the legal obligation imposed upon them of keeping a register of—

(*a.*) Common Lodging Houses.
(*b.*) Slaughter-houses.
(*c.*) Dairies, Cowsheds, and Milkshops.

And the Authority are reminded of the necessity of enforcing in their entirety the bye-laws relating to them.

8. *Houses let in Lodgings.*—With a view to the proper regulation of the tenemented houses within the district, the Sanitary Authority should draw up bye-laws for the registration and regulation of such houses, or parts of houses, not being Common Lodging Houses, which are let in lodgings or are occupied by the members of more than one fami y.

No. 11.

REPORT on the WATER SUPPLY of the BOROUGH of ALDEBURGH-ON-SEA; by Dr. DEANE SWEETING.

App. A. No. On the Water Supply of Aldeburgh-on-Sea; by Dr. Sweeting.

Complaints having been received from residents at Aldeburgh-on-Sea, in Suffolk, as to the quality of the water supplied to them, and extensive correspondence on this subject having ensued between the Town Council of the Borough of Aldeburgh and the Board, it was finally determined by the Board to order an investigation as to the nature and circumstances of the water supply of the borough. I therefore visited Aldeburgh for this purpose in the early part of December, 1898.

Before describing the Aldeburgh Water Company's works and their surroundings, it will be well to refer to the powers under which the Company are constituted and the relation of the Corporation of Aldeburgh to the Water Company.

The Aldeburgh Water Company, formed by Memorandum and Articles of Association in 1870, pursuant to the Companies Acts, 1862–1867, obtained in April, 1871, from the Board of Trade a Provisional Order, under the "Gas and Water-works Facilities Act," 1870, which was confirmed by the "Gas and Water Orders Confirmation Act," 34 and 35 Vic. ch. cxliv. The Order incorporates the "Water-works Clauses Acts, 1847 and 1863," except when they are expressly varied by the Order. It consists of 32 sections and 7 sub-sections. The most important of these are the following, here quoted *in extenso*:—

Section 10 (*a*).—Under the heading of "works authorised" by the Order:—

"An abstraction reservoir or reservoirs, with all necessary embankments, cuttings, filtering beds, sluices, culverts, pipes, roads, and other works connected therewith, situate on a certain piece of pasture land belonging to the Corporation or trustees of the town lands of the town of Aldeburgh, otherwise Aldeborough, and which said piece of pasture land abuts upon the western boundary wall of the Aldeburgh Park Estate."

Section 14.—Under the heading of "quality of water":—

"The water supplied by the undertakers shall be effectually filtered, and shall be as pure as, having regard to the source and nature of supply, circumstances will admit."

Section 25.—Under the heading of "penalties":—

"If on any day the water supplied by the undertakers is of less purity than it ought to be, according to the provisions of this Order, the undertakers shall in every such case be liable to a penalty not exceeding ten pounds. Provided that no penalty shall be incurred in any case in which it is proved that the defect in purity was occasioned by an unavoidable cause or accident."

Now, first with regard to Section 10 (*a*), this refers to the old source of supply, analysed by Sir E. Frankland on October 28th,

. A. No. 11
the Water
ply of
eburgh-on-
: by Dr.
eeting.

1896 (*see* Appendix A). In effect, the "abstraction reservoir" in question was a shallow well, fed by an elongated ditch impounding water on the borders of Aldeburgh Marsh. This marsh has been and is extensively used for grazing purposes, and has been consequently from time to time manured. Water was piped from the "abstraction reservoir" by "Conduit No. 1," specified in Section 10 (*b*) of the Order (not quoted), and which consisted of ordinary field tile drains, to a well in the engine shed at the pumping station, whence the water was pumped to a water-tower for distribution by gravitation. In addition, water from the above-mentioned ditch was, I understand, directly piped to the well at the pumping station, it having been found that two supplemental shallow surface wells, sunk close to the "abstraction reservoir" (*see* map), furnished water that proved brackish. These supplemental wells were in fact never used for purposes of supply.

After the severe and unmitigated condemnation of this marsh-collected water in 1896 by Sir E. Frankland, the Company sought to improve their supply. They therefore shifted their "abstraction reservoir" to the *inner* side of "the western boundary wall of the Aldeburgh Park Estate," whereas the old source of supply abutted upon the *outer* side of that wall.

The Water Company in doing this encountered no objection from the Corporation of Aldeburgh, who were in each case the ground landlords. By a lease, dated March 18th, 1871, the Corporation granted the Company one acre of land on their grazing marshes, for 150 years from March 25th, 1860, for the purpose of a reservoir. This was in addition to two other pieces of land amounting to 1a. 17r. 23p. on the Aldeburgh Park Estate, granted by a similar lease for a pumping station and a water tower. No part of the above estate had then been sub-let by the Corporation. The old "abstraction reservoir" was contained in the one-acre site above-mentioned. As to the site of the new "abstraction reservoir" the Water Company have acquired the interest in a plot of 1a. 1r. 25p. adjacent to their pumping station on the Aldeburgh Park Estate from the Aldeburgh Land Company, who hold it from the Corporation at a nominal (peppercorn) rental. According to the Town Clerk, the Company shifted their "abstraction reservoir" to the new site without consulting or informing the Corporation, and the works were in progress before the Corporation were even aware of the change that was being effected. The pumping-station and the water-tower remain as before.

The present works of the Aldeburgh Water Company consist of :—

(*A.*) Collecting Reservoir.
(*B.*) Filtration Chamber.
(*C.*) Pumping Well.
(*D.*) Engine House.
(*E.*) Water-tower.

These are all close to each other and situated on rising ground west of and behind the main portion of the town of Aldeburgh, about 30 chains distant from the sea and 60 from the tidal river Alde. From the water-works the ground slopes somewhat abruptly westward towards the marsh which intervenes between them and the river. Northward the ground gradually rises. On the east the ground rises rather precipitately and forms a plateau overlooking the main part of the town.

(*A.*) *Collecting reservoir.*—This is a receptacle loosely boarded over with planks, the sides being formed of concrete and the base of the white or Coralline Crag. The base of the reservoir, which is 0·15 feet below Ordnance Datum, is 13 feet from the surface of the ground at its easterly end and 8 feet from the surface of the ground at its westerly end; the ground, as I have said, here sloping from east to west. Its dimensions are: 62 feet long, 33 feet wide, and the walls are of the same height all round, viz., 7 feet. It is capable of containing some 88,795 gallons; but this maximum water capacity is rarely if ever realised. At my visit the depth of water was 2 feet 8 inches, and I am informed that the normal depth of water is 2 feet.

(*B.*) *Filtration chamber.*—This is a small concrete receptacle, covered by a stone slab, intervening between the collecting reservoir and the pumping well. It forms duplicate channels, each of which, measuring 11 feet × 4 feet, is furnished with a filtering screen, through which a portion of the water passes horizontally. Each screen is 2 feet wide × 2 feet 6 inches long × 2 feet deep, perforated by small openings, about ¼ inch in diameter, and containing three layers of gravelly material, the first (nearest to the collecting reservoir) coarse, the second medium, and the third fine. The amount of material in each section of filter is nearly equal, that in the coarse section somewhat predominating.

The partial filtration thus described, and which cannot be regarded as in any sense adequate, was an after-thought, not having been established until the spring of 1898, some months after the rest of the works had been inaugurated.

(*C.*) *Pumping well.*—This, which is circular in shape and nine feet in diameter, has its bottom composed of cement concrete, and its sides of loosely placed bricks, which are neither cemented internally or puddled externally.

(*D.*) *Engine house.*—This contains a modern and apparently efficient six-horse-power steam engine.

(*E.*) *Water-tower.*—The water is pumped from the circular well to a brick water-tower on the summit of adjacent high ground. This tower contains a cast-iron tank, 21 feet × 21 feet × 4 feet, and holding therefore some 10,000 to 11,000 gallons. From this tank the water descends by gravitation to the town.

It is computed that three-fourths of the inhabitants of Aldeburgh, which are estimated at about 2,300 in the non-visitors' season, consume the water supplied by the Company. I am informed that pumping does not take place regularly, but only as required. When the level of the water in the service tank falls to within 18 inches of the bottom, pumping is practised until the tank is again full. The collecting reservoir is said to be never emptied by the pumping that is practised. But I learned that, during the construction of the pumping well, this reservoir was on one occasion entirely emptied by twenty minutes' continuous

A. No. 11.

The Water Supply of Aldeburgh-on-Sea, by Dr. Darling.

working of a 6-inch portable centrifugal pump. Since, therefore, this reservoir holds, at its maximum, some 88,000 gallons, and since some 1,700 people habitually consume the water, its maximum contents are equal to less than two days' consumption (at 30 gallons per head per day). When, as during the summer season, visitors to Aldeburgh double the population, pumping must—seeing that the service tank holds only some 10,000 gallons—be frequently resorted to. At such times it is highly probable that the collecting reservoir must be pumped to exhaustion, unless, of course, the pumping well receives much soil water directly through its pervious walls.

Coming now to the question of the actual source of this public water supply, it may first be mentioned that locally the water in the collecting reservoir is spoken of as impounded "spring water," and that the source is vaguely indicated as "the high land" north of the water-works. There seems to be no particular reason for the view that the water is "spring water" beyond the fact that it has been noticed to "bubble up" in the collecting reservoir. And the view as to its supposed source from "high land" was not supported by any observed facts. There seems no reason to regard this water as other than "Crag water," *i.e.*, the ordinary subsoil or ground water of the Coralline Crag formation. This view is supported by observations of the levels of water in certain wells in the Coralline Crag in the neighbourhood of the waterworks. Passing from north to south, for instance, the level above Ordnance Datum of the water in wells sunk into the Crag tended to diminish as the site of the water-works was approached. Thus, at Brick-kiln Farm (*see* map), where the ground is 26·77 feet above Ordnance Datum, the well-water level is 3 feet above Ordnance Datum; whilst at Station Row, 20·50 feet above Ordnance Datum, the well-water level was 1 foot above Ordnance Datum. At the site of the waterworks the surface is some 10 feet above, and the bottom of the collecting reservoir 0·15 feet below, Ordnance Datum. Dr. Thresh, too, in his report of the analysis of the water on June 27th, 1898 (Appendix D), speaks of the water as "derived from the Coralline Crag," and compares it with other "subsoil waters."

The Coralline Crag is found at or near the surface in several patches in Suffolk, but the "main mass" of it is in the vicinity of Aldeburgh. This formation rests directly upon the London Clay, whether horizontally or with a progressive dip N.N.E. is disputed by geologists. As to its composition, the older view of Prestwich regarded it as made up of some eight zones, but later field observation has restricted these to two, viz., an upper zone of "ferruginous rock" and a lower of "shelly sand" (Harmer).* Both the depth at which it is reached and the thickness of the Coralline Crag are variable. It is met with at depths below the superficial soil varying from 5 feet to 20 feet, though the average depth appears to be about 8 feet, as at Ipswich (Prestwich),† and near Aldeburgh (Reid).‡ As to its thickness, the accounts given by authorities that I consulted vary considerably. The range appears to be between 30 feet and 76 feet, though Mr. Whitaker

* "Proc. Geolog. Assoc.," November, 1898. † Quart. Journ. Geol. Soc., 1871.
‡ "Pliocene deposits of Britain."

places it as from 40 feet to 60 feet. But the most recent observation is that of Mr. Harmer, who, as a result of an excursion organised by Mr. Whitaker and himself in the summer of 1898, attributes to the section of the Coralline Crag at Aldeburgh a thickness of some 75 feet. There seems little doubt, therefore, that the Coralline Crag is of considerable thickness at the site of the collecting reservoir, and that the bottom of this, being only some 10 feet from the surface of the ground, is at a great distance above the London Clay. In the circumstances there can hardly be here any question of springs in the Coralline Crag, thrown up by the underlying London Clay. The water, as said before, must be regarded as subsoil water, and nothing else. The "bubbling up," upon which so much local stress was laid, in favour of the "spring water" view, is to be referred, not to the uprising of spring water, but to the inflow of the subsoil water to the collecting reservoir, more or less accelerated during depletion of this reservoir by the operations of pumping.

It will be noticed that the bulk of the Coralline Crag at Aldeburgh is below sea level. It extends underneath the whole of the shingle, and forms the bed of the foreshore; an important fact, as will be seen later. The structure of the Coralline Crag may be briefly defined as a "shelly sand, with thin bands of crystalline limestone, overlaid by soft yellow rock containing bryozoa or corallines" (*see* Memoirs of Geological Survey).

As far as I could gather from local evidence as to borings, the Coralline Crag at the site of the waterworks is found at a depth of about 6 feet to 7 feet below the surface, a red loamy sand intervening. There can be no doubt, therefore, that the superficial soil is, like this formation, distinctly porous, and that both would allow the passage of liquid matter downward through them with great facility.

As to the surroundings of the Aldeburgh waterworks, and to the circumstances of the gathering ground from which the water is derived. This gathering ground may be regarded, since the water is derived from the Coralline Crag, as to all intents and purposes co-extensive with the portion of the "main mass" of this formation situated at Aldeburgh. The Coralline Crag here is of an irregular, straggling shape, its longest axis, from north-east to south-west, being about one mile and a quarter in length. The new collecting reservoir is placed almost at its southern extremity. Most of the Coralline Crag here is devoid of superincumbent strata, but on the west, north, and east it is covered by patches of Red Crag, which in turn are partly overlaid in these directions by strips of Chillesford Clay (*see* map annexed, where the formations are coloured). Its eastern portion is partly overlaid by glacial drift.

The inhabited portions of Aldeburgh are situated geologically upon strata as follows:—

(1.) Coralline Crag uncovered by other strata (coloured brown in map).

(2.) Coralline Crag covered by Red Crag (coloured red in map).

(3.) Coralline Crag covered by Red Crag and Chillesford Clay (coloured grey in map).

(4.) Coralline Crag covered by Glacial Drift (coloured pink in map).

(a.) At the extreme no th and west.

(b.) At the east, comprising the main part of the town facing the sea.

It will now be convenient to refer to the circumstances of excrement disposal and drainage obtaining at inhabited dwellings situated upon each of the above formations.

First, then, as to (1) above, Coralline Crag coloured brown in map. Within a few yards of the pumping well and the collecting reservoir is the *Waterworks Cottage.* Here is a garden manured by the contents of a pail-closet. Until recently, household slops were thrown on the garden, too; but now these are piped from a slopstone gulley to an adjoining ditch. A little north is the *Stableman's Cottage.* Here there is a manure heap, and the garden is manured by the refuse of pig-styes and the solid matter from a large brick cesspool. This is not emptied more frequently than twice a year, and the liquid contents meanwhile soak away into the ground. North-west of this are *Workhouse Cottages*, a group of several old and dilapidated cottages, the property of Aldeburgh Corporation. I found here a huge ash heap containing decomposing animal and vegetable matter. Domestic slops are thrown by hand on the ground around the cottages. Excrement is disposed of in rude privy pits, the liquid contents of which find their way, after traversing a small portion of garden ground, to a ditch which feeds the dykes in the neighbouring Aldeburgh Marsh. The above cottages and the ditch which receives liquid matter from the privies are a few yards from the elongated ditch spoken of earlier as feeding the old "abstraction reservoir" (*see* p.). North-east of these cottages is *Forthampton Place*, a row of nine houses, from which the contents of hand-flushed water-closets and domestic sewage from slopstone gulleys are delivered into large deep brick cesspools, most of them uncemented and unventilated. The bricks composing these cesspools are laid dry, and not rendered in mortar. There are in all five such cesspools serving the nine houses, situated at various distances behind them. Their construction allows soakage of liquid filth into the surrounding soil; as matter of fact, they appear to be emptied at irregular and wide intervals. *Station Row*, a little north of Forthampton Place, comprises in all 23 houses. There is no drainage here at all. Slop water is thrown upon garden allotments on the opposite side of the road. Ordinary cesspit privies are provided, one for each house at the end of each back garden. These privy pits are rude uncemented receptacles, constructed usually of brick. They are emptied at irregular intervals, and their contents are allowed to soak away into the soil. West of Station Row is *Aldeburgh Hall Farm*, where there are three large fold-yards containing much liquid filth and manure. A stagnant duck-pond at this farm receives the drainage from pig-styes, as well as much surface washing and liquid percolation from adjacent manure heaps. House refuse is stored here in an uncemented large open brick ash-pit, the bottom of which is below the level of the ground. Excrement is stored in two large, deep, uncemented and unventilated cesspools within 30 feet of the farmhouse; these receive also much slop *water directly* piped to them. Removal of their contents is infrequent, and soakage of liquid from them into the soil is very probable.

Other habitations on this geological area are *New Cottage* at the northern part, and *Convalescent Cottage* and *Kersey's folly* near the centre. At the first two of these there are cesspools of the kind mentioned above, whilst at the latter the old rude cesspit obtains. These receptacles are all unsatisfactory in construction and are imperfectly scavenged.

Next, as to (2) or the Red Crag overlying the Coralline Crag (coloured red on map). It will be convenient to consider here the houses on the *Aldeburgh Park Estate*, some of which are placed on this formation, though others are on the remaining formations (3) and (4), viz., the Chillesford Clay and the Drift. These houses consist of nine superior class private residences, two of them semi-detached, the others detached, placed north, east, and south of the Aldeburgh Waterworks, and within some 300 yards of those works. They are provided with cesspools varying from one to three in number, and altogether there are some thirteen of these on the estate. The general character of these cesspools is that of large and deep pervious receptacles. They are constructed of ordinary bricks laid dry, not rendered in cement or mortar. Most of them are unventilated. Some are provided, on the drains delivering to them, with cemented catch-pits, which tend to retain the more solid matters. I had one cesspool opened, viz., that serving Nos. 3 and 4 Westfields ; it was 26 feet 9 inches deep and 4 feet in diameter. The "catch-pit" belonging to this cesspool measured 6 feet in all directions and was full of solid matter, whilst the cesspool which received its overflow contained 13 feet of reeking and offensive liquid filth. I had another cesspool at No. 6 Westfields opened which was 18 feet deep. This also was provided with a catch-pit, the overflow from which passed to the cesspool. At my visit this cesspool contained very foul-smelling liquid matter. The depth of another of the cesspools on the Aldeburgh Park Estate is said to be 16 feet. As regards levels, the bottom of the cesspool at Nos. 3 and 4 Westfields is about one foot above Ordnance Datum, and the bottom of that at No. 6 Westfields is some 3 feet above Ordnance Datum. As stated before, the base of the collecting reservoir is 0·15 feet below Ordnance Datum.

It is thus seen that the bottoms of these cesspools are generally at an appreciably higher level than the bottom of the collecting reservoir. Consequently, liquid filth from these and similar cesspools on the Aldeburgh Park Estate must pass into the water of the Coralline Crag, and may subsequently be impounded in the Water Company's collecting reservoir. Though details are given above of only two cesspools on the Aldeburgh Park Estate, the conditions surrounding the others are not materially different. Only one cesspool on the estate, viz., that at a new house south of the waterworks, can be said to be constructed in accordance with modern requirements as to imperviousness.

The only other habitations on formation (2), excluding present consideration of the southernmost part of the town proper and a few outlying houses, are *Brick-kiln Farm* in the north and *Red House* in the north-west. Much the same general conditions obtain at these farms as at Aldeburgh Hall Farm, previously

APP. A. No. 11.
On the Water Supply of Aldeburgh-on-Sea; by Dr. Sweeting.

described, with respect to the state of the fold-yards and as to sewage disposal generally. At *Red House* there are two cesspools, which communicate with each other, one of which retains the solid matters, whilst the other receives liquid overflow from the first, which is now and again pumped directly on to the land.

Coming now to (3), or the Chillesford Clay overlying the Red Crag (coloured grey in map), and omitting for the present a few houses at the southern portion of the town proper, the only habitation directly situated on the Chillesford Clay, in addition to some of the Aldeburgh Park Estate houses already mentioned, is the *Brickworks Cottage*, adjacent to the Hall Farm Brickworks. Here the old privy pit obtains, which consists of a simple hole in the ground covered over by wooden planks. Its contents are applied from time to time to the garden, which also receives domestic slops which are thrown on to it.

Finally, (4), as to the houses on the Glacial Drift overlying the Coralline Crag (coloured pink in map):—Those comprised in 4 (*a*) are, on the north, *Warren House* and *Telegraph Cottage*, and on the west, *Watering House*. Excrement disposal is effected here by the unsatisfactory cesspits of the sort previously described in connection with the cottages placed on the other formations mentioned.

As for 4 (*b*), viz., the Glacial Drift overlying the Coralline Crag east of the waterworks, among possible sources of pollution of the water of the subsoil are the following:—*Churchyard's Farm*, where there are fold-yards containing a large amount of liquid filth, where the slop water is piped to the vault of a badly-constructed cesspit, and where a duck-pond receives abundance of surface washings; the *Brewery*, close to this, where there are three brick cesspools, all of them uncemented and unventilated, and only emptied at irregular intervals, and where two open "soak-pits" receive slop water and urine respectively; *Lee Road* and *The Terrace*, where of a total of some fifteen cesspools only three, according to the Borough Surveyor, have any pretensions whatever to proper construction, the others being so constructed and arranged and emptied so irregularly as to facilitate soakage of their contents into the soil.

The above and other places similar to them may be looked upon as sensibly contributing to pollution of the Coralline Crag subsoil water. But there is yet another and perhaps more important source of pollution of the water, which is impounded in the new collecting reservoir of the Aldeburgh Water Company.

It has been before stated that the Coralline Crag extends underneath the whole of the shingle at Aldeburgh, and forms the bed of the foreshore. *Reid* (op. cit.) says that this is the only place on the coast where the Coralline Crag has been found exposed. Now, although a scheme of main drainage has just lately been sanctioned by the Board, after local inquiry, and a loan granted, Aldeburgh is at present altogether a cesspool town. The Surveyor informs me that there are in all 375 cesspools there, and that only 14 are properly constructed and ventilated. Nearly the whole of Aldeburgh, comprising the main part fronting the sea, is built upon the shingle, and the majority of the cesspools of the place, which are almost entirely of brick construction at the sides and without any bottom, are sunk directly into this shingle.

These Aldeburgh cesspools are from 14 feet to 23 feet deep, and the bottom of most of them is only from 4 feet to 5 feet above Ordnance Datum. They are in a manner flushed out by the tide, which first rises in them and then falls, taking the cesspool contents with it. This method of sewage disposal has hitherto received considerable local approbation, especially amongst those ratepayers who object to the cost of main drainage. However apparently convenient such a system may have been, there seems little doubt that complete transport of cesspool contents into the sea is not readily effected. High-water mark being only 8 feet above Ordnance Datum, and the cesspools being bottomless, their contents must constantly be mingling with the shingle. Since, therefore, the Coralline Crag is immediately below the shingle and is porous, that formation must in turn become contaminated with cesspool filth. The natural trend of the subsoil water in the Coralline Crag hereabouts is, it is true, generally towards the sea. But it is very probable that this subsoil water becomes tide-locked and dammed back at high tides, and its exit to the sea prevented. In this case, the bottom of the collecting reservoir at the waterworks being below Ordnance Datum, the frequently repeated pumping operations, such as are apt to be carried on at the works, must tend to produce a cone of exhaustion, and so encourage passage of Crag water, polluted by soakage of filth from the shingle, to the Water Company's pumping well.

In a word, in view of the facts tending to indicate that the contents of the new collecting reservoir must be often well nigh exhausted, especially during the visitors' season at Aldeburgh, it is not improbable that this reservoir and the pumping well, at such times of exhaustion, receive polluted subsoil water of the Coralline Crag, passing towards them from the region between the reservoir and the sea.

The above argument will equally apply to the cesspools at the south of the main part of the town, belonging to houses placed immediately on the Red Crag (2), and the Chillesford Clay (3). (*See* map for colourings.) For, as before stated, these are superimposed directly upon the Coralline Crag, the subsoil water of which is liable to pollution from the town cesspools generally.

[Similarly, the surface water in Aldeburgh Marsh, south and west of the waterworks, is dammed back and held up by the high tides of the tidal river Alde. This is shewn by the increased height of the water in the Marsh dykes during the high tides, which occur as a rule two hours later than the high tides on the coast, the Alde having a circuitous course of some nine miles from the back of Aldeburgh to its entrance into the sea at Shingle Street. Inasmuch as the Aldeburgh Marsh is composed of alluvium covering London clay, no question arises of pollution of the new reservoir of the Water Company directly from the river Alde. But there is no doubt that the old "abstraction reservoir" of the Company, which, as before related, was a shallow well fed from an elongated ditch in the Marsh, received during high tides a perhaps not insignificant amount of surface impurities and filth from the manured grazing grounds of Aldeburgh Marsh. Indeed, during floods at the latter end of 1897, the whole marsh was covered by 4 feet of water.]

App. A. No. 11.
On the Water Supply of Aldeburgh-on-Sea; by Dr. Sweeting.

The foregoing view as to risk of pollution of the Company's collecting reservoir and pumping well by cesspool filth, whether from the receptacles of the class found in the main part of the town—namely, those cesspools sunk in the shingle of the foreshore—or from cesspools of outlying houses or groups of houses closer to the waterworks, as the Aldeburgh Park Estate, was locally challenged on the ground that the water in the Crag is effectually cut off from these and other sources of pollution by an extensive and horizontal "band of clay" underlying all these cesspools. On examining into this, and consulting authorities at the Geological Survey Office, I found that the clay referred to was the "Chillesford Clay." At best this is here a thin and non-persistent layer; the Chillesford Clay around Aldeburgh having been eroded and superseded by later gravels. It is altogether a localised and superficial patch, or series of patches, resting directly here on the Red Crag. It can therefore in no sense be considered as affording any protection to the Crag water, which is gathered from an area of considerably larger extent than that locally covered by the clay in question. Moreover, where it exists in the inhabited area, as at a small portion at the south of the main part of the town, it is doubtless pierced by some of the above described cesspools.

It should be added that many of the fields to the north of and at higher level than the waterworks are heavily manured, especially where root-crops are grown, such as turnips and mangel-wurzel. And a field, north-west of and within a mile of the waterworks, used as the town "tipping place," receives the contents of the town cesspools, which are deposited there at irregular intervals. At my visit there was a considerable accumulation of semi-liquid filth at this place. Up to a recent date, the contents of the town ash-pits were also tipped there. Now, however, these are used by the Surveyor for road-making purposes.

To briefly sum up the facts elicited during the inquiry upon which I have reported above, it may be said that the old source of the water supply, which was condemned in 1896 by Sir E. Frankland, was subject to manifold pollutions, and was, moreover, unguarded by any filtration whatever. And that the new supply, whilst professedly "spring water," is really nothing more than water soaking through the superficial soil. Its gathering ground, coterminous with a particular geological formation of high porosity, is a much polluted subsoil, contaminated by filth products from cesspools, cesspit-privies, fold-yards, manured fields, manure heaps, pig-styes, and the like. It is, moreover, not improbable that cesspool impurities of considerable aggregate quantity find entrance to the waterworks from the polluted shingle of the foreshore, owing to holding back of the subsoil water by high tides, aided by extra pumping operations, especially during the visitors' season. It may indeed be suspected that the Aldeburgh inhabitants are, in a manner and on occasion, forced into drinking their own diluted sewage. Undoubtedly, when main drainage is established throughout Aldeburgh and the cesspools are done away with, there will be less risk of dangerous pollution of the water-supply. But there will then be even greater need than now for an adequate supply, in view of the demand there will be for proper flushing of closets and of sewers. That the present supply is inadequate has been shewn, and the evidence bearing on this may be supplemented by facts brought to my notice as to frequent complaints by the Corporation officials of the lack of sufficient water for watering the streets during the summer of 1898.

The inefficient filtration of the water has been already referred to, with note that the method adopted was but an after-thought of the Water Company. Confirmatory evidence of inefficient filtration was afforded by the relation to me of instances where "centipedes" and other low forms of animal life had appeared in water drawn from house taps.

Other evidence, pointing to serious animal organic pollution of the water, was afforded by numerous complaints made to me by inhabitants that the water, especially when heated, gave forth a distinct odour of sewage. This odour had at times proved so nauseating that people stated to me that they had abandoned the practice of washing in warm water or taking hot baths.

In addition to the analysis of the old water supply by Sir E. Frankland in 1896, already referred to and reproduced in Addendum A, there is given in Addendum B a copy of an analysis by the same chemist of the new supply in January, 1898; and in Addendum C a copy of an analysis by Mr. Cassal of the new supply in July, 1898. A copy of Dr. Thresh's analysis of the new supply in June, 1898, is reproduced in Addendum D.

In comparing the conclusions arrived at by these three chemists after analysis of the new supply (*see* Addenda B., C., D.), it will be noticed that Dr. Thresh differs entirely from Sir E. Frankland and Mr. Cassal in the interpretation he places upon the results of his analysis. He ascribes the high proportion of chlorides that he found to "free diffussion of certain of the saline constituents of sea-water, especially of the chloride of sodium (common salt)"; though, from the absence of chloride of magnesium from the saline constituents of the water, he denies actual admixture with sea-water. However this may be, it is to be noted that in Dr. Thresh's analysis not only the chlorides but also the total solids and the nitrates, when reduced to the same standard, are very high in amount, and practically the same as in the other two chemists' analyses; that the large quantity of organic ammonia in his analysis is identical with the average of Mr. Cassal's three samples; and that the high proportion of oxygen absorbed from permanganate in his analysis is almost identical with Mr. Cassal's average (*see* note to Addendum D.). To my mind, these positive facts outweigh the mere negative value of the absence of the bacillus coli, referred to by Dr. Thresh, and point rather to an uniformly suspicious, if not dangerous, water.

Dr. Thresh gives no details of the "examination" of the "source" of the water which lead him to the not unfavourable opinion he gave of it; whether or not, therefore, he submitted the gathering ground to the "adequate inspection" which Mr. Cassal rightly regarded as necessary to determine the question of the safety of the Aldeburgh water, is not clear.

The facts that I have set out above embody the result of such actual physical inspection. It is in reference to these facts that the chemical data of the several analysts must be interpreted.

Addendum E *comprises a brief* note as to rainfall.

App. A. No. 11.

On the Water Supply of Aldeburgh-on-Sea; by Dr. Sweeting.

ADDENDUM A.

Water Analysis Laboratory,
"The Yews," Reigate, Surrey.
28th October, 1896.

DEAR SIR,

HEREWITH I enclose results of analysis, chemical and bacterioscopic, of the sample of water collected by my assistant, Mr. Burgess, at Aldeburgh, on the 16th instant. This water, besides being highly saline, contains a large quantity of organic matter, and exhibits strong evidence of previous pollution by sewage or other animal matter. It is quite unfit for dietetic purposes, and, being very hard, is useless for washing. The bacterioscopic examination is confirmatory. The sample specially collected for this purpose contained 450 microbes per c.c., whilst 100 is the maximum number allowable.

I am,
Yours very truly,
E. FRANKLAND.

HENRY C. CASLEY, Esq.

Results of analysis expressed in parts per 100,000.

No. of sample 9,988, Aldeburgh, October 16th, 1896.

Total solid matters	... ...	95·00
Organic carbon	... ...	·210
Organic nitrogen	... ...	·034
Ammonia	... ...	·002
Nitrogen as nitrates and nitrites	... ...	1·456
Total combined nitrogen	... ...	1·492
Previous sewage or animal contamination	... ...	14·210
Chlorine	... ...	26·5
Hardness { Temporary	... ...	20·3
Hardness { Permanent	... ...	0
Total	... ...	18·3 38

Remarks :—Highly turbid.

ADDENDUM B.

BOROUGH OF ALDEBURGH.

Water Analysis Laboratory,
"The Yews," Reigate,
January 22nd, 1898.

DEAR SIR,

HEREWITH I enclose results of complete analysis of a sample of water collected on the 12th instant, by my chief assistant (Mr. W. T. Burgess), from a tap supplying Ald cottages. Although this water does not contain an excess of organic matter it is derived from very impure sources, and a small proportion of the organic matter is of animal origin. Indeed about one-eighth of the water has been in the condition of average London sewage. It is hard but not excessively so. Although the water is clear the filtration had not been bacterially efficient, for the sample contained the enormous number of 20,660 microbes per c.c. (about 20 drops), 100 per c.c. being the maximum number usually found in efficiently filtered water.

I am,
Yours very truly,
E. FRANKLAND.

HENRY C. CASLEY, Esq.,
Town Clerk of Aldeburgh.

No. of Sample.	Description.	Total solid matters.	Organic carbon.	Organic nitrogen.	Ammonia.	Nitrogen as nitrates and nitrites.	Total combined nitrogen.	Previous sewage contamination.	Chlorine.	Temporary hardness.	Permanent hardness.	Total hardness.	Remarks.
10,640	12th January, 1898. Tap supplying Ald Cottages.	94·00	·117	·023	0	1·312	1·335	12,800	28·9	22·9	6·8	29·7	Clear.

ADDENDUM C.

PUBLIC ANALYST'S DEPARTMENT.

CHARLES E. CASSAL, F.I.C., Public Analyst.
Town Hall, Mount Street, W., and
101 Leadenhall Street, E.C.

Town Hall, Kensington, W.,
London, 15th July, 1898.

Report on the analysis of three samples of water received on the 25th June, 1898, from C. T. Walrond, Esq., C.E., and respectively labelled :—

I. Aldeburgh Water Supply. Sample from reservoir, 24th June, 1898.

II. Aldeburgh Water Supply. Sample from garden tap at Mr. Padd's house, 24th June, 1898.

III. Aldeburgh Water Supply. Sample from scullery tap, Sunnymeade, 24th June, 1898.

I certify that I have analysed the three samples of water above referred to, and that the results of my analysis were as stated in the appended analytical reports, which are intended for the information of experts should necessity arise. From a consideration of the analytical results and of the information with which I have been supplied, I have arrived at the following conclusions :—

The three samples are closely similar in certain respects, but in others they exhibit differences which are to some extent accounted for by the facts as to the sampling. The amounts of total solid matter in Nos. I. and II. are practically the same, being slightly lower in No. II. than in No. I.; while in No. III. the amount is a little higher than in Nos. I. and II. The degrees of hardness are the same in all three samples, and the amounts of oxidised nitrogen (nitrates) are also the same. The amounts of organic matter contained in Nos. I. and III. are very similar, and are higher than the amount contained in No. II. The presence of the microscopic eels, the living infusoria, and the finely divided mineral suspended matter in No. III. appears to be accounted for by the fact that this sample was taken from a point where the water is likely to have remained stagnant. No. II. would appear to represent the water which was being delivered at the time of sampling to the greater part of the town, and the fact that it is of rather better quality than No. I. is no doubt due to the effects produced by the settlement and slight filtration to which I understand that the water is subjected.

No. I., the sample from the reservoir, which, I understand, represents the untreated supply at the time of sampling, is a water containing a high amount of total solid matter, an amount decidedly higher than is desirable in a water supplied for drinking and domestic purposes.

The organic matter present is of vegetable origin, but it is higher in amount than is desirable. The water, however, cannot be condemned as unfit for use upon these grounds.

The amount of "nitrogen as nitrates" is very high : but while this fact, with respect to many waters, would justify grave suspicion, I am not prepared to assert, in view of the evidence before me, that in this case the high nitrates afford serious ground for alarm. I consider, however, that the point is one which calls for searching investigation by adequate inspection and analytical examination. No evidence is afforded of the existence of pollution or contamination with matters partaking of the nature of sewage at present. While the results in the case of No. II. appear to show that an improvement is brought about by the filtration arrangement, it is probable that more satisfactory results could be attained by the introduction of a more effective system of purification. In any case the supply is one which should be subjected to periodical analytical examination.

(Signed) CHARLES E. CASSAL, F.I.C.,
Public Analyst for Kensington, St. George's Hanover Square, Battersea, High Wycombe, and the County of Lincoln (Administrative Counties of Kesteven and Holland, Lincolnshire).

APP. A. No. 11.

On the Water Supply of Aldeburgh-on-Sea; by Dr. Sweeting.

Analytical Report.

ALDEBURGH WATER SUPPLY.

Analytical Report upon three samples of water received on the 25th Ju
1898, from C. T. Walrond, Esq., C.E., and labelled as stated below :—

Number and Description of Samples.	(I.) "Sample from Reservoir."	(II.) "Garden tap, Mr. Padd's house."	(III.) "Scullery tap, Sunnymeade."
Appearance in a 2-ft. tube—No. I.	Fairly clean.	Slight greenish tint.	
Appearance in a 2-ft. tube—No. II.	"	"	
Appearance in a 2-ft. tube—No. III.	Cloudy.	"	
Odour	—	Not abnormal (in each case).	
Taste	—	"	
Reaction	Neutral.	Neutral.	Neutral.
Total solid matters	106·4	106·4	108·0 pts. per 100,00
" hardness..	26·0	26·0	26·0 " "
(a) Permanent hardness	8·6	8·6	8·6 " "
(b) Temporary "	17·4	17·4	17·4 " "
Chlorine as chlorides	25·5	27·4	27·0 " "
= Chloride of sodium ..	42·02	45·15	44·49 " "
Nitrogen as nitrates	1·4	1·4	1·4 " "
Oxygen absorbed from Permanganate (at 30° C. in 4 hours).	·714	·571	·743 " per millio
Saline ammonia	·05	·022	·046 " "
Organic	·096	058	·072 " "
Lead, copper, iron	Absent.	Absent.	Absent.
Appearance of solids on ignition.	Slight browning in each.		

Microscopic Examination of the Suspended Matters.

No. I. A little flocculent vegetable matter. A few diatoms. Fragments o
Algæ (water weed). Mineral matter.

No. II. Same as No. I.

No III. Flocculent vegetable matter. Diatoms. Much finely divided miner
matter. Living infusoria. Living anguillulæ.

Remarks.

Nitrites. Absent in each case.

Phosphates. Marked traces in each case.

(Signed) CHARLES E. CASSAL, F.I.C.

Town Hall, Kensington, W.; Town Hall,
Mount Street, W.; 101, Leadenhall Street, E.C.
London, 15th July, 1898.

ADDENDUM D.

EXTRACTS from REPORT and ANALYSIS on the ALDEBURGH PUBLIC WATER SUPPLY by JOHN C. THRESH, Esq., M.D., D.Sc., D.P.H. Cambridge M.O.H. Essex County Council, Fellow of the Institute of Chemistry, Membe of the Society of Public Analysts, Associate Member of the Society o Waterworks Engineers, Editor of "Journal of State Medicine," Author o "Water and Water Supplies," "Protection of Underground Waters," &c., &c.

County Public Health Laboratory, Chelmsford,
June 27, 1898.

On the 22nd June I visited Aldeburgh, and examined the works belonging t the Aldeburgh Water Company. The water supplying Aldeburgh is derive from the Coralline Crag. The exposed Crag, not having a very large area, th flow of water through it cannot be large. This permits of the free diffusion o certain of the saline constituents of sea water, especially of the chloride o sodium (common salt). Hence, the chlorides in the Aldeburgh water are high

The water rises through the bottom of a shallow tank or reservoir in the Crag. From this tank it passes through a second or smaller tank, in which are vertical sand screens or filters, into the pumping well, from which it is forced through a charcoal filter* into a tank situated on a tower near the top of the hill. From this tank the water flows by gravitation to supply the town. A sample of water was taken directly from the collecting reservoirs for analysis, and at the same time gelatine plates were prepared for bacteriological examination after due incubation. The result of the chemical analysis is appended. A special examination was made for the bacillus coli in both the filtered and unfiltered waters. There was no growth at the expiration of 48 hours. As no bacillus coli could be found, there was no indication of sewage contamination. The nitrates present were not excessive compared with many other undoubtedly good gravel waters. As the result of my examination of the source of the water itself, I am of opinion that it compares very favourably with the best subsoil waters of the Eastern Counties in organic purity. The composition of the saline constituents of the water, especially the absence of chloride of magnesium, indicates that there is no actual admixture with sea water at the site of the waterworks.

(Signed) JOHN C. THRESH.

Analysis of sample of water taken from the reservoir over the springs at Aldeburgh Water Company's works by myself on June 22nd. 1898:—

	Physical character	Clear and colourless.	
(a)	Total solids	62·8	Results expressed in grains per gallon.
(b)	Nitric nitrogen...	·78	
(c)	Chlorine	17·9	
	Nitrites	Nil.	
	Hardness—temporary...	9°	
	Hardness—total	18°	
	Sodium carbonate	Nil.	
	Lead and iron	Nil.	
	Free ammonia	·00	In parts per million.
	Organic ammonia	·07	
	Oxygen absorbed in 4 hours. ...	·56	

Result of bacteriological examination—Bacillus coli sought for but not found.

(Signed) JOHN C. THRESH.

NOTE by the Reporter (R. D. Sweeting).—The above figures of (a), (b), and (c) reduced to "parts per 100,000," and compared with Cassal's and Frankland's are as follows:—

——	Thresh.	Cassal.	Frankland.
(a) Solids	89·7	107·2 (average)	94·0
(b) Nitrates	1·1	1·4 "	1·3
(c) Chlorine	25·5	26·6 "	28·9

The organic ammonia is identical with the average of Cassal's three samples.

The oxygen absorbed from permanganate in four hours is ·56 parts per million compared with ·67 (average of Cassal's three samples).

R. D. S.

ADDENDUM E.

NOTE AS TO RAINFALL.

No data as to rainfall at Aldeburgh were available. But my late colleague, Dr. Airy, of Woodbridge, fixes the average annual rainfall at this place at 23·4 inches, according to his own records. He regards "Rendlesham Hall" as the Suffolk station most nearly comparable with Aldeburgh. The average annual rainfall here is 24·83, compared with 24·75 at Greenwich (Symons and Biddell).

R. D. S.

---

* This has since been abandoned.—R.D.S.

On the circumstance of the Sources of Water Supply for the Borough of Carnarvon; by Dr. Wheaton.

No. 12.

REPORT on the CIRCUMSTANCE of the SOURCES of WATER SUPPLY for the BOROUGH of CARNARVON; by Dr. S. W. WHEATON.

For some years past the unsatisfactory nature of the water supplied to the Borough of Carnarvon, more particularly its liability to pollution by liquid refuse from the village of Rhyd-ddu, has been before the Board. Attention was called to this liability to pollution of the Carnarvon water supply by the Medical Officer of Health for the Carnarvonshire combined districts in his report on the health of the Borough of Carnarvon in the year 1893. In 1895, Dr. Bruce Low, one of the Board's Medical Inspectors, visited the Carnarvonshire combined districts and reported upon sanitary progress and administration in the districts. In his report, dated August 19th, 1895, a short account was given of the Carnarvon water supply, and it was shown that the water was liable to pollution from several sources, especially by liquid refuse from the village of Rhyd-ddu, by fæcal matter from privies, and by liquid matter from cesspools pertaining to various dwellings situated in the upper part of the valley of the Gwyrfai, from which river the water supply is derived. Since the issue of Dr. Low's report the Board have received complaints as to the character of the Carnarvon water supply from private individuals, and the matter has also from time to time been referred to in the reports of Dr. Fraser, the present Medical Officer of Health for the Carnarvonshire combined districts. The Board have been in correspondence with the Gwyrfai and Glaslyn Rural District Councils, in whose districts the village of Rhyd-ddu is situated, and have urged them to take measures for dealing with the sewerage of this village so as to prevent the pollution of the river Gwyrfai, but hitherto without result. On December 28th, 1897, the Carnarvon Town Council wrote to the Board stating that nothing had been done by the Gwyrfai or Glaslyn Rural District Councils for the sewerage of Rhyd-ddu, and that they were not aware that they could themselves do anything more in the matter.

Under these circumstances, I was instructed to make an inspection of the gathering ground from which the water supply in question is derived, and to report in detail as to the exact conditions under which the water is obtained.

The Water supply of the Borough of Carnarvon, as stated above, is derived from the river Gwyrfai. The valley of the Gwyrfai forms one of a number of valleys which run in a westerly direction from the Snowdonian range towards the Menai Straits. These valleys are nearly parallel to one another, and each contains one or more lakes. On the north, the Gwyrfai Valley is separated from the valley of the Seiont, containing lakes Padarn and Peris, by a range of mountains which attains a height of from 1,000 to 2,000 feet. On the south, the Gwyrfai Valley is bounded by a range of mountains of nearly equal height, by which it is separated from the valley of the Llyfni and the two lakes Llyniau Nantyllef. At the head of the Gwyrfai Valley are the steep western slopes of Snowdon, forming a deep hollow, Cwm Clogwyn, and a prolongation of the central ridge of Snowdon in a south-westerly direction. There is thus formed a watershed which is sharply defined, and which is situated midway between the villages of Rhyd-ddu and Beddgelert. The valley of the Gwyrfai contains two lakes. That nearest the head of the valley is Llyn-y-Gader; that situated at a lower level is known as Llyn Quellyn. It is with the water issuing from the latter lake that Carnarvon is supplied. The head of the Gwyrfai Valley is about three miles wide, and Lyn-y-Gader is situate nearly in the centre of it. The lake is nearly circular in shape and about a quarter of a mile broad. From this lake the Gwyrfai issues as a large stream, and after a course of nearly three miles, passing the village of Rhyd-ddu which is situated on its banks, falls into Lake Quellyn. At this point the valley narrows considerably and is not more than a mile and a half in width. Lake Quellyn is about one mile in length, and a third of a mile in greatest breadth. From this lake the Gwyrfai issues as a fair sized river, and after a course of about 12 miles, falls into the sea at the southern extremity of the Menai Straits, at a point four miles south of the town of Carnarvon. The intake for the supply of the Borough is situate at a point half a mile below Lake Quellyn, and 6¼ miles distant from the town of Carnarvon. At the intake a dam has been constructed in the course of the stream, and the water thus held up flows through 12 copper gauze screens, or strainers, directly into an iron pipe by which it is conveyed to a reservoir situated at Yspytty, 2½ miles from the town. This reservoir is of 2,780,000 gallons capacity, and from it the town is supplied by gravitation. There is no filtration of the water. The supply is a constant one; there are very few storage cisterns in the Borough. The town of Carnarvon contained in 1891 a population of 9,804 persons living in 2,154 houses. There are no large manufactories or businesses requiring a large quantity of water in the town. During the summer months the supply of water to Carnarvon has been at times inadequate. This inadequacy would appear to be in part due to corrosion and consequent narrowing of the calibre of the smaller mains, in part to leakages from mains and to waste from taps in dwellings and from stand-pipes in courts. The Town Council have decided therefore to provide an increased storage by means of an additional reservoir. The present water supply was inaugurated in 1865, the Town Council having obtained power to take water from the Gwyrfai by an Act of Parliament, the Carnarvon Waterworks Act, 1865. At that time the water was obtained from the river at a point much nearer to Carnarvon than the present intake. The Town Council finding that the water was polluted by the drainage of the village of Waenfawr, consulted Mr. Baldwin Latham, who advised the removal of the intake to its present position, and a Provisional Order was obtained for this purpose in 1879. The Town Council do not own Lake Quellyn nor any portion of the area from which the water is derived. The whole of the area is under the control for sanitary purposes of the Gwyrfai and Glaslyn Rural District Councils. That portion of the valley which is in the Gwyrfai Rural District comprises both sides of the river above the intake, includes Lake Quellyn, and extends to a short distance above the upper end of that lake. From this point the river upward forms the boundary between the districts of the two Rural

APP. A. No. 12.

On the circumstance of the Sources of Water Supply for the Borough of Carnarvon; by Dr. Wheaton.

Councils, the area on the north of the stream being in the Glas Rural District, that on the south in the area of the Gwyrfai R District. The village of Rhyd-ddu is situate on both sides of river at a distance of about three-fourths of a mile above L Quellyn, and is consequently in two different Rural Districts.

The whole of the valley above the intake is occupied by far At nearly all the farms cows and pigs are kept, and at some them there are several cow-houses. The land of these farm chiefly pasture, but each farm has a few fields of arable la Towards the summit of the high land bordering the valley on e side the ground is rough and uncultivated, and used entirely sheep walks. Each farmhouse, with one exception, is situated the banks of a stream falling into Gwyrfai. Such stream is quently employed to turn a waterwheel by which the churn and chaff cutting for the farm are performed. No doubt convenience of having a supply of water constantly at ha determined the situation of these buildings. Unfortunately, h ever, where farm buildings are situated upon streams, as in th instances, there is a temptation to discharge all refuse, whet liquid or solid, into the streams with consequent fouling of water furnished by them.

No recent analysis of the water furnished by the Gwyrfai r are available, the latest test of it having been made by Dr. Fr in 1893.

On proceeding to examine the Gwyrfai, its tributaries, and lakes in its course, with a view to the detection of source pollution of the water, I commenced at the intake of the Carnar water supply. From this point to Lake Quellyn is a distance half a mile. The river flows through fields which are chi pasture, and at the time of my visit there was no arable l directly bordering on the stream. On the south side of the ri between the intake and the lake, is one farmhouse, with two c sheds. None of these buildings are less than 200 yards from river, and drainage from them does not flow directly into river, but passes over grass land sloping towards it. The fa house, which has four inmates, is unprovided with a privy, the occupants "use the fields." Returning to the north bank the river, it was seen that a considerable quantity of water ente the river from the roadside channel at this point. The road question is the main road between Carnarvon and Portma which passes through Beddgelert. Along this road many tra pass daily, also numbers of quarrymen going to and fro to th work, and their was abundant evidence that the roadside char was much used for purposes of nature by persons passing al the road. About 500 yards above the intake is a farmhouse, yards from the river, situate on a small tributary stream. T farmhouse has eight inmates. At the time of Dr. Low's visit 1895 there was no privy accommodation here, but a pail privy since been provided, the contents of which are buried in garden. The feeder of the river passing this farmhouse does appear to receive any liquid refuse from the farm buildings, s refuse having been diverted into a field; but before joining Gwyrfai it receives drainage from arable land which was litte with farmyard manure at the time of my visit. Above dwelling is another farmhouse, about 200 yards distant from L

Quellyn, through the yard of which another feeder of the river passes. The greater part of the farmyard drainage is kept out of the stream by a bank of earth; but a rubble drain, taking liquid refuse from a portion of the farmyard, appears to join it. At the time of Dr. Low's visit there was no privy at this house, but a pail privy has since been provided. A little further on are three cottages which have six occupants; one of them, which was empty, is used as a summer residence only. For these three dwellings a pit privy has been provided, the receptacle of which is merely a rough hole in the ground situate about one yard distant from a rubble drain. There would appear to be reason to fear the washing during heavy rainfall of the contents of this privy into the small stream which passes the cottages on its way to the river. The next dwelling is a villa residence, situate at the side of the lake, close to the point at which the Gwyrfai issues from it. This house is provided with an indoor watercloset, the discharge from which is conducted to a cesspool in a field. The contents of this cesspool are from time to time distributed upon the surface of the field. Pertaining to this house also is a pit privy. The pit of this privy is of loose stone, and is situated about 30 yards from the margin of the lake.

With regard to the surroundings of Lake Quellyn, from which the Gwyrfai flows.—Its southern side is bounded by hills which are for the most part steep, and there are no dwellings on this side of the valley. On the northern side of the valley are a number of farmhouses situated on small streams falling into the lake. There is also a large hotel, the "Snowdon Ranger," situate about 60 yards from the lake side. A railway also passes up this side of the valley, and there is a station, Quellyn, near the hotel. The main road borders on the lower part of the lake here for a distance of half a mile, and there is no fence between it and the lake. The drainage water from the road enters the lake, and at the time of my visit there was sufficient evidence to show that the shore of the lake itself was not unfrequently used for purposes of nature by passers by. I visited the farms on this side of the valley. The first of these, Cae-au-Gwynion, has been provided with a pail privy. The farmyard drainage has also been diverted, so that it does not fall directly into the watercourse which passes by the farmhouse; but I noted a collection of manure placed on the bank of the stream, liquid matter from which made its way to the water. The next farm, Llwyn Onn, has, since Dr. Low's visit, been provided with a pail privy, and the manure heap has been removed from the position which it formerly occupied, and which was so situated that liquid refuse escaped into the stream flowing by the farmhouse. The next farm, Bron-y-fedw Uchaf, has been supplied with a pail privy, also since the date of Dr. Low's report, and the liquid drainage from the manure heap has also been diverted from entering the stream. The next farm, Bron-y-fedw Isaf, is a large one. Here, formerly the house drainage and the liquid drainage from the farm buildings entered the stream flowing by the farmhouse, but recently a pipe drain has been constructed to convey the drainage to the surface of a field. At this farm two privies formerly discharged into the stream over which they are built. After Dr. Low's report, these privies were provided with pails; but at the time of my visit the

APP. A. No. 12.

On the circumstance of the Sources of Water Supply for the Borough of Carnarvon; by Dr. Wheaton.

pails were disused, and the privy contents were again escaping into the stream. The next farm is Glan-r-afon, situated on the side of a large stream, falling into the Gwyrfai at a point just above Quellyn lake. There are four occupants, but no privy accommodation, and the inmates "use the fields" in the neighbourhood of the house. All liquid matters from the house and farm buildings are discharged into the stream. Above this farm is a small cottage provided with a midden privy which is not placed near any watercourse.

The Snowdon Ranger hotel at the time of Dr. Low's visit was provided with three privies, which were placed directly over a small stream falling into the lake. Since his visit, however, these privies have been replaced by pail privies. There is a watercloset in connection with this hotel, and the discharge from this, together with liquid refuse from the premises, is conducted to a cesspool, which is roughly constructed of stone, and is about 40 yards distant from the margin of the lake. At the time of my visit this cesspool was overflowing on the surface of a field sloping to the shore of the lake. Opposite the upper extremity of the lake, and distant about one-and-a-half miles from it, is a large slate quarry, Glan-r-afon quarry. A large stream passes through the quarry precincts, and is used for the working of the machinery by means of a large water-wheel. This stream is formed by the effluent water from four small lakes lying on the western slope of Snowdon, and, after passing through the quarry, it crosses Glan-r-afon farm, as already described. In its course through the quarry the water turns a large wheel which drives machinery for cutting the slate rock. There is sufficient head of water to provide all the motive power required at the quarry. This quarry is in the Glaslyn Rural District. At the quarry 250 men are employed; of these 60 live on the premises during the week, returning to their homes on Saturday night and coming back to work on Monday morning. For these, dwellings, known as "barracks," are provided. The liquid refuse from these dwellings escapes into small channels formed by feeders passing through the quarry to join the larger stream. Three privies were seen, placed over the course of streams of this description. These privies are rudely constructed troughs, formed of slabs of slate, and there could be no doubt that in time of heavy rainfall the greater part of their contents would be washed away into the stream. One bucket privy was also seen upon the premises, and another trough privy, which is disused. The neighbourhood of two of the trough privies which were used by the workmen was thickly littered with excrement, and several places were noted in the course of my visit which were evidently used by the men instead of the privies. Several collections of excrement were noted in the neighbourhood of small surface channels, down which fæcal matter could not fail to be carried in times of heavy rainfall. The whole of the drainage of the area covered by the quarry premises and workings falls into the stream passing Glan-r-afon farm. It is evident that this quarry forms a very serious source of pollution of the water of Lake Quellyn, and in times of heavy rainfall a large amount of fæcal matter cannot fail to be conveyed from the quarry to the lake and thence into the river. Should at any time a person employed in the quarry suffer from enteric fever, the risk to the health of the persons consuming this water would become very great. The fact that the workmen come from numerous different towns and villages to their work at the quarry increases this risk of conveyance of specific infection to the water of the lake and river.

Returning to the upper end of Lake Quellyn, I proceeded to examine that portion of the river which intervenes between Lakes Quellyn and Llyn-y-Gader. The main road here crosses the river by a bridge a short distance above Lake Quellyn. The village of Rhyd-ddu is situated one mile above this bridge. Close to the bridge on the southern side of the river is a farmhouse, Planwydd. Here a pail privy has been provided, but the drainage from pigsties and from a portion of the farmyard finds its way into a ditch, discharging into a small stream passing through the farm on its way to the Gwyrfai. On the northern side of the river is another farm, Quellyn, which is so situated that there is no danger of liquid refuse from it entering the river.

The village of Rhyd-ddu consists of 39 dwellings and has a population of about 150 persons. The dwellings are situated on each side of the Gwyrfai, which is here crossed by a bridge. Owing to the opening of a station here, generally known as Snowdon, and the development of the quarries at Glan-r-afon, a certain amount of building has taken place and is in progress. I noted six new dwellings on the southern side of the river, and about a similar number on the northern side. That portion of the village which is on the south side of the river is in the Gwyrfai Rural District; that on the northern side, which includes about three-quarters of the dwellings, is in the Glaslyn Rural District.

The whole of the liquid refuse from this village is discharged into the river, either directly—the refuse being thrown from the dwellings on the river banks—or indirectly by means of rubble drains or surface channels on the roadside. The village is not sewered. Formerly a large part of the fæcal matter from dwellings was also thrown into the stream, but of late steps have been taken to prevent this as far as possible by posting a large notice in Welsh warning persons of the consequences, and also by verbal warnings to householders. The result of my observations was to show that a certain amount of fæcal matter was still thrown into the stream from the bucket privies which are in use in the village. In the Gwyrfai section of the village there is a row of seven dwellings, which are served by four bucket privies. These dwellings have no gardens; consequently the occupiers have no means for disposal of the contents of the buckets from the privies. In the Glaslyn section of the village there are nine dwellings, having a yard in common which slopes directly to the stream. Here there was evidence that the contents of pail privies are thrown into the stream and that the bank of the stream had been used for purposes of defæcation. The majority of the dwellings on this side of the stream have sufficient garden ground for the disposal of the contents of the pails from privies. On this side of the river the drainage from a cowshed passes directly into it. At Rhyd-ddu station also the overflow of a roughly-constructed cesspool in connection with the station watercloset escapes into a ditch. This ditch communicates with the Gwyrfai. Formerly the whole of the refuse from the village was thrown into the stream, but three open

.A. No. 12

he circumstances of the ...ces of ...er Supply ...he ...ugh of ...arvon; by Wheaton.

receptacles of slate have been constructed by the Glaslyn Rural District Council for the storage of household refuse and ashes, the contents of which are removed by a farmer for use on his land. The majority of dwellings on the south side of the stream, however, have no such receptacles. Here a large heap of house refuse and ashes was seen on the bank of the stream, the centre of which heap was hollowed out to form a receptacle for liquid refuse from the dwellings. There can be no doubt that the sanitary circumstances of this village are such as to form a serious danger to the consumers of the water from the Gwyrfai river, and there would be grave risk to such consumers should enteric fever or cholera at any time break out in this village.

The two authorities in whose districts this village is situated, have failed up to the present to come to any arrangement for a joint system of sewerage, although there can be no doubt that combination for this purpose is indicated by the local circumstances. The Glaslyn and Gwyrfai Rural District Councils have each caused plans to be prepared for the sewerage of that portion of the village, which is under their jurisdiction. But nothing further has been done. It has been suggested that the want of a proper supply of water for the Glaslyn portion of the village, forms a serious obstacle to the efficient working of any sewers; but it appears to me that there are several water-courses which might be impounded and the water used for flushing the sewers. Negotiations have been in progress for some time between the Glaslyn Rural District Council and neighbouring land-owners for the obtaining of a supply of water for the village, but so far without result.

Above the village of Rhyd-ddu there is a villa which is used as a summer residence only. There is a watercloset here, the pipe from which is conducted to a ditch which ultimately joins the river. Above the village also there is a woollen mill, which is situated upon the bank of the river, and the liquid refuse from it is discharged by a rubble drain directly into the stream. There is no privy accommodation at this mill. Situated at a higher level than the village there is a farm, Fridd-isaf, close to a small stream falling into the Gwyrfai. The liquid drainage from the farmyard and manure-heaps enters this stream, and a privy discharges on the surface of the ground near it. An examination of the surroundings of Lake Llyn-y-Gader, from which the Gwyrfai issues, and which is situated at the head of the valley, showed that the land in its immediate vicinity is almost entirely uncultivated marsh. This lake is fed by numerous small streams, and appears to be shallow. The drainage from a cowshed and liquid refuse from farm premises of Fridd-uchaf farm enters a watercourse which falls into the lake, and the pasture land at some distance from the lake side appeared to have been littered with ashpit refuse.

From the foregoing account, it will be seen that the work of preventing pollution of the Carnarvon Water supply is a very difficult undertaking, and requires constant vigilance to be exerted over a very large area. The steps necessary for this purpose appear to be :—

(1.) The thorough sewerage of the village of Rhyd-ddu, and the disposal of the sewage in such a manner as to avoid any risk of pollution of the river.

(2.) The scavenging of this village at frequent and regular intervals, in order to put an end to the throwing of the contents of pail closets, of privies, or of household refuse, into the stream.

(8.) Frequent inspection of all farms and dwellings upon the gathering ground. The disposition of all liquid refuse from dwellings, farmyards, manure heaps, and the like, in such fashion that these matters shall not foul streams. The prevention, as far as possible, of manuring of land in the immediate neighbourhood of the lakes and streams.

(4.) The provision of moveable receptacles for fæcal matter for all dwellings upon the gathering ground, in order to prevent the direct fouling of streams by fæcal matter. Strict attention to be directed to the method of disposal of contents of such moveable receptacles.

(5.) The diversion of all water passing through Glan-r-afon quarry from the river above the intake.

(6.) The diversion of all surface water from the main roads from the river.

The last two measures require works of engineering of considerable difficulty.

Filtration of the Carnarvon Water Supply has been recommended as a precautionary measure. It is doubtful, however, whether filtration, as ordinarily carried out, can, in the present circumstances of that supply, be trusted to secure a uniformly wholesome water.

No. 13.

No. 13. ...ni- and ...stra- ...the ...n-the- ...ural ; by

REPORT on the SANITARY CIRCUMSTANCES AND ADMINISTRATION of the STOW-ON-THE-WOLD RURAL DISTRICT; by DR. R. DEANE SWEETING.

Stow-on-the-Wold Rural District, situated chiefly in Gloucestershire, and in part in a detached portion of Worcestershire, is the most easterly in the former county, and is placed approximately midway between the cities of Gloucester and Oxford. It is bounded on the north, west, and south, respectively, by the Unions of Shipston-on-Stour, Winchcomb, and Northleach; whilst on the east it is conterminous with the county boundaries of Oxford and Warwick. The Worcester portions of the rural district are the two detached parishes of Evenlode and Daylesford at the extreme north-east of the Union. The Stow-on-the-Wold Urban District, comprising the Parish of Stow-on-the-Wold, is the only other sanitary district in the Registration District of Stow-on-the-Wold, which is, therefore, made up of the Urban and of the Rural District. The Registration District is sub-divided into two Registration Sub-districts, viz.—(A) *Stow-on-the-Wold* Sub-district, containing the Stow-on-the-Wold Urban District and 13 parishes in the Rural District; (B) *Bourton-on-the-Water* Sub-district, containing 14 parishes, all in the Rural District. A list of these 27 parishes is given as an appendix to this report, particulars as to the acreage, number of inhabited houses, and population being added in each case. The area of the whole Rural District is 44,504 acres, the number of inhabited houses 1,782, and the population (1891 census) 7,800. The rateable value is £51,207; the annual assessable value is £36,843. Its shape is an irregular quadrilateral, the greatest length, from north to south, being about 12 miles, and the greatest breadth, from east to west, about 11 miles.

The Rural District is, generally speaking, hilly on the south and west, where spurs of the Cotswold Hills extend into it. On the north, the centre, and the east, the country is flatter, and is watered by various named and unnamed tributaries and feeders of the River Thames. Between these two tracts of hills and vale is a belt of undulating country.

Geologically, many formations are met with. In the valleys and the flat portions of the district, Lower Lias clay is the usual formation. As the ground rises the Upper Lias clay and the Oolites appear, until at the highest points, as, for example, the hills around the village of Bourton-on the-Water, sections show, first, Marlstone lying on Lower Lias clay, then Upper Lias clay, above that "Midford Sands"; then, in order, Inferior Oolite, "Fuller's Earth," and at the top Great Oolite. The latter contains in the west of the district a flaggy bed called "Stonesfield Slate," which is quarried and used locally for roofing purposes. These strata are much intersected by faults. Considerable beds of porous gravel are found in the flat portions of the district. These are especially well marked at Bourton-on-the-Water village, where the underground water is only a few feet from the surface.

The industry of the district is almost purely agricultural, the only exception being the quarries of Stonesfield slate abovementioned, which are situated at Eyford. There is a little trading done at Bourton-on-the-Water, the most considerable place in the district, but this, again, is wholly connected with the cultivation of the land.

The district has not been inspected by the Medical Department of the Local Government Board since the visit of the late Dr. Ballard to the village of Bourton-on-the-Water in 1874 on account of an epidemic of enteric fever. Briefly, Dr. Ballard found that this fever, which had been originally imported, had been associated with and fostered by defective drainage, excremental nuisances, and contaminated water supply. The conditions of the latter have remained practically unaltered since Dr. Ballard's visit 24 years ago; and, as at that date, the responsibility for the health of the district is divided between two Medical Officers of Health. Hence this inquiry.

After preliminary conference with the Clerk on September 28th, I visited on this and subsequent days, along with the Medical Officers of Health and the Inspector of Nuisances, 23 of the 27 parishes of the district, omitting only those which contained inconsiderable populations.

Before proceeding to set out the result of my inspection of the sanitary circumstances of the district, it should be mentioned that this is divided, for sanitary purposes, into two sub-divisions, viz., Stow-on-the-Wold and Bourton-on-the-Water. The following are the area, number of inhabited houses, and population of each of these sub-divisions, which are co-extensive with the Registration Sub-districts so far as these are contained in the Rural District:—

Sub-division of Rural District.	Area in Acres.	Number of Inhabited Houses.	Population (1891 Census).	Estimated Population (1898).
Stow-on-the-Wold	21,139	842	3,605	3,284
Bourton-on-the-Water	23,365	940	4,195	4,266
Stow-on-the-Wold Rural District	44,504	1,782	7,800	7,550

Inasmuch as the sanitary circumstances of these two subdivisions are generally alike, it will be convenient to describe them together, noting at the same time any differences. Particular reference will be made to places in either sub-division, the sanitary arrangements of which have in the past engaged the attention of the Board.

I. Sanitary Circumstances.

1. *Condition of Dwellings.*—These are, in the majority of cases, substantial cottages, built of local stone and covered with local slates (Stonesfield slate from the Great Oolite beds). They are as a rule in good repair. The usual arrangement consists of a sitting-room and kitchen (combined) on the ground floor, and two bedrooms on the first floor. Overcrowding is uncommon,

though not unknown. When discovered, it has been at once dealt with by the Council. I found no instance of it during my inspection. Good ample garden space is attached to each cottage or group of cottages. There are, however, a few places, *e.g.*, Bourton-on-the-Water and Church Westcote, where groups of cottage property are huddled together on insufficient space. No action under the Housing of the Working Classes Act of 1890 has yet been taken with regard to any of these groups of cottages. There is need for better paving of the ground around cottages, and also for more extended provision of eave-spouting. The absence of these two requirements has in many instances rendered both cottage walls and the ground surrounding cottages unduly damp.

2. *Water Supply.*—This is obtained from—
(*a*) Springs.
(*b*) Shallow wells.
(*c*) Open watercourses.

(*a*) *Springs* are in many cases exposed to pollution from surface washings and manurial and slop nuisances in their neighbourhood. One of them, at Broadwell, issues from a bank only some 50 yards below a churchyard. The springs are piped either to open dipping places (Donnington and Notgrove), roadside spouts (Oddington and Little Rissington), standpipes and private taps (Longborough and Naunton), or underground tanks fitted with pumps (Condicote and Great Barrington). In several parts of the district, two or more of the above methods of distribution are in vogue at the same time. At some of them, the pumps serving the underground tanks were out of repair at my inspection. Most of the underground tanks are cement lined.

For the Bourton-on-the-Water sub-division the spring water supplied to the villages is generally less adequate in quantity and less accessible in position than for the Stow-on-the-Wold sub-division.

(*b*) The *shallow wells* vary in depth from 10 feet (Donnington) to 20 feet (Bledington). They are usually dry-steined with stone, and provided with pumps. Some, however, are mere draw wells. Each well supplies a single cottage or small group of cottages. Many of them are placed very close to houses, and nearly all are in proximity to privy vaults. There is no external puddling or internal cementing of these shallow wells, which consequently allow passage of surface impurities into them. Shallow wells constitute a larger proportion of the supply in Bourton-on-the-Water sub-division, since the village of that name is wholly served by them.

Inasmuch as the water supply of Bourton-on-the-Water village has already formed the subject of inquiry by the Board, it will be well to refer to it in more detail.

The wells are all shallow at Bourton-on-the-Water and are computed to number at least 200. As the number of inhabited houses is now 254, it is reckoned that each house practically has its own well. I met, however, with one instance where five houses were dependent upon one well.

The wells are fed by water from the superficial gravel bed before mentioned, and held up by the subjacent Lower Lias clay. Most

APP. A. No. 18.

On the Sanitary circumstances and Administration of the Stow-on-the-Wold Rural District; by Dr. Sweeting.

of them are fitted with a pump, but there are in addition a few draw wells. In depth none are said to exceed 10 feet, and many are even shallower than this. They are all dry-steined, and very few are puddled or cemented to prevent entrance of surface soakage. Though Bourton is now provided with a sewerage system, not all the houses in the village are as yet connected with the main sewers. There still remain some 50 houses which have the old objectionable privy vault; and I understand that when the new sewerage system was adopted, many of these old privy vaults and many of the old rubble drains were not dug out, so that the wells of Bourton-on-the-Water are still liable to contamination by impurities soaking into the bed of porous gravel from which the water is obtained. Such contamination is, in fact, constantly going on, and the Medical Officer of Health informed me that he had lately seen water which was "thick with sewage." In addition, many of the wells have run dry during the recent drought, which is not to be wondered at, in view of the fact that the bed of gravel which forms the water-bearing stratum is of no great thickness.

A public inquiry was held at Bourton-on-the-Water, on February 16th, 1894, by Mr. F. H. Tulloch, one of the Board's Engineer Inspectors, into an application of the Stow-on-the-Wold Rural District Council, for permission to borrow £1,500 for works of water supply. The source of supply proposed was what is known as the "Glebe Land Spring," which crops out at the foot of the Inferior Oolite, on rising ground about a mile west of the village. An intake chamber, a gravitation reservoir, and distributing mains were proposed. It transpired at the inquiry that the reservoir was not placed sufficiently high to allow of the whole of the village being supplied. The pressure would only have been sufficient to extend to about the middle of the village, and would have been inadequate to force water to the top floors of houses, and sanction to the scheme was withheld. Since that time, nothing has been done towards procuring a suitable public supply of water.

Dr. Corser in his annual report for the following year (1895), pressed for this, urging that water was imperatively required for flushing the newly-established sewerage system, and that the latter would prove of little value unless supplemented by a proper water supply. Since then, continuous correspondence has ensued between the Board and the Rural District Council. Thus, in reply to a letter from the Board of August 5th, 1896, the clerk stated that the Council "undertook to provide a pure water supply in cases where the wells had been condemned."

The Board, on September 9th, rejoined that this was not sufficient, suggesting that the Council's engineer should advise them on the matter.

The Board again wrote on November 9th, drawing attention to the need for flushing water-closets and house-drains, and pointing out that nuisances would arise from the absence of this requirement.

In a special report to the Council, dated November 26th, 1896, Dr. Corser drew attention to the need for flushing house drains, and to the fact that out of 67 water-closets in the village 48 had no flushing arrangements whatever, and alluded to the circumstance [illegible] that in 189[illegible] he found two out of every three samples of water [illegible] for [illegible] purposes: whereupon the Board wrote on [illegible] 1896, again urging the services of an engineer to [illegible] Council.

[illegible] Rural Council forwarded a Report of the Bourton-on-the-Water Parish Council dated January 6th, 1897, stating that there was no need for a fresh water supply, that the Medical Officer of Health had selected samples of water for analysis in a [illegible] and partial manner, and that increased flushing would [illegible] universal adoption of water-closets. At the same time, [illegible] Council stated that they had forwarded 12 samples of [illegible] the County Analyst. The report of this analyst, Mr. G. Embrey, F.C.S., dated February 17th, 1897, was subsequently forwarded to the Board. Briefly, he condemned four samples as "bad," other four were held to be "not quite satisfactory," and four he regarded as "good." Bacteriological examination confirmed the chemical analysis of the four "bad" waters as "sewage-polluted." The covering letter from the Clerk, dated February 20th, 1897, and enclosing Mr. Embrey's report, informed the Board that the Rural Council had ordered four wells to be closed, and four to be cleaned out. The Board, on March 9th, 1897, adverting to Mr. Embrey's analysis, informed the Council that they looked upon only three of the 12 well waters as free from evidence of excessive organic matter, and added that in their opinion the mere cleaning out of a foul well was not alone sufficient to render it a safe source of supply. Dr Corser's Annual Report for 1896, received on March 8th, 1897, contained bold and severe strictures on the water supply of the village and drew attention to the grave responsibility of the Council. This outspoken attitude gave great offence to certain members of the Council, and an unsuccessful attempt was made to get rid of the Medical Officer of Health. Early in April, 1897, the Council forwarded another report of Mr. Embrey, dated March 25th, 1897, advocating a supply which should be "above suspicion." A reminder from the Board on May 12th, 1897, had no effect except to evoke a request for sanction to employ a water diviner, expenditure for whose services was refused by the Board, who again, on July 27th, 1897, advised the services of a competent engineer. In November, 1897, the Council forwarded a report from Mr. G. B. Witts, their engineer, in which he advised the acquisition of a spring known as "Nicholl's Barrow." He regarded this as likely to furnish an adequate supply of water for Bourton-on-the-Water, and at such a height as to provide sufficient pressure.

Negotiations with the owner of this spring, and with certain riparian owners of land abutting on ditches and streams which receive its overflow, were then actively entered upon. The riparian owners proved accommodating; but the owner of this spring has given a definite and final refusal to the Council to sell the water rights to them. The Council have therefore been obliged to finally abandon the "Nicholl's Barrow" spring scheme, and to turn their attention again to the "Glebe land" spring, which formed the subject of the 1894 inquiry.

In his Annual Report for 1897, Dr. Corser again reverted to the subject, and after referring to the need of a water supply of sufficient pressure to reach upstairs water-closets, advised pumping of the water from the "Glebe land" spring to a raised reservoir by means of water power, to be furnished by the River Windrush, less than 100 yards distant. Evidently, this advice has had some effect upon the Council, for in August, 1898, they gave instructions to their engineer, Mr. Witts, to formulate a scheme of pumping, including the probable cost of a water wheel.

Mr. Witts has not yet submitted a report on this matter, but the Clerk informs me that it is daily expected, and that on its receipt he will at once forward a copy to the Board.

Such is a *précis* of the history of the Bourton-on-the-Water water question, which has occupied attention for many years, both before and after the public inquiry in 1894. As will be seen under the next heading (3), the need for an improved supply is more urgent than ever since the introduction of main drainage into the village. Perhaps, however, the requirement that the water shall be laid on at a pressure sufficient to force it to the top floors of houses need not be insisted upon, since very few houses in the village have water-closets or draw-taps on the upper floors.

(*c*) *Open water courses.*—Some of these, as at Longborough, are used for drinking purposes, the water being conveyed to spouts at the roadside. They receive, especially at flood-time, surface washings and manurial soakage from fields draining into them above. One of these streams was said to be quite muddy in wet weather, and even at times to contain soapsuds and other evidence of domestic sewage.

As regards the village of Upper Slaughter, the only public water supply is an open dipping place or trough in the upper part of the village, which is fed by a private spring. This trough is so inaccessible for the inhabitants of the lower part of the village that they dip water for drinking purposes out of a stream which passes through this part of the village. This stream is a feeder of the Dickler, which is a tributary of the Thames. It receives the unpurified sewage of Upper Slaughter. In December, 1897, a public inquiry was held by Mr. Meade-King, one of the Board's Engineering Inspectors, into an application of the Rural Council for sanction to borrow £375 for purposes of water supply at Upper Slaughter. It appeared from the evidence given there that the water of the private spring above alluded to as supplying the trough in the village was partly piped to it direct by gravitation, and partly pumped by two water wheels to a tank at the top of the rectory, the overflow from which reached the trough. The spring itself is situated on private property, 10 or 12 feet below the surface of the ground, and is said to rise from the Midford Sands. The proposition at the inquiry was to erect another wheel, to be worked by the water proceeding from one of the existing wheels, and to pump the water of the spring from the neighbourhood of the present spring to a rising main, and thence to stand-pipes. Sanction to the application was not given by the Board for the following reasons, viz. :—That it was not proposed to take water from the spring itself, but from an unknown source, which might be either a leakage from that spring, or a leakage passing from the adjacent stream through the porous Marlstone over the Lower

Lias clay; and that water from such a source was liable to serious pollution in its passage through an unsewered village like Upper Slaughter.

Since the Board's refusal of assent to the above scheme, the Rural Council have been in negociation with the Ecclesiastical Commissioners as to terms of acquiring the lease of another spring in the village. The Board have intimated to the Council that there would be no objection to the incurring of certain preliminary professional expenses in connection with this new scheme. The Clerk informed me that an application for sanction to a loan will be made to the Board very shortly.

In some instances, water has been supplied by private munificence, and not by the Sanitary Authority. Thus, at Lower Swell a deep spring has been tapped by a private owner, and the water raised to the village by three hydraulic rams. Again, at Upper Swell a spring has been piped by a property owner to a roadside dipping place and to two stand-pipes in the village.

It will be gathered from the above description that the water from the shallow wells and water-courses in the district is distinctly dangerous, whilst the surroundings of some of the springs render their water open to suspicion. The mode of distribution of the water of some of the springs, too, viz., at open dipping places or roadside spouts, is also open to grave objection.

3. *Drainage, Sewerage, and Sewage Disposal.*—In many parts of the district there is no drainage at all. Slop-water in these places is usually thrown over garden ground, and often in such a way as to cause nuisance (Notgrove). Where house drains are met with, they are usually square stone rubble drains (Lower Swell); but a few short lengths of socketted pipes are found (Bledington, Little Rissington). Inlets to the house-drains are earthenware gullies, which are sometimes trapped (Condicote), but more frequently are large untrapped receptacles resembling open cesspools (Broadwell). These are in many cases placed quite close to the back doors of houses, and constitute distinct nuisances (Naunton, Upper Swell).

Where there is no sewer, the house-drains discharge into a "slop cesspit" or "dumb-well," which is placed often at some unknown spot in the garden, and is consequently never or seldom emptied.

In not a few villages of the district there is no sewerage whatever (*e.g.*, Longborough, Naunton).

Where there is sewerage, this is represented often by old highway rubble drains, usually constructed of loose stone slabs at the sides and top, and having no solid bottom (Upper Slaughter and Lower Swell); but a few short lengths of 6-inch and 9-inch socketted pipes are found acting as sewers (Evenlode, Bledington). The highway drains which act as sewers get flushed only by storm and surface water. The pipe sewers have no special means of flushing, and are not ventilated at all.

As for sewage disposal—in addition to direct disposal of slop sewage on garden ground and its soakage away from the rude cesspits placed in the gardens, the chief methods of disposal are (*a*) ditches which empty into feeders and tributaries of the River Thames; (*b*) ditches which overflow into pasture fields.

At Bledington, 67 of the 80 houses in the village drain directly to the stream in the centre, which joins the Evenlode, a tributary of the Thames, about a quarter of a mile below the village. Both this village stream and the Evenlode, where joined by it, are sluggish, and almost stagnant. The flat land adjacent to the Evenlode is often submerged in flood-time. A scheme of main drainage for Bledington, propounded under pressure from the Thames Conservancy Board, which was before the Local Government Board in 1897, was rejected without public inquiry, owing to absence of any proposal of land purification of the sewage effluent, after treatment through coke filters. Another application is now before the Board, dealing only with slop-water sewage, earth-closets being proposed for excrement disposal. Public inquiry has not yet been ordered by the Board.

The drainage of Evenlode village also reaches the River Evenlode. This is effected by means of certain ditches which empty into that river.

Similarly, all the drainage of Broadwell village reaches the same tributary of the Thames by means of ditches which feed that stream; and the Dickler, another tributary of the Thames, receives sewage from Lower Swell by way of a ditch which finds its way to that tributary. Some slop-water, too, is directly thrown into it.

Again, the sewage of Upper and Lower Slaughter reaches a feeder of the Dickler by highway drains, and that of Little Rissington enters another feeder of the same tributary of the Thames in like manner. Pressure has been put upon the Council by the Thames Conservancy Board with reference to the pollution of the Dickler by Upper Slaughter sewage. The latter feeder of the Dickler receives also at Little Rissington overflows from cesspits, which enter it directly.

The river Windrush, in its passage through Naunton, receives much surface drainage, and some slop-water from that village.

In addition to the above modes of sewage disposal, certain village ponds (Wyck Rissington) receives sewage directly from farm yards; and in a few small villages (Addlestrop and Daylesford) slop sewage is conveyed by jointed pipes to cemented cesspits having an overflow to neighbouring ditches.

The village of Bourton-on-the-Water has been recently provided with a system of main drainage, and disposes of its sewage on land by irrigation.

A public inquiry was held by Mr. Tulloch on February 16th, 1884, into an application of the Stow-on-the-Wold Rural District Council for sanction to borrow money for the sewerage and sewage disposal of the village of Bourton-on-the-Water. The application was sanctioned, after some modification of the original plans, and assent was given by the Board to a loan of £2,432. A further sum of £250, in excess of the original estimate, was afterwards agreed to by the Board, making £2,682 in all.

The scheme of main drainage was completed at the end of 1895. There is a main 12" outfall sewer, and this receives two 9" branch sewers. Ventilation of the sewers is effected by open manholes at regular intervals. They are provided with inlets at the manholes, and with cast-iron ventilating pipes, which pass from the manholes to be carried up the sides of adjacent houses. Flushing

App. A. No. 13.

On the Sanitary circumstances and Administration of the Stow-on-the-Wold Rural District; by Dr. Sweeting.

is accomplished by running into a 700-gallon tank, for the whole of one day in each week, water from an adjoining mill-sluice, by arrangement with the mill-owner. This tank is placed at the head of the sewer, and discharges automatically when full.

House drains, 6″ main and 4″ branch, are disconnected from the sewers by syphon traps, provided in most cases with gullies, over which sink pipes discharge. But these gullies have been found liable to get blocked by sewage and choked by sand, thus showing the need for better flushing. Soil-pipes are usually ventilated by pipes of equal bore led above the eaves of houses. Waste pipes are cut off from the drains, and empty over channels leading to trapped gullies. Of the 254 inhabited houses at Bourton-on-the-Water, 201 are at present connected with the main sewer. This work of connection, under section 23 of the Public Health Act, 1875, has not been as rapid as it might have been; but the Inspector of Nuisances claims to have cut off nearly 200 old rubble drains that formerly discharged sewage directly into the river Windrush, which passes through the centre of the village. Little or no domestic sewage now reaches this stream, in marked contrast to what obtained at the date of the report of Dr. Ballard, who described the Windrush as virtually an open sewer.

After a course of about 1½ miles, during which it is led across the river Windrush, the outfall sewer discharges into two alternating subsidence tanks. From thence, the liquid portion passes into a field of about four acres, where it is conveyed along some dozen parallel carriers. The result of this treatment is that the effluent from the irrigation-area, which passes into the river Windrush, appears remarkably clear. The sludge from the tanks is applied directly to land. There is no chemical treatment. At my visit, the sewage farm appeared altogether well managed. It is leased at a low rent to a farmer who makes what he can out of the sludge and the grass grown on the farm.

Inasmuch as the sewers are laid at very flat gradients, which appears to have been unavoidable, it cannot be said that the flushing arrangement above mentioned is at all adequate. The tank at the head of the sewer should be supplied at more frequent intervals than one day a week.

4. *Excrement Disposal.*—This is effected for most part in the usual village privy vault, often a mere hole in ground, covered by a wooden plank or planks, or by a stone slab. The contents, being infrequently removed, soak away into the ground, endangering neighbouring water supplies. I am told that 12 months is the usual period which elapses between the removals of privy contents. In some instances, however, I found that they had not been removed for four years. Some of these privy vaults I found very offensive and foul smelling. Some effort has been made to introduce "dry" methods, and a few pail closets are to be met with (Evenlode, Notgrove). In a few better class private houses, there are water-closets, which drain to cemented cesspools, having overflow to ditches. A few midden privies were met with (Great Barrington) and seemed well kept.

There is no public system of scavenging in the district.

In Bourton-on-the-Water, since the establishment of a sewerage scheme, the introduction of water-closets has proceeded gradually. Of the 201 houses connected to the main sewer 92 are now provided with water-closets, or nearly one-half. Of these 80 are hand-flushed and 12 are provided with a two-gallon flush, pumped from the well to a cistern in the water-closet. Of the 80 hand-flushed closets, 76 are situated out of doors and four indoors, all on the first floor. Most of the hand-flushed closets that I saw were very foul, many of them being full to the top with excrement and urine. The pans of many of these are dirty in the extreme. Some of the water-closets are badly situated, viz., inside houses and opening directly into living rooms.

The above facts point strongly to the need of a proper public water service. The choked condition of many of the gullies, the filthiness of most of the hand-flushed closets, and the inconvenience of having no water supply to upstairs closets, all point to this urgent necessity. It is, indeed, an indispensable corollary to the system of public drainage already established in the village.

5. *Refuse Disposal.*—Very few properly constructed ash-pits are to be found in the district. Refuse is, as a rule, either placed in rude holes dug in the ground or thrown on to ash heaps in gardens. The latter sometimes receive in addition to various kinds of animal and vegetable refuse, slop water, which is hand thrown upon them by occupiers of the cottages. It is probable that at Burton-on-the-Water village some refuse is still discharged into the river Windrush, which passes through it.

6. *Slaughter-houses.*—There are four in Bourton-on-the-Water village, of which I visited three. Two of these I found very dirty and badly kept. Bye-laws are urgently required to deal with them.

Pig-keeping and pig-slaughtering on private premises are often so carried out as to become grave nuisances.

7. *Cowsheds, Dairies, and Milkshops.*—Milk is brought direct to the consumers from the farms. Those cowsheds and dairies that I saw seemed fairly clean.

8. *Common Lodging-houses.*—There are none in the district.

II. Sanitary Administration.

The Rural District Council of Stow-on-the-Wold, the successors of the Stow-on-the-Wold Rural Sanitary Authority, consists of 27 councillors, each representing a parish of the district. They meet every fortnight at the workhouse of Stow-on-the-Wold Union.

The Infectious Disease (Notification) Act, 1889, was adopted in June, 1893. Part III. of the Public Health Act Amendment Act (1890) was adopted in July, 1890. The Infectious Disease (Prevention) Act, 1890, has not been adopted. No action has been taken under the Housing of the Working Classes Act (1890).

There is no hospital for infectious diseases in the district. A tent was temporarily prepared in 1896 for small-pox cases, owing to the scare produced by the epidemic of that disease then raging at Gloucester. This tent was not used, no small-pox cases having occurred, a fact largely due, in local medical opinion, to the abundant vaccination and re-vaccination that was then carried out. In reply to a communication from the Board in July of this

.A. No. 13. — he Sani- circums- ces and ministra- of the w-on-the- ld Rural trict; by Sweeting.

present year, the Council replied that they were considering the question of a permanent hospital.

No proper disinfection of articles of clothing is carried out, owing to the absence of any apparatus suitable for the purpose. Fumigation of rooms by sulphur candles is the only process followed out in houses where infectious disease has occurred.

There are no bye-laws of any kind under the Public Health Act, 1875, though these are urgently needed, especially for scavenging.

As regards the sanitary work done by the Council, with the exception of the protracted consideration of the Bourton-on-the-Water water-supply question, already set out above, and of schemes for the drainage of Bledington and water supply of Upper Slaughter, none of which have reached maturity, it cannot be said that they have paid much attention to the urgent sanitary requirements of their district as a whole. I refer especially to the need for water-supplies free from danger of pollution, for main drainage in the larger villages, for improved methods of sewage disposal and protection of streams from contamination by sewage, for better modes of excrement and refuse disposal, and of scavenging, and for the adoption of a good code of bye-laws.

They do not usually require the attendance of the Medical Officers of Health at their meetings. These officers very rarely attend, only, in fact, if requested to do so by the Council. Instead of attending, they send their journals by the Inspector of Nuisances, who attends the Council meetings on every occasion. Their advisory functions, under Art. 18 (4) of the Board's Order of March 23rd, 1891, are therefore reduced to a minimum, the more so since these journals, as will be related later, are not kept in the full and complete manner that they should be.

A Medical Officer of Health is appointed for each sub-division of the district. The arrangement under which the duties of Medical Officer of Health are divided between a plurality of officers is not one which, in the Board's experience, generally conduces to efficient discharge of duty. It is their practice, therefore, as occasion arises, to direct special inquiry into the methods and efficiency of sanitary administration in districts thus officered; and, on the last occasion on which the Board assented to the re-appointment for one year of the two Medical Officers of Health for the Stow-on-the-Wold Rural District, they intimated to the Council that the district would be officially inspected before the expiration of the period for which they were re-appointed.

The two officers are Mr. E. Dening for the Stow-on-the-Wold division, and Dr. F. R. S. Corser for the Bourton-on-the-water division of the district. Each officer is paid only £22 10*s.* 0*d.* a year for his services, half of which is repaid from County funds. Mr. Dening lives at Stow-on-the-Wold, and Dr. Corser at Bourton-on-the-Water. Both places are on the Cheltenham and Banbury branch of the Great Western Railway, and each of these stations serves a large agricultural district. Each officer is conveniently situated for his district, especially Mr. Dening, whose district is more compact and less scattered than that of Dr. Corser.

Mr. E. Dening, L.R.C.P., M.R.C.S., is District Medical Officer and Public Vaccinator, and Medical Officer to the Workhouse, as

APP. A. N

On the ...tary cir...stances ... Admini...tion of ... Stow-o... Wold R... District; Dr. Sweet...

well as Medical Officer of Health to the Stow-on-the-Wold Urban District Council. His annual reports have been so meagre, and have contained so little information, that a Supplemental Report for 1897 was asked for. No record of systematic periodical inspection, or of the exact sanitary state of the different parts of the district, appears to have been made. He has paid attention to little else than questions of water supply, as a rule omitting reference to other important matters, such as drainage and excrement disposal. With regard to water supply he has sometimes, as in his 1895 Report, chronicled the state of things in his district in more favourable terms than the actual circumstances appeared to warrant, and he has generally trusted more to the results of chemical analysis than to circumstances of physical environment. I find from Mr. Dening's journal, and from the official correspondence that he has on several occasions neglected to inform the Board of the grounds on which he has advised the Council to require the closure of schools, as laid down in Art. 18 (15) of the Board's Order. This journal has been very inadequately kept. There is seldom any record in it of the "sanitary condition of premises" (col. 5), or of any "action taken" (col. 6), on his representations to the Council.

Col. 5 is, indeed, usually taken up with a brief general account of the "health of the district," which is nearly always reported to be "good," and col. 6 is invariably blank. It is only just to add that, since the Supplemental Report was called for this year, Mr. Dening appears to have been distinctly more systematic in the inspections of his district.

Dr. F. R. S. Corser, M.B., C.M., is District Medical Officer and Public Vaccinator, as well as Medical Officer of Health. His action on the Bourton-on-the-Water water question has been already alluded to. He certainly braved unpopularity and displeasure by his conscientious attitude with regard to that question. The trouble in which the performance of his duty involved him at this time seems to have constituted a dividing line in Dr. Corser's work. Previous to 1897, his Annual Reports were distinctly good. He showed that he had systematically inspected his district, his reports covered a wide range of subjects, and he gave excellent advice to the Council. Moreover, his journal, up to this period, was very fairly kept, though col. 6, relating to "action taken," was seldom filled up. But since 1897, to judge by the Report for that year, and by his recent entries in the journal, this officer's work has shown signs of distinct retrogression. His 1897 report is the most meagre report that he has ever presented; whilst, in his journal, he has of late emulated his colleague by filling up the important col. 5 by such irrelevant entries as "Health of the district good," and the journal has been badly kept.

The two Medical Officers of Health never meet at all officially, and, indeed, seldom see each other. Each is engaged in large private practice. Their practices are so independent and distinct that they have little opportunity of meeting. Consequently, no joint or concerted action is ever taken by them. As before said, they rarely, if ever, attend the meetings of the Council. No comprehensive scheme for the sanitary amelioration of the district has ever been brought to the notice of the Council. Indeed, no representation is ever made by them. The sanitary circumstances of the two divisions of the district being practically identical, there would be no difficulty in the supervision of the whole district as an administrative entity. In regard of topography and means of transit, there would be no inconvenience in one Medical Officer of Health undertaking such supervision. As matter of fact, a single Inspector of Nuisances is appointed for the whole Rural District. It would undoubtedly conduce to more efficient sanitary administration in this district if the action of the District Council were guided, to a greater extent than it has been in the past, by expert advice. Such advice may be expected to be more consistent in character, and to carry more weight than at present, if it were to come from a single Medical Officer of Health in constant personal relation with the Council.

There is, as just stated, only one Inspector of Nuisances for the whole district, viz., *Mr. A. E. Clifford*, who receives £70 a year for his services, part of which is repaid, and resides at Bourton-on-the-Water. Though not technically trained, and having no certificate from the Sanitary Institute or other corporation, yet Mr. Clifford struck me as an active and zealous officer. His early training in the building trade has given him facilities in acquiring details of construction and the like, which have been useful to the Council, especially during the sewerage of Bourton-on-the-Water. His journal shows that he makes frequent inspection, systematic and occasional, of the district, and keeps himself well informed as to nuisances. This book is on the whole very well kept. Sometimes, however, he has omitted to note the "Date of inspection" in column 1, and the "Result of action" in column 7, of the journal. He appears, as regards cases of infectious sickness, to follow out the instructions of the Medical Officers of Health, and now and again to confer with them as to the repression of nuisances.

Before leaving the district, I conferred with the Clerk to the Council as to certain points in the sanitary administration of the district. In particular, I advised the adoption of a code of bye-laws, and the attendance of the Medical Officers of Health at the Council's meetings, in order to advise that body. In regard of the question of bye-laws, I pointed out that the Council could at once make bye-laws for scavenging, and that it was desirable that they should seek Urban powers to frame regulations for new streets and buildings, and for the control of slaughter-houses. I strongly advised the adoption of the Board's model code. With regard to the post of Medical Officer of Health I suggested to him the abvisability of determining the present dual arrangement, and of concentrating the appointment in the hands of one officer, at a more liberal salary.

No. 14.

Report on persistence of Enteric Fever in the Swinton and Pendlebury Urban District, and on the Sanitary Circumstances and Administration of the District: by Dr. Theodore Thomson.

App. A. No. 14
On Enteric Fever in, and the Sanitary Circumstances and Administration of, the Swinton and Pendlebury Urban District by Dr. Thomson.

In the spring of the present year the Board's attention was directed to a statement by the Medical Officer of Health of the Swinton and Pendlebury Urban District, in his Report on the health of Swinton and Pendlebury during 1897, that there had been unusual prevalence of enteric fever in the district in the latter part of that year. Investigation regarding recent years anterior to 1897 showed that there had been marked prevalence of enteric fever in the district in those years also. Thereupon the Board wrote to the Urban District Council directing their attention to these facts and to the existence of unsanitary conditions in their district referred to in the Medical Officer of Health's Report for 1897, more especially in association with the old privy-midden system still in vogue in Swinton and Pendlebury. The Board requested the Urban District Council to give these matters their serious consideration, and expressed a hope that the Council would take such steps as might be necessary to substitute a proper system of excrement and refuse disposal for that in use in Swinton and Pendlebury, and also that they would do their utmost to remove all unwholesome conditions tending to foster the existence of enteric fever in the district. Correspondence ensued between the Board and the Urban District Council, from which it appeared that the latter regarded the Board as taking too unfavourable a view of the health and the sanitary circumstances of the district. I was accordingly instructed by the Board to visit Swinton and Pendlebury, and to make inquiry as to prevalence of enteric fever there, and also as to the sanitary circumstances and administration of the district. This I did on several occasions in the months of June and July, ascertaining the facts now about to be set forth.

The Urban District of Swinton and Pendlebury, some four miles to the north-west of Manchester, had, at the census of 1891, a population of 21,637, with an area of 2,196 acres. Its rateable value for the present year is £96,198; and the current rates are 2*s.* 8*d.* in the £ for the general district rate and 1*s.* 10*d.* in the £ for the poor rate. The chief industries are collieries, cotton spinning, and weaving; while there are also bleach works and breweries in the district. Geologically, Swinton and Pendlebury are on the Coal Measures; which, however, are here overlaid by a considerable amount of Drift, chiefly in the form of Glacial Sand and Gravel and of River Valley Gravel. The soil on which the dwellings in the more populous parts of the district stand is locally stated to be for the most part dry sand of uncertain depth.

The condition of the public health in Swinton and Pendlebury may be gathered from the following tables (Tables A and B), in which are shown the deaths and death rates from all causes and from certain particular causes during the 10 years 1888–97; while, for purposes of comparison, like data are given for England and Wales, and also for the 33 Great Towns in Table C.

TABLE A.

Swinton and Pendlebury Urban District.

Years.	Number of Deaths from					Estimated Population.	Death-rate per 1,000 living from all causes.
	All Causes.	Seven principal Zymotics.	Enteric Fever.	Enteric *plus* "Continued" Fever.	Diarrhœa.		
1888	387	57	11	11	13	20,511	18·9
1889	397	75	6	16	13	20,880	19·0
1890	370	36	5	5	11	21,255	17·4
1891	395	87	7	7	5	21,645	18·2
1892	394	34	4	4	12	22,135	17·8
1893	426	52	8	8	26	22,615	18·8
1894	449	108	8	8	11	23,165	19·4
1895	440	85	12	12	18	23,735	18·5
1896	439	69	12	12	19	24,570	17·9
1897	452	65	12	12	23	25,230	17·9
Total	4,149	668	85	95	151	225 741	18·4

The estimates of the population of Swinton and Pendlebury during the years 1888–97 are based partly on the 1891 census returns and partly on local information derived from the rate-books. These estimated populations do not include either the Industrial Schools or the Children's Hospital, both situated within the district but mainly comprising children from without the district. Similarly, deaths in these institutions of children admitted from other districts are omitted from the Table. Deaths of persons admitted from Swinton and Pendlebury to the Monsall Fever Hospital, to the Salford Workhouse, and to the Barton-on-Irwell Workhouse, are included in the Table.

TABLE B.

—	The death-rates per 1,000 living for the whole period, 1888–97, from					The Mean Population.
	All Causes.	Seven principal Zymotics.	Enteric Fever.	Enteric *plus* "Continued" Fever.	Diarrhœa.	
Swinton and Pendlebury Urban District.	18·38	2·96	·38	·42	·67	22,574

TABLE C.

—	The death-rates per 1,000 living for the whole period, 1888-97, from				The Mean Population.
	All Causes.	Seven principal Zymotics.	"Fever."*	Diarrhœa.	
England and Wales...	18·41	2·10	·17	·63	29,576,528
The 33 Great Towns†	20·13	—	·19	·86	—

* "Fever" includes enteric fever, typhus fever, and continued fever.
† During the years 1888–91 (inclusive) there are only 28 Great Towns.

From these Tables it appears that the death-rates in Swinton and Pendlebury from all causes, and from diarrhœa, are about the same as the corresponding rates in England and Wales as a whole. The death-rate, however, from the seven principal zymotic diseases is materially higher in Swinton and Pendlebury than in England and Wales; while the death-rate from enteric *plus* continued fever in the former is more than double the "fever" rate in the 33 great towns, and two-and-a-half times that of England and Wales. The undue prevalence of enteric fever in Swinton and Pendlebury, already indicated by the death-rate from this cause, is further illustrated in the following Table. This gives the number of *cases* of enteric fever notified to the Local Authority as having occurred in their district during the years 1890-97, together with the annual attack rates from this cause during that period. In a subsidiary table are added the attack rates per 1,000 persons living in certain towns in England and Wales during the five years 1893-97. The towns selected are those in which the Infectious Disease (Notification) Act, 1889, is in force; and the attack rate in each instance is based on the number of cases of enteric fever notified under that Act to the Local Authority.

TABLE D.

Showing the number of cases of enteric fever notified, and the annual attack rate from this cause per 1,000 persons, in Swinton and Pendlebury during the period 1890-97; as also the annual attack rates from enteric fever per 1,000 persons in certain towns during the five years 1893-97.

—	Number of cases of Enteric Fever notified in period 1890-97.	Annual Attack Rate per 1,000 from Enteric Fever in period 1890-97.
Swinton and Pendlebury	393	2·13

TOWNS.	Annual Attack Rates per 1,000 from Enteric Fever during the five years 1893-97.	TOWNS.	Annual Attack Rates per 1,000 from Enteric Fever during the five years 1893-97.
Aston Manor	·94	London	·77
Barnsley	3·03	Macclesfield	·67
Barrow in-Furness ...	·81	Merthyr Tydfil ...	2·21
Bath	·32	Middlesbrough ...	2·41
Birmingham	1·00	Newport (Mon.) ...	·98
Bournemouth	·19	Northampton	·41
Bradford...	·73	Oxford	·47
Bristol and St. George ...	·62	Plymouth	·45
Burton-on-Trent ...	1·07	Salford	1·80
Cambridge	1·15	St. Helens (Lancs.) ...	2·68
Cardiff	·56	Sheffield	1·54
Carlisle	·42	Smethwick	·77
Colchester	·53	Southampton	·95
Coventry	·65	Southport	·81
Darlington	1·61	South Shields	1·25
Devonport	·64	Stockport	1·37
Dover	·36	Tottenham	1·52
Eastbourne	·69	Tynemouth	·93
Exeter	1·65	Walsall	1·01
Gloucester	·72	Walthamstow	1·59
Grimsby	2·96	West Bromwich ...	1·42
Hornsey	·60	West Hartlepool ...	·95
Hull	1·60	Wolverhampton ...	1·44
*Leeds	1·09	Worcester	·76
Leyton	1·29		
Liverpool	2·17	Fifty towns ...	1·03

* Leeds from 1895-97 only.

From the figures in Table D it appears that, of the 50 towns included in the table, only six exhibit a higher attack rate from enteric fever than the Swinton and Pendlebury District: and that the mean attack rate from enteric fever in these towns during the five years 1893-97 was 1·03 per thousand living, while that of Swinton and Pendlebury during the eight years 1890-97 was 2·13—more than double the rate of these towns.

GENERAL SANITARY CONDITION.

Dwellings.—For the most part houses in the Swinton and Pendlebury Urban District are inhabited by persons of the working class. They are usually two-storied brick buildings with two rooms on each floor. In the main they are in fair condition: although there would seem to have been in the past a good deal of "jerry" building in the district, with result that not a few houses of apparently good exterior are damp, and also admit rain through defective roofs. Houses of this description, it may be anticipated, will rapidly deteriorate; and will be fruitful sources of defective sanitary conditions in the near future. There are also houses in the district which are old and dilapidated: and these are usually damp and badly lighted. Of crowding of dwellings upon insufficient area there is but little, although occasional instances of this sort came under my observation.

5, *Bold Street:* an old and dilapidated building. Is very damp: and, according to the tenant, rain comes in through the roof freely.

17, *Melbourn Street:* a two-roomed dwelling. The ground floor room is 12 ft. square and 7 ft. 3 in. high. Is very damp: and the tenant states that rain comes through the roof.

144, *Bolton Road:* a house with six rooms and a scullery. The gable end wall is damp; rain comes in through roof into first floor front room, and also into a back room situated over the scullery. The latter room is so damp as to be uninhabitable, and the tenant does not use it.

24, *Cobden Street:* is damp. According to the tenant rain comes through the roof, notwithstanding that it has been repaired more than once.

12, *Gate Street:* is damp. Tenant states that, in winter time, water can be "scraped off the wall."

40 *and* 42, *Partington Lane:* both houses are old, dilapidated, damp, and badly lighted.

Alice Street and Ellen Street: crowding of houses upon insufficient area. These two streets are parallel to one another: and the back walls of eight houses in the one street are distant only 11 feet from the back walls of eight houses in the other. The narrow space thus provided between the two streets is further diminished by five blocks of privy middens.

House and Court Yards.—The condition of house and court yards leaves much to be desired. In a considerable number of instances, indeed, house and court yards are well paved and clean. But many are paved with cobbles, which allow accumulation of filth between their interstices. As regards not a few, paved with bricks or with stone flags, the paving material has become broken and uneven. Yet other yards are unpaved save for a narrow strip of cobbles or other material skirting the house wall. Some yards are entirely unpaved. The channelling leading to the yard gully is frequently uneven, with the result that slop water tends to stagnate in the channel instead of flowing freely to the yard gully. Many yards are in a very dirty condition owing to vegetable and other objectionable refuse cast on the yard surface: and in not a few instances, also, owing to the keeping of fowls, which are allowed to roam about the yard, which they foul with their droppings.

1–33, *Worsley's Buildings:* a large yard common to the occupants of Nos. 1–33. Is unpaved, save for a strip of brick paving, 3 feet in width, which skirts the houses. This strip is old, dilapidated, and uneven. Uneven channelling along the centre of the yard, containing stagnant pools of slop-water. Fowls kept. The yard surface is littered with ashes, vegetable refuse, rags, scraps of paper, and fowl droppings.

474–482, *Bolton Road:* a large yard partly paved with cobbles. This paving is very uneven, and allows slop-water to remain on the yard surface. The yard is littered with ashes, other refuse, and fowl droppings.

36–38, *Worsley Street:* the yards of these houses are paved with brick which has become uneven and broken, allowing slop-water to form pools on the yard surface.

1–13, *Bridge Street:* the yard common to these houses is unpaved save for a narrow strip of cobbles skirting the houses. The yard surface has a downward slope away from the houses, and slop-water runs over this surface, which is unpaved for the most part. The yard is littered with ashes, other refuse, and fowl droppings.

50–62, *Station Road:* yard unpaved save for a narrow strip of flagstones. Ashes, bricks, cans, bones, vegetable refuse, scraps of paper, littering the surface of the yard; which is also fouled by fowl droppings.

15, *Arthur Street:* yard unpaved. Tenant states that it becomes very dirty in wet weather.

323–337, *Chorley Road:* the yard is paved, in the vicinity of the houses, with cobbles and bricks in a very dilapidated state. There are pools of slop-water on the yard surface.

59, *Eaton Street:* the yard is paved with flagstones, bricks, and cobbles, but the paving is broken and uneven.

Roadways and Passages.—Most of the streets in Swinton and Pendlebury are paved: the material employed for the greater number being, as regards the roadways, either granite setts or gritstone setts, while the footpaths are generally paved with flagstones. Cobbles, however, still form the paving material in some instances; while not a few of the streets paved with setts are in bad repair. Other streets are entirely unpaved, and in wet weather some of them become, it is said, unfit for traffic. There are also, in this district, many passages serving as back streets These are usually about 12 feet in width, and are for the most part unpaved. They become very dirty in wet weather. As regard cleanness, the streets leave something to be desired; dried horse manure, dust, scraps of paper, and the like may at times be observed scattered on their surface and blowing about with every gust of wind. The passages that serve as back streets are not infrequently littered with ashes, vegetable refuse, and other objectionable matters.

Partington Lane: in the greater part of its length the roadway of this street is paved with setts in one half and with cobbles in the other. The paving is uneven in many places. The footways are paved with flagstones in some parts, with cobbles in others; while in yet other parts there is no paving save the kerbstone. The channelling is commonly rough and uneven. There are accumulations of horse refuse, straws, and scraps of paper in the gutters.

Birkdale Road: the roadway is unpaved, very uneven, and deeply channelled by rain-storms.

Back passage between Stafford Road and Pendlebury Road: is about 12 feet wide, unpaved, surface uneven, channelled by rain-storms, and littered with ashes and other refuse.

Disposal and Removal of Excrement and House-refuse.—The chief method of disposal of excrement and house-refuse is the privy-midden system. There are also some water-closets, with dry ash-pits; but these are few in comparison with the privy-middens. The middens are usually capacious structures, capable of containing the excrement and refuse of three or four dwellings contributed during several months. The older middens are constructed of bricks set in cement with a floor of flagstones or of bricks; the newer middens have cement-concrete floors and a layer of cement lining the interior of their walls. The floor of the midden is below the ground-level, save in those that have been constructed during the last two or three years, and not infrequently liquid may be observed standing in them. Nearly half the middens in the district are drained into the sewers. Some of them are roofed over, others are uncovered: probably the latter are the more numerous. Into these receptacles are cast ashes, vegetable refuse of all sorts, and not uncommonly animal refuse also. They are usually offensive nuisances. Dry ash-pits, where they exist, are usually structures capable of containing several months' accumulations of house refuse. They are not infrequently without roof: while the floor of some is below the ground level. Middens, as well as ashpits, are cleansed by contractors employed for this purpose by the Urban District Council. For the most part these receptacles are cleansed once every two or three months: although, in many instances, a longer period than this elapses between successive cleansings. The contents of middens are disposed of to farmers: while dry refuse is taken to a "tip" near the northern boundary of the district and there deposited.

1–7, *Moor Street:* two privy middens. One of these measures 8 ft. by 3 ft.; its floor is 2 ft. below ground level. It is uncovered and smells offensively. It is stated to be cleansed twice a year.

178–184, *Worsley Road:* midden measures 22 ft. by 5 ft.; its floor is 1 ft. below ground level. It is roofed over. Is very offensive. Has not been cleansed, it is said, for a year.

121–123, *Partington Lane:* midden is uncovered, and measures 5 ft. by 5 ft.; its floor is 1½ ft. below ground level. It smells offensively.

Police Station, Arthur Street: covered midden, measuring 3 ft. by 5 ft.; its floor is on the ground level. Is very offensive.

54, *Wellington Road:* covered midden, measuring 3 ft. by 4½ ft., with floor 6 in. below the ground-level. Smells very offensively. The tenant of the dwelling states that he uses "disinfectant" with a view to lessening the smell. He also complains that the midden is sometimes allowed to become too full before it is cleansed.

Yard at Corner of Grosvenor Street and Little Cross Lane: 11 houses in yard, with common privy midden. The midden measures 30 ft. by 2¾ ft., and its floor is 6 in. below ground level. It has no roof: and liquid stands in its bottom. Is extremely offensive.

7, *Whitley Street:* uncovered and very offensive midden measuring 6 ft. by 2½ ft., with floor 6 in. below ground level. This midden is within 4 ft. of the back-door and window of the house.

Potter's Square: uncovered, very offensive midden, measuring 12 ft. by 5 ft., with floor 1½ feet below ground level.

48–50, *Stafford Road:* uncovered midden, measuring 5¼ ft. by 3½ ft., with floor 9 in. below ground level. Is full; and very offensive. Its liquid contents are leaking through the midden wall on to the surface of the passage on which it abuts.

House-drainage.—House drains are, for the most part, earthenware pipes jointed with clay or with cement. Prior to 1895 cement was seldom used for this purpose. House drains are seldom disconnected from the sewers to which they discharge.

The interiors of dwellings are usually disconnected from the drains serving them; although, in those instances where there is an indoor water-closet, the soil-pipe is seldom ventilated by a full-bore upcast shaft. In the course of my inspection I observed instances of nuisance arising from defective conditions of house-drainage.

3, *Partington Lane:* house-drain blocked; tenant complains of offensive smells proceeding from it.

179–189, *Worsley Road:* complaint made of offensive smells from a yard gully, to which the privy-midden serving No. 189 is said to be drained.

325, *Chorley Road:* tenant complains that during heavy rains the drain of the yard-gully opposite her door "backs up," and, in consequence, sewage makes its way into her house, through which it flows, by reason of the slope of the floors, from the back-room to the front-room and thence into the street.

Sewerage and Sewage Disposal.—With the exception of two small groups of dwellings, one situated toward the western and the other towards the eastern extremity of the district, all houses in those parts of Swinton and Pendlebury that are not of a rural character, drain to sewers. For the most part the sewers are earthenware pipes, jointed in some instances with clay, in other instances with cement. There is no provision for ventilation of sewers otherwise than by perforated manhole covers. Many manhole covers, however, are not perforated; while the holes in those that are perforated are usually blocked by refuse from the street surface. The flushing of sewers is effected by the introduction of a stream of water from a 4-inch hose pipe by way of a manhole; the water being retained at the point of introduction by means of a lowered penstock until sufficient head has been attained, whereupon the penstock is raised and the accumulated water flushes the length of sewer below. There are two systems of sewers in this district; of which one serves a population of about 7,000, and conveys the sewage to the Pendlebury sewage works, while the other serves a population of about 18,000 and conveys the sewage to the Swinton works. The two systems are not connected with one another; and the Swinton Sewage Outfall Works are on the south side of the district, while the Pendlebury Sewage Outfall Works are on the north side.

The Pendlebury Sewage Outfall Works deal with an estimated dry-weather flow of 150,000 gallons. They comprise 16 acres of land, of which 1¾ acres belong to the Urban District Council, while other 14¼ acres are rented for the purposes of these works. All sewage brought to these works goes through the following stages of treatment:—

(1.) Straining off the grosser solids;

(2.) Treatment by the alumino-ferric process;

(3.) Either (*a*) treatment by polarite filter beds, or (*b*) treatment by being passed through "cinder and earth" beds.

These latter consist of two beds of boiler-cinders one foot in depth, beneath which are 3 ft. 6 in. of soil. The beds cover a total area of ⅝ acre, and are under-drained. The polarite filter-beds are four in number, with a total area of 150 square yards. The effluent from these works is discharged into the river Irwell.

The Swinton Sewage Outfall Works comprise 37 acres of land, of which 32 are the property of the Urban District Council, while the additional five acres are rented. The daily dry-weather flow of sewage to these works is estimated at 500,000 gallons. This sewage is dealt with as follows:—

(1.) Straining off the grosser solids;

(2.) Treatment with lime and copperas;

(3.) Either (*a*) treatment by intermittent downward filtration through land; or (*b*) treatment by means of coke beds; or (*c*) treatment by means of cinder-beds; or (*d*) treatment by means of "cinder and earth" beds.

As regards (*a*) it may be noted that only 10½ acres of land regarded as suitable for sewage disposal, are at a sufficiently low level to permit sewage being passed over them. These 10½ acres are under-drained at a depth of 3 ft. to 3½ ft.; the soil is clay. As regards (*b*), these beds are four in number with a total area of 500 square yards, and consist of a layer of broken coke, 4 ft. in depth, As regards (*c*), these beds are two in number, with a surface area of 1,320 square yards each, and consist of a layer of cinders, 3 ft. 6 in. in depth. As regards (*d*), these beds are four in number, with a total area of 1¾ acres, and consist of a layer of cinders on top, about 1 ft. in depth, beneath which come nearly 3 ft. of soil. The coke-beds, the cinder-beds, and the "cinder and earth" beds are all under-drained. The effluent from the Swinton Sewage Outfall Works is discharged into a neighbouring stream, known as the Folly Brook, and not obviously liable to other pollution. I examined samples of the effluent flowing from the cinder-beds, as also from the "cinder and earth" beds. They were opalescent and had very little smell. At the point where the effluent from these beds discharged into the brook, the bed of the stream was tolerably clean and patches of green weed were noticeable here and there. On following the downward course of the stream, however, increasing evidences of pollution appeared: the green weed disappeared and deposits of black mud became more and more noticeable in the bed and on the banks of the brook. About a quarter of a mile below the highest point at which the sewage effluent passes into the brook, its waters were in a foul condition.

Water Supply.—The water supplying the whole district is obtained from the mains of the Manchester Corporation, and is derived from Thirlmere. Only one house in the district is known to be supplied from a local well.

Slaughterhouses: Dairies and Cowsheds: Bakehouses.—There are 21 slaughterhouses, 29 cowsheds, and 10 milkshops in the district. The slaughterhouses seen by me were, for the most part, fairly good as regards their structural conditions; in the matter of cleanness, however, some of them left a good deal to be desired. The structural conditions of the cowsheds that came under my notice fell short of what is desirable in several respects. Defective paving of their floors and inadequate means of ventilation were the shortcomings mainly noted by me. The bakehouses inspected by me were satisfactory as regards construction and cleanness.

Sanitary Administration.

Swinton and Pendlebury is an Urban District, with a Council of twelve members, meeting once a month.

Adoptive Acts, Bye-laws and Regulations in force in the District.—The Public Health Acts Amendment Act has been in force in the district since January 8th, 1891; and the Infectious Disease (Notification) Act, 1889, has been in force in the district since December 3, 1889. The following list shows the matters concerning which bye-laws have been made, and also the dates on which these several sets of bye-laws were allowed:—

Prevention of Nuisances and Removal of Refuse: January 11th, 1869.
Regulation of Slaughterhouses: January 11th, 1869.
Regulation of Common Lodging-houses: January 11th, 1869.
Regulation of Markets: May 26th, 1877.
Keeping of Animals: May 13th, 1885.
New Streets and Buildings: December 5th, 1894.
Paving of Yards: December 10th, 1897.

Regulations regarding Dairies, Cowsheds and Milkshops have been in force in the district since May 26th, 1887.

The three series of bye-laws approved in 1869 are, in some respects, not in accordance with modern requirements. As regards common lodging-houses, however, it should be noted that there are no longer any of these in the district. The bye-laws for New Streets and Buildings follow, in a general way, the Board's Model Series on this subject, but with an important difference as regards privies and ashpits, the maximum cubic capacity of these receptacles being put at 20 cubic ft. instead of 8 cubic ft., as almost invariably recommended by the Board. The bye-laws concerning paving of yards and open spaces in connection with dwelling-houses are under Section 23 of the Public Health Acts Amendment Act, 1890. Owing to the recent date of their approval, the powers obtained under these bye-laws have not as yet been much utilised.

No action has been taken by the Urban District Council under the Housing of the Working Classes Act, 1890.

Scavenging.—The cleansing of "earth-closets, pails, privies, ashpits, and cesspools" within the district is undertaken by the Urban District Council, who employ two contractors to carry out this business. Each of these two contractors has an area allotted to him for this purpose: one contractor receiving £410 per year, the other £470 per year for the work. Each contractor is bound to cleanse the receptacles above specified within three days of notice to do so given him by the Inspector of Nuisances, and within twenty-four hours when there is infectious disease in the dwelling. The usual method of cleansing ashpits and middens is to shovel out their contents on to the surface of a back passage, whence they are transferred to night-soil carts and removed. When there is no back passage the midden contents are conveyed in wheelbarrows to the street, on the surface of which they are deposited to await removal. The contractor undertakes to "disinfect all earth-closets, pails, privies, ashpits, and cesspools at the time of being emptied," and to "sprinkle disinfectants along any court, passage, or street in which the night soil may have been deposited for the purpose of removal," the Urban District Council supplying him with disinfectants. These receptacles are to be cleansed during the day time from September to May inclusive, and during the night time in June, July, and August.

Hospital Accommodation for cases of Infectious Disease: and Disinfection of Infected Articles.—There is no hospital for infectious diseases within the district. Cases of small-pox from Swinton and Pendlebury are admitted by the Manchester City Council to their hospital at Monsall; payment for this accommodation being made to the latter body by the Urban District Council under agreement. The agreement expires on June 24, 1899. Cases of infectious disease, other than small-pox, notifiable under the Infectious Disease (Notification) Act, 1889, are admitted to the Salford Town Council's hospital at Ladywell; the Urban District Council paying the Salford Town Council for this accommodation, according to a scale fixed by agreement. This agreement has been made for a period of ten years dating from May of the present year. Prior to that time cases of fever not treated at their homes were for the most part sent to Monsall Hospital. The Urban District Council have no disinfecting apparatus. The means of disinfection resorted to are fumigation of infected rooms and articles with sulphur; with subsequent washing of these, where this measure is applicable. In some cases, infected articles are disinfected by the Salford Town Council at a fixed scale of charges. Articles infected by small-pox are destroyed.

Sanitary Staff.—The sanitary staff for the district consists of a Medical Officer of Health, who is engaged in private practice; and of an Inspector of Nuisances, who is also Inspector of Markets. The Inspector of Nuisances receives occasional assistance in the disinfection of houses invaded by infectious disease. This

App. A. No. 14.

On Enteric Fever in, and the Sanitary Circumstances and Administration of, the Swinton and Pendlebury Urban District; by Dr. Thomson.

assistance is afforded by a man who is ordinarily employed on the district roads: he receives no special emoluments for help rendered to the Inspector of Nuisances.

The Medical Officer of Health is Samuel Hosegood, M.R.C.S., L.S.A., who receives a salary of £90 per year, half repaid from county funds. The Inspector of Nuisances is Albert Bleakley, who receives £110 per year, half repaid from county funds. He also receives £10 per year as Inspector of Markets. In addition to the ordinary duties of Inspector of Nuisances, he has to see that the undertakings of the contractors to cleanse ashpits, privies, &c., are properly performed. He also keeps the Health Office books. He is a careful and painstaking officer; but the duties devolving upon him cannot adequately be discharged by a single Inspector of Nuisances, provided with but trifling assistance, in a district of the size and character of Swinton and Pendlebury.

Nuisances.—Nuisances abound in the district. Mainly these arise in connection with the offensive privy-middens that form the chief method of excrement-disposal in the district. But unpaved and defectively paved yards and passages, accumulations of refuse in yards, defective sink-pipes and rainfall pipes causing dampness of house walls, defective drains, and other unsanitary conditions are also fruitful sources of nuisance. The annexed returns made by the Inspector of Nuisances in regard of the two years 1896 and 1897 afford some indication of the condition of the district as regards nuisances.

Summary of Sanitary Work attended to during the year 1896:—

NUISANCES AND REPAIRS, &c., NECESSARY TO PROPERTY.

Defective soil pipes and water-closet arrangements...	11
Dilapidated state of closet and ashpit walls and wet ashpits	44
Defective sinkstone pipes, and paving around gullies and the pipes being directly connected with the drains	17
Blocked gullies and drains, and defective condition of same	58
Defective structural condition of houses—chimney flues, spouting, roofings and ceilings, cellar floors	15
Allowing waste water to lodge on surface of yards, streets, and passages	21
Dirty state of houses, premises, and yards... ...	10
Accumulation of, and deposit of, refuse, vegetables, and decayed fish, &c., in prohibited places ...	28
Overcrowding	5
Defective state of urinals and manure receptacles ...	12
Unsound food offered for sale (onions)	1
Inefficient emptying of ashpits and clearing afterwards	15
Firing of house chimneys	23
Keeping poultry in house (ducks)	1
Keeping pigs in contravention of bye-laws ...	4

Summary of Sanitary Work attended to during the year 1897 :—

NUISANCES AND REPAIRS, &C., NECESSARY TO PROPERTY.

Closets and ashpit walls requiring repairs... ...	95
Wet ashpits, "owing chiefly to blocked drains and tenants throwing waste water into covered ashpits"	88
Depositing ashpit refuse in prohibited places ...	11
Ashpits without doors	63
Broken and short slopstone pipes, "allowing waste water to drip down housewall"	407
Defective pavement around gullies in yards, "allowing waste water to lodge"	366
Untrapped drains. "Slopstone pipes directly connected with drains and untrapped grids in yards"	47
Blocked gullies. "Chiefly tenants' fault for not clearing same"	29
Defective drains and broken gullies	42
Dirty houses and accumulation of rubbish, &c., in yards	14
Defective structural condition of houses : "Walls, roofs, and spouting"	32
Inefficient draining of yards and passages... ...	6
Firing of house chimneys	8
Defective manure receptacles	5
Overcrowding	2

It will be observed that the nuisances "attended to" in 1897 far out-number those attended to in 1896. This is, in part at least, to be attributed to an investigation made by a Committee of the Urban District Council appointed to conduct a "street-to-street visitation" of Swinton and Pendlebury.

In searching for an explanation of the persistence of enteric fever in Swinton and Pendlebury, no evidence was forthcoming which would justify its being referred either to the water supply or to the sewerage of the district. For the water supply, as stated, is derived from the Manchester mains, and is supplied to Manchester and to other districts which are not marked by a like persistence of fever. The sewerage of the Swinton and Pendlebury district comprises two entirely separate systems; and there was no evidence of heavier incidence of the fever on dwellings on one system as compared with dwellings on the other.

But although broad contrasts of this kind did not appear, there were indications that some parts of the district had suffered more from enteric fever than others. The data available in this regard are, unfortunately, of very limited amount, extending as they do over a period of three years only; and a corresponding limitation therefore attaches to their value as bases of induction. They may, however, probably be taken as affording some ground for suggestion as to the conditions mainly responsible for the fever. These data are set out in the following table (Table E.), in which are given the death rates and attack rates from enteric fever per 1,000 living in each of the four wards which constitute the Swinton and Pendlebury Urban District. As matter of interest, death rates from all causes, as well as from certain other causes, in these wards are also given.

TABLE E.

Showing for each of the Four Wards of the Swinton and Pendlebury Urban District the mean Annual Death Rates per 1,000 living from all causes, as well as from certain particular causes, during the Three Years 1895-97; and also the mean Annual Enteric Fever Attack Rates per 1,000 living in each Ward during the same period.

Ward.	Area in Acres.	Estimated Population.	Mean Annual Death Rate per 1,000 living from				Mean Annual Attack Rate per 1,000 living from Enteric Fever.
			All Causes.	Seven Principal Zymotics.	Diarrhœa.	Enteric Fever.	
North ...	265	6,960	17·2	3·6	1·4	0·57	2·4
South ...	975	7,350	12·3	1·8	0·41	0·13	1·2
West ...	236	6,620	20·2	3·9	1·1	0·60	3·3
East ...	746	4,230	26·1	1·9	0·07	0·47	3·1

The most striking feature in the above table is the comparatively small incidence of enteric fever, both as regards attacks and deaths, upon the South Ward. This ward, which has also the lowest death-rate from all causes, contains a materially smaller proportion of dwellings of the poorer classes than any of the other three wards. Pendlebury West, much of Pendlebury North, and the populous part of Pendlebury East, on the other hand, mainly comprise dwellings inhabited by the working classes and by the poorest classes; and it will be observed that these three wards have suffered heavily from enteric fever.

So far as these figures afford guidance they tend to suggest that the fever has been associated with neighbourhoods inhabited by the working and the poorer classes; and to this extent they suggest also association of the disease with the filthy conditions too often found in connection with the surroundings of this class of dwelling in town districts. That these filthy conditions do abound in Swinton and Pendlebury appears with sufficient clearness from the account that has been given of the general sanitary circumstances of the district. These conditions comprise: Unpaved and badly paved yards, allowing refuse, both liquid and solid, to pollute the ground; defective house and yard drains, bringing about a like result; and, above all, large privy middens containing the accumulated excreta of months, much of the more liquid part of which soaks into and fouls the neighbouring soil. In addition to the fouling of the soil thus brought about by these middens, pollution of the ground-surface takes place each time that they are cleansed, by reason of their contents being cast upon passages and streets and left there pending removal of these matters by the night-soil cart. In these various ways the soil on which much of Swinton and Pendlebury stands is liable to serious fouling; and in view of the suitability of soils thus befouled as a medium for the growth and multiplication of the bacillus of enteric fever, it may well be that here is to be found an explanation of the persistence of that fever in the district.

But, however this may be, the old privy middens that abound in the more populous parts of Swinton and Pendlebury are fertile sources of offensive nuisance; and for this reason the Urban District Council should take steps to procure their abolition and to substitute in their stead a proper system of excrement and refuse disposal.

There are other matters that also require the attention of the Council. Prominent among these is the proper paving of house-yards and streets. As regards paving of house-yards the Council have recently acquired bye-law powers, as already noted, of which they have been making some use and will, it is to be trusted, make further use. As regards paving of streets the Council have endeavoured to obtain the Board's sanction of a loan for this purpose; but sanction was refused on the ground that some of the works proposed would involve the construction of sewers that would convey sewage to outfall works not in a condition to receive a further volume of sewage. The present methods of disposal of sewage, whether at the Swinton outfall works or at the Pendlebury outfall works, are unsatisfactory; and this question should have the Council's immediate and careful attention, as also should the need for efficient ventilation of the sewers throughout the district. Dwellings and house drainage also require constant and careful supervision; and regular visitation of all parts of the district is necessary for the detection and remedy of nuisances. For this purpose the Council should add to their sanitary staff, which is at present inadequate to the proper supervision and control of the sanitary circumstances of the district.

The Council should also make new bye-laws in place of such of those now in force in the district as are no longer in accordance with modern requirements.

No. [illegible]

REPORT upon MEASLES in the BOROUGH of BURTON-ON-TRENT; by DR. THEODORE THOMSON.

[illegible] Burton-on-Trent; by Dr. Thomson.

Towards the end of February, 1898, Dr. Robinson, the Medical Officer of Health for the Borough of Burton-on-Trent, reported the occurrence of an outbreak of measles in that town. From subsequent reports made periodically by Dr. Robinson, it appeared that the disease continued to prevail in Burton. I was accordingly instructed by the Board, in the latter part of May, to visit the locality and to make inquiry as to the circumstances associated with this continued prevalence of measles. To this end I visited Burton on June 1st, and on subsequent occasions, making inquiry in the sense indicated. The facts ascertained by me in this connection are as follows:—

The Borough of Burton-on-Trent had, at the census of 1891, a population of [illegible], on an area of [illegible] acres. This population is estimated to have increased to [illegible] by the middle of 1898. The greater part of the town stands on flat ground to the west of the Trent; the remainder is on rising ground skirting the eastern banks of the river. The population is mainly of the working class, employed in connection with the breweries that form the staple industry of the town. Most dwellings are of the artisan class, and are usually two-storied buildings, with two rooms on the ground floor and two rooms on the floor above. Save in a few instances in the older part of the town, there is no overcrowding of dwellings upon area. The prevailing method of excrement disposal is the pail system; but there is still a considerable number of privy-middens, while there are also a good many water-closets. Nearly all parts of the town are drained to sewers. There is a public water supply furnished by the South Staffordshire Waterworks Company, and said to be derived from the New Red Sandstone.

The recent history of Burton as regards measles may be gathered from the following data, furnished by Mr. Perks, the former Medical Officer of Health for Burton, in a report made by him in 1897. He states in that report that in Burton the mean annual death rate per 1,000 of the population from this cause was 0·29 during the quinquennium 1887–91; and rose to 0·74 per 1,000 in the following quinquennium, 1892–96. This increase was in considerable part due to the occurrence of a severe epidemic of the disease in 1896, which gave rise to a mortality from this cause of 1·7 per 1,000 in that year alone. In 1897 there were 274 known cases of measles in the district, while seven deaths were referred to this cause. These cases occurred mainly in the first half of the year, and were due to the continuance of the 1896 epidemic over the early months of 1897.

The outbreak of measles which forms the subject of this Report appears to have commenced in the month of January, 1898; the few cases occurring in this month being scattered somewhat widely apart over the town. In the course of the four weeks ending January 29th, five cases of the disease in Burton came to the knowledge of the Local Authority. By the latter part of February cases had become numerous, and in early March as many as 118 persons were notified to the Authority in the course of a single week as having measles. During April there was some decrease of prevalence of the disease, but in May and early June the number of cases reached a higher limit than had yet been attained. In the course of the week ending May 21st no fewer than 288 cases came to the knowledge of the Local Authority. Towards the end of June the outbreak began to wane, and continued to abate throughout the following month of July; until, in mid-August, the disease had ceased to be regarded as epidemic in the district. The course of the epidemic, from week to week, will be gathered from the following Table (Table A):—

APP. A. No. 15.

On Measles in the Borough of Burton-on-Trent: by Dr. Thomson.

TABLE A.

Showing, week by week, the number of cases of Measles in Burton that came to the knowledge of the Local Authority during the period January 2—August 13, 1898.

Week ending	Cases of Measles.	Week ending	Cases of Measles.
January 8	1	April 30	33
" 15	—	May 7	94
" 22	2	" 14	92
" 29	2	" 21	288
February 5	5	" 28	152
" 12	6	June 4	209
" 19	12	" 11	109
" 26	39	" 18	91
March 5	25	" 25	65
" 12	118	July 2	57
" 19	37	" 9	36
" 26	115	" 16	35
April 2	60	" 23	28
" 9	113	" 30	17
" 16	86	August 6	10
" 23	75	" 13	3
		January 2—August 13	2,015

From the foregoing Table it will be observed that during the period January 2–August 13 the total number of cases that came to the knowledge of the Local Authority was 2,015. Of these 2,015 persons attacked, 29 died, giving the low fatality rate of 14 per 1,000.

In the following Table (Table B) the incidence of attacks as well as deaths on certain age-groups is shown; and also the fatality rates as regards those age-groups.

App. A. No. 15.

On Measles in the Borough of Burton-on-Trent; by Dr. Thomson.

TABLE B.

Showing the number of persons known to have been attacked by Measles at each of several age-groups, and also, in each instance, the number of deaths referred to this cause; with the attack rate and the death rate from this cause in each group per 1,000 living, and the fatality rate per 1,000 persons attacked by Measles.

Age-groups.	Estimated Population.	Attacked by Measles.	Deaths from Measles.	Measles Attack rate per 1,000 living.	Measles Death rate per 1,000 living.	Measles Fatality rate per 1,000 attacked.
At all ages ...	51,664	2,015	29	39	0·56	14
0—1	1,430	148	6	103	4·2	40
1—2	1,425	217	11	152	7·7	50
2—3	1,495	235	9	157	6·0	38
3—4	1,356	309	1	228	0·7	3
4—5	1,445	315	1	218	0·7	3
Under 5	7,151	1,224	28	171	3·9	23
5—10	7,150	680	1	95	0·1	1
10 and upwards...	37,363	111	—	3	—	—

From this table it appears that as regards measles the chief incidence of *attacks* was on children in the 2nd, 3rd, 4th, and 5th years of life, and more especially on children in their 4th and 5th years; while the chief incidence of *deaths* was on children in the 1st, 2nd, and 3rd years, and more especially on those in their 2nd year. Thus the fatality rate was much higher on children in the 1st, 2nd, and 3rd years than on children in the 4th and later years of life, and was highest in the case of children in the 2nd year of life. Much the same facts appeared from certain data on this subject furnished by me in my Report on Measles in England and Wales.

ACTION OF THE LOCAL AUTHORITY.

1. *Measures adopted with a view to obtaining information as regards occurrence of Measles in the district.*

(*a.*) *Compulsory Notification.*—In December, 1893, the Burton Town Council scheduled measles as one of the diseases to be notified under the Infectious Disease (Notification) Act, 1889, which had already been for some time in force in the district. Measles thus became, and still is, compulsorily notifiable in Burton. The form of notification under the Infectious Disease (Notification) Act is dual: the duty of notification, that is to say, devolving upon the householder and upon the medical attendant. In Burton, however, only a very small proportion of cases of measles is notified by the householder. But according to the Medical Officer of Health there are few cases of measles which do not receive medical attention; so that, in his belief, there are few cases of measles occurring in the district that do not come to the knowledge of the Authority by means of notification on the part of the medical attendant. Inquiry is made by a visiting officer of the Local Authority as to the probable source of infection of cases notified as suffering with measles; and, with a view to detecting unrecognised sources of infection, house to house visitation is made in the invaded neighbourhood. This customary procedure, however, was not found practicable during the height of the epidemic by reason of inadequacy of sanitary staff for this purpose. Effort was from the first made to induce householders to be on the alert to detect symptoms of the disease and to notify it to the Local Authority, by placarding neighbourhoods known to be invaded by measles with bills briefly setting out the main symptoms of the malady, and calling the attention of the householder to his duty in regard of notification. Handbills of a like tenour were also distributed from house to house in invaded neighbourhoods.

(b.) Information from School Authorities.—Information as to children absent from school on account of illness suspected to be measles is not systematically furnished to the Local Authority by the Authorities of either public elementary, private, or Sunday schools. Occasional information has, however, been derived from these sources. This was the case in the early stages of the outbreak of measles dealt with in this Report, the Medical Officer of Health having made inquiries of School Authorities as to whether there were any cases of suspicious illness among children attending schools in the invaded neighbourhoods.

2. *Measures adopted with a view to preventing spread of Measles in invaded dwellings.*

Save during the height of the epidemic, when the Sanitary Staff proved inadequate for the purpose, each household notified to the Local Authority as invaded by measles was visited by an officer of the Health Department on the same day as that on which the notification was received. This officer gave verbal instructions with a view to securing within the dwelling as great a measure of isolation of the sick from the healthy as was practicable. He also left with the householder a handbill containing advice on this subject, and conveying a warning as to the penalties liable to be incurred by the exposure of infected persons or articles. Subsequent visits were made in newly invaded neighbourhoods, with a view to securing observance of these instructions as to isolation of the sick. But when the outbreak had passed beyond this early stage, second visits were rarely made to an invaded house, save on the occurrence of an additional case in the family. Burton is provided with a hospital for cases of infectious disease, but the accommodation is regarded as insufficient to allow the reception there of cases of measles, and, accordingly, no cases of measles were removed to the hospital.

In the earlier stages of the outbreak, disinfection of invaded dwellings, after recovery or death of the patient or patients, was in all instances carried out. The measures adopted were fumigation of infected rooms with sulphur, followed by washing of their floors and woodwork, and by whitewashing of the walls and ceilings. Infected articles of clothing and bedding were disinfected by high-pressure steam. But, in the latter stages of the outbreak, the amount of work cast on the officials of the health department was such that in many instances it was found impracticable to perform disinfection of invaded dwellings.

3. *Measures adopted with a view to checking extension of Measles throughout the Borough.*

(*a.*) *Information furnished to School Authorities.*—No information was furnished by the Local Authority to the Authorities of Sunday schools or of private schools as to occurrence of measles in families of which members attended these schools. The Local Authority, however, furnished information of this sort daily to the Clerk of the Burton School Board as regards families of which members attended a public elementary school.

(*b.*) *Exclusion from School of members of households invaded by Measles.*—Children in households invaded by measles were forbidden to attend any school, whether public elementary, Sunday, or private. They were prohibited from returning to school until 28 days after the occurrence of the last case in the house. In this connection it is noteworthy that in several instances children in neighbourhoods invaded by measles, although not themselves living in an invaded household, were forbidden to attend a school situated in a neighbourhood in which measles had not become prevalent.

(*c.*) *School Closure.*—In a large number of instances, public elementary schools in Burton were closed at the instance of the Local Authority, with a view to checking spread of the disease. The managers of a great many Sunday schools, and of two private schools, were advised that it was desirable that their schools should be closed for a like reason, and for the most part this advice was acted on.

(*d.*) *Other Measures.*—There is a public library in Burton, and the Local Authority furnished the librarian daily with a list of households known to be invaded by measles. The librarian thereupon ceased to give out books to members of these households until the children from these families had resumed attendance at school. On this point the statement of the householder was accepted as sufficient evidence. All library books returned from houses notified to the librarian as invaded by measles were, it is stated, sent to the Local Authority with a view to their disinfection.

Origin and Propagation of the Outbreak.

Apprehension of the manner in which the outbreak of measles dealt with in this Report was propagated in Burton will be facilitated by a description of the topography of the town.

Across Burton, from north-east to south-west, run the Trent river, the Midland Railway, and the Trent and Mersey Canal, in such fashion that to the north-west of the canal lie two groups of suburbs, while to the south-east of the river lie two other groups of suburbs. Between the river and the canal is the central and chief portion of the town, and this central portion is itself sub-divided by the Midland Railway, which runs nearly parallel to the canal on one side and the river on the other. In this way Burton is divided into north-western suburbs, south-eastern suburbs, and two central areas; one central area between the railway and the canal, which may be termed the north-west central area, and the other between the railway and the river, which may be termed the south-east central area. The central areas, more particularly the south-east central area, are yet further sub-divided into sections by great brewery buildings, with their inter-communicating lines of private railways. Between these various parts of Burton there are communicating roads; but the facilities for communication are materially lessened by the conditions described.

During the first three weeks of January two houses in the north-west central area, and one in a south-eastern suburb, are known to have been invaded by measles. As to the source of origin of the infection in these cases, no evidence is forthcoming. In the following fortnight, that ending on February 5th, seven other cases of this disease came to the knowledge of the Local Authority. These seven cases occurred in five houses, of which three were in the north-west central area, while the remaining two were in the south-eastern suburbs, one in each suburb. By February 19th the disease had markedly increased in the north-west central area; and by March 5th had become epidemic there. By the latter part of April the outbreak in this area was on the wane. Meanwhile, however, cases of measles had occurred here and there in the other parts of Burton; and during the first fortnight of May the disease suddenly assumed epidemic proportions in one of the north-western suburbs and in one of the south-eastern suburbs. In the following fortnight measles became epidemic in the other south-eastern suburb, while the cases in the south-east central area, where the disease had been persisting in some degree for several weeks, began to increase markedly. In early June measles had become epidemic in this latter part of Burton also, while in the invaded north-west suburb and in one of the invaded south-east suburbs the disease was declining. By the latter part of June measles was epidemic only in the south-east central area. In this area the disease had by mid-July begun to decline, and was, in early August, all but extinct. No further extension of measles occurred in Burton.

Measles, therefore, in epidemic proportions, did not, in this instance, display that rapid invasion of the whole district so often characteristic of this disease. Nor did the disease always, on becoming epidemic in a fresh section of the place, select the section nearest to that in which it was already most prevalent. Thus, measles assumed epidemic form in the south-east suburbs at a time when it was elsewhere epidemic only in a north-west suburb, and when it had not as yet become epidemic in the south-east central area, which immediately abuts on the south-eastern suburbs. One of the north-western suburbs escaped the disease, save for a few scattered cases. Considerable part of the south-east central area was similarly exempted : this part being separated from the rest of that area by a large extent of brewery buildings. Nevertheless, notwithstanding these two exceptions, Burton suffered heavily with measles ; more than 2,000 cases having occurred in the district, as already stated, in the eight months January to August, 1898.

As regards the manner of propagation of this epidemic, investigation led to the conclusion that schools had materially contributed thereto. As to this relationship between schools and extension of measles, and also as to what may have been the influence of closure of certain schools on the course of the epidemic, account will be found in an appendix to this Report. Communication, however, between sick and healthy at school was not the only factor concerned in extension of the disease. Attainment of a sufficient degree of isolation of the sick in their own homes was often impracticable : and even where this was attainable, the work cast upon them by an increasing epidemic soon rendered it impossible for the visiting officers of the Sanitary Authority to pay invaded households the repeated visits necessary to ensure that their instructions on this point were being carried out. It is much to be regretted that the Local Authority's hospital for infectious diseases does not provide accommodation regarded as sufficient to permit the admission of persons suffering from measles. For the application of this measure at a time when cases of measles have not yet become numerous, and while there is still hope of preventing the disease from assuming epidemic proportions, no large amount of hospital accommodation is necessary. Removal of insufficiently isolated cases of measles to hospital at a time when cases of the disease are few, may, combined with other appropriate measures, suffice to check what might otherwise prove to be a serious outbreak.

The measures adopted by the Local Authority with a view to checking extension of measles were deficient, also, in certain other respects. This is notably the case as regards information from School Authorities as to occurrence of cases of disease suspected to be measles among children attending school. Systematic information of this kind should, if possible, be obtained from all schools in the district. No arrangement of this kind, however, obtains in Burton, either with public elementary, Sunday, or private schools. Indeed, the only information obtained from School Authorities in Burton is by the Medical Officer of Health making occasional inquiries of School Authorities on this subject. Such inquiries, it may be anticipated, would be most likely to be made at a time when measles had already gained a foothold in the district ; whereas the advantage of systematic information from School Authorities is that it betters the chance of the Local Authority getting to know of those cases which occur when the disease is but little thought of, but which may serve to originate a serious outbreak. Systematic information should be furnished to all School Authorities by the Local Authority as to cases of measles in families of which children attend school.

In some respects, however, the action of the Local Authority, in its endeavour to obtain control of the measles epidemic that forms the subject of this Report, was commendable. Thus, the inquiries made at invaded houses by the Visiting Officer of the Local Authority as to the probable source of infection; house-to-house visitation of invaded neighbourhoods with the same object; the exclusion from school not only of members of invaded families, but also of members of families not so invaded but residing in an invaded neighbourhood, when the school happened to be in a neighbourhood not yet invaded; and the performance of disinfection of infected premises and articles, are all measures worthy of commendation. But even these measures ceased to be carried out during the periods of greatest prevalence of measles in the district; since the staff on whom these duties devolved proved inadequate to their discharge. The Sanitary Staff in Burton consists of a Medical Officer of Health, who gives his whole time to his duties, an Inspector of Nuisances, an Assistant-Inspector of Nuisances, and a Clerk. It is doubtful whether this staff suffices to cope with the ordinary work falling on the Health Department of a town of the size of Burton; it is quite insufficient for the added labour of measures undertaken with a view to control an epidemic of measles.

The Burton Town Council, therefore, would do well, in future attempts to control measles in their district, to provide a staff sufficient to properly carry out the necessary measures. They should also endeavour to make arrangements for the systematic exchange of information as to the occurrence of measles with the Authorities of all public elementary, Sunday, and private schools in their district. And they should take into consideration the advisability of providing hospital accommodation adequate at least for the reception of cases of measles, insufficiently isolated at their homes, during inter-epidemic periods.

APPENDIX.

The Relationship between Schools and extension of Measles in Burton; and the amount of influence of school-closure on the progress of the epidemic.

There are in Burton 18 public elementary schools, 55 Sunday schools, and two large private schools. The two private schools referred to are the Grammar School and the Girls' High School. At one or another time in the course of the epidemic of measles dealt with in this Report, both these private schools, 9 public elementary schools (in whole or in part), and 34 Sunday schools, were closed at the instance of the Local Authority, with a view to obtaining control of the disease. The public elementary schools were not in all instances entirely closed, as noted: the closure not infrequently being applied to the infant department only. The following table shows the relationship in time between closures of these public elementary and private schools and the course of the epidemic. This measure, as applied to Sunday schools, does not appear in the table. As a rule, however, the managers of Sunday schools in neighbourhoods where public elementary schools had been closed consented to close their schools also. To this rule there were, indeed, exceptions; but these exceptions were not numerous.

TABLE C.

?. A. No. 15.
Measles in Borough of ton-on- at; by Thomson.

Table showing the number of cases of measles in Burton that came to the knowledge of the Local Authority, week by week, during the period January 2nd to August 13th, 1898: together with the public elementary and private day schools closed during that period, and the time during which they were closed.

Week ending		Number of known cases of Measles.	The Public Elementary and the Private Schools closed; with the periods over which closure extended.			
January	8	1				
"	15	—				
"	22	2				
"	29	2				
February	5	5				
"	12	6				
"	19	12				
"	26	39				
March	5	25				
"	12	118	Two private schools closed March 2nd—March 28th.			
"	19	37				
"	26	115		Victoria Road School closed March 10th—April 18th.		
April	2	60			Wellington Street and Grange Street Schools closed March 30th—April 18th.	
"	9	113	All public elementary day schools closed for Easter holidays April 7th—17th.			
"	16	86			Wellington Street School (infant department) kept closed until April 23rd.	
"	23	75			Goodman Street Infant School closed April 18th—30th.	
"	30	33				
May	7	94				
"	14	92				
"	21	288	Winshill Infant School closed May 9th—June 11th.			
"	28	152		Horninglow and Stapenhill Infant Schools closed May 19th—June 18th.		
June	4	209				
"	11	109			Broadway Street Infant School closed June 4th—July 9th.	
"	18	91				
"	25	65				Uxbridge Street Infant School closed June 14th—July 9th.
July	2	57				
"	9	36				
"	16	35	All public elementary day schools closed for summer holidays July 8th—Aug. 2nd.			
"	23	28				
"	30	17				
August	6	10				
"	13	3				

From this table it appears that measles began to assume serious proportions in the last week of February. On March 2nd, two private schools (the Grammar School and the Girls' High School) were closed; while a public elementary school was closed on March 10th, and two other public elementary schools were closed on March 30th. At this time, and up to the middle of April, measles was mainly prevalent in the north-west central district; and with the exception of the Grammar School, which is in the south-east central district, all these schools are in the north-west central district. No reduction of the disease, however, would seem to have followed on the closure of these schools: nor, indeed, does any decrease of the epidemic appear until the latter part of April. It is possible that this decrease had some connection with the closure of all public elementary day schools during the Easter holidays, which lasted from April 7th to 17th. The decrease, at all events, manifested itself about a fortnight after the Easter holidays commenced. Up to the commencement of these holidays one public elementary school in the north-west central district had remained open, while two of the remaining three public elementary schools in that district did not close, as already stated, until March 30th. All the Sunday schools in the district, save one, were closed in March. It will be seen, however, that the remission of the disease after Easter was of brief duration. There was marked renewal of measles prevalence in the week ending May 7th, and in weeks following. This renewal is mainly attributable to the disease appearing in epidemic proportions in a north-western and in a south-eastern suburb. Its increase in these areas was sudden and great, and occurred, it will be observed, about a fortnight after schools re-opened on the termination of the Easter holidays. This fact is strongly suggestive of schools in these areas having aided in extension of the disease. Two infant departments of schools were kept closed, notwithstanding expiry of the Easter holidays; one of these (Wellington Street School) is in the north-west central district, the other (Goodman Street School), is in the north-western suburb that had become invaded by measles. A Sunday school in the latter area, however, was not closed. In the south eastern suburb that had become invaded, school closure was not resorted to until May 9th, when the infant department of the public elementary school in that neighbourhood was closed. About a fortnight after this measure had been adopted, the disease began to decline in the south-eastern suburb referred to; but, in view of the previous duration and intensity of the outbreak in that neighbourhood, this may have been due to exhaustion of susceptible material rather than to the action taken. Meanwhile, and also subsequently, measles invaded other parts of the town, and the closure of infant departments of certain schools had been resorted to with a view to checking its course. But, for the reason given in referring to the south-eastern suburb already mentioned, it is difficult to feel assurance that the desired limitation of the epidemic was attained in this way. From the table it appears that the epidemic began to wane about the middle of June, and rapidly decreased during July. Whether school closure in individual instances in May and June, and the closure of all schools for the summer holidays on July 8th, had any share in this, is not clear. Exhaustion of susceptible material, and the decline in measles prevalence that is usually associated with this season of the year, may have been the determining factors in the course taken by the epidemic in its later stages.

On the whole, therefore, these facts, while not inconsistent with benefit having been derived from school closure, lend but little active support to this thesis. It is, however, open to doubt whether this action on the part of the Local Authority might not have been taken earlier than it was, in at least some instances. Burton is, as stated in the body of this Report, cut up into several areas, with less inter-communication than is usual in town districts; and it may be that closure of all schools, in an area that had become invaded, at an earlier stage than that at which this measure was usually adopted, would have given better results. Thus, in the north-west central area, one school was closed on March 2nd, another on March 10th, and two on March 30th. The three last referred to are public elementary schools, and at no great distance from each other; and the closure of two of them so late as three weeks after the closure of the third seems a tardy measure. Furthermore, in two of these schools the attendance in the infant department had, I ascertained, dropped, at the date of closure, to 20 per cent. below what it had been during the weeks immediately preceding the appearance of measles in the neighbourhood; while, in the third, this drop amounted to 15 per cent. It may be questioned whether it was wise to so long postpone the closure of these schools. In other instances the falling off of attendance in the infants' department was considerably greater than at the above-mentioned three schools before the school or its infant department was closed. In view of the partial nature of the check to education involved in closure of only the infant departments of schools, this measure might well have been resorted to at an earlier stage than was usually the case during the epidemic.

Inference regarding relationship between schools and extension of the epidemic may also, in some degree, be drawn from the data given in the following table:—

TABLE D.

Week ending		Measles attacks among: Children attending Public Elementary Schools	Children attending Private Schools	Children attending Sunday School only	Children having Relatives at School	Children entirely Unconnected with School	TOTAL
January	8 ..	—	—		—	1	1
"	15 ..	—	—		—	—	—
"	22 ..	—	—		—	2	2
"	29 ..	1	—		1	—	2
February	5 ..	1	—		—	4	5
"	12 ..	1	1		—	4	6
"	19 ..	3	1		—	8	12
"	26 ..	7	21		5	6	39
March	5 ..	8	22		—	5	35
"	12 ..	68	36	5	17	2	128
"	19 ..	21	4	—	11	1	37
"	26 ..	56	3	4	37	35	135
April	2 ..	37	2	—	15	6	60
"	9 ..	51	7	—	36	19	113
"	16 ..	42	4	—	33	7	86
"	23 ..	22	3	6	27	17	75
"	30 ..	12	—	—	13	8	33
May	7 ..	78	—	2	7	7	94
"	14 ..	70	1	1	11	9	92
"	21 ..	225	—	1	49	13	288
"	28 ..	94	—	—	43	15	152
June	4 ..	69	1	—	88	45	209
"	11 ..	45	—	—	47	17	109
"	18 ..	33	2	1	28	27	91
"	25 ..	31	2	1	26	5	65
July	2 ..	27	1	—	16	13	57
"	9 ..	16	1	—	12	7	36
"	16 ..	14	1	—	12	8	35
"	23 ..	10	—	1	9	8	28
"	30 ..	3	—	1	8	5	17
August	6 ..	1	—	—	7	2	10
"	13 ..	1	—	—	1	1	3
Jan. 2 to Aug. 13		1,066	95	23	554	268	2,015

NOTE.—By "children having relatives at school" is meant children who do not themselves attend any school, but belong to a family of which other members attend school. "Children entirely unconnected with school" are children that neither attend school themselves nor belong to a family of which any member attends school.

Information as to the number of "children attending Sunday school only," attacked by measles prior to March 7th, is wanting. It is, therefore, possible that some of the children entered prior to March 7th as "entirely unconnected with school," or as "having relatives at school," may have attended Sunday school.

In considering what inferences may with reason be based upon the data in this table, it has to be borne in mind that information is lacking as to the number of children in each class, and also as to the ages of these children. But it is noteworthy that, while the earliest cases of measles appeared in the class of children entirely unconnected with school, yet there was no great extension of the disease until it had gained a foothold among children attending school. This suggests that schools had a large share in dissemination of measles. The same inference is suggested by consideration of the relationship in time of the maximum incidence of the disease on the several classes of children. This, it will be noted, was first attained among children attending school; the time being, as regards private schools, in late February and early March, and as regards public elementary schools, during the month of May. On the other hand, the maximum incidence of attack on children not at school themselves, but having relatives at school, took place from about the middle of May to the middle of June, that is, about a fortnight later than the period of maximum incidence on children attending public elementary schools. This condition of things is consistent with in the first place, direct infection of children by one another at school, and in the second place, children thus infected subsequently giving the infection to relatives at home. A similar culminating of the maximum incidence of measles in these two classes of children was observed by me in an epidemic of measles in Lancaster in 1897. The period of maximum incidence of the disease on children entirely unconnected with school is, as shown in the table, rather later than on children having relatives at school.

It appears, also, from the table, that the decrease in measles prevalence which occurred about a fortnight after the commencement of the Easter holidays on April 7th, is more marked among children attending school than among other children. It is, in fact, to lessened number of attacks among children attending school that this decrease is mainly to be attributed—a fact which strengthens the suggestion that the Easter school closure may have been beneficial. On the other hand, it will be noted that the recrudescence of the epidemic at the beginning of May, a fortnight after re-opening of the schools at the termination of the Easter holidays, is chiefly due to increase of cases among children attending school.

No. 16.

A. No. 16.
Plague on "Caledonia";
Bulstrode.

MEMORANDUM on the occurrence of cases of **BUBONIC PLAGUE** on board the P. and O. R.M.S. "**CALEDONIA**" from **BOMBAY**, and on the precautionary measures taken in relation thereto; by **DR. BULSTRODE.**

In accordance with my instructions I travelled to Plymouth on the night of December 6th, and on my arrival there on the morning of December 7th, at 7 a.m., I was met by Dr. Williams, the Port Medical Officer of Health, who informed me that the "Caledonia" had arrived early that morning and that he was awaiting daylight in order to commence his inspection. We were at once conveyed by tender to the "Caledonia," which was at anchor in the Sound, and on boarding this vessel we conferred with the captain and also with the ship's surgeon (Dr. Charles Mitchell), from whom, and from the report with which he subsequently furnished me, I obtained the following information:—

The "Caledonia," which has a gross tonnage of 7,558 tons, and is the fastest vessel employed in the Indian service, left Bombay with a cargo of general goods on November 19th. She carried 166 passengers, and her crew was made up of 128 Europeans and 212 natives.

On November 18th, the day prior to sailing, all the crew were examined on shore by the Bombay Port Medical Officer of Health and his staff in accordance with the provisions of the Venice Convention.

At this examination a "paniwalla," or greaser, who was found to be in a debilitated condition and with a high temperature, was rejected. The crew then returned on board the "Caledonia," and the vessel put out into "the stream" so as to avoid all clandestine communication with the shore.

On the following day (November 19th) the passengers, who had been medically examined on shore, came on board, and on the same day the crew were mustered—this time on board—for further inspection. It was then discovered that the native above referred to, who had been rejected at the first examination by the port authority, had in some unexplained way found his way on board again.

There had clearly been some oversight, neglect, or actual disobedience of orders in the matter, and, whoever was responsible, no one was desirous of fixing the blame.

The "paniwalla" on being a second time detected was at once removed in custody to one of the Bombay hospitals, and all the bedding and clothing of the native crew was passed through the "Equifex" disinfector on board the "Caledonia." After leaving Bombay the crew was examined every other day by the surgeon, and on November 24th, shortly after leaving Aden (where four marines from a British war vessel were taken on board), it was found that a coal-trimmer named Shaik Jarvodeen was suffering pain and tenderness due to swelling in the right axilla. Hereabouts there was a "slight abrasion," which "abrasion" had, he said, occurred while shaving his axilla, a practice common amongst natives. Apparently the surgeon was inclined to regard this injury as the cause of the swollen glands, and the man was not therefore isolated on suspicion.

The glands increased in size day by day, but the temperature did not rise above 99°. On November 27th, however, the condition of the patient was apparently such that the surgeon asked one of the passengers, a medical man, who had had experience of plague, to see the patient. He suspected the case to be one of plague, but was unable to give a definite opinion.

At 4 p.m. on the same day (November 27th) the "Caledonia" arrived at Suez, and it seems that the authorities here were in possession of information—by telegram—to the effect that the native who had been sent ashore to the hospital at Bombay on November 19th had developed "symptoms of plague."

Jarvodeen, the patient with the axillary swelling, was, after examination, declared to the Sanitary Authority at Suez as a suspicious case of plague, and the Port Medical Officer there gave instructions for him to be landed at Moses' Wells, along with two firemen who had been in contact with him on board. On the examination of the crew another lascar was discovered with an enlarged gland in his right axilla, and he, too, was landed at Moses' Wells. This latter patient had, it appears, been under the notice of the ship's surgeon for a wounded finger. All the clothes and effects of these four men were also landed.

The quarters of the native crew were disinfected under the immediate superintendence of the Port Sanitary Authority at Suez, and the whole of the effects of the native crew were subjected to steam disinfection; the contents of their chests were also turned out and similarly dealt with. After these precautions had been carried out to the satisfaction of the Port Medical Officer of Health, he endorsed the Bill of Health.

No specimen of the blood of either patient had been procured by the ship's surgeon, and I pointed out to him that the collection of a specimen of blood or pus in a sterilised vaccine tube would be useful under similar circumstances on another occasion. The Suez authorities stated their intention of examining the blood of the patients, but no evidence was, at the time of my visit to the vessel, forthcoming as to the result of such examination.*

The "Caledonia" then passed through the Suez Canal under "quarantine" conditions; she also coaled in quarantine, landed her mails at Port Said, and then steamed direct to Plymouth.

After leaving Suez the crew were frequently examined by the ship's surgeon, but no case suspicious of plague was discovered. The "Caledonia" arrived at Plymouth on the morning of December 7th, and she was therefore, in the terms of the Venice Convention, a "suspected" vessel, *i.e.*, one on which there had been a case of plague within twelve days.

After obtaining the above information it was decided by Dr. Williams and myself that none of the passengers' luggage should leave the "Caledonia" until after examination by us of all persons on board, as were any suspicious cases detected, it would be necessary to detain their baggage.

---

* It appears from a communication which reached the Local Government Board on July 7, 1899, *i.e.*, long after the above account was written, that the President of the Quarantine Board (Dr. Bernand Ruffer) stated, in reply to a question put to him by the Bombay Port Medical Officer of Health, that in his opinion neither of the cases removed from the "Caledonia" developed plague. Clearly, however, the circumstances were such that it was incumbent upon all persons having administrative concern with those cases to regard them as plague.

The passengers, each of whom was inspected separately in the surgeon's cabin by Dr. Williams, were asked from what part of India he, or she, came; how long passers through had remained in Bombay, and whether the servants or themselves had had any communication with the native quarters.

Each passenger received from Dr. Williams an initialed card, which was to be presented to the Sanitary Inspector in the saloon, and the passenger, after giving this latter official his name and address, received a pass to leave the vessel with his luggage.

The whole of the crew, European and native, were then mustered on deck and inspected, the ship's surgeon explaining any conditions as regards the natives which seemed to call for comment. As a result of this detailed inspection, no case suspicious of plague was discovered, and hence the "Caledonia" was allowed to proceed at about noon on her journey to London.

Source of Infection.—As regards the origin of this case, or these cases, it is clear that the presence of what was apparently a case of plague on board during the stay of the ship at Bombay would afford an explanation of the infection.

It is also to be remembered that the action taken at Bombay with a view to the exclusion of any chance case of plague in the crew was, in the case of the "Caledonia," rendered nugatory by the re-shipment of the once rejected native. Had not this man been discovered before the ship sailed, the consequences might have been far more serious than they actually were. The occurrence points to the necessity for a better method of controlling the movements of the natives after inspection than was exercised in this instance.

No. 17.

MEMORANDUM as to PLAGUE on BOARD the S.S. "GOLCONDA"; by DR. BULSTRODE.

App. A. No. 17. On Plague on S.S. "Golconda"; by Dr. Bulstrode.

At about 1.30 a.m. on Christmas Day (December 25th) I received a telegram at my private address from Dr. Williams, the Port Medical Officer of Health of Plymouth. The telegram stated that the S.S. "Golconda" had arrived at Plymouth on Saturday evening (December 24th), and that she had on board a case of plague, which had been removed to the hospital ship "Pique." Dr. Williams also added that he had telegraphed to the Board's Medical Officer, but that he had some doubts as to the telegram reaching its destination.

On receipt of this information I deemed it desirable, having regard to the hour and to the holiday period, to communicate in the first instance direct with Dr. Williams, and simultaneously to notify the Medical Officer by telegram of my action. I therefore drove to Victoria Station at 3 a.m. on Sunday morning, and despatched a telegram to Dr. Williams asking for full particulars, and also whether, under all the circumstances, he would wish for a conference with one of the Board's Medical Inspectors. Dr. Williams replied that the patient had been landed at Plymouth, and that the "Golconda," after disinfection, &c., had been carried out, had proceeded to her destination at London.

After conferring with the Medical Officer, I was instructed to proceed to Gravesend to ascertain the facts as to the case. As, however, Dr. Collingridge informed me, in response to a telegram, that the "Golconda" had already left Gravesend, I proceeded to the Albert Docks to await her arrival there.

Before the vessel reached her berth an opportunity was afforded me of hearing from Dr. Finlay, the Medical Officer of the British India Steam Navigation Company, a history of the occurrence, and Dr. Finlay also informed me that some important information as regards the possible sources of infection had but quite recently come to the knowledge of the officers of the Company. When the "Golconda" came alongside, I conferred with the ship's surgeon (Dr. Stride) as to the additional information received, and took certain precautionary steps which seemed called for under the circumstances. These will be referred to later on.

HISTORY OF THE VOYAGE OF THE "GOLCONDA," AND OF THE CASE OF PLAGUE.

The "Golconda," whose tonnage is 5,874 tons, belongs to the British India Steam Navigation Company, and she is at present engaged in trading with passengers and cargo between Calcutta and London, calling at Madras, Colombo, Aden, Suez, Port Said, Naples, and Marseilles.

The vessel left Calcutta on November 19th, having on board 77 passengers, amongst whom was a saloon passenger, Mr. W——, who subsequently developed plague, and also a crew, chiefly native, of 151 (Europeans 25, natives 126). On November 23rd the "Golconda" arrived at Madras, where 29 passengers were landed and 15 embarked, amongst these latter being Mr. B——, to whom further reference will be made.

The "Golconda" left Madras the same day, and on November 25th she arrived at and left Colombo, where five passengers were landed and one embarked. On December 3rd Aden was reached, where five passengers were taken on; and on the following day the vessel left for Suez, where she arrived on December 9th. The Egyptian Quarantine Board's Officers were apparently satisfied as to the health of the ship, and, after disembarking four passengers for Egypt, she proceeded through the canal to Port Said, which place was reached on December 10th. Here three passengers were embarked, and the vessel sailed for Naples, where she arrived on December 14th, and where seven passengers were landed. On December 15th the "Golconda" left Naples, and on December 17th arrived at Marseilles, where 18 passengers were landed, amongst them being Mr. B——, to whom brief reference has already been made. The vessel left Marseilles on December 18th, reaching Plymouth on the evening of December 24th, and the Albert Docks (London) on December 26th, at about 11.30 a.m.

The passenger who had developed plague before touching at Plymouth was Mr. W——, an officer in the service of the British India Company, who had been employed as second officer on board a vessel trading between Indian ports, the Straits Settlements, Rangoon, and Burmah; the patient's last voyage before his return to England on the "Golconda" having been from Rangoon to Calcutta, which port he reached on November 8th. Antecedent to embarking on board the "Golconda" at Calcutta, Mr. W—— had been engaged in superintending the loading of a coasting vessel, the "India" (B. I. Co.), trading between Bombay and Calcutta. The crew of this vessel, which consisted almost entirely of Natives, had, I was informed, been shipped at Bombay.

Mr. W—— was a saloon passenger on board the "Golconda" and he occupied a cabin in that part of the vessel allotted to such passengers. When the "Golconda" arrived at Madras on November 23rd, Mr. W—— went ashore for a time, and, as has been said, Mr. B—— there joined the vessel. Shortly after leaving Colombo on November 25th, Mr. W——, according to the information given by him at a later date to the ship's surgeon, suffered from febrile symptoms and from tenderness of the inguinal glands in both groins. This condition continued, and the glands commenced to swell, until Suez was reached, when the febrile condition is stated to have abated, but the glands continued to increase in size. The patient, however, did not regard himself as ill enough to consult the ship's surgeon. The day after leaving Marseilles (*i.e.*, on December 19th), the glands became so painful that he at last sought the assistance of the ship's surgeon, who at once diagnosed the case as one of *pestis ambulans*, and promptly dealt with it as such in the matter of isolation, etc. At this time the patient had an indurated swelling about the size of a small hen's egg in each groin, there being no obvious lesion either of the genital organs or of the legs to account for the inguinal swelling. T = 100·8, P = 100.

Next day the patient's general condition was little altered, but the buboes were exhibiting a tendency to suppurate. On the arrival of the "Golconda" at Plymouth, Dr. Williams, who had already been apprised by a telegram from Ushant that the vessel would, on reaching Plymouth, require immediate medical assistance, went on board. Upon examining the patient, who was dressed and sitting up in his cabin, Dr. Williams found ovoid-shaped swellings in either groin measuring some 3½—4 inches in length and about 2 inches in breadth. The rounded apex of each bubo was inflamed and showed a tendency to suppurate, there being considerable induration at the periphery of the swellings.

The patient expressed himself as feeling much better though still slightly feverish, but, apart from this and the inconvenience occasioned by the buboes, his condition was satisfactory. Dr. Williams, after a review of all the facts, had no hesitation in concluding that he had to do with a case of *pestis ambulans*, and he at once had the case isolated on board the "Pique," the "Golconda" being dealt with as "infected" under the provisions of the Board's Order as to Plague, &c. Ten passengers were allowed to land at Plymouth after the usual precautions as to medical examination and as to disinfection had been carried out. The names and addresses of the ten passengers were also taken, and the medical officers of health of the districts to which they were travelling were communicated with. The "Golconda" then proceeded on her journey. On arrival at Gravesend the vessel was finally dealt with by the medical staff of the London Port Sanitary Authority, and was allowed to proceed to her berth in the Albert Docks, instructions having been given for further precautionary measures to be taken there.

Before the vessel arrived at the docks, I was informed by Dr. Finlay that it had just come to his knowledge that at Madras there came on board the "Golconda" a Mr. B——, who had apparently given certain of his friends amongst the passengers to understand that he had quite recently been discharged from a hospital either at Madras or Bangalore, and that he had there suffered from an attack of plague. Had this information been correct, it afforded a possible explanation of Mr. W——'s infection. Mr. B—— had, however, left the "Golconda" before she reached Plymouth, and hence, I had no opportunity of questioning him as to the actual facts. I ascertained, however, which cabin he had occupied on board, and pending the receipt of further information concerning him, as a matter of precaution I arranged that his cabin should be dealt with as infected. I also telegraphed to Dr. Williams at Plymouth telling him of the allegations as to Mr. B——, and stating that certain Plymouth agents held Mr. B——'s luggage. Although I conferred with two of the passengers who had been friends of Mr. B——, I could not ascertain from them any really trustworthy information. Mr. B—— had apparently left an impression that he had been recently in hospital in India with plague, but as to whether an interval of a few days or a few months had elapsed since his attack I was unable to determine.

Dr. Williams has, however, since seen Mr. B—— in Cornwall, and from Mr. B——'s statement it would appear that he entered the "Mine" hospital of the Mysore Gold Mines on October 18th, 1898, that he had vomiting and diarrhœa, and that there was

P. A. No. 17.

Plague S.S. "Golconda"; Bul-e.

some suspicion of the attack having been cholera. There was, however, doubt as to the exact nature of the illness, and specimens of blood had been examined with negative results. Plague was epidemic at Bangalore, 50 miles away, but no cases were known of in the mining districts, and Mr. B—— suffered no glandular enlargement whatever. His attack, too, would not appear to have been severe enough to be regarded as one of "non-bubonic plague"; and, on the whole, it must be admitted that there is no sufficient evidence as to his having had plague.

Any attempt to determine the real source of Mr. W——'s infection must obviously be attended with much speculation. It would certainly seem that immediately prior to embarking, Mr. W—— was brought in the course of his duties into association with natives who had come recently from Bombay on board the "India." We have no evidence as to the presence of plague amongst these natives, but, keeping in view the obvious difficulty of ascertaining the existence of certain cases of *pestis minor*, and our previous experience as to cases of plague amongst this class of persons, it is not altogether unjustifiable to assume that in the native crew aboard the "India" we have a possible source of Mr. W——'s infection.

With regard to Calcutta, we have no official or other information of the presence of plague in that city at the time in question, but the general diffusion of plague over India renders caution necessary in accepting the absence of definite information as evidence that plague was then non-existent in Calcutta, and Mr. W—— may possibly have acquired infection there.

So, too, with regard to Madras, where we know of the existence of plague, Mr. W——'s trip ashore may have been responsible for his infection.

Although the period of incubation is fairly constant, too much reliance must not be placed upon it. The "Golconda" left Calcutta on November 19th, and apparently Mr. W—— felt quite well until November 24th, after the vessel left Colombo. This interval of six days would be not inconsistent with his having been infected at Calcutta, and the interval of three days after leaving Madras would fit in with a theory that he was infected at this latter place.

[*N.B.*—Plague has since been announced at Calcutta, and the city has been notified to foreign Governments as infected under the terms of the Venice Convention. January 14th, 1899.]

No. 18.

REPORT upon the PROGRESS and DIFFUSION of BUBONIC PLAGUE from 1879 to 1898; by Dr. BRUCE LOW.

INTRODUCTORY.

The responsible duty which devolves upon the Medical Department of watching the progress of foreign epidemics, with a view of preventing their extension to this country, has of late years been largely limited to one disease, Asiatic cholera. Simultaneously, however, with the rapid subsidence of the cholera prevalences in Europe after 1895, there appeared in Southern Asia another dangerous epidemic disease, BUBONIC PLAGUE, which has since seriously threatened the safety of Western nations.

An outbreak of plague at Hong Kong in 1894 attracted general attention to the disease, and since then the foci of infection have rapidly multiplied in China, in India, and elsewhere.

Since 1879, when the late Mr. Netten Radcliffe reported to the Board "on the recent progress of Levantine plague,"* it has not been found necessary by the Medical Department to issue any account of prevalences of plague in one or another country, for the reason that the outbreaks were apparently limited to certain recognised endemic areas, and did not, at the time, threaten to extend beyond these areas to Europe.

But with the recent alarming activity of plague in Asia, and the remarkably rapid increase of foci of infection in China and India, it has become necessary once again to watch closely the behaviour of the disease, and to report upon its extensions.

With a view of bringing the history of plague outbreaks up to date, I have collected information from various sources showing shortly the course which the disease has followed from 1880 onwards in each of the countries hereafter named, that is, since 1879, the date of the last report on plague by Mr. Netten Radcliffe.

The chief areas in which bubonic plague has long been regarded as endemic are (1) the district of Assyr in Arabia, which lies on the Eastern shores of the Red Sea; (2) portions of Persia and Mesopotamia; (3) Southern China, and especially the province of Yunnan, on the frontiers of Thibet. To these Professor R. Koch† has recently added another, viz., the district of Kisiba, to the extreme north-west of German East Africa. This African endemic area extends beyond the German territory into British Central Africa, and includes that part of the province of Uganda which lies to the south-west of Lake Victoria Nyanza. Plague has been proved to be endemic in this Central African area for the last eight years, and reports point to its existence in the locality 40 years ago. (*See* page .) Dr. Zupitza, a German military surgeon, has shown, by clinical, pathological, and

---

* Report and Papers on the Recent Progress of Levantine Plague, Supplement to the 9th Annual Report of the Local Government Board, 1879–80.

† Deutsche Medicinische Wochenschrift, July 14, 1898.

P. A. No. 18. — *the ... fusion of bubonic plague from ... to 1898; by Bruce Low.*

bacteriological evidence, that this African endemic disease is true bubonic plague.*

Indian plague which is known as "Mahamari" and is endemic in Upper India, in Kumaon, and Garhwal, is believed to be identical with bubonic plague. It has shown itself in these districts in small localised outbreaks for a number of years. It is possible that this Indian plague area receives its infection from Yunnan, through Thibet. In past times Syria and Asia Minor were also looked upon as plague endemic areas, but there is no evidence to show that they are centres of infection at the present time.

In collecting data with regard to plague outbreaks, a difficulty is encountered through the laxity with which the term "plague" is applied. In the less civilised countries plague is used as a synonym for pestilence, and hence some reported plague epidemics have proved, upon investigation, to have been in reality outbreaks of cholera or small-pox, and in one or two instances relapsing fever.

The narrative in the following pages shows that until the last few years, the outbreaks of plague which were heard of, were confined to a great extent, to the Arabs in Assyr, to the Bedouins in Mesopotamia, and to the nomadic tribes in Northern Persia and Central Asia. Since, however, the well marked outbreak in Southern China, in 1894, plague has been prevalent, not as before among wandering tribes, but in large and populous cities like Canton, Hong Kong, and Bombay. Instead of haunting thinly populated districts like Assyr and Mesopotamia, it is now prevalent in the densely populated districts of Southern China and Western India. This circumstance is not without serious import, and suggests the possibility of further extension of the disease from East to West. Former epidemics of plague have been observed to move from East to West.

Though plague has been spreading in Western Asia for the last four or five years, it had not, at the close of 1898, spread to Europe, notwithstanding the frequent and almost constant communications that take place between the infected ports in the East and the healthy ports in the West. It is true that three fatal cases of plague occurred in Vienna (*see* page) in October, 1898, but these were due to infection received in a bacteriological laboratory from cultures which had been brought from India to Vienna some eighteen months before. The first of the three cases, an attendant, infected himself through neglect of the rules laid down for his guidance, and subsequently the disease was communicated to his medical attendant and to one of his nurses.

It has also to be recorded that in England, in 1896, two Portuguese native Goanese steward's helpers, from a steamer arriving in the London docks from Bombay, developed a disease believed to have been plague, of which they died in the Seamen's Hospital a fortnight or more after their arrival in England. In this instance

* More recently yet another endemic area has been reported by Dr. Thiroux (pupil of the Pasteur Institute, Director of the Bacteriological Institute at Antananarivo), who was employed on plague investigation in Madagascar. He was sent from there to Réunion to inquire into a suspicious outbreak, which he proved to be true bubonic plague. His researches go to establish the fact that the disease has been appearing from time to time in the island of Réunion for the last 60 years under the name of "Lymphangite infectieuse."—*Journal Officiel de la Réunion*, 4th July, 1899.

the infection was thought to have been contained in clothing bought at Bombay, but not unpacked till after this country was reached. Another suspicious case occurred, just before these other two, in the person of a native of Bombay, who was brought to the Seamen's Hospital in a moribund state with pneumonic symptoms, two days after his arrival in the London Docks from Calcutta. (*See* Dr. Buchanan's report, 26th Annual Report of the Medical Officer to the Local Government Board, 1896-97.) Four instances have occurred since 1896 in which illness resembling plague has developed on board a ship after it has left India *en route* for this country. These are (1) the "Dilwara," in 1896, on board which an English girl died of plague eight days after leaving Bombay (*see* page); (2) the "Carthage," in 1897, on board which two Lascar coal trimmers developed plague, one on the fourth and the other on the twelfth day after leaving Bombay (*see* page); (3) the "Caledonia," in 1898, on board which two Lascars were reported to have been taken ill with what was at the time believed to be plague after the ship sailed from Bombay (*see* memorandum by Dr. Bulstrode, page); (4) the "Golconda," in 1898, on which a passenger from Calcutta was found to have mild plague symptoms after the ship had passed through the Suez Canal and had touched at Marseilles. (*See* extract from memorandum by Dr. Bulstrode, p. .)

The present indications are not in favour of the disease being likely to become epidemic in this country after being brought here from abroad by shipping. There appear to be conditions on board ship which are unfavourable to the development of virulence in plague. If infection reaches this country, it would seem most likely to be conveyed either by means of mild and unrecognised cases marked in the main by indolent forms of bubo, or perhaps through infected articles of clothing.

At the time of writing, plague is still, in 1899, prevalent in India and China. It has appeared at Jeddah and in Formosa; Mauritius and Egypt have yielded cases, and some anxiety has been caused in South Africa by the occurrence of suspicious cases there.*

An endeavour has been made to bring this account of plague prevalences up to the end of 1898, but in certain instances official figures have not yet been received as to the number of attacks and deaths, and in this respect the present report will probably require to be supplemented at a future date.

With regard to the serious epidemic of plague in India much has already been written by medical officers, sanitary commissioners, and others. A full report on Plague in India up to August, 1897, has been compiled by Mr. R. Nathan, of the Indian Civil Service, in four volumes. Another report has been issued by the Bombay Plague Committee, under General Gatacre, which contains much useful information. There are also others which call for no special mention here. A Royal Commission on Plague was sent from this country to India in the latter part of 1898, under Professor T. R. Fraser, M.D., F.R.S., and the result of their investigations is expected to be published shortly.

Since the epidemic began, nearly all the great European nations have sent medical delegates to India to study plague, so that, in

---

* Since this was written plague has appeared in Portugal at Oporto. Outbreaks have also been reported from South America, in Brazil and in Paraguay. Suspicious outbursts, too, have occurred in Russia, in Europe, in the Governments of Astrakhan and Samara.

APP. A. No.
On the Diffusion of Bubonic Plague from 1879 to 1898; by Dr. Bruce Low.

the event of the disease establishing a foothold in Europe, th knowledge thus obtained will be of great service in devisin measures to repress any outbreaks.

The Indian Government has supplemented its permanent medic staff by a number of temporary medical officers, female as well a male, and trained nurses, who are employed on railway inspection and on other duty connected with the prevention of plague. Th present severe and deplorable visitation of plague in India ma not improbably lead to important reforms in sanitary administr tion in that country, and it is to be hoped that medical officers health will ultimately be appointed to supervise the sanita condition of districts, and with power to enforce their reco mendations, in order that a recurrence of this devastation plague, or other epidemic disease, may in future be prevented.

The following summary shows briefly that from 1879 to 18 not a single year has passed without the development of plag in at least one country, and in later years the disease has be present in several countries at one and the same time.

In 1880, plague was reported to be present in Mesopotamia.
„ 1881, it was present in Mesopotamia, Persia and China.
„ 1882, in Persia and in China.
„ 1883, in China.
„ 1884, in China and in India (as "Mahamari").
„ 1885, in Persia.
„ 1886, in India (as "Mahamari").
„ 1887, in India (as "Mahamari").
„ 1888, in India (as "Mahamari").
„ 1889, in Arabia, Persia and China.
„ 1890, in Arabia, Persia and China.
„ 1891, in Arabia, China and India (as "Mahamari").
„ 1892, in Mesopotamia, Persia, China, Russia and ? Tripoli.
„ 1893, in Arabia, China, Russia, and India (as "Mahamari"
„ 1894, in Arabia, China, and India as ("Mahamari").
„ 1895, in Arabia and China.
„ 1896, in Arabia, Asia Minor, China, Japan, Russia and Indi
„ 1897, in Arabia, China, Japan, India, Russia, and East Afric
„ 1898, in Arabia, Persia, China, Japan, India, Russia, E Africa, and Madagascar.†

It has also to be remembered that since 1890, and probably lo before that time, plague has been occurring in Central Afri whence it might have been carried northwards by Arab travell to Egypt and Tripoli in past years, or eastwards towards the por in communication with Jeddah and the Red Sea Coast.

ARABIA.

The Assyr district, which lies between the provinces of th Hedjaz and Yemen in Western Arabia (*see* map), is a centre which plague has long been endemic; its population in 1879 wa estimated at 60,000. The people are a warlike race, wh habitually go about armed to the teeth. A part of the peop leads a nomadic life in tents, while the remainder has more per manent habitation in villages. The country has three well-marke

* Also two suspected cases in the London Docks
† Also three fatal cases in Vienna.

† Das österreichische Sanitätswesen, October 17, 1889.
‡ Veröffentlichungen d. K. Gesundheitsamtes, November 18, 1890.

* Also two suspected cases in the London Docks
† Also three fatal cases in Vienna.

divisions: (1) a plain lying parallel to the Red Sea Coast, on which the three ports of Lith (or Leet), Confuda, and Hali are situated; (2) a mountainous region intersected by deep valleys, and (3) an elevated plateau or table land. The interior is seldom visited by strangers and food supplies are often scarce as a result of periods of drought. The people for the most part live in poverty, and exhibit a disregard for personal hygiene.

Though plague epidemics are frequent in Assyr, the inhabitants have the firm conviction that plague, like the rain and sunshine, is sent from Heaven, and therefore cannot be prevented. When persons die of plague their clothing is commonly purchased by friends and relatives, and carried away to be worn without washing or other previous precaution. Since 1815 plague has been known in Assyr. At that time the disease was raging in the adjoining Hedjaz district, into which it is alleged to have been brought by the expedition under Mehemet Ali. The plague infection was again imported, so it is alleged, by Egyptian troops in 1825–26. As communication with the interior is difficult and infrequent, definite reports on the plague outbreaks are seldom obtained; but information as to serious prevalences of this malady came to hand in 1853–54, and again in 1862. The disease reappeared in 1868 at Namas, and again in 1871, as well as in 1873–74. In 1879 there was a widespread epidemic which continued three months or more, and was attended with great fatality.*

Certain caravan routes to Mecca pass through Assyr, and this circumstance lends importance to the occurrence of any plague outbreaks in that country. As to plague outbreaks from 1879 to 1888 no information has been received, but in January, 1889, plague was again reported to have broken out in Assyr, and continued to show itself in various localities till the following October. Active measures were taken in July by the Ottoman sanitary officials to establish a cordon along the borders between the Hedjaz and Assyr, extending from Lith on the coast to Taif on the borders of the Desert (see map appended); in addition, a ten days' quarantine was imposed on all arrivals from the coast of the Red Sea from Lith to Hodeida. These restrictions were withdrawn in October. The plague epidemic is stated† to have raged in 500 small villages or hamlets in the districts of Ebha and Ben Sheïr. The mortality was 75 per cent. of those attacked. The intensity of the outbreak was rendered greater through the personal uncleanliness of the people. One reporter observes it was not uncommon for families of five or six persons, sick and healthy, to live crowded together in one small apartment. The duration of the disease was from three to fifteen days, and the convalescence in those who recovered was protracted. The symptoms observed were high fever, great prostration, severe pain in the axilla or groin, with swellings which inflamed and suppurated. The epidemic was associated with famine and poverty, the result of a drought.

In September, 1890, in a report from Jeddah by Dr. Baum,‡ it is stated that plague had broken out again in Assyr, and that a caravan composed of 2,500 persons from Samaa arrived at Arafat,

---

* Recueil des Travaux du Comité Consultatif d'Hygiène Publique de France, Vol. IX.

† Das Oesterreichische Sanitätswesen, October 17, 1889.

‡ Veröffentlichungen d. K. Gesundheitsamtes, November 18, 1890.

APP. A. No. 18.

On the Diffusion of Bubonic Plague from 1879 to 1898; by Dr. Bruce Low.

after the Hedjaz, with scarcely 1,000 survivors, the suggestion being that many had died from plague; but no confirmation of this statement was forthcoming from other sources. An official report from Constantinople* in April, 1891, stated that plague had again broken out on the Red Sea Coast between Lith and Lohaja in Assyr, and that a physician had been despatched to inquire into the matter.

No outbreak of plague in Assyr was reported in 1892, although an epidemic did occur on the Eastern borders of Arabia in the Turkish district of Bassorah, or Basra, in Mesopotamia. In 1893 a report from Alexandria† stated that plague had again appeared in Assyr, but no details were given. This report was repeated in December of the same year, and it was then alleged that plague was appearing on the Red Sea Coast in Assyr between Lith and Lohaja.

In the thirty-fourth Annual Report of the Sanitary Commissioner with the Government of Bombay, the Acting Port Medical Officer, Major Crimmins, V.C., I.M.S., says that: "On reference to records it will be found that plague was epidemic in Assyr in October, 1893, July, 1894, August, 1895, and in October, 1896." He adds that plague is endemic along the Red Sea Coast, and that "quarantine is imposed by Egypt and Aden almost yearly against that portion of the Red Sea Coast between Lith and Lohaja."

As has been said, communication with Assyr by Europeans is infrequent, and information as to outbreaks is often vague and wanting in confirmation. Outbreaks on the other hand occur which are never reported. Current literature is silent as to plague in Assyr during 1894, except the mention by Major Crimmins quoted above. But mention is made in a continental medical journal‡ that plague had appeared in Assyr in April and in July, 1895, and that eight villages had been attacked in the Beni-Sheir district. A despatch from Her Majesty's Consul at Jeddah, dated May 8, 1895, states that "the epidemic of plague in the canton of Assyr continues; it is said that eighty deaths per week occur from the disease." The same report alludes to the habits of the people as being "remarkably uncleanly." Later advices mention that the infection had spread to Taif, a town on the borders of the Hedjaz. In September, 1896, it was reported§ that plague had shown itself once again in Assyr. In the annual report of the British Consul at Jeddah it is mentioned that owing to cases of plague, "which is endemic in Assyr," the ports of Lohaja and Assyr were quarantined during the latter part of 1896.

In June, 1897, plague was officially declared to exist at Jeddah, forty-six deaths having been certified from that cause during the month. The first cases observed were at the end of May; the persons attacked were Hadramauts, Arabs, and Abyssinians. No cases occurred among the British Indian pilgrims, the first case being a pilgrim from Yemen.

On the other hand, there is evidence supplied by the British Vice-Consul that cases of fever attended with buboes had been occurring in Jeddah, and also at Mecca, as early as February, 1897. Cases of the same kind had been observed early in the year at Taif,

* United States Public Health Reports, 1891, p. 184.
† Veröffentlichungen d. K. Gesundheitsamtes, 1893.
‡ Veröffentlichungen d. K. Gesundheitsamtes, 1896.
§ Das Oesterreichische Sanitätswesen, 1896.

which, as has been mentioned, is situated close to the border of the Hedjaz. Though only fifty deaths had been declared due to plague up to the end of June, 1897, the mortality was in all probability far greater, the statistics of plague issued by the Health Office at Jeddah being almost certainly incorrect. The Consul gives the following reasons for doubting the correctness of the official figures :—

"(1) The mortality returned under other causes than plague during June is much higher than usual.

"(2) Apart from plague there was no other prevalent malady in the town to cause this rise in the mortality.

"(3) In 1897 the Holy Pilgrimage was not a heavy one owing to prohibition of the departure of pilgrims from India and the French colonies, and for this reason Jeddah has been less crowded than in former years.

"(4) On the perusal of the statement showing the deaths from plague, with sexes and nationalities, it will be seen that among the forty-six victims only four were females and one a child. This proves that the return is far from being correct. The apparent immunity of females and children from plague is very suspicious. It is known that female corpses are never inspected owing to the absence of female doctors. Female victims from plague were probably buried without inspection of the corpses and therefore the disease escaped notice. Women ill of plague are known to have been concealed from the authorities. Several of the fatal cases occurred among Yemens and Hadramauts on their way back from Mecca. Some pilgrims taken ill on the journey complained of swollen and painful inguinal glands, but the condition of the glands, as well as the illness, was attributed to 'fatigue from riding.'"

On June 29, two cases of plague occurred at the quarantine station at El Tor. These cases occurred among a batch of Egyptian pilgrims who had left Jeddah on June 12. Several other mild cases were also detected among the pilgrims, and these were at once isolated. No extension of the disease took place on this occasion.

The measures taken at Jeddah comprised active scavenging of the town; infected houses were whitewashed, an isolation hospital with accommodation for fifty patients was established, and 1,500 poor pilgrims were sent to undergo quarantine on the islands of Abu Said and Wasta. There was found to be great local objection to the segregation of patients. The scavenging undertaken by the authorities is said to have changed the aspect of the town, which before was "a mass of stinking filth," into a "clean and decent place." In March, 1898, three deaths from suspected plague were reported at Jeddah among Hadramaut pilgrims. Dr. Xanthopulides reported on the outbreak to the International Sanitary Commission. He said that previous to the 1898 epidemic he had noticed deaths among goats and mice from a pulmonary affection. He further states that at Nucla, a village three-quarters of an hour distant from Jeddah, and of which the inhabitants are exclusively Bedouins, a death from suspected plague had occurred in March. It was believed that at Jeddah the plague cases were concealed. Only twenty-one cases of plague were reported between March 22 and April 16, and thirty-four plague deaths were registered. Sick women were not permitted to see a doctor, nor after death were their bodies examined. In one group of cases the relations of the patients testified to their having recently purchased some second-hand garments. As it was known that clothing belonging to persons who had died of plague had not been disinfected or burnt, it is suggested that the disease was set agoing again by this means. The people of Jeddah object to the destruction of their clothing by the authorities, and, in view of this, they hide their illness from the sanitary officials.

The plague epidemic of 1898 at Jeddah was reported upon by Dr. Nowry Bey,* who states that the outbreak began on March 21st, and he expresses the opinion that the infection must have been imported from Bombay through the medium of rice-bags,† or by means of rats or mice from infected ships. At the beginning of the outbreak there were seen in the streets, where the sick persons lived, numerous mice in a dazed or dying condition, and unable to run away. Plague bacilli were found in the dead mice. A further official report was issued by Dr. Cozzonis Effendi,‡ who mentions 43 cases, 35 of which proved fatal in the period from March 4th to April 13th. But these statistics are incomplete, for Dr. Mohammed Husain, who resides at Jeddah, reported that illness characterised by fever and buboes existed in the town some time after it was declared free from infection. He mentions for example the case of a man, Haji Ishay, who died on May 28th, whose death was registered by the Health Office as having occurred from dysentery. But as this man's daughter had fever and buboes just before her father was taken ill, and as the person who nursed the father was also taken ill with identical symptoms, as also were two servants in the house, Dr. Husain concludes that all these cases were in reality bubonic plague. He further adds that the Health Officials never visited the house to ascertain the cause of Haji Ishay's death, but issued a certificate that it was due to dysentery. Such instances as the above show the lax manner in which mortality statistics are compiled, and suggest that the officials wilfully concealed at the time the existence of plague in the town.

One of the mail steamers, the "Mehallah," sailing fortnightly between Suakim, Jeddah and Suez, was reported, on May 3rd, to have had on board two cases of plague. The "Mehallah" left Suez on April 19th, left El Tor on the 20th, and reached the outer harbour of Jeddah on April 23rd, where she stayed 2¾ hours to land mails, but took no passengers or cargo on board. The same day she called at Ras el Aswed, the temporary port for Mecca, and here she disembarked 67 passengers. She then proceeded to Suakim, and then returned to Suez, reaching that port on April 30th. The vessel was disinfected as soon as possible at the lazaret station at Moses Wells, but the baggage, it is stated, was not disembarked. The ship's Egyptian doctor, who remained on board, fell ill of plague on May 22nd, and died on the 26th; and during the next few days two fresh plague cases occurred on board. The baggage was now disembarked and the ship thoroughly

* Annales de l'Institut Pasteur, T. XII., No. 9.

† The "rice-bags from Bombay" theory is a favourite with some French writers.

‡ Rapport sur la manifestation pestilentielle à Djeddah en 1898 par le Docteur Cozzonis Effendi, Constantinople.

re-disinfected. It is said that on this occasion a number of dead rats were found on board, but the quarantine doctor at Suez said they were not discovered till after the process of disinfection, which, probably, killed them.

How the "Mehallah" became infected in the first instance has not been discovered. The ship's crew and officers were all Egyptians. It is denied that the stoker who was the first to show signs of plague, or anyone else, had landed at Jeddah on April 23rd. Such denial may be taken for what it is worth. It has to be remembered that the Governor of Jeddah denied that plague had ever existed in the town, though it was well known that a number of fatal cases had already occurred. It is strongly suspected that plague was introduced into the "Mehallah" by infected clothing purchased at Jeddah.

It has been asserted that plague was introduced into Jeddah in 1898, not from the adjoining district of Assyr, where the disease had been occurring annually, but that it had been landed from small sailing boats ("sambooks" or "hutras") which swarm in the Red Sea and Indian Ocean (*see* also p). These boats are subject to no supervision, and it is stated that a good deal of so-called "sanitary smuggling" of persons from infected districts into the Hedjaz goes on along the coast during the pilgrimage. It is suggested, especially by the French Authorities, that the infection of Jeddah was due to importation of plague from India, and not from Assyr. But apart from the fact that no evidence or proof has been put forward in support of this view, it is only reasonable to suppose that plague was brought by pilgrims from Assyr, or by Yemeni or Hadramauti pilgrims who got infected in passing *en route* through Assyr, an admitted endemic centre of plague. It may be added that in February, 1899, plague once more appeared in Jeddah, the first recognised case being a native of Assyr. From February to May, 120 deaths from plague had occurred at Jeddah.*

Mesopotamia.

Mesopotamia has been for many years one of the best-known endemic areas of plague. In the basin of the Tigris and Euphrates, plague was epidemic in 1834, and again in 1866–67. It reappeared in 1873, and continued to show itself during 1874–75, as also in 1876–77. No marked epidemic appears to have occurred in the province of Bagdad from 1877 till 1880, but in September, 1880, plague attacked the El Zayed tribe. From this centre the disease spread among the neighbouring population and reached Nedjef, a much-frequented centre for pilgrims from Persia. The malady appeared towards the end of the annual pilgrimage. Nedjef had a resident population estimated at from 5,000 to 6,000. In the first five days after plague had appeared, 50 deaths from it were reported, and in this number were not included deaths of females. Owing to the virulence of the epidemic, the disease was called by the Turkish physicians sent to the spot "the black death," the name given to a rapidly fatal form of Oriental or bubonic plague. In most cases the illness ran a short and fatal course; the epidemic

* "La Peste à Djeddah et le Pelerinage de 1899." Rapport présenté au Conseil Superieur de Santé par le Dr. Cozzonis Effendi, Inspecteur Général du Service Sanitaire Ottoman.

lasted from September, 1880, to June, 1881. It carried off a considerable number of the population of the district; some 17 towns or villages were invaded. The precise numbers of attacks and deaths are not stated in the reports by the Turkish physicians.

From 1881 till 1892 there are no reliable accounts of any plague epidemics in Mesopotamia. The absence of reports, however, does not prove that no outbreaks occurred. In May, 1892, plague broke out at Hascham and raged among the Arab tribes of Elden and Ferishan, on both banks ot the Schatt el Arab. The disease spread among the Bedouins in the neighbourhood of Hai, and to other districts in the Government of Bassorah. The Turkish doctors give the mortality as 50 per cent. of those attacked. The tents of the Bedouins were burnt and the population were "dislodged"; after this the outbreak subsided.

There is a very dangerous system of transportation of corpses going on between India, Southern Turkestan, Persia, and Mesopotamia, in connexion with the Mohammedan sect of the Shiahs, whose religious creed makes them desirous of being buried in Nedjef or Kerbela, where are situated the tombs of Ali and Husein, who are held in great veneration.

There is a possibility that this transportation of corpses and the communications it entails may be the means of conveying infection to or from Mesopotamia. In this connection it is interesting to note that Roux asserts that at least 4,000 Persian corpses are imported into Nedjef annually, and that in 1894, after the Persian famine, the number was as high as 12,000.

According to the sanitary regulations in force, corpses cannot be imported into Turkey till three years after they have been buried in Persia, *i.e.*, only bones are allowed to be imported. But these regulations are evaded, and there is evidence that corpses of those recently dead, and others in a putrefactive state, are brought across the Persian frontier to be buried at Nedjef according to the religious tenets of the Moslem Shiahs. There is, however, a bacteriological side to this question, for the experiments of Klein and others tend to show that as a rule the organisms accompanying putrefaction destroy those of specific infection within the dead body in a short period after death; and this may to some extent reduce the danger alluded to above.

In February, 1897, the Turkish Government issued orders applying strict measures against the Shiah pilgrims and the transportation of corpses into Turkey, arriving from Baluchistan, Persia, and India.

It is reported that on February 6th, 1897, an Indian woman died in the lazaretto at Bassorah. She had been landed from the steamer *Khandallah*, which had come from Karachi. No other cases supervened. The deceased was asserted to have died of plague, but no evidence was adduced in support of this assertion to show that the case was one of true bubonic plague.

Asia Minor.

Rumours of plague in various parts of Turkey have not been infrequent, but, for various reasons, definite reports have not always been obtainable. The most recent outbreak heard of is one that was reported in 1896 from Bitlis on the Lake Van, where plague was asserted to have broken out among the men of the Hamedieh Regiment, the Kurdish Imperial Cavalry. At the time of the dispatch, May 22nd, 1896, some 15 deaths had taken place. It is well known that the sanitary condition of most of the towns in Asia Minor is far from satisfactory. A recent Consular report speaks of the people of these towns of Khurdistan as "living in filth." The town of Erzeroum, for example, is described as being one of the filthiest in Asia Minor, and it is seldom free from epidemic disease of one or another kind.

Persia.

In March, 1881, plague broke out in some villages in the northern part of the Province of Khorassan (*see* map), the disease having been active in this same locality in 1876–77 and 1878. One authority, Dr. Mahé,* regards this repeated appearance of plague in the N.E. of Persia as pointing to a possible relation between the plague centres of Mesopotamia and Khurdistan on the one hand, and those to the N.E. of the Himalayas on the other. In 1882, plague appeared at the village of Ouzoundéré, the population of which numbered 524; of these no less than 259 were attacked and 155 of them died. Other villages in the neighbourhood also suffered; at Hadje Hassan, for example, out of 130 inhabitants, 63 were attacked, and 47 died.

In 1885, during January and February, plague was prevalent in the neighbourhood of Hamadan, and in April, 1889, the disease broke out in the Bana and Soudzi Bulah districts of Persian Kurdistan.

In January, 1890, several villages in the Mahidest district, 14 hours distant from Kermanshah, were invaded by plague. Various accounts were given of the cases, some calling it Oriental plague, others regarding it as a virulent typhus fever. But the outbreak was officially reported as bubonic plague to the English Foreign Office. In one small village there were 51 cases and 30 deaths. In 1892, there were many fatal plague cases in Astarabad and Axdbil. No definite records of plague incidence in Persia have been forthcoming during more recent years. It has to be borne in mind that Persia is not a country well supplied with railways or telegraphs, and for this reason news travels slowly, or not at all. The extensive plague epidemics in India and China have occupied public attention and possibly tended to divert notice from lesser outbreaks in Persia and elsewhere. In 1898, news came from Teheran that Persia was free from plague. At Meshed, however, many persons, among whom were children, were suffering from enlargement of the lymphatic glands (? pestis minor). It must not be forgotten that Meshed is a town at which there is a continuous gathering of Moslem pilgrims from India, Afghanistan, Baluchistan, as well as from distant parts of Persia. Some critics regarded the enlargement of the glands, spoken of above, as due to prevalence of venereal disease,† but, on the other hand, the fact that many of the sufferers were little children is against this view.

---

* Recueil de la Comité Consultatif d'Hygiène Publique, Vol. XI., p. 242.

† *Vide* report by Dr. Zavitziano, U.S. Sanitary Commissioner at Constantinople. Public Health Reports, January 18th, 1899.

China.

Epidemiologists have been aware for some time that in certain localities in China, particularly in the mountain-valleys of Yunnan, there existed endemic centres of bubonic plague. Records of outbreaks of this disease can be traced back some 50 years or more. Owing, perhaps, to the hitherto scanty means of communication with the interior of China, we are wanting in definite records of virulent epidemics of plague which are said to have ravaged extensive districts in China, without the knowledge of their existence at the time coming to Europe.

In Southern China plague has broken out with considerable regularity during the last 20 years; for example, it appears regularly in January of each year in the province of Kwangsi, and in the districts of Lienchou and Leichou. The disease was reported as occurring from 1881 to 1884 at Pakhoi (*see* map), an open Chinese port to the north of the Gulf of Tonking. In 1889 it was epidemic at Lungchow, and in 1890 it broke out at Wu-Chu on the coast between Pakhoi and Canton. In 1891 it appeared at Kao-Chau, the prefecture adjoining Lienchou, in which Pakhoi is situated. Thousands are said to have perished of the pestilence in the towns and villages round Kao-Chau.

Pakhoi was a few years ago stated, on good authority, to be "a most filthy town; in hot weather the streets become loathsome, and the stench arising from fermenting garbage on every side and from stagnant sewage is past description." More recent reports, however, say that since 1891 the sanitary condition of Pakhoi has been improved.

In 1892 plague was epidemic at An-pu, about 100 miles east of Pakhoi, and carried off a large number of the inhabitants during the months of March and April. The French missionaries, who have resided for many years in the district, say that plague is endemic in the neighbourhood of An-pu, and that the disease recurs almost annually in epidemic form during the spring.

In 1893 plague was again reported to be prevalent at Pakhoi. It was also appearing in Yunnan, more especially at Mengtsz, in the south of the province, where at least 1,000 deaths occurred in a population of about 12,000. Plague also was rife at Lungchou, and in many towns in the Kwang-Si province. There is continual commercial communication between Yunnan and Kwang-Si.

In February, 1894, a terrible epidemic of plague began to devastate Canton; this outbreak reached its height in May, and by July had nearly disappeared. The total number of deaths from plague during this epidemic at Canton cannot be definitely ascertained, as there is no system of death registration in operation. But it has been calculated by missionaries, resident in the locality, that upwards of 100,000 persons out of an estimated population of 1,600,000 must have then lost their lives.* Other authorities estimate the number of plague deaths at 60,000. One reporter bases his statements on the fact that in four months no fewer than 90,000 coffins were sold in Canton, 75 per cent. of which were believed to have been for the burial of persons dead of plague.

* Proust; "Defense de l'Europe contre la Peste." Recueil des Travaux du Comité Consultatif d'Hygiène Publique de France, Vol. XXVII., p. 273.

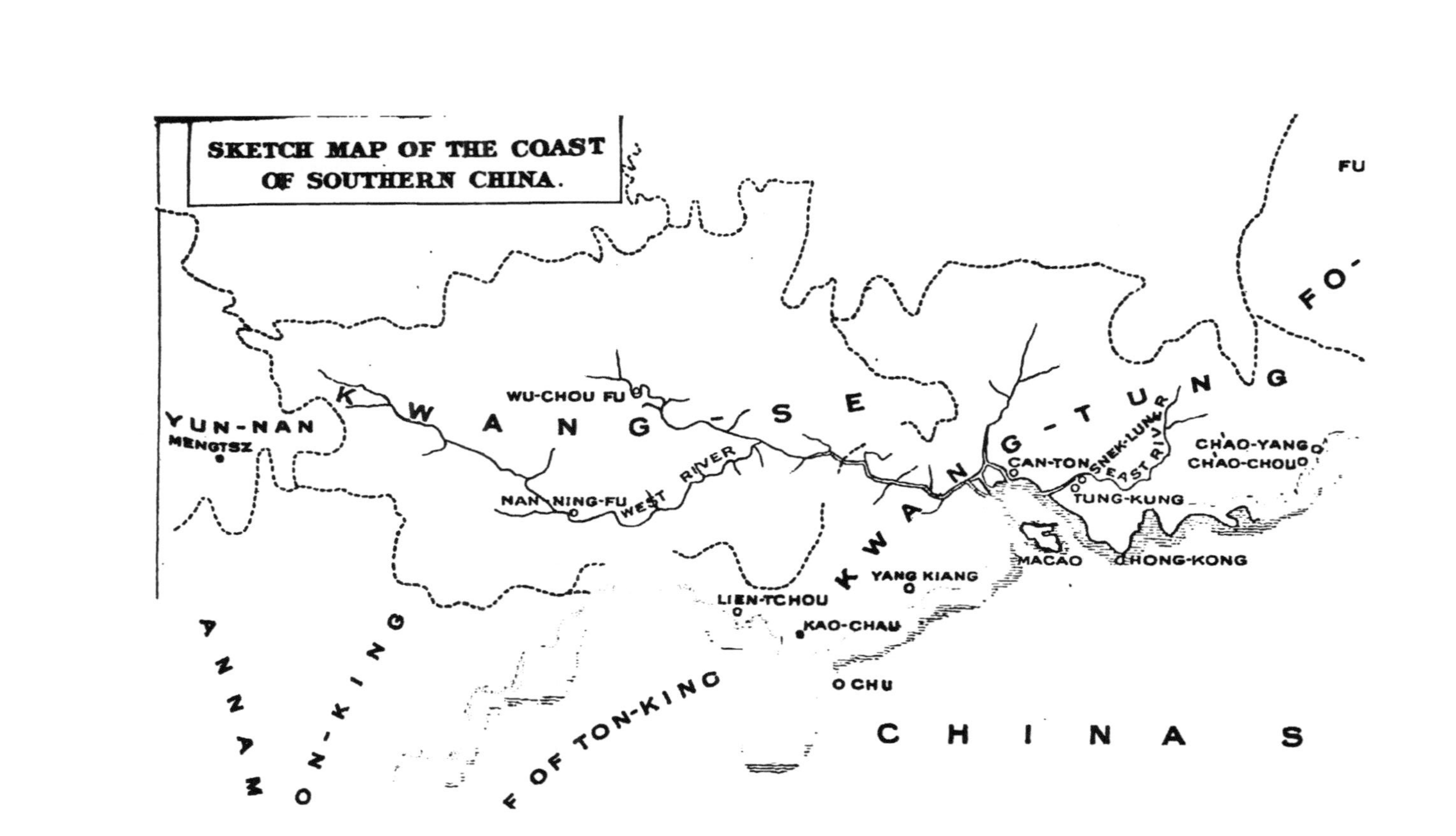
SKETCH MAP OF THE COAST OF SOUTHERN CHINA.
YUN-NAN
MENGTSZ
KWANG-SE
WU-CHOU FU
NAN NING-FU
WEST RIVER
KWANG-TUNG
CAN-TON
SNEK-LUN
EAST RIVER
TUNG-KUNG
CHAO-YANG
CHAO-CHOU
MACAO
HONG-KONG
YANG KIANG
LIEN-TCHOU
KAO-CHAU
O CHU
FO-
FU
CHINA S
ANNAM
ON-KING
F OF TON-KING

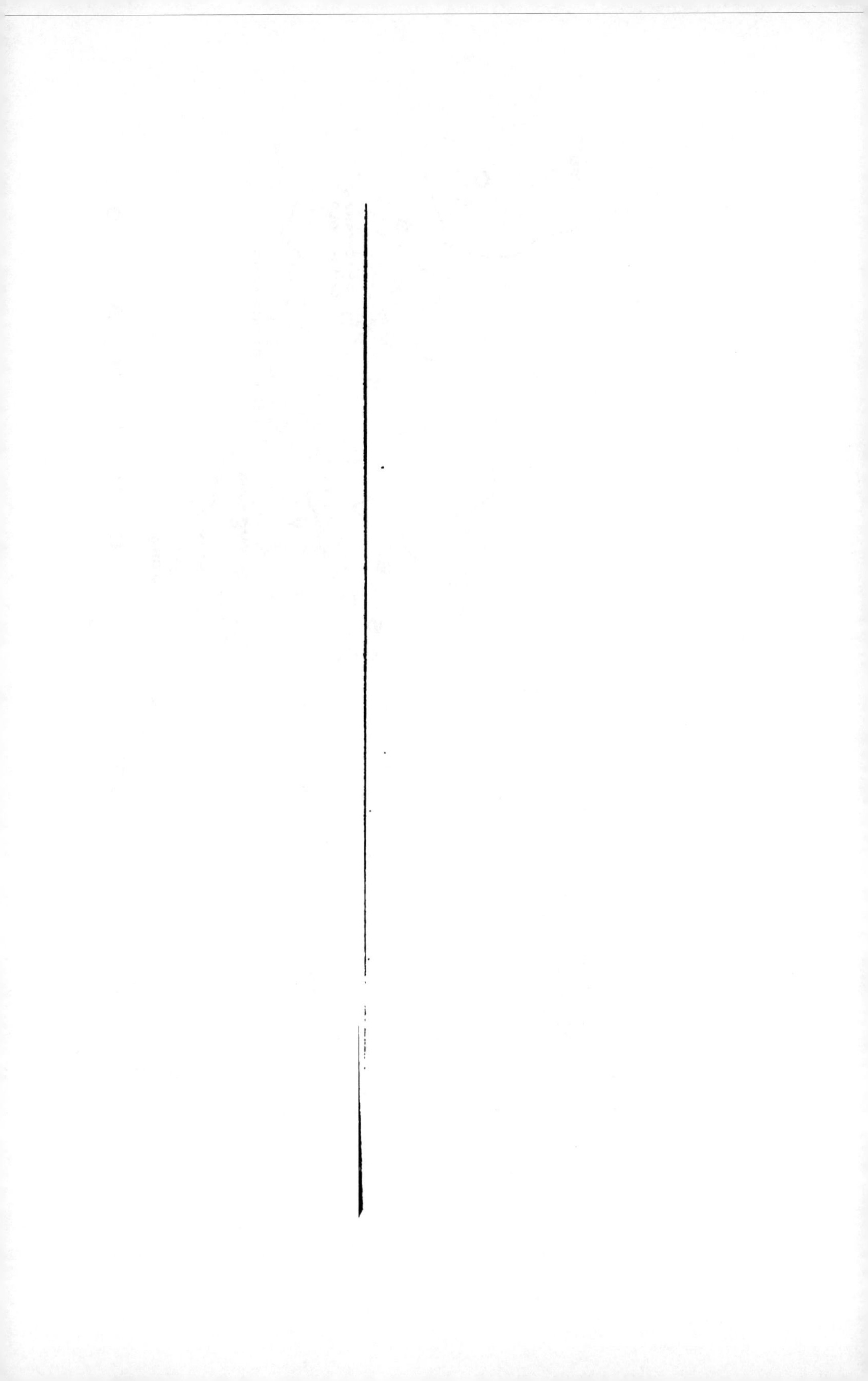

Hong Kong was attacked in May, 1894, by plague. The reported deaths are stated to have been 2,550, but this does not in all probability represent the true amount of the fatality. It is asserted that 90 per cent. of those attacked died. The Chinese, as soon as they felt ill, at once left Hong Kong for their native villages on the mainland, since they had great fear of leaving their bones upon a foreign shore. Of a native population of about 210,000, it is said that 80,000 left Hong Kong during May and June, many of them being already ill, or developing the malady. At one period during this exodus there were as many as 15 deaths a night on board the British steamers which leave Hong Kong at six in the evening and arrive at Canton at six next morning. All these persons were apparently well when they walked on board the steamer at Hong Hong, otherwise they would not have been permitted to embark. As is well known, serious riots occurred in Hong Kong in consequence of the Chinese objection to Western methods of isolation and treatment of the infected sick in special hospitals.

At Canton, as well as at Hong Kong, in 1894, the appearance of plague was preceded by a remarkable mortality among rats. Indeed, in some parts of China plague is spoken of as "the rat disease." The superstitious Chinamen regarded these animals as the messengers of the devil, and endeavoured to drive them away. In certain quarters of Hong Kong as many as 20,000 dead rats were collected in a short space of time, 1,500 being obtained alone from one street. The 1894 outbreak in Hong Kong gradually subsided in July and August. The outbreak had appeared in the cool season, when the houses of the Chinese are much more overcrowded than they are in the hot season. The malady prevailed most in the districts inhabited by the poorest classes of Chinese, among whom bad hygienic conditions were common.

There can be little doubt but that the plague infection was brought to Hong Kong from Canton or Pakhoi, probably by ships during the ordinary operations of commercial intercourse. Dr. Wilm,* in a report upon this outbreak, mentions as carriers of contagion (1) persons arriving in Hong Kong suffering from plague in the early stages; (2) articles of clothing contaminated by fæcal matter, urine, or expectoration of plague patients; (3) rats, mice, or even pigs coming from plague infected places; (4) flies were also regarded as carriers of the disease from one infected house to another. Other observers have drawn attention since then to fleas and bugs as probable conveyors and inoculators of contagion from the infected to the healthy. Dr. Wilm gives it as his opinion that plague bacilli may be taken into the system with food, and he adds that food of various kinds, and fruit, brought from markets and places which were in a most filthy condition, were habitually hawked about the poorer streets, and he believed they had helped to diffuse the disease.†

* Report on the epidemic of bubonic plague in Hong Kong, by Staff-Surgeon Wilm, M.D., of the Imperial German Navy.

† The Bombay Plague Research Committee state that there was no autopsy which went to show that the plague bacillus had reached the stomach or intestines in food, and had thence infected the messenteric glands. The German, Russian, and Austrian Plague Commissions confirm this view that plague is not spread by means of articles of food among human beings, though rats are readily infected by feeding them with food purposely mixed with plague cultures.

P. A. No. 18.

the ...usion of ...onic ...gue from ... to 1898; by Bruce Low.

After a period of six months apparent freedom from plague the malady re-appeared in Hong Kong in 1895. Cases began to be heard of in April, but the disease did not on this occasion assume epidemic proportions. Only 44 cases were officially reported.

Places in the vicinity of Hong Kong were also invaded in 1895. In November and December cases were reported from Canton, Fatsham, Sheklung, Fungkun and elsewhere. There was also an outbreak at Amoy in 1895, but no details have been given showing its extent.

Foochow also suffered, as well as the native villages situated on the island of Nantai.

Swatow, where a few cases had been heard of in 1894, was also invaded by plague in 1895.

From April to June, 1895, this disease caused considerable havoc among the Chinese population of the Portuguese Colony of Macao. Dr. Da Silva,* the Chief Medical Officor to the Colony, reports that 2,592 deaths from plague occurred in 1895, of whom 2,559 were Chinese.

During 1895 plague re-appeared at Mengtsz, in Yunnan, near the borders of Tonking. It is estimated that between 1,200 and 1,500 persons lost their lives in the outbreak. Early in the same year (1895) plague made its appearance in the junk-port of Pochin, which is about 20 miles distant from Hoihow (Kiung-chow). For about five months the malady raged in Hoihow, and in the adjacent villages. It is estimated that 3,000 persons died of plague during the epidemic out of a population of 30,000. From Hoihow an infected person carried the disease to Kiung-chow, where, however, only 30 cases resulted at the time. But early in January, 1896, plague became epidemic at Kiungchow, where it is stated that "people died in great numbers"; but no details are given in the official reports.

In the spring of 1896 plague appeared in the town of Ch'ao-yang, whence it spread to neighbouring villages. The epidemic raged with great virulence during April and May, gradually subsiding in June, and by the middle of July the outbreak had all but disappeared. A modest estimate of the plague mortality in Ch'ao-yang would be 2,000 deaths. At Haimên, at least 1,200 persons died of plague.

Hong Kong, early in 1896, again suffered from a plague outbreak, which reached its height at the end of May, though cases continued to be reported up to November; 1,204 cases were notified during the year, and of these 1,078 proved fatal, a case mortality of 89·5 per cent. A considerable number of cases occurred upon the native boats in the harbour. The distribution of the cases in Hong Kong throughout the year is given in the subjoined Table.

1896.	January.	February.	March.	April.	May.	June.	July.	August.	September.	October.	November.	December.	Total.
Plague cases reported.	49	125	168	316	344	113	52	25	9	2	1	...	1,204

* *Medical Reports of the Chinese Imperial Maritime Customs.*

Plague also reappeared in Canton in 1896. Some sporadic cases were reported in January, in April the disease had assumed epidemic proportions, and in May it began to abate. No details as to the probable number of cases are obtainable. Amoy was also invaded by plague in May, and Swatow in June, but in the latter place the cases were never in sufficient numbers to amount to an epidemic.

In April, 1897, the Portuguese settlements of Lappa and Macao suffered from plague. The first case was observed in March, at Macao, in the person of a Sister of Mercy, and this case was followed by a number of others in the Convent to which she belonged. The outbreak in Macao lasted till June. The full records of this epidemic are not available. Dr. Da Silva,* who reported upon it, does not give the total number of cases and deaths, but he gives a table showing that in the three months, April, May and June, 251 deaths occurred from plague, and of these, 13 were among inmates of the Convent of the Sisters of Mercy.

During 1897 plague reappeared at Amoy, where in June it was stated upon the authority of the Chinese interpreter of the United States Consulate, that as many as 100 cases were occurring daily. Swatow and Foochow also suffered from epidemic plague during the year, but no statistical information is available as to amount and fatality of the disease. At the beginning of May plague broke out on the frontier of Kwangsi in four villages, three kilometres south of Biha, in Tonking. There was a reported occurrence of plague in several Chinese villages near the French frontier about the same time. A marked mortality among rats preceded this outbreak, as has been observed elsewhere.

At Hong Kong, in 1897, scattered cases of plague were observed, but the disease did not assume epidemic proportions, only 17 cases having been officially reported. These were said to have been imported from the mainland. But in the first quarter of 1898 the malady again began to prevail, and continued till the end of June. The total number of cases officially reported numbered 1,315, of which 1,160 proved fatal, a case-mortality of 88·2 per cent. During the first quarter of the year 213 cases came under observation, and during the second quarter 1,094 cases; while in the third quarter only 8 cases were reported. Of the total 1,315 cases 1,240 were Chinese, and of these 1,111 died, a mortality of 89·6 per cent., while of 75 non-Chinese 49 died, a mortality of 65·3 per cent. Certain villages on the mainland also suffered from plague during 1898. At the close of the year sporadic cases were again reported from Hong Kong, and 1899 unfortunately saw a recrudescence there of the disease.

In 1898 there was again a serious outbreak at Amoy, but, owing to the fact that the Chinese Authorities collected no statistics upon the subject, no figures can be quoted showing the extent of the epidemic. This is said to have prevailed from May to September, the maximum being reached in July. The disease was confined to natives, the foreign community entirely escaping infection. In some of the towns and villages near Amoy plague also prevailed, and caused considerable mortality.

---

* Medical Reports of the Chinese Imperial Maritime Customs.

P. A. No. 18.
—
the
'ssion of
onic
gue from
to 1898 : by
Bruce Low.

During the first half of 1898 plague was rife in parts of the Consular district of Swatow, more especially in the city of Ch'ao Yang. During April and May it appeared in some of the villages near Ch'aochow Fu, and in the immediate vicinity of Swatow. By the middle of July it had practically disappeared. The occurrence of plague in the Swatow district was reported by the British Consul, but no statistics as to the number of attacks and deaths have been transmitted up to the present.

Plague outbreaks were also reported from Annam in June and July 1898, and cases continued to occur up to November. The infection is stated to have been carried to Annam by Chinese Junks, trading from Canton and Hainam. The first cases occurred in March, at Culao, where the Junks were in the habit of calling for the purpose of purchasing pork. Culao is situated on one side of an estuary, on the other side of which is the Institute of Dr. Yersin, whose name is well known in association with the preparation of a serum for the cure of plague. It was at first alleged that the plague infection had been introduced into the neighbourhood from Dr. Yersin's laboratory,* as at Vienna (see page), but this was completely disproved upon investigation. The chief centres of infection were Xuong-Huan, Phuong Can and Nhatrang. The complete returns of recorded cases are not given, but Dr. Yersin reports that of 72 cases observed by him 53 died. Of the 72 cases, 33 were treated by the serum method, and of these 14 died, a mortality of 42 per cent. The epidemic was arrested by burning the infected and adjoining houses. In the infected zone an antecedent mortality among rats had been observed.

Mongolia:—Since 1888 bubonic plague has appeared year by year in the valley of So-len-ko (or Selenga), which lies in the mountain range on the eastern border of the great Mongolian plateau. The district known to be infected is called Toung-Kia-Yng-Tze, and is stated to be "about 12 days ride from Pekin." The district is sparsely populated by a people who originally came from the province of Chang-Tong, on the frontier of Manchuria. Labourers from this province occasionally go to So-len-ko Valley to assist in agricultural work, and it is believed that plague was carried there from Chang-Tong, which has trade communication along its coast, by means of shipping with Amoy, Canton, and other southern Chinese ports where plague has been occurring for a number of years.

The people of the Toung-Kia-Yng-Tze district live in mud huts, under unfavourable conditions of dirt, overcrowding and want of ventilation. Dr. Maignon,† who visited the locality in 1897, says that the majority of the inhabitants do not wash themselves once a year. Soap, he says, is a thing absolutely unknown among the people. Their clothes, which are caked with dirt, are worn till they fall to pieces. The houses have no closet accommodation, the custom being to defecate in the open air, near to the huts. When plague breaks out in a family, the sick person remains in the living apartment common to the rest of the inmates.

---

* Journal de l'Institut Pasteur.

† La Peste Bubonique in Mongolia by Surgeon Major Matignon (Attaché to the French Legation in China).—Annales d'Hygiène Publique. T. XXXIX., p. [illegible]

Patients suffering from the pneumonic form of plague expectorate upon the walls or floor, or when they become weaker, upon the pillows or bed covering. Sometimes the relatives remove the sputum from the sufferers lips by means of their fingers, which are wiped upon their clothing, and neither washed nor cleansed in any way. Under such conditions it is not surprising that multiple cases in families are common, and that the type of the disease is usually of a severe kind. During periods of epidemic prevalence of plague it is stated that the dead are not buried, but are thrown down a neighbouring ravine, where the corpses are devoured at night by wolves. No precise records as to numbers of attacks and deaths are obtainable, but in 1896 Dr. Matignon heard of at least 160 fatal cases, occurring in ten small villages. In 1897, seven villages suffered, but the number of cases is not known.

Towards the end of 1898, news reached St. Petersburg, through a Russian doctor, that plague was prevalent in the So-len-ko valley. The Russian reporter said he had seen 30 persons ill of plague during his visit to the valley. They were living in abject misery, among filthy surroundings. In addition to the cases which the Russian doctor himself saw, there had been, he stated, a considerable amount of the disease in the district, attended with a heavy mortality.

Japan.

The information from Japan as to plague is not very definite; but it is known that isolated cases were imported into Japanese ports by ships from Hong Kong and other infected Chinese centres during and after the outbreaks of 1894. The disease, notwithstanding these repeated importations, does not appear to have hitherto secured any foothold in Japan itself. The island of Formosa, however, was the chief place that suffered.

Formosa:—In July, 1896, bubonic plague was reported to have appeared at Anpei, where 43 cases came under observation. Later in the year, other localities became infected by the disease.

Plague was officially declared in the town of Taihoku on October 28th, 1896, and the Japanese Government despatched a Commission of medical men, including Professor Ogato, of the University of Tokio, to investigate the outbreak. In the town of Taihoku and in the adjoining districts of Kirung and Tamsui, from November 28th to December 31st, 132 cases were recorded. The mortality among the recognised attacks was 56·1 per cent. Previous to the recognition of the disease, there had been a fatal prevalence of illness characterised by fever, which was regarded as "virulent malaria" by the local Chinese and Japanese medical practitioners. Such cases were known to have occurred from June, 1896, up to November, when plague was at last recognised. These earlier cases are now regarded as having been true bubonic plague. It is believed that the infection was brought to Taihoku from Anpei, which has trade communications with southern Chinese ports. About the time of the Taihoku outbreak, a considerable mortality among rats and mice had been observed there. Subsequently it was proved by bacteriological examination that these animals had died from bubonic plague. Professor Ogato states that he, as well as other medical men, picked up dead rats in the public streets, and he was able to affirm after examination that they had been suffering from plague. Some dead rats were sent to him wrapped up in newspapers, and when the parcels were opened the bodies of the rats which had died of plague were found to be infested by fleas. He therefore believes that there is a danger of plague infection being transferred by fleas from plague stricken rats to human beings.*

The actual numbers of cases and deaths which occurred in Formosa, from July to December 1896, are not forthcoming, but they are stated to have been considerable. Plague continued to occur in Formosa in the early months of 1897, the number of reported plague attacks from January to July being given as 541, and the deaths as 418. But an authority resident in the island says that these numbers ought to have been doubled to arrive at an approximation of the actual incidence of the disease. Towards the latter part of the year the outbreak seems to have subsided, but it reappeared with increased severity in 1898. The cases during 1898 were almost exclusively confined to the shipping ports of Taipeh, Tainan, and Taiwan. In Taiwan, with a population of about 135,000, the epidemic reached its height towards the beginning of May; in the three weeks ended May 19th, no fewer than 2,223 persons were attacked, of whom 1,421 died, a fatality of 63·9 per cent. The outbreak subsided in June, and seemed to disappear during the autumn months, but on December 25th, 10 fresh cases, two of them fatal, were officially reported. In the first quarter of 1899 plague was found to be epidemic in Tainan, 683 cases having been reported, with 477 deaths.

INDIA.

It is asserted by Indian medical authorities that plague, under the name of "*Mahamari*," is endemic in the districts of Kumaon and Garhwal, situated on the south western slopes of the Himalayas, on the frontiers of India and Thibet. A form of plague was epidemic at Pali, in Rajputana, in 1836–37, and this is usually referred to as the "Pali Plague," which is believed also to have been true Bubonic plague.

Some contend that "*Mahamari*" is imported from time to time from Thibet into Kumaon and Garhwal, and is not, strictly speaking, endemic in these districts. This "*Mahamari*," or "Indian Plague," which is said by those who have had opportunities of studying it, to be identical with the Bubonic plague of late years so prevalent in India, has broken out frequently in the above localities during the last 20 or 30 years. Such outbreaks were reported in 1884, 1886, 1887, 1888, 1891, 1893, 1894, and 1897. All these reported outbreaks were on a comparatively small scale. For example, the most recent outbreak, which occurred in 1897,

* "Ueber die Pestepidemie in Formosa," by Professor Ogato, Centralblatt für Bacteriologie, &c., Vol. XXI., p. 769.

at a small and remote village among the hills, 20 miles from Kedarnath, comprised only 17 fatal cases, and the disease did not spread beyond the village first attacked. This "Himalayan Plague" had, apparently, no connection with the recent outbreak of plague in western India, which began in September 1896 at Bombay.

The City of Bombay was declared to be infected by Bubonic plague on September 23rd, 1896. From reliable sources, it is stated that fatal cases of the disease were recognised there during August, and it is further alleged that cases of fever with glandular swellings occurred during the summer months of 1896, and, according to the testimony of some, had even been noticed in previous years. From Bombay the disease spread rapidly during the last three months of 1896 throughout the Presidency of Bombay. During December a few suspicious cases were observed in Calcutta, the first being a traveller from Bombay; and at the end of the year cases were imported into Rajputana, but no epidemic then followed. During 1897 plague continued to prevail in Bombay City and Presidency, and extension of the epidemic took place to some of the native states in Central and Southern India. Cases also occurred in the Hardwar Union Municipality in the North-West Provinces, but the malady did not spread very widely, though it lingered on till December. In February plague appeared in the Jullundur district in the Punjaub. A few imported cases were also reported from Madras. In 1898 the disease continued in Bombay City and in the Presidency. Cases, too, were reported from the North-West Provinces, the Punjaub, the Central Provinces, the native states of Hyderabad and Mysore, and in the cities of Madras and Calcutta, as well as in parts of the Presidencies of Bengal and Madras.

At the beginning of 1899 plague was still epidemic in Bombay, and was increasing in Bengal and Calcutta, though declining elsewhere. It is estimated that from September 1896 to January 1899 more than 250,000 persons died of plague in India. This number is probably below the truth, since cases and deaths were commonly concealed from the authorities in many places, and, moreover, the registration of deaths is defective throughout most of India.

THE CITY OF BOMBAY.

How the infection was brought to Bombay in the first instance is still unascertained. Several theories have been advanced, the more important being as follows:—

1. The disease may have been brought to Bombay by sea from Hong Kong, or from some other infected port in Southern China, where plague had been epidemic for some years previous to 1896.

2. It may have been conveyed from the Persian Gulf, the disease being endemic in Mesopotamia, and there being continual communication by shipping between the ports in the Persian Gulf and Bombay.

3. The infection may have been carried from the Kamaon and Garhwal hills to Bombay by pilgrims. In support of this last theory it is affirmed that in May, June and July of 1896 a large

under conditions of warmth, moisture, and darkness. It is evident that in the winter months all these conditions are to be found in the small, overcrowded filthy rooms of the lower class population of Bombay.

The chawls, or tenements, are high buildings, sometimes of seven storeys, and each shelters as many as 500 to 1,000 individuals. These tenements are crowded, too, upon area. It is stated that 70 per cent of the people of Bombay live in such tenements. The sub-soil of the city is sodden with sewage, and many of the drains are choked. Basements of houses are damp, and in the wet season the cellars are flooded. From the table on page it will be seen that in Bombay the first epidemic, which reached its height in February, was practically over in June, 1897. The number of plague deaths recorded during the period, September 1896, to the beginning of June 1897, was 11,577, and the reported attacks 13,823. These numbers are probably far below the truth.

With the subsidence of the epidemic in June 1897, public confidence began to return, and those who had fled in panic early in the year flocked back to their homes in the City.

During 1897 famine existed in the Bombay Presidency, and in the summer months many starving families came to Bombay. In August cholera broke out, and 220 deaths from it were recorded during the first seven days of the month. Relapsing fever also appeared, and ultimately became epidemic. The lowest number of plague cases and deaths recorded in any month in Bombay during 1897 was in July. The following table gives the gradual development of the *second* epidemic in the City :—

—	1897.						1898.					
	July.	August.	September.	October.	November.	December.	January.	February.	March.	April.	May.	June.
Plague cases	72	124	247	290	378	924	2910	4812	4709	2372	?	?
Deaths ...	44	70	168	178	252	644	2934	4498	4973	2171	524	150

It will be seen that in January and again in March the recorded number of plague deaths is given as in excess of the number of reported plague cases.

The curve of this second epidemic follows that of the first as regards the period of the year in which it began, reached its height, and subsided. The complete figures as to the cases during the last two months of the second epidemic have not been obtained. It appears, however, that more than 17,000 persons were reported as having been attacked, and no fewer than 16,606 were certified as having died from plague between July, 1897, and the end of June, 1898. This second outbreak, therefore, it will be seen, was on a more serious scale than the first.

The *third* epidemic, which extended into the spring of 1899, occurred practically in the same months of the year as the first and second epidemics, and reached its maximum in March.

to Novr. 1898.

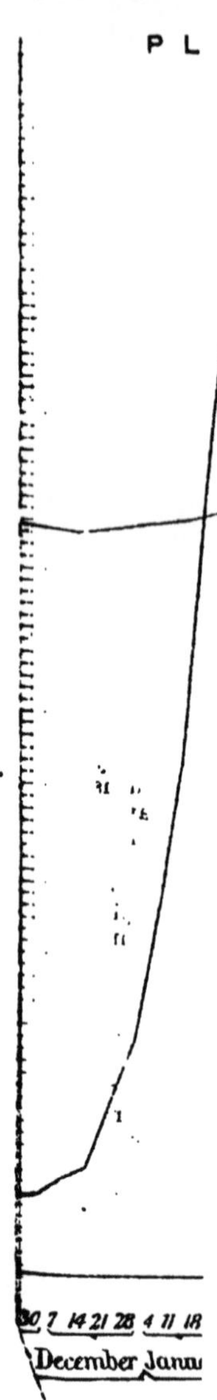
P L
30 7 14 21 28 4 11 18
December Janu

table shows the number of ascertained plague
e *third* epidemic :—

1898.		1899.						Total from Nov., 1898, to June, 1899.
Nov.	Dec.	Jan.	Feb.	Mar.	April.	May.	June.	
198	388	1678	3045	4262	2856	1092	249	13,768

sidence of the epidemic in each of the three outbreaks
coincided with the advent of the rainy season. The
ying chart, for which I am indebted to Dr. A. H. Smee,
e weekly plague mortality from December, 1896, to
9, together with the recorded data of rainfall and
The weekly number of plague deaths in this chart is
m the Weekly Return of the Registrar-General. The
are no doubt understated, but such error is probably
hroughout, and need not, therefore, affect the comparison.
asures at first employed in dealing with the outbreak were
al character. They included the closing of insanitary
the abatement of overcrowding, the evacuation of
while disinfection and cleansing were carried on, and
washing of dirty houses. Early in March, 1897, a
mmittee was appointed, under General Gatacre, to take
s for suppressing the epidemic. It was found necessary
special efforts for the discovery of cases, and for the
of all those attacked to isolation hospitals, together
segregation for a time of those who had been in
ith the sick. Considerable opposition of a fanatical
met with in carrying out these measures. Both
dans and Hindus objected to isolation of their sick, and
ith intense indignation any attempt to intrude upon the
their homes by the house-to-house visitors. The Moham-
owed active opposition, and in public meetings urged
preferable to isolation. As showing the popular feeling
spital treatment it may be mentioned that on several
mobs threatened to demolish the hospital buildings. On
ions a mob of mill hands threatened to wreck one of the
and had to be dispersed by the police. The Plague
e issued a document explaining the objects of isola-
and replying to the misrepresentations that had been
. After this, active opposition became less marked, and
ittee carried out a very elaborate system of preventive
by aid of European officers and native soldiers. The
divided into 10 districts, each with a medical officer,
l steps were taken to detect cases of plague and remove
ublic or private hospitals, the latter being provided by the
ects for their own people. Fifteen Government hospitals
al of 680 beds were opened, and 29 private hospitals for
ects and castes were established throughout the city.
sal of corpses, and the disinfection of houses and articles
g, etc., were carefully attended to. General measures for
the sanitary condition of the city were also adopted.
nbay is the principal port of entry and departure for
s travelling *to and from* India, it was necessary to prevent

diffusion of plague by passengers whether on their way through Bombay to other Indian districts or leaving the country by this seaport. A system of railway inspection was organised on the Great Indian Peninsula and East Indian Railways.

When plague subsided in Bombay, it was found necessary to provide for inspection of all persons entering Bombay by land from infected districts outside. The rules issued for the principal inspection stations in Bombay provided that the train should be detained till the Medical Officer certified that all persons travelling by the train were free from plague. The Medical Officers were authorised to detain suspicious cases, as well as those actually attacked, and to keep them under observation for a time. Hospital accommodation was provided close to the stations, and also separate camps for suspected cases. For the examination of female passengers, female doctors or trained nurses were appointed. Dirty clothing and articles believed to be infected were disinfected. Measures were also taken for the disinfection of railway carriages.

Arrangements were also made for inspecting persons entering or leaving Bombay by road or by river. Native craft were brought with some difficulty under inspection. They often carried passengers who had come from infected places, or who were leaving Bombay when the epidemic was raging.

The proper medical examination of passengers leaving Bombay, on board ship, was also held to be a matter of considerable importance to Europe.

During 1896 the number of vessels inspected was 570, and the number disinfected 59. Only two plague cases were found on board vessels during the last three months of 1896.

During 1897, 60 vessels entered the harbour having on board cases of suspicious illness, 65 of which ultimately proved to be cases of plague after the patients had been sent to hospital for observation. Eleven other vessels with 14 fully developed plague cases on board also arrived. These 71 vessels came from ports in proximity to Bombay. With regard to outward bound vessels, up to February 6th, 1897, only those sailing for European or other ports where inspection and other restrictions are imposed were inspected; but subsequent to this date all vessels were examined before leaving port.

From the report of the acting Port Medical Officer (Major Crimmin, V.C., I.M.S.) I have taken the following figures as regards the vessels which left or entered Bombay harbour :—

TABLE giving a SUMMARY of the SHIPPING and of the PLAGUE CASES DISCOVERED on VESSELS ENTERING or LEAVING BOMBAY HABOUR during 1897.

Vessels, including Native Craft.	Number of Vessels Inspected.	Number of Persons Inspected, including the Crew.	Number of Vessels Disinfected.	Number of Persons Rejected or Segregated.	Number of Persons with Plague.
Entered the harbour ...	32,187	602,401	13	39,230	79
Left the harbour	40,565	710,716	—	3,322	71
In harbour during the year	56	—	56	—	19
	72,808	1,313,117	69	42,552	169

Owing to the outbreak of plague in Bombay, only two ships, with 1106 pilgrims, left the port of Bombay for Jeddah in January, 1897. After that date all pilgrims were forbidden to leave Indian ports for the Hedjaz.* No cases of plague or deaths occurred on board these two pilgrim vessels between leaving Bombay and arriving at the Turkish Quarantine Station at Camaran. Four ships returned from Jeddah with pilgrims, and a plague had meanwhile been reported at Jeddah, the baggage of the pilgrims in the last two ships was disinfected, but no plague cases developed. The returning pilgrims from the four ships were not allowed to enter the town of Bombay, but were sent direct to their homes at the expense of the Government.

The success of the system of inspection of departing passengers and crews at Bombay is shown by the fact that from September, 1896, to December, 1897, only two ocean-sailing steamers are known to have developed plague on board after leaving Bombay. These two vessels were (1) the "Pekin," which left Bombay on December 28, 1896, with 1,045 pilgrims on board for Jeddah. Two cases of plague developed, but no subsequent extension of the disease took place on the ship, notwithstanding the fact that more than 1,000 pilgrims, usually a class by no means remarkable for personal cleanliness, were on board; and (2) the transport "Dilwara," which sailed for Southampton on March 10, 1897, with 1,235 souls on board. On the evening after leaving Bombay a child of a non-commissioned British officer became ill, and she died of plague on March 18, in the Red Sea, where the body was consigned to the deep. The family had been living prior to departure at Kolaba, a suburb of Bombay, where plague was then prevalent. No further cases followed on board the "Dilwara."†

During 1898 two ocean steamers developed plague on board after leaving Bombay: (1) The "Carthage" left Bombay on July 2, 1898, with 111 passengers and 220 officers and crew. Four days later a coal-trimmer, a Lascar, was found to be ill with plague. There was evidence that he was ill at starting, but had evaded the examination in a surreptitious way. He was isolated, with two attendants, in a boat swung on the quarter. All three persons were landed at Aden, and the ship was disinfected. A second case, another Lascar coal-trimmer, was discovered on July 14. He was isolated as in the former case, and made a good recovery. The men's quarters were again disinfected. There was no further extension of the disease. (2) The "Caledonia" left Bombay on November 19, 1898, and reached Suez on the 27th. On that date one of the crew, a Lascar, was found with an illness which had some of the clinical symptoms of plague, and another of the crew also had suspicious symptoms. They were landed at the quarantine station at Moses' Wells. No further cases followed. Later advices from the President of the Quarantine Board in Egypt point to these cases not having been plague. (*See* memorandum on the "Caledonia," by my colleague, Dr. Bulstrode, page   of present volume.)

* At the end of this report will be found a copy of the Orders issued by the Government of India under the Epidemic Diseases Act, 1897, dealing with the pilgrimage to the Hedjaz and emigration from India (*see* page   ).

† As already mentioned, there occurred in September, 1896, three suspicious cases, believed to have been plague, who died in the Seamen's Hospital, London. Two of the patients were Goanese steward's helpers on board a vessel from Bombay, and they are believed to have been infected in the Port of London, after their arrival, by the wearing of infected clothing bought at Bombay, but never worn during the voyage. (*See Report* by Dr. Buchanan in the Twenty-sixth Annual *Report of the Medical Officer* to the Local Government Board, page 139.)

Bombay Presidency.

PP. A. No. 18. [...] the [...]ffusion of [...]bonic [...]gue from [...]9 to 1898; by [...]. Bruce Low.

From Bombay city plague was rapidly diffused in 1896 by the panic-stricken people who fled from the city in fear of the disease or of the preventive measures adopted to arrest it. The first cases recognised outside Bombay were at Ahmedabad, in October, and during the same month cases were reported from Poona, Thana, Khandesh, Kolaba, and Dharwar districts. In November Bijapur and Surat districts became infected. A wider extension of the epidemic occurred during December, when seven districts in the Presidency were declared to be infected, viz., Ahmednagar, Sholapur, Broach, Kaira, Karachi, Hyderabad, and Shikapur. Up to the end of December the official returns state that 367 plague attacks, with 283 deaths, had been recorded in the above districts during October, November, and December. Table I. shows the districts infected from, and including, Bombay city during the last quarter of 1896, the population and area of the districts, the date of infection, along with number of plague cases and deaths.

Table I.—Plague in 24 Circles or Collectorates of the Presidency of Bombay in 1896.

Collectorates or Circles.	Population.	Area in Square Miles.	Date of first reported case of plague.	Plague.		Deaths per cent. of Attacks.	Plague Death-rate per 10,000 population.
				Attacks.	Deaths.		
			1896.				
1. Kándesh	1,460,319	9,944	Oct.	6	5		
2. Nasik	841,087	5,940	...	...	...		...
3. Thana	818,967	3,936	Oct.	129	95		1·16
4. Bombay City ...	806,144	22	Sept.	2,544	1,936		
5. Kolába	594,779	1,811	Oct.	14	13		
6. Ahmednagar ...	887,656	6,666	Dec.	6	4		
7. Poona	1,061,449	5,348	Oct.	53	39		
8. Sholápur	750,255	4,521	Dec.	1	1		
9. Sátára	1,225,511	4,988	...	...	...		
10. Ratnágiri	1,105,862	3,922	...	...	...		
11. Belgaum	1,011,453	4,657	...	...	...		
12. Dhárwár	1,050,533	4,535	...	...	...		
13. Bijápur	796,286	5,757	...	...	...		
14. Kanara	446,156	3,911	...	...	...	...	
15. Surat	649,824	1,662	Nov.	23	21	91·30	
16. Broach	341,450	1,453	Dec.	5	5		
17. Kaira	871,529	1,609	Dec.	11	10		
18. Panch Mahâls ...	313,381	1,613		...	...		
19 Ahmedabad ...	920,928	3,821		54	29		
20. Karáchi	561,013	14,115		63	59		
21. Hyderabad... ...	883,836	9,030		1	1		
22. Shikapur	915,058	9,277		1	1		
23. Thar and Parkar ...	332,401	12,729		...	...		
24. Upper Sind Frontier	174,469	2,863		...	..		
Total for the Presidency.	18,820,346	124,130	...	2,911	2,219	76·23	1

During January, 1897, four more districts were invaded by plague, viz., Nasik, Satara, Ratnagiri, and Belgaum. In February Thar and Parkar, as well as Kathiawar, became infected, and cases also began to appear in Lower Damaun (Portuguese India). In March there were a few imported cases in the Upper Sind frontier, and some in the Palunpur and Baroda native states, which are in close association with the Presidency. The total recorded plague cases and attacks in the Presidency of Bombay during 1897 are given in Table II.

TABLE II.—PLAGUE in the PRESIDENCY of BOMBAY in 1897.

Collectorate.	Population (1891).	Area in Square Miles.	Date of first reported case of Plague.	Plague.			Plague Death-rate per 10,000 population.
				Attacks.	Deaths.	Deaths per cent. of Attacks.	
Kándesh... ...	1,460,319	9,944	15 Oct. 96	103	76	73·79	·52
Nasik	841,087	5,940	1 Jan. 97	1,389	1,075	77·39	12·78
Thana... ...	818,967	3,936	8 Oct. 96	6,352	4,156	65·43	50·75
Bombay City...	806,144	22	Sept. 96	13,314	11,003	82·64	136·49
Kolába ...	594,779	1,811	29 Oct. 96	1,517	1,309	86·29	22·01
Ahmadnagar ...	887,656	6,666	12 Dec. 96	551	386	70·05	4·35
Poona	1,061,449	5,348	2 Oct. 96	11,971	8,871	74·10	83·57
Sholápur ...	750,255	4,521	17 Dec. 96	2,845	2,353	82·71	31·36
Sátára... ...	1,225,511	4,988	10 Jan. 97	13,242	10,031	75·75	81·85
Ratnágiri ...	1,105,862	3,922	"	27	26	96·30	·24
Belgaum ...	1,011,453	4,657	24 Jan. 97	569	444	78·03	4·39
Dhárwár ...	1,050,533	4,535	20 Oct. 96	46	38	82·61	·36
Bijápur ...	796,286	5,755	29 Nov. 96	3	3	100·00	·4
Kanara ...	446,156	3,911	...	...	...	...	...
Surat	649,824	1,662	21 Nov. 96	4,316	3,070	71·13	47·24
Broach ..	341,450	1,453	17 Dec. 96	16	9	56·25	·26
Kaira... ...	871,529	1,609	10 Dec. 96	9	3	33·33	·3
Panch Mahals	313,381	1,613	...	...	...	...	...
Ahmedabad ...	920,928	3,821	3 Oct. 96	91	65	71·43	·71
Karáchi	561,013	14,115	10 Dec. 96	4,442	3,572	80·42	63·67
Hyderabad ...	883,836	9,030	27 Dec. 96	643	499	77·60	5·65
Shikapur ...	915,058	9,277	30 Dec. 96	987	716	72·54	7·82
Thar and Parkar.	332,401	12,729	15 Feb. 97	3	2	66·67	...
Upper Sind Frontier.	174,469	2,863	13 Mar. 97	3	3	100·00	...
Total for the Presidency	18,820,346	124,130	...	62,439	47,710	76·41	25·35

In the next Table (III.) is shown the number of attacks and deaths, month by month, in the Presidency (not including Bombay city). From this it will be seen that the 1897 epidemic outside the city was rising in January, reached its height in March, and then declined up to June, from which point there commenced a fresh accession, which attained its climax in November, there being a distinct fall in December. I am, unfortunately, unable to continue the table throughout 1898, as at the time of writing I have not obtained trustworthy figures. The further consideration of the 1898

A. No. 18. — he ...sion of ...onic ...ue from ...to 1898; by Bruce Low.

epidemic must, therefore, be deferred for a future report. To some extent the behaviour of the disease in the Presidency districts was not unlike that which occurred in Bombay city, that is to say, there were periods in which the malady increased in severity, then for a time diminished, only to burst forth again in epidemic form after a short interval.

TABLE III.—Showing reported PLAGUE ATTACKS and DEATHS in the PRESIDENCY of BOMBAY (*not* including the CITY of BOMBAY) Month by Month during 1897.

—	1897.												Total for 1897.
	January.	February.	March.	April.	May.	June.	July.	August.	September.	October.	November.	December.	
Plague Cases ...	1633	3098	5818	5218	1355	228	590	2660	4150	8002	9699	6656	49,107
„ Deaths ...	1375	2508	4374	3966	998	129	294	1853	2947	5596	7266	5401	36,707

In Table IV. is given a summary of the incidence of plague upon the whole Presidency of Bombay from the fourth quarter of 1896 to the end of December 1897, showing, month by month, the reported attacks and deaths in each of the 24 Collectorates comprised in the five registration districts into which the Presidency is divided. From this table it will be seen that the number of recorded cases during the 16 months in question was 65,350, and the deaths 49,748, but, as has elsewhere been said, these numbers do not show the full incidence of the disease. It is stated that up to the end of 1898 no fewer than 134,000 persons had died from plague in the Bombay Presidency, not including the City of Bombay. A later report* gives the total number of plague deaths, in the Presidency from September, 1896, to March, 1899, as 191,000.

* "Statement exhibiting the moral and material progress and condition of India during the year 1897–98." Printed by Order of the House of Commons, June 7, 1899.

TABLE IV.—Showing, Month by Month, the number of PLAGUE ATTACKS and DEATHS in the BOMBAY PRESIDENCY for 1897, and the total PLAGUE ATTACKS and DEATHS from September, 1896, to December, 1897, for each District:—

Registration Districts.	Collectorates.	1896. September to December.		1897. January.		February.		March.		April.		May.		June.		July.		August.		September.		October.		November.		December.		TOTAL. Sept. 1896 to Dec. 1897.	
		Cases.	Deaths.	Cases.	Deaths.	Cases.	Deaths.	Cases.	Deaths.	Cases.	Deaths.	Cases.	Deaths.	Cases.	Deaths.	Cases.	Deaths.	Cases.	Deaths.	Cases.	Deaths.	Cases.	Deaths.	Cases.	Deaths.	Cases.	Deaths.	Cases.	Deaths.
Western.	Kándesh	6	5	6	1	8	4	7	1	2	2	3	3	5	5	..	..	2	1	1	1	..	..	1	..	68	58	109	81
	Nasik	..	..	24	16	10	10	6	3	4	2	1	..	..	..	5	4	132	92	314	271	227	175	268	212	398	290	1,380	1,075
	Thana	129	95	423	324	1,208	948	1,709	1,350	1,373	990	429	262	107	17	88	4	149	47	249	77	363	46	161	26	93	56	6,481	4,251
	City of Bombay	2,544	1,936	2,587	2,110	3,532	3,349	2,694	2,242	1,866	1,309	448	345	202	104	72	44	124	70	247	171	290	180	378	255	924	643	15,858	12,758
	Kolába	14	13	19	19	114	84	212	158	639	608	226	188	36	35	22	16	38	28	81	62	51	38	48	40	39	33	1,531	1,222
Central.	Ahmednagar ..	6	4	5	3	7	4	8	4	..	1	..	..	..	..	..	..	12	9	44	32	117	77	138	80	220	107	567	390
	Poona	53	39	205	159	300	238	1,284	910	730	510	98	78	21	15	196	118	954	619	1,147	825	2,272	1,640	2,951	2,239	1,811	1,520	12,024	8,910
	Sholapur.. ..	1	1	1	..	1	..	..	..	..	..	..	..	..	..	..	..	..	..	15	9	260	109	1,211	983	1,357	1,162	2,846	2,354
	Satara	..	..	3	3	3	3	3	2	1	1	..	..	..	..	198	107	1,316	1,024	2,118	1,543	4,214	3,127	3,092	2,858	1,694	1,363	13,342	10,031
	Ratnágiri ..	..	..	5	5	6	6	..	..	..	..	..	..	..	..	..	..	9	9	..	..	2	1	5	5	..	..	27	26
Southern.	Belgaum	..	..	1	..	1	1	..	..	..	..	..	..	..	..	..	..	..	..	..	..	14	6	222	162	331	275	569	444
	Dhárwár.. ..	..	..	..	..	..	..	..	..	..	..	..	..	..	..	..	..	..	..	..	..	13	10	27	23	6	5	46	38
	Bijápur	..	..	..	..	..	..	..	..	..	..	..	..	..	..	..	..	..	..	..	..	..	..	1	1	2	2	3	3
	Kanara	..	..	..	..	..	..	..	..	..	..	..	..	..	..	..	..	..	..	..	..	..	..	..	..	..	..	..	..
Guzerat.	Surat	23	21	34	27	174	118	903	703	793	615	120	96	19	19	66	36	46	22	178	126	459	274	907	579	617	453	4,339	3,091
	Broach	5	5	4	4	5	2	6	2	1	1	..	..	..	..	..	..	..	..	..	..	..	..	..	..	..	..	21	14
	Kaira	11	10	3	..	2	..	1	1	..	..	..	..	1	1	..	..	..	..	1	..	..	..	..	..	1	1	20	13
	Panch Mahals ..	..	..	..	..	..	..	..	..	..	..	..	..	..	..	..	..	..	..	..	..	..	..	..	..	..	..	..	..
	Ahmedabad ..	54	29	24	19	20	10	22	15	9	6	7	6	3	3	..	..	2	2	1	..	1	2	1	1	1	1	145	94
Sind.	Karáchi	63	59	844	768	1,202	1,052	1,232	905	787	574	244	179	26	20	13	9	..	..	1	1	9	1	66	48	18	15	4,505	3,631
	Hyderabad ..	1	1	30	26	24	19	134	98	389	295	65	60	1	1	..	..	..	..	..	..	..	..	..	..	..	..	644	500
	Thar and Parkar	..	..	..	..	1	1	2	1	..	..	..	..	..	..	..	..	..	..	..	..	..	..	..	..	..	..	3	2
	Shirkarpur ..	1	1	2	1	12	8	287	219	499	351	160	134	27	13	..	..	..	..	..	..	..	..	..	..	..	..	988	717
	Jacobabad ..	..	..	..	..	..	..	2	2	1	1	..	..	..	..	..	..	..	..	..	..	..	..	..	..	..	..	3	3
Total in the whole Presidency ..		2,911	2,219	4,170	3,485	6,630	5,857	8,512	6,616	7,084	5,275	1,803	1,343	448	233	662	338	2,784	1,923	4,397	3,118	8,292	5,776	10,077	7,521	7,580	6,044	65,350	49,748

The figures in this Table are mainly based upon the Report for 1897 of the Sanitary Commissioners with the Government of Bombay.

APP. A. N[illegible]

On the Diffusion [of] Bubonic Plague fr[om] 1879 to 189[illegible] Dr. Bruce

APP. A. No. 18.

On the Diffusion of Bubonic Plague from 1879 to 1898: by Dr. Bruce Low.

The appended map shows the topographical relations of the different collectorates or districts, and indicates, by the colouring, the time at which each district is believed to have become infected by plague.

The subjoined Table (V.) gives the total recorded plague attacks and deaths in the following localities in the Native States adjoining the Bombay Presidency and in Portuguese India, from September 1896 to end of December 1897, compiled from various official sources.

TABLE V.—Showing Chief Localities invaded by BUBONIC PLAGUE from September, 1896, to the end of December, 1897, in the NATIVE STATES adjoining the BOMBAY PRESIDENCY, and in PORTUGUESE TERRITORY.

City or District.	Population.	Attacks.	Deaths.	Plague Rates per 10,000 of Population.		Percentage of Deaths to Attacks.
				Attack Rate.	Death Rate.	Mortality.
Native States within Bombay Presidency.						
*Baroda Territory ...	2,415,396	755	612	3·1	2·5	81
Kathiawar ...	2,761,879	263	168	1·0	·6	64
*Kolhapur	913,131	252	175	2·8	1·9	69
*Kutch	558,415	6,689	5,720	119·8	102·4	86
Mahi Kantha ...	279,434	4	2	...	...	..
*Palanpur	645,526	1,219	793	18·9	12·3	65
Savan Toadi ...	192,948	26	25	1·3	1·3	96
Janjiri State ...	81,780	349	213	42·7	26·1	61
Akalkot State ...	75,774	71	50	9·4	6·6	70
Bhor State ...	155,699	2	2	...	...	...
Portuguese Territory.						
Lower Damaun ...	11,000	?	2,352	...	2138·2	...
Goa	420,868	62	16	1·5	·4	26

As illustrations of the spread of plague in the Bombay Presidency, the following short accounts of the disease in Karachi and Poona are given.

KARACHI.

The first appearance of plague in the port of Karachi was on December 10th, 1896, the infection having been presumably brought from Bombay. The epidemic reached its height in February, and afterwards declined slowly in March and April, cases continuing to occur, however, up to 27th July 1897. During that period 4,181 plague cases and 3,398 deaths were recorded, a case mortality of 82·4 per cent. There was a recrudescence in March 1898, which after continuing towards the end of the year

* In those districts marked with an asterisk, plague was still epidemic during the early part of 1898.

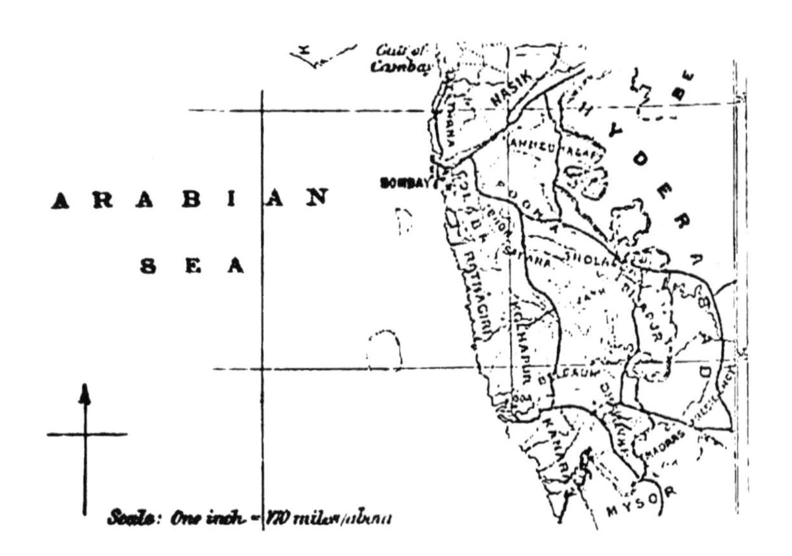

Scale: One inch = 170 miles/about

subsided, to reappear again in February 1899. I am unable to give the precise figures for the year 1898, as they have not yet been received, but further reference will be made to them in a future report.

There is a very intimate and constant communication between Bombay and Karachi by means of shipping. In the period between November 1st, 1896, and 31st July, 1897, 42 ships arrived at Karachi from Bombay and intermediate ports, bringing 6,697 passengers who had submitted to a medical inspection before going on board. Only two plague cases were discovered among this large number of travellers. But out of 4,929 of the passengers who were detained under observation at Karachi, 13 subsequently developed plague. On the other hand 19,623 persons were medically inspected before leaving Karachi by ship, and 11 of them were found to be suffering from plague; of 50 other passengers who had been kept under observation before being allowed to depart, 13 developed plague before the period for their departure had arrived.

The malady raged in the older portions of the town which were in an insanitary condition, and where overcrowding was common. No European or soldier suffered at Karachi during the outbreak of 1896–97.

The appended table shows the course of the epidemic in Karachi, from December 1896 to August 1897, inclusive :—

Plague.	December 1896.	1897.								Total from Dec. 1896 to August 1897.
		January.	February.	March.	April.	May.	June.	July.	August.	
Reported cases	38	758	1,141	1,057	920	212	40	15	—	4,181
„ deaths	37	698	1,006	785	673	152	36	11	—	3,398

Haffkine's preventive treatment by inoculation was carried out in a number of persons, apparently with success.

Segregation camps were found of the greatest service and disinfection of houses and clothing was largely practised. Difficulty was at first experienced in inducing patients to enter the isolation hospitals provided for the purpose, although the fatality rate of cases treated in hospital was only 61 per cent. as compared with 82 per cent. for all cases.

In the district around Karachi, plague was carried to a number of villages; but no epidemic on a large scale followed in 1897; for only 238 cases with 178 deaths were reported between December 18th and June 11th.

POONA.

This City became infected by plague early in December 1896, and the disease spread among the inhabitants during January and February 1897, culminating in a severe epidemic in March. The number of cases continued high till the middle of April and thereafter subsided rapidly. The population of the city at the census of 1891 was 118,790. After plague had appeared in Bombay, a number of people flocked from that place to Poona. The first imported case was recognised on December 19th, 1896, but it is more than probable that other plague stricken persons had arrived at Poona before this. The following table shows the course of the epidemic from December to May in the city of Poona :—

	1896.	1897.					
	December.	Jan.	Feb.	March.	April.	May.	Total for 6 months.
Plague cases ...	4	101	203	1,098	593	50	2,049
„ deaths	4	95	172	756	416	38	1,481

Although the number of plague deaths is given above as 1,481 it is estimated that at least 2,900 persons lost their lives in Poona from the epidemic between December and May. It is evident, therefore, that if the deaths are reckoned at 2,900, the plague seizures must have been considerably more than given above. Unfortunately there is no means of arriving at a correct estimation of the number. Although the epidemic subsided to a large extent in May, and in June only 11 cases were reported, there was soon a recrudescence of the disease. This is seen in the table below :—

1897.	July.	August.	Sept.	October.	Nov.	Dec.
Reported Plague Cases ...	60	77	296	1,446	2,534	1,648

—a total of 6,061 cases in the six months. In January 1898, the reported plague seizures had fallen to 462.

From Poona city the infection extended to the suburbs, the cantonment, and beyond these into the surrounding district.

The Poona suburbs have a population of about 6,706. During the epidemic in the city, 275 cases were reported in the suburbs with 164 deaths up to May 20th, 1897. The cantonment had a population at the 1891 census of 35,094, and among these some 188 cases and 130 deaths were recorded between January and May. The cantonment is chiefly occupied by military lines and bungalows where plague was not likely to secure a hold. The majority of the above 188 cases occurred in the bazaar portions of the cantonment, which are densely populated.

The Poona district beyond the city, suburbs and cantonment, had 12 localities infected.

The total amount of reported plague which occurred in Poona city, together with its suburbs and cantonment, during the period January to December 1897 was 8,538 attacks. and 6,278 deaths ; and in the Poona district outside the city and cantonment during the same period there were 3,705 cases and 2,737 deaths.

The plague measures taken in Poona comprised the removal of the plague-stricken to hospitals, the isolation, in segegration camps, of persons who had been exposed to the infection, and the disinfection of invaded houses, and of clothing, &c., which had been exposed to infection. It was also found necessary to institute a new system for obtaining a more accurate registration of deaths than had hitherto existed. Registration clerks were appointed to be on duty the whole day, and officials were posted at the principal burning and burial grounds to obtain from funeral parties the death certificates given by the registrars. Policemen were instructed to examine the certificates of all funeral parties that passed them, and to take steps for detaining parties not having the requisite certificates. These new arrangements were duly made public. The military were employed to assist in enforcing these measures, which excited much ill-feeling among the natives. In addition, a system of medical inspection of passengers at railway stations, and at certain points on the roads was instituted. It was found that large numbers of plague-stricken patients were removed from Poona by night, and accordingly military picquets were called on to prevent this. These picquets found 103 plague cases, and 23 plague corpses on the road during the time they were employed, and the inspections at the railway stations yielded 799 plague cases. The booking of third-class passengers from Poona and neighbouring stations was suspended, except where formal permits had been obtained from the district magistrate or from the Plague Committee. Arrangements also were made for preventive inoculation.

To lessen the chances of disturbance in the city by the carrying out of these measures, native cavalry patrolled the streets when military search parties, fumigators, or limewashers were at work. The search work was carried out by British soldiers, each party being accompanied by a native of position; ladies too accompanied the parties to search the rooms occupied by "purdah" or covered women.

The Brahmin community was the only section of the community in Poona which maintained an unfriendly attitude towards the valuable work carried on by the military; frequent obstruction was offered to the operations of the workers. False and malicious statements were circulated by agitators against those carrying out the plague preventive measures, with the result of inflaming the minds of ignorant and fanatical persons against those performing this difficult and disagreeable work. The vernacular press too gave publicity to the wildest of false rumours, and the excitement thus fomented ended in the assassination of Mr. Rand, I.C.S., and Lieutenant Ayerst, on the 22nd of June, at a time when the plague measures were proving successful. The murderers, though they escaped justice for a time, have since been taken and have paid the full penalty of their crime.

The following is an example of the behaviour of the epidemic in a native state.

The Kutch State.

The Kutch State is practically upon an island, south of Sind, at the mouth of the Indus. Plague has ravaged Kutch before; namely, in 1812–17, when it is said to have destroyed half of the

P. A. No. 18. — the ...usion of ...bonic ...gue from ... to 1898; by Bruce Low.

people in the State. About the beginning of October 1896 imported cases of plague began to be brought to the port town of Kutch Mandyi by the natives fleeing from Bombay. But no epidemic prevalence was reported until the middle of the following April. Between April and July the disease raged, reaching its maximum in May, diminishing in June, and practically disappearing in July and August. With regard to the beginning of the epidemic it is stated that a large number of cases had already occurred before the outbreak became known to the authorities. From October 1896 to the end of March 43 plague cases, and 25 deaths had come under observation, more than half of them having been reported at the middle of February. The following table shows the course of the disease during the epidemic :—

Plague.	For the 4-weeks period ended							Total.
	March 12.	April 9.	May 7.	June 4.	July 2.	July 30.	August 27.	
Reported cases ...	2	12	1,824	2,115	324	33	8	4,318

Of the 4,318 attacks between March and August 3,828 proved fatal, a fatality rate of 88·7 per cent.

From Kutch Mandvi town the disease was carried into other parts of the State. Twenty eight localities were ultimately known to have become infected, yielding 840 cases of which 641 died, a case mortality of 76·3 per cent. During September, 1897, the disease recrudesced in the Kutch State, but was declining again in December. The cases reported in July were 107, in August 237, in September 522, October 464, November 336, and December 151, making a total of 1,817 reported cases from July to December inclusive in the State of Kutch. The people as soon as it was known that the plague was becoming epidemic were panic stricken, and deserted the sick members of their families, leaving them to die unattended. Many of the persons left to perish in the locked up houses were women. Some of these deserted or concealed patients were discovered by the search parties which were organised to seek out plague cases. Those people who remained with their sick, were very reluctant to allow them to be removed to hospital for treatment. Five hospitals were established for the different sects, the Hindus forming the largest number needing treatment. In the five hospitals 1,220 persons were treated between April and June, and of these 790 died, a case mortality of 64·8 per cent. Many people, estimated at about 10,000, at least, had left their homes in terror when plague was known to be raging in the town ; it was therefore easier than might have been expected to cleanse and disinfect the dwellings since many were already empty. A staff of officers and subordinates had been obtained from Bombay for this purpose. The roofs were taken off in many instances, and the walls cleaned with lime mixed with carbolic acid. Chloride of lime was employed for a preparatory fumigation before the whitewashers began. Dead rats and dead cats in the houses added to the stenches encountered and it is stated that half of the houses that were entered for cleansing purposes had dead rats or cats in them.

During 1897 the recognised cases of plague in Kutch amounted to 6,653, and the deaths to 5,709.

Rajputana.

During the last few weeks of 1896 plague was imported into several localities in Rajputana, and sixteen deaths resulted, all of them of travellers from infected localities outside the Province. At the end of December 1896 cases of plague were heard of at Abu Road in Rajputana close to the borders of the Bombay Presidency. From December 1896 to March 1897 some seven cases were reported here but the disease failed to spread. In December 1896 also, plague was reported from Marwar, which lies not far to the north of Abu Road, and up to February 1897, seven cases came under observation. In December 1896, three mendicants came from Bombay to Jeypur by rail. They were refused admittance to the city, but took refuge in some cenotaphs. All three are stated to have died from a disease which had a resemblance to plague, but the cases were not officially reported as cases of bubonic plague, though the precautions adopted were taken as if the diagnosis had been confirmed. There were no other cases here. Single cases were reported in January 1897 from Jodhpur and Nadbai; and from Jowalia four cases were recorded in May and June 1897. No extension of the disease followed at this time. But in December 1897 suspicious cases and deaths began to be reported from four adjacent villages in the Sirohi State in the south of Rajputana. The first cases broke out among refugees from Poona and its neighbourhood, and the diagnosis of plague was subsequently confirmed. In January 1898, the villages of Kalinderi, Shurdial and Tewri all situated near the border of the Palanpur State were declared infected. Among other measures for stopping the spread of the infection, evacuation of the villages was tried with satisfactory results. About 128 cases were reported in these villages and some 107 of them proved fatal.

The Sirohi State in which these cases occurred is about 40 miles from Pali, where a form of plague prevailed in 1836. This Pali-plague is, as has been said, generally believed to have been true bubonic plague.

Plague in the Punjaub.

A single imported case of plague was discovered at Rewari in January 1897; another isolated case came under observation at Sialkot in February, and two other imported cases were reported at Shershah in March. In April plague broke out at Khatkar-Kalan, a village of 1,200 inhabitants in the Jullundur district. The infection was imported by a native who went on a pilgrimage to Hardwar. He returned home on April 29th suffering from an illness which speedily killed him. Soon after this, a number of his relatives who had been with him during his fatal illness were

No. 18. — n of from 898; by ne Low.

attacked by the same malady which afterwards was proved to be bubonic plague. About the end of July and beginning of August it was observed that rats were dying in considerable numbers in some cattle sheds in the village. Gradually the disease spread among the people, and up to October, 79 had been attacked and 45 of them had died. At the end of October the village was evacuated with the view of staying the epidemic. Meanwhile some neighbouring villages became infected; Jhandher Khurd, 6 miles from Khatkar-Kalan, was found to be invaded on November 2nd, and yielded 29 cases and 18 deaths up to December 19th. Khan Khana, with a population of 2,500, was attacked on November 24th, and in a fortnight 96 persons had been struck down by the disease. Berampur in the Hoshiapur district, 10 miles distant from Khatkar-Kalan, was invaded on December 11th. Up to the end of 1897 plague had infected four villages in the Jullundur district and one village in the Hoshiapur district, yielding a total of 302 attacks, and 175 deaths.

During the early months of 1898, plague unfortunately continued to spread in both of the above mentioned districts and in the adjoining localities. By the end of April 62 villages in Jullundur and 14 in Hoshiapur had been invaded. Subsequently further extension of the epidemic malady took place, so that by the end of the year 1898 the total number of persons attacked by plague reached 3,528, while the deaths amounted to 2,103. The plague epidemic of 1898 in the Punjaub involved 86 villages comprising a population of 91,892.

A measure which was regarded as highly successful in staying the progress of the village epidemics, was that of evacuation and disinfection of all the houses. It was found possible to remove for a time from their villages into temporary encampments, populations of 4,000 or more; and it was observed that after this was done if plague was re-introduced it showed no tendency to spread. In the earlier months of the epidemic serious riots took place, and several lives were lost, owing to the native objection to segregation of the sick members of the family, and also owing to their refusal to accept the sanitary measures ordered by the local authorities. It was found that in consequence of these objections sick persons were concealed, and that the dead were even sometimes buried inside the dwellings.

The North-West Provinces and Oudh.

In February 1897 three plague cases imported from Bombay were discovered in Delhi, but no epidemic developed. Single cases were also detected at Rai-Bareilly in January, and at Unao in February, and Bareilly in March.

In April bubonic plague was reported to have appeared in the Hardwar Union Municipality. Hardwar is situated on the Ganges, and is a place of great importance to the Hindoos. Several times a year religious bathing fairs are held at Hardwar, and are frequented by pilgrims from all parts of India. The usual population of the Municipality is about 30,000, but during fair times as many as 200,000 persons are resident in the town and neighbourhood. On April 8th, 1897, the body of a woman was found at Hardwar, and subsequent examination proved her death to have been caused by plague. A search was instituted, and a female fellow pilgrim was found in a lodging house suffering from plague. The next day three more cases were discovered. Prompt and appropriate measures were taken, and though a few cases were afterwards reported no epidemic occurred. The total number of reported cases at Hardwar was 18, of which 15 died. Eight of the cases were in April, seven in May, and three in June. The Plague Commissioner, Major D. S. Reade, R.A.M.C., regarded it as conclusively proved that the infection was brought to Hardwar by pilgrims from Sind, probably from Karachi. He further believed that in the first instance the infection was brought by means of contaminated clothing, "the evidence pointing to the person or persons who brought the clothes remaining unaffected."

Within a mile and a half of Hardwar lies the town of Kankhal, with a population of between 5,000 and 6,000. There is constant and intimate inter-communication between the two towns. About the middle of June (1897) an excessive mortality among rats was noted at Kankhal, and the plague bacillus was discovered in the dead bodies of some of the animals which were examined. This excited considerable alarm, and many rats were destroyed; houses were temporarily evacuated so that they might be cleansed, and a reward was given for every rat, dead or living, brought in from Kankhal. The first plague cases observed among the inhabitants of Kankhal occurred in September. In October attention was drawn to the danger arising from the monkeys which swarm in the town and frequent the houses, picking up food. Dead monkeys* had been found in the streets, and in some of them Mr. Hankin found plague bacilli. As the monkeys are regarded by the Hindoos with a feeling something like religious veneration, it was found impossible to deal with these animals by wholesale destruction, as had been done with rats. It was decided to trap as many of them as possible and keep them under observation in cages for a time. A large number were trapped and kept in quarantine for 10 days or more. A few died of plague, but the disease did not spread among the monkeys, and they were liberated in the jungle when the 10 days isolation period had passed. The number of plague cases among the people of Kankhal, from September to December, amounted to 64, and of these 43 died.

In the village of Jagitpur, with a population of 1,500, situated 2½ miles from Kankhal, plague appeared in December 1897, the infection having been derived from that town. Twenty-four cases were reported, with 19 deaths. The town of Jawalapur has a population of about 17,000; it lies two miles from Kankhal, from which it is separated by the Ganges Canal and the railway. A case of plague was reported at Jawalapur on January 9th 1898. The disease spread in the town during February and March, causing some panic among the inhabitants, some 3,000 or 4,000 fleeing from the place. Between January and April 113 cases

* During the 1898 epidemic of plague in the Bombay Presidency, monkeys were found dead, and others were seen to fall from trees in a dying state at Gadag, in the Dharwar District, by Miss Alice Corthorne, M.D. Bacteriological examination showed that the monkeys had died of plague; this also was proved as regards squirrels in the same locality (*Indian Medical Gazette*, March 1899).

were reported, and of these 87 proved fatal. The village of Jamalpur became infected in February, and during February, March, and April 35 cases were reported, with 23 deaths. The infection had been brought from Jawalapur. In eight other villages within a radius of 10 miles of Hardwar, some 13 further plague cases, all fatal, were found during the first three months of 1898. It is probable that other cases escaped observation, though a careful and systematic examination of all the villages in an area of 400 square miles round Hardwar was made.

In the period from January 1897 to the end of April 1898, plague was restricted to this one region in the North-West Provinces. And in a resident population estimated at 34,000 there occurred 271 plague attacks, and 217 deaths, a case mortality of 80 per cent. Owing to the careful measures planned and taken by the authorities, the outbreak of bubonic plague in the centre of the Hindu Pilgrimage has been limited within comparatively insignificant dimensions. Much of this success is due to the counsel rendered to the authorities by the Sanitary Commissioner, Colonel Thomson, C.I.E., I.M.S., and his assistants.

BENGAL.

Calcutta seems to have escaped to a large extent the serious epidemics of plague which affected Bombay during 1896, 1897 and 1898. A few cases of glandular swellings with fever were reported to the Calcutta Municipal Authorities at the end of 1896 and beginning of 1897; but, though the presence of the plague bacillus was said to have been demonstrated in five out of the eleven cases brought under notice of the Authorities, and although the first case was known to have travelled direct from Bombay, the diagnosis of plague was not upheld. A medical Board examined the evidence on which the diagnosis was established, and expressed themselves satisfied that none of the persons affected were suffering from bubonic plague. Precautions by strict isolation and disinfection were nevertheless carried out. But in the spring of 1898 the presence of plague in Calcutta could not be disputed. Cases continued to be reported from April to October, though not in any large numbers. During 1898 some 230 recognised cases, with 192 deaths, were recorded between April 17th and October 10th. In the autumn the disease seemed to have subsided, but there was another and more serious recrudescence in the early part of 1899.

The infection in 1898 is said to have been introduced into Calcutta through the agency of infected articles of clothing coming by sea or by rail from infected localities. But, as has been said, cases of fever with glandular enlargement are stated to have been not uncommon in the city for some few months, if not longer, previous to plague being officially declared.* Some

* These antecedent, mild or suspicious cases (Pestis Minor?) were stated by Major Evans, I.M.S., Physician to the Calcutta Medical College Hospital (who afterwards himself died from plague in Calcutta), to have had nothing in common with plague either clinically or pathologically.

persons thought that plague had been brought ashore by rats from ships in the harbour. In support of this contention it was pointed out that the first cases occurred in houses close to the jetties. In April, 1898, rats were noticed to be dying in the sheds, near the jetties, in considerable numbers. The persons first recognised as attacked had been resident in Calcutta for considerable periods of time. They could not in all probability have acquired the disease elsewhere than in Calcutta. That the disease was introduced by the agency of rats or of infected articles arriving by sea or by land, seems the only feasible solution; unless there had been, as is alleged, during a period of some months a succession of mild or unrecognised cases following upon the importation of the disease from Bombay. Dr. Griffiths, the principal medical officer of the East Indian Railway, noticed that at the height of the Bombay epidemic a large percentage of the poorer class passengers from that city were, upon examination, found to be suffering from enlarged glands, while travellers from non-infected districts had no such ailment.

How far the development of the Calcutta outbreak was connected with the previous cases of fever with glandular swellings it is not possible here to say, but it has to be borne in mind that the history of plague epidemics shows often that the outbreak has been preceded by the occurrence of mild and unrecognised cases; and that when the epidemic has raged for a time there occur periods of decline, followed by periods of recrudescence, the intervals being filled up, seemingly, and the continuity of the disease carried on, by cases of a mild, non-typical, or not easily recognisable character, described by some writers as Pestis Ambulans, or Pestis Minor. On the other hand some residents have stated their opinion that the Calcutta outbreak was sudden in its development and that there was no dropping fire of cases before the epidemic burst out. This, if true, might be accounted for by a failure to recognise or diagnose an unfamiliar disease, or by the difficulty of detecting plague cases when the prevalent type is the pneumonic form. It is known, as has already been mentioned, that prior also to the Bombay outbreak mild cases of glandular swellings with fever had been observed in that city.

Considerable precautions had been taken during 1897 and 1898 to prevent the introduction of plague into Calcutta. Major Dyson, I.M.S., Sanitary Commissioner, states that 1,800,000 travelling persons had been inspected; that over 40,000 had been detained as suspects; but that of these only six subsequently developed plague. There is evidence to show that upon occasion travellers deceived the inspecting staff, and that some of them escaped observation; and such persons may, therefore, have introduced the infection into Calcutta. It is also stated that plague cultures were obtained from clothing that was supposed to have been sufficiently disinfected. However this may be, it is fortunate that the malady had not, up to the end of 1898, displayed any large epidemic proportions in Calcutta. Some authorities have alleged that there is something in the circumstances of Calcutta which renders it unfavourable for the development of a plague epidemic. There had been, it must be said, much cleaning and scavenging done during 1897, in view of the possibilities of an outbreak, and some persons believe that the mild proportions of the plague epidemic in Calcutta during 1898 were due to this.

The nature of the houses in which the poorer classes live in Calcutta is different from what prevails at Bombay. There is less aggregation of population in tenements; and the dwellings themselves are of a less permanent kind. The conditions therefore are, on the whole, more favourable for ventilation, and for the admission of light as well as air.

There was a recommencing recrudescence of the disease at the end of 1898 and beginning of 1899.

In 1898, when plague was declared to exist in the city, there was great alarm among the inhabitants, a large number of whom fled from the city into adjoining districts in the Bengal Presidency. It is calculated that 250,000 persons, the majority of them women and children, left the city in fear of plague and to escape the plague-preventive measures. A number of the natives submitted to inoculation with Professor Haffkine's preventive material. In May the panic began to subside, but the natives showed, as elsewhere, their aversion to segregation by active and determined opposition. In one of the suburbs of Calcutta, for example, a doctor who was known to be engaged in selecting a site for an isolation hospital was mobbed and pursued. The doctor took refuge in a house, and in self-defence, to keep the mob at bay, fired his pistol at his assailants, one of whom was killed and two were wounded. The doctor was subsequently rescued by the police from his perilous position.

The disease was spread in the country districts outside Calcutta during 1898 by those who fled from the city. One instance of this is related below. In the winter outbreak of 1898–99 a number of places in the Bengal Presidency are known to have become infected by plague carried from Calcutta; these local outbreaks have been, fortunately, of small dimensions. Backergunge (mentioned further on), Faridpur, Dacca, Saran, and Darbhanga are among the places known to have been infected by fugitives from Calcutta.

A suspicion that plague had been carried from Calcutta to Europe by a passenger on board the ship *Golconda* arose in 1898, under circumstances of a somewhat peculiar kind. The case has been reported upon separately by my colleague, Dr. Bulstrode. (*See* page .)

A small, but virulent, outbreak of plague occurred about 120 miles from Calcutta, affecting the villages of Siddhakati and Abhoynil, in Thana Nalchihi, of the Backergunge district, during September, 1898. Eleven persons were attacked by plague between September 4th and 24th, and all of them died. The infection was imported from Calcutta by a pleader, who, accompanied by some members of his family and a servant, left Calcutta on August 30th. The party reached Siddhakati on August 31st. It appears that on August 27th a student living with the pleader fell ill, and died on August 30th. On the same day the medical man attending the student was also attacked by plague of the pneumonic kind, and he too died on September 1st. The pleader, suspecting these cases to have been plague, fled with the members of his family to his native village, where he himself developed plague and died on September 6th; his wife was attacked there on September 9th and died on the 12th. A nephew, who had fled with them, was taken ill and died on September 4th; a servant, who showed symptoms of plague on September 2nd, died on the 5th. The remaining persons who suffered were all relatives or neighbours of this family. Energetic measures were taken as soon as the disease was recognised, and fortunately no extension of the outbreak followed in the two villages infected by these people.

Later information states that up to April 1899, 953 deaths from plague had been recorded in the Bengal Presidency.

THE CENTRAL PROVINCES.

Itarsi is a railway inspection station for plague prevention purposes on the main line from Bombay. A case of plague was recognised there on December 31st, 1896, and between that date and February 13th, 1897, nine cases of plague were reported. On January 8th, at another station, Jubblepore, not far from Itarsi, a single plague case was reported. The infection in both of the above instances was carried by travellers from Bombay. During January, 1897, cases of plague were observed in the Chanda district. Seven recognised cases, with five deaths, occurred during January and February. Later in the year there was a larger outbreak in the Wardha district, upwards of 163 cases having been noted. The Wardha district lies to the north-west of Chanda. In December, 1898, plague was said to be still rife in the Wardha district.

CENTRAL INDIA.

Gwalior State.—Plague was reported in the village of Khandraoni, in the Gwalior State, in January, 1897. The village had a population of 558 (in 1896), and it has frequent communication with Bombay. Two Brahmins came to their homes in the village in January from Bombay. One of them died there five days after his arrival, and the other sickened and died a few days later. Two native doctors who had attended these patients contracted the disease, and both of them died. The plague gradually spread through the village, and ultimately 74 of the inhabitants were attacked and 51 died.

Between February 22nd and March 4th, 1897, four plague cases occurred at Ujjain. There is a large religious fair known as the Singhest Mela, which is held once in 12 years at Ujjain, in the Gwalior State, and is usually attended by three or four hundred thousand persons, chiefly religious mendicants from all parts of India. During 1897 the fair fell due on April 17th to May 16th; but owing to the famine in the autumn of 1896, the fair was forbidden by the Maharajah of Gwalior, and notices to this effect were sent in November to all local governments and authorities, as well as to the railway companies throughout India. Notwithstanding this notice some 15,000 pilgrims arrived at Ujjain on

No. 18. — ... of ... rom ... 98; by ... e Low.

April 17th, and great alarm was felt by the authorities lest plague should break out amongst them. With the assistance of the religious heads of the different sects, stringent measures were taken as to the sanitary arrangements of the encampments outside the city. On April 23rd it was found that 22,000 devotees had bathed at Ujjain, but that they had dispersed at once. Fortunately no development of plague followed, and the fair passed off without any injury to the district or to the pilgrims themselves.

In June, 1897, a case of plague was taken out of the Bombay train at Rutlam, not far from the borders of the Bombay Presidency. Up to July 30th five more cases occurred, but no epidemic developed.

Lower Damaun (Portuguese India).

In February, 1897, some cases of plague occurred in Lower Damaun amongst new arrivals from Bulaar, a neighbouring town which was then badly affected by the disease; and cases were noticed also among sailors arriving from Karáchi in native craft. Before the end of February the fishermen living close to the shore became affected. As soon as this happened the public began to be alarmed, and many people hurried way from Lower Damaun. It is believed that about 2,000 people quitted the locality in March. The plague outbreak reached its climax in the middle of April. The population of the affected part of Lower Damaun was estimated before the epidemic began at about 11,000. Some difficulty was experienced in obtaining accurate figures regarding the mortality of this epidemic, as no proper records are kept. By careful inquiry, however, it was calculated that 2,352 persons died of plague in Lower Damaun between February and July, 1897. During the height of the epidemic in April, the largest number of deaths from plague that occurred in one day was 80. Some use of Dr. Haffkine's protective inoculation was made here, and the result is stated to have been satisfactory. The outbreak was of an exceptionally virulent nature.

Hyderabad State.

At the end of 1897 an epidemic developed in the Naldrug district of the Hyderabad State and spread to the neighbouring localities. The villages first attacked were Itkal, Kashgaon, Dhotri, Warligaon, Sarali, Raviamba, Ajagu, and Gungoti, all of which are situated near the Sholapur border of the Bombay Presidency, where at the time plague was epidemic. On February 21st, 1898, an official telegram declared that plague existed in all the villages within a radius of 30 miles of Gungoti, and that 450 deaths had been recorded from plague up to February 17th. On the same date 220 cases were said to be under treatment at Wadi, a town with a railway station on the main line from Bombay to Madras. It is believed that the first three or four cases were imported by rail from Bombay, and that later the disease spread by human agency from village to village. Ultimately, at least,

68 localities in the Hyderabad State are known to have been invaded by plague during 1898; the total attacks it is said amounted to about 5,000. No less than 3,200 of these were reported from the Lingsugar district. During the hot months of 1898 plague subsided, but in the winter months it again showed renewed activity. At the close of the year plague had again become epidemic in some districts of the Hyderabad State. Up to the beginning of April, 1899, the recorded plague deaths had numbered 4,764, but the actual number is in all probability considerably more than this.

Mysore State.

The State of Mysore did not suffer from epidemic plague until April, 1898. The chief locality in Mysore to feel the full force of the plague epidemic was Bangalore (population about 84,000), where it is stated that no fewer than 3,300 cases occurred in the city, and 3,837 more in the Cantonment; 1,963 cases were received into hospital, and of these 1,261 died. Altogether about 9,000 cases of plague were heard of up to January, 1899. It is believed that 50 per cent. of the plague deaths were undetected. Bangalore was in the first instance infected by a man who travelled in April by rail from Hubli where there was a serious epidemic at the time. The infection did not spread at Bangalore on this occasion, for it was not till another month had passed that the next case was recognised. It is possible that there was a second importation of plague infection. The real epidemic in Bangalore began about August, 1898. At the first intimation of the presence of plague in the city there was a panic, and it is estimated that as many as 25,000 persons fled from the city. Some of these carried infection with them, and thus spread the disease to the rural villages and districts in the Mysore State. Bangalore had some overcrowded areas, and in these plague is said to have exhibited an aggravated type. Attempts to segregate the infected persons led to active opposition and ultimately to rioting. When segregation was given up at Bangalore the people became quiet and amenable to reason. At the end of 1898 plague was still prevalent in Bangalore; several Europeans had been attacked, and one British soldier died of it in December. In one bungalow three of the family were attacked by plague. It had been noticed that previous to this dead rats and dead squirrels had been found in the bungalow gardens. Further reports state that there had been exceptional mortality among monkeys, as well as among rats and squirrels, in Bangalore and its vicinity.

Next to Bangalore the chief city to be attacked by plague in Mysore during 1898 was Seringapatam. Here the populace exhibited the same violent objection to isolation of their sick. This attitude interfered greatly with the prevention of the spread of the disease. A riot occurred, and attempts were made to rescue prisoners arrested for their violence in a previous plague riot. The prisoners had to be removed from the gaol to the fort, which the mob attempted to take by storm, and a number of

No. 18. [illegible]n of [illegible] from [illegible]: by [illegible] Low.

rioters were shot in the fight that ensued. The town of Chikbullapore also became infected in the early winter of 1898. Riots also developed here in connection with plague preventive measures, and in one of these riots forty-nine persons were taken prisoners and brought to Bangalore where they were sentenced to six months imprisonment. All of the forty-nine prisoners were inoculated with preventive material before removal to jail. During the next week fresh riots occurred; a mob of 800 Mohammedans sacked the segregation camp, and, after removing every movable effect, burned the camp to the ground. By December the epidemic in the town was waning, but the infection was spreading in the rural districts. This epidemic continued well into the spring of 1899, and the scene of its greatest virulence was the Kholar gold fields where it attacked the coolies employed in the mines.

It is recorded that from the first appearance of the disease up to July 1st, 1899, no fewer than 18,862 cases of plague had occurred in the Mysore State, of which 12,364 were in Bangalore City, Cantonment and District. Later information gives the total plague deaths in the Mysore State up to July 1st, 1899, as 15,597.

The following are the localities in the Mysore State which had been invaded by plague up to July 1st, 1899, with the number of cases and deaths in each district or municipality* :—

District or City.	Plague cases.		Total cases.	Deaths.
	Imported.	Indigenous.		
Bangalore City	3,346		3,346	2,665
Bangalore Cantonment	4,033		4.033	3,324
Bangalore District...	238	4.747	4,985	4,467
Mysore City...	14	2,653	2,667	2,171
Mysore District	49	619	668	509
Kholar District	170	1,990	2,160	1,706
Tumkur District	116	872	988	745
Hassan District	2	...	2	1
Kadur District	5	1	6	5
Shimoga District	...	...	...	...
Chittaldrug District	7	...	7	4
Totals	...	...	18,862	15,597

MADRAS CITY.

Plague infection was said to have been carried by rail to Madras from Poona; but the disease did not spread. The first recognised case was reported on October 7, 1897, and the first death from plague on October 1[illegible]. According to the returns of the Registrar-General there were, in all, three deaths from plague during the last quarter of 1897 in the City of Madras. During 1898 no information as to any plague outbreak is forthcoming; for, though seven plague deaths are recorded by the Registrar-General, there

* For these figures I am indebted to Dr. [illegible] Low, late Medical Officer in charge of the [illegible] Camp, Jalarpet.

was apparently no epidemic prevalence. The occurrence of the cases is believed to have been due to importation of infection from districts outside the city. Although there was no epidemic, some 5,360 persons or more submitted to inoculation with Haffkine's material.

Later reports state that up to July 1st, 1899, only 17 plague cases, with 11 deaths, had been recorded in Madras City.

Madras Presidency.

In common with the rest of Southern India, including the Native States of Hyderabad and Mysore, some of the districts in the Madras Presidency suffered from localised outbreaks of plague. Since the two above-named Native States are in constant communication by railway with adjoining districts of the Presidency, it was not to be wondered at that outbreaks were reported in a number of places during the latter months of 1898. The inland parts of the Presidency, as was to be expected, were the most affected, the towns and villages on the Carnatic shore suffering but little in comparison. Cases in the seaport towns were in all instances imported ones. Nowhere, however, did the mortality show that the epidemic was severe. During November the town of Namakal was attacked, and six districts in the Presidency were reported after this as having plague. In the Anantapur district the spread of infection by human agency was traced from railway stations to several villages. Unusual mortality among rats was also observed here, as in other parts of India, to coincide with occurrence of plague among human beings. The early months of 1899 showed a distinct diminution in the amount of plague in the Presidency. A series of railway inspection stations were established during 1897 and 1898 for the purposes of examining passengers passing from infected districts into the Madras Presidency. It was also deemed advisable to forbid the sale of railway tickets to pilgrims desirous of proceeding to the religious festival which is held at Vaikanta, near Trichinopoly, from December 24, 1897, to January 3, 1898. This festival is attended usually by pilgrims from all parts of India. A number of additional medical men for plague and for general duty were engaged. The result of these measures was that, notwithstanding the numerous foci of the disease adjoining the whole of the western frontier of the Presidency, there had been up to the end of 1898 nothing like a wide spread epidemic. It is estimated, however, that throughout the whole of the Madras Presidency up to July 1st, 1899, there had occurred upwards of 2,092 plague deaths. The Madras Presidency had a clean bill of health for the week ended July 8th, 1899.

The areas chiefly affected by plague were the districts of Anantapur, Bellary, Kurnool, North Arcot and Salem, along with the municipalities of Vellore and Wallajapet. The districts of Anantapur, Bellary, North Arcot and Salem border on the Mysore frontier, and Kurnool on the Hyderabad frontier. The municipalities of Vellore and Wallajapet are situated a few miles apart in the North Arcot district. Of the remaining 17 districts in the Madras Presidency seven had isolated cases, all imported, and 10 districts were completely free from plague. Of 59 municipalities in the Presidency 38 remained entirely free from plague infection. The subjoined table shows the distribution of plague in the Presidency to July, 1st, 1899.*

Districts and Municipalities.	Plague cases.		Total cases.	Deaths.
	Imported.	Indigenous.		
Districts.				
Anantapur	62	625	687	552
Bellary	21	282	303	260
Kurnool	4	170	174	143
North Arcot	115	436	551	413
Salem	143	467	610	444
7 other districts	15	...	15	11
Municipalities.				
Madras	17	...	17	11
Vellore	6	182	188	170
Vaniyambadi	8	1	9	6
Wallajapet	8	59	67	49
Tirupati	1	3	4	4
16 other municipalities	38	...	38	29
Total	...	...	2,663	2,092

BALUCHISTAN.

In a telegram to his Government, Dr. Campassampiero, the Ottoman Sanitary Delegate at Teheran, under date February 1st, 1897, stated that "bubonic plague was announced at Djewadir, the maritime port of Baluchistan." But the only instances of plague which came under the notice of the Indian Government were two imported cases, one stopped by the Inspectors on 30th March, 1897, at Sharigh, and the other on April 12th at Sibi.

AFGHANISTAN.

No direct information is to hand from Afghanistan in recent years as to the occurrence of plague in that country. The Russian Consul-General at Khorassan (Eastern Persia) states that "several Indians who had come to Afghanistan, died at Khandahar, in December 1896, of bubonic plague." But no confirmation of this statement has as yet been made public. It is probable that if such an occurrence had taken place, the officials of the Indian Government would have been made acquainted with it.

SINGAPORE.

Singapore is in incessant communication through shipping with Hong Kong as well as with Indian ports; nevertheless no epidemic of plague has during the last five years developed at this port. It is true that there have been several importations of

* For these figures I am indebted to Dr. J. Spencer Low.

the disease, but owing, it is stated, to the measures carried out at Singapore, the imported infection has not been able to establish itself.*

In June 1894, a fireman, ill of plague, was landed at the isolation station on St. John's Island, from the s.s. *Pakshan* on its arrival from Hong Kong. The patient after a tedious illness recovered, and no subsequent cases occurred.

In March 1896, a coolie was landed at St. John's Island from the s.s. *Wing Sang* suffering from plague of which he died next day. Five other "doubtful" cases were also landed from this ship and were isolated as a precautionary measure, but all recovered and no extension of the disease took place.

During 1898, two plague cases were landed from the *Keong Wai*, and one case from the *Lightning*. All three cases proved fatal. There were also three other "doubtful" cases taken to St. John's Island, all of which recovered.

In 1899, up to August, 10 plague cases had been treated at the isolation hospital, having been landed from vessels arriving from infected ports; eight of these patients died. Five other "doubtful" cases were isolated and recovered.

In all the so-called "doubtful" cases mentioned above, fever was present, and there were enlarged glands for which there existed no obvious cause. In several of the patients pneumonic symptoms were observed.

In addition to these, numerous cases of illness with feverish symptoms were found among passengers arriving at the port. All such cases were isolated and kept under observation till a correct diagnosis had been established.

The measures carried out for guarding against the importation of plague at Singapore have been the following:—

Ships arriving from plague infected ports have been compelled to proceed to a mooring station where the passengers and crew, one by one, were medically inspected. The passengers were then landed at St. John's Island, where they remained under "observation" while completing a period of nine days' probation dating from the time of leaving the infected port. At the end of this period the passengers were allowed to proceed to their destinations in the colony. The passengers having been landed, "the ship was disinfected," and was then allowed free pratique. No attempt was made to disinfect the cargo, no matter whether it was of the kind classed as "susceptible" or otherwise. Not a single instance occurred, during the five years under consideration, where plague was caused by the landing, or handling of cargoes.

The work of medically inspecting large numbers of persons, many of whom were coolies, was heavy, but the result must be regarded as highly satisfactory to those responsible for the arrangements.

The subjoined table shows the number of passengers and crews arriving at Singapore in ships from ports declared infected, from 1895 to 1898 inclusive, and who underwent medical inspection;

---

* Certain rumours were current in 1897 that "Pestis Minor" was prevalent in Singapore, but no confirmation of these rumours has been forthcoming up to the present.

the actual number of detected plague cases occurring among those inspected is also given, as well as the number of suspicious cases classed as "doubtful."

Year.	Passengers.	Crew.	Total inspected.	Cases.	
				Plague.	"Doubtful."
1895	196,635	11,438	208,073	...	...
1896	145,814	13,457	159,271	1	5
1897	115,548	12,515	128,063	1	...
1898	144,686	12,571	157,257	3	3
	602,683	49,981	652,664	5	8

In 1899, up to August, there were, as has been said, 10 plague cases, and five others classed as "doubtful."

During 1898 some 200 cases of feverish illness were detected among the passengers, but after detention under "observation," the illness was not found to be plague. In 1899, up to August, similarly, over 300 of these "fever" cases were detained, but none of them turned out to be plague.

Russia.

The last recorded outbreak of plague in past years in European Russia occurred in 1878–79 in the district of the Lower Volga in the province of Astrakhan;* since that time, so far at least as can be judged from the absence of reports upon the subject, the disease had not re-appeared in epidemic form in Russia till 1892. In June, 1892, the reappearance of plague in Russia was reported from Olti in the Government of Kars. Reports from Tiflis stated that the disease had been smuggled into the district from Turkey. It was also announced at the time that cases of plague had occurred in the neighbourhood of Batoum among the Mohammedan population.

In September, 1892, bubonic plague suddenly appeared in Turkestan. On September 10th it had broken out at Askabad, which is not far from the Persian frontier, and in six days, out of a total population of about 30,000, it had killed some 1,303 persons. The outbreak described by the Governor of the Province in his report as "the Black Death," followed shortly upon an outbreak of cholera, but as regards the persons attacked by plague it is mentioned distinctly that none of them had diarrhoea or vomiting. An official report of March 13th, 1893, says that there were several cases of Siberian plague near Kiew, and that one case was in the City itself. No further particulars are given. In 1896, plague

* *Vide* Mr. Netten Radcliffe's account in his memorandum on the progress of Levantine Plague in 1876–79. Supplement to the Ninth Annual Report of the Local Government Board, 1879–80. *See* also Dr. J. F. Payne's report on this epidemic, Transactions of the Epidemiological Society, Vol. IV., p. 262. *See*, likewise, summary of Report of the Imperial German Medical Commission on the Plague which prevailed in the Province of Astrakhan during the Winter of 1878–79, communicated to the Epidemiological Society by Dr. J. Lawrence Hamilton on May 5, 1880, Vol. IV., p. 376.

is said to have appeared at Merv, but no particulars were forthcoming as to the outbreak. A newspaper telegram, however, went so far as to allege that some 10,000 inhabitants of Merv and neighbourhood had been killed during September and October, 1896, by some malignant fever, the suggestion being that plague was in question. It was not "cholera, small-pox or typhus, for these would have been easily recognised." On September 17, 1897, a newspaper correspondent telegraphed from Odessa that plague had broken out in the Northern Caucasus, and that several deaths had already occurred ; it was added that measures were being taken to arrest the spread of the malady. No further confirmation of this outbreak was forthcoming.

In the Weekly Public Health Reports, issued by the supervising Surgeon-General of United States Marine Service, it is stated that during the week ended November 28th, 1896, two cases and one death from plague occurred in St. Petersburg; but no details of these cases have been obtainable. Again, in the same publication, it is mentioned that another death from plague occurred at St. Petersburg during the week ended April 17, 1897. Details were also given of a case of plague on board the British ship *Baldwin* arriving at the port of Theodosia (or Kaffa) in the Black Sea. The patient was isolated but bacteriological investigation did not support the diagnosis. As the ship had sailed from Cardiff to Theodosia and had touched at no infected port, it is unlikely that the man (a member of the crew) could have contracted the disease. No doubt, in times of alarm, cases such as this are liable to be reported as plague, but, on further inquiry, they turn out not to be plague at all.

In November, 1898, information was officially received that plague had, early in October, invaded the village of Anzob in the district of Iskander in the Hissar Range, 167 miles south-east of Samarkand. The population of the village numbered 357, and of these there died during October no fewer than 233. The infection came from the village of Marzin in the same canton. A native woman of Anzob waited upon a sick relative there. On her return to Anzob she fell sick and died within three days. The first extension of the disease took place among this woman's relatives, and from them the disease rapidly spread throughout the village. The local measures taken by the Government Authorities comprised the establishment of an isolation hospital, the clothes worn by persons who died, and their bedding, were burnt, the village was thoroughly scavenged; precautions were taken as to nursing of the sick and speedy burial of the dead. Later, the infected houses were destroyed by fire. A cordon made up of the inhabitants of neighbouring villages was organised. Bodies of Cossacks were employed to watch the posts of observation, and medical men and nurses, with the necessary medicines and disinfectants, were despatched to the locality. A special bacteriologist, who had studied plague in India, accompanied the party. No further extension of the disease appears to have occurred. The Russian Government assert that the infection was imported to Anzob through Baluchistan and Afghanistan from India by pilgrims who had returned from the pilgrimage to Mecca by way of Karachi, a port which was known to be infected. But no facts in support of this assertion have been brought forward.

..No. 13.

...on of
...ic
...from
1898; by
...nce Low.

Anzob, it may be mentioned, is about a thousand miles from Karachi, as the crow flies. It is not unlikely that in the villages like Anzob in that district there may be an endemic infection like the "Mahamari" of Damaun and Garhwal, which now and again comes under observation in the remote mountain villages.

In a recent paper published in the *Zeitschrift für Hygiene und Infectionskrankheiten* for May, 1899, Dr. Favre of Charkow describes a disease that sometimes breaks out amongthe wan dering Mongolian tribes which inhabit the steppes in the district of Akscha in the Northern part of Eastern Siberia. This illness is characterised by high fever and painful glandular swellings in the axilla or groin; those who have no buboes have lung affection and a bloody sputum. The disease runs a rapid course, and has a very high mortality, death taking place in from two to four days from the onset. The illness is very infectious, and Dr. Favre relates an instance where a Russian doctor and his assistant made a postmortem examination upon a victim of the epidemic; both doctor and assistant contracted the disease within four days and both died from it. Notes of five outbreaks in the Akscha district are given, viz., one in 1888, one in 1889, two in 1891, and one in 1894. The wandering tribesmen hunt, for purposes of food, a rodent, a kind of Marmot (Arctomys Bobae) which is plentiful in the locality. These men know that the animals suffer at times from an epidemic disease which is very fatal. It is also known that if they handle or cut animals showing signs of the disease they themselves are very liable to suffer also. Dr. Favre does not commit himself to saying that this disease which he calls "Sarbaganpest" is true plague, but he points out that it resembles it closely.

TRIPOLI.

Plague was epidemic in Tripoli from 1856 to 1859, it devastated the town of Benghasi in 1858–59, and appeared at Merge near Benghasi in 1874. No definite origin of these outbreaks was traced.

Towards the end of 1892, after a long interval during which no plague epidemic had been reported, news came that plague had again broken out in the town of Benghasi. The appearance of the disease coincided with the arrival of a number of Bedouins from the interior. Some 20,000 of these, it is said, arrived in the town during the course of three months. This influx of Arabs was the result of the failure of their crops for three successive seasons through drought, and through the destruction caused by a visitation of locusts. The Bedouins were in a famine-stricken condition. They crowded into the town and subsisted as best they could in hunger and dirt, in overcrowding and misery; these conditions being aggravated by much fouling of the ground by filth, through the unclean habits of these people. As has been said, the disease was at first regarded as bubonic plague, but a Medical Commission from Malta reported that the outbreak was really "spotted typhus," though it is admitted by the Commission that the cases they saw and others which were described to them were not characteristic. Some had "boils," and even axillary abscess was noted. In 1892, our knowledge of plague was hardly so advanced as it is at the present time, and one cannot refrain from entertaining suspicion when "typhus" associated with boils or buboes is mentioned. It is possible, therefore, that after all, the disease was plague, since it is no uncommon affliction among the Arabs on the Red Sea Coast, as well as among the Bedouins in Mesopotamia. As the disease is also endemic in Central Africa, it could obviously travel under certain circumstances northwards as far as Tripoli.

At Benghasi there were in 1892 a large number of deaths, but owing to the peculiar nature of Turkish rule, and owing also to local ignorance, it is no easy matter to obtain even rough statistics. The Arabs neither register deaths nor do they call in qualified physicians. It is said that about a fourth part of the population caught the disease.

East Africa.

For a number of years a virulent disease, the clinical characters of which were said to resemble those of bubonic plague, has been epidemic from time to time in parts of Central Africa, notably, it is said, in the Buddu District in the territory of King Mitsa, in the Province of Uganda.* There are some who believe that the occurrence of plague in Uganda is due to the increasing communication which that province has with India, whose natives are now employed as soldiers or as labourers in East and Central Africa.

During the last eight or nine years outbreaks of this epidemic disease, locally known as "Rubwunga," have occurred in districts on the north-west borders of the German East African Protectorate, the infection having been brought in the first instance, so it is suggested, from Uganda.

Referring to the depopulation of Uganda by plague epidemics, one observer expresses the opinion that the scourge is undoubtedly fostered by the filthy and uncleanly habits of the natives, and in addition by the overcrowding in their dwellings. The ordinary Uganda house has no other opening for purposes of ventilation than the door. The interior of the dwelling is dim and gloomy and is divided into partitions by curtains of bark cloth, which add to the general stuffiness by impeding the circulation of air and the entrance of light. The floor, which is strewn with fine grass, becomes the receptacle for all kinds of filth. This is not swept away or removed, but as often as its condition reaches a state which is too appalling even for the Waganda to put up with, an additional layer of fresh grass is introduced, "so that the ultimate condition of things may be better imagined than described." It is only natural that such miserable and dirty habits should foster disease and pestilence.†

German East Africa.—In the Autumn of 1897 the German Imperial Government was informed that "*Rubwunga*" was prevalent in the Kisiba district situated to the west of Bukoba, which lies on the Western shore of Lake Victoria Nyanza. On the advice

* Among others, Captain Lugard, D.S.O., in his book on "The Rise of our East African Empire," published in 1893, refers to the depopulation of certain districts in Uganda by plague.

† Letter in the "Times" of July 6th, 1893, from a special correspondent in Uganda.

A. No. 18.

ue
ision of
mic
ue from
o 1898; by
ruce Low.

of Professor R. Koch, who was at that time in the Protectorate pursuing his scientific researches at Dar-es-Salaam, an expedition was dispatched under Staff Surgeon Dr. Zupitza to Kisiba, to study the nature of the epidemic malady. The expedition reached Bukoba in November, and although the epidemic was for the time subsiding, ample opportunities were available for studying the disease in the infected localities. Dr. Zupitza found that the clinical characters of *Rubwunga* were indistinguishable from those of bubonic plague. He sent preparations of blood taken from persons suffering from the malady, and also portions of the spleen and other viscera taken from fatal cases, to Professor Koch, for bacteriological examination; with the result that bacilli were found in these preparations identical with those with which Koch had become familiar in India while studying plague at Bombay.

It was also noted that rats had died in large numbers during the epidemic at Kisiba. The natives had already recognised that the spread of the disease was in some way associated with the rats, and as soon as the dead bodies of these animals began to be found in or near their huts, these were at once abandoned. Some of the dead rats picked up at Kisiba were forwarded for examination to Koch, who found in them extraordinary numbers of plague bacilli, and in pure culture. Artificial inoculation was also practised. Rats, inoculated with blood taken from an incised bubo, died in two or three days: the blood, and in particular the spleen, was found to contain swarms of plague bacilli. The investigations of Koch and Zupitza place beyond all dispute the fact that *Rubwunga* is true bubonic plague, and that it is endemic in Central Africa.

Towards the end of 1898 *Rubwunga* was prevalent in at least two localities in German East Africa:—(1) in the district under the Sultan of Seissawala, north and north-west of Bukoba; (2) in the district in and around Kitengule.

According to statements made by the Sultan of Seissawala, the outbreak in 1898 had not, in his opinion, been so extensive as in some former years. Only seven "Shambas" or plantations in the northern part of his territory had been attacked during the year: but of a population of 715 in these seven "Shambas," 467 fell victims to the epidemic.*

In Kitengule district the outbreak began about the middle of August, 1898, and from that date to the middle of December some 60 deaths were known to have been due to the disease. In two small hamlets it is stated that half of the inhabitants died from the malady.

Early in 1898 measures were taken to divert the trading caravan routes, so that travellers should not pass through the infected districts; and precautions were taken to limit intercommunication between the invaded and uninvaded localities in the German Protectorate.

Lorenzo Marques.—A report from the British Consul at Lorenzo Marques stated that at the beginning of December 1898, the steamer *Gironde* of the *Messageries Maritimes* arrived from Madagascar and entered Mozambique, having a case of plague on

* Veröffentlichungen des Kaiserlichen Gesundheitsamtes, 1899, page 589.

board, a coloured man, employed in the engine room. As the ship was not allowed to land the sick man, or to communicate with the shore, she sailed again for Madagascar. In January 1899, three Indians were striken with what was regarded at first as bubonic plague, but the diagnosis was not definitely confirmed. The United States Consul reports to his goverment that the sanitary condition of Lorenzo Marques is bad. He says that for years "Asiatic traders have been living worse than pigs. As many as 20 persons sleep in a little room 12 feet square." There is no drainage system, and slops, urine, and excrement, are deposited in the tiny back yards.

Mombassa.—Late in December 1898, the British India Steam Navigation Company's Steamer, the *Bhundara*, arrived at Kilindini, near Mombassa (British East Africa), with 1,000 coolies on board, brought from Karachi, to work on the Uganda railway which is in process of construction. The ship had been 12 days on the voyage, and during that time seven deaths, six of them from plague, had occurred among the coolies. On arrival the authorities refused at first to allow the coolies to disembark; but ultimately the ship landed her passengers on an island near Zanzibar, where the ship was disinfected and fresh water taken on board. No other cases developed subsequently among the coolies or the crew.

MADAGASCAR.

On November 23rd, 1898, plague was medically recognised to be present at Tamatave; and, this diagnosis was shortly afterwards confirmed by bacteriological examination. It was admitted, however, that the disease must have been present in Tamatave for a month, or even two months prior to November 24th. Here, as elsewhere, it was noticed that rats died in considerable numbers just before plague was recognised at Tamatave. The first plague cases were found to have occurred in a certain quarter where quantities of merchandise, rice and other grain, were stored. The first sufferers were natives who were in the habit of working at the discharge of vessels bringing merchandise to Tamatave; and the theory adopted by the French officials was that the disease was brought from Bombay by a ship laden with rice. Another theory was that the infection had been smuggled into the island by one or other of the numerous Arab boats which frequent the Indian Ocean from the Straits of Malacca to the Mascarenha islands, doing an active trade on the shores of Malacca, India, Socotra and other islands. These sailing boats carry no papers and, it is alleged, can land cargoes or passengers in a clandestine way at any point on the coast that they may choose. But opposed to this view is the fact that Zanzibar, which is the favourite port for such craft, had not, up to end of 1898, become infected by plague. Another suggestion as to the introduction of infection into Tamatave, was that the French steamer the *Gironde* which trades between Diego Suarez, a port to the extreme north of Madagascar, and Lorenzo Marques, had been in contact at Beira with an English vessel and had from it contracted plague by which *eight* of the crew of the *Gironde* were

P. A. No. 18. the [...]usion of [...]bonic [...]gue from [...] to 1898; by Bruce Low.

attacked in November 1898. M. Proust, in an article published in the *Bulletin de l'Académie de Medicine* (No. 2, 1899, page 67), makes the definite statement that the *Gironde* became infected while lying in port at Beira side by side with an English ship the captain of which had omitted to notify plague from which several persons on board had been suffering. The sailors of the *Gironde*, it was alleged, had come into close contact with the English ship's crew in the process of unloading the cargoes which had to be delivered from both ships into the same lighter. M. Proust is, however, careful to add that these "facts" require further confirmation. So far as I can ascertain, no such British plague-stricken ship has been heard of. Another hypothesis was that the *Gironde* became infected through meeting an English ship at sea after leaving Diego Saurez. Finally, some maintained that the plague was brought by emigrants from Mauritius which, it was pointed out, had frequent communications with India. But, as a matter of fact, Mauritius did not, so far at least as is known, become infected till January 25th, 1899, *i.e.*, not till after plague had been established some two months at Tamatave.

The total number of reported cases from November 24th, 1898, to January 5th, 1899, a period of six weeks, at Tamatave, in a population estimated at 8,000, was 239 of which 158 proved fatal. During January, 1899, some 48 fresh cases were reported and of these 38 proved fatal. By February, the official reports state that the disease had almost disappeared. These reported cases do not, however, represent the full incidence of the disease at Tamatave and its surrounding villages. The native population "did not lend themselves to observation;" they concealed their sick and buried their dead clandestinely. Many persons it is alleged never sought any medical assistance at all, and died from plague without any record of the fact being made. Even medical men refrained from reporting any but the typical cases, fearing to "exaggerate the epidemic." An official report on this outbreak was published in the "Journal Officiel de Madagascar" on January 10th, 1899, by Dr. Lidin (Director of the Health Service of the Corps of Occupation), who gives an outline of the facts, and sums up his report by several conclusions of which the following are the chief, viz.:—

"1. The disease which occurred at Tamatave was undoubtedly bubonic plague.

2. The ailment was diagnosed first on November 23rd, 1898, but had existed previous to this date for a month if not longer.

3. The disease was originated by means of infected grain from India.

4. It attacked first and above all others the natives employed in the handling of rice in certain houses of business.

5. It attacked the coloured races to the exclusion, almost altogether, of the pure whites. Only three Europeans were attacked and but one of them died.

6. At first the conveyance of contagion from person to person seemed feeble, but it became stronger as time went on.

This official report, however, only deals with the epidemic up to December 14th, 1898. But from consular and other sources particulars have been received giving information up to the end of February, 1899, at which time the outbreak had all but ceased.

The total reported cases from November 24th, 1898 to February 2nd, 1899, were 305, of which 206 were fatal, a case mortality of 67·5 per cent."

Vienna.

The Royal Academy of Science in Vienna sent four medical delegates to India early in 1897, to study plague. Drs. Müller and Poech concerned themselves with the clinical characters of the disease, while Drs. Albrecht and Ghon investigated its pathology and bacteriology.

On the return of these delegates further investigations were instituted by Drs. Albrecht and Ghon of the materials they had brought with them from Bombay. A laboratory for the work was assigned to them in the anatomical and pathological department of the Vienna General Hospital. A laboratory attendant of some experience, named Barisch, was appointed, mainly to look after the experimental animals, to keep the instruments and apparatus clean, to disinfect, and to destroy the bodies of dead animals. This man received minute instructions as to the manner in which he was to carry out his several duties, and he was warned of the risks to which he was exposed in the event of his neglecting the precautions. The investigations were undertaken with a view to ascertain how plague was propagated, by what means the bacilli penetrate into the system, and also as to how far immunity against plague could be established. The work was begun in August, 1897, and was continued up to October, 1898, when the events I am about to relate occurred.

During the investigations the plague cultures were kept in the exclusive possession of Dr. Ghon, and were propagated by him alone. A special dress was worn over their other clothing by the two doctors and by the attendant. Elaborate disinfection by perchloride of mercury solution was practised daily as soon as the work was done.

At the beginning of October, 1898, the labours of the experimenters were approaching a conclusion. The work had lasted upwards of thirteen months, and all that remained to be done was the testing of immunised animals, and infecting those used for control purposes. The last plague inoculation was made on October 4th, 1898.

Barisch, the attendant, had a rigor on October 14th. He had indulged in a debauch on the night of October 8th–9th, and had returned to his home at 5 o'clock in the morning, as he expressed it, "frozen." On October 15th Dr. Ghon and Dr. Stejskal saw the patient and regarded him as suffering from influenza. Dr. Ghon, however, examined the man's sputum, and found alongside pneumococci some bacilli which he thought might possibly be regarded as degenerate forms of the plague organism. He showed his microscopic preparations to Dr. Albrecht, his fellow worker, who agreed with this view. Dr. Ghon therefore communicated his suspicions to Dr. Müller, who had had large opportunities for gaining practical experience of the clinical features of plague while acting as a delegate in India. The patient the same day, October 15th, was seen by Dr. Müller, who declared most decidedly that, judging from the clinical symptoms,

P. A. No. 18.

the ... fusion of ... ponic ... gue from ... to 1898; by Bruce Low.

the case was not one of plague, but was a commencing croupous pneumonia. Dr. Ghon, however, thought it advisable to make cultures from the man's sputum, and injected some of it into a rat. On October 16th Dr. Müller again made a careful and prolonged examination of the patient, and once more declared his inability to believe, from the clinical characters of the illness, that the man was suffering from plague infection. Dr. Ghon, still suspicious, had the man removed to the General Hospital, where he was isolated in a detached ward, with two nurses to wait upon him. Further cultures were made from the sputum, and a second rat was injected on the 16th.

On October 17th the new cultures showed no very definite result. The rat injected on the 15th remained alive and apparently well. Dr. Müller reported that, after once more examining the sick man, he could find "nothing to contra-indicate an ordinary pneumonia." Meanwhile a quantity of plague serum was supplied to Dr. Müller, but he declined to make use of it in the treatment of the patient, or as a preventive upon himself or the nurses.

The same evening (October 17th) the second rat died, and from a small mass of bloody exudation in its peritoneum Dr. Ghon found some diplo-bacilli which showed bi-polar staining.

On October 18th Dr. Müller, though admitting that the bacteriological evidence was stronger in favour of a diagnosis of plague, at the same time refused to assent to this on the basis of the clinical characters of the illness. The same afternoon, October 18th, Barisch died. The condition of the patient a few hours before his death was such as to convince Dr. Müller that after all he had to do with a case of plague pneumonia, and he sent a communication in this sense to Dr. Ghon.

On October 19th the first rat, which was injected on the 15th, died, and subsequent examination showed that it had died of plague.

On October 20th one of the nurses became ill, and on the 21st Dr. Müller was attacked. Both of these cases were strictly isolated, with Dr. Poech (who had studied plague in Bombay) to attend them, and two Sisters of Mercy to wait upon them. Dr. Müller died on October 23rd, and the nurse on October 29th, both of them from pneumonic plague. The second nurse who had waited upon Barisch had a feverish attack, as also had one of the Sisters of Mercy who attended to Dr. Müller, but there was no evidence to show that the illness in either case was due to plague infection.

As to the precise way in which Barisch became infected it is not possible to speak with certainty, but it is believed that he had of late become less scrupulously careful about his work than before, and that he had been handling, in the absence of the two bacteriologists, the experimental animals, one of which, inoculated on October 4th, had an abscess at the place of inoculation. It is also known that Barisch, who was a confirmed smoker, had, notwithstanding a regulation forbidding smoking in the laboratory, smoked there frequently of late; and it is surmised that his fingers, infected by handling the experimental animals, had touched his mouth in the act of replacing his cigarette between his lips. He may thus have practically infected himself by disobedience of the orders issued by his employers.

Dr. Müller and the nurse in all probability were infected by the sprayed expectoration discharged into the air of the sick room by the patient in the process of coughing; the medical attendant having special opportunities for infection while engaged in auscultating the patient's chest.

The occurrence of this localised plague outbreak created a profound sensation, not only in Vienna, but also in the rest of Europe. In Vienna the wildest rumours as to extension of the infection received credence for a time; but a statement of the facts made in Parliament by the Austrian Prime Minister, and repeated in the press, re-assured the public mind, and the excitement gradually subsided. The Vienna outbreak is noteworthy from the fact that the disease developed unrecognised directly under the eyes of those who had made a special study of plague in India; and that even the clinical characters of the disease passed undetected by the medical man, Dr. Müller, who was presumably the best qualified in Vienna to diagnose the disease. It is also worthy of note that the bacteriologist, Dr. Ghon, who had for 18 months or more been studying the bacillus of plague, in India and afterwards in Vienna, was not able at first to say definitely that the organisms in the sputum of Barisch were really those of bubonic plague. Such circumstances show the difficulties in the way of recognition of the disease by the clinician and the bacteriologist, and it is not surprising that at first, at least, in epidemics of plague, the cases pass unrecognised and unrecorded, no measures of prevention, therefore, being applied to check the spread of the disease until it has got a firm footing in a locality.

In connection with the publicity given to the possibility of plague infection spreading from laboratories where cultures are kept, some colonial governments, after this event, forbade medical men to retain plague cultures any longer in their possession. One bacteriologist in Australia was called upon to give up, to be destroyed, certain plague cultures with which he was working. He refused to do so unless he were paid a certain sum in compensation for his loss. This sum his Government declined to pay, and as he still refused to give up his plague cultures, the police entered his laboratory and then destroyed the cultures in question.

Mention has been already made of the groundless accusation against M. Yersin that he brought plague infection into his laboratory in Annam, and that the disease spread from the laboratory into the neighbouring villages.

. A. No. 19.

pilation of
of
fied Infec-
Diseases,
registered
there-

No. 19A.

TABLE showing Quarter by Quarter, during the Year 1898, for
CASES and of REGISTERED DEATHS from the under-menti
each of the URBAN SANITARY DISTRICTS in question.

[The Cases are a Summary of the Weekly Returns of Notifiable Diseases re
Quarterly Retu

In Registration Divisions.	Urban Sanitary Districts.	Population (1891).	Small-Pox.								
			1st Quarter.		2nd Quarter.		3rd Quarter.		4th Quarter.		T fo
			Cases.	Deaths.	Cases.	Deaths.	Cases.	Deaths.	Cases.	Deaths.	Cases.
I.	London .. .. .. (*Administrative Co.*)	4,232,118	10	..	12	..	..	..	..	1	22
II. South-Eastern.	Croydon	102,695	..	..	..	..	..	..	..	..	..
	Dover	33,300	..	..	..	..	..	..	..	..	..
	Eastbourne	34,969	..	..	..	..	..	..	..	..	..
	Portsmouth	159,278	..	..	..	..	..	..	..	..	..
	Bournemouth	37,781	..	..	..	..	..	..	..	..	..
	Southampton	65,325	2	..	9	..	5	..	1	..	17
	Reading	60,054	..	..	..	..	..	..	..	..	..
III. South Midland.	Willesden	61,265	..	..	1	..	..	..	..	..	1
	Hornsey	44,205	..	..	..	..	..	..	..	..	..
	Tottenham	71,343	..	..	..	..	..	..	..	..	..
	Oxford	45,742	..	..	..	..	..	..	..	..	..
	Northampton	61,012	..	..	..	..	..	..	..	..	..
	Cambridge	36,983	..	..	..	..	..	..	..	..	..
IV. Eastern.	Leyton	63,056	1	..	..	..	..	..	..	..	1
	Walthamstow	46,346	..	..	..	..	..	..	..	..	..
	West Ham	204,903	..	..	..	..	..	..	..	..	..
	Colchester	34,559	..	..	..	..	..	..	..	..	..
	Norwich	100,970	..	..	..	..	..	..	..	..	..
V. South-Western.	Exeter	37,404	..	..	..	..	..	..	..	..	..
	Plymouth	84,248	..	..	..	..	..	..	..	..	..
	Devonport	54,803	..	..	..	..	..	..	..	..	..
	Bath	51,844	..	..	..	..	..	..	..	..	..
VI. West Midland.	Bristol	258,296	1	..	..	..	1	..	..	..	2
	Gloucester	39,444	..	..	..	..	..	..	..	..	..
	Cheltenham	47,514	..	..	..	..	..	..	..	..	..
	Hanley	54,946	..	..	..	..	..	..	..	..	..
	Longton	34,327	..	..	..	..	..	..	..	..	..
	Burton-on-Trent ..	46,047	..	..	..	..	..	..	..	..	..
	Wolverhampton ..	82,662	..	..	..	..	..	..	..	..	..
	Walsall	71,789	..	..	..	..	..	..	..	..	..
	West Bromwich ..	59,474	..	..	..	..	..	..	..	..	..
	Worcester..	42,908	..	..	..	..	..	..	..	..	..
	Smethwick	36,170	..	..	..	..	..	..	..	..	..
	Birmingham	478,113	..	..	..	..	..	..	..	..	..
	Aston Manor	68,639	..	..	..	..	..	..	..	..	..
	Coventry	52,724	..	..	..	..	..	..	..	..	..
VII. North Midland.	Leicester	174,624	..	..	..	..	..	..	..	..	..
	Grimsby	51,934	1	..	..	..	..	..	..	..	1
	Nottingham	213,877	..	..	..	..	..	..	..	..	..
	Derby	94,146	..	..	..	..	..	..	..	..	..

The Governor General in Council is further pleased to direct as follows :—

APP. A. No. 18.

On the Diffusion of Bubonic Plague from 1879 to 1898; by Dr. Bruce Low

(1) That no person shall be permitted to embark with the object of making a pilgrimage to Mecca, except at the ports of Karachi in Sind and Chittagong in the Lower Provinces of Bengal, and unless and until such person has been kept under observation in a place appointed for the purpose until such time as the Medical Officer in charge shall be satisfied that all risk of such person conveying the infection of plague has been completely abated ; and unless such person is conveyed from the place of observation to the place of embarkation under the orders of an officer appointed for the purpose.

(2) That no passage tickets for the Hedjaz shall be sold except at the places appointed for the detention of intending pilgrims under observation.

(3) That all persons who have entered the Bombay Presidency (excluding Sind), with the object of proceeding to the Hedjaz, shall be placed in a camp of observation until such time as the Medical Officer in charge shall satisfy himself that all risk of the occurrence of plague among them has been completely abated. They shall then be sent to their homes.

Although the Government of India have decided that the pilgrimage to the Hedjaz may be permitted subject to the safeguards enumerated above, they are nevertheless convinced that intending pilgrims will be well advised in deferring their purpose until another season, in view specially of the stringent quarantine rules imposed by the Turkish Government and the inconvenience and harassment to which it is likely they will be subjected on arriving in Arabia.

The latest information received by the Government of India shows that Hedjaz is still in a most disturbed condition, and that the journey to Mecca is fraught with danger to life and property.

EMIGRATION FROM INDIA.

Home Department, No. 836, dated the 6th March 1897.

No. 3.

In exercise of the powers conferred by section 2, sub-section (*1*), of the Epidemic Diseases Act (III. of 1897), the Governor General in Council is pleased to direct that no person who has, since the 1st January 1897, resided, or been in, or passed through, the territories (other than Sind) administered by the Governor of Bombay in Council, or the State of Baroda, or the territories of any Native Prince or State under the suzerainty of Her Majesty exercised through the Governor of Bombay in Council, shall, until further orders, be permitted to embark on any ship at any port in British India with the object of proceeding as an emigrant or as a labourer to any port out of British India.

Home Department, No. 1038, dated the 28th April 1898.

In exercise of the powers conferred by section 2, sub-section (*1*), of the Epidemic Diseases Act (III. of 1897), the Governor General in Council is pleased to direct that no person who has, since the 25th March 1898, resided or been in, or passed through, the province of Sind, shall, until further orders, be permitted to embark on any ship at any port in British India with the object of proceeding as an emigrant or as a labourer to any port out of British India.

No. 19A.

COMPILATION of RETURNS of NOTIFIED CASES of CERTAIN INFECTIOUS DISEASES, and REGISTERED DEATHS therefrom.

1898.

APP. A. N

Compilati Returns of Notified I tious Dise and regist Deaths th from.

It should be noted, with reference to these returns, that some of the districts to which they apply administer compulsory notification under local Acts; and that not only are some of the diseases named in the Infectious Disease (Notification) Act of 1889 not notifiable under some of the local Acts, but that under certain of these Acts no payment is made for the notification of multiple attacks of the same disease occurring in the same house within a specified date of the first attack notified.

The Urban Districts finding place in these returns in which compulsory notification has been adopted under local Acts are—

Accrington.	Newcastle-on-Tyne.
Ashton-under-Lyne.	Norwich.
Birkenhead.	Nottingham.
Blackburn.	Oldham.
Bolton.	Portsmouth.
Burnley.	Preston.
Bury.	Reading.
Cheltenham.	Rotherham.
Chester.	Scarborough.
Croydon.	Stockton-on-Tees.
Darwen.	Sunderland.
Derby.	Swansea.
Grimsby.	Wakefield.
Halifax.	Warrington.
Hartlepool.	West Ham.
Huddersfield.	Wigan.
Leicester.	Willesden.
Manchester.	York.

. A. No. 19.
of
of
Infec-
iseases,
gistered
there-

In Registration Divisions.	Urban Sanitary Districts.	Population (1891).	DIPHTHERIA.								
			1st Quarter.		2nd Quarter.		3rd Quarter.		4th Quarter.		fc[...]
			Cases.	Deaths.	Cases.	Deaths.	Cases.	Deaths.	Cases.	Deaths.	Cases.
I.	London .. .. .. (*Administrative Co.*)	4,232,118	2,834	534	2,529	367	2,704	380	3,493	511	11,5
II. South-Eastern.	Croydon	102,695	50	6	27	5	49	4	29	2	1
	Dover	33,300	8	2	4	1	2	..	5	1	1
	Eastbourne	34,969	19	3	9	2	3	1	11	5	4
	Portsmouth	159,278	40	4	42	8	43	9	159	34	2
	Bournemouth	37,781	8	..	4	..	3	..	1	2	1
	Southampton	65,325	10	2	21	5	25	10	88	40	1
	Reading	60,054	6	1	1	..	2	1	10	3	1
III. South Midland.	Willesden	61,265	83	21	81	19	99	13	69	10	3
	Hornsey	44,205	32	2	18	2	16	4	33	4	[illegible]
	Tottenham	71,343	28	5	50	12	44	5	78	7	1
	Oxford	45,742	4	1	3	1	3	..	18	1	[illegible]
	Northampton	61,012	4	1	2	..	9	2	1	1	1
	Cambridge	36,983	3	1	3	1	4	2	20	2	[illegible]
IV. Eastern.	Leyton	63,056	51	13	36	7	37	4	38	4	1
	Walthamstow	46,346	76	14	44	12	65	13	41	6	2
	West Ham	204,903	183	43	200	32	191	36	393	70	9
	Colchester	34,559	8	5	5	1	24	11	28	7	[illegible]
	Norwich	100,970	17	1	17	6	2	..	15	7	[illegible]
V. South-Western.	Exeter	37,404	6	3	2	1	1	1	3	1	1
	Plymouth	84,248	8	4	11	2	19	2	13	2	[illegible]
	Devonport	54,803	2	1	6	2	5	..	2	..	1
	Bath	51,844	15	..	7	1	4	1	16	2	4
VI. West Midland.	Bristol	258,206	81	17	41	13	48	10	44	4	21
	Gloucester	39,444	24	9	12	2	2	..	19	6	[illegible]
	Cheltenham	47,514	15	1	7	..	8	1	20	4	[illegible]
	Hanley	54,946	14	5	3	2	13	3	11	1	4
	Longton	34,327	220	34	115	20	94	26	125	25	5
	Burton-on-Trent ..	46,047	12	4	13	..	7	3	19	3	[illegible]
	Wolverhampton ..	82,662	57	18	25	5	63	8	51	7	1
	Wallsall	71,789	3	2	4	5	4	3	7	1	1
	West Bromwich ..	59,474	19	4	5	2	12	3	7	2	4
	Worcester	42,908	90	14	63	4	15	2	18	5	1
	Smethwick	36,170	14	8	9	2	14	6	18	4	[illegible]
	Birmingham	478,113	202	39	118	33	167	27	169	32	6
	Aston Manor	68,639	30	1	9	1	16	2	18	4	[illegible]
	Coventry	52,724	..	2	3	1	7	2	11	7	[illegible]
VII. North Midland.	Leicester	174,624	45	11	85	13	39	8	105	30	2
	Grimsby	51,934	38	4	18	4	15	..	19	3	[illegible]
	Nottingham	213,877	16	5	16	8	19	3	32	8	[illegible]
	Derby	94,146	20	3	16	8	14	..	24	3	[illegible]

…of 93 URBAN SANITARY DISTRICTS, the Number of NOTIFIED DISEASES, together with an ANNUAL SUMMARY of these data for

by the Board from Medical Officers of Health. The Deaths are extracted from the the Registrar-General.]

Compila Returns Notified tions and regi Deaths from.

In Registration Divisions.	Urban Sanitary Districts.	Population (1891).	SMALL-POX—*continued.*									
			1st Quarter.		2nd Quarter.		3rd Quarter.		4th Quarter.		Total for 1898.	
			Cases.	Deaths.	Cases.	Deaths.	Cases.	Deaths.	Cases.	Deaths.	Cases.	Deaths.
VIII. North-Western.	Stockport	70,263	..	..	..	..	..	..	..	..	..	..
	Macclesfield	36,000	..	..	..	..	..	..	..	..	..	..
	Chester	37,105	..	..	..	..	..	..	..	..	..	..
	Birkenhead	99,857	1	..	..	..	..	..	..	..	1	..
	Wallasey	33,229	..	..	..	..	..	..	..	..	..	..
	Bootle	49,217	..	..	..	..	..	1	..	..	..	1
	Liverpool	629,443	9	..	2	..	1	1	5	1	17	2
	St. Helens (Lancs.)	71,288	..	..	..	..	..	..	..	..	..	..
	Southport	41,406	..	..	..	..	..	..	..	..	..	..
	Wigan	55,013	..	..	..	..	..	..	..	..	..	..
	Warrington	52,743	..	..	..	..	..	..	..	..	..	..
	Bolton	115,002	..	..	..	..	..	..	..	..	..	..
	Bury (Lancs.)	57,212	..	..	..	..	..	..	..	..	..	..
	Salford	198,139	..	..	..	..	..	..	..	..	..	..
	Manchester	505,368	..	..	..	..	..	..	..	..	..	..
	Ashton-under-Lyne	40,463	..	..	..	..	..	..	..	..	..	..
	Oldham	131,463	1	..	..	..	..	..	..	..	1	..
	Rochdale	71,401	..	..	..	..	..	..	..	..	..	..
	Accrington	38,603	..	..	..	..	..	..	..	..	..	..
	Burnley	87,016	..	..	..	..	..	..	..	..	..	..
	Blackburn	120,064	..	..	..	..	..	..	..	..	..	..
	Darwen	34,192	..	..	..	..	..	..	..	..	..	..
	Preston	107,573	..	..	..	..	..	..	..	..	..	..
	Barrow-in-Furness	51,712	..	..	..	..	..	..	..	..	..	..
IX. Yorkshire.	Keighley	30,810	..	..	..	..	..	..	..	..	..	..
	Huddersfield	95,420	1	..	1	..	..	..	..	..	2	..
	Halifax	89,832	..	..	..	..	..	..	..	..	..	..
	Bradford	216,361	1	..	..	..	..	..	..	..	1	..
	Leeds	367,505	3	1	..	1	..	..	..	..	3	2
	Wakefield	33,146	..	..	..	..	..	..	..	..	..	..
	Barnsley	35,427	..	..	..	..	..	..	..	..	..	..
	Sheffield	324,243	..	..	..	..	..	..	..	..	..	..
	Rotherham	42,061	..	..	..	..	..	..	..	..	..	..
	York	67,004	3	1	1	..	..	..	..	..	4	1
	Hull	200,044	1	..	3	..	..	..	..	..	4	..
	Scarborough	33,776	..	..	..	..	..	..	..	..	..	..
	Middlesbrough	75,532	1,200	153	197	44	4	1	..	1	1,401	199
X. Northern.	Darlington	38,060	3	1	1	..	..	..	..	..	4	1
	Stockton-on-Tees	49,705	4	..	12	..	..	..	..	..	16	..
	West Hartlepool	42,710	8	1	2	..	..	..	..	..	10	1
	Sunderland	131,015	4	1	1	1	..	..	..	..	5	2
	Jarrow	33,675	..	..	..	..	..	..	..	..	..	..
	South Shields	78,391	..	..	6	..	..	..	..	..	6	..
	Gateshead	85,092	4	..	10	2	3	1	..	..	17	3
	Newcastle-on-Tyne	186,300	1	..	16	3	..	..	..	..	17	3
	Tynemouth	46,588	3	..	..	..	..	..	..	..	3	..
	Carlisle	39,176	..	..	5	..	..	..	..	..	5	..
XI. Welsh.	Newport (Mon.)	54,707	..	..	..	..	..	..	..	..	..	..
	Cardiff	128,915	..	..	..	..	..	..	..	..	..	..
	Rhondda	88,351	..	..	..	..	..	..	..	..	..	..
	Merthyr Tydfil	58,080	..	..	..	..	..	..	..	..	..	..
	Swansea	90,349	..	..	..	..	..	..	..	..	..	..
	Totals, 93 Districts	13,128,484	1,262	158	279	51	14	4	6	3	1,561	216

App. A. No. 19.

Compilation of Returns of Notified Infectious Diseases, and registered Deaths therefrom.

In Registration Divisions.	Urban Sanitary Districts.	Population (1891).	Scarlet Fever.								
			1st Quarter.		2nd Quarter.		3rd Quarter.		4th Quarter.		T[otal] fo[r]
			Cases.	Deaths.	Cases.	Deaths.	Cases.	Deaths.	Cases.	Deaths.	Cases.
I.	London (*Administrative Co.*)	4,232,118	3,856	182	3,907	154	4,001	108	5,064	130	16,[illegible]
II. South-Eastern.	Croydon	102,095	83	4	74	2	43	..	100	3	[illegible]
	Dover	33,300	4	1	5	..	18	1	84	2	111
	Eastbourne	34,969	7	1	8	..	18	..	21	..	54
	Portsmouth	159,278	140	8	147	7	145	4	283	11	714
	Bournemouth	37,781	22	..	13	..	5	..	24	..	64
	Southampton	65,325	49	..	76	..	47	3	61	..	[illegible]
	Reading	60,054	8	..	9	..	38	..	36	..	91
III. South Midland.	Willesden	61,265	93	4	97	2	78	2	94	1	[illegible]
	Hornsey	44,205	39	..	38	..	42	..	59	..	[illegible]
	Tottenham	71,343	67	1	80	1	81	1	107	1	[illegible]
	Oxford	45,742	10	..	8	..	3	..	5	..	[illegible]
	Northampton	61,012	86	3	72	..	236	5	323	7	717
	Cambridge	36,983	13	..	17	..	22	..	57	..	[illegible]
IV. Eastern.	Leyton	63,056	107	1	115	3	86	1	73	1	381
	Walthamstow	46,346	59	..	72	..	74	1	91	1	[illegible]
	West Ham	204,[illegible]	251	6	232	6	200	4	276	7	959
	Colchester	34,559	28	..	11	..	10	..	16	..	63
	Norwich	100,970	165	2	174	5	168	7	285	9	792
V. South-Western.	Exeter	37,404	4	..	1	..	5	..	13	..	23
	Plymouth	84,248	23	..	38	..	107	3	195	1	363
	Devonport	54,8[illegible]3	26	1	5	1	41	1	63	1	135
	Bath	51,844	24	..	16	1	15	..	13	..	68
VI. West Midland.	Bristol	[illegible]	120	5	83	2	67	4	137	3	407
	Gloucester	39,444	54	3	43	2	27	..	38	..	162
	Cheltenham	47,5[illegible]	63	..	53	..	42	..	140	5	[illegible]
	Hanley	54,946	38	1	27	2	52	1	159	1	276
	Longton	34,327	6	..	5	..	25	..	101	2	137
	Burton-on-Trent	46,[illegible]	74	2	61	1	65	1	104	1	[illegible]
	Wolverhampton	82,662	85	8	102	5	74	5	107	5	368
	Walsall	71,789	91	3	69	5	107	3	86	5	353
	West Bromwich	59,474	126	2	58	2	87	3	73	2	[illegible]
	Worcester	42,908	28	..	66	3	97	2	125	5	316
	Smethwick	36,170	58	1	27	..	59	2	32	..	176
	Birmingham	478,113	330	16	311	15	361	9	362	6	1,3[illegible]
	Aston Manor	[illegible]	63	1	36	1	25	..	39	2	163
	Coventry	52,724	44	..	26	3	90	3	119	5	279
VII. North Midland.	Leicester	174,624	2.7	14	199	10	235	8	280	12	971
	Grimsby	51,934	21	1	21	..	30	1	87	1	159
	Nottingham	213,877	111	5	112	5	263	[illegible]	413	17	899
	Derby	94,146	101	5	80	5	119	5	182	5	482

"FEVER."														
[3r]d Quarter.				4th Quarter.					Totals for 1898.					Urban Sanitary Districts.
[Noti]fications.				Notifications.					Notifications.					
[Enteric]	Continued.	Total.	Deaths.	Typhus.	Enteric.	Continued.	Total.	Deaths.	Typhus.	Enteric.	Continued.	Grand Total.	Deaths.	
7	18	715	122	7	1,487	14	1,508	286	18	3,030	55	3,105	548	London. (Administrative Co.)
8	..	8	..	..	42	1	43	6	..	64	2	66	11	Croydon.
2	..	2	..	..	1	2	3	1	..	11	3	14	2	Dover.
7	..	7	1	..	9	..	9	..	..	25	..	25	4	Eastbourne.
6	13	111	10	..	134	16	150	23	..	307	44	351	43	Portsmouth.
1	..	1	..	..	4	..	4	..	..	6	..	6	..	Bournemouth.
6	..	76	8	..	77	4	81	15	..	155	4	159	34	Southampton.
7	..	7	..	..	6	..	6	1	..	38	..	38	9	Reading.
7	..	7	1	..	45	..	45	6	..	67	..	67	8	Willesden.
0	..	10	2	..	13	..	13	1	..	51	..	51	4	Hornsey.
5	..	25	6	..	48	..	48	7	..	101	..	114	17	Tottenham.
8	..	8	..	..	17	2	19	2	..	30	2	32	3	Oxford.
7	..	7	..	..	14	..	14	14	..	36	..	36	18	Northampton.
7	..	7	1	..	18	..	18	3	..	59	..	59	9	Cambridge.
7	1	28	6	..	39	6	45	5	..	89	8	97	18	Leyton.
8	..	18	4	..	46	1	47	5	..	75	1	76	9	Walthamstow.
2	3	115	17	..	260	3	263	37	..	461	8	469	72	West Ham.
8	..	8	3	..	7	..	7	2	..	20	..	20	7	Colchester.
5	..	95	9	..	108	..	108	23	..	252	..	252	45	Norwich.
3	3	25	2	..	22	2	24	1	..	93	9	102	7	Exeter.
8	8	26	2	..	13	..	13	2	..	40	9	49	6	Plymouth.
6	..	26	6	..	10	..	10	4	..	41	..	41	13	Devonport.
1	..	1	1	..	6	..	6	..	..	17	..	17	3	Bath.
9	..	29	5	..	88	..	88	10	..	113	..	113	25	Bristol.
5	..	5	1	..	..	..	..	1	..	8	..	8	3	Gloucester.
3	..	3	1	..	17	..	17	3	..	23	..	23	6	Cheltenham.
0	1	22	7	..	14	2	16	1	..	60	11	71	15	Hanley.
8	..	16	2	..	25	..	25	9	..	61	..	61	16	Longton.
4	..	11	2	..	6	2	8	..	..	24	2	26	3	Burton-on-Trent
8	..	28	4	..	43	1	44	6	..	121	1	122	20	Wolverhampton.
5	..	15	3	..	10	..	10	5	..	98	..	98	22	Walsall.
11	1	12	3	..	27	..	27	6	..	72	1	73	18	West Bromwich.
5	..	5	..	..	3	..	3	..	..	17	2	19	2	Worcester.
3	..	3	1	..	19	..	19	2	..	46	..	46	10	Smethwick.
65	3	168	18	..	162	2	164	26	..	660	12	672	111	Birmingham.
16	1	17	2	..	35	..	35	4	..	107	2	109	15	Aston Manor.
11	..	11	1	..	12	1	13	1	..	50	3	53	6	Coventry.
75	..	75	7	..	113	..	113	16	..	241	..	241	30	Leicester.
190	10	209	5	..	56	2	58	15	..	286	13	299	25	Grimsby.
160	..	160	10	..	123	..	123	16	..	421	..	421	56	Nottingham.
25	..	25	5	..	71	..	71	12	..	160	..	160	28	Derby.

In Registration Divisions.	Urban Sanitary Districts.	Population (1891).	"FEVER." 1st Quarter. Notifications. Typhus.	Enteric.	Continued.	Total.	Deaths.	2nd Quart[er] Notifications Typhus.	Enteric.	Continued.
VIII. North-Western.	Stockport	70,263	..	24	..	24	3	..	12	1
	Macclesfield	36,000	..	21	..	21	3	..	5	..
	Chester..	37,105	..	9	..	9	1	..	3	..
	Birkenhead	99,857	..	42	1	43	8	..	35	1
	Wallasey	33,229	..	12	..	12	..	..	12	..
	Bootle	49,217	..	23	..	23	7	28	24	..
	Liverpool	629,443	11	146	11	168	25	34	170	14
	St. Helens (Lancs.)	71,288	..	20	..	20	5	..	15	..
	Southport	41,406	..	15	..	15	..	..	4	..
	Wigan	55,013	..	22	..	22	3	..	7	..
	Warrington	52,743	..	18	..	18	5	..	7	..
	Bolton	115,002	..	45	..	45	7	..	28	..
	Bury (Lancs.)	57,212	..	9	2	11	3	..	7	..
	Salford	198,139	1	89	4	94	17	..	27	5
	Manchester	505,368	..	167	..	167	34	..	80	..
	Ashton-under-Lyne ..	40,463	..	8	1	9	2	..	6	..
	Oldham	131,463	..	19	..	19	7	..	12	..
	Rochdale	71,401	..	9	2	11	2	..	6	..
	Accrington	38,603	..	23	..	23	4	..	4	..
	Burnley	87,016	..	19	..	19	9	..	20	..
	Blackburn	120,064	..	61	..	61	9	..	38	..
	Darwen	34,192	..	17	1	18	4	..	7	..
	Preston..	107,573	..	83	..	83	16	..	23	..
	Barrow-in-Furness ..	51,712	..	3	2	5	1	..	5	3
IX. Yorkshire.	Keighley	30,810	..	22	..	22	3	..	9	..
	Huddersfield	95,420	..	11	..	11	2	..	6	..
	Halifax..	89,832	..	9	..	9	5	..	13	..
	Bradford	216,361	4	48	..	52	10	..	20	..
	Leeds	367,505	..	80	2	82	16	..	45	2
	Wakefield	33,146	..	3	1	4	..	..	5	..
	Barnsley	35,427	..	4	..	4	1	..	6	1
	Sheffield	324,213	..	152	3	155	22	..	58	..
	Rotherham	42,061	..	17	..	17	3	..	11	..
	York	67,004	..	9	1	10	3	..	8	..
	Hull	200,044	..	29	2	31	10	..	19	12
	Scarborough	33,776	..	5	..	5	1	..	3	..
	Middlesbrough	75,532	..	66	..	66	12	1	42	..
X. Northern.	Darlington	38,060	..	18	..	18	3	..	7	..
	Stockton-on-Tees	49,705	..	16	5	21	3	..	26	2
	West Hartlepool	42,710	..	6	..	6	..	..	2	..
	Sunderland	131,015	..	41	7	48	10	..	35	7
	Jarrow	33,675	..	5	..	5	2	..	16	..
	South Shields	78,391	..	27	1	28	6	..	27	1
	Gateshead	85,692	..	39	2	41	7	..	19	..
	Newcastle-on-Tyne ..	186,300	..	41	3	47	10	..	36	6
	Tynemouth	46,588	..	9	..	9	1	..	13	..
	Carlisle..	39,176	..	4	..	4	..	..	7	..
XI. Welsh.	Newport (Mon.)	54,707	..	8	..	8	1	..	6	1
	Cardiff	128,915	..	25	..	25	4	..	12	..
	Rhondda	88,351	..	43	3	46	3	..	43	5
	Merthyr Tydfil	58,080	..	12	1	13	2	..	9	..
	Swansea	90,349	..	8	..	8	1	..	21	..
	Totals, 93 Districts ..	13,128,484	23	3,038	88	3,149	505	67	2,097	97

"FEVER."

3rd Quarter. Notifications. Enteric.	Continued.	Total.	Deaths.	4th Quarter. Notifications. Typhus.	Enteric.	Continued.	Total.	Deaths.	Totals for 1898. Notifications. Typhus.	Enteric.	Continued.	Grand Total.	Deaths.	Urban Sanitary Districts.
19	..	19	1	..	67	..	67	12	..	122	1	123	19	Stockport.
3	..	3	2	..	16	..	16	..	..	45	..	45	5	Macclesfield.
28	..	28	..	..	25	..	25	3	..	65	..	65	4	Chester.
76	1	77	13	..	92	..	92	14	..	235	3	238	38	Birkenhead.
36	..	36	2	..	27	..	27	3	..	87	..	87	9	Wallasey.
41	..	45	10	..	40	..	40	7	32	128	..	160	32	Bootle.
246	16	262	42	23	318	27	368	57	88	880	68	1,036	166	Liverpool.
48	1	49	11	..	51	..	51	10	..	134	1	135	30	St. Helens (Lancs.)
18	..	18	3	..	19	1	20	1	..	56	1	57	4	Southport.
29	..	29	4	..	35	..	35	8	..	93	..	93	16	Wigan.
7	..	7	3	..	10	..	10	2	..	42	..	42	13	Warrington.
35	..	35	7	..	155	..	155	21	..	263	..	263	38	Bolton.
17	6	23	4	..	21	8	29	7	1	54	15	70	18	Bury (Lancs.)
84	..	84	16	12	190	13	215	30	13	390	22	425	80	Salford.
109	..	109	12	..	290	..	290	50	..	646	..	646	125	Manchester.
16	1	17	2	..	28	10	38	7	..	58	12	70	13	Ashton - under - Lyne.
9	..	9	1	..	28	..	28	10	..	68	..	68	22	Oldham.
24	1	25	7	6	32	3	41	8	6	71	6	83	18	Rochdale.
8	..	10	3	..	18	..	18	5	2	53	..	55	13	Accrington.
32	..	32	8	..	48	..	48	5	..	119	..	119	27	Burnley.
23	..	23	1	..	114	..	114	16	..	236	..	236	32	Blackburn.
8	1	9	..	..	11	1	12	..	..	43	3	46	6	Darwen.
33	..	33	8	..	94	..	94	17	..	233	..	233	43	Preston.
20	3	23	4	..	29	8	37	5	..	57	16	73	10	Barrow - in - Furness.
21	..	21	6	..	22	..	22	5	..	74	..	74	15	Keighley.
29	..	29	4	..	28	..	28	3	..	77	..	77	10	Huddersfield.
12	1	13	..	..	42	..	42	10	..	76	1	77	18	Halifax.
6[illegible]	1	[illegible]2	7	..	152	1	153	28	7	288	2	297	49	Bradford.
115	5	120	12	..	290	4	294	55	..	530	13	543	92	Leeds.
28	..	28	3	..	24	..	24	2	..	60	1	61	20	Wakefield.
21	..	21	2	..	102	1	103	17	..	133	2	135	21	Barnsley.
150	2	152	20	..	558	13	571	85	..	918	18	936	127	Sheffield.
8	..	8	1	..	26	..	26	6	..	62	..	62	12	Rotherham.
61	1	62	7	..	51	..	51	6	..	129	2	131	16	York.
80	6	86	13	..	184	9	193	32	..	312	29	341	58	Hull.
11	..	11	4	..	10	1	11	1	..	29	1	30	7	Scarborough.
47	..	47	4	..	86	..	86	11	1	241	..	242	33	Middlesbrough.
30	1	31	..	..	38	..	38	5	..	93	1	94	8	Darlington.
21	6	27	2	..	34	..	34	4	..	100	13	113	10	Stockton-on-Tees.
6	..	6	2	..	11	..	11	2	..	25	..	25	4	West Hartlepool.
136	14	150	19	1	240	22	263	31	1	452	50	503	69	Sunderland.
10	..	10	1	..	9	..	9	3	..	40	..	40	8	Jarrow.
48	4	52	12	..	62	1	63	11	..	164	7	171	37	South Shields.
31	1	32	2	..	31	2	33	5	..	120	5	125	18	Gateshead.
104	7	111	17	..	118	2	120	32	..	302	18	330	67	Newcastle - on - Tyne.
7	..	7	1	..	5	..	5	..	..	34	..	34	3	Tynemouth.
6	..	6	..	..	5	1	6	..	..	22	1	23	1	Carlisle.
13	1	14	1	..	10	..	10	3	..	37	2	39	5	Newport (Mon.)
26	3	29	7	..	23	3	26	4	..	86	6	92	17	Cardiff.
39	3	4[illegible]	8	..	232	9	241	29	..	357	20	377	41	Rhondda.
19	..	19	2	..	37	2	39	5	..	77	3	80	9	Merthyr Tydfil.
39	1	40	6	..	40	..	40	4	..	108	1	109	13	Swansea.
183	148	4,361	604	49	7,422	203	7,674	1,294	109	16,740	536	17,445	2,903	Totals, 93 Districts.

No. 19B.

TABLE showing, Week by Week, during the Year 1898, for each of the SANITARY AREAS w
NOTIFIED CASES of and REGISTERED DEATHS from the following DISEASES, together
[The Cases are copied from the Weekly Returns of Notifiable Diseases received by the Board from the Met

	Sanitary Areas.	Population (1891).	SMALL-POX.										
			Weekly Statement, 1st Quarter, 1898.										
			Jan. 8.		Jan. 15.		Jan. 22.		Jan. 29.		Feb. 5.		Feb
			Cases.	Deaths.	Cases.	Deaths.	Cases.	Deaths.	Cases.	Deaths.	Cases.	Deaths.	Cases.
	London .. .. (Administrative County)	4,232,118	..	..	..	..	3	..	1	..	..	..	..
W. District.	Kensington	166,308	..	..	..	..	..	..	..	..	..	..	..
	Fulham	91,639	..	..	..	..	..	..	1	..	..	..	..
	Hammersmith	97,239	..	..	..	..	..	..	..	..	..	..	..
	Paddington	117,846	..	..	..	..	..	..	..	..	..	..	..
	Chelsea..	96,253	..	..	..	..	..	..	..	..	..	..	..
	St. George, Hanover Sq.* ..	78,599	..	..	..	..	..	..	..	..	..	..	..
	Westminster	55,539	..	..	..	..	..	..	..	..	..	..	..
	St. James, Westminster ..	24,995	..	..	..	..	..	..	..	..	..	..	..
N. District.	St. Marylebone	142,404	..	..	..	..	..	..	..	..	..	..	..
	Hampstead	68,416	..	..	..	..	..	..	..	..	..	..	..
	St. Pancras	234,379	..	..	..	..	..	..	..	..	..	..	..
	Islington	319,143	..	..	..	..	..	..	..	..	..	..	..
	St. Mary, Stoke Newington	30,936	..	..	..	..	..	..	..	..	..	..	..
	Hackney	198,606	..	..	..	..	1	..	..	..	..	..	..
Central District.	St. Giles and St. George, Bloomsbury.	39,782	..	..	..	..	..	..	..	..	..	..	..
	St. Martin-in-the-Fields ..	14,616	..	..	..	..	..	..	..	..	..	..	..
	Strand†	25,217	..	..	..	..	..	..	..	..	..	..	..
	Holborn‡	34,043	..	..	..	..	..	..	..	..	..	..	..
	Clerkenwell	66,216	..	..	..	..	..	..	..	..	..	..	..
	St. Luke, Middlesex ..	42,440	..	..	..	..	..	..	..	..	..	..	..
	London, City of§	37,583	..	..	..	..	..	..	..	..	..	..	..
E. District.	Shoreditch	124,009	..	..	..	..	..	..	..	..	..	..	..
	Bethnal Green	129,132	..	..	..	..	..	..	..	..	..	..	..
	Whitechapel	74,420	..	..	..	..	..	..	..	..	..	..	..
	St George in the East ..	45,795	..	..	..	..	..	..	..	..	..	..	..
	Limehouse	57,376	..	..	..	..	..	..	..	..	..	..	..
	Mile End Old Town ..	107,592	..	..	..	..	..	..	..	..	..	..	..
	Poplar	166,748	..	..	..	..	2	..	..	..	..	..	..
S. District.	St Saviour, Southwark ..	27,177	..	..	..	..	..	..	..	..	..	..	..
	St. George Southwark ..	59,712	..	..	..	..	..	..	..	..	..	..	..
	Newington	115,804	..	..	..	..	..	..	..	..	..	..	..
	St. Olave, Southwark ..	12,723	..	..	..	..	..	..	..	..	..	..	..
	Bermondsey	84,682	..	..	..	..	..	..	..	..	..	..	..
	Rotherhithe	39,255	..	..	..	..	..	..	..	..	..	..	..
	Lambeth	275,203	..	..	..	..	..	..	..	..	..	..	..
	Battersea	150,558	..	..	..	..	..	..	..	..	..	..	..
	Wandsworth	156,942	..	..	..	..	..	..	..	..	..	..	..
	Camberwell	235,344	..	..	..	..	..	..	..	..	..	..	..
	Greenwich	165,413	..	..	..	..	..	..	..	..	..	..	..
	Lewisham	92,647	..	..	..	..	..	..	..	..	..	..	..
	Woolwich	40,848	..	..	..	..	..	..	..	..	..	..	..
	Plumstead	52,436	..	..	..	..	..	..	..	..	..	..	..
	Lee	36,103	..	..	..	..	..	..	..	..	..	..	..
	Port of London	..	..	..	..	..	..	..	..	..	..	..	..

* *Including St. Peter's, Westminster (popula*
‡ *Including Gray's Inn (popu*
Staple Inn (population,

† Including Middle Temple (popula
lation, 71), Charterhouse (popula

"Fever."														Urban Sanitary Districts.
3rd Quarter.				4th Quarter.					Totals for 1898.					
Notifications.				Notifications.					Notifications.					
Enteric.	Continued.	Total.	Deaths.	Typhus.	Enteric.	Continued.	Total.	Deaths.	Typhus.	Enteric.	Continued.	Grand Total.	Deaths.	
697	18	715	122	7	1,487	14	1,508	206	18	3,032	55	3,105	598	London. (*Administrative Co.*)
8	..	8	..	..	42	1	43	8	..	64	2	66	11	Croydon.
2	..	2	..	..	1	2	3	1	..	11	3	14	2	Dover.
7	..	7	1	..	9	..	9	..	..	25	..	25	4	Eastbourne.
98	13	111	10	..	134	16	150	22	..	307	44	351	43	Portsmouth.
1	..	1	..	..	4	..	4	..	..	6	..	6	..	Bournemouth.
76	..	76	8	..	77	4	81	15	..	165	4	169	24	Southampton.
7	..	7	..	..	6	..	6	1	..	23	..	23	2	Reading.
7	..	7	1	..	45	..	45	6	..	67	..	67	8	Willesden.
10	..	10	2	..	13	..	13	1	..	31	..	31	4	Hornsey.
25	..	25	6	..	48	..	48	7	..	101	..	114	17	Tottenham.
8	..	8	..	..	17	2	19	2	..	30	2	32	2	Oxford.
7	..	7	..	..	14	..	14	14	..	36	..	36	16	Northampton.
7	..	7	1	..	18	..	18	3	..	59	..	59	9	Cambridge.
27	1	28	5	..	30	6	45	5	..	89	8	97	18	Leyton.
18	..	18	4	..	46	1	47	5	..	75	1	76	9	Walthamstow.
112	3	115	17	..	260	3	263	37	..	461	8	469	72	West Ham.
8	..	8	3	..	7	..	7	2	..	20	..	20	7	Colchester.
95	..	95	9	..	108	..	108	23	..	252	..	252	45	Norwich.
23	3	26	2	..	22	2	24	1	..	93	9	102	7	Exeter.
18	8	26	2	..	13	..	13	2	..	40	9	49	6	Plymouth.
26	..	26	6	..	10	..	10	4	..	41	..	41	13	Devonport.
1	..	1	1	..	6	..	6	..	..	17	..	17	3	Bath.
29	..	29	6	..	38	..	38	10	..	113	..	113	25	Bristol.
5	..	5	1	..	..	..	..	1	..	8	..	8	3	Gloucester.
3	..	3	1	..	17	..	17	3	..	23	..	23	6	Cheltenham.
21	1	22	7	..	14	2	16	1	..	60	11	71	15	Hanley.
16	..	16	2	..	25	..	25	9	..	61	..	61	16	Longton.
14	..	14	2	..	6	2	8	..	..	24	2	26	3	Burton-on-Trent
28	..	28	4	..	43	1	44	6	..	121	1	122	20	Wolverhampton.
15	..	15	3	..	10	..	10	3	..	98	..	98	22	Walsall.
11	1	12	3	..	27	..	27	6	..	72	1	73	18	West Bromwich.
5	..	5	..	..	3	..	3	..	..	17	2	19	2	Worcester.
3	..	3	1	..	10	..	10	2	..	46	..	46	10	Smethwick.
165	3	168	18	..	162	2	164	28	..	660	12	672	111	Birmingham.
16	1	17	2	..	38	..	38	4	..	107	2	109	15	Aston Manor.
11	..	11	1	..	12	1	13	1	..	50	3	53	6	Coventry.
75	..	75	7	..	113	..	113	16	..	241	..	241	30	Leicester.
199	10	209	5	..	56	2	58	15	..	286	13	299	35	Grimsby.
100	.	100	19	..	123	..	123	16	..	421	..	421	56	Nottingham.
25	..	25	5	..	71	..	71	12	..	160	..	160	28	Derby.

Sanitary Areas.	Population (1891).	SMALL-POX—*continued.*											
		Weekly Statement, 2nd Quarter, 1898.											
		April 9.		April 16.		April 23.		April 30.		May 7.		May 14.	
		Cases.	Deaths.	Cases.	Deaths.	Cases.	Deaths.	Cases.	Deaths.	Cases.	Deaths.	Cases.	Deaths.
London (*Administrative County*)	4,232,118	..	..	..	..	2	..	3	..	..	..	..	..
W. District. Kensington	166,308	..	..	..	..	..	..	..	..	..	..	..	..
Fulham	91,639	..	..	..	..	..	..	..	..	..	..	..	..
Hammersmith	97,239	..	..	..	..	..	..	..	..	..	..	..	..
Paddington	117,846	..	..	..	..	..	..	..	..	..	..	..	..
Chelsea	96,253	..	..	..	..	..	..	..	..	..	..	..	..
St. George, Hanover Sq.*	78,590	..	..	..	..	..	..	..	..	..	..	..	..
Westminster	55,534	..	..	..	..	1	..	..	..	..	..	..	..
St. James, Westminster	24,995	..	..	..	..	..	..	..	..	..	..	..	..
N. District. St. Marylebone	142,404	..	..	..	..	..	..	1	..	..	..	..	..
Hampstead	68,416	..	..	..	..	..	..	..	..	..	..	..	..
St. Pancras	234,379	..	..	..	..	..	..	..	..	..	..	..	..
Islington	319,143	..	..	..	..	..	..	..	..	..	..	..	..
St. Mary, Stoke Newington	30,936	..	..	..	..	..	..	..	..	..	..	..	..
Hackney	198,606	..	..	..	..	..	..	..	..	..	..	..	..
Central District. St. Giles and St. George, Bloomsbury.	39,782	..	..	..	..	..	..	..	..	..	..	..	..
St. Martin-in-the-Fields	14,616	..	..	..	..	..	..	..	..	..	..	..	..
Strand†	25,217	..	..	..	..	..	..	..	..	..	..	..	..
Holborn‡	31,043	..	..	..	..	..	..	..	..	..	..	..	..
Clerkenwell	66,216	..	..	..	..	..	..	..	..	..	..	..	..
St. Luke, Middlesex	42,440	..	..	..	..	..	..	..	..	..	..	..	..
London, City of§	37,583	..	..	..	..	..	..	..	..	..	..	..	..
E. District. Shoreditch	124,009	..	..	..	..	..	..	..	..	..	..	..	..
Bethnal Green	129,132	..	..	..	..	..	..	..	..	..	..	..	..
Whitechapel‖	74,420	..	..	..	..	..	..	..	..	..	..	..	..
St. George-in the East	45,795	..	..	..	..	..	..	..	..	..	..	..	..
Limehouse	57,376	..	..	..	..	..	..	..	..	..	..	..	..
Mile End Old Town	107,592	..	..	..	..	..	..	1	..	..	..	..	..
Poplar	166,748	..	..	..	..	..	..	..	..	..	..	..	..
S. District. St. Saviour, Southwark	27,177	..	..	..	..	..	..	..	..	..	..	..	..
St. George, Southwark	59,712	..	..	..	..	..	..	..	..	..	..	..	..
Newington	115,804	..	..	..	..	..	..	..	..	..	..	..	..
St. Olave, Southwark	12,723	..	..	..	..	..	..	..	..	..	..	..	..
Bermondsey	84,682	..	..	..	..	..	..	..	..	..	..	..	..
Rotherhithe	39,255	..	..	..	..	..	..	..	..	..	..	..	..
Lambeth	275,203	..	..	..	..	..	..	..	..	..	..	..	..
Battersea	150,558	..	..	..	..	..	..	..	..	..	..	..	..
Wandsworth	156,942	..	..	..	..	..	..	..	..	..	..	..	..
Camberwell	235,344	..	..	..	..	1	..	1	..	..	..	..	
Greenwich	165,413	..	..	..	..	..	..	..	..	..	..	..	
Lewisham	92,647	..	..	..	..	..	..	..	..	..	..	..	
Woolwich	40,848	..	..	..	..	..	..	..	..	..	..	..	
Plumstead	52,436	..	..	..	..	..	..	..	..	..	..	..	
Lee	36,103	..	..	..	..	..	..	..	..	..	..	..	
Port of London	..	..	..	..	..	..	..	..	..	..	..	..	

* Including St. Peters, Westminster (population, 235). † Including Middle Temple (population, 95)
‡ Including Gray's Inn (population, 253), Lincoln's Inn (population, 27), Charterhouse (population, 126) Staple Inn (population, 21), and Furnival's Inn (population, 19).

SMALL-POX—*continued.*

Weekly Statement, 2nd Quarter, 1898—*continued.*

[21].	May 28.		June 4.		June 11.		June 18.		June 25.		July 2.		Totals for 2nd Quarter, 1898.		Sanitary Areas.
Deaths.	Cases.	Deaths.	Cases.	Deaths.	Cases.	Deaths.	Cases.	Deaths.	Cases.	Deaths.	Cases.	Deaths.	Cases.	Deaths.	
.	1	..	2	..	2	..	..	..	2	..	..	..	12	..	London. (*Administrative County.*)
	..	..	..	..	..	..	..	..	..	..	..	..	..	..	Kensington.
	..	..	..	..	..	..	..	..	..	..	..	..	..	..	Fulham.
	..	..	..	..	..	..	..	..	..	..	..	..	..	..	Hammersmith.
	..	..	..	..	..	..	..	..	..	..	..	..	..	..	Paddington.
	..	..	..	..	..	..	..	..	..	..	..	..	..	..	Chelsea.
	..	..	1	..	..	..	..	..	..	..	..	..	1	..	St. George, Hanover Sq.*
	..	..	..	..	..	..	..	..	..	..	..	..	1	..	Westminster.
	..	..	..	..	..	..	..	..	..	..	..	..	..	..	St. James, Westminster.
	..	..	..	.	..	..	..	..	..	..	..	..	1	..	St. Marylebone.
	..	..	..	..	..	..	..	..	..	..	..	..	..	..	Hampstead.
	..	..	..	..	..	..	..	..	..	..	..	..	..	..	St. Pancras.
	..	..	..	..	..	..	..	..	..	..	..	..	..	..	Islington.
	..	..	..	..	..	..	..	..	..	..	..	..	..	..	St. Mary, Stoke Newington.
	..	..	..	..	..	..	..	..	..	..	..	..	..	..	Hackney.
	..	..	..	..	..	..	..	..	..	..	..	..	..	..	St. Giles and St. George Bloomsbury.
	..	..	..	..	..	..	..	..	..	..	..	..	..	..	St. Martin-in-the Fields.
	..	..	..	..	..	..	..	..	..	..	..	..	..	..	Strand.†
	..	..	..	..	..	..	..	..	..	..	..	..	..	..	Holborn.‡
	..	..	..	..	..	..	..	..	..	..	..	..	..	..	Clerkenwell.
	..	..	..	..	..	..	..	..	..	..	..	..	..	..	St. Luke, Middlesex.
	..	..	..	..	..	..	..	..	..	..	..	..	..	..	London, City of.§
	..	..	..	..	..	..	..	..	..	..	..	..	..	..	Shoreditch.
	..	..	..	..	..	..	..	..	..	..	..	..	..	..	Bethnal Green.
	..	..	1	..	..	..	..	..	..	..	..	..	1	..	Whitechapel.‖
	..	..	..	..	..	..	..	..	..	..	..	..	..	..	St. George-in-the-East.
	.	..	..	..	..	..	..	..	..	..	..	..	..	..	Limehouse.
	..	..	..	..	..	..	..	..	..	..	..	..	1	..	Mile End Old Town.
	..	..	..	..	..	..	..	..	..	..	..	..	..	..	Poplar.
	..	..	..	..	..	..	..	..	..	..	..	..	..	..	St. Saviour, Southwark.
	..	..	..	..	..	..	..	..	..	..	..	..	..	..	St. George Southwark.
	..	..	..	..	..	..	..	..	..	..	..	..	..	..	Newington.
	..	..	..	..	..	..	..	..	..	..	..	..	..	..	St Olave, Southwark.
	1	..	..	..	..	..	..	..	..	..	..	..	1	..	Bermondsey.
	..	..	..	..	..	..	..	..	..	..	..	..	..	..	Rotherhithe.
.	..	..	..	..	..	..	..	..	..	..	..	..	..	..	Lambeth.
.	..	..	..	..	..	..	..	..	..	..	..	..	..	..	Battersea.
.	..	..	..	..	..	..	..	..	..	..	..	..	..	..	Wandsworth.
.	..	..	..	..	..	..	..	..	..	..	..	..	2	..	Camberwell.
..	..	..	..	..	..	..	..	..	..	..	..	..	..	..	Greenwich.
..	..	..	..	..	..	..	..	..	..	..	..	..	..	..	Lewisham.
..	..	..	..	..	..	..	..	..	..	..	..	..	..	..	Woolwich.
..	..	..	..	..	..	..	..	..	..	..	..	..	..	..	Plumstead.
..	..	..	..	..	2	..	..	..	..	..	..	..	2	..	Lee.
..	..	..	..	..	..	..	..	..	2	..	..	..	2	..	Port of London.

…uding Inner Temple (population, [illegible]).

‖ Including Tower of London (population, [illegible]).

District	Sanitary Areas.	Population (1891).	Small-pox—continued. Weekly Statement, 3rd Quarter, 1898. July 9. Cases.	July 9. Deaths.	July 16. Cases.	July 16. Deaths.	July 23. Cases.	July 23. Deaths.	July 30. Cases.	July 30. Deaths.	Aug. 6. Cases.	Aug. 6. Deaths.	Aug. Cases.
	London .. .. (*Administrative County*)	4,232,118	..	..	2	..	..	..	..	..	3	..	..
W. District.	Kensington	166,308	..	..	..	..	..	..	..	..	..	..	..
W. District.	Fulham	91,639	..	..	..	..	..	..	..	..	..	..	..
W. District.	Hammersmith	97,239	..	..	..	..	..	..	..	..	..	..	..
W. District.	Paddington	117,846	..	..	..	..	..	..	..	..	..	..	..
W. District.	Chelsea	96,253	..	..	..	..	..	..	..	..	..	..	..
W. District.	St. George, Hanover Sq.* ..	78,599	..	..	..	..	..	..	..	..	..	..	..
W. District.	Westminster	56,539	..	..	..	..	..	..	..	..	..	..	..
W. District.	St. James, Westminster ..	24,995	..	..	..	..	..	..	..	..	..	..	..
N. District.	St. Marylebone	142,404	..	..	..	..	..	..	..	..	..	..	..
N. District.	Hampstead	68,416	..	..	..	..	..	..	..	..	..	..	..
N. District.	St. Pancras	234,379	..	..	..	..	..	..	..	..	1	..	..
N. District.	Islington	319,143	..	..	..	..	..	..	..	..	..	..	..
N. District.	St. Mary, Stoke Newington	30,936	..	..	..	..	..	..	..	..	..	..	..
N. District.	Hackney	198,606	..	..	..	..	..	..	..	..	..	..	..
Central District.	St. Giles and St. George, Bloomsbury.	39,782	..	..	..	..	..	..	..	..	..	..	..
Central District.	St. Martin-in-the-Fields ..	14,616	..	..	..	..	..	..	..	..	..	..	..
Central District.	Strand†	25,217	..	..	..	..	..	..	..	..	..	..	..
Central District.	Holborn‡	34,043	..	..	..	..	..	..	..	..	..	..	..
Central District.	Clerkenwell	66,216	..	..	..	..	..	..	..	..	..	..	..
Central District.	St. Luke, Middlesex ..	42,440	..	..	..	..	..	..	..	..	..	..	..
Central District.	London, City of§	37,583	..	..	..	..	..	..	..	..	..	..	..
E. District.	Shoreditch	124,009	..	..	..	..	..	..	..	..	..	..	..
E. District.	Bethnal Green	129,132	..	..	..	..	..	..	..	..	..	..	..
E. District.	Whitechapel‖	74,420	..	..	..	..	..	..	..	..	..	..	..
E. District.	St. George-in-the-East ..	45,795	..	..	..	..	..	..	..	..	..	..	..
E. District.	Limehouse	57,376	..	..	..	..	..	..	..	..	..	..	..
E. District.	Mile End Old Town ..	107,592	..	..	..	..	..	..	..	..	..	..	..
E. District.	Poplar	166,748	..	..	..	..	..	..	..	..	..	..	..
S. District.	St. Saviour, Southwark ..	27,177	..	..	..	..	..	..	..	..	..	..	..
S. District.	St. George, Southwark ..	59,712	..	..	..	..	..	..	..	..	1	..	..
S. District.	Newington	115,804	..	..	..	..	..	..	..	..	..	..	..
S. District.	St. Olave, Southwark ..	12,723	..	..	..	..	..	..	..	..	..	..	..
S. District.	Bermondsey	84,682	..	..	..	..	..	..	..	..	..	..	..
S. District.	Rotherhithe	39,255	..	..	..	..	..	..	..	..	..	..	..
S. District.	Lambeth	275,203	..	..	..	..	..	..	..	..	..	..	..
S. District.	Battersea	150,558	..	..	..	..	..	..	..	..	1	..	..
S. District.	Wandsworth	156,942	..	..	..	..	..	..	..	..	..	..	..
S. District.	Camberwell	235,344	..	..	2	..	..	..	..	..	..	..	..
S. District.	Greenwich	165,413	..	..	..	..	..	..	..	..	..	..	..
S. District.	Lewisham	92,617	..	..	..	..	..	..	..	..	..	..	..
S. District.	Woolwich	40,848	..	..	..	..	..	..	..	..	..	..	..
S. District.	Plumstead	52,436	..	..	..	..	..	..	..	..	..	..	..
S. District.	Lee	36,103	..	..	..	..	..	..	..	..	..	..	..
S. District.	Port of London	..	..	..	..	..	..	..	..	..	..	..	..

* Including St. Peters, Westminster (population, 235). † Including Middle Temple (populat
‡ Including Gray's Inn (population, 253), Lincoln's Inn (population, 27), Charterhouse (populati
Staple Inn (population, 21), and Furnival's Inn (population, 121).

SMALL-POX—*continued.*													
Weekly Statement, 3rd Quarter, 1898.											Totals for 3rd Quarter 1898.		Sanitary Areas.
ug. 27.	Sept. 3.		Sept. 10.		Sept. 17.		Sept. 24.		Oct. 1.				
Deaths.	Cases.	Deaths.	Cases.	Deaths.	Cases.	Deaths.	Cases.	Deaths.	Cases.	Deaths.	Cases.	Deaths.	
..	..	..	1	..	1	..	..	..	..	..	10	..	London. (*Administrative County.*)
..	..	..	..	..	..	..	..	..	..	..	..	..	Kensington.
..	..	..	..	..	..	..	..	..	..	..	..	..	Fulham.
..	..	..	..	..	..	..	..	..	..	..	..	..	Hammersmith.
..	..	..	..	..	..	..	..	..	..	..	..	..	Paddington.
..	..	..	..	..	..	..	..	..	..	..	..	..	Chelsea.
..	..	..	..	..	..	..	..	..	..	..	..	..	St. George, Hanover Sq.*
..	..	..	..	..	..	..	..	..	..	..	..	..	Westminster.
..	..	..	..	..	..	..	..	..	..	..	..	..	St. James, Westminster.
..	..	..	..	..	..	..	..	..	..	..	..	..	St. Marylebone.
..	..	..	..	..	..	..	..	..	..	..	..	..	Hampstead.
..	..	..	..	..	..	..	..	..	..	..	2	..	St. Pancras.
..	..	..	..	..	..	..	..	..	..	..	..	..	Islington.
..	..	..	..	..	..	..	..	..	..	..	..	..	St. Mary, Stoke Newington.
..	..	..	..	..	..	..	..	..	..	..	1	..	Hackney.
..	..	..	..	..	..	..	..	..	..	..	..	..	St. Giles and St. George, Bloomsbury.
..	..	..	..	..	..	..	..	..	..	..	..	..	St. Martin-in-the-Fields.
..	..	..	..	..	..	..	..	..	..	..	..	..	Strand.†
..	..	..	..	..	..	..	..	..	..	..	..	..	Holborn.‡
..	..	..	..	..	..	..	..	..	..	..	..	..	Clerkenwell.
..	..	..	..	..	..	..	..	..	..	..	..	..	St. Luke, Middlesex.
..	..	..	..	..	..	..	..	..	..	..	1	..	London, City of.§
..	..	..	..	..	..	..	..	..	..	..	..	..	Shoreditch.
..	..	..	..	..	..	..	..	..	..	..	..	..	Bethnal Green.
..	..	..	1	..	..	..	..	..	..	..	1	..	Whitechapel.‖
..	..	..	..	..	..	..	..	..	..	..	..	..	St. George-in-the-East.
..	..	..	..	..	..	..	..	..	..	..	..	..	Limehouse.
..	..	..	..	..	..	..	..	..	..	..	..	..	Mile End Old Town.
..	..	..	..	..	..	..	..	..	..	..	..	..	Poplar.
..	..	..	..	..	..	..	..	..	..	..	..	..	St. Saviour, Southwark.
..	..	..	..	..	..	..	..	..	..	..	1	..	St. George, Southwark.
..	..	..	..	..	..	..	..	..	..	..	..	..	Newington.
..	..	..	..	..	..	..	..	..	..	..	..	..	St. Olave, Southwark.
..	..	..	..	..	..	..	..	..	..	..	..	..	Bermondsey.
..	..	..	..	..	..	..	..	..	..	..	..	..	Rotherhithe.
..	..	..	..	..	..	..	..	..	..	..	..	..	Lambeth.
..	..	..	..	..	..	..	..	..	..	..	1	..	Battersea.
..	..	..	..	..	1	..	..	..	..	..	1	..	Wandsworth.
..	..	..	..	..	..	..	..	..	..	..	2	..	Camberwell.
..	..	..	..	..	..	..	..	..	..	..	..	..	Greenwich.
..	..	..	..	..	..	..	..	..	..	..	..	..	Lewisham.
..	..	..	..	..	..	..	..	..	..	..	..	..	Woolwich.
..	..	..	..	..	..	..	..	..	..	..	..	..	Plumstead.
..	..	..	..	..	..	..	..	..	..	..	..	..	Lee.
..	..	..	..	..	..	..	..	..	..	..	..	..	Port of London.

ner Temple (population, 96).

‖ Including Tower of London (population, [illegible]).

District	Sanitary Areas	Population (1891).	SMALL-POX—*continued.* Weekly Statement, 4th Quarter, 1898. Oct. 8. Cases.	Oct. 8. Deaths.	Oct. 15. Cases.	Oct. 15. Deaths.	Oct. 22. Cases.	Oct. 22. Deaths.	Oct. 29. Cases.	Oct. 29. Deaths.	Nov. 5. Cases.	Nov. 5. Deaths.	Nov. 12. Cases.	Nov. 12. Deaths.	Nov. 1[illegible] Cases.	Nov. 1[illegible] Deaths.
	London .. .. (*Administrative County*)	4,232,118	..	..	..	1	..	..	..	..	1	..	..	..	..	..
W. District.	Kensington	166,308	..	..	..	..	..	..	..	..	..	..	..	..	..	..
W. District.	Fulham	91,639	..	..	..	..	..	..	..	..	..	..	..	..	..	..
W. District.	Hammersmith	97,239	..	..	..	..	..	..	..	..	..	..	..	..	..	..
W. District.	Paddington	117,846	..	..	..	..	..	..	..	..	..	..	..	..	..	..
W. District.	Chelsea..	96,253	..	..	..	..	..	..	..	..	..	..	..	..	..	..
W. District.	St. George, Hanover Sq.*..	78,599	..	..	..	1	..	..	..	..	..	..	..	..	..	..
W. District.	Westminster	55,539	..	..	..	..	..	..	..	..	..	..	..	..	..	..
W. District.	St. James, Westminster ..	24,995	..	..	..	..	..	..	..	..	..	..	..	..	..	..
N. District.	St. Marylebone	142,404	..	..	..	..	..	..	..	..	..	..	..	..	..	..
N. District.	Hampstead	68,416	..	..	..	..	..	..	..	..	..	..	..	..	..	..
N. District.	St. Pancras	234,379	..	..	..	..	..	..	..	..	..	..	..	..	..	..
N. District.	Islington	319,143	..	..	..	..	..	..	..	..	..	..	..	..	..	..
N. District.	St. Mary, Stoke Newington	30,936	..	..	..	..	..	..	..	..	..	..	..	..	..	..
N. District.	Hackney	198,606	..	..	..	..	..	..	..	..	..	..	..	..	..	..
Central District.	St. Giles and St. George, Bloomsbury.	39,782	..	..	..	..	..	..	..	..	..	..	..	..	..	..
Central District.	St. Martin-in-the-Fields ..	14,616	..	..	..	..	..	..	..	..	..	..	..	..	..	..
Central District.	Strand†..	25,217	..	..	..	..	..	..	..	..	..	..	..	..	..	..
Central District.	Holborn‡	34,043	..	..	..	..	..	..	..	..	..	..	..	..	..	..
Central District.	Clerkenwell	66,216	..	..	..	..	..	..	..	..	..	..	..	..	..	..
Central District.	St. Luke, Middlesex ..	42,440	..	..	..	..	..	..	..	..	..	..	..	..	..	..
Central District.	London, City of§	37,583	..	..	..	..	..	..	..	..	..	..	..	..	..	..
E. District.	Shoreditch	124,009	..	..	..	..	..	..	..	..	..	..	..	..	..	..
E. District.	Bethnal Green	129,132	..	..	..	..	..	..	..	..	..	..	..	..	..	..
E. District.	Whitechapel‖	74,420	..	..	..	..	..	..	..	..	..	..	..	..	..	..
E. District.	St. George-in-the-East ..	45,795	..	..	..	..	..	..	..	..	..	..	..	..	..	..
E. District.	Limehouse	57,376	..	..	..	..	..	..	..	..	..	..	..	..	..	..
E. District.	Mile End Old Town ..	107,592	..	..	..	..	..	..	..	..	..	..	..	..	..	..
E. District.	Poplar	166,748	..	..	..	..	..	..	..	..	..	..	..	..	..	..
S. District.	St. Saviour, Southwark ..	27,177	..	..	..	..	..	..	..	..	..	..	..	..	..	..
S. District.	St. George, Southwark ..	59,712	..	..	..	..	..	..	..	..	..	..	..	..	..	..
S. District.	Newington	115,804	..	..	..	..	..	..	..	..	..	..	..	..	..	..
S. District.	St. Olave, Southwark ..	12,723	..	..	..	..	..	..	..	..	..	..	..	..	..	..
S. District.	Bermondsey	84,682	..	..	..	..	..	..	..	..	..	..	..	..	..	..
S. District.	Rotherhithe	39,255	..	..	..	..	..	..	..	..	..	..	..	..	..	..
S. District.	Lambeth	275,203	..	..	..	..	..	..	..	..	..	..	..	..	..	..
S. District.	Battersea	150,558	..	..	..	..	..	..	..	..	1	..	..	..	..	..
S. District.	Wandsworth..	156,942	..	..	..	..	..	..	..	..	..	..	..	..	..	..
S. District.	Camberwell	235,344	..	..	..	..	..	..	..	..	..	..	..	..	..	..
S. District.	Greenwich	165,413	..	..	..	..	..	..	..	..	..	..	..	..	..	..
S. District.	Lewisham	92,647	..	..	..	..	..	..	..	..	..	..	..	..	..	..
S. District.	Woolwich	40,848	..	..	..	..	..	..	..	..	..	..	..	..	..	..
S. District.	Plumstead	52,436	..	..	..	..	..	..	..	..	..	..	..	..	..	..
S. District.	Lee	36,103	..	..	..	..	..	..	..	..	..	..	..	..	[illegible]	..
S. District.	Port of London	..	..	..	..	..	..	..	..	..	..	..	..	..	..	..

* Including St. Peters, Westminster (population, 235). † Including Middle Temple (population, [illegible]
‡ *Including* Gray's Inn (population, 253), Lincoln's Inn (population, 27), Charterhouse (population, 1[illegible]
Staple Inn (population, 21), and Furnival's Inn (population, 131).

SMALL-POX—*continued.*																
Weekly Statement, 4th Quarter, 1898—*continued.*												Totals for 4th Quarter, 1898.		Grand Totals for Year 1898.		Sanitary Areas.
Nov. 26.		Dec. 3.		Dec. 10.		Dec. 17.		Dec. 24.		Dec. 31.						
Cases.	Deaths.	Cases.	Deaths.	Cases.	Deaths.	Cases.	Deaths.	Cases.	Deaths.	Cases.	Deaths.	Cases.	Deaths.	Cases.	Deaths.	
..	..	..	..	..	..	1	..	1	..	..	..	3	1	35	1	London. (*Administrative County.*)
..	..	..	..	..	..	..	..	..	..	..	..	..	..	..	..	Kensington.
..	..	..	..	..	..	..	..	..	..	..	..	..	..	1	..	Fulham.
..	..	..	..	..	..	..	..	..	..	..	..	..	..	..	..	Hammersmith.
..	..	..	..	..	..	..	..	..	..	..	..	..	..	..	..	Paddington.
..	..	..	..	..	..	..	..	..	..	..	..	..	..	..	..	Chelsea.
..	..	..	..	..	..	..	..	..	..	..	..	..	1	1	1	St. George, Hanover Sq.*
..	..	..	..	..	..	..	..	..	..	..	..	..	..	1	..	Westminster.
..	..	..	..	..	..	..	..	..	..	..	..	..	..	..	..	St. James, Westminster.
..	..	..	..	..	..	..	..	..	..	..	..	..	..	1	..	St. Marylebone.
..	..	..	..	..	..	..	..	..	..	..	..	..	..	..	..	Hampstead.
..	..	..	..	..	..	..	..	..	..	..	..	..	..	2	..	St. Pancras.
..	..	..	..	..	..	..	..	..	..	..	..	..	..	..	..	Islington.
..	..	..	..	..	..	..	..	..	..	..	..	..	..	..	..	St Mary, Stoke Newington.
..	..	..	..	..	..	..	..	..	..	..	..	..	..	2	..	Hackney.
..	..	..	..	..	..	..	..	..	..	..	..	..	..	..	..	St. Giles and St. George, Bloomsbury.
..	..	..	..	..	..	..	..	..	..	..	..	..	..	..	..	St. Martin-in-the-Fields.
..	..	..	..	..	..	..	..	..	..	..	..	..	..	..	..	Strand.†
..	..	..	..	..	..	..	..	..	..	..	..	..	..	1	..	Holborn.‡
..	..	..	..	..	..	..	..	..	..	..	..	..	..	..	..	Clerkenwell.
..	..	..	..	..	..	..	..	..	..	..	..	..	..	..	..	St. Luke, Middlesex.
..	..	..	..	..	..	..	..	..	..	..	..	..	..	1	..	London, City of.§
..	..	..	..	..	..	..	..	..	..	..	..	..	..	..	..	Shoreditch.
..	..	..	..	..	..	..	..	..	..	..	..	..	..	..	..	Bethnal Green
..	..	..	..	..	..	..	..	..	..	..	..	..	..	3	..	Whitechapel.‖
..	..	..	..	..	..	..	..	..	..	..	..	..	..	..	..	St. George in-the-East.
..	..	..	..	..	..	..	..	..	..	..	..	..	..	..	..	Limehouse.
..	..	..	..	..	..	..	..	..	..	..	..	..	..	1	..	Mile End Old Town.
..	..	..	..	..	..	..	..	..	..	..	..	..	..	4	..	Poplar.
..	..	..	..	..	..	..	..	..	..	..	..	..	..	..	..	St. Saviour, Southwark.
..	..	..	..	..	..	..	..	..	..	..	..	..	..	1	..	St. George, Southwark.
..	..	..	..	..	..	..	..	..	..	..	..	..	..	..	..	Newington.
..	..	..	..	..	..	..	..	..	..	..	..	..	..	..	..	St. Olave, Southwark.
..	..	..	..	..	..	..	..	..	..	..	..	..	..	1	..	Bermondsey.
..	..	..	..	..	..	..	..	..	..	..	..	..	..	..	..	Rotherhithe.
..	..	..	..	..	..	1	..	..	..	..	..	1	..	1	..	Lambeth.
..	..	..	..	..	..	..	..	1	..	..	..	2	..	4	..	Battersea.
..	..	..	..	..	..	..	..	..	..	..	..	..	..	1	..	Wandsworth.
..	..	..	..	..	..	..	..	..	..	..	..	..	..	5	..	Camberwell.
..	..	..	..	..	..	..	..	..	..	..	..	..	..	..	..	Greenwich.
..	..	..	..	..	..	..	..	..	..	..	..	..	..	..	..	Lewisham.
..	..	..	..	..	..	..	..	..	..	..	..	..	..	..	..	Woolwich.
..	..	..	..	..	..	..	..	..	..	..	..	..	..	..	..	Plumstead.
..	..	..	..	..	..	..	..	..	..	..	..	..	..	3	..	Lee.
..	..	..	..	..	..	..	..	..	..	..	..	..	..	2	..	Port of London.

...cluding Inner Temple (population, 86).

‖ Including Tower of London (population, [illegible]).

	Sanitary Areas	Population (1891).	SMALL-POX—*continued.* Weekly Statement, 4th Quarter, 18										
			Oct. 8.		Oct. 15.		Oct. 22.		Oct. 29.		Nov. 5.		No
			Cases.	Deaths.	Cases.	Deaths.	Cases.	Deaths.	Cases.	Deaths.	Cases.	Deaths.	Cases.
	London (*Administrative County*)	4,232,118	..	..	..	1	..	..	..	..	1	..	..
W. District.	Kensington	166,308	..	..	..	..	..	..	..	..	..	..	..
	Fulham	91,639	..	..	..	..	..	..	..	..	..	..	..
	Hammersmith	97,239	..	..	..	..	..	..	..	..	..	..	..
	Paddington	117,846	..	..	..	..	..	..	..	..	..	..	..
	Chelsea	96,253	..	..	..	..	..	..	..	..	..	..	..
	St. George, Hanover Sq.*	78,599	..	..	..	1	..	..	..	..	..	..	..
	Westminster	55,539	..	..	..	..	..	..	..	..	..	..	..
	St. James, Westminster	24,995	..	..	..	..	..	..	..	..	..	..	..
N. District.	St. Marylebone	142,404	..	..	..	..	..	..	..	..	..	..	..
	Hampstead	68,416	..	..	..	..	..	..	..	..	..	..	..
	St. Pancras	234,379	..	..	..	..	..	..	..	..	..	..	..
	Islington	319,143	..	..	..	..	..	..	..	..	..	..	..
	St. Mary, Stoke Newington	30,936	..	..	..	..	..	..	..	..	..	..	..
	Hackney	198,606	..	..	..	..	..	..	..	..	..	..	..
Central District.	St. Giles and St. George, Bloomsbury.	39,782	..	..	..	..	..	..	..	..	..	..	..
	St. Martin-in-the-Fields	14,616	..	..	..	..	..	..	..	..	..	..	..
	Strand†	25,217	..	..	..	..	..	..	..	..	..	..	..
	Holborn‡	34,043	..	..	..	..	..	..	..	..	..	..	..
	Clerkenwell	66,216	..	..	..	..	..	..	..	..	..	..	..
	St. Luke, Middlesex	42,440	..	..	..	..	..	..	..	..	..	..	..
	London, City of§	37,583	..	..	..	..	..	..	..	..	..	..	..
E. District.	Shoreditch	124,009	..	..	..	..	..	..	..	..	..	..	..
	Bethnal Green	129,132	..	..	..	..	..	..	..	..	..	..	..
	Whitechapel‖	74,420	..	..	..	..	..	..	..	..	..	..	..
	St. George-in-the-East	45,795	..	..	..	..	..	..	..	..	..	..	..
	Limehouse	57,376	..	..	..	..	..	..	..	..	..	..	..
	Mile End Old Town	107,592	..	..	..	..	..	..	..	..	..	..	..
	Poplar	166,748	..	..	..	..	..	..	..	..	..	..	..
S. District.	St. Saviour, Southwark	27,177	..	..	..	..	..	..	..	..	..	..	..
	St. George, Southwark	59,712	..	..	..	..	..	..	..	..	..	..	..
	Newington	115,804	..	..	..	..	..	..	..	..	..	..	..
	St. Olave, Southwark	12,723	..	..	..	..	..	..	..	..	..	..	..
	Bermondsey	84,682	..	..	..	..	..	..	..	..	..	..	..
	Rotherhithe	39,255	..	..	..	..	..	..	..	..	..	..	..
	Lambeth	275,203	..	..	..	..	..	..	..	..	..	..	..
	Battersea	150,558	..	..	..	..	..	..	..	..	1	..	..
	Wandsworth	156,942	..	..	..	..	..	..	..	..	..	..	..
	Camberwell	235,344	..	..	..	..	..	..	..	..	..	..	..
	Greenwich	165,413	..	..	..	..	..	..	..	..	..	..	..
	Lewisham	92,647	..	..	..	..	..	..	..	..	..	..	..
	Woolwich	40,848	..	..	..	..	..	..	..	..	..	..	..
	Plumstead	52,436	..	..	..	..	..	..	..	..	..	..	..
	Lee	36,108	..	..	..	..	..	..	..	..	..	..	..
	Port of London	..	..	..	..	..	..	..	..	..	..	..	..

* Including St. Peters, Westminster (population, 235). † Including Middle Temple
‡ Including Gray's Inn (population, 253), Lincoln's Inn (population, 27), Charterhouse
Staple Inn (population, 21), and Furnival's Inn (population, 121).

SMALL-POX—*continued*.

g. 20.	Aug. 27.		Sept. 3.		Sept. 10.		Sept. 17.		Sept. 24.		Oct. 1.		Totals for 3rd Quarter 1898.		Sanitary Areas.
Deaths.	Cases.	Deaths.	Cases.	Deaths.	Cases.	Deaths.	Cases.	Deaths.	Cases.	Deaths.	Cases.	Deaths.	Cases.	Deaths.	
..	2	..	..	..	1	..	1	..	..	..	..	..	10	..	London. (*Administrative County.*)
..	..	..	..	..	..	..	..	..	..	..	..	..	..	..	Kensington.
..	..	..	..	..	..	..	..	..	..	..	..	..	..	..	Fulham.
..	..	..	..	..	..	..	..	..	..	..	..	..	..	..	Hammersmith.
..	..	..	..	..	..	..	..	..	..	..	..	..	..	..	Paddington.
..	..	..	..	..	..	..	..	..	..	..	..	..	..	..	Chelsea.
..	..	..	..	..	..	..	..	..	..	..	..	..	..	..	St. George, Hanover Sq.*
..	..	..	..	..	..	..	..	..	..	..	..	..	..	..	Westminster.
..	..	..	..	..	..	..	..	..	..	..	..	..	..	..	St. James, Westminster.
..	..	..	..	..	..	..	..	..	..	..	..	..	..	..	St. Marylebone.
..	..	..	..	..	..	..	..	..	..	..	..	..	..	..	Hampstead.
..	..	..	..	..	..	..	..	..	..	..	..	..	2	..	St. Pancras.
..	..	..	..	..	..	..	..	..	..	..	..	..	..	..	Islington.
..	..	..	..	..	..	..	..	..	..	..	..	..	..	..	St. Mary, Stoke Newington.
..	1	..	..	..	..	..	..	..	..	..	..	..	1	..	Hackney.
..	..	..	..	..	..	..	..	..	..	..	..	..	..	..	St. Giles and St. George, Bloomsbury.
..	..	..	..	..	..	..	..	..	..	..	..	..	..	..	St. Martin-in-the-Fields.
..	..	..	..	..	..	..	..	..	..	..	..	..	..	..	Strand.†
..	..	..	..	..	..	..	..	..	..	..	..	..	..	..	Holborn.‡
..	..	..	..	..	..	..	..	..	..	..	..	..	..	..	Clerkenwell.
..	..	..	..	..	..	..	..	..	..	..	..	..	..	..	St. Luke, Middlesex.
..	1	..	..	..	..	..	..	..	..	..	..	..	1	..	London, City of.§
..	..	..	..	..	..	..	..	..	..	..	..	..	..	..	Shoreditch.
..	..	..	..	..	..	..	..	..	..	..	..	..	..	..	Bethnal Green.
..	..	..	..	..	1	..	..	..	..	..	..	..	1	..	Whitechapel.‖
..	..	..	..	..	..	..	..	..	..	..	..	..	..	..	St. George-in-the-East.
..	..	..	..	..	..	..	..	..	..	..	..	..	..	..	Limehouse.
..	..	..	..	..	..	..	..	..	..	..	..	..	..	..	Mile End Old Town.
..	..	..	..	..	..	..	..	..	..	..	..	..	..	..	Poplar.
..	..	..	..	..	..	..	..	..	..	..	..	..	..	..	St. Saviour, Southwark.
..	..	..	..	..	..	..	..	..	..	..	..	..	1	..	St. George, Southwark.
..	..	..	..	..	..	..	..	..	..	..	..	..	..	..	Newington.
..	..	..	..	..	..	..	..	..	..	..	..	..	..	..	St. Olave, Southwark.
..	..	..	..	..	..	..	..	..	..	..	..	..	..	..	Bermondsey.
..	..	..	..	..	..	..	..	..	..	..	..	..	..	..	Rotherhithe.
..	..	..	..	..	..	..	..	..	..	..	..	..	..	..	Lambeth.
..	..	..	..	..	..	..	..	..	..	..	..	..	1	..	Battersea.
..	..	..	..	..	..	..	1	..	..	..	..	..	1	..	Wandsworth.
..	..	..	..	..	..	..	..	..	..	..	..	..	2	..	Camberwell.
..	..	..	..	..	..	..	..	..	..	..	..	..	..	..	Greenwich.
..	..	..	..	..	..	..	..	..	..	..	..	..	..	..	Lewisham.
..	..	..	..	..	..	..	..	..	..	..	..	..	..	..	Woolwich.
..	..	..	..	..	..	..	..	..	..	..	..	..	..	..	Plumstead.
..	..	..	..	..	..	..	..	..	..	..	..	..	..	..	Lee.
..	..	..	..	..	..	..	..	..	..	..	..	..	..	..	Port of London.

luding Inner Temple (population, 86). ‖ Including Tower of London (population, 868).

	Sanitary Areas.	Population (1891).	SCARLET FEVER. Weekly Statement, 1st Quarter, 1898.											
			Jan. 8.		Jan. 15.		Jan. 22.		Jan. 29.		Feb. 5.		Feb. 12.	
			Cases.	Deaths.	Cases.	Deaths.	Cases.	Deaths.	Cases.	Deaths.	Cases.	Deaths.	Cases.	Deaths.
	London .. .. (*Administrative County*)	4,232,118	350	19	274	15	344	16	334	17	294	12	[illegible]	19
W. District.	Kensington	166,308	6	..	6	2	9	..	13	2	8	..	6	2
	Fulham	91,639	13	..	7	4	20	..	18	2	14	2	16	..
	Hammersmith	97,239	6	..	2	..	9	..	9	..	4	1	9	..
	Paddington	117,846	6	..	7	1	9	..	6	..	3	..	9	1
	Chelsea	96,258	10	..	1	..	7	..	9	..	6	..	7	1
	St. George, Hanover Sq.* ..	78,599	4	..	1	..	5	1	2	..	2	..	2	..
	Westminster	55,539	4	..	3	..	5	1	5	1	4	..	2	1
	St. James, Westminster ..	24,995	1	..	..	..	1	..	..	..	..	..	..	..
N. District.	St. Marylebone	142,404	5	1	5	..	8	..	9	..	3	..	5	1
	Hampstead	68,416	1	..	2	..	5	..	3	..	5	..	2	..
	St. Pancras	234,379	10	1	14	1	10	..	16	..	9	..	15	3
	Islington	319,143	17	1	14	..	17	1	22	2	17	1	18	1
	St. Mary, Stoke Newington	30,936	2	..	1	..	2	..	..	..	..	..	..	..
	Hackney	198,606	24	1	15	1	11	..	23	..	20	2	18	1
Central District.	St. Giles and St. George, Bloomsbury.	39,782	6	..	1	..	3	..	2	..	..	..	..	..
	St. Martin-in-the-Fields ..	14,616	..	..	1	..	1	1	..	..	..	..	..	..
	Strand†	25,217	..	..	..	..	1	..	2	..	1	..	..	..
	Holborn‡	34,043	1	..	2	..	3	..	..	..	2	..	4	..
	Clerkenwell	66,216	5	..	2	..	6	1	5	..	2	1	3	..
	St. Luke, Middlesex ..	42,440	3	..	7	..	3	..	1	..	3	..	4	1
	London, City of§	37,583	2	..	..	..	2	..	1	..	2	..	1	..
E. District.	Shoreditch	124,009	13	1	5	..	5	..	4	1	10	..	8	..
	Bethnal Green	129,132	18	..	6	..	14	2	15	1	13	1	13	..
	Whitechapel‖	74,420	4	..	3	..	2	..	3	1	4	..	7	..
	St. George-in-the-East ..	45,795	5	..	5	..	2	..	6	..	5	..	6	..
	Limehouse	57,376	6	..	6	..	4	1	2	..	5	..	2	..
	Mile End Old Town ..	107,592	17	1	13	..	15	..	7	1	6	..	7	..
	Poplar	166,748	28	1	13	..	22	1	18	1	15	..	21	..
S. District.	St. Saviour, Southwark ..	27,177	1	..	..	..	3	..	2	..	..	..	3	..
	St. George, Southwark ..	59,712	1	1	1	..	4	..	9	..	3	..	5	..
	Newington	115,804	7	2	18	..	7	1	9	1	10	1	9	..
	St. Olave, Southwark ..	12,723	2	..	1	..	3	..	..	..	3	..	6	..
	Bermondsey	84,682	4	1	5	..	8	1	4	..	3	..	5	1
	Rotherhithe	39,255	6	..	4	..	1	..	6	1	3	..	1	..
	Lambeth	275,203	19	1	21	2	19	2	16	..	15	..	15	..
	Battersea	150,558	19	4	15	2	26	..	31	1	21	..	16	1
	Wandsworth	156,942	26	1	15	2	21	..	11	..	18	2	18	2
	Camberwell	235,344	10	..	20	..	22	1	21	2	13	..	12	..
	Greenwich	165,413	15	1	19	..	13	1	8	..	16	..	7	..
	Lewisham	92,647	1	..	4	..	..	..	4	..	7	..	2	..
	Woolwich	40,848	12	..	1	..	3	1	4	..	8	..	6	2
	Plumstead	52,436	4	1	5	..	8	..	7	..	9	..	6	..
	Lee	36,103	6	..	3	..	5	..	1	..	2	1	3	..
	Port of London	..	..	..	..	..	..	..	..	..	..	..	..	..

* Including St. Peters, Westminster, (population, 235). † Including Middle Temple (population, [illegible]

SCARLET FEVER—*continued.*

Feb. 19.		Feb. 26.		Mar. 5.		Mar. 12.		Mar. 19.		Mar. 26.		April 2.		Totals for 1st Quarter, 1898.		Sanitary Areas.
Cases.	Deaths.	Cases.	Deaths.	Cases.	Deaths.	Cases.	Deaths.	Cases.	Deaths.	Cases.	Deaths.	Cases.	Deaths.	Cases.	Deaths.	
183	9	261	11	262	8	250	8	328	10	269	16	331	22	3,958	181	London. (*Administrative County.*)
11	..	15	..	6	..	11	..	6	..	8	..	16	3	130	9	Kensington.
15	..	20	1	13	..	16	1	10	1	15	..	13	..	190	11	Fulham.
5	..	3	..	4	..	2	1	7	..	5	..	6	1	71	3	Hammersmith.
3	1	8	..	6	..	4	..	6	..	3	2	5	..	74	5	Paddington.
7	..	4	1	15	..	10	..	5	..	1	..	5	..	87	2	Chelsea.
6	..	1	..	7	..	2	1	2	..	6	1	3	..	44	3	St. George, Hanover Sq.*
4	..	..	..	..	..	..	..	4	..	2	..	4	..	37	3	Westminster.
1	..	..	..	..	..	..	..	..	..	..	..	1	..	4	..	St. James, Westminster.
..	..	2	..	8	..	2	..	7	..	6	1	3	..	66	3	St. Marylebone.
1	..	7	..	4	..	5	..	4	..	1	..	4	..	44	..	Hampstead.
12	2	13	..	7	..	17	..	14	..	17	2	12	2	166	11	St. Pancras.
23	..	22	1	14	1	12	1	25	..	21	..	25	1	263	10	Islington.
1	..	2	..	3	..	4	..	..	..	1	..	..	..	16	..	St. Mary, Stoke Newington.
27	..	14	1	26	1	22	..	23	..	11	1	16	..	260	8	Hackney.
2	..	..	..	..	..	1	..	1	..	1	..	1	..	18	..	St. Giles and St. George, Bloomsbury.
1	..	..	..	..	..	..	..	..	..	..	..	1	..	4	1	St. Martin-in-the-Fields.
2	..	..	..	2	..	1	..	..	..	1	..	..	..	10	..	Strand †
1	..	..	..	1	..	2	..	..	..	2	..	..	..	18	..	Holborn.‡
9	.	5	.	2	..	6	..	5	..	7	..	10	..	67	2	Clerkenwell.
5	..	2	1	..	..	..	..	1	..	1	..	5	1	35	3	St. Luke, Middlesex.
1	..	1	..	1	..	..	..	1	..	..	..	1	..	13	..	London, City of.§
7	..	1	..	5	1	6	..	14	1	8	..	17	..	103	4	Shoreditch.
8	..	10	1	10	..	9	..	11	..	11	..	6	..	144	5	Bethnal Green.
7	..	7	..	2	..	6	1	8	..	3	1	3	..	89	3	Whitechapel ¶
3	..	1	..	3	..	1	..	1	..	3	..	4	3	45	3	St. George-in-the-East.
6	..	4	..	7	..	4	..	11	1	1	..	9	..	67	2	Limehouse.
10	1	5	1	3	..	6	..	8	..	7	..	9	1	111	5	Mile End Old Town.
13	..	18	1	6	2	7	..	18	..	9	1	17	1	210	8	Poplar.
2	..	1	..	..	..	1	..	2	..	4	..	..	..	19	..	St. Saviour, Southwark.
1	..	6	..	6	..	6	..	8	..	5	..	2	..	55	1	St. George, Southwark.
12	1	17	1	15	..	26	..	10	1	4	..	7	..	141	8	Newington.
.	..	..	..	1	..	3	..	..	..	1	..	..	..	20	..	St. Olave, Southwark.
3	1	6	..	3	..	1	1	5	..	7	..	6	..	60	5	Bermondsey.
1	..	3	..	5	1	3	..	6	1	1	1	4	1	44	5	Rotherhithe.
23	..	20	..	16	..	11	1	34	1	14	..	29	1	252	8	Lambeth.
16	1	20	..	19	..	20	..	26	2	20	2	25	3	273	16	Battersea.
15	..	13	1	17	..	14	..	6	1	12	..	17	..	203	10	Wandsworth.
26	..	10	..	4	1	10	..	9	..	17	1	10	..	184	5	Camberwell.
11	1	6	1	10	..	7	..	14	1	9	..	12	..	147	5	Greenwich.
2	..	4	..	3	..	3	..	5	..	15	..	13	2	63	2	Lewisham.
6	..	5	..	4	..	2	..	4	..	2	..	4	..	61	3	Woolwich.
10	1	2	..	3	1	4	..	6	..	4	2	5	1	73	6	Plumstead.
6	..	2	..	2	..	1	1	2	..	3	..	2	1	38	3	Lee.
.	..	..	..	..	..	..	..	..	..	..	..	..	..	..	..	Port of London.

luding Gray's Inn (population, 253), Lincoln's Inn (population, 27), Charterhouse (population, 136), Staple pulation, 21), and Furnival's Inn (*population, 121*).
luding Inner *Temple (population, 96).* § Including Tower of London (population, 989).

Sanitary Areas.	Population (1891).	Scarlet Fever—continued. Weekly Statement, 2nd Quarter, 1896. April 9. Cases.	April 9. Deaths.	April 16. Cases.	April 16. Deaths.	April 23. Cases.	April 23. Deaths.	April 30. Cases.	April 30. Deaths.	May 7. Cases.	May 7. Deaths.	May 14. Cases.	May 14. Deaths.
London.. .. .. .. (*Administrative County*)	4,232,118	307	19	300	12	306	10	314	13	288	10	312	[illegible]
W. District. Kensington	166,308	10	..	9	..	6	..	3	2	10	..	11	[illegible]
Fulham	91,639	18	1	14	..	18	..	11	..	17	2	13	[illegible]
Hammersmith	97,239	6	..	9	..	8	..	12	..	7	..	10	[illegible]
Paddington	117,846	6	..	4	..	4	..	5	..	6	..	6	[illegible]
Chelsea	96,253	7	1	10	..	4	1	5	..	3	..	4	[illegible]
St. George, Hanover Sq.* ..	78,599	3	..	4	..	4	..	1	..	6	..	9	[illegible]
Westminster	55,539	1	..	2	..	..	..	3	..	2	..	3	[illegible]
St. James, Westminster ..	24,995	1	..	1	..	..	..	2	..	..	..	2	[illegible]
N. District. St. Marylebone	142,404	6	..	5	..	12	..	5	..	6	..	3	[illegible]
Hampstead	68,416	1	1	6	..	4	..	3	..	8	..	9	[illegible]
St. Pancras	234,379	17	1	15	1	13	..	14	..	15	..	16	..
Islington	319,143	20	1	18	..	33	1	29	1	18	..	30	1
St. Mary, Stoke Newington	30,934	..	..	5	..	3	..	5	..	6	..	1	..
Hackney	198,606	14	..	19	1	19	1	17	..	16	..	23	..
Central District. St. Giles and St. George, Bloomsbury.	39,782	1	..	..	..	3	..	1	..	1	..	..	..
St. Martin-in-the-Fields ..	14,616	..	..	..	..	1	..	..	..	..	..	..	..
Strand†	25,217	1	..	2	..	1	..	1	1	..	..	1	..
Holborn‡	34,943	1	..	1	..	1	..	..	..	..	..	2	..
Clerkenwell	66,216	3	2	4	..	5	..	6	..	10	..	6	..
St. Luke, Middlesex ..	42,440	3	..	1	..	8	..	5	1	3	..	1	..
London, City of§	37,583	4	..	1	..	1	..	1	..	1	..	3	..
E. District. Shoreditch	124,009	10	1	14	1	12	1	9	1	10	2	9	..
Bethnal Green	129,132	11	..	7	..	4	..	8	..	3	..	2	1
Whitechapel‖	74,420	2	..	1	1	..	..	5	..	2	..	11	..
St. George-in-the-East ..	45,795	2	..	5	1	..	..	3	..	2	..	2	..
Limehouse	57,376	7	..	2	..	4	..	3	..	6	..	1	..
Mile End Old Town ..	107,592	8	..	5	..	7	..	6	1	7	..	6	[illegible]
Poplar	166,748	22	1	11	..	14	..	19	2	5	..	13	[illegible]
S. District. St. Saviour, Southwark ..	27,177	2	..	2	..	..	..	1	..	1	..	1	[illegible]
St. George, Southwark ..	59,712	6	1	7	2	3	..	7	..	4	1	9	
Newington	115,804	3	..	11	..	7	..	11	..	11	..	14	
St. Olave, Southwark ..	12,723	..	..	2	..	1	..	..	..	..	..	..	
Bermondsey	84,682	3	..	1	1	11	1	1	1	7	1	10	
Rotherhithe	39,255	2	..	1	1	2	..	3	1	5	1	1	
Lambeth	275,203	21	1	33	1	25	2	23	..	21	1	19	
Battersea	150,558	22	1	17	..	17	..	20	1	17	..	16	
Wandsworth	156,942	19	..	12	..	13	..	14	..	12	..	14	
Camberwell	235,344	9	1	13	1	6	..	26	..	14	..	11	
Greenwich	165,413	10	1	8	1	10	1	14	..	11	1	8	
Lewisham	92,647	8	5	2	..	7	..	3	..	2	1	4	
Woolwich	40,848	2	..	3	..	3	..	5	..	..	..	3	
Plumstead	52,436	7	..	8	..	6	2	9	..	4	..	2	
Lee	36,108	8	..	5	..	5	..	1	1	4	..	4	
Port of London	..	..	..	..	..	..	..	..	..	..	..	-	

* Including St. Peters, Westminster (population, 235). † Including Middle Temple (population, [illegible]
‡ *Including Gray's* Inn (population, 253), Lincoln's Inn (population, 71), Charterhouse (population, [illegible]
Staple Inn (population, 21), and Furnival's Inn, 121).

SCARLET FEVER—*continued.*

Weekly Statement, 2nd Quarter, 18[illegible]—*continued.*														Totals for 2nd Quarter, 18[illegible].		Sanitary Areas.
May 21.		May 28.		June 4.		June 11.		June 18.		June 25.		July 2.				
Cases.	Deaths.	Cases.	Deaths.	Cases.	Deaths.	Cases.	Deaths.	Cases.	Deaths.	Cases.	Deaths.	Cases.	Deaths.	Cases.	Deaths.	
288	18	325	6	301	17	304	10	290	10	295	8	287	12	3,905	154	London.
																(*Administrative County*
7	3	11	2	5	..	13	..	8	..	10	1	8	..	106	8	Kensington.
16	1	15	..	16	2	9	1	17	1	9	..	5	1	177	9	Fulham.
7	..	9	..	5	..	5	..	11	..	11	1	10	2	110	4	Hammersmith.
5	..	9	..	8	..	10	..	10	..	12	..	4	..	89	..	Paddington.
9	..	6	2	8	..	8	..	8	..	5	..	3	..	60	4	Chelsea.
7	..	3	..	6	..	2	..	4	..	9	..	7	..	65	..	St. George, Hanover Sq
2	..	3	..	4	..	1	..	2	..	..	..	..	..	23	..	Westminster.
..	..	2	..	1	..	..	..	1	..	..	..	..	..	10	..	St. James, Westminster
3	..	3	..	7	..	8	..	8	2	9	..	5	..	80	2	St. Marylebone.
15	..	9	..	11	..	3	1	2	..	3	..	5	..	79	2	Hampstead.
9	..	11	..	14	4	9	1	13	..	14	1	11	2	171	10	St. Pancras.
28	1	21	..	28	..	24	..	17	..	25	1	24	..	315	6	Islington.
5	1	1	..	2	..	3	..	3	..	7	..	1	..	42	1	St. Mary, Stoke Newingt
18	4	34	..	18	..	29	4	28	1	20	..	23	1	272	12	Hackney.
2	..	1	..	1	..	..	..	1	..	..	..	1	..	12	..	St. Giles and St. Geor Bloomsbury.
..	..	1	..	1	..	..	..	..	..	..	..	..	..	3	..	St. Martin-in-the-Field
..	..	2	..	1	..	..	..	..	..	3	..	..	..	13	1	Strand.
2	1	1	..	..	..	1	..	5	..	1	..	2	..	17	1	Holborn.†
6	1	2	..	6	..	8	..	4	1	9	1	6	..	76	5	Clerkenwell.
2	..	4	1	3	..	9	..	6	..	2	..	3	1	50	3	St. Luke, Middlesex.
3	..	2	..	2	1	3	..	2	..	3	..	..	..	24	1	London, City of.§
6	1	14	1	18	1	6	..	2	..	10	..	4	1	118	10	Shoreditch.
5	..	6	..	3	..	9	..	8	..	7	..	3	..	81	1	Bethnal Green.
6	..	7	..	6	..	3	..	6	..	1	1	1	..	51	2	Whitechapel.‡
7	..	2	..	1	..	1	..	6	1	1	.	1	..	33	2	St. George-in-the-East.
4	..	4	..	4	..	4	..	4	..	2	..	5	..	59	..	Limehouse.
12	..	8	..	14	..	12	..	11	..	6	..	14	..	118	1	Mile End Old Town.
14	2	9	..	8	1	21	..	9	1	11	..	11	..	168	8	Poplar.
3	..	..	..	..	..	1	..	1	..	..	..	..	..	11	..	St. Saviour, Southwark
7	..	2	..	5	1	5	..	9	..	8	1	7	..	79	9	St. George, Southwark.
18	1	19	..	13	1	4	1	15	..	7	1	9	..	140	5	Newington.
..	..	..	..	2	1	1	..	4	..	3	..	..	..	13	1	St. Olave, Southwark.
6	..	4	..	12	..	9	..	5	..	6	..	7	..	92	4	Bermondsey.
..	..	3	..	6	..	2	..	3	..	5	..	1	..	34	3	Rotherhithe.
14	..	24	..	15	2	18	..	12	1	16	..	15	2	[illegible]	10	Lambeth.
14	..	22	..	12	2	14	1	12	..	10	..	9	1	202	6	Battersea.
12	..	16	..	6	..	9	..	13	..	7	..	11	..	177	1	Wandsworth.
14	..	17	..	16	..	20	1	9	1	18	..	13	..	[illegible]	4	Camberwell.
12	..	7	..	6	..	7	..	13	1	10	..	9	1	125	6	Greenwich.
..	..	4	..	..	1	7	..	1	..	4	..	7	..	49	7	Lewisham.
..	..	1	..	10	..	5	..	2	..	1	..	3	..	38	..	Woolwich.
8	..	2	..	4	..	2	..	3	..	7	..	7	..	64	3	Plumstead.
..	..	4	..	5	..	..	..	6	..	4	..	2	..	48	2	Lee.
..	..	..	..	..	..	..	..	..	..	..	..	1	..	1	..	Port of London.

† Including Inner Temple (population, [illegible]).

§ Including Tower of London (population, [illegible]).

District	Sanitary Areas.	Population (1891).	SCARLET FEVER—*continued.* Weekly Statement, 3rd Quarter, 1898. July 9.		July 16.		July 23.		July 30.		Aug. 6.		Aug
			Cases.	Deaths.	Cases.	Deaths.	Cases.	Deaths.	Cases.	Deaths.	Cases.	Deaths.	Cases.
	London .. .. (*Administrative County*)	4,232,118	326	8	336	8	389	8	321	7	278	8	250
W. District.	Kensington	166,208	12	..	10	..	13	1	16	..	11	2	15
W. District.	Fulham	91,639	13	..	13	..	23	..	12	..	15	..	17
W. District.	Hammersmith	97,239	8	..	11	1	12	..	14	..	19	2	17
W. District.	Paddington	117,846	7	..	8	1	8	..	6	..	8	..	1
W. District.	Chelsea	96,253	8	..	5	1	13	..	9	..	8	..	4
W. District.	St. George, Hanover Sq.* ..	78,599	3	..	5	..	10	..	18	..	6	1	3
W. District.	Westminster	55,539	3	..	1	..	4	..	..	..	3	..	..
W. District.	St. James, Westminster ..	24,995	..	..	..	..	2	..	1	..	1	..	2
N. District.	St. Marylebone	142,404	7	1	14	..	10	..	11	..	5	1	6
N. District.	Hampstead	68,416	2	..	5	..	2	..	4	..	2	..	5
N. District.	St. Pancras	234,379	18	3	28	1	23	..	25	1	17	1	12
N. District.	Islington	319,143	13	..	25	..	34	..	21	..	16	..	9
N. District.	St. Mary, Stoke Newington	30,936	2	..	2	..	6	..	2	..	1	..	1
N. District.	Hackney	198,606	20	..	21	..	16	2	15	..	14	..	14
Central District.	St. Giles and St. George, Bloomsbury.	39,782	..	..	1	..	..	..	1	..	..	..	1
Central District.	St. Martin-in-the-Fields ..	14,616	..	..	..	..	..	..	1	..	..	..	..
Central District.	Strand†	25,217	3	..	..	..	..	..	..	..	..	..	..
Central District.	Holborn‡	34,043	2	..	..	1	2	..	2	..	1	..	1
Central District.	Clerkenwell	66,216	3	..	2	..	8	1	2	..	8	..	7
Central District.	St. Luke, Middlesex ..	42,440	3	..	1	..	4	..	5	..	3	..	1
Central District.	London, City of§	37,583	3	..	1	..	2	..	2	..	2	..	1
E. District.	Shoreditch	124,009	5	..	3	..	5	..	4	..	7	..	3
E. District.	Bethnal Green	129,132	9	..	8	..	10	..	5	..	13	..	6
E. District.	Whitechapel‖	74,420	6	..	4	1	8	..	8	..	..	..	3
E. District.	St. George-in-the-East ..	45,795	5	..	2	..	2	..	..	..	1	..	4
E. District.	Limehouse	57,376	7	..	7	..	3	1	2	..	2	..	4
E. District.	Mile End Old Town ..	107,592	10	..	14	..	7	..	5	..	9	..	6
E. District.	Poplar	166,748	15	1	14	..	7	..	13	..	14	..	14
S. District.	St. Saviour, Southwark ..	27,177	2	..	2	..	2	..	3	1	..	..	2
S. District.	St. George, Southwark ..	59,712	10	..	9	..	14	1	4	1	12	1	6
S. District.	Newington	115,804	27	2	11	..	9	1	15	2	7	..	5
S. District.	St. Olave, Southwark ..	12,723	2	..	3	..	4	..	3	..	1	..	1
S. District.	Bermondsey	84,682	7	..	11	..	17	..	18	..	5	..	18
S. District.	Rotherhithe	39,255	2	..	1	..	2	..	3	1	3	..	1
S. District.	Lambeth	275,203	19	..	23	..	17	..	18	..	14	..	10
S. District.	Battersea	150,558	9	..	6	..	18	1	11	..	10	..	11
S. District.	Wandsworth	156,942	16	..	10	..	17	..	11	1	7	..	1
S. District.	Camberwell	235,344	24	1	27	..	20	..	15	..	16	..	17
S. District.	Greenwich	165,413	8	..	12	1	14	..	7	..	4	..	10
S. District.	Lewisham	92,647	1	..	4	..	3	..	2	..	3	..	5
S. District.	Woolwich	40,848	1	..	2	..	2	..	2	..	1	..	1
S. District.	Plumstead	52,436	8	..	9	1	11	..	4	..	8	..	3
S. District.	Lee	36,103	3	..	..	..	5	..	1	..	1	..	2
S. District.	Port of London	..	..	..	1	..	..	..	..	..	..	..	..

* Including St Peters, Westminster (population, 235). † Including Middle Temple (populatio
‡ Including Gray's Inn (population, 253), Lincoln's Inn (population, 27), Charterhouse (population
Staple Inn (population, 21), and Furnival's Inn (population, 121).

Scarlet Fever—*continued.*														
Weekly Statement, 3rd Quarter, 1898—*continued.*												Totals for 3rd Quarter, 1898.		Sanitary Areas.
Aug. 27.		Sept. 3.		Sept. 10.		Sept. 17.		Sept. 24.		Oct. 1.				
Cases.	Deaths.	Cases.	Deaths.	Cases.	Deaths.	Cases.	Deaths.	Cases.	Deaths.	Cases.	Deaths.	Cases.	Deaths.	
229	7	283	5	346	8	281	3	337	9	398	10	4001	108	London. (*Administrative County.*)
7	..	12	..	13	..	6	..	11	..	9	..	142	4	Kensington.
8	1	10	1	8	..	6	..	7	..	22	1	163	3	Fulham.
4	..	4	..	10	..	5	..	16	1	9	..	134	5	Hammersmith.
7	..	5	..	7	..	1	..	4	..	6	..	71	2	Paddington.
1	..	5	..	8	..	7	1	6	1	5	1	83	4	Chelsea.
5	..	5	..	7	..	2	..	7	..	2	..	74	1	St. George, Hanover Sq.*
1	..	..	..	1	..	..	..	3	..	5	..	21	..	Westminster.
..	..	..	..	1	..	..	..	..	..	2	..	9	2	St. James, Westminster.
5	1	2	..	3	..	5	..	8	..	11	1	88	4	St. Marylebone.
6	..	7	..	12	..	10	..	3	..	..	..	61	..	Hampstead.
21	1	18	..	27	..	26	..	14	2	25	1	276	15	St. Pancras.
17	..	17	1	24	..	31	..	30	2	34	1	291	5	Islington.
2	..	3	..	9	..	4	..	6	..	4	..	42	..	St. Mary, Stoke Newington.
8	..	10	1	14	..	17	..	29	..	25	..	214	4	Hackney.
..	..	..	..	1	..	1	..	1	..	..	..	7	1	St. Giles and St. George, Bloomsbury.
..	..	..	..	..	..	..	..	..	..	..	..	2	..	St. Martin-in-the-Fields.
..	..	..	..	1	..	2	1	..	..	4	..	11	1	Strand.†
4	..	1	..	1	..	1	..	3	..	5	..	24	1	Holborn.‡
8	..	10	..	4	1	3	..	2	..	8	..	75	2	Clerkenwell.
1	..	1	..	..	..	6	..	5	..	4	..	39	..	St. Luke, Middlesex.
..	..	..	..	1	..	2	..	2	..	3	..	22	1	London, City of.§
7	1	9	..	9	..	18	..	9	..	6	..	95	2	Shoreditch.
1	..	9	..	8	..	9	..	6	..	9	..	98	1	Bethnal Green.
6	..	7	..	8	..	2	..	9	1	3	..	68	2	Whitechapel.‖
3	..	2	1	3	..	..	..	1	..	1	..	29	1	St. George-in-the-East.
2	..	3	..	4	..	2	..	1	..	6	..	45	1	Limehouse.
7	..	9	..	3	..	8	..	6	..	10	..	103	..	Mile End Old Town.
7	..	22	..	11	..	6	1	6	..	13	..	150	3	Poplar.
1	..	1	..	3	..	3	..	4	..	1	..	27	1	St. Saviour, Southwark.
10	..	13	..	9	..	7	..	6	..	9	1	114	8	St. George, Southwark.
7	..	6	..	20	1	14	..	13	..	26	1	170	8	Newington.
..	..	3	..	1	..	..	..	2	..	3	..	24	..	St. Olave, Southwark.
16	..	5	..	12	3	11	..	23	1	14	1	168	8	Bermondsey.
5	..	4	..	4	..	..	..	1	..	1	..	27	1	Rotherhithe.
8	1	19	..	20	..	10	..	14	..	30	..	215	1	Lambeth.
10	..	10	..	15	..	10	..	9	..	12	..	142	1	Battersea.
9	..	10	..	8	..	7	..	11	..	7	..	130	2	Wandsworth.
9	..	19	1	31	..	22	..	15	1	26	..	250	4	Camberwell.
6	..	6	..	12	3	5	..	7	..	11	..	109	4	Greenwich.
..	..	2	..	2	..	3	..	9	..	4	1	42	1	Lewisham.
4	1	3	..	4	..	1	..	4	..	8	..	34	1	Woolwich.
4	1	9	..	6	..	6	..	9	..	11	..	92	2	Plumstead.
2	..	1	..	1	..	3	..	6	..	4	1	30	1	Lee.
..	..	..	..	..	..	..	..	..	..	..	..	1	..	Port of London.

…g Inner Temple (population, 96). ‖ Including Tower of London (population, 868).

District	Sanitary Areas.	Population (1891).	SCARLET FEVER—*continued.* Weekly Statement, 3rd Quarter, 1898. July 9. Cases.	July 9. Deaths.	July 16. Cases.	July 16. Deaths.	July 23. Cases.	July 23. Deaths.	July 30. Cases.	July 30. Deaths.	Aug. 6. Cases.	Aug. 6. Deaths.	Aug. 13. Cases.	Aug. 13. Deaths.
	London .. .. (*Administrative County*)	4,232,118	326	8	336	8	389	8	321	7	278	8	250	10
W. District.	Kensington	166,208	12	..	10	..	13	1	16	..	11	2	15	..
	Fulham	91,639	13	..	13	..	23	..	12	..	15	..	17	..
	Hammersmith	97,239	8	..	11	1	12	..	14	..	19	2	17	1
	Paddington	117,846	7	..	8	1	8	..	6	..	8	..	1	1
	Chelsea	96,253	8	..	5	1	13	..	9	..	8	..	4	..
	St. George, Hanover Sq.* ..	78,599	3	..	5	..	10	..	18	..	6	1	3	..
	Westminster	55,539	3	..	1	..	4	..	..	..	3	..	..	..
	St. James, Westminster ..	24,995	..	..	..	..	2	..	1	..	1	..	2	2
N. District.	St. Marylebone	142,404	7	1	14	..	10	..	11	..	5	1	6	..
	Hampstead	68,416	2	..	5	..	2	..	4	..	2	..	5	..
	St. Pancras	234,379	18	3	28	1	23	..	25	1	17	1	12	2
	Islington	319,143	13	..	25	..	34	..	31	..	16	..	9	1
	St. Mary, Stoke Newington	30,936	2	..	2	..	6	..	2	..	1	..	1	..
	Hackney	198,606	20	..	21	..	16	2	15	..	14	..	14	..
Central District.	St. Giles and St. George, Bloomsbury.	39,782	..	..	1	..	..	..	1	..	..	..	1	..
	St. Martin-in-the-Fields ..	14,616	..	..	..	..	..	..	1	..	..	..	..	..
	Strand†	25,217	3	..	..	..	..	..	..	..	..	..	..	..
	Holborn‡	31,043	2	..	..	1	2	..	2	..	1	..	1	..
	Clerkenwell	66,216	3	..	2	..	8	1	2	..	8	..	7	..
	St. Luke, Middlesex ..	42,440	3	..	1	..	4	..	5	..	3	..	1	..
	London, City of§	37,583	3	..	1	..	2	..	2	..	2	..	1	..
E. District.	Shoreditch	124,009	5	..	3	..	5	..	4	..	7	..	3	1
	Bethnal Green	129,132	9	..	8	..	10	..	5	..	13	..	6	1
	Whitechapel‖	74,420	6	..	4	1	8	..	8	..	..	..	3	..
	St. George-in-the-East ..	45,795	5	..	2	..	2	..	..	..	1	..	4	..
	Limehouse	57,376	7	..	7	..	3	1	2	..	2	..	4	..
	Mile End Old Town ..	107,592	10	..	14	..	7	..	5	..	9	..	6	..
	Poplar	166,748	15	1	14	..	7	..	13	..	14	..	14	..
S. District.	St. Saviour, Southwark ..	27,177	2	..	2	..	2	..	3	1	..	..	2	..
	St. George, Southwark ..	59,712	10	..	9	..	14	1	4	1	12	1	6	1
	Newington	115,804	27	2	11	..	9	1	15	2	7	..	5	..
	St. Olave, Southwark ..	12,723	2	..	3	..	4	..	3	..	1	..	1	..
	Bermondsey	84,682	7	..	11	..	17	..	18	..	5	..	18	..
	Rotherhithe	39,255	2	..	1	..	2	..	3	1	3	..	1	..
	Lambeth	275,203	19	..	23	..	17	..	18	..	14	..	10	..
	Battersea	150,558	9	..	6	..	18	1	11	..	10	..	11	..
	Wandsworth	156,942	16	..	10	..	17	..	11	1	7	..	1	..
	Camberwell	235,344	24	1	27	..	20	..	15	..	16	..	17	..
	Greenwich	165,413	8	..	13	1	14	..	7	..	4	..	10	..
	Lewisham	92,647	1	..	4	..	3	..	2	..	3	..	5	..
	Woolwich	40,848	1	..	2	..	2	..	2	..	1	..	1	..
	Plumstead	52,436	8	..	9	1	11	..	4	..	8	..	3	..
	Lee	36,103	3	..	..	..	5	..	1	..	1	..	2	..
	Port of London	..	..	..	1	..	..	..	..	..	..	..	..	..

* Including St. Peters, Westminster (population, 235). † Including Middle Temple (population, 95).
‡ Inglnding Gray's Inn (population, 253), Lincoln's Inn (population, 27), Charterhouse (population, 136), [illegible] (population, 21), and Furnival's Inn (population, 121).

SCARLET FEVER—*continued.*																
Weekly Statement, 4th Quarter, 1898—*continued.*												Totals for 4th Quarter, 1898.		Grand Totals for Year 1898.		Sanitary Areas.
Nov. 26.		Dec. 3.		Dec. 10.		Dec. 17.		Dec. 24.		Dec. 31.						
Cases.	Deaths.	Cases.	Deaths.	Cases.	Deaths.	Cases.	Deaths.	Cases.	Deaths.	Cases.	Deaths.	Cases.	Deaths.	Cases.	Deaths.	
[illegible]	7	227	9	329	11	265	5	279	14	329	12	5,068	138	16,917	581	London. (*Administrative County.*)
7	..	7	1	10	1	10	..	6	..	8	..	106	8	474	24	Kensington.
26	1	19	2	15	..	18	1	18	..	14	1	281	12	811	35	Fulham.
8	1	6	..	14	..	4	..	13	3	5	1	148	10	468	22	Hammersmith.
3	..	8	..	1	..	4	1	3	..	4	..	70	1	304	8	Paddington.
3	..	3	..	5	1	5	..	..	..	6	..	86	5	336	15	Chelsea.
4	..	3	..	4	..	4	..	..	..	1	..	33	..	221	4	St. George, Hanover Sq.*
1	..	..	..	1	..	2	..	..	1	..	..	18	2	90	5	Westminster.
..	..	1	..	..	..	1	..	..	..	..	..	8	..	31	2	St. James, Westminster.
6	..	6	..	13	1	4	..	4	..	6	..	79	4	312	13	St. Marylebone.
5	..	1	..	4	..	5	..	1	..	4	..	60	1	244	3	Hampstead.
35	1	27	..	14	3	34	..	31	1	21	2	367	17	960	55	St. Pancras.
29	..	36	2	26	..	32	..	29	..	18	2	477	5	1,336	26	Islington.
4	..	2	..	1	..	10	..	3	..	2	..	49	..	149	1	St. Mary, Stoke Newington.
16	..	20	..	17	..	31	..	10	3	10	..	290	6	1,026	30	Hackney.
..	..	2	..	2	..	3	1	1	..	..	..	26	1	65	2	St. Giles and St. George, Bloomsbury.
..	..	..	..	..	..	..	..	..	..	..	..	2	..	11	1	St. Martin-in-the-Fields.
3	..	3	..	..	..	1	..	..	..	2	..	14	..	47	2	Strand.†
1	..	2	..	4	..	1	..	..	..	2	..	38	..	97	2	Holborn.‡
7	..	3	..	3	..	8	..	5	..	5	..	114	4	331	13	Clerkenwell.
4	..	7	..	2	..	2	..	1	..	3	..	73	9	197	15	St. Luke, Middlesex.
1	..	..	..	1	..	3	..	1	..	..	..	20	..	79	2	London, City of.§
8	..	6	..	4	1	4	..	8	1	8	..	106	5	422	21	Shoreditch.
4	..	6	..	5	..	8	..	6	1	4	..	110	4	433	11	Bethnal Green.
5	..	6	..	9	..	7	..	5	..	4	..	74	1	263	8	Whitechapel.‖
1	..	2	..	4	..	2	..	1	..	8	..	36	..	143	6	St. George-in-the-East.
3	..	2	1	1	..	..	..	1	..	1	..	28	2	190	5	Limehouse.
2	..	6	..	12	..	9	..	4	..	1	..	96	2	435	8	Mile End Old Town.
9	..	6	1	8	1	10	..	5	..	..	..	119	3	645	22	Poplar.
2	..	1	..	..	..	1	..	8	..	2	..	22	..	79	1	St. Saviour, Southwark.
3	..	6	..	3	..	1	..	1	..	5	1	62	2	310	20	St. George, Southwark.
26	..	9	..	8	1	12	..	8	..	5	..	170	3	621	24	Newington.
..	..	..	..	1	..	2	..	..	..	..	..	10	..	67	1	St. Olave, Southwark.
7	..	8	..	10	..	5	..	4	..	10	..	148	2	458	19	Bermondsey.
4	..	3	..	..	..	5	..	4	1	10	..	40	1	145	10	Rotherhithe.
16	..	14	1	15	1	21	1	20	2	20	3	280	11	1,002	30	Lambeth.
10	..	12	..	13	..	9	..	9	..	13	2	192	5	809	28	Battersea.
11	1	9	..	14	1	14	..	11	..	55	..	236	3	706	16	Wandsworth.
21	..	33	1	26	..	19	..	19	..	19	..	341	5	965	18	Camberwell.
14	1	18	..	28	..	19	1	12	..	25	..	271	6	652	21	Greenwich.
1	..	..	..	7	..	2	..	1	..	6	..	49	..	208	10	Lewisham.
1	1	5	..	2	..	7	..	14	..	9	..	71	1	204	5	Woolwich.
17	..	14	..	18	..	24	..	21	1	10	..	194	1	423	12	Plumstead.
3	1	6	..	7	..	3	..	4	..	3	..	42	1	156	7	Lee.
..	..	..	..	..	..	..	..	..	..	..	..	..	..	2	..	Port of London.

§ Including Inner Temple (population, 96).

‖ Including Tower of London (population, 868).

	Sanitary Areas.	Population (1891).	DIPHTHERIA. Weekly Statement, 1st Quarter, 1898.											
			Jan. 8.		Jan. 15.		Jan. 22.		Jan. 29.		Feb. 5.		Feb. 12.	
			Cases.	Deaths.	Cases.	Deaths.	Cases.	Deaths.	Cases.	Deaths.	Cases.	Deaths.	Cases.	Deaths.
	London (*Administrative County*)	4,232,118	231	50	192	43	200	33	278	35	240	41	239	45
W. District.	Kensington	166,308	2	..	2	2	3	..	5	4	6	1	..	..
	Fulham	91,639	7	2	5	1	10	3	14	3	11	3	14	4
	Hammersmith	97,239	4	..	1	..	2	..	2	..	4	..	3	..
	Paddington	117,846	5	4	4	1	2	1	7	1	4	1	2	..
	Chelsea	96,253	7	1	4	1	3	..	8	3	4	..	5	1
	St. George, Hanover Sq.*	78,599	1	..	1	..	1	..	2	..	2	1	1	1
	Westminster	55,539	1	..	4	2	2	2	..	..	5	1	3	1
	St. James, Westminster	24,995	1	..	2	..	..	..	1	..	..	..	..	..
N. District.	St. Marylebone	142,404	3	2	5	1	3	1	3	1	4	1	6	3
	Hampstead	68,416	4	2	3	..	1	1	3	..	2	2	1	..
	St. Pancras	234,379	11	2	8	2	9	1	15	..	7	2	9	3
	Islington	319,143	14	6	10	10	10	1	13	..	19	1	19	3
	St. Mary, Stoke Newington	30,936	..	..	..	..	2	..	1	1	..	..	1	..
	Hackney	198,606	18	4	16	4	17	1	19	1	25	5	24	6
Central District.	St. Giles and St. George, Bloomsbury.	39,782	..	..	1	..	..	..	1	1	3	..	..	..
	St. Martin-in-the-Fields	14,616	..	..	2	1	..	..	..	..	..	..	..	..
	Strand†	25,217	..	..	..	..	1	..	1	..	..	..	..	1
	Holborn‡	34,043	4	..	3	1	4	2	3	..	6	..	3	1
	Clerkenwell	66,216	3	2	7	..	6	..	14	..	7	..	4	..
	St. Luke, Middlesex	42,440	1	..	2	..	5	1	4	..	3	..	4	..
	London, City of §	37,583	4	1	..	1	1	..	2	..	1	..	..	..
E. District.	Shoreditch	124,009	5	1	4	..	6	3	5	2	3	..	6	..
	Bethnal Green	129,132	7	3	4	3	11	2	14	..	5	..	5	2
	Whitechapel ‖	74,420	4	..	..	..	5	1	5	..	1	1	2	..
	St. George-in-the-East	45,795	2	..	1	..	..	..	1	..	2	1	1	..
	Limehouse	57,376	4	..	2	..	3	2	7	..	4	..	2	1
	Mile End Old Town	107,592	3	..	6	1	3	..	3	..	4	3	3	1
	Poplar	166,748	8	1	7	2	13	2	10	2	8	3	7	1
S. District.	St. Saviour, Southwark	27,177	5	1	1	..	1	1	4	..	1	..	4	..
	St. George, Southwark	59,712	3	3	1	..	6	1	9	1	7	2	4	..
	Newington	115,804	10	3	6	1	2	1	7	1	13	1	5	..
	St. Olave, Southwark	12,723	..	1	1	..	2	..	..	..	1	..	1	1
	Bermondsey	84,683	1	1	1	1	4	..	3	..	2	1	2	1
	Rotherhithe	39,255	..	1	1	..	..	..	1	..	1	1	2	1
	Lambeth	275,203	11	3	7	1	11	1	11	2	11	2	18	1
	Battersea	150,558	20	1	13	1	10	2	20	5	9	2	17	[illegible]
	Wandsworth	156,942	12	2	12	2	10	1	15	1	16	2	24	[illegible]
	Camberwell	235,344	14	3	16	1	8	..	11	4	11	2	12	[illegible]
	Greenwich	165,413	17	..	16	1	14	2	21	1	9	1	8	[illegible]
	Lewisham	92,647	4	..	5	..	..	..	4	1	6	..	4	[illegible]
	Woolwich	40,848	1	1	2	1	2	..	2	..	1	..	4	[illegible]
	Plumstead	52,436	2	..	3	..	5	..	4	..	3	1	6	[illegible]
	Lee	36,103	3	..	3	1	2	..	3	..	9	..	..	[illegible]
	Port of London	..	..	..	..	..	..	..	..	..	..	..	..	[illegible]

* Including St. Peter's, Westminster (population, 235). † Including Middle Temple (population, [illegible]
‡ Including Gray's Inn (population, [illegible]), Lincoln's Inn (population, 37), Charterhouse (population, 136), [illegible] Inn (population, 21), and Furnival's Inn (population, 121).

Feb. 26. Cases.	Feb. 26. Deaths.	Mar. 5. Cases.	Mar. 5. Deaths.	Mar. 12. Cases.	Mar. 12. Deaths.	Mar. 19. Cases.	Mar. 19. Deaths.	Mar. 26. Cases.	Mar. 26. Deaths.	April 2. Cases.	April 2. Deaths.	Totals for 1st Quarter, 18[illegible]. Cases.	Totals for 1st Quarter, 18[illegible]. Deaths.	Sanitary Areas.
DIPHTHERIA—*continued.* Weekly Statement, 1st Quarter, 18[illegible]—*continued.*														
209	46	202	38	174	33	221	36	218	53	182	45	2,685	531	London. (*Administrative County.*)
3	1	3	..	5	1	9	2	3	1	5	..	58	13	Kensington.
10	..	4	1	6	..	9	1	10	3	3	1	110	22	Fulham.
4	1	2	2	3	1	6	..	5	..	2	2	41	6	Hammersmith.
2	1	3	2	1	..	5	1	1	..	5	1	45	14	Paddington.
4	1	7	..	..	..	6	..	3	2	4	1	60	10	Chelsea.
..	..	1	..	..	..	1	..	1	..	1	..	16	2	St. George, Hanover Sq.*
3	2	3	..	1	..	1	1	2	..	5	2	35	11	Westminster.
2	1	1	..	2	..	1	..	1	..	1	..	15	1	St. James, Westminster.
5	1	3	..	5	2	1	2	5	1	6	1	55	16	St. Marylebone.
6	1	3	..	4	..	4	..	1	..	..	..	31	6	Hampstead.
9	..	8	4	7	1	12	1	12	5	8	5	127	28	St. Pancras.
7	4	15	3	8	..	18	2	15	3	7	2	182	40	Islington.
3	..	..	1	..	1	..	..	1	..	..	..	19	3	St. Mary Stoke Newington.
22	10	24	5	19	2	18	2	23	7	24	3	277	52	Hackney.
1	..	..	..	1	..	..	..	1	1	1	..	9	2	St. Giles and St. George, Bloomsbury.
..	..	..	..	..	..	..	1	..	..	..	..	2	2	St. Martin-in-the-Fields.
..	..	1	..	1	..	..	..	5	..	..	..	15	1	Strand.
4	1	4	..	3	1	2	1	5	1	3	1	46	9	Holborn.
4	3	4	1	5	1	6	..	6	3	2	1	78	11	Clerkenwell.
3	..	4	1	2	..	2	..	6	..	5	..	45	2	St. Luke, Middlesex.
..	..	..	..	1	..	1	..	1	..	..	..	12	2	London, City of.‖
3	..	5	..	4	..	5	5	2	2	4	2	57	15	Shoreditch.
5	1	8	2	5	1	5	..	5	4	5	1	82	20	Bethnal Green.
4	1	3	2	..	1	3	1	4	1	2	..	41	6	Whitechapel.
2	..	3	..	3	..	3	..	2	..	1	..	25	1	St. George-in-the-East.
4	1	1	..	1	..	4	..	3	..	1	..	39	4	Limehouse.
4	1	4	1	9	..	3	2	6	1	3	1	58	12	Mile End Old Town.
6	3	6	1	4	1	14	5	9	2	4	..	114	28	Poplar.
2	..	3	..	2	2	..	..	3	1	6	2	34	7	St. Saviour, Southwark.
..	1	7	..	6	3	4	..	7	..	3	1	62	12	St. George, Southwark.
6	1	5	1	6	1	8	1	6	2	9	..	91	16	Newington.
..	..	1	..	..	..	1	..	..	..	..	..	7	2	St. Olave, Southwark.
..	..	1	1	1	..	2	..	..	..	1	..	19	5	Bermondsey.
1	2	..	..	1	..	2	..	1	..	1	1	12	6	Rotherhithe.
11	..	11	2	7	..	8	1	13	1	15	4	143	19	Lambeth.
16	3	14	3	14	2	8	2	11	5	2	2	169	33	Battersea.
17	3	12	2	11	5	13	..	11	3	3	5	171	34	Wandsworth.
9	1	14	1	12	4	11	1	2	1	17	..	153	21	Camberwell.
8	..	8	1	5	1	9	2	14	2	8	2	144	16	Greenwich.
3	..	1	..	2	1	6	..	3	..	..	..	41	3	Lewisham.
2	1	..	..	2	..	1	..	..	..	3	..	21	4	Woolwich.
5	..	2	1	1	1	5	1	2	1	10	4	48	11	Plumstead.
3	..	3	..	4	..	4	1	3	..	..	..	42	3	Lee.
..	..	..	..	..	..	..	..	..	..	..	..	..	..	Port of London.

…ing Inner Temple (population, 96).

‖ Including Tower of London (population, [illegible]).

District	Sanitary Areas.	Population (1891).	DIPHTHERIA—continued. Weekly Statement, 2nd Quarter, 1898. April 9. Cases.	April 9. Deaths.	April 16. Cases.	April 16. Deaths.	April 23. Cases.	April 23. Deaths.	April 30. Cases.	April 30. Deaths.	May 7. Cases.	May 7. Deaths.	May 14. Cases.	May 14. Deaths.
	London .. .. (*Administrative County*)	4,232,118	174	29	208	34	170	31	179	38	196	27	191	29
W. District.	Kensington	166,308	7	..	4	1	7	..	..	..	3	1	11	..
	Fulham	91,639	7	2	7	2	3	1	8	..	6	1	6	1
	Hammersmith	97,239	2	..	2	2	2	..	..	..	3	..	2	..
	Paddington	117,816	2	1	5	..	4	2	2	..	6	1	3	..
	Chelsea	96,253	4	..	6	..	5	1	5	4	6	2	8	..
	St. George, Hanover Sq.* ..	78,599	3	..	5	..	3	..	3	..	..	..	1	..
	Westminster	55,539	3	1	3	1	9	..	2	2	4	..	7	..
	St. James, Westminster ..	24,995	..	..	..	..	..	..	..	..	1	..	..	..
N. District.	St. Marylebone	142,404	3	1	7	2	4	2	4	..	5	1	9	1
	Hampstead	68,416	3	..	1	..	..	..	5	..	2	..	3	2
	St. Pancras	234,379	11	..	8	1	7	1	7	4	6	1	6	1
	Islington	319,143	11	2	20	4	17	2	12	4	14	1	12	2
	St. Mary, Stoke Newington	30,936	..	..	2	1	1	..	..	..	1	..	1	..
	Hackney	198,606	6	5	16	5	20	3	11	3	15	2	20	3
Central District.	St. Giles and St. George, Bloomsbury.	39,782	1	..	1	..	1	..	..	..	1	..	..	..
	St. Martin-in-the-Fields ..	14,016	1	..	..	..	..	..	..	..	..	..	1	..
	Strand†	25,217	1	1	2	..	2	..	9	..	..	..	..	..
	Holborn‡	34,043	2	..	3	..	2	1	2	1	2	..	..	..
	Clerkenwell	66,216	4	2	1	1	4	2	2	..	6	..	2	..
	St. Luke, Middlesex ..	42,440	1	1	3	..	2	..	2	..	8	2	3	1
	London, City of§	37,583	..	..	1	..	1	..	1	..	3	1	1	..
E. District.	Shoreditch	124,009	4	..	4	..	3	..	8	1	2	1	3	1
	Bethnal Green	129,132	3	1	3	1	6	..	8	4	4	1	4	..
	Whitechapel‖	74,420	2	..	4	1	2	..	4	..	4	2	2	..
	St. George-in-the-East ..	45,795	1	..	2	..	5	1	2	..	5	2	1	..
	Limehouse	57,376	1	..	1	1	..	..	..	..	..	..	3	..
	Mile End Old Town ..	107,592	2	..	7	..	3	..	7	2	3	..	5	..
	Poplar	166,748	10	1	2	2	9	1	6	3	12	..	11	1
S. District.	St. Saviour, Southwark ..	27,177	1	..	1	..	2	..	..	..	4	1	..	..
	St. George, Southwark ..	59,712	2	..	3	1	1	..	6	..	2	..	3	1
	Newington	115,804	2	1	10	2	3	..	4	1	6	..	4	.
	St. Olave, Southwark ..	12,723	1	..	..	..	1	..	..	..	..	..	..	.
	Bermondsey	84,682	..	..	..	..	3	..	2	1	1	..	..	.
	Rotherhithe	39,255	3	..	2	..	1	..	3	1	1	..	..	.
	Lambeth	275,203	8	1	11	1	5	2	11	3	15	2	9	
	Battersea	150,558	18	2	9	1	3	2	11	1	6	1	13	
	Wandsworth	156,942	13	1	12	2	8	6	6	2	10	2	6	
	Camberwell	235,344	10	4	17	2	4	3	6	1	12	2	10	
	Greenwich	165,113	11	..	8	..	8	..	7	..	3	..	10	
	Lewisham	92,647	2	..	1	..	7	..	3	..	8	..	7	
	Woolwich	40,848	1	..	5	..	..	..	2	..	1	..	1	
	Plumstead	52,436	1	..	..	..	1	..	1	..	4	..	1	
	Lee	36,108	6	2	4	..	1	1	3	..	2	..	2	
	Port of London	..	..	..	..	..	..	..	..	..	..	..	..	

* Including St. Peters, Westminster (population, 235). † Including Middle Temple (population,
‡ Including Gray's Inn (population, 253), Lincoln's Inn (population, 27), Charterhouse (population,
Staple Inn (population, 21), and Furnival's Inn (population, 121).

[illegible]A—*continued.*

Weekly Statement, 2nd Quarter, 1898—*continued.*														Totals for 2nd Quarter, 1898.		Sanitary Areas.
May 21.		May 28.		June 4.		June 11.		June 18.		June 25.		July 2.				
Cases.	Deaths.	Cases.	Deaths.	Cases.	Deaths.	Cases.	Deaths.	Cases.	Deaths.	Cases.	Deaths.	Cases.	Deaths.	Cases.	Deaths.	
187	31	197	26	205	30	210	23	216	26	185	21	208	27	2,581	363	London. (*Administrative County.*)
3	..	5	2	9	1	3	2	1	..	5	..	7	1	79	8	Kensington.
6	..	9	1	9	1	12	1	19	2	15	..	7	1	109	13	Fulham.
3	..	2	..	..	..	1	..	3	..	1	..	1	..	22	2	Hammersmith.
2	..	10	1	4	1	6	1	6	1	2	..	4	1	56	9	Paddington.
17	1	7	2	7	1	8	1	12	1	4	4	2	1	91	18	Chelsea.
4	..	3	..	1	..	3	..	3	1	3	..	1	1	33	3	St. George, Hanover Sq.*
7	..	4	2	1	..	2	..	1	..	1	..	2	..	46	6	Westminster.
1	..	2	..	1	..	2	1	..	..	..	..	..	..	7	1	St. James, Westminster.
4	..	2	..	5	1	8	..	8	1	2	..	8	2	69	11	St. Marylebone.
1	..	3	1	5	1	4	1	2	..	6	..	4	..	59	5	Hampstead.
6	2	7	2	9	..	8	..	8	..	5	..	12	1	100	18	St. Pancras.
5	3	4	..	10	3	7	2	10	..	7	1	6	2	135	26	Islington.
..	..	2	..	1	..	1	..	2	..	1	..	..	..	12	1	St. Mary, Stoke Newington.
19	1	25	..	20	5	10	1	16	3	12	1	14	..	210	21	Hackney.
..	..	1	..	1	..	1	..	1	..	2	..	..	..	10	..	St. Giles and St. George, Bloomsbury.
..	..	..	..	..	..	..	..	..	..	..	..	..	..	2	..	St. Martin-in-the-Fields
..	..	1	..	1	1	..	..	1	..	..	..	..	..	27	3	Strand.†
1	..	1	..	1	..	2	..	3	..	..	..	1	..	20	3	Holborn.‡
5	1	1	..	6	..	5	1	3	..	8	2	12	..	69	9	Clerkenwell.
1	3	..	..	2	..	6	..	1	..	4	..	5	..	36	7	St. Luke, Middlesex.
1	..	1	..	1	..	..	..	..	..	..	..	2	..	12	1	London, City of.§
4	..	3	1	5	1	6	..	4	1	6	1	6	1	58	8	Shoreditch.
6	1	10	2	13	3	7	1	9	3	5	..	2	..	86	17	Bethnal Green.
..	1	6	..	2	..	2	..	2	..	4	..	5	..	39	4	Whitechapel.‖
1	..	4	..	1	1	3	..	1	..	1	..	4	1	31	5	St. George-in-the-East.
3	1	1	1	2	..	2	..	2	1	2	1	1	1	18	6	Limehouse.
6	1	4	..	6	1	7	1	3	1	5	1	3	..	61	7	Mile End Old Town.
7	3	4	1	5	..	8	1	6	2	5	1	15	4	108	22	Poplar.
1	..	2	..	1	..	..	..	..	1	3	..	1	..	16	2	St. Saviour, Southwark.
1	..	3	..	5	1	2	2	4	1	..	1	5	..	37	7	St. George, Southwark.
7	2	6	1	4	..	7	1	5	..	2	..	7	1	67	9	Newington.
..	..	..	..	..	..	..	..	..	..	..	..	..	..	2	..	St. Olave, Southwark.
1	..	3	..	4	..	8	..	10	2	8	..	5	..	40	3	Bermondsey.
..	..	..	..	1	1	2	..	..	..	..	..	1	..	14	2	Rotherhithe.
6	..	9	1	10	2	10	..	16	..	17	3	8	..	135	16	Lambeth.
11	3	11	3	9	..	15	2	13	2	13	1	12	1	144	21	Battersea.
9	..	9	1	12	2	13	2	10	2	12	..	8	1	128	22	Wandsworth.
20	2	13	1	9	3	9	1	12	1	15	4	19	2	156	26	Camberwell.
14	..	8	..	14	..	8	..	8	..	7	..	8	1	114	2	Greenwich.
12	4	9	2	7	..	7	1	5	..	4	..	6	1	78	8	Lewisham.
3	1	1	1	1	..	..	..	3	..	1	..	..	1	19	3	Woolwich.
1	1	..	..	..	..	2	..	1	..	..	..	3	2	15	3	Plumstead.
3	..	1	..	..	..	2	..	4	..	2	..	1	..	21	3	Lee.
..	..	..	..	..	..	..	..	1	..	..	..	..	..	1	..	Port of London.

‡ Including Inner Temple (population, 93).

§ Including Tower of London (population, 888).

	Sanitary Areas.	Population (1891).	DIPHTHERIA—continued. Weekly Statement, 3rd Quarter, 1893. July 8. Cases.	July 8. Deaths.	July 15. Cases.	July 15. Deaths.	July 22. Cases.	July 22. Deaths.	July 29. Cases.	July 29. Deaths.	Aug. 5. Cases.	Aug. 5. Deaths.	A[…] Cases.
	London .. .. (Administrative County)	4,232,118	214	27	242	38	245	29	198	21	187	20	15
W. District.	Kensington	166,308	1	..	3	..	1	..	3	..	3	..	
	Fulham	91,639	14	..	16	2	7	..	9	1	9	..	1
	Hammersmith	97,239	4	..	4	3	3	1	4	..	4	..	
	Paddington	117,846	5	2	7	3	6	1	5	2	2	1	
	Chelsea..	96,253	13	1	4	1	4	..	6	1	7	1	
	St. George, Hanover Sq.* ..	78,599	3	..	1	..	2	..	2	..	1	..	
	Westminster	55,539	2	..	1	..	1	..	2	3	..	..	
	St. James, Westminster ..	24,995	4	..	1	..	1	..	2	1	..	..	..
N. District.	St. Marylebone	142,404	5	..	2	..	7	..	4	2	1	..	..
	Hampstead	68,416	2	1	3	..	2	..	3	..	1	2	
	St. Pancras	234,379	5	1	7	1	9	..	8	1	9	3	
	Islington	319,143	10	..	6	..	15	1	9	..	9	2	1
	St. Mary, Stoke Newington	30,936	1	..	..	..	..	..	3	..	2	..	..
	Hackney	198,606	11	2	8	..	5	2	15	2	17	..	1
Central District.	St. Giles and St. George, Bloomsbury.	39,782	..	..	1	..	2	..	1	1	3	1	
	St. Martin-in-the-Fields ..	14,616	..	..	1	1	..	..	..	..	..	..	..
	Strand†	25,217	..	..	1	..	1	1	..	..	..	..	..
	Holborn‡	34,043	3	..	1	..	2	..	1	..	1	..	..
	Clerkenwell	66,216	12	2	7	2	14	2	3	2	6	1	
	St. Luke, Middlesex ..	42,440	2	2	6	1	7	..	3	1	4	..	
	London, City of§	37,583	1	..	3	..	..	..	..	..	..	..	..
E. District.	Shoreditch	124,009	3	3	1	2	4	..	5	..	2	..	
	Bethnal Green	129,132	7	..	5	..	4	..	3	..	7	..	
	Whitechapel‖	74,420	8	1	3	1	4	1	6	..	4	..	
	St. George-in-the-East ..	45,795	4	1	6	..	1	1	2	..	1	1	
	Limehouse	57,376	2	..	..	1	3	..	3	..	4	..	
	Mile End Old Town ..	107,592	5	..	8	..	7	1	6	1	7	2	
	Poplar	166,748	8	5	28	3	15	3	9	..	5	1	
S. District.	St. Saviour, Southwark ..	27,177	..	..	5	1	1	..	1	..	..	..	
	St. George, Southwark ..	59,712	3	1	5	..	1	..	6	1	3	..	
	Newington	115,804	7	..	7	1	16	1	9	..	6	2	1
	St. Olave, Southwark ..	12,723	..	..	1	..	1	1	..	..	..	..	
	Bermondsey	84,682	1	..	3	..	4	..	4	..	6	2	
	Rotherhithe	39,255	1	..	..	..	1	1	..	..	1	..	
	Lambeth	275,203	16	2	26	7	22	4	17	..	13	2	
	Battersea	150,558	20	1	14	1	14	..	9	..	11	2	1
	Wandsworth	156,942	8	..	12	2	17	1	9	..	2	..	
	Camberwell	235,344	5	..	18	2	18	2	7	1	13	5	
	Greenwich	165,413	3	1	7	1	14	2	8	1	11	1	1
	Lewisham	92,647	12	1	9	1	6	2	6	..	5	1	
	Woolwich	40,848	..	..	1	..	2	1	..	..	3	..	
	Plumstead	52,436	1	..	..	..	1	..	2	..	3	..	..
	Lee	36,103	3	..	1	1	1	..	3	..	1	..	
	Port of London	..	..	..	..	..	..	..	..	..	..	..	..

* Including St. Peters, Westminster (population, 235). † Including Middle Temple (popula[…]
‡ Including Gray's Inn (population, 2[…]), Lincoln's Inn (population, 27), Charterhouse (populati[…]
Staple Inn (population, 21), and […]'s Inn (population, 121).

DIPHTHERIA—*continued.*																
Weekly Statement, 3rd Quarter, 1898—*continued.*														Totals for 3rd Quarter, 1898.		Sanitary Areas.
Aug. 20.		Aug. 27.		Sept. 3.		Sept. 10.		Sept. 17.		Sept. 24.		Oct. 1.				
Cases.	Deaths.	Cases.	Deaths.	Cases.	Deaths.	Cases.	Deaths.	Cases.	Deaths.	Cases.	Deaths.	Cases.	Deaths.	Cases.	Deaths.	
[illegible]	[illegible]	162	20	174	18	203	20	240	26	217	20	200	23	2,704	356	London. (*Administrative County.*)
3	1	3	..	1	..	5	..	8	1	2	..	7	..	42	2	Kensington.
20	..	12	4	8	1	4	1	9	1	5	..	11	..	127	13	Fulham.
6	1	3	..	1	..	4	..	5	1	4	1	10	..	54	8	Hammersmith.
1	1	4	..	4	1	6	3	2	..	7	2	5	1	57	16	Paddington.
4	2	2	..	5	..	8	1	4	..	1	..	5	..	60	9	Chelsea.
5	..	3	..	..	..	..	..	4	1	1	..	6	1	20	3	St. George, Hanover Sq.*
2	..	..	..	1	..	1	..	1	..	2	..	4	1	19	4	Westminster.
..	..	1	..	..	..	1	..	1	..	..	..	..	..	11	1	St. James, Westminster.
2	..	4	..	8	..	5	1	9	2	7	2	10	1	63	8	St. Marylebone.
..	..	3	..	2	1	2	..	1	..	2	..	2	..	27	5	Hampstead.
7	..	8	1	8	..	7	3	9	..	4	1	10	1	84	13	St. Pancras.
3	..	7	1	5	..	6	..	10	2	16	2	10	2	118	12	Islington.
..	..	..	..	1	..	..	..	1	..	..	..	..	..	8	..	St. Mary, Stoke Newington.
6	..	14	2	19	1	13	..	18	..	10	..	18	4	148	16	Hackney.
2	..	..	..	1	..	..	..	..	..	1	..	2	..	14	9	St. Giles and St. George, Bloomsbury.
1	..	..	..	..	..	..	..	..	..	..	..	..	..	2	1	St. Martin-in-the-Fields.
..	..	..	..	..	..	1	..	..	..	..	..	1	..	4	1	Strand.†
5	1	3	..	3	..	7	3	10	2	4	1	1	2	40	8	Holborn.‡
7	..	3	1	9	..	5	3	3	..	4	..	8	..	81	13	Clerkenwell.
3	..	2	..	3	3	1	1	5	..	7	..	6	1	50	8	St. Luke, Middlesex.
..	..	..	..	..	..	1	..	..	..	..	1	1	..	5	1	London, City of.§
5	..	5	..	1	1	4	2	6	1	4	..	9	..	40	10	Shoreditch.
6	2	3	2	1	..	2	..	6	..	3	..	6	..	38	4	Bethnal Green.
3	..	..	..	1	..	4	..	6	..	4	1	1	..	42	4	Whitechapel.‖
2	..	1	..	4	..	..	..	1	..	3	..	1	..	30	3	St. George-in-the-East.
4	2	3	..	1	..	..	..	5	..	2	1	3	..	21	4	Limehouse.
4	3	3	2	6	..	14	1	11	5	9	1	10	2	86	21	Mile End Old Town.
2	..	6	2	12	1	12	1	11	1	7	4	13	3	131	26	Poplar.
1	..	3	..	1	..	1	..	1	..	2	..	..	..	17	1	St. Saviour, Southwark.
5	..	5	..	1	1	3	..	8	1	6	..	9	1	68	6	St. George, Southwark.
21	4	12	..	8	2	14	2	7	2	15	3	8	..	131	21	Newington.
..	..	1	..	1	..	1	..	..	..	..	..	..	..	6	1	St. Olave, Southwark.
1	..	6	..	7	1	5	3	4	1	10	1	10	4	62	11	Bermondsey.
..	..	..	..	..	..	1	..	..	..	..	..	..	..	5	1	Rotherhithe.
7	1	9	1	12	4	14	4	27	2	20	4	22	1	211	34	Lambeth.
20	6	5	..	16	1	9	3	12	1	17	3	25	3	178	21	Battersea.
5	..	11	3	5	1	5	..	12	2	9	..	21	4	131	13	Wandsworth.
20	..	8	2	9	..	13	..	18	..	16	1	15	..	180	14	Camberwell.
8	2	9	..	10	1	11	1	17	1	12	..	6	1	122	13	Greenwich.
4	..	1	..	3	..	3	..	3	..	1	..	4	..	62	6	Lewisham.
2	..	1	..	..	..	2	..	3	..	1	..	1	..	17	1	Woolwich.
3	..	5	..	2	..	..	..	2	..	3	..	2	..	24	..	Plumstead.
3	..	..	..	1	..	7	..	3	1	4	..	6	..	24	2	Lee.
..	..	..	..	..	..	..	..	..	..	..	..	..	..	..	..	Port of London.

§ Including Inner Temple (population, 96). ‖ Including Tower of London (population, 888).

		Sanitary Areas.	Population (1891).	DIPHTHERIA—*continued.* Weekly Statement, 4th Quarter, 1898. Oct. 8.		Oct. 15.		Oct. 22.		Oct. 29.		Nov. 5.		Nov. 12.		[illegible]
				Cases.	Deaths.	Cases.	Deaths.	Cases.	Deaths.	Cases.	Deaths.	Cases.	Deaths.	Cases.	Deaths.	Cases.
London (*Administrative County*)		.. ..	4,232,118	275	33	266	41	305	30	345	46	371	28	283	30	[illegible]
		Kensington	166,208	6	1	1	..	8	..	5	..	3	..	4	..	[illegible]
W. District.		Fulham	91,639	7	1	17	1	15	1	10	3	5	1	12	1	[illegible]
		Hammersmith	97,239	4	1	4	..	10	..	2	1	4	1	1	1	[illegible]
		Paddington	117,846	7	1	5	2	13	3	3	4	4	1	3	1	[illegible]
		Chelsea	96,253	1	1	3	..	4	..	4	1	5	..	..	..	[illegible]
		St. George, Hanover Sq.* ..	78,599	5	..	3	1	..	..	1	1	4	..	5	..	..
		Westminster	55,539	..	..	2	..	..	..	4	1	6	..	1	..	
		St. James, Westminster ..	24,995	..	..	2	..	..	..	..	..	..	..	2	..	
N. District.		St. Marylebone	142,404	4	..	5	..	2	1	3	..	5	..	2	1	
		Hampstead	68,416	1	..	1	..	1	..	5	..	..	..	1	..	
		St. Pancras	234,379	7	..	9	4	12	3	13	2	12	3	15	5	1
		Islington	319,143	12	4	9	1	12	..	3	1	6	..	5	..	
		St. Mary, Stoke Newington	30,936	2	..	..	..	1	..	..	..	..	..	3	1	
		Hackney	198,606	26	2	16	3	16	1	18	2	15	3	26	1	
Central District.		St. Giles and St. George, Bloomsbury.	39,782	2	..	..	..	1	1	..	..	..	..	2	1	1
		St. Martin-in-the-Fields ..	14,616	..	..	..	..	1	..	..	..	1	..	..	..	..
		Strand†	25,217	.	..	..	..	..	..	..	..	..	..	..	..	..
		Holborn‡	34,043	5	..	2	..	4	..	2	1	5	..	3	..	
		Clerkenwell	66,216	5	1	..	..	5	..	1	..	1	..	3	..	
		St. Luke, Middlesex ..	42,440	4	1	4	..	2	..	7	..	6	1	3	..	
		London, City of§	37,583	..	..	..	..	2	..	1	..	2	1	2	1	
E. District.		Shoreditch	124,009	6	1	2	..	6	..	5	1	7	..	10	1	[illegible]
		Bethnal Green	129,132	7	1	9	3	9	..	6	..	11	2	16	1	
		Whitechapel‖	74,420	6	..	5	..	..	..	4	..	3	..	3	..	
		St. George-in-the-East ..	45,795	1	1	..	..	3	..	2	1	2	1	1	1	
		Limehouse	57,376	4	3	3	1	4	1	3	1	4	..	6	..	
		Mile End Old Town ..	107,592	9	2	6	..	4	1	3	4	6	..	..	..	
		Poplar	166,748	12	1	18	3	7	2	5	1	7	1	7	..	[illegible]
S. District.		St. Saviour, Southwark ..	27,177	2	..	2	..	..	..	2	..	7	1	3	1	
		St. George, Southwark ..	59,712	5	..	4	2	5	..	3	1	5	1	3	..	
		Newington	115,804	12	1	13	1	7	3	11	..	7	1	21	2	[illegible]
		St. Olave, Southwark ..	12,723	..	..	..	..	1	..	..	..	..	..	..	..	[illegible]
		Bermondsey	84,682	12	1	6	1	15	1	15	4	14	3	16	1	[illegible]
		Rotherhithe	39,255	3	..	..	1	1	..	..	..	..	..	..	..	
		Lambeth	275,203	25	2	26	2	24	3	23	3	27	2	20	3	[illegible]
		Battersea	150,558	13	1	20	5	36	6	29	0	27	..	22	3	
		Wandsworth	156,942	14	3	15	3	15	1	15	..	13	1	11	..	
		Camberwell	235,344	14	1	12	3	21	1	14	3	17	2	17	2	
		Greenwich	165,413	19	..	19	1	13	1	13	1	7	2	13	..	
		Lewisham	92,647	4	1	8	1	10	..	4	2	7	..	8	..	
		Woolwich	40,848	2	..	2	1	2	..	1	1	6	..	8	..	
		Plumstead	52,436	2	..	1	..	3	..	3	..	2	..	3	..	
		Lee	36,103	5	1	7	1	9	..	2	..	7	..	1	2	
		Port of London	..	..	..	..	..	1	..	..	..	1	..	..	..	.

* Including St. Peters, Westminster (population, 235). † Including Middle Temple (populat
‡ Including Gray's Inn (population, 253), Lincoln's Inn (population, 27), Charterhouse (populati
Staple Inn (population, 21), and Furnival's Inn (population, 131).

Nov. 26. Cases.	Nov. 26. Deaths.	Dec. 3. Cases.	Dec. 3. Deaths.	Dec. 10. Cases.	Dec. 10. Deaths.	Dec. 17. Cases.	Dec. 17. Deaths.	Dec. 24. Cases.	Dec. 24. Deaths.	Dec. 31. Cases.	Dec. 31. Deaths.	Totals for 4th Quarter, 1898. Cases.	Totals for 4th Quarter, 1898. Deaths.	Grand Totals for Year 1898. Cases.	Grand Totals for Year 1898. Deaths.	Sanitary Areas.
DIPHTHERIA—*continued.* Weekly Statement, 4th Quarter, 1898—*continued.*																
241	37	269	43	276	43	273	39	248	46	248	46	3,491	506	11,561	1,756	London. (*Administrative County.*)
2	..	3	..	2	..	..	1	4	..	3	..	45	3	215	26	Kensington.
12	..	4	..	8	1	7	2	5	1	8	1	119	14	465	61	Fulham.
9	..	3	2	4	..	..	1	..	..	..	..	43	7	160	23	Hammersmith.
6	1	8	2	14	1	13	3	10	2	6	1	98	22	256	61	Paddington.
7	..	7	..	4	..	6	..	2	..	2	..	48	4	268	41	Chelsea.
1	..	3	..	4	..	1	..	5	1	..	..	32	4	110	10	St. George, Hanover Sq.*
4	..	..	..	3	..	2	..	..	..	..	..	25	1	125	22	Westminster.
..	..	..	..	..	..	1	..	1	1	..	..	8	2	41	5	St. James, Westminster.
6	2	2	1	6	1	6	..	4	1	4	3	56	10	232	45	St. Marylebone.
3	..	6	..	3	1	2	..	5	2	6	..	36	3	133	19	Hampstead.
13	4	18	6	20	6	9	3	9	5	10	1	164	43	485	96	St. Pancras.
5	1	1	..	9	1	9	1	18	1	4	1	100	12	533	90	Islington.
1	..	5	..	1	..	1	..	4	..	3	..	23	1	53	5	St. Mary, Stoke Newington.
9	1	15	..	16	1	24	1	14	2	16	2	220	21	855	120	Hackney.
..	..	..	..	1	..	..	..	..	..	..	..	7	2	40	6	St. Giles and St. George, Bloomsbury.
..	..	..	..	..	..	..	..	..	..	..	..	2	..	8	3	St. Martin-in-the-Fields.
1	..	1	..	2	..	1	..	3	..	1	..	9	..	45	4	Strand.†
5	1	3	..	..	..	1	..	2	..	1	..	34	3	144	22	Holborn.‡
6	..	1	..	2	..	5	1	7	1	5	..	48	3	264	36	Clerkenwell.
9	2	3	1	2	..	1	1	6	..	2	..	55	7	186	24	St. Luke, Middlesex.
3	1	2	1	1	1	..	..	..	..	..	..	14	6	44	10	London, City of.§
5	1	5	2	8	1	4	..	3	1	8	1	82	11	246	44	Shoreditch.
4	3	7	2	9	2	12	1	9	2	5	2	107	21	321	62	Bethnal Green.
4	..	6	..	..	..	5	2	7	2	6	..	51	4	174	20	Whitechapel.‖
1	..	..	..	3	1	3	..	1	..	1	..	19	5	105	14	St. George-in-the-East.
3	..	2	..	5	2	3	1	7	1	2	1	46	12	134	26	Limehouse.
7	1	4	2	6	1	8	3	3	1	8	2	70	23	279	63	Mile End Old Town.
6	..	7	2	4	..	15	2	5	..	8	2	111	14	458	89	Poplar.
12	..	2	..	3	1	1	..	3	1	6	..	47	5	114	15	St. Saviour, Southwark.
5	1	8	1	8	2	9	2	10	1	5	..	75	11	237	36	St. George, Southwark.
13	3	13	1	12	2	18	..	10	4	11	6	162	26	451	72	Newington.
..	..	..	..	..	..	..	..	..	..	..	..	1	..	16	3	St. Olave, Southwark.
8	3	13	1	10	2	4	..	6	..	9	6	138	25	259	44	Bermondsey.
1	..	2	1	1	..	..	..	..	..	1	..	10	2	41	11	Rotherhithe.
18	4	23	5	21	3	31	6	22	5	32	3	323	45	812	114	Lambeth.
14	4	24	1	22	3	26	3	27	6	14	5	295	47	786	120	Battersea.
11	..	20	4	13	3	11	..	6	..	14	3	173	20	593	89	Wandsworth.
9	1	14	3	29	3	5	2	13	..	28	3	201	26	670	87	Camberwell.
10	..	15	2	7	1	6	1	9	3	8	2	158	14	538	44	Greenwich.
4	1	11	1	3	1	16	1	4	1	9	..	99	9	280	26	Lewisham.
3	3	6	2	5	1	3	1	1	..	1	..	52	11	109	19	Woolwich.
3	..	..	..	..	..	1	..	..	..	4	1	24	1	111	15	Plumstead.
1	..	3	..	5	1	4	..	3	1	3	..	54	6	161	14	Lee.
1	..	..	..	..	..	..	..	..	..	..	..	3	..	4	..	Port of London.

§ Including Inner Temple (population, 86).

‖ Including Tower of London (population, [illegible]).

	Sanitary Areas.	Population (1891).	Jan. 8.		Jan. 15.		Jan. 22.		Jan. 29.		Feb. 5.		Feb. 12.	
			Cases.	Deaths.	Cases.	Deaths.	Cases.	Deaths.	Cases.	Deaths.	Cases.	Deaths.	Cases.	Deaths.
	London .. .. (*Administrative County*)	4,232,118	52	10	47	9	34	7	48	9	28	8	34	[illegible]
W. District.	Kensington	166,308	1	1	1	..	1	..	..	..	..	..	2	..
	Fulham	91,639	..	1	..	..	1	..	5	..	1	..	..	..
	Hammersmith	97,239	1	..	..	..	1	..	1	..	..	..	..	..
	Paddington	117,846	1	..	1	1	1	..	1	2	4	1	..	..
	Chelsea	96,253	1	..	..	..	2	..	..	..	1	..	..	..
	St. George, Hanover Sq.* ..	78,599	..	..	1	..	1	..	1	1	..	..	..	..
	Westminster	55,539	1	..	..	..	..	..	..	..	..	..	..	..
	St. James, Westminster ..	24,995	..	..	1	..	..	..	..	..	..	..	..	..
N. District.	St. Marylebone	142,404	2	..	1	..	..	1	1	..	..	..	..	..
	Hampstead	68,416	..	..	..	..	..	..	..	..	..	..	..	..
	St. Pancras	234,379	5	1	5	..	2	..	4	2	1	1	2	1
	Islington	319,143	11	1	3	..	3	1	..	..	3	..	5	..
	St. Mary, Stoke Newington	30,936	..	..	..	..	..	..	..	..	..	..	..	..
	Hackney	198,606	9	..	4	2	2	1	2	..	..	..	1	..
Central District.	St. Giles and St. George, Bloomsbury.	39,782	..	..	..	..	..	..	..	..	..	..	..	..
	St. Martin-in-the-Fields ..	14,616	..	..	..	..	..	..	..	..	..	..	..	..
	Strand†	25,217	..	..	..	..	..	..	..	..	..	..	..	..
	Holborn‡	34,043	..	..	..	..	..	..	1	..	..	..	..	1
	Clerkenwell	66,216	1	..	1	1	..	..	1	..	1	..	1	..
	St. Luke, Middlesex ..	42,440	..	1	..	..	..	..	1	..	..	..	..	..
	London, City of§	37,583	2	..	1	..	..	..	2	..	..	1	1	..
E. District.	Shoreditch	124,009	..	..	3	..	1	..	2	1	1	..	1	..
	Bethnal Green	129,132	..	1	1	..	3	..	2	..	2	..	1	..
	Whitechapel‖	74,420	..	..	1	..	..	..	1	..	2	1	3	..
	St. George-in-the-East ..	45,795	1	..	1	..	1	1	..	..	1	..	1	..
	Limehouse	57,376	..	..	..	..	1	..	..	..	..	..	..	..
	Mile End Old Town ..	107,592	..	1	..	..	..	..	2	..	..	..	2	..
	Poplar	166,748	3	..	7	1	1	2	1	..	1	..	3	3
S. District.	St. Saviour, Southwark ..	27,177	..	..	..	..	..	..	..	..	..	..	..	..
	St. George, Southwark ..	59,712	..	..	2	..	..	..	2	..	..	1	..	..
	Newington	115,804	2	..	..	..	1	..	..	..	2	..	1	..
	St. Olave, Southwark ..	12,723	..	..	..	..	..	..	..	..	..	..	..	..
	Bermondsey	84,682	2	1	..	..	1	1	2	..	..	..	..	..
	Rotherhithe	39,255	1	..	1	..	..	..	..	..	..	..	1	
	Lambeth	275,203	2	..	1	3	2	..	4	..	1	..	1	
	Battersea	150,558	..	..	1	1	1	..	3	1	..	..	1	
	Wandsworth	156,942	1	1	2	..	1	..	3	..	3	1	2	
	Camberwell	235,344	..	..	2	..	3	..	1	1	..	..	2	
	Greenwich	165,413	..	..	2	..	4	..	5	1	3	2	2	
	Lewisham	92,647	3	1	2	..	..	..	..	..	..	..	1	
	Woolwich	40,848	1	..	1	..	..	..	..	..	..	..	..	
	Plumstead	52,436	1	..	..	..	..	..	..	..	1	..	..	
	Lee	36,103	..	..	..	..	..	..	..	..	..	..	1	
	Port of London	..	..	..	1	..	..	..	..	..	..	..	..	

* Including St. Peter, Westminster (population, 235). † Including Middle Temple (population, [illegible]
‡ Including [illegible] Inn (population, [illegible]), Lincoln's Inn (population, 27), Charterhouse (population, [illegible]
[illegible] Inn (population, [illegible]), and Furnival's Inn (population, 121).

Enteric Fever—continued.

Feb. 19. Cases.	Feb. 19. Deaths.	Feb. 26. Cases.	Feb. 26. Deaths.	Mar. 5. Cases.	Mar. 5. Deaths.	Mar. 12. Cases.	Mar. 12. Deaths.	Mar. 19. Cases.	Mar. 19. Deaths.	Mar. 26. Cases.	Mar. 26. Deaths.	April 2. Cases.	April 2. Deaths.	Totals for 1st Quarter, 1898. Cases.	Totals for 1st Quarter, 1898. Deaths.	Sanitary Areas.
						Weekly Statement, 1st Quarter, 1898—continued.										
35	8	32	9	40	11	41	6	30	10	30	1	30	7	481	101	London. (*Administrative County.*)
..	..	2	..	..	..	2	..	1	..	2	..	1	..	13	1	Kensington.
1	..	..	..	..	..	..	..	..	..	3	..	2	..	13	1	Fulham.
..	..	2	..	1	..	..	..	2	1	1	..	..	..	9	1	Hammersmith.
1	..	1	..	1	..	1	..	2	2	1	..	1	..	16	6	Paddington.
1	..	1	..	3	..	1	2	..	..	1	..	..	1	11	3	Chelsea.
1	..	..	..	1	..	..	1	..	..	1	..	..	..	6	2	St. George, Hanover Sq.*
..	..	..	..	..	..	..	..	..	..	1	..	..	..	2	..	Westminster.
..	..	..	..	..	..	..	..	..	..	..	..	..	..	1	..	St. James, Westminster.
1	..	2	1	1	1	2	..	1	1	1	..	2	1	14	5	St. Marylebone.
..	..	..	..	..	..	..	..	..	..	..	..	1	..	1	..	Hampstead.
2	1	1	1	4	1	5	..	..	..	2	..	3	1	36	9	St. Pancras.
5	1	2	..	4	..	1	..	3	1	..	..	1	..	41	4	Islington.
..	..	..	..	..	..	..	..	..	..	..	..	..	..	..	..	St.Mary,StokeNewington.
1	..	1	..	6	3	2	..	..	..	4	..	2	2	34	8	Hackney.
..	..	..	..	1	..	..	..	1	1	..	..	..	..	2	1	St. Giles and St. George, Bloomsbury.
..	..	..	..	..	..	1	..	..	..	..	..	..	1	1	1	St. Martin-in-the-Fields.
1	1	..	..	..	..	..	..	..	..	1	..	..	..	2	1	Strand.†
..	..	1	..	..	..	1	..	..	..	..	..	..	..	3	1	Holborn.‡
..	..	..	1	..	..	1	..	..	..	..	..	..	..	6	2	Clerkenwell.
..	..	..	..	1	..	..	..	1	..	..	..	..	..	3	1	St. Luke, Middlesex.
..	..	..	..	..	..	..	..	1	..	..	..	..	..	7	1	London, City of.§
..	2	..	..	1	..	1	..	1	..	..	..	1	..	12	3	Shoreditch.
..	..	1	..	..	..	..	..	2	..	..	..	1	..	13	1	Bethnal Green.
1	..	2	..	..	..	1	..	1	..	1	..	..	..	12	1	Whitechapel.‖
1	..	..	..	1	..	..	..	..	..	2	..	2	..	11	1	St. George-in-the-East.
..	..	..	..	1	..	..	..	1	..	..	..	1	..	4	..	Limehouse.
..	..	4	..	..	1	..	..	..	1	1	..	..	..	9	3	Mile End Old Town.
3	..	..	..	3	..	4	..	1	..	2	..	..	..	29	5	Poplar.
..	..	..	..	..	..	..	..	..	..	..	..	..	..	..	..	St. Saviour, Southwark.
..	1	..	..	1	..	..	..	..	..	..	..	..	..	5	2	St. George, Southwark.
2	..	..	..	..	..	3	..	3	..	1	1	2	1	17	2	Newington.
..	..	..	..	..	..	..	..	..	..	..	..	1	..	1	..	St. Olave, Southwark.
1	..	1	1	..	..	1	1	..	..	..	..	..	..	8	4	Bermondsey.
1	..	..	..	..	..	..	..	..	..	..	..	..	..	4	..	Rotherhithe.
3	..	1	1	3	1	3	1	3	1	2	..	2	..	28	8	Lambeth.
2	..	1	..	1	..	..	..	3	1	..	..	1	..	14	3	Battersea.
1	..	1	..	..	..	2	..	..	..	..	..	2	..	18	2	Wandsworth.
2	..	2	2	4	2	2	..	..	..	1	..	1	..	20	5	Camberwell.
2	1	2	1	1	1	5	..	1	1	2	..	2	..	31	8	Greenwich.
..	..	..	..	..	..	1	..	2	..	..	..	1	..	10	1	Lewisham.
..	..	..	..	..	..	..	..	..	..	..	..	..	..	2	..	Woolwich.
1	..	2	1	..	..	1	..	..	..	..	..	..	..	6	1	Plumstead.
1	1	2	..	1	1	..	1	..	..	..	..	..	..	5	3	Lee.
..	..	..	..	..	..	..	..	..	..	..	..	..	..	1	..	Port of London.

† Including Inner Temple (population, 98).

‡ Including Tower of London (population, 888).

District	Sanitary Areas.	Population (1891).	Enteric Fever—continued. Weekly Statement, 2nd Quarter, 1892. April 9. Cases.	April 9. Deaths.	April 16. Cases.	April 16. Deaths.	April 23. Cases.	April 23. Deaths.	April 30. Cases.	April 30. Deaths.	May 7. Cases.	May 7. Deaths.	May [cut off] Cases.
	London .. .. (*Administrative County*) ..	4,232,118	19	9	22	6	29	5	32	7	28	4	24
W. District.	Kensington	166,308	..	..	2	1	2	..	1	..	3	..	4
	Fulham	91,639	..	..	..	..	..	..	..	..	1	..	..
	Hammersmith	97,239	..	..	..	..	..	..	..	..	2	..	..
	Paddington	117,846	..	1	..	..	..	..	1	1	..	..	1
	Chelsea	96,253	..	1	..	..	1	..	..	1	..	..	..
	St. George, Hanover Sq.* ..	78,599	1	..	1	1	1	..	1	..	..	..	..
	Westminster	55,539	1	..	..	..	..	..	..	..	1	..	..
	St. James, Westminster ..	24,995	1	..	..	..	..	..	..	..	..	..	1
N. District.	St. Marylebone	142,404	..	..	2	..	2	..	..	..	..	..	1
	Hampstead	68,416	1	..	..	..	..	1	..	..	..	..	..
	St. Pancras	234,379	..	..	2	..	2	..	2	..	1	..	2
	Islington	319,143	2	..	7	..	1	1	2	1	..	1	..
	St. Mary, Stoke Newington	30,936	..	..	..	..	..	..	..	..	..	..	..
	Hackney	198,606	1	..	3	1	5	..	1	..	1	..	..
Central District.	St. Giles and St. George, Bloomsbury.	39,783	..	1	..	..	1	..	..	..	1	..	..
	St. Martin-in-the-Fields ..	14,616	..	..	..	..	..	..	2	..	1	..	1
	Strand†	25,217	..	..	..	..	..	..	..	..	..	..	..
	Holborn‡	34,043	..	..	..	..	..	..	1	..	..	..	..
	Clerkenwell	66,216	..	..	..	..	..	..	..	..	..	..	..
	St. Luke, Middlesex ..	42,440	..	..	..	..	..	..	..	..	..	..	..
	London, City of§	37,583	..	..	..	..	1	..	..	..	..	..	..
E. District.	Shoreditch	124,009	1	1	..	..	2	..	1	..	..	..	2
	Bethnal Green	129,132	..	..	3	..	2	..	3	2	1	1	1
	Whitechapel‖	74,420	..	1	..	..	1	..	3	..	..	..	1
	St. George-in-the-East ..	45,795	..	..	..	..	..	..	..	..	4	..	2
	Limehouse	57,376	..	1	1	..	..	..	..	..	1	..	..
	Mile End Old Town ..	107,592	..	..	..	..	..	..	2	..	..	..	..
	Poplar	166,748	3	1	2	1	3	..	..	..	1	..	1
S. District.	St. Saviour, Southwark ..	27,177	..	..	..	..	1	..	1	..	..	..	..
	St. George, Southwark ..	59,712	..	..	1	..	..	..	..	..	..	..	..
	Newington	115,804	..	1	..	..	..	..	1	..	..	..	1
	St. Olave, Southwark ..	12,723	..	..	..	..	..	..	..	..	..	..	..
	Bermondsey	84,682	..	..	..	..	..	..	1	..	..	..	..
	Rotherhithe	39,255	..	..	1	..	..	..	..	..	..	..	..
	Lambeth	275,203	3	1	1	1	2	1	3	2	2	..	3
	Battersea	150,558	..	..	..	..	2	..	..	..	..	..	..
	Wandsworth	156,942	2	..	..	..	..	1	1	..	2	..	1
	Camberwell	235,344	..	..	1	1	2	..	3	..	2	1	..
	Greenwich	165,413	2	..	5	..	6	1	3	..	4	1	2
	Lewisham	92,647	1	..	..	..	1	..	..	..	..	..	..
	Woolwich	40,818	..	..	..	..	1	..	..	..	..	..	..
	Plumstead	52,436	..	..	..	..	..	..	..	..	..	..	..
	Lee	26,108	..	..	..	..	..	..	..	..	..	..	..
	Port of London	..	..	..	..	..	..	..	..	..	..	..	..

* Including St. Peters, Westminster (population, 295). † Including Middle Temple (populatio[cut off]
‡ Including Gray's Inn (population, 253), L[illegible] (population, 27), Charterhouse (population,
Staple Inn (population, 21), and Furn[illegible] [illegible]on, 133).

Enteric Fever—continued.

Weekly Statement, 1st Quarter, 1898—continued.													Totals for 1st Quarter, 1898.		Sanitary Areas.
18.	Feb. 26.		Mar. 5.		Mar. 12.		Mar. 19.		Mar. 26.		April 2.				
Deaths.	Cases.	Deaths.	Cases.	Deaths.	Cases.	Deaths.	Cases.	Deaths.	Cases.	Deaths.	Cases.	Deaths.	Cases.	Deaths.	
8	33	9	40	11	41	6	30	10	30	1	30	7	481	101	London. (*Administrative County.*)
..	2	..	..	..	2	..	1	..	2	..	1	..	13	1	Kensington.
..	..	..	..	..	..	..	..	..	3	..	2	..	13	1	Fulham.
..	2	..	1	..	..	..	2	1	1	..	..	..	9	1	Hammersmith.
..	1	..	1	..	1	..	2	2	1	..	1	..	16	6	Paddington.
..	1	..	3	..	1	2	..	..	1	..	..	1	11	3	Chelsea.
..	..	..	1	..	..	1	..	..	1	..	..	..	6	2	St. George, Hanover Sq.*
..	..	..	..	..	..	..	..	..	1	..	..	..	2	..	Westminster.
..	..	..	..	..	..	..	..	..	..	..	..	..	1	..	St. James, Westminster.
..	2	1	1	1	2	..	1	1	1	..	2	1	14	5	St. Marylebone.
..	..	..	..	..	..	..	..	..	..	..	1	..	1	..	Hampstead.
1	1	1	4	1	5	..	..	..	2	..	3	1	36	9	St. Pancras.
1	2	..	4	..	1	..	3	1	..	..	1	..	41	4	Islington.
..	..	..	..	..	..	..	..	..	..	..	..	..	..	..	St.Mary,StokeNewington.
..	1	..	6	3	2	..	..	..	4	..	2	2	34	8	Hackney.
..	..	..	1	..	..	..	1	1	..	..	..	..	2	1	St. Giles and St. George Bloomsbury.
..	..	..	..	..	1	..	..	..	..	..	..	1	1	1	St. Martin-in-the-Fields.
1	..	..	..	..	..	..	..	..	1	..	..	..	2	1	Strand.†
..	1	..	..	..	1	..	..	..	..	..	..	..	3	1	Holborn.‡
..	..	1	..	..	1	..	..	..	..	..	..	..	6	2	Clerkenwell.
..	..	..	1	..	..	..	1	..	..	..	..	..	3	1	St. Luke, Middlesex.
..	..	..	..	..	..	..	1	..	..	..	..	..	7	1	London, City of.§
2	..	..	1	..	1	..	1	..	..	..	1	..	12	3	Shoreditch.
..	1	..	..	..	..	..	2	..	..	..	1	..	13	1	Bethnal Green.
..	2	..	..	..	1	..	1	..	1	..	..	..	12	1	Whitechapel.‖
..	..	..	1	..	..	..	..	..	2	..	2	..	11	1	St. George-in-the-East.
..	..	..	1	..	..	..	1	..	..	..	1	..	4	..	Limehouse.
..	4	..	..	1	..	..	..	1	1	..	..	..	9	3	Mile End Old Town.
4.	..	..	3	..	4	..	1	..	2	..	..	..	29	5	Poplar.
..	..	..	..	..	..	..	..	..	..	..	..	..	..	..	St. Saviour, Southwark.
1	..	..	1	..	..	..	..	..	..	..	..	..	5	2	St. George, Southwark.
..	..	..	..	..	3	..	3	..	1	1	2	1	17	2	Newington.
..	..	..	..	..	..	..	..	..	..	..	1	..	1	..	St. Olave, Southwark.
..	1	1	..	..	1	1	..	..	..	..	..	..	8	4	Bermondsey.
..	..	..	..	..	..	..	..	..	..	..	..	..	4	..	Rotherhithe.
..	1	1	3	1	3	1	3	1	2	..	2	..	28	8	Lambeth.
..	1	..	1	..	..	..	3	1	..	..	1	..	14	3	Battersea.
..	1	..	..	..	2	..	..	..	..	..	2	..	18	2	Wandsworth.
..	2	2	4	2	2	..	..	..	1	..	1	..	20	5	Camberwell.
1	3	1	1	1	5	..	1	1	2	..	2	..	31	8	Greenwich.
..	..	..	..	..	1	..	2	..	..	..	1	..	10	1	Lewisham.
..	..	..	..	..	..	..	..	..	..	..	..	..	2	..	Woolwich.
..	3	1	..	..	1	..	..	..	..	..	..	..	6	1	Plumstead.
1	3	..	1	1	..	1	..	..	..	..	..	..	5	3	Lee.
..	..	..	..	..	..	..	..	..	..	..	..	..	1	..	Port of London.

Including Inner Temple (population, 96). § Including Tower of London (population, 869).

Sanitary Areas.	Population (1891).	ENTERIC FEVER—continued. Weekly Statement, 3rd Quarter, 1898. July 9.		July 16.		July 23.		July 30.		Aug. 6.	
		Cases.	Deaths.	Cases.	Deaths.	Cases.	Deaths.	Cases.	Deaths.	Cases.	Deaths.
London .. .. (*Administrative County*)	4,232,118	35	7	29	4	46	8	44	6	49	10
W. District. Kensington	166,308	..	..	..	..	1	..	2	1	..	..
Fulham	91,639	1	..	1	..	..	1	1	..	1	1
Hammersmith	97,239	..	..	..	..	1	..	2	..	2	..
Paddington	117,846	..	..	..	..	3	..	..	..	..	..
Chelsea	96,253	1	..	2	..	1	..	..	1	..	1
St. George, Hanover Sq.* ..	78,590	..	..	..	..	2	..	..	..	1	..
Westminster	55,539	1	..	3	..	..	..	..	..	3	..
St. James, Westminster ..	24,995	..	..	..	..	..	..	..	..	..	..
N. District. St. Marylebone	142,404	3	..	3	1	1	1	2	..	1	..
Hampstead	68,416	3	..	..	..	2	..	1	..	1	..
St. Pancras	234,379	3	..	2	..	1	1	5	..	6	1
Islington	319,143	3	..	2	1	6	1	2	1	7	..
St. Mary, Stoke Newington	30,936	..	..	..	..	..	..	..	..	..	..
Hackney	198,606	2	..	5	1	4	..	4	..	1	2
Central District. St. Giles and St. George, Bloomsbury.	39,782	..	..	..	..	1	..	1	..	..	..
St. Martin-in-the-Fields ..	14,616	..	..	..	..	1	..	..	..	..	..
Strand†	25,217	..	..	..	..	..	..	..	..	..	..
Holborn‡	34,043	..	..	..	..	..	..	..	..	..	..
Clerkenwell	66,216	..	..	1	..	1	..	2	..	2	..
St. Luke, Middlesex ..	42,440	..	..	..	..	..	..	..	..	..	..
London, City of§	37,583	..	..	..	..	..	..	..	..	..	..
E. District. Shoreditch	124,009	..	..	1	..	1	..	1	..	4	..
Bethnal Green	129,132	..	..	..	..	2	1	1	..	1	..
Whitechapel‖	74,420	..	1	..	..	..	..	2	..	1	..
St. George-in-the-East ..	45,795	..	..	..	..	2	..	1	..	..	..
Limehouse	57,376	1	..	..	..	..	..	1	..	1	..
Mile End Old Town ..	107,592	1	1	1	..	..	..	2	..	..	..
Poplar	166,748	2	..	1	..	4	..	2	2	3	1
S. District. St. Saviour, Southwark ..	27,177	1	..	1	..	..	..	..	..	..	..
St. George, Southwark ..	59,712	1	1	..	..	..	..	..	..	..	..
Newington	115,804	..	..	1	..	..	..	1	1	..	..
St. Olave, Southwark ..	12,723	..	..	..	..	..	..	..	..	..	..
Bermondsey	84,682	2	..	1	..	..	1	..	..	..	..
Rotherhithe	39,255	..	..	..	..	..	..	1	..	1	..
Lambeth	275,203	2	..	..	..	4	..	2	..	4	1
Battersea	150,558	1	1	2	..	2	..	2	..	2	1
Wandsworth	156,942	3	..	1	..	2	1	..	..	1	..
Camberwell	235,344	1	..	..	..	..	..	2	..	2	2
Greenwich	165,413	2	1	..	..	4	..	2	..	3	..
Lewisham	92,647	..	..	..	..	..	..	1	..	..	..
Woolwich	40,848	..	1	..	..	..	..	..	..	1	..
Plumstead	52,436	..	..	1	..	..	..	..	..	..	..
Lee	36,108	1	1	..	1	..	1	..	..	..	..
Port of London	..	..	..	..	..	..	..	1	..	1	..

* Including St. Peters, Westminster (population, 235). † Including Middle Temple (popu

‡ *Including* Gray's Inn (population, 253), Lincoln's Inn (population, 27), Charterhouse (popul *Staple* Inn (population, 21), and Furnival's Inn (population, 131).

Enteric Fever—*continued*.														
Weekly Statement, 3rd Quarter, 1898—*continued*.												Totals for 3rd Quarter, 1898.		Sanitary Areas.
Aug. 27.		Sept. 3.		Sept. 10.		Sept. 17.		Sept. 24.		Oct. 1.				
Cases.	Deaths.	Cases.	Deaths.	Cases.	Deaths.	Cases.	Deaths.	Cases.	Deaths.	Cases.	Deaths.	Cases.	Deaths.	
29	8	53	14	67	19	34	7	60	7	82	7	697	113	London. (*Administrative County.*)
1	..	..	..	2	..	6	..	2	..	3	..	22	1	Kensington.
3	1	2	..	..	..	..	..	..	..	..	..	11	3	Fulham.
..	..	1	..	1	1	3	..	2	..	1	1	14	2	Hammersmith.
1	..	3	1	5	..	1	..	1	..	2	..	18	1	Paddington.
1	..	1	1	2	1	1	..	..	..	1	..	10	4	Chelsea.
..	..	1	..	..	1	..	..	..	..	..	..	5	2	St. George, Hanover Sq.*
..	..	..	..	2	..	2	..	..	1	3	..	16	1	Westminster.
..	..	..	..	..	..	..	..	..	..	..	..	..	..	St. James, Westminster.
1	..	1	..	..	1	1	..	..	..	3	..	20	3	St. Marylebone.
1	..	..	1	..	..	3	..	2	..	2	..	18	1	Hampstead.
4	..	3	..	7	2	4	..	3	..	6	2	50	8	St. Pancras.
..	1	2	2	5	1	6	..	4	..	9	1	60	11	Islington.
..	..	..	..	1	..	..	..	..	1	..	..	1	1	St. Mary, Stoke Newingto
2	2	5	3	3	1	3	1	5	..	8	..	53	10	Hackney.
1	..	..	..	..	..	..	..	..	..	1	..	5	..	St. Giles and St. George, Bloomsbury.
..	..	..	..	..	..	..	..	1	..	..	..	2	..	St. Martin-in-the-Fields.
..	..	..	..	..	..	..	..	1	..	1	..	2	..	Strand.†
..	..	..	..	1	..	..	..	..	..	..	..	1	..	Holborn.‡
1	1	..	..	1	1	2	..	..	1	..	..	11	3	Clerkenwell.
..	..	..	..	..	..	1	..	..	..	2	..	3	..	St. Luke, Middlesex.
..	..	1	..	..	..	1	1	..	..	..	..	2	1	London, City of.§
..	..	4	..	3	2	4	..	3	2	1	..	26	5	Shoreditch.
..	..	..	..	3	..	3	1	3	..	2	..	23	2	Bethnal Green.
..	..	2	1	1	1	..	..	2	..	2	..	12	4	Whitechapel.‖
2	..	1	1	..	..	..	..	1	..	2	..	10	2	St. George-in-the-East.
1	1	..	..	3	..	2	..	1	..	..	..	10	2	Limehouse.
..	..	3	..	1	1	1	1	3	..	1	..	16	3	Mile End Old Town.
1	..	3	1	6	..	6	..	1	..	5	..	36	5	Poplar.
..	..	1	..	..	..	..	..	..	..	1	1	4	1	St. Saviour, Southwark.
2	..	1	..	1	..	2	..	..	..	..	..	9	2	St. George, Southwark.
1	1	2	..	6	..	4	1	5	..	3	..	24	3	Newington.
..	..	1	..	..	..	..	..	..	..	..	..	1	..	St. Olave, Southwark.
..	..	3	..	..	1	2	..	..	..	1	..	14	2	Bermondsey.
..	..	1	..	..	..	..	..	1	..	..	..	4	..	Rotherhithe.
3	2	1	1	4	..	1	..	4	1	..	..	31	5	Lambeth.
1	..	..	..	1	..	4	..	1	..	3	..	20	2	Battersea.
4	..	2	1	3	1	2	..	3	..	2	..	29	4	Wandsworth.
2	..	4	..	2	1	2	..	1	..	3	1	19	4	Camberwell.
5	..	2	2	1	3	13	2	5	..	8	1	53	9	Greenwich.
..	..	3	..	..	..	1	..	1	..	2	..	10	..	Lewisham.
..	..	1	..	1	..	1	..	..	1	..	..	5	2	Woolwich.
..	..	..	..	..	..	..	..	2	..	1	..	5	..	Plumstead.
1	..	1	..	..	..	..	..	..	..	2	..	6	2	Lee.
..	..	1	..	2	..	..	..	2	..	2	..	9	..	Port of London.

…g Inner Temple (population, 96).

§ Including Tower of London (population, 868).

	Sanitary Areas.	Population (1891).	Enteric Fever—continued. Weekly Statement, 4th Quarter, 189[illegible].										
			Oct. 8.		Oct. 15.		Oct. 22.		Oct. 29.		Nov. 5.		Nov. 1[illegible]
			Cases.	Deaths.	Cases.	Deaths.	Cases.	Deaths.	Cases.	Deaths.	Cases.	Deaths.	Cases.
	London .. .. (Administrative County)	4,232,118	85	6	97	18	110	23	148	19	150	24	131
W. District.	Kensington	166,308	1	1	..	..	1	..	7	..	6	1	4
	Fulham	91,639	1	..	4	..	1	1	6	1	4	1	5
	Hammersmith	97,239	4	..	2	..	2	2	2	..	3	..	4
	Paddington	117,846	2	..	..	..	6	..	4	..	2	2	3
	Chelsea..	96,253	2	..	..	..	2	1	4	..	3	..	2
	St. George, Hanover Sq.* ..	78,599	..	..	1	1	1	..	2	..	1	..	1
	Westminster	55,539	1	..	3	..	1	..	4	1	4	..	1
	St. James, Westminster ..	24,995	..	..	..	..	..	..	..	..	..	..	..
N. District.	St. Marylebone	142,404	3	..	4	1	4	1	3	..	5	1	8
	Hampstead	68,416	2	..	2	1	..	2	1	..	..	1	4
	St. Pancras	234,379	6	1	8	5	7	1	12	1	10	..	18
	Islington	319,143	5	1	7	1	9	2	13	1	13	2	11
	St. Mary, Stoke Newington	30,936	..	..	1	..	2	..	..	..	2	..	1
	Hackney	198,606	9	..	13	2	16	1	7	3	8	..	14
Central District.	St. Giles and St. George, Bloomsbury.	39,782	1	..	3	..	..	..	1	..	..	1	5
	St. Martin-in-the-Fields ..	14,616	..	..	..	..	..	..	3	..	..	..	..
	Strand†	25,217	1	..	..	..	..	..	..	1	..	..	1
	Holborn‡	34,043	1	..	..	..	..	..	1	..	..	..	..
	Clerkenwell	66,216	..	..	2	..	..	..	5	..	1	..	1
	St. Luke, Middlesex ..	42,440	..	..	..	..	..	..	..	..	4	1	2
	London, City of§	37,583	..	..	..	..	..	..	1	..	1	..	1
E. District.	Shoreditch	124,009	3	1	2	..	4	1	4	..	4	..	6
	Bethnal Green	129,132	1	..	2	1	4	..	10	1	5	2	5
	Whitechapel‖	74,420	1	..	1	..	..	1	3	..	2	..	1
	St. George-in-the-East ..	45,795	4	..	..	..	..	..	2	1	8	1	..
	Limehouse	57,376	..	..	3	..	2	..	5	1	..	1	3
	Mile End Old Town ..	107,592	1	1	1	..	1	..	3	1	5	..	1
	Poplar	166,748	5	..	4	1	8	1	10	1	10	3	2
S. District.	St. Saviour, Southwark ..	27,177	1	..	1	..	..	..	..	..	..	..	1
	St. George, Southwark ..	59,712	2	..	..	1	2	..	5	..	..	..	1
	Newington	115,804	4	..	3	..	2	..	6	..	7	..	1
	St. Olave, Southwark ..	12,723	..	..	..	..	..	..	..	..	..	..	2
	Bermondsey	84.682	1	..	1	..	3	..	2	..	2	1	1
	Rotherhithe	39,255	1	..	..	..	1	..	..	..	3	..	..
	Lambeth	275,203	5	..	4	1	10	2	5	1	7	2	8
	Battersea	150,558	1	..	4	..	2	1	3	1	3	1	4
	Wandsworth	156,942	3	..	6	..	3	2	2	1	3	1	3
	Camberwell	235,344	4	..	1	1	2	..	4	..	8	..	4
	Greenwich	165,413	7	1	9	1	6	3	6	2	6	..	2
	Lewisham	92,647	..	..	2	..	2	..	..	..	2	1	1
	Woolwich	40,848	..	..	1	..	1	..	..	1	3	1	1
	Plumstead	52,436	..	..	2	1	2	..	2	..	1	..	2
	Lee	36,108	..	..	..	..	2	1	..	..	3	..	1
	Port of London	..	2	..	1	..	1	..	..	..	1	..	..

* Including St. Peters, Westminster (population, 236). † Including Middle Temple (po
‡ Including Gray's Inn (population, 265), Lincoln's Inn (population, 27), Charterhouse (popu
Inn (population, 21), and Furnival's Inn (population, 131).

Enteric Fever—*continued.*

Weekly Statement, 3rd Quarter, 1898—*continued.*

Aug. 20.		Aug. 27.		Sept. 3.		Sept. 10.		Sept. 17.		Sept. 24.		Oct. 1.		Totals for 3rd Quarter, 1898.		Sanitary Areas.
Cases.	Deaths.	Cases.	Deaths.	Cases.	Deaths.	Cases.	Deaths.	Cases.	Deaths.	Cases.	Deaths.	Cases.	Deaths.	Cases.	Deaths.	
57	13	39	9	58	14	67	19	84	7	60	7	82	7	697	113	London. (*Administrative County.*)
3	..	1	..	..	..	3	..	6	..	2	..	3	..	22	1	Kensington.
2	..	3	1	2	..	..	..	..	..	..	..	..	..	11	3	Fulham.
..	..	..	..	1	..	1	1	3	..	2	..	1	1	14	2	Hammersmith.
..	..	1	..	3	1	5	..	1	..	1	..	2	..	16	1	Paddington.
..	..	1	..	1	1	2	1	1	..	..	..	1	..	10	4	Chelsea.
1	1	..	..	1	..	..	1	..	..	..	..	..	..	5	2	St. George, Hanover Sq.*
3	..	..	..	..	..	2	..	2	..	..	1	2	..	16	1	Westminster.
..	..	..	..	..	..	..	..	..	..	..	..	..	..	..	..	St. James, Westminster.
2	..	1	..	1	..	..	1	1	..	..	..	3	..	20	3	St. Marylebone.
3	..	1	..	..	1	..	..	3	..	2	..	2	..	18	1	Hampstead.
4	2	4	..	3	..	7	2	4	..	3	..	6	2	50	8	St. Pancras.
4	2	..	1	8	2	5	1	8	..	4	..	9	1	60	11	Islington.
..	..	..	..	..	..	1	..	..	..	..	1	..	..	1	1	St.Mary,StokeNewingto
5	1	2	2	5	2	3	1	3	1	5	..	8	..	53	10	Hackney.
1	..	1	..	..	..	..	..	..	..	..	..	1	..	5	..	St. Giles and St. George, Bloomsbury.
..	..	..	..	..	..	..	..	..	..	1	..	..	..	2	..	St. Martin-in-the-Fields.
..	..	..	..	..	..	..	..	..	..	1	..	1	..	2	..	Strand.†
..	..	..	..	..	..	1	..	..	..	..	..	..	..	1	..	Holborn.‡
1	..	1	1	..	..	1	1	2	..	..	1	..	..	11	3	Clerkenwell.
..	..	..	..	..	..	..	..	1	..	..	..	2	..	3	..	St. Luke, Middlesex.
..	..	..	..	1	..	..	..	1	1	..	..	..	..	2	1	London, City of.§
2	1	..	..	4	..	3	2	4	..	3	2	1	..	26	5	Shoreditch.
2	..	..	..	..	..	3	..	3	1	3	..	2	..	22	2	Bethnal Green.
3	1	..	..	2	1	1	1	..	..	2	..	2	..	12	4	Whitechapel.‖
1	1	2	..	1	1	..	..	..	..	1	..	2	..	10	2	St. George-in-the-East.
1	..	1	1	..	..	2	..	2	..	1	..	..	..	10	2	Limehouse.
3	..	..	..	3	..	1	1	1	1	3	..	1	..	16	3	Mile End Old Town.
..	1	1	..	3	1	5	..	6	..	1	..	5	..	36	5	Poplar.
..	..	..	..	1	..	..	..	..	..	..	..	1	1	4	1	St. Saviour, Southwark.
3	1	2	..	1	..	1	..	2	..	..	..	..	..	9	2	St. George, Southwark.
..	..	1	1	2	..	6	..	4	1	5	..	3	..	24	3	Newington.
..	..	..	..	1	..	..	..	..	..	..	..	..	..	1	..	St. Olave, Southwark.
4	1	..	..	3	..	..	1	2	..	..	..	1	..	14	3	Bermondsey.
..	..	..	..	1	..	..	..	..	..	1	..	..	..	4	..	Rotherhithe.
2	..	3	2	1	1	4	..	1	..	4	1	..	..	31	5	Lambeth.
1	..	1	..	..	..	1	..	4	..	1	..	3	..	20	2	Battersea.
3	1	4	..	2	1	3	1	2	..	3	..	2	..	29	4	Wandsworth.
..	..	2	..	4	..	2	1	2	..	1	..	3	1	19	4	Camberwell.
4	..	5	..	2	2	1	3	13	2	5	..	8	1	53	9	Greenwich.
1	..	..	..	3	..	..	..	1	..	1	..	2	..	10	..	Lewisham.
..	..	..	..	1	..	1	..	1	..	..	1	..	..	5	2	Woolwich.
1	..	..	..	..	..	..	..	..	..	2	..	1	..	5	..	Plumstead.
1	..	1	..	1	..	..	..	..	..	..	..	2	..	6	3	Lee.
..	..	..	..	1	..	2	..	..	..	2	..	2	..	9	..	Port of London.

…cluding Inner Temple (population, 96).

‖ Including Tower of London (population, 868).

District	Sanitary Areas.	Population (1891).	Enteric Fever—continued. Weekly Statement, 4th Quarter, 1892. Oct. 8. Cases.	Oct. 8. Deaths.	Oct. 15. Cases.	Oct. 15. Deaths.	Oct. 22. Cases.	Oct. 22. Deaths.	Oct. 29. Cases.	Oct. 29. Deaths.	Nov. 5. Cases.	Nov. 5. Deaths.	Nov. 12. Cases.	Nov. 12. Deaths.	Nov. 19. Cases.	Nov. 19. Deaths.
	London (*Administrative County*)	4,232,118	85	6	97	18	110	23	148	19	150	24	131	26	118	27
W. District.	Kensington	166,308	1	1	..	..	1	..	7	..	6	1	4	..	2	..
	Fulham	91,639	1	..	4	..	1	1	6	1	4	1	5	1	3	1
	Hammersmith	97,239	4	..	2	..	2	2	2	..	3	..	4	1	4	1
	Paddington	117,846	2	..	..	..	6	..	4	..	2	2	3	2	3	1
	Chelsea	96,253	2	..	..	..	2	1	4	..	3	..	2	..	4	1
	St. George, Hanover Sq.*	78,599	..	..	1	1	1	..	2	..	1	..	1	..	1	..
	Westminster	55,539	1	..	3	..	1	..	4	1	4	..	1	..	2	..
	St. James, Westminster	24,995	..	..	..	..	..	..	..	..	..	..	..	..	1	..
N. District.	St. Marylebone	142,404	3	..	4	1	4	1	3	..	5	1	8	2	6	1
	Hampstead	68,416	2	..	2	1	..	2	1	..	..	1	4	..	3	..
	St. Pancras	234,379	6	1	8	5	7	1	12	1	10	..	13	2	6	3
	Islington	310,143	5	1	7	1	9	2	13	1	13	2	11	1	8	..
	St. Mary, Stoke Newington	30,936	..	..	1	..	2	..	..	..	2	..	1	..	1	..
	Hackney	198,606	9	..	13	2	16	1	7	3	8	..	14	3	6	4
Central District.	St. Giles and St. George, Bloomsbury.	39,782	1	..	3	..	..	..	1	..	..	1	5	1	..	..
	St. Martin-in-the-Fields	14,616	..	..	..	..	..	..	3	..	..	..	..	..	..	..
	Strand†	25,217	1	..	..	..	..	..	..	1	..	..	1	..	..	..
	Holborn‡	34,043	1	..	..	..	..	..	1	..	..	..	..	..	..	..
	Clerkenwell	66,216	..	..	2	..	..	..	5	..	1	..	1	1	2	..
	St. Luke, Middlesex	42,440	..	..	..	..	..	..	..	..	4	1	2	..	1	..
	London, City of§	37,583	..	..	..	..	..	..	1	..	1	..	1	1	1	..
E. District.	Shoreditch	124,009	3	1	2	..	4	1	4	..	4	..	6	2	3	..
	Bethnal Green	129,132	1	..	2	1	4	..	10	1	5	2	5	1	2	2
	Whitechapel‖	74,420	1	..	1	..	..	1	3	..	2	..	1	..	..	..
	St. George-in-the-East	45,795	4	..	..	..	..	..	2	1	8	1	..	1	1	..
	Limehouse	57,376	..	..	2	..	2	..	5	1	..	1	3	..	2	..
	Mile End Old Town	107,592	1	1	1	..	1	..	3	1	5	..	1	..	4	..
	Poplar	166,748	5	..	4	1	8	1	10	1	10	3	2	..	6	4
S. District.	St. Saviour, Southwark	27,177	1	..	1	..	..	..	..	..	..	..	1	..	..	1
	St. George, Southwark	59,712	2	..	..	1	2	..	5	..	..	..	1	..	3	..
	Newington	115,804	4	..	3	..	2	..	6	..	7	..	1	..	4	..
	St. Olave, Southwark	12,723	..	..	..	..	..	..	..	..	..	..	2	..	..	..
	Bermondsey	84,682	1	..	1	..	3	..	2	..	2	1	1	..	4	..
	Rotherhithe	39,255	1	..	..	..	1	..	..	..	3	..	..	..	..	..
	Lambeth	275,203	5	..	4	1	10	2	5	1	7	2	8	1	8	1
	Battersea	150,558	1	..	4	..	2	1	3	1	3	1	4	..	5	..
	Wandsworth	156,942	3	..	6	..	3	2	2	1	3	1	3	..	5	..
	Camberwell	235,344	4	..	1	1	2	..	4	..	8	..	4	2	5	2
	Greenwich	165,413	7	1	9	1	6	3	6	2	6	..	2	2	4	1
	Lewisham	92,647	..	..	2	..	2	..	..	..	2	1	1	..	5	1
	Woolwich	40,848	..	..	1	..	1	..	..	1	3	1	1	2	1	..
	Plumstead	52,436	..	..	2	1	2	..	2	..	1	..	2	..	1	..
	Lee	36,103	..	..	..	..	2	1	..	..	3	..	1	..	..	..
	Port of London	..	2	..	1	..	1	..	..	..	1	..	..	..	1	..

* Including St. Peters, Westminster (population, 235). † Including Middle Temple (population, 85).
‡ Including Gray's Inn (population, 253), Lincoln's Inn (population, 27), Charterhouse (population, 136), S[illegible] Inn (population, 21), and Furnival's Inn (population, 121).

	SIMPLE CONTINUED FEVER—*continued.*															
	Weekly Statement, 2nd Quarter, 1898—*continued.*												Totals for 2nd Quarter, 1898.		Sanitary Areas.
21.	May 28.		June 4.		June 11.		June 18.		June 25.		July 2.				
Deaths.	Cases.	Deaths.	Cases.	Deaths.	Cases.	Deaths.	Cases.	Deaths.	Cases.	Deaths.	Cases.	Deaths.	Cases.	Deaths.	
..	2	..	1	..	..	..	1	..	1	..	1	..	12	1	London. (*Administrative County.*)
..	..	..	..	..	..	..	..	..	..	..	..	..	1	..	Kensington.
..	..	..	..	..	..	..	..	..	..	..	..	..	..	..	Fulham.
..	..	..	..	..	..	..	..	..	..	..	..	..	1	..	Hammersmith.
..	..	..	..	..	..	..	..	..	..	..	..	..	..	..	Paddington.
..	..	..	..	..	..	..	..	..	..	..	..	..	1	..	Chelsea.
..	..	..	..	..	..	..	..	..	..	..	..	..	..	..	St. George, Hanover Sq.*
..	..	..	..	..	..	..	..	..	..	..	..	..	..	..	Westminster.
..	..	..	..	..	..	..	..	..	..	..	..	..	..	..	St. James, Westminster.
..	..	..	..	..	..	..	..	..	..	..	..	..	..	..	St. Marylebone.
..	..	..	..	..	..	..	..	..	..	..	..	..	..	..	Hampstead.
..	..	..	..	..	..	..	..	..	..	..	..	..	..	..	St. Pancras.
..	..	..	..	..	..	..	..	..	..	..	..	..	..	..	Islington.
..	..	..	..	..	..	..	..	..	..	..	..	..	..	..	St. Mary, Stoke Newington
..	..	..	..	..	..	..	..	..	..	..	..	..	..	..	Hackney.
..	..	..	..	..	..	..	..	..	..	..	..	..	..	..	St. Giles and St. George, Bloomsbury.
..	..	..	..	..	..	..	..	..	..	..	..	..	..	..	St. Martin-in-the-Fields.
..	..	..	..	..	..	..	..	..	..	..	..	..	..	..	Strand.†
..	..	..	..	..	..	..	..	..	..	..	..	..	..	..	Holborn.‡
..	..	..	..	..	..	..	..	..	..	..	..	..	..	..	Clerkenwell.
..	..	..	..	..	..	..	..	..	..	..	..	..	..	..	St. Luke, Middlesex.
..	..	..	..	..	..	..	..	..	..	..	..	..	..	..	London, City of.§
..	..	..	..	..	..	..	..	..	..	..	..	..	..	..	Shoreditch.
..	..	..	..	..	..	..	..	..	..	..	..	..	..	..	Bethnal Green.
..	..	..	..	..	..	..	..	..	..	..	..	..	..	..	Whitechapel.‖
..	..	..	..	..	..	..	..	..	..	..	..	..	..	..	St. George-in-the-East.
..	..	..	..	..	..	..	..	..	..	..	..	..	..	..	Limehouse.
..	..	..	..	..	..	..	..	..	..	..	..	..	1	..	Mile End Old Town.
..	2	..	..	..	..	..	..	..	..	..	..	..	2	..	Poplar.
..	..	..	..	..	..	..	..	..	..	..	..	..	..	..	St. Saviour, Southwark.
..	..	..	..	..	..	..	..	..	..	..	..	..	1	..	St. George, Southwark.
..	..	..	..	..	..	..	..	..	..	..	..	..	..	..	Newington.
..	..	..	..	..	..	..	..	..	..	..	..	..	..	..	St. Olave, Southwark.
..	..	..	..	..	..	..	..	..	..	..	..	..	..	..	Bermondsey.
..	..	..	..	..	..	..	..	..	..	..	..	..	..	..	Rotherhithe.
..	..	..	1	..	..	..	1	..	1	..	1	..	5	..	Lambeth.
..	..	..	..	..	..	..	..	..	..	..	..	..	..	1	Battersea.
..	..	..	..	..	..	..	..	..	..	..	..	..	..	..	Wandsworth.
..	..	..	..	..	..	..	..	..	..	..	..	..	..	..	Camberwell.
..	..	..	..	..	..	..	..	..	..	..	..	..	..	..	Greenwich.
..	..	..	..	..	..	..	..	..	..	..	..	..	..	..	Lewisham
..	..	..	..	..	..	..	..	..	..	..	..	..	..	..	Woolwich.
..	..	..	..	..	..	..	..	..	..	..	..	..	..	..	Plumstead.
..	..	..	..	..	..	..	..	..	..	..	..	..	..	..	Lee.
..	..	..	..	..	..	..	..	..	..	..	..	..	..	..	Port of London.

…uding Inner Temple (population, 86). ‖ Including Tower of London (population, [illegible]).

	Sanitary Areas.	Population (1891).	SIMPLE CONTINUED FEVER—continued.											
			Weekly Statement, 3rd Quarter, 1898.											
			July 9.		July 16.		July 23.		July 30.		Aug. 6.		Au	
			Cases.	Deaths.	Cases.	Deaths.	Cases.	Deaths.	Cases.	Deaths.	Cases.	Deaths.	Cases.	
	London .. .. (*Administrative County*)	4,232,118	..	..	..	..	2	..	4	..	1	..	1	
W. District.	Kensington	166,308	..	..	..	..	..	..	..	..	..	..	..	
	Fulham	91,639	..	..	..	..	..	..	..	..	..	..	..	
	Hammersmith	97,239	..	..	..	..	..	..	..	..	..	..	..	
	Paddington	117,846	..	..	..	..	..	..	..	..	..	..	..	
	Chelsea	96,253	..	..	..	..	..	..	..	..	..	..	..	
	St. George, Hanover Sq * ..	78,599	..	..	..	..	..	..	..	..	..	..	..	
	Westminster	55,539	..	..	..	..	..	..	..	..	..	..	..	
	St. James, Westminster ..	24,995	..	..	..	..	..	..	..	..	..	..	..	
N. District.	St. Marylebone	142,404	..	..	..	..	1	..	..	..	..	..	..	
	Hampstead	68,416	..	..	..	..	..	..	..	..	..	..	..	
	St. Pancras	234,379	..	..	..	..	..	..	..	..	1	..	..	
	Islington	319,143	..	..	..	..	..	..	..	..	..	..	..	
	St. Mary, Stoke Newington	30,936	..	..	..	..	..	..	..	..	..	..	..	
	Hackney	198,606	..	..	..	..	..	..	..	..	..	..	..	
Central District.	St. Giles and St. George, Bloomsbury.	39,782	..	..	..	..	..	..	..	..	..	..	..	
	St. Martin-in-the-Fields ..	14,616	..	..	..	..	..	..	..	..	..	..	..	
	Strand†	25,217	..	..	..	..	..	..	..	..	..	..	..	
	Holborn‡	34,043	..	..	..	..	..	..	..	..	..	..	..	
	Clerkenwell	66,216	..	..	..	..	..	..	1	..	..	..	..	
	St. Luke, Middlesex ..	42,440	..	..	..	..	..	..	..	..	..	..	..	
	London, City of§	37,583	..	..	..	..	..	..	..	..	..	..	..	
E. District.	Shoreditch	124,009	..	..	..	..	..	..	..	..	..	..	..	
	Bethnal Green	129,132	..	..	..	..	..	..	..	..	..	..	..	
	Whitechapel‖	74,420	..	..	..	..	..	..	..	..	..	..	..	
	St. George-in-the-East ..	45,795	..	..	..	..	..	..	..	..	..	..	..	
	Limehouse	57,376	..	..	..	..	..	..	..	..	..	..	..	
	Mile End Old Town ..	107,592	..	..	..	..	..	..	..	..	..	..	..	
	Poplar	166,748	..	..	..	..	..	..	..	..	..	..	..	
S. District.	St. Saviour, Southwark ..	27,177	..	..	..	..	..	..	..	..	..	..	..	
	St. George, Southwark ..	59,712	..	..	..	..	..	..	..	..	..	..	..	
	Newington ..	115,804	..	..	..	..	..	..	1	..	..	..	..	
	St. Olave, Southwark ..	12,723	..	..	..	..	..	..	..	..	..	..	..	
	Bermondsey	84,682	..	..	..	..	..	..	..	..	..	..	..	
	Rotherhithe	39,255	..	..	..	..	..	..	..	..	..	..	..	
	Lambeth	275,203	..	..	..	..	..	..	..	..	..	..	1	
	Battersea	150,558	..	..	..	..	..	..	..	..	..	..	..	
	Wandsworth	156,942	..	..	..	..	..	..	..	..	..	..	..	
	Camberwell	235,344	..	..	..	..	..	..	2	..	..	..	..	
	Greenwich	165,413	..	..	..	..	..	..	..	..	..	..	..	
	Lewisham	92,647	..	..	..	..	..	..	..	..	..	..	..	
	Woolwich	40,848	..	..	..	..	..	..	..	..	..	..	..	
	Plumstead	52,436	..	..	..	..	1	..	..	..	..	..	..	
	Lee	36,103	..	..	..	..	..	..	..	..	..	..	..	
	Port of London	..	..	..	..	..	..	..	..	..	..	..	..	

* Including St. Peters, Westminster (population, 235). † Including Middle Temple (popula
‡ Including Gray's Inn (population, 263), Lincoln's Inn (population, 27), Charterhouse (populat
Staple Inn (population, 21), and Furnival's Inn (population, 121).

SIMPLE CONTINUED FEVER—*continued.*

Weekly Statement, 3rd Quarter, 1898—*continued.*

[...].	Aug. 27.		Sept. 3.		Sept. 10.		Sept. 17.		Sept. 24.		Oct. 1.		Totals for 3rd Quarter 1898.		Sanitary Areas.
Deaths.	Cases.	Deaths.	Cases.	Deaths.	Cases.	Deaths.	Cases.	Deaths.	Cases.	Deaths.	Cases.	Deaths.	Cases.	Deaths.	
1	2	..	2	1	2	..	1	..	1	..	..	..	18	3	London. (*Administrative County.*)
..	..	..	..	..	..	..	..	..	..	..	..	..	..	..	Kensington.
..	..	..	..	1	..	..	..	..	..	..	..	..	..	1	Fulham.
..	..	..	1	..	..	..	..	..	..	..	..	..	1	..	Hammersmith.
..	..	..	..	..	1	..	..	..	..	..	..	..	1	..	Paddington.
..	..	..	..	..	..	..	..	..	..	..	..	..	..	..	Chelsea.
..	..	..	1	..	..	..	..	..	..	..	..	..	1	..	St. George, Hanover Sq.*
..	..	..	..	..	..	..	..	..	..	..	..	..	..	..	Westminster.
..	..	..	..	..	..	..	..	..	..	..	..	..	..	..	St. James, Westminster.
..	..	..	..	..	1	..	..	..	..	..	..	..	2	..	St. Marylebone.
1	..	..	..	..	..	..	..	..	..	..	..	..	..	1	Hampstead.
..	..	..	..	..	..	..	..	..	..	..	..	..	1	..	St. Pancras.
..	..	..	..	..	..	..	..	..	..	..	..	..	..	..	Islington.
..	..	..	..	..	..	..	..	..	..	..	..	..	..	..	St.Mary,StokeNewington.
..	..	..	..	..	..	..	..	..	1	..	..	..	1	..	Hackney.
..	..	..	..	..	..	..	..	..	..	..	..	..	..	..	St. Giles and St. George, Bloomsbury.
..	..	..	..	..	..	..	..	..	..	..	..	..	..	..	St. Martin-in-the-Fields.
..	..	..	..	..	..	..	..	..	..	..	..	..	..	..	Strand.†
..	..	..	..	..	..	..	..	..	..	..	..	..	..	..	Holborn.
..	..	..	..	..	..	..	..	..	..	..	..	..	1	..	Clerkenwell.
..	..	..	..	..	..	..	..	..	..	..	..	..	..	..	St. Luke, Middlesex.
..	..	..	..	..	..	..	..	..	..	..	..	..	.	..	London, City of.§
..	1	..	..	..	..	..	..	..	..	..	..	..	1	..	Shoreditch.
..	1	..	..	..	..	..	..	..	..	..	..	..	1	..	Bethnal Green.
..	..	..	..	..	..	..	..	..	..	..	..	..	..	..	Whitechapel.‖
..	..	..	..	..	..	..	..	..	..	..	..	..	..	..	St. George-in-the-East
..	..	..	..	..	..	..	..	..	..	..	..	..	..	..	Limehouse.
..	..	..	..	..	..	..	..	..	..	..	..	..	..	..	Mile End Old Town.
..	..	..	..	..	..	..	..	..	..	..	..	..	..	..	Poplar.
..	..	..	..	..	..	..	..	..	..	..	..	..	..	..	St. Saviour, Southwark.
..	..	..	..	..	..	..	..	..	..	..	..	..	..	..	St. George, Southwark.
..	..	..	..	..	..	..	..	..	..	..	..	..	1	..	Newington.
..	..	..	..	..	..	..	..	..	..	..	..	..	..	..	St. Olave, Southwark.
..	..	..	..	..	..	..	..	..	..	..	..	..	..	..	Bermondsey.
..	..	..	..	..	..	..	..	..	..	..	..	..	..	..	Rotherhithe.
..	..	..	..	..	..	..	..	..	..	..	..	..	2	1	Lambeth.
..	..	..	..	..	..	..	..	..	..	..	..	..	..	..	Battersea.
..	..	..	..	..	..	..	..	..	..	..	..	..	..	..	Wandsworth.
..	..	..	..	..	..	..	..	..	..	..	..	..	2	..	Camberwell.
..	..	..	..	..	..	..	1	..	..	..	..	..	2	..	Greenwich.
..	..	..	..	..	..	..	..	..	..	..	..	..	..	..	Lewisham.
..	..	..	..	..	..	..	..	..	..	..	..	..	..	..	Woolwich.
..	..	..	..	..	..	..	..	..	..	..	..	..	1	..	Plumstead.
..	..	..	..	..	..	..	..	..	..	..	..	..	..	..	Lee.
..	..	..	..	..	..	..	..	..	..	..	..	..	..	..	Port of London.

...uding Inner Temple (population, 96). ‖ Including Tower of London (population, 898)

	Sanitary Areas.	Population (1891).	Oct. 8. Cases.	Oct. 8. Deaths.	Oct. 15. Cases.	Oct. 15. Deaths.	Oct. 22. Cases.	Oct. 22. Deaths.	Oct. 29. Cases.	Oct. 29. Deaths.	Nov. 5. Cases.	Nov. 5. Deaths.	Nov. 12. Cases.	Nov. 12. Deaths.
			Simple Continued Fever—continued. Weekly Statement, 4th Quarter, 1898.											
	London (*Administrative County*)	4,232,118	1	..	1	..	1	..	2	2	1	..	1	..
W. District.	Kensington	166,308	..	..	1	..	..	..	..	..	..	..	..	..
W. District.	Fulham	91,639	..	..	..	..	..	..	..	..	..	..	..	..
W. District.	Hammersmith	97,239	1	..	..	..	..	..	..	..	..	..	..	..
W. District.	Paddington	117,846	..	..	..	..	..	..	..	1	..	..	..	..
W. District.	Chelsea	96,253	..	..	..	..	..	..	..	..	..	..	..	..
W. District.	St. George, Hanover Sq.*	78,599	..	..	..	..	..	..	..	..	..	..	..	..
W. District.	Westminster	55,539	..	..	..	..	..	..	..	..	..	..	..	..
W. District.	St. James, Westminster	24,995	..	..	..	..	..	..	..	..	..	..	..	..
N. District.	St. Marylebone	142,404	..	..	..	..	..	..	..	..	..	..	..	..
N. District.	Hampstead	68,416	..	..	..	..	..	..	..	..	..	..	..	..
N. District.	St. Pancras	234,379	..	..	..	..	..	..	..	..	..	..	..	..
N. District.	Islington	319,143	..	..	..	..	..	..	..	..	1	..	..	..
N. District.	St. Mary, Stoke Newington	30,936	..	..	..	..	..	..	..	..	..	..	..	..
N. District.	Hackney	198,606	..	..	..	..	..	..	..	..	..	..	..	..
Central District.	St. Giles and St. George, Bloomsbury.	39,782	..	..	..	..	..	..	..	..	..	..	..	..
Central District.	St. Martin-in-the-Fields	14,616	..	..	..	..	..	..	..	..	..	..	..	..
Central District.	Strand†	25,217	..	..	..	..	..	..	..	..	..	..	..	..
Central District.	Holborn‡	34,043	..	..	..	..	..	..	..	..	..	..	..	..
Central District.	Clerkenwell	66,216	..	..	..	..	..	..	..	..	..	..	..	..
Central District.	St. Luke, Middlesex	42,440	..	..	..	..	..	..	..	..	..	..	..	..
Central District.	London, City of§	37,583	..	..	..	..	..	..	..	..	..	..	..	..
E. District.	Shoreditch	124,009	..	..	..	..	..	..	..	..	..	..	..	..
E. District.	Bethnal Green	129,132	..	..	..	..	1	..	..	..	..	..	..	..
E. District.	Whitechapel‖	74,420	..	..	..	..	..	..	..	..	..	..	..	..
E. District.	St. George-in-the-East	45,795	..	..	..	..	..	..	..	..	..	..	..	..
E. District.	Limehouse	57,376	..	..	..	..	..	..	1	1	..	..	..	..
E. District.	Mile End Old Town	107,592	..	..	..	..	..	..	..	..	..	..	..	..
E. District.	Poplar	166,748	..	..	..	..	..	..	..	..	..	..	..	..
S. District.	St. Saviour, Southwark	27,177	..	..	..	..	..	..	..	..	..	..	..	..
S. District.	St. George, Southwark	59,712	..	..	..	..	..	..	..	..	..	..	..	..
S. District.	Newington	115,804	..	..	..	..	..	..	..	..	..	..	..	..
S. District.	St. Olave, Southwark	12,723	..	..	..	..	..	..	..	..	..	..	..	..
S. District.	Bermondsey	84,682	..	..	..	..	..	..	..	..	..	..	..	..
S. District.	Rotherhithe	39,255	..	..	..	..	..	..	..	..	..	..	..	..
S. District.	Lambeth	275,203	..	..	..	..	..	..	1	..	..	..	..	..
S. District.	Battersea	150,558	..	..	..	..	..	..	..	..	..	..	..	..
S. District.	Wandsworth	156,942	..	..	..	..	..	..	..	..	..	..	1	..
S. District.	Camberwell	235,344	..	..	..	..	..	..	..	..	..	..	..	..
S. District.	Greenwich	165,413	..	..	..	..	..	..	..	..	..	..	..	..
S. District.	Lewisham	92,647	..	..	..	..	..	..	..	..	..	..	..	..
S. District.	Woolwich	40,848	..	..	..	..	..	..	..	..	..	..	..	..
S. District.	Plumstead	52,436	..	..	..	..	..	..	..	..	..	..	..	..
S. District.	Lee	36,103	..	..	..	..	..	..	..	..	..	..	..	..
S. District.	Port of London	..	..	..	..	..	..	..	..	..	..	..	..	..

* Including St. Peters, Westminster (population, 286). † Including Middle Temple (popul

SIMPLE CONTINUED FEVER—*continued.*

Weekly Statement, 4th Quarter, 1898—*continued.*												Totals for 4th Quarter, 1898.		Grand Totals for Year 1898.		Sanitary Areas.
Nov. 26.		Dec. 3.		Dec. 10.		Dec. 17.		Dec. 24.		Dec. 31.						
Cases.	Deaths.	Cases.	Deaths.	Cases.	Deaths.	Cases.	Deaths.	Cases.	Deaths.	Cases.	Deaths.	Cases.	Deaths.	Cases.	Deaths.	
[illegible]	..	1	..	1	..	..	..	3	..	..	1	14	3	55	9	London.
																(Administrative County.)
[illegible]	..	..	..	1	..	..	..	..	..	..	..	3	..	4	..	Kensington.
..	..	..	..	..	..	..	..	..	..	..	..	..	..	1	1	Fulham.
..	..	..	..	..	..	..	..	..	..	..	..	1	..	5	..	Hammersmith.
[illegible]	..	..	..	..	..	..	..	..	..	..	..	1	1	2	1	Paddington.
..	..	..	..	..	..	..	..	..	..	..	..	..	..	1	..	Chelsea.
..	..	..	..	..	..	..	..	..	..	..	..	..	..	1	..	St. George, Hanover Sq.*
..	..	..	..	..	..	..	..	..	..	..	..	..	..	1	..	Westminster.
..	..	..	..	..	..	..	..	..	..	..	..	..	..	..	..	St. James, Westminster.
..	..	..	..	..	..	..	..	..	..	..	..	..	..	2	..	St. Marylebone.
..	..	..	..	..	..	..	..	..	..	..	..	..	..	..	1	Hampstead.
..	..	..	..	..	..	..	..	..	..	..	..	..	..	2	..	St. Pancras.
..	..	..	..	..	..	..	..	..	..	..	..	1	..	1	..	Islington.
..	..	..	..	..	..	..	..	..	..	..	..	..	..	..	..	St. Mary, Stoke Newington.
..	..	1	..	..	..	..	..	1	..	..	..	2	..	3	..	Hackney.
..	..	..	..	..	..	..	..	..	..	..	..	..	..	..	..	St. Giles and St. George, Bloomsbury.
..	..	..	..	..	..	..	..	..	..	..	..	..	..	..	..	St. Martin-in-the-Fields.
..	..	..	..	..	..	..	..	..	..	..	..	..	..	..	..	Strand.†
..	..	..	..	..	..	..	..	..	..	..	..	..	..	..	..	Holborn.‡
..	..	..	..	..	..	..	..	..	..	..	..	..	..	1	..	Clerkenwell.
..	..	..	..	..	..	..	..	..	..	..	..	..	..	..	..	St. Luke, Middlesex.
..	..	..	..	..	..	..	..	..	..	..	1	..	1	..	1	London, City of.§
..	..	..	..	..	..	..	..	..	..	..	..	..	..	1	..	Shoreditch.
..	..	..	..	..	..	..	..	..	..	..	..	1	..	2	..	Bethnal Green.
..	..	..	..	..	..	..	..	..	..	..	..	..	..	..	..	Whitechapel.‖
..	..	..	..	..	..	..	..	..	..	..	..	..	..	..	..	St. George-in-the-East.
..	..	..	..	..	..	..	..	..	..	..	..	1	1	1	1	Limehouse.
..	..	..	..	..	..	..	..	..	..	..	..	..	..	1	..	Mile End Old Town.
..	..	..	..	..	..	..	..	..	..	..	..	..	..	3	1	Poplar.
..	..	..	..	..	..	..	..	..	..	..	..	..	..	..	..	St. Saviour, Southwark.
..	..	..	..	..	..	..	..	..	..	..	..	..	..	1	..	St. George, Southwark.
..	..	..	..	..	..	..	..	..	..	..	..	..	..	1	..	Newington.
..	..	..	..	..	..	..	..	..	..	..	..	..	..	..	..	St. Olave, Southwark.
..	..	..	..	..	..	..	..	..	..	..	..	..	..	..	..	Bermondsey.
..	..	..	..	..	..	..	..	..	..	..	..	..	..	..	..	Rotherhithe.
..	..	..	..	..	..	..	..	1	..	..	..	2	..	13	1	Lambeth.
..	..	..	..	..	..	..	..	..	..	..	..	..	..	..	2	Battersea.
..	..	..	..	..	..	..	..	..	..	..	..	1	..	1	..	Wandsworth.
..	..	..	..	..	..	..	..	1	..	..	..	1	..	3	..	Camberwell.
..	..	..	..	..	..	..	..	..	..	..	..	..	..	3	..	Greenwich.
..	..	..	..	..	..	..	..	..	..	..	..	..	..	..	..	Lewisham.
..	..	..	..	..	..	..	..	..	..	..	..	..	..	..	..	Woolwich.
..	..	..	..	..	..	..	..	..	..	..	..	..	..	1	..	Plumstead.
..	..	..	..	..	..	..	..	..	..	..	..	..	..	..	..	Lee.
..	..	..	..	..	..	..	..	..	..	..	..	..	..	..	..	Port of London.

Including Gray's Inn (population, 258), Lincoln's Inn (population, 27), Charterhouse (population, 136), Staple Inn (population, 21), and Furnival's Inn (population, 121).

Including Inner Temple (population, 80).

‖ Including Tower of London (population, [illegible]).

Sanitary Areas.		Jan. 8.	Mar. 19.	Mar. 26.		April 2.	Totals for 1st Qr. 1896.		April 9.	April 16.	April 23.	May 14.	Totals for 2nd Qr. 1896.	
		Cases.	Cases.	Cases.	Deaths.	Deaths.	Cases.	Deaths.	Cases.	Cases.	Cases.	Cases.	Cases.	Deaths.
London .. .. (*Administrative County*)	4,232,118	1	1	5	1	1	7	2	1	1	1	1	4	..
W. District — Kensington	166,308	..	..	5	1	1	5	2	1	..	1	..	2	..
Fulham	91,639	..	..	..	..	..	..	..	..	..	..	..	..	..
Hammersmith	97,239	..	..	..	..	..	..	..	..	..	..	..	..	..
Paddington	117,846	..	..	..	..	..	..	..	..	..	..	..	..	..
Chelsea..	96,253	..	..	..	..	..	..	..	..	..	..	..	..	..
St. George, Hanover Sq.* ..		..	..	..	..	..	..	..	..	..	..	..	..	..
Westminster		..	..	..	..	..	..	..	..	..	..	..	..	..
St. James, Westminster ..		..	..	..	..	..	..	..	..	..	..	..	..	..
N. District — St. Marylebone		..	..	..	..	..	..	..	..	..	..	..	..	..
Hampstead		..	..	..	..	..	..	..	..	..	..	..	..	..
St. Pancras		..	..	..	..	..	..	..	..	..	..	..	..	..
Islington		..	..	..	..	..	..	..	..	..	..	..	..	..
St. Mary, Stoke Newington		..	..	..	..	..	..	..	..	..	..	..	..	..
Hackney		..	..	..	..	..	..	..	..	..	..	..	..	..
Central District — St. Giles and St. George, Bloomsbury.		..	..	..	..	..	..	..	..	..	..	..	..	..
St. Martin-in-the-Fields ..		..	..	..	..	..	..	..	..	..	..	..	..	..
Strand†		..	..	..	..	..	..	..	..	..	..	..	..	..
Holborn‡		..	..	..	..	..	..	..	..	..	..	..	..	..
Clerkenwell		..	..	..	..	..	..	..	..	..	..	..	..	..
St. Luke, Middlesex ..		..	..	..	..	..	..	..	..	..	..	..	..	..
London, City of§		..	..	..	..	..	..	..	..	..	..	..	..	..
E. District — Shoreditch		..	..	..	..	..	..	..	..	..	..	..	..	..
Bethnal Green		..	..	..	..	..	..	..	..	..	..	..	..	..
Whitechapel‖		..	..	..	..	..	..	..	..	..	..	..	..	..
St. George-in-the-East ..		..	..	..	..	..	..	..	..	..	..	..	..	..
Limehouse		..	..	..	..	..	..	..	..	..	..	..	..	..
Mile End Old Town ..		..	..	..	..	..	..	..	..	..	..	..	..	..
Poplar		..	..	..	..	..	..	..	..	..	..	..	..	..
S. District — St. Saviour, Southwark ..		..	..	..	..	..	..	..	..	..	..	..	..	..
St. George, Southwark ..		..	..	..	..	..	..	..	..	1	..	..	1	..
Newington		..	..	..	..	..	..	..	..	..	..	..	..	..
St. Olave, Southwark ..		..	..	..	..	..	..	..	..	..	..	..	..	..
Bermondsey		..	..	..	..	..	..	..	..	..	..	..	..	..
Rotherhithe		..	..	..	..	..	..	..	..	..	..	..	..	..
Lambeth		..	1	..	..	..	1	..	..	..	..	..	..	..
Battersea		1	..	..	..	..	1	..	..	..	..	..	..	..
Wandsworth		..	..	..	..	..	..	..	..	..	..	..	..	..
Camberwell		..	..	..	..	..	..	..	..	..	..	..	..	..
Greenwich		..	..	..	..	..	..	..	..	..	..	1	1	..
Lewisham		..	..	..	..	..	..	..	..	..	..	..	..	..
Woolwich		..	..	..	..	..	..	..	..	..	..	..	..	..
Plumstead		..	..	..	..	..	..	..	..	..	..	..	..	..
Lee		..	..	..	..	..	..	..	..	..	..	..	..	..
Port of London		..	..	..	..	..	..	..	..	..	..	..	..	..

* Including St. Peters, Westminster (population, 285).
† Including Middle Temple (population, 85).

Typhus Fever—continued.										
Weekly Statement, 1898—continued.								Grand Totals for Year 1898.		Sanitary Areas.
...s for 3rd ...ter, 1898.	Oct. 22.	Nov. 5.	Nov. 12.	Dec 10.		Totals for 4th Quarter, 1898.				
Deaths.	Cases.	Cases.	Cases.	Cases.	Deaths.	Cases.	Deaths.	Cases.	Deaths.	
..	2	3	1	1	1	7	1	18	3	London. (*Administrative County.*)
..	..	2	..	..	..	2	..	9	2	Kensington.
..	..	..	..	..	..	..	..	..	..	Fulham.
..	..	..	..	..	..	..	..	..	..	Hammersmith
..	..	..	..	..	..	..	..	..	..	Paddington.
..	..	..	..	..	..	..	..	..	..	Chelsea.
..	..	..	..	..	..	..	..	..	..	St. George, Hanover Sq.*
..	..	..	..	..	..	..	..	..	..	Westminster.
..	..	..	..	..	..	..	..	..	..	St. James, Westminster.
..	..	..	..	..	..	..	..	..	..	St. Marylebone.
..	..	..	..	..	..	..	..	..	..	Hampstead.
..	..	..	..	..	..	..	..	..	..	St. Pancras.
..	..	..	1	1	1	2	1	2	1	Islington.
..	..	..	..	..	..	..	..	..	..	St. Mary, Stoke Newington.
..	..	..	..	..	..	..	..	..	..	Hackney.
..	..	..	..	..	..	..	..	..	..	St. Giles and St. George, Bloomsbury.
..	..	..	..	..	..	..	..	..	..	St. Martin-in-the-Fields.
..	..	..	..	..	..	..	..	..	..	Strand.†
..	..	..	..	..	..	..	..	..	..	Holborn.‡
..	..	..	..	..	..	..	..	..	..	Clerkenwell.
..	..	..	..	..	..	..	..	..	..	St. Luke, Middlesex.
..	..	..	..	..	..	..	..	..	..	London, City of.§
..	..	..	..	..	..	..	..	..	..	Shoreditch.
..	..	..	..	..	..	..	..	..	..	Bethnal Green.
..	..	..	..	..	..	..	..	..	..	Whitechapel.‖
..	..	..	..	..	..	..	..	..	..	St. George-in-the-East.
..	..	..	..	..	..	..	..	..	..	Limehouse.
..	..	..	..	..	..	..	..	..	..	Mile End Old Town.
..	..	..	..	..	..	..	..	..	..	Poplar.
..	..	..	..	..	..	..	..	..	..	St. Saviour, Southwark.
..	1	1	..	..	..	2	..	3	..	St. George, Southwark.
..	..	..	..	..	..	..	..	..	..	Newington.
..	..	..	..	..	..	..	..	..	..	St. Olave, Southwark.
..	..	..	..	..	..	..	..	..	..	Bermondsey.
..	..	..	..	..	..	..	..	..	..	Rotherhithe.
..	1	..	..	..	..	1	..	2	..	Lambeth.
..	..	..	..	..	..	..	..	1	..	Battersea.
..	..	..	..	..	..	..	..	..	..	Wandsworth.
..	..	..	..	..	..	..	..	..	..	Camberwell.
..	..	..	..	..	..	..	..	1	..	Greenwich.
..	..	..	..	..	..	..	..	..	..	Lewisham.
..	..	..	..	..	..	..	..	..	..	Woolwich.
..	..	..	..	..	..	..	..	..	..	Plumstead.
..	..	..	..	..	..	..	..	..	..	Lee.
..	..	..	..	..	..	..	..	..	..	Port of London.

...ding Gray's Inn (population, 253), Lincoln's Inn (population, 27), Charterhouse (popula-
...), Staple Inn (population, 21), and Furnival's Inn (population, 121).
...ding Inner Temple (population, [illegible]). ‖ Including Tower of London (population, [illegible]).

District	Sanitary Areas.	Population (1891).	SIMPLE CONTINUED FEVER—*continued.* Weekly Statement, 4th Quarter, 1898. Oct. 8.		Oct. 15.		Oct. 22.		Oct. 29.		Nov. 5.		Nov. 12.	
			Cases.	Deaths.	Cases.	Deaths.	Cases.	Deaths.	Cases.	Deaths.	Cases.	Deaths.	Cases.	Deaths.
	London .. .. (*Administrative County*)	4,232,118	1	..	1	..	1	..	2	2	1	..	1	..
W. District.	Kensington	166,308	..	..	1	..	..	..	..	..	..	..	..	..
W. District.	Fulham	91,639	..	..	..	..	..	..	..	..	..	..	..	..
W. District.	Hammersmith	97,239	1	..	..	..	..	..	..	..	..	..	..	..
W. District.	Paddington	117,846	..	..	..	..	..	..	..	1	..	..	..	..
W. District.	Chelsea	96,253	..	..	..	..	..	..	..	..	..	..	..	..
W. District.	St. George, Hanover Sq.* ..	78,599	..	..	..	..	..	..	..	..	..	..	..	..
W. District.	Westminster	55,539	..	..	..	..	..	..	..	..	..	..	..	..
W. District.	St. James, Westminster ..	24,995	..	..	..	..	..	..	..	..	..	..	..	..
N. District.	St. Marylebone	142,404	..	..	..	..	..	..	..	..	..	..	..	..
N. District.	Hampstead	68,416	..	..	..	..	..	..	..	..	..	..	..	..
N. District.	St. Pancras	234,379	..	..	..	..	..	..	..	..	..	..	..	..
N. District.	Islington	319,143	..	..	..	..	..	..	..	..	1	..	..	..
N. District.	St. Mary, Stoke Newington	30,936	..	..	..	..	..	..	..	..	..	..	..	..
N. District.	Hackney	198,606	..	..	..	..	..	..	..	..	..	..	..	..
Central District.	St. [illegible] and St. George, Bloomsbury.	39,783	..	..	..	..	..	..	..	..	..	..	..	..
Central District.	St. Martin-in-the-Fields ..	14,616	..	..	..	..	..	..	..	..	..	..	..	..
Central District.	Strand†	25,217	..	..	..	..	..	..	..	..	..	..	..	..
Central District.	Holborn‡	34,043	..	..	..	..	..	..	..	..	..	..	..	..
Central District.	Clerkenwell	66,216	..	..	..	..	..	..	..	..	..	..	..	..
Central District.	St. Luke, Middlesex ..	42,440	..	..	..	..	..	..	..	..	..	..	..	..
Central District.	London, City of§	37,583	..	..	..	..	..	..	..	..	..	..	..	..
E. District.	Shoreditch	124,009	..	..	..	..	..	..	..	..	..	..	..	..
E. District.	Bethnal Green	129,132	..	..	..	..	1	..	..	..	..	..	..	..
E. District.	Whitechapel‖	74,420	..	..	..	..	..	..	..	..	..	..	..	..
E. District.	St. George-in-the-East ..	45,795	..	..	..	..	..	..	..	..	..	..	..	..
E. District.	Limehouse	57,376	..	..	..	..	..	..	1	1	..	..	..	..
E. District.	Mile End Old Town ..	107,592	..	..	..	..	..	..	..	..	..	..	..	..
E. District.	Poplar	166,748	..	..	..	..	..	..	..	..	..	..	..	..
S. District.	St. Saviour, Southwark ..	27,177	..	..	..	..	..	..	..	..	..	..	..	..
S. District.	St. George, Southwark ..	59,712	..	..	..	..	..	..	..	..	..	..	..	..
S. District.	Newington	115,804	..	..	..	..	..	..	..	..	..	..	..	..
S. District.	St. Olave, Southwark ..	12,723	..	..	..	..	..	..	..	..	..	..	..	..
S. District.	Bermondsey	84,682	..	..	..	..	..	..	..	..	..	..	..	..
S. District.	Rotherhithe	39,255	..	..	..	..	..	..	..	..	..	..	..	..
S. District.	Lambeth	275,208	..	..	..	..	..	..	1	..	..	..	..	..
S. District.	Battersea	150,558	..	..	..	..	..	..	..	..	..	..	..	..
S. District.	Wandsworth	156,942	..	..	..	..	..	..	..	..	..	..	1	..
S. District.	Camberwell	235,344	..	..	..	..	..	..	..	..	..	..	..	..
S. District.	Greenwich	165,413	..	..	..	..	—	..	..	..	..	..	..	..
S. District.	Lewisham	92,647	..	..	..	..	..	..	..	..	..	..	..	..
S. District.	Woolwich	40,848	..	..	..	..	..	..	..	..	..	..	..	..
S. District.	Plumstead	52,436	..	..	..	..	..	..	..	..	..	..	..	..
S. District.	Lee	36,103	..	..	..	..	..	..	..	..	..	..	..	..
S. District.	Port of London	..	..	..	..	..	..	..	..	..	..	..	..	..

* Including St. Peters, Westminster (population, 285). † Including Middle Temple (popul

SIMPLE CONTINUED FEVER—*continued.*														
Weekly Statement, 4th Quarter, 1898—*continued.*										Totals for 4th Quarter, 1898.		Grand Totals for Year 1898.		Sanitary Areas.
Dec. 3.		Dec. 10.		Dec. 17.		Dec. 24.		Dec. 31.						
Cases.	Deaths.	Cases.	Deaths.	Cases.	Deaths.	Cases.	Deaths.	Cases.	Deaths.	Cases.	Deaths.	Cases.	Deaths.	
1	..	1	..	..	..	3	..	..	1	14	3	55	9	London. (*Administrative County.*)
..	..	1	..	..	..	..	..	..	..	3	..	4	..	Kensington.
..	..	..	..	..	..	..	..	..	..	..	..	1	1	Fulham.
..	..	..	..	..	..	..	..	..	..	1	..	5	..	Hammersmith.
..	..	..	..	..	..	..	..	..	..	1	1	2	1	Paddington.
..	..	..	..	..	..	..	..	..	..	..	..	1	..	Chelsea.
..	..	..	..	..	..	..	..	..	..	..	..	1	..	St. George, Hanover Sq.*
..	..	..	..	..	..	..	..	..	..	..	..	1	..	Westminster.
..	..	..	..	..	..	..	..	..	..	..	..	..	..	St. James, Westminster.
..	..	..	..	..	..	..	..	..	..	..	..	2	..	St. Marylebone.
..	..	..	..	..	..	..	..	..	..	..	..	..	1	Hampstead.
..	..	..	..	..	..	..	..	..	..	..	..	2	..	St. Pancras.
..	..	..	..	..	..	..	..	..	..	1	..	1	..	Islington.
..	..	..	..	..	..	..	..	..	..	..	..	..	..	St.Mary,StokeNewington.
1	..	..	..	..	..	1	..	..	..	2	..	3	..	Hackney.
..	..	..	..	..	..	..	..	..	..	..	..	..	..	St. Giles and St. George, Bloomsbury.
..	..	..	..	..	..	..	..	..	..	..	..	..	..	St. Martin-in-the-Fields.
..	..	..	..	..	..	..	..	..	..	..	..	..	..	Strand.†
..	..	..	..	..	..	..	..	..	..	..	..	..	..	Holborn.‡
..	..	..	..	..	..	..	..	..	..	..	..	1	..	Clerkenwell.
..	..	..	..	..	..	..	..	..	..	..	..	..	..	St. Luke, Middlesex.
..	..	..	..	..	..	..	..	..	1	..	1	..	1	London, City of.§
..	..	..	..	..	..	..	..	..	..	..	..	1	..	Shoreditch.
..	..	..	..	..	..	..	..	..	..	1	..	2	..	Bethnal Green.
..	..	..	..	..	..	..	..	..	..	..	..	..	..	Whitechapel.‖
..	..	..	..	..	..	..	..	..	..	..	..	..	..	St. George-in-the-East.
..	..	..	..	..	..	..	..	..	..	1	1	1	1	Limehouse.
..	..	..	..	..	..	..	..	..	..	..	..	1	..	Mile End Old Town.
..	..	..	..	..	..	..	..	..	..	..	..	3	1	Poplar.
..	..	..	..	..	..	..	..	..	..	..	..	..	..	St. Saviour, Southwark.
..	..	..	..	..	..	..	..	..	..	..	..	1	..	St. George, Southwark.
..	..	..	..	..	..	..	..	..	..	..	..	1	..	Newington.
..	..	..	..	..	..	..	..	..	..	..	..	..	..	St. Olave, Southwark.
..	..	..	..	..	..	..	..	..	..	..	..	..	..	Bermondsey.
..	..	..	..	..	..	..	..	..	..	..	..	..	..	Rotherhithe.
..	..	..	..	..	..	1	..	..	..	2	..	13	1	Lambeth.
..	..	..	..	..	..	..	..	..	..	..	..	..	2	Battersea.
..	..	..	..	..	..	..	..	..	..	1	..	1	..	Wandsworth.
..	..	..	..	..	..	1	..	..	..	1	..	3	..	Camberwell.
..	..	..	..	..	..	..	..	..	..	..	..	3	..	Greenwich.
..	..	..	..	..	..	..	..	..	..	..	..	..	..	Lewisham.
..	..	..	..	..	..	..	..	..	..	..	..	..	..	Woolwich.
..	..	..	..	..	..	..	..	..	..	..	..	1	..	Plumstead.
..	..	..	..	..	..	..	..	..	..	..	..	..	..	Lee.
..	..	..	..	..	..	..	..	..	..	..	..	..	..	Port of London.

ng Gray's Inn (population, 253), Lincoln's Inn (population, 27), Charterhouse (population, 186),
e Inn (population, 21), and Furnival's Inn (population, 121).
ng Inner Temple (population, [illegible]).
‖ Including Tower of London (population, [illegible])

Sanitary Areas.	Population (1891).	Jan. 8. Cases.	Mar. 12. Cases.	Mar. 26. Cases.	Mar. 26. Deaths.	April 2. Deaths.	Totals for 1st Qr. 1898. Cases.	Totals for 1st Qr. 1898. Deaths.	April 9. Cases.	April 16. Cases.	April 23. Cases.	May 14. Cases.	Totals for 2nd Qr. 1898. Cases.	Totals for 2nd Qr. 1898. Deaths.
London (*Administrative County*)	4,232,118	1	1	5	1	1	7	2	1	1	1	1	4	..
W. District: Kensington	166,308	..	..	5	1	1	5	2	1	..	1	..	2	..
Fulham	91,639	..	..	..	..	..	..	..	..	..	..	..	..	..
Hammersmith	97,239	..	..	..	..	..	..	..	..	..	..	..	..	..
Paddington	117,846	..	..	..	..	..	..	..	..	..	..	..	..	..
Chelsea	96,253	..	..	..	..	..	..	..	..	..	..	..	..	..
St. George, Hanover Sq.*	78,599	..	..	..	..	..	..	..	..	..	..	..	..	..
Westminster	55,539	..	..	..	..	..	..	..	..	..	..	..	..	..
St. James, Westminster	24,995	..	..	..	..	..	..	..	..	..	..	..	..	..
N. District: St. Marylebone	142,404	..	..	..	..	..	..	..	..	..	..	..	..	..
Hampstead	68,416	..	..	..	..	..	..	..	..	..	..	..	..	..
St. Pancras	234,379	..	..	..	..	..	..	..	..	..	..	..	..	..
Islington	319,143	..	..	..	..	..	..	..	..	..	..	..	..	..
St. Mary, Stoke Newington	30,936	..	..	..	..	..	..	..	..	..	..	..	..	..
Hackney	198,606	..	..	..	..	..	..	..	..	..	..	..	..	..
Central District: St. Giles and St. George, Bloomsbury.	39,782	..	..	..	..	..	..	..	..	..	..	..	..	..
St. Martin-in-the-Fields	14,616	..	..	..	..	..	..	..	..	..	..	..	..	..
Strand†	25,217	..	..	..	..	..	..	..	..	..	..	..	..	..
Holborn‡	34,043	..	..	..	..	..	..	..	..	..	..	..	..	..
Clerkenwell	66,216	..	..	..	..	..	..	..	..	..	..	..	..	..
St. Luke, Middlesex	42,440	..	..	..	..	..	..	..	..	..	..	..	..	..
London, City of§	37,583	..	..	..	..	..	..	..	..	..	..	..	..	..
E. District: Shoreditch	124,009	..	..	..	..	..	..	..	..	..	..	..	..	..
Bethnal Green	129,132	..	..	..	..	..	..	..	..	..	..	..	..	..
Whitechapel‖	74,420	..	..	..	..	..	..	..	..	..	..	..	..	..
St. George-in-the-East	45,795	..	..	..	..	..	..	..	..	..	..	..	..	..
Limehouse	57,376	..	..	..	..	..	..	..	..	..	..	..	..	..
Mile End Old Town	107,592	..	..	..	..	..	..	..	..	..	..	..	..	..
Poplar	166,748	..	..	..	..	..	..	..	..	..	..	..	..	..
S. District: St. Saviour, Southwark	27,177	..	..	..	..	..	..	..	..	..	..	..	..	..
St. George, Southwark	59,712	..	..	..	..	..	..	..	..	1	..	..	1	..
Newington	115,804	..	..	..	..	..	..	..	..	..	..	..	..	..
St. Olave, Southwark	12,723	..	..	..	..	..	..	..	..	..	..	..	..	..
Bermondsey	84,682	..	..	..	..	..	..	..	..	..	..	..	..	..
Rotherhithe	39,255	..	..	..	..	..	..	..	..	..	..	..	..	..
Lambeth	275,203	..	1	..	..	..	1	..	..	..	..	..	..	..
Battersea	150,558	1	..	..	..	..	1	..	..	..	..	..	..	..
Wandsworth	156,942	..	..	..	..	..	..	..	..	..	..	..	..	..
Camberwell	235,344	..	..	..	..	..	..	..	..	..	..	..	..	..
Greenwich	165,413	..	..	..	..	..	..	..	..	..	..	1	1	..
Lewisham	92,647	..	..	..	..	..	..	..	..	..	..	..	..	..
Woolwich	40,848	..	..	..	..	..	..	..	..	..	..	..	..	..
Plumstead	52,436	..	..	..	..	..	..	..	..	..	..	..	..	..
Lee	36,103	..	..	..	..	..	..	..	..	..	..	..	..	..
Port of London	..	..	..	..	..	..	..	..	..	..	..	..	..	..

* Including St. Peters, Westminster (population, 235).
† Including Middle Temple (population, 95).

Typhus Fever—continued.										
Weekly Statement, 1898—continued.								Grand Totals for Year 1898.		Sanitary Areas.
...s for 3rd ...ter, 1898.	Oct. 22.	Nov. 5.	Nov. 12.	Dec 10.		Totals for 4th Quarter, 1898.				
Deaths.	Cases.	Cases.	Cases.	Cases.	Deaths.	Cases.	Deaths.	Cases.	Deaths.	
..	2	3	1	1	1	7	1	18	3	London. (*Administrative County.*)
..	..	2	..	..	..	2	..	9	2	Kensington.
..	..	..	..	..	..	..	..	..	..	Fulham.
..	..	..	..	..	..	..	..	..	..	Hammersmith
..	..	..	..	..	..	..	..	..	..	Paddington.
..	..	..	..	..	..	..	..	..	..	Chelsea.
..	..	..	..	..	..	..	..	..	..	St. George, Hanover Sq.*
..	..	..	..	..	..	..	..	..	..	Westminster.
..	..	..	..	..	..	..	..	..	..	St. James, Westminster.
..	..	..	..	..	..	..	..	..	..	St. Marylebone.
..	..	..	..	..	..	..	..	..	..	Hampstead.
..	..	..	..	..	..	..	..	..	..	St. Pancras.
..	..	..	1	1	1	2	1	2	1	Islington.
..	..	..	..	..	..	..	..	..	..	St. Mary, Stoke Newington.
..	..	..	..	..	..	..	..	..	..	Hackney.
..	..	..	..	..	..	..	..	..	..	St. Giles and St. George, Bloomsbury.
..	..	..	..	..	..	..	..	..	..	St. Martin-in-the-Fields.
..	..	..	..	..	..	..	..	..	..	Strand.†
..	..	..	..	..	..	..	..	..	..	Holborn.‡
..	..	..	..	..	..	..	..	..	..	Clerkenwell.
..	..	..	..	..	..	..	..	..	..	St. Luke, Middlesex.
..	..	..	..	..	..	..	..	..	..	London, City of.§
..	..	..	..	..	..	..	..	..	..	Shoreditch.
..	..	..	..	..	..	..	..	..	..	Bethnal Green.
..	..	..	..	..	..	..	..	..	..	Whitechapel.‖
..	..	..	..	..	..	..	..	..	..	St. George-in-the-East.
..	..	..	..	..	..	..	..	..	..	Limehouse.
..	..	..	..	..	..	..	..	..	..	Mile End Old Town.
..	..	..	..	..	..	..	..	..	..	Poplar.
..	..	..	..	..	..	..	..	..	..	St. Saviour, Southwark.
..	1	1	..	..	..	2	..	3	..	St. George, Southwark.
..	..	..	..	..	..	..	..	..	..	Newington.
..	..	..	..	..	..	..	..	..	..	St. Olave, Southwark.
..	..	..	..	..	..	..	..	..	..	Bermondsey.
..	..	..	..	..	..	..	..	..	..	Rotherhithe.
..	1	..	..	..	..	1	..	2	..	Lambeth.
..	..	..	..	..	..	..	..	1	..	Battersea.
..	..	..	..	..	..	..	..	..	..	Wandsworth.
..	..	..	..	..	..	..	..	..	..	Camberwell.
..	..	..	..	..	..	..	..	1	..	Greenwich.
..	..	..	..	..	..	..	..	..	..	Lewisham.
..	..	..	..	..	..	..	..	..	..	Woolwich.
..	..	..	..	..	..	..	..	..	..	Plumstead.
..	..	..	..	..	..	..	..	..	..	Lee.
..	..	..	..	..	..	..	..	..	..	Port of London.

...ding Gray's Inn (population, 253), Lincoln's Inn (population, 27), Charterhouse (popula-...), Staple Inn (population, 21), and Furnival's Inn (population, 121).
...ding Inner Temple (population, 96). ‖ Including Tower of London (population, 886).

No. 1.

FURTHER REPORT on the BACILLUS ENTERITIDIS SPOROGENES; by DR. KLEIN, F.R.S.

P. B. No. 1. the Bacillus teritidis rogenes; Dr. Klein.

In last year's Report (27th Annual Report of the Medical Officer of the Local Government Board, 1897–1898, pp. 210–250), I described the Morphology and Biology of the bacillus enteritidis sporogenes; its distribution in milk, in sewage, in sewage polluted water, and in manure; and its relations with diarrhœa and cholera nostras.

During the past year, I have made investigation of this bacillus enteritidis as to its relations with the diarrhœa associated with typhoid fever, and have taken opportunity of examining in similar sense some additional cases of fatal cholera nostras, and an extensive epidemic of diarrhœa that occurred in the wards of St. Bartholomew's Hospital. This diarrhœa outbreak is the third of like sort that has occurred in this Hospital within two years. The two previous outbreaks have been fully described by me in the Reports of the Medical Officer, one in 1896–1897, the other in 1897–1898. This third outbreak occurred during the early part of August, 1898. It was investigated by Dr. F. W. Andrewes on exactly the same lines as the two previous outbreaks, and he has published an account of it, to which I shall refer in due course.

The occurrence and distribution of the spores of Bacillus enteritidis in typhoid fever and other intestinal diseases.

During the autumn and winter, of the typhoid fever occurring in London, numerous cases are taken in at St. Bartholomew's Hospital. I have therefore had ample opportunity of investigating typhoid stools with regard to the bacillus enteritidis. In obtaining such stools, with notes of the cases furnishing them, I have been greatly assisted by Dr. T. G. Forbes, of that hospital. The cases investigated were not in any way selected, but were taken just as they came into hospital. All were diagnosed as typhoid fever by the physicians of the wards; and in addition and as a matter of routine, the serum test of Widal was in all instances made in my colleague Dr. Andrewes' laboratory; so that confirmatory evidence as to their being true typhoid was in each instance forthcoming. As will be seen the number of such cases investigated was considerable; so that general conclusions may be justly drawn from the results of this investigation. In several instances, the stools of a case were examined on more than one occasion; in other instances, only one stool was subjected to examination.

The following were the cases dealt with :—

Case 1.—Edward W., age 33; duration of illness five days. Diarrhœa commenced on fourth day, stools typical "typhoid." Temperature on sixth day 103° F. Widal reaction positive.

Of a typical fluid typhoid stool (*a*) obtained on sixth day, $\frac{1}{100}$th of a cubic centimetre (10 cubic millimetres) was added to milk, which was then heated for 15 minutes at 80° C. After cooling, the vessel containing the inoculated milk was placed in a Buchner tube which was closed and incubated anaërobically at 37° C. This is the method which I described fully in last year's Report and which proved not only simple but perfectly efficient in demonstrating the presence of spores of bacillus enteritidis.

The above milk after 24 hours' incubation was typically changed in the fashion recorded in last year's Report, and the whey was full of the typical bacilli enteritidis. Of this whey 1 cc. was injected subcutaneously into the groin of a guinea-pig of about 200 grammes weight. The animal was found dead next morning, with typical gangrene and copious sanguineous malodorous exudation full of the typical bacilli.

On the ninth day of this patient's illness (temperature 103° F.) a further stool, (*b*), was obtained, which was not typical; it was fairly solid, loam coloured, with a little fluid. Of this fluid 10 cmm., 50 cmm., and 100 cmm., respectively, were used in anaërobic milk culture. The first two cultures remained unchanged, the third (100 cmm.) exhibited fairly typical change. Of its whey 2 cc. were injected into a guinea-pig, which had next day a big swelling on the abdomen, but did not die till after 48 hours.

This experiment shows then that the atypical stool of the ninth day of illness contained only a tenth part as many spores as the earlier typical stool; and further, that the milk culture of this stool proved considerably less virulent than that of stool (*a*).

On the twelfth day of illness the patient voided again a typical fluid typhoid stool (*c*). Temperature 102·5° F. Of this stool anaërobic milk cultures were made with 10 cmm., 100 cmm., and 250 cmm., respectively. The first remained unchanged; the other two showed the typical enteritidis change, and proved on injection into guinea-pigs of the normal virulence.

On the eighteenth day of illness (temperature 102·4° F.) a further stool (*d*) was obtained. This was of the typical fluid character, and the whey of an anaërobic milk culture therefrom, on injection into a guinea-pig, proved that the stool contained at least one spore per 50 cmm., and that the microbe was of the normal virulence.

On the thirty-sixth day the temperature of the patient was subnormal; but a stool (*e*) obtained at this date was still of the typical fluid character and contained at least one spore per 50 cmm. The whey of this culture was of the normal virulence.

On the forty-sixth day the patient was considered convalescent, temperature subnormal; a stool was atypical, solid, with a little fluid at the bottom of the vessel. From this fluid part anaërobic milk cultures were made, and in this way the stool was found to contain one spore per 100 cmm. This milk culture proved of the normal virulence.

It appears then that in the early stage of typhoid fever, when the temperature was high (103° F.) and the evacuation had all the characters of a typical typhoid stool, the result was most pronounced. Even 10 cmm. of the fluid from a stool at this stage sufficed to set up typical and virulent culture of bacillus enteritidis. Further, it is seen that on subsequent occasions the number of spores was greater in the typical stool than in stools which were atypical, though even during convalescence the fæces still contained virulent spores.

Case 2.—George W. M., aged 16. A stool (*a*) was obtained from this patient on the thirteenth day of his illness; temperature on that day was 99·6° F. Widal reaction positive. The patient had relaxed bowels, but no definite diarrhœa, since the commencement of his illness; the stool obtained on the thirteenth day of his illness was semi-solid, with a little fluid at the bottom of the vessel. Milk culture made anaërobically with 50 cmm. of this fluid yielded positive result, showing that the fluid part of the stool contained spores of bacillus enteritidis to the number of at least 20 per 1 cc. Experiment on a guinea-pig showed that this milk culture was of the normal virulence.

On the seventeenth day a further stool (*b*) of this case was obtained. The stool was of the same semi-solid consistency as the other, and there was a little fluid at the bottom of the vessel. The temperature of the patient on that day was 98° to 99·8° F. Anaërobic milk cultures made with the fluid part of the stool showed that it contained the spores of the bacillus enteritidis to the number of 10 per 1 cc. The action of the whey on the guinea-pig was, however, in this instance of a decidedly less virulent nature; 1·5 cc. of a typically changed milk culture, per 200 grammes body weight, produced a tumour and made the animal ill; but the animal recovered, with ulceration of the skin which ultimately healed.

On the twenty-sixth day a further stool (*c*) of this case was obtained. The temperature of the patient was 99° to 100° F. The stool consisted of solid fæces. Using for an anaërobic milk culture as much as $\frac{1}{10}$th of a gramme of the fæces, a positive result was obtained. But the whey of this milk culture, although in aspect, reaction, and in its containing numerous bacilli it was typical, proved when injected into a guinea-pig without any effect whatever; no swelling occurred, and the animal remained quite well. From this experiment I feel justified in not considering the culture in question as one of bacillus enteritidis, but (for reasons which I pointed out in last year's Report) one of the non-pathogenic bacillus butyricus. The patient died on the thirty-first day from exhaustion.

In this case then the stools were not typical typhoid stools; and although at first (thirteenth day of illness) the stools contained the virulent spores of bacillus enteritidis, later (on the seventeenth day) they were greatly reduced in number, and their action decidedly was less virulent. On the twenty-sixth day they could not be demonstrated.

Case 3.—Florence B., aged 19. A typical fluid typhoid stool (*a*) was obtained on the eighth day of illness. The patient's temperature on that day was 101·2° to 103·6° F. Widal reaction positive. Anaërobic milk cultures proved that the stool contained the spores of bacillus enteritidis to the number of at least 100 per 1 cc.; milk culture infected with 10 cubic millimetres of the stool becoming, after 24 hours' incubation, typically changed. One cc. of the whey of this culture subcutaneously injected into the groin of a guinea-pig produced death, with typical extensive gangrene and copious sanguineous malodorous exudation full of the bacilli enteritidis.

A further stool (*b*) of this case was obtained on the twentieth day. This stool was not typical; it was solid. The temperature of the patient was normal. A particle of the solid fæces (about $\frac{1}{10}$th to $\frac{1}{12}$th of a gramme) was used for making an anaërobic milk culture after the usual method. The result was positive; the milk culture was typically changed after 27 hours' incubation at 37° C. With 1 cc. of the whey of this milk culture one guinea-pig was injected; a second guinea-pig received 2 cc. The first guinea-pig showed after 27 hours a mere trace of swelling, and was lively; the second guinea-pig had a distinct swelling, but was also quite lively. The swelling led in this instance to ulceration of the skin, the animal remaining meanwhile seemingly quite well and lively.

A further stool (*c*) was obtained on the thirty-fourth day; this stool was solid fæces. The temperature was subnormal and the patient convalescent. Anaërobic milk cultures of the fæces were made, each with a fair sized particle (about $\frac{1}{10}$th to $\frac{1}{12}$th of a gramme) in the usual manner; but they remained quite unchanged, and no bacillus enteritidis could be demonstrated.

This case is so far instructive, as it shows that at the early stage of enteric fever, when the temperature was high and the stool of the nature of the typical typhoid stool, the evacuations contained abundance of virulent spores of the bacillus enteritidis; and that later, when the stools became atypical and the temperature normal, the number of spores had decreased and their virulence also had abated. And, further, that during convalescence, even a fair sized quantity of the solid fæces ($\frac{1}{10}$th to $\frac{1}{12}$th of a gramme) did not contain the spores of bacillus enteritidis. Conclusions must not, however, be hastily drawn, for other cases, to be presently described, yielded results different from those described above in regard of this Case 3.

Case 4.—Charles C. This case came first under observation* on the twenty-fourth day of illness, the temperature being then 103° to 104° F. Widal reaction was positive. The patient had diarrhœa during the first two weeks, but then became constipated. The stool (*a*) obtained on the twenty-fourth day was solid. About $\frac{1}{10}$th or $\frac{1}{8}$th of a gramme of it in anaërobic milk culture yielded positive result, and the whey of the typically changed culture proved of the normal virulence.

A further stool (*b*) was obtained on the thirty-fifth day of this patient's illness; it was solid. The temperature on this day was 98° F., and the patient was regarded as convalescent. Of the solid fæces two anaërobic milk cultures were made; one with about $\frac{1}{8}$th of a gramme, the other with about $\frac{1}{40}$th of a gramme. Only the first became typically changed, the second remained unchanged. With whey from the former one guinea-pig was injected. This was dead within 20 hours with typical symptoms, extensive gangrene and copious malodorous sanguineous exudation full of the bacilli.

---

* It will be understood that this expression is used here and subsequently in this Report to indicate that this was the first opportunity that I had of obtaining a stool of the case.

Case 5.—John B., sewerman, came under observation on the eleventh day of illness. Temperature 101° to 103° F. Widal reaction positive. Stool (*a*), obtained at this date was fluid, dark greenish brown. Anaërobic milk cultures were made with it, with the result that the stool was found to contain the spores of the bacillus enteritidis to the number of at least 100 per 1 cc. But their virulence was not of the usual standard; the injection of 1 cc. into a guinea-pig of about 200 grammes weight produced tumour, but the animal remained alive although the tumour led to extensive ulceration of the skin. Only when 1·5 cc. of whey was used per guinea-pig (*i.e.*, an increased dose) did the animal die, but even then not until after three days' illness.

A further stool (*b*) of this patient was obtained on the twenty-second day of illness. The stool was not typical, but solid. The temperature of the patient was subnormal. A fair-sized lump of fæcal matter (about $\frac{1}{10}$th to $\frac{1}{8}$th of a gramme) was used for anaërobic milk culture. The result was positive, and the whey proved of normal virulence.

A further stool (*c*) was obtained from this case on the thirty-first day of attack, the temperature still being subnormal. The stool was solid, but there was a little fluid fæcal matter at the bottom of the vessel. Of this fluid part of the stool, 50 cubic millimetres were used in anaërobic milk culture. The result was positive, and the whey of the culture proved of normal virulence.

In this case then, although the spores, at first (temp. 101° to 103°) very abundant, were not of the normal virulence; but later on (twenty-second day), when the temperature had become subnormal, and the number of the spores had decreased, they were of normal virulence. Still later, on the thirty-first day, the temperature of the patient being still subnormal, the number of spores had slightly increased again, and they proved of normal virulence.

Case 6.—Sidney W., aged 22. Came under observation on the thirteenth day of illness. Temperature 101° to 102° F. Widal reaction positive. A stool (*a*), obtained on this day, was a typical fluid typhoid stool. Anaërobic milk culture of this stool proved that it contained the spores of bacillus enteritidis to the number of at least 100 per 1 cc.; and experiment on the guinea-pig proved the whey of the culture of normal virulence.

A further stool (*b*) was obtained on the twenty-first day of illness, the temperature being subnormal. This stool was atypical—solid fæces. Anaërobic milk cultures of the stool proved the presence of the spores of bacillus enteritidis, but in small number; $\frac{1}{10}$th of a gramme did not yield any result in culture; $\frac{1}{5}$th of a gramme proved, however, positive in this respect, and the whey of this culture proved of normal virulence.

The patient had a relapse on the thirty-first day, temperature being 100·6° F.; and on the thirty-second day a stool (*c*) was obtained which was fairly typical. With fluid from this stool an anaërobic milk culture was set going, with the result of proving the presence of spores to the number of about 20 per 1 cc. The whey of this culture proved of normal virulence.

The patient was convalescent after the thirty-eighth day. On the forty-second day a further stool (*c*) was obtained; it was semi-fluid. Anaërobic milk culture of the stool showed spores to the number of ten per 1 cc., and they were of normal virulence.

Case 7.—Albert T., age 36. This case came under observation on the twenty-third day of his illness. Temperature 101° to 103° F. Widal reaction positive. Stool (*a*) obtained at this date was a typical fluid typhoid stool. Anaërobic milk culture of it showed the presence of spores of bacillus enteritidis to the number of about 50 per 1 cc. The milk culture was, however, of subnormal virulence; it took three to four days to kill a guinea-pig with the usual dose.

A further stool (*b*) was obtained on the thirty-eighth day, the temperature being 102·6° F. The stool was of the typical fluid typhoid character. Anaërobic milk culture of the stool showed the presence of spores of bacillus enteritidis to the number of 100 per 1 cc., and experiment on the guinea-pig with the whey of the culture proved them to be of the normal virulence.

Case 8.—Sidney W. This case came under observation in the convalescent stage on the twenty-ninth day; the temperature was normal. The stool (*a*) was of a semi-solid character. Anaërobic milk culture made with $\frac{1}{8}$th of a gramme of the stool proved negative. A further stool (*b*) was obtained on the thirty-eighth day, the patient's temperature being normal. This stool was normal, solid. With it anaërobic milk cultures were made, the amounts of fæces used being considerable. A milk tube that received about $\frac{1}{4}$th to $\frac{1}{5}$th of a gramme proved positive; the whey injected into a guinea-pig was of normal virulence.

Case 9.—Ada S., age 26. This case came under observation on the twentieth day, the temperature being [illegible] F. Widal reaction positive. The stool obtained at this time was of the typical fluid typhoid character. A platinum loop full yielded in milk culture positive result, and the whey proved of normal virulence.

[In this case, as also in several subsequent cases, the stool had been carbolised in the wards immediately after it was voided: so that no exact measurement of the amount of stool added to the milk in making anaërobic milk culture could be made. But it turned out that the "carbolising" of this stool, as also of several other similar stools of typhoid cases (a large volume of a 5 per cent. solution of carbolic acid being added to the typhoid stool in each instance), had no deleterious effect on the spores of the bacillus enteritidis contained in it.]

A further stool (b) was obtained on the thirty-eighth day, the temperature being normal and the patient convalescent. This stool was atypical, solid. With a fair-sized lump of the stool, an anaërobic milk culture was made, but this proved negative.

Case 10.—[illegible] C., age 26. Came under observation on the twenty-fifth day of illness. Temperature [illegible] to [illegible] F. Widal reaction positive. The stool obtained, a typical fluid typhoid stool, had been carbolised: but it produced in anaërobic milk culture the typical change of bacillus enteritidis, and the whey, on injection into a guinea-pig, proved of normal virulence.

Case 11.—Ellen H., age 14. Came under observation on the twenty-third day of illness, temperature being [illegible] F. Widal reaction positive. The stool, which was of typical character, had, when obtained in the laboratory, been carbolised. Anaërobic milk culture, with a loop full, gave positive result; virulence of whey normal.

A further stool (b) was obtained on the fortieth day of illness, the temperature being subnormal, and the patient convalescent. The stool, which was solid, had, when received in the laboratory, been carbolised. A good-sized lump of the solid part of the stool was employed in anaërobic milk cultures. These showed after three days' incubation only slight changes; gas was produced, but there was much curdled casein and not the normal amount of whey, though this contained a fair number of bacilli. Injection into the guinea pig of a normal dose of whey produced swelling and transitory quietness; but the animal recovered, after the skin over the abdomen had ulcerated. This shows that the number of spores in the fæces could not have been large, and that these possessed decidedly subnormal virulence.

Case 12.—William B., age 15. Came under notice on the forty-second day of illness, the temperature being 104·2 F. Widal reaction positive. The stool obtained on that day was semi-fluid. Anaërobic milk culture, made with a platinum loop full of the stool, was, after 24 hours' incubation, typically changed, and the whey proved of normal virulence.

Case 13.—Alfred C., age 21. This case came under notice on the ninth day of illness, the temperature being 101·6 F. Widal reaction positive. The stool was atypical, semi-solid, the patient, however, had had diarrhœa during the first four days. With a platinum loop full of the solid part of the stool an anaërobic milk culture was made. This, after 24 hours' incubation, was typically changed, and the whey, when injected into a guinea-pig, proved of normal virulence. The patient had normal temperature on the twentieth day, and was then considered convalescent. Owing to a misunderstanding no further stool was obtainable.

Case 14.—Ellen S., age 16. Came under observation on the fourteenth day of illness, temperature being 103·8 F. Widal reaction positive. The stools were typical fluid typhoid. A sample (*a*) that was brought to the laboratory had been carbolised. With a loop full of the stool an anaërobic milk culture was, however, made, and this became typically changed, while its whey proved of normal virulence.

A further stool (*b*) was obtained on the twenty-fifth day of the patient's illness, her temperature being 101·4° F. The stool was solid. With a fair sized lump of it (about $\frac{1}{12}$th of a gramme) an anaërobic milk culture was made, which became typically changed, and the whey of which proved of normal virulence.

A further stool (*c*) was obtained on the thirty-second day, the temperature being normal and the patient convalescent. The stool was normal, solid. With about $\frac{1}{10}$th and $\frac{1}{4}$th of a gramme respectively, anaërobic milk cultures were made. These did not undergo any change.

Case 15.—Mary D., aged 3½. This case came first under notice on the fourteenth day of illness, temperature being 100·6° F. Widal reaction positive. The patient had had diarrhœa and typical typhoid stools in the earlier stages of the disease. The patient's bowels on the fourteenth day were constipated, and the stool (*a*) obtained was solid. It had been carbolised. Anaërobic milk cultures made with a loop full of the solid material became typically changed, and the whey proved of normal virulence.

A further stool (*b*) was obtained on the twenty-fourth day, temperature being normal and the patient convalescent. The stool was solid. With a fair-sized lump (about 1/10th of a gramme) an anérobic milk culture was made, which became typically changed after twenty-four hours' incubation, and the whey of which proved of normal virulence.

Case 16.—William W., age 16. This case came first under notice on the fourteenth day of illness. Temperature 102·8° F. Widal reaction positive. The patient had had no diarrhœa at any time during his fortnight of illness. The stool (*a*) obtained on this day was solid. With a loop full of the solid fæces an anaérobic milk culture was made, which became typically changed in due time, and the whey of which proved of the normal virulence.

A further stool (*b*) was obtained on the twenty-first day, temperature being 102·6° F. This stool was of a semi-fluid character. Anaerobic milk culture of the stool proved negative. No further stool was obtainable from this case.

Case 17.—William S., aged 5. This case came under notice on the fourteenth day of illness; temperature then 101·4° F. Widal reaction positive. A stool (*a*) obtained at this date was a typical fluid typhoid stool. With about 10 cmm. (1/100th of a cc.) of it an anaërobic milk culture was made, which became atypically changed, and the whey proved devoid of pathogenic action.

A further stool (*b*) was obtained on the twenty-second day, temperature being still 101·4° F. This stool was semi-fluid; and a loop full in anaërobic milk culture produced the typical change, the whey proving of normal virulence.

No further stool was obtainable in this case.

Case 18.—Eliza T., aged 9. Came first under notice on the fifteenth day of illness, the temperature then being 104° F. Widal reaction positive. Stool (*a*) was typical typhoid. With a loop full of it an anaërobic milk culture was made which proved negative.

A further stool (*b*) was obtained on the twenty-seventh day of illness, the temperature being still high, 103·4° F. The stool was semi-solid. Anaërobic milk culture proved negative. The patient died on the thirty-first day.

Case 19.—Eliza S., aged 4. This case came under observation on the twelfth day of illness, the temperature being 99·4° F. Widal reaction positive. Stool (*a*) was typical, fluid. With a platinum loop of this fluid stool an anaërobic milk culture was made which became typically changed in due time, and the whey of which proved of normal virulence.

A further stool (*b*) was obtained on the thirty-first day, the temperature being 100° F. This stool also was typical. With a platinum loop an anaërobic milk culture was made, which became typically changed in due time, and the whey of which proved of normal virulence. No further stool of this case was obtainable.

Case 20.—William S., aged 16. Came under notice on the thirty-first day of his illness; the temperature was normal, but the stool was of the typical character. Widal reaction positive. With a platinum loop an anaërobic milk culture was made, which became changed in an atypical manner, though the whey of it proved of normal virulence.

No further stool obtainable from this case.

Case 21.—Ellen K., aged 5. This case came under observation on the fifteenth day of illness, the temperature being 100·6° F. Widal reaction positive. The patient had been throughout constipated. On the fifteenth day a carbolised stool was obtained. With a solid particle of it an anaërobic culture was made which yielded positive result, and the whey proved of normal virulence.

Case 22.—Fred. G., aged 6. This case after eighteen days' illness had for eight days normal temperature; then on the twenty-sixth day the temperature suddenly rose to 102° F. and the patient "relapsed." Widal reaction positive. All through the first illness and until relapse the bowels were in this case constipated. With the relapse the patient voided a typical typhoid stool, one platinum loop of which was used for anaërobic milk culture. The culture became changed in a typical manner, and the whey proved pathogenic, inasmuch as it produced great swelling in the groin and on the abdomen. The animal, however, recovered after extensive ulceration of the skin of the abdomen.

[The last six cases (Case 17 to Case 22) formed an interesting group comprised in a localised outbreak of typhoid fever in a court off Red Lion Street, Clerkenwell. This outbreak began with James S., aged 4½, on November 20th, 1898. His case, which did not come under my observation, was followed by that of Fred G. (Case 22) on November 28; by that of Eliza T. (Case 18) on December 3rd; by Eliza S. (Case 19) on December 5th; by William S. (Case 17) on December 6th; by William S. (Case 20) on December 8th; and lastly by Ellen K. (Case 21) on December 13th. All these patients lived in the same court and, with the exception of William S. (Case 20), whose age was 16, were children between 4 and 9 years of age, accustomed to play together in the court. No definite cause of the fever could be traced. There was no uniformity in the symptoms as regards diarrhœa; some had typical fluid typhoid stools from the first, while others were constipated. As mentioned above, in most instances (except Case 18) the spores of the bacillus enteritidis were present in the bowel evacuations in sufficient abundance to be easily detected by using small amounts (a platinum loop, *i.e.*, about 10 cm.) of the fluid stool for anaërobic milk culture.]

Case 23.—John W., aged 23. Came under observation on the thirteenth day of illness. Temperature had been on the eighth day 102° F.; on the thirteenth day it had fallen to 100° F. Widal reaction was positive. The patient had diarrhœa from the fourth day of illness till the eighth day; on the thirteenth day the stool was of semi-solid nature, pale yellow in colour. Anaërobic milk culture made with a platinum loopful of stool became typically changed, and the whey proved of normal virulence. No further stool was obtainable.

Case 24.—Rachel R., aged 23. First came under notice on the twenty-fourth day of illness; temperature 104·2° F. Widal reaction positive. A stool, typical fluid typhoid stool, in anaërobic milk culture produced typical change, and the whey of the culture proved of normal virulence.

Case 25.—Emeline B., aged 32. Came first under notice on the twenty-first day of illness, when the temperature was varying between 103·6° and 105° F. The patient was in a bad state, having rigors almost daily. Widal reaction positive. Stools typical, fluid. A platinum loop from such stool yielded positive result in anaërobic milk culture, and the whey proved of normal virulence.

Case 26.—James S., aged 7. This case came first under notice on the eighth day of illness, the temperature being 104° F. Bowels constipated. Widal reaction positive. A small portion of semi-solid stool in anaërobic milk culture gave typical change. The virulence, however, of the whey was sub-normal; it produced a big tumour, but this broke and led to ulceration of the skin, which ultimately quite healed up.

Case 27.—John C., aged 21. Came under notice on the twenty-eight day of his illness; temperature on the fourteenth day 102° F. Widal reaction positive. Had diarrhœa for ten days from the eleventh till twenty-first day; stools typical, fluid. On the twenty-eighth day patient's temperature was normal, and the stool voided was solid. Anaërobic milk culture made with a fair sized lump of it became atypically changed; such culture contained a few slim bacilli which were not bacillus enteritidis and which were not pathogenic to the guinea-pig.

Case 28.—Osmond N., aged 18. Came under observation on the fourteenth day of illness, the temperature being 101°F. Widal reaction positive. Had no diarrhœa until the eighth day of illness. The stool obtained was typical of typhoid. Anaërobic milk culture with one loopful of it proved positive; virulence of the whey normal.

Case 29.—George B., aged 18. Came under observation on the twelfth day of illness; temperature 100·2° F. Widal reaction positive. Bowels constipated. Obtained (by enema) a solid stool, which did not yield positive result in anaërobic milk culture; the culture became atypically changed, and though it contained few slim bacilli, these were not bacilli enteritidis and were non-virulent.

Case 30.—Felix K., aged 21. Came under observation on the fourteenth day of illness, temperature 100° F. Widal reaction positive. Patient had been constipated; solid stool obtained by enema. With a fair-sized lump of fæces an anaërobic milk culture was made, which gave negative result.

[These three cases (28, 29 and 30) were employés at the same establishment in Holborn. Case 28 and case 30 were taken ill with typhoid fever on January 16th, 1899; case 29 on January 18th. A fourth person, William D., aged 23 (of whom no stool was obtainable), was taken ill with typhoid fever also on January 16th, at this establishment. This case also was free from diarrhœa, his bowels being constipated throughout. There was failure, it will be noted, to demonstrate the spores of bacillus enteritidis in cases 29 and 30 (constipated), but positive results were obtained in case 28. Now in this latter case, in which the spores of bacillus enteritidis were abundant in the stool, and of normal virulence, the stool voided was a typical fluid one; whereas in the other two cases (constipated), from which solid fæces only were obtainable, the culture test made with a fair-sized lump of material failed to yield the spores of bacillus enteritidis.]

Case 31.—James M., aged 24. Came first under observation on the eighteenth day of illness, the temperature being then 103° F. Widal reaction positive. Had had diarrhœa throughout, and this was still present on the eighteenth day. The stool when received in the laboratory had been carbolised. With a few drops of this carbolised stool an anaërobic milk culture was made, which yielded positive result, and the whey proved of normal virulence.

Case 32.—Arthur S., aged 14. This case came first under notice on the seventeenth day of illness, the temperature being then 102·9° F. Widal reaction positive. The patient had had diarrhœa throughout, and the stool received on the 17th day was a typical fluid typhoid stool. Anaërobic milk culture made with a loopful of stool yielded positive result, and the whey proved of normal virulence.

Case 33.—Kate H., aged 47. Came first under notice on the twenty-seventh day of illness, the temperature being then 101° F. Widal reaction positive. Had had diarrhœa throughout; the stool received on the tweny-seventh day was typical and fluid. With a loopful of this an anaërobic milk culture was made, which proved positive; but the whey was of slightly subnormal virulence, the animal recovering after the fifth day.

Case 34.—Sophy L., aged 7. Came under observation on the twenty-eighth day of illness, temperature on that day being 99–102·2° F. Widal reaction positive. The patient had had no diarrhœa up to that date, and the stool obtained was solid fæces. The result of anaërobic milk culture, made with a fair-sized lump of it, proved negative as to bacillus enteritidis.

Case 35.—Nurse B., age 24. Came under observation on the fifteenth day of illness, temperature being then normal. Widal reaction positive. No diarrhœa till the thirteenth day, when the bowels became loose. The stool obtained on the twenty seventh day was semi-fluid, and from it anaërobic milk culture was made, which proved negative. This same patient suffered a relapse on the forty-fifth day from the onset of illness; her temperature rose to 103·2° F., but there was no diarrhœa. A stool was obtained on the fifty-second day, at which date her temperature was still 102° F. This stool was semi-solid. Anaërobic milk culture of it yielded negative result.

Case 36.—Charles R., age 32. Came first under notice on the twenty-third day of illness, temperature being 100° F. Widal reaction positive. The stool received was fluid, typical typhoid. An anaërobic milk culture, made with a platinum loopful, became typically changed, and the whey proved of normal virulence.

Case 37.—Walter P., age 7. Came under notice on the twenty-second day of illness, temperature being 100·2° to 102·8° F. Widal reaction doubtful. A stool received on that day was solid. Anaërobic milk culture, made with a fair-sized lump of it, yielded positive result, and the whey was of normal virulence.

Case 38.—Nelly H., age 15. Came under notice on the twenty-first day of her illness, temperature being 97·8° to 99·8° F. Widal reaction negative. No diarrhœa; solid dark stools. Anaërobic milk culture of these proved negative qua bacillus enteritidis.

Case 39.—John H., age 29. Came under observation on the twenty third day of his illness, temperature being normal. Widal reaction positive. Had had no diarrhœa during his illness. A stool on the twenty-third day was semi-fluid. With a fair-sized lump of it an anaërobic milk culture was made, which proved positive, and the whey was of normal virulence.

Case 40. James C., age 32. Came under notice on the [illegible] day of his illness, temperature being [illegible] F. Widal reaction positive. A stool received at this date was typical, fluid typhoid. Anaerobic milk culture, with a [illegible] of the stool, proved positive and the whey was of normal virulence.

In addition to certain cases above described, which were examined by me in the early stage of their illness but also in the [illegible] stage, I had opportunity of subjecting to examination other cases that came under my notice after convalescence had already set in.

Case 41. Amelia V., age 25. Came first under notice at the end of [illegible] eight weeks from the commencement of illness. Temperature normal. Stools [illegible]. The patient had until towards the end of the [illegible] week typical fluid stools; afterwards her bowels were constipated. Widal reaction positive. With a fair-sized particle of a stool anaerobic milk culture was made, and proved positive; the whey, too, was of normal virulence.

Case 42. Edward D., age 27. Came under notice at the end of five weeks from commencement of illness. Temperature normal. Stools semi-solid. The patient had diarrhœa during the first three weeks, after which the stools became of normal consistency. Widal reaction positive. With fair-sized particle of a stool an anaerobic milk culture was made, which proved negative (the bacillus absent).

Case 43. Eliza H., age 27. Came under notice on the thirty-ninth day from commencement of illness, her temperature being then subnormal and her stools solid [illegible]. This patient had diarrhœa for the first two weeks of attack. Widal reaction positive. With a fair-sized particle of a solid stool an anaerobic milk culture was made, which proved positive, and the whey was of normal virulence.

Case 44. Christopher M., age 14. Came first under notice on the thirty-ninth day from commencement of illness. Temperature had been normal for two weeks, and the patient convalescent. Widal reaction positive. Had had diarrhœa in the earlier stages of his illness. Stool obtained for examination solid. With a fair-sized lump of this stool an anaërobic milk culture was made, which proved positive, and the whey was of normal virulence.

I proceed now to place in tabular form all the cases of typhoid fever in regard of which opportunity was afforded me for examining the stools. All had been, in the first instance, diagnosed as instances of this disease, and in regard of two only did doubt subsequently arise: Namely, *Case* 38 (Nelly H.), which had been diagnosed, by the general nature of her illness and the raised temperature in the early stages, as typhoid fever, but which failed to give Widal reaction and was a case without diarrhœa at any time during illness; and *Case* 37 (Walter P.), whose blood behaved doubtfully in respect of Widal reaction. This case had, however, characteristic temperature during the first part of his illness, and had diarrhœa also during the first twelve or fourteen days.

Some preliminary comments are necessary as to the several columns in the table. The third column gives the temperature of the body of the patient on the day that a stool was obtained—the two figures mean: left morning, right evening temperature. The fifth column states the condition of the bowels on the day that a stool was obtained, and the sixth column notes the nature of the stool actually received at the laboratory. It is to be understood that "No Diarrhœa" is not retrospective; for, as was pointed out in the description of the cases on previous pages, in many instances diarrhœa had been present for a longer or shorter period antecedent to the day on which a stool was obtained for examination. The seventh and eighth columns require somewhat more detailed explanation.

As to column 7 "Number of Spores" (of bacillus enteritidis) found in the stool on a particular day. Wherever it was possible—*e.g.*, in the case of the typical thin fluid stool—the amounts of the stool that were added in each instance to 10–12 c.c. of milk had been actually and accurately measured; so that in these cases the number of spores that produced the typical enteritidis change of the anaërobic milk culture could be fairly accurately estimated. But since the number of milk cultures that could be devoted to each sample of stool were of necessity limited, the several amounts of a given fluid stool added to separate milk cultures differed within considerable limits. Each sample of fluid stool was thus dealt with: There was added to one milk culture 10 cmm., to another 50 cmm., and to a third 100 cmm. of the stool; and if it were found that the first culture remained unaffected, while the other two showed the typical enteritidis change, it was inferred that the particular stool contained *at least* 20 spores per 1 cc. I say "at least," because it is possible that like change might have been also brought about by any amount of stool less than 50 cmm. and more than 10 cmm.

Where the stool was typical, though not so thin a fluid as the above, a platinum loopful was dealt with in lieu of an actually measured quantity. Such a loopful amounts, as far as can be approximately determined, to nearly 10 cmm., *i.e.*, $\frac{1}{100}$th of 1 cc., certainly not more; and in many instances, the statement "Numerous" refers to the result of this method, for in such case *at least* 100 spores must have been present per 1 cc. of stool.

In those cases in which the stool was semi-solid or solid, a "fair-sized particle" of the material was added to the milk. I have in several instances attempted to get an idea of the amount of stool thus dealt with as compared with a sample of typical thin fluid typhoid stool obtained in the way above noted. This I have done by taking up with the same loop a similarly sized particle of the solid stool, and distributing it in water so as to render the mixture as like as possible (so far as can be judged by the eye) as regards turbidity and consistency, to a thin fluid stool; and I have found that such a "fair-sized particle" corresponds to from $\frac{1}{4}$ to $\frac{1}{2}$ a cc. of the fluid stool. A positive result after inoculation of the milk, therefore, with a "fair-sized particle or lump" would mean that about 2–4 spores were present per 1 cc. of the fluid.

It has been stated in the notes of the cases in several instances, that the amount of solid fæces was weighed out as nearly as could be done; that $\frac{1}{10}$th of a gramme of a solid stool was then distributed in water, so as to make an emulsion that approximately corresponded in turbidity and consistency to a typical thin fluid typhoid stool; and that it was found that $\frac{1}{10}$th of a gramme of the solid material would correspond to about 1 cc.

of a typhoid stool. In cases, therefore, in which $\frac{1}{10}$th of a gramme of solid stool had to be added to milk to give a positive result, only about 1 spore was present per 1 cc. of stool; whereas, as regards the typical fluid typhoid stool, in many instances the number of spores present was at least 100 per 1 cc. of stool.

When the milk (heated after inoculation to 80°C. for 12–15 minutes in order to kill everything except spores), having been anaërobically incubated at 37°C., showed no change after three days, such milk certainly did not contain the spores of bacillus enteritidis. For if these latter be present the change is certainly well under way, if not fully developed, after 24–36 hours. In several instances, the milk culture changed only after lapse of 2–3 days, or even later; but in such case the change was quite different from that produced by bacillus enteritidis. The latter grows very rapidly, and in 24–36 hours the milk in which it is growing exhibits copious clear (or more or less turbid) whey, with casein flocculi adhering here and there to the wall and bottom of the tube, and particularly floating as slightly pinkish threads buoyed, on the top, up by copious gas bubbles, which by this time have so completely disarranged and disintegrated the superficial layer of cream as practically to destroy it. The whey under the microscope is full of the characteristic rod-shaped bacilli enteritidis.

When, however, the milk underwent change only after 2–3 days, and when this change consisted in a complete solidification of the milk with a few cracks in it containing clear fluid, the cream layer remaining undisturbed and separated from the subjacent clot by a thin layer of clear liquid, the clear fluid contained under the microscope a few thin long bacilli. As I have pointed out in my former reports, the only microbe besides bacillus enteritidis that produces the rapid and characteristic change of the anaërobic milk is the bacillus butyricus of Botkin. But this microbe is without any pathogenic action. In the case of bacillus enteritidis the pathogenic action is very typical and rapid. But, as was also pointed out in a former report (1897–1898), the degree of its pathogenicity, as regards the guinea-pig, varies in different samples of spores. In most cases of diarrhœa (as in summer diarrhœa, epidemic diarrhœa, cholera nostras, typhoid stools) a typically changed anaërobic milk culture forms a whey 1 c.c. of which, when injected subcutaneously into the groin of a guinea-pig weighing 200-300 grammes, causes in a few hours swelling about the seat of inoculation. This swelling by the next morning (within 20 hours) extends over the groin, the abdomen, and even chest, and feels to the touch like a big bag of gas or fluid. The animal, as a rule, is dead in this space of time, or dies in 24 or 30 hours. On post-mortem examination the skin of the groin, thigh, abdomen, and even chest is separated from the subjacent muscles by gas or fluid, the latter being a sanguineous (but showing under the microscope no blood discs), malodorous, thin or viscid fluid crammed with the characteristic rod-shaped bacilli. The subcutaneous and muscular tissues are partly dissolved into shreds, and are gangrenous. A milk culture the whey of which

APP. B. No. 1. On the Bacillus Enteritidis Sporogenes; by Dr. Klein.

in the above amount causes this typical fatal result with extensive gangrene I consider as of "normal virulence," and in the foregoing description of the cases and in the tabular statement the expression "normal virulence" is to be taken in this sense.

In those instances when the normal dose of 1 c.c. of the whey of a typically changed milk culture* fails to kill a guinea-pig of 200–300 grammes weight in 24–36 hours, although it has produced extensive swelling in the region of the groin and abdomen, and even chest, such culture is spoken of as "of decreased virulence." And there are various degrees of such decreased virulence, ranging from one which produces distinct more or less gelatinous tumour, leaving the animal lively and otherwise well, up to a virulence rendering the animal distinctly ill and quiet, with definite tumour, and killing it in two to three days or later, with post-mortem appearance of fairly well marked gangrenous change in the tumour and surrounding parts. A more than normal dose—*i.e.*, 1·5 to 2 c.c. of the whey of a typically changed milk culture per guinea-pig of 200–300 grammes—which does no more than cause tumour and retarded death, is of course also a culture of "decreased virulence."

But whatever the degree of virulence may be, and whether the experimental animal appears lively and well or is more or less quiet and off its feed, there always is after 24 hours a tumour produced; and by this the pathogenicity of the culture is established. This local effect, in the case of a culture of "decreased virulence" as a rule leads to sloughing of the skin, which ultimately, after a week or two, completely heals. In a few instances the tumour (fluid) after 24–48 hours was of great size and extent and led rapidly to sloughing, becoming open and discharging copious thin malodorous fluid.

Although I have not met in the above typhoid fever experiments with cultures of the spores of bacillus enteritidis possessing any greater virulence than is expressed by 0·5 c.c. of the whey producing, in a guinea-pig of 200–300 grammes weight, typical tumour, extensive gangrene, and death in 20–24 hours, I have occasionally obtained milk cultures of the spores of bacillus enteritidis from raw sewage, as also from sewage-polluted water, exhibiting virulence considerably greater. For instance, from Hendon sewage and from the deposit of the water of the River Lea and of the River Thames, I have had milk cultures of which 0·25 c.c. of the whey caused the above extensive gangrene and death in 16 hours; the spores of the crude Hendon sewage being, indeed, those possessed of the highest degree of virulence that I have yet met with. I consider these facts, viz., as to especial virulence of spores of sewage (crude or water diluted), of importance and worthy of being placed on record.

* It will be understood that invariably, for these cultures, quite recently sterilized milk was used; for, as I have pointed out in last year's report stale milk is not so well suited for these cultivation experiments.

P. B. No 1

the Bacillus
teritidis
orogenes;
Dr. Klein

Case. (1.)	Day of Illness at which Stools Examined. (2.)	Body Temperature. (3.)	Widal reaction. (4.)
No. 1.—E. W.	6th day ...	103	+
	9th „ ...	103	
	12th „ ...	102·5	
	18th „ ...	102·4	
	36th „ ...	Subnormal ...	
	46th „ ...	„ ...	
No. 2.—G. W. M.	13th „ ...	99·6 to 98	+
	17th „ ...	98 to 99·8	
	26th „ ...	99·4 to 100	
No. 3.—F. B.	8th „ ...	101·2 to 103·6	+
	20th „ ...	Normal	
	34th „ ...	Subnormal, convalescent.	
No. 4.—C. C.	24th „ ...	103 to 104	+
	35th „ ...	98	
No. 5.—J. B.	11th „ ...	101 to 103	
	22nd „ ...	Subnormal ...	
	31st „ ..	„ ...	
No. 6.—S. W.	13th „ ...	100 to 102	+
	21st „ ...	Subnormal ...	
	32nd „ ...	Relapse 100·6 ...	
No. 7.—A. J.	23rd „ ...	101 to 103	+
	38th „ ...	102·6	
No. 8.—S. W.	29th „ ...	Convalescent ...	Not notec
	38th „ ...	„ ...	
No. 9.—A. S.	20th „ ...	101 to 102	+
	38th „ ...	Convalescent ...	
No. 10.—B. C.	25th „ ...	100 to 102	+
No. 11.—E. H.	23rd „ ...	104·2	+
	40th „ ...	Subnormal, convalescent.	
No. 12.—W. B.	42nd „ ...	104·2	+
No. 13.—A. C.	9th „ ...	101·6	+
No. 14.—E. S.	14th „ ..	103·8	+
	25th „ ...	101·4	
	32nd „ ...	Convalescent ...	
No. 15.—M. D.	14th „ ...	100·6	+
	24th „ ...	Convalescent ...	
No. 16.—W. W.	14th „ ...	102·8	+
	21st „ ...	102·6	
No. 17.—W. S.	14th „ ...	101·4	+
	22nd „ ...	101·4	

State of Bowels on occasion of Examination of Stool. (5.)	Character of Stool Examined. (6.)	Number of Spores in Stool. (7.)	Action of Whey of Milk Culture on Guinea-pig. (8.)
Diarrhœa ...	Typical	At least 100 per 1 cc.	Normal virulence.
No diarrhœa	Not typical	Decreased	Decreased „
Diarrhœa ...	Typical	„	Normal „
Loose ...	Fairly typical ...	Increased	„ „
Diarrhœa ...	Typical	„	„ „
Loose ...	Not typical	Decreased	„ „
„ ...	Not typical	At least 20 per 1 cc.	„ „
„ ...	„	Decreased	Decreased „
„ ...	Solid	Negative	Not pathogenic.
Diarrhœa ...	Typical	At least 100 per 1 cc.	Normal virulence.
No diarrhœa	Solid	Decreased	Greatly decreased virulence.
„	„	Negative	Negative.
No diarrhœa	Solid	Small number ...	Normal virulence.
„	„	„ ...	„ „
Diarrhœa ...	Fluid...	At least 100 per 1 cc.	Decreased „
No diarrhœa	Solid	Decreased	Normal „
„	„	Increased	„ „
Diarrhœa ...	Typical	At least 100 per 1 cc.	„ „
No diarrhœa	Solid	Decreased	„ „
Loose ...	Not typical, but fluid	Increased	„ „
Diarrhœa ...	Typical	About 50 per 1 cc. ...	Decreased „
„ ...	„	At least 100 per 1 cc.	Normal „
No diarrhœa	Semi-solid	Negative	—
„	Solid	Positive, few spores	Normal virulence.
Diarrhœa ...	Typical	Numerous	„ „
No diarrhœa	Solid	Negative	—
Diarrhœa ...	Typical	Numerous	Normal virulence.
„ ...	Typical	Numerous	„ „
No diarrhœa	Solid	Much decreased ...	Decreased „
No diarrhœa, loose.	Semi-solid	Numerous	Normal „
No diarrhœa	„	Positive	„ „
Diarrhœa ...	Typical	Numerous	„ „
No diarrhœa	Solid	Decreased	„ „
„	„	Negative	—
Constipated	Solid	Positive	Normal virulence.
„	„	„	„ „
No diarrhœa	Solid	„	„ „
„	Semi-fluid	Negative	—
Diarrhœa ...	Typical	„	—
No diarrhœa	Semi-fluid	Positive	Normal virulence.

.B. No. 1.

he Bacillus tidis

Dr. Klein.

Case. (1.)	Day of Illness at which Stool is Examined. (2.)	Body Temperature. (3.)	Widal reaction. (4.)
No. 18.—E. J.	16th day ... 27th ,, ...	104 103·4	+
No. 19.—E. S.	12th ,, ... 31st ,, ...	99·4 100	+
No. 20.—W. S....	31st ,, ...	Normal	+
No. 21.—E. K....	15th ,, ...	100·6	+
No. 22.—F. G.	28th ,, ... Relapse.	102	+
No. 23.—T. W.	13th day ...	100	+
No. 24.—R. R.	24th ,, ...	104·2	+
No. 25.—E. B....	21st ,, ...	103·6 to 105	+
No. 26.—J. S.	8th ,, ..	104	+
No. 27.—T. C.	28th ,, ...	Normal	+
No. 28.—O. N....	14th ,, ...	101	+
No. 29.—G. B....	12th ,, ...	100·2	+
No. 30.—F. K....	14th ,, ...	100	+
No. 31.—T. M.	18th ,, ...	103	+
No. 32.—A. S.	17th ,, ...	102·9	+
No. 33.—K. H.	27th ,, ...	101	+
No. 34.—S. L.	28th ,, ...	99 to 102·2	+
No. 35.—Nurse	15th ,, ... 52nd ,, ...	Normal ... ... Relapse 102·0 ...	+
No. 36.—C. R....	23rd ,, ...	100	+
No. 37.—W. P.	22nd ,, ...	100·2 to 102·8	?
No. 38.—N. H.	21st ,, ...	97 8 to 89 8	Negative
No. 39.—T. H....	23rd ,, ...	Normal	+
No. 40.—T. C.	17th ,, ...	100	+
No. 41.—A. V....	8 weeks ...	Normal convalescent.	+
No. 42.—E. D....	5 ,, ...	Normal convalescent.	+
No. 43.—E. H....	39th day ...	Subnormal ...	+
No. 44.—C. M....	39th ,, ...	Normal convalescent.	+

State of Bowels on occasion of Examination of Stool. (5.)	Character of Stool Examined. (6.)	Number of Spores in Stool. (7.)	Action of Whey of Milk Culture on Guinea-pig. (8.)
Diarrhœa ...	Typical	Negative	—
No diarrhœa	Semi-solid	"	—
Diarrhœa ...	Typical	Numerous	Normal virulence.
" ...	"	"	" "
Diarrhœa ...	Typical	"	" "
Constipated	Solid	Positive	" "
Diarrhœa ...	Typical	"	Decreased "
No diarrhœa	Semi-solid	"	Normal virulence.
Diarrhœa ...	Typical	Numerous	" "
" ...	"	"	" "
Constipated	Semi-solid	"	Decreased "
No diarrhœa	Solid	Negative	—
Diarrhœa ...	Typical	Numerous	Normal virulence.
Constipated	Solid	Negative	—
"	"	"	—
Diarrhœa ...	Typical	Numerous	Normal virulence.
" ...	"	"	" "
" ...	"	"	Decreased "
No diarrhœa	Solid	Negative	—
Loose ...	Semi-fluid	"	
No diarrhœa	Semi-solid	"	—
Diarrhœa ...	Typical	Numerous	Normal virulence.
No diarrhœa	Solid	Positive	" "
	"	Negative	—
"	Semi-solid	Positive	Normal virulence.
Diarrhœa ...	Typical	Numerous	" "
Constipated	Semi-solid	Positive	" "
Normal ...	Solid	Negative	—
No diarrhœa	"	Positive	Normal virulence.
"	"	"	" "

B. No. 1.

e Bacillus
ritidis
genes;
r. Klein.

Proceeding now to summarise and compare the results obtained by examination of stools in the several cases :—

In 25 instances there was diarrhœa and typical typhoid stools at the time of experiment; and in 23 of the 25 the spores of bacillus enteritidis were numerous enough to be detected in small quantities of the stool (in 10 cubic millimetres more or less), in two instances only did such small quantity fail to give evidence of the presence of these spores. A negative result in this sense does not of course mean that no such spores were present in the stool; all that such experiment shows is that they were not numerously present—that they were in quantity less than 100 per 1 cc. In other 26 instances the stool at the date of examination was solid. Of these 26, ten yielded negative results quá spores of bacillus enteritidis, 0·25–1 cc. emulsion comparable to fluid typhoid stools being employed in each instance; that is to say, large quantities of the solid material were in each instance tested.

Whereas, therefore, in the phases of enteric fever associated with typical fluid typhoid stools the spores of bacillus enteritidis are as a rule *numerously present*, in a considerable percentage of instances (about 39 per cent.) of the typhoid cases wherein the stools had become, owing to constipation or to convalescence, normally formed and solid, the spores were altogether *absent* in relatively considerable amounts of the fæcal matter. And even in those instances of formed solid stools in which the spores were found to be present, they could only be demonstrated by using large amounts of the fæcal matters. So too with stools which were of a semi-solid or semi-fluid nature; in 12 such cases the test with considerable quantities of stool proved negative in four instances (or 33 per cent.).

It may be affirmed then that though in the diarrhœa stage of typhoid fever the number of spores of bacillus enteritidis is, as a rule, great and easily demonstrable, in the non-diarrhœal condition or in the convalescent stage of that malady, their number so greatly and conspicuously diminishes as to render it very difficult if not practically impossible to demonstrate them.

Having thus, during the progress of these experiments, ascertained that the spores of bacillus enteritidis are occasionally so limited in number in stools as to require examination of considerable amounts of fæcal matter in order to demonstrate them, I considered it necessary to reconsider the statement made in previous reports by myself and Dr. Andrewes (Reports for 1896–1897 and 1897–1898) as to the absence of spores of bacillus enteritidis in the stools of non-diarrhœa patients and in those of healthy persons. The statements in question referred to investigations which made comparison in this sense between the fluid stools of cases of epidemic diarrhœa and cholera nostras on the one hand, and ordinary diarrhœa and healthy evacuations on the other. In the fluid stools of epidemic diarrhœa, demonstration of the spores of bacillus enteritidis by anaërobic milk culture was, by using a small quantity of the fluid for inoculation of the milk, in most instances easily achieved. In some cases it will be remembered that the mucus flakes were crowded with these spore-bearing bacilli. But the experiment failed in some (a minority of) instances of fluid stools of diarrhœa, and failed, too, in normal stools; the method of using small quantities of stool being in these cases also adopted. I am now inclined to think that the negative results in the case of summer diarrhœa and in the case of healthy persons were due to not using sufficient amount of the stool for inoculation of the milk. However that may be, I have now devoted further attention to this subject, and am, consequently, in a position to supplement the statements above referred to by the results of a number of new observations in the same subject-matter, as follows:—

1. A girl (A. P.) was suddenly taken ill with severe diarrhœa on September 17th, and died in fourteen hours. Stools, typical rice-water character. They did not yield any cholera vibrios, but the spores of bacillus enteritidis of normal virulence were easily demonstrated, a loopful of the stool being used for the anaërobic milk culture.

2. Similarly, in all the cases of the third epidemic outbreak of diarrhœa in the wards of St. Bartholomew's Hospital on August 6th, 1898, investigated and described by Dr. Andrewes (*The Lancet*, January 7th, 1899), the presence of numerous spores of bacillus enteritidis had been microscopically and culturally easily demonstrated by Dr. Andrewes.

3. Alfred F., aged 16, taken ill with pneumonia, had also diarrhœa; the stool was semi-fluid; widal reaction negative: crisis on January 30th. With a platinum loop a sample of the top part of the fluid stool was transferred to an anaërobic milk culture. Result positive, and whey of normal virulence.

4. Of the following cases the contents of the colon transversum were obtained in the post-mortem room. These were of more or less fluid character, and fair amounts of them (one to three platinum loops) were used for anaërobic milk cultures, which proved positive quà spores of bacillus enteritidis, and which were found, by experiment, of the normal virulence.

(*a.*) Harriett D., age 26. Died from gastric ulcer, complicated with perforation and peritonitis.

(*b.*) Abraham L. Empyema, perforation of diaphragm, peritonitis. Contents of ileum fluid.

(*c.*) William E., age 47. Œsophageal stricture; broncho-pneumonia; septicæmia.

(*d.*) Catherine J., age 58. Ovariotomy, peritonitis; both ileum and colon distended with gas; contents of colon pale thin fluid.

(*e.*) Samuel S., age 22. Otitis media: cerebral abscess. Ileum contained a little fluid, with mucus flakes.

(*f.*) Moses Y., age 55. Epilepsy. Contents of colon black thin fluid.

5. In the following cases neither the history nor the actual state of the bowels and contents of the colon on the post-mortem table suggested diarrhœa.

(*a.*) Maria H. Ankylosis of jaw; failure of heart. Contents of colon normal. With a good-sized lump taken up with a platinum loop made anaërobic milk cultures. Result was negative.

(*b.*) James L., age 18. Otitis media. Contents of colon normal. With good-sized lumps made anaërobic milk cultures. Result was negative.

(*c*.) Charles E , age 40. Fractured skull; fracture of left tibia and fibula. Contents of colon semi-solid. With one, two and three loops of solid parts, respectively, made anaërobic milk cultures. The former two remained unchanged, the last became typically changed and proved of normal virulence.

(*d*.) Catherine H., age 35. Fractured skull. Contents of colon normal. With a fair-sized-lump of fæces made an anaërobic milk culture: this yielded positive result and proved of normal virulence.

(*e*.) J. B. W. was brought dead into the Hospital, a marble slab having fallen on him and killed him. Contents of colon normal. With-fair-sized lumps of the solid parts made anaërobic milk cultures, three in number; two remained unchanged, one became typically changed and proved of normal virulence.

The last two series of cases require further consideration and comparison. The cases of Series 4 might perhaps be added in a general way to the group in which diarrhœa and loose stools are associated with the presence of numerous or fairly numerous spores of bacillus enteritidis. The cases of Series 5 could not, however, be so added, since in them the contents of the ileum were of the normal character, and, so far as could be recognised on the post-mortem table, or could be ascertained from the history of the patient, no abnormal state of the bowels was or had been present. As regards the five cases in this series, it is seen that in two the culture test made with comparatively large amounts of the contents of the colon yielded negative results; but that in the three others, by the same method, viz., using comparatively large amounts of fæcal matter, positive results were obtained. It is, therefore, necessary to infer that in a considerable percentage of instances of persons in the normal state, spores of bacillus enteritidis are present in their large intestines, and that these can be demonstrated by using for culture considerable amounts of the fæcal matter. And in this way may be accounted for the negative results previously obtained by Dr. Andrewes and myself when we were working with only small amounts of the normal stool.

At first sight the facts now to hand might appear to call in question not only previous conclusions as to the importance of the demonstration in diarrhœa stools of various origin of the spores of bacillus enteritidis, but also, and perhaps in a higher degree, conclusion as to the pathogenic relation of bacillis enteritidis to diarrhœa. On a careful consideration, however, of all the circumstances, and of the different series of cases, such inference would appear a hasty one, for reasons which will be discussed presently.

Before doing so I wish again to insist on what has been in some detail pointed out in last year's report, viz., that the anaërobic bacillus enteritidis sporogenes is an entirely different species from the anaërobic bacillus of malignant œdema of Koch, Gaffky, Fraenkel and others. In morphological, biological, and physiological respects; as regards size, flagella, terminal spores, staining by Gram's method, mode of growth in milk, symptoms of the disease induced in rodents, and the character of the post-mortem appearances in these animals, the bacillus enteritidis is as different as can be from the classical bacillus of malignant œdema. I mention this once again, because it might be thought by a superficial observer that since Gaffky has stated* that the spores of the bacillus of malignant œdema occur under normal conditions in the large intestine, the results I have obtained go to confirm those of Gaffky ; and that perhaps after all the bacillus enteritidis sporogenes is no other than the bacillus of malignant œdema. Such inference would be altogether wrong, and on several grounds :—

(*a*) In the first place, as stated above, the bacillus enteritidis is a definite and distinct species. The only points of similarity between it and the classical bacillus of malignant œdema is that both form gas, both are anaërobic, both are motile, both form spores, and both liquify sugar gelatine. But these are characters which are common to a considerable series of well-defined species of spore-bearing anaërobes. To accentuate the difference between the above two anaërobes, I need go no further than point out again the utterly different action of the two on the guinea pig. A pure culture of the bacillus enteritidis, when of normal virulence, causes in this animal well-marked extensive gangrene, with malodorous copious sanguineous exudation filled with bacilli of the shape of rods and short cylinders. And when the virulence of this bacillus is subnormal, a tumour is produced, though it does not kill the animal. In the case of a pure culture of malignant œdema bacillus of normal virulence, the condition produced in the guinea pig is a gelatinous tumour, with infiltration of the tissues and death of the animal. No gangrene, however, results, and the thin exudation oozing out of the infiltrated tissue contains numerous longer or shorter threadlike and filamentous bacilli.

(*b*) In the second place, Gaffky, in his memoir, makes no mention of examination by him of the contents of the human intestine, but refers to that of animals in general—the guineapig in particular ; and he arrives at his conclusion on the ground of the following experiment. A normal guinea pig, dead, is placed in the incubator at 37° C. for 12–24 hours, after which interval it is found, on microscopic examination, that numerous filamentous anaërobic motile bacilli (of the size and thickness and aspect of the bacillus of malignant œdema) have grown out from the large intestines and have penetrated the surfaces of the abdominal viscera. This experiment of Gaffky's is easily verified. But these thread-like anaërobic bacilli which grow out from the large intestine on to and into the abdominal viscera are not bacilli of malignant œdema, but belong to a species of motile anaërobe, which forms exquisite terminal spores and drumsticks, and which, moreover, are barren of pathogenic action. In another report in this volume I deal in detail with this microbe (which I have called bacillus cadaveris sporogenes) as regards its morphology and biology, and as to its connection with the anaërobic destruction of the viscera such as occurs in buried bodies ; and I there show the important role it plays in the destruction of dead tissues.

The question now to be considered is : Can the bacillus enteritidis, after all, be thought of as having causative relation to a pathological state of the intestine such as obtains in typhoid

* Mittheilungen aus d. Kais Gesundheitsamte I. p. 81.

No. 1. acillus lis ion; lein.

fever, in epidemic diarrhœa, in cholera nostras, and in other sporadic cases of intestinal disease associated with diarrhœa; seeing that the spores of bacillus enteritidis are (limited in number it is true) now found present in apparently a large percentage of instances in the normal large intestine?

I think from the results which I have recorded of the examination of the numerous cases of the fluid stools in epidemic diarrhœa, in cholera nostras, and in typhoid fever, that it is obvious not only that in these states of the bowel the spores of the bacillus enteritidis are, as a matter of fact, present in the evacuations in considerable numbers, but also that they must be present in the fluid contents of the small intestines. For in previous reports, as also in the present one, the cases dealt with have often been cases in which the fluid stools voided were, and had been so for some time, copious, and were of a strong alkaline reaction, and, therefore, undoubtedly to a large extent referable to the small intestine. What, therefore, may justly be at least concluded is that in such diseased conditions the bacillus enteritidis is present abundantly in the small intestine, and forms there abundantly the spores which are so readily demonstrated by the method of milk culture.

Strictly parallel cases are to hand as regards bacillus coli and proteus vulgaris. Both these microbes are normal inhabitants of the intestine; in the large intestine they are generally found in great numbers. Both microbes are capable of causing severe congestion and exudation within the small intestine, if perchance they undergo there rapid and copious multiplication. In my report for 1895-96 I have mentioned cases of sporadic cholera, in which the mucus flakes of the rice water evacuations contained one or the other of those microbes in amount and arrangement—"fish in stream arrangement"—strictly comparable to that of the cholera vibrio in the epithelial flakes of typical cholera rice water stools.

Now both bacillus coli and proteus vulgaris produce on occasion powerful toxins, as is now proved by numerous experiments. Neither of these microbes is quite of the harmless nature that many persons seem to think; the experimental evidence of clinical sort (cystitis when proteus or bacillus coli is introduced into the bladder; abscesses, necroses and gangrenes in various viscera associated with the copious presence of one or the other of these microbes) shows that by their action they are capable of setting up severe disorders. In a report in this volume on the microbes found in the viscera of buried bodies, I describe in detail certain experiments showing the intense virulence which proteus vulgaris, derived most probably from the intestine and having after death penetrated into the spleen, is capable of exerting when inoculated into rodents. Then there is the particular variety of bacillus coli, known as Gärtner's bacillus enteritidis. In various outbreaks due to consumption of particular meat (pork and beef) it has been shown that the gastro-enteritis of the sufferers is due to the copious multiplication in the intestine of this microbe, and to a general infection of the system, as a result of the bacillus being taken with food stuffs. Dr. Durham* has recently given a summary of the more recent literature of these outbreaks. I myself have described one which occurred at Portsmouth (*see* the Medical Officer's

* British Medical Journal, December 17th, 1898.

Report for 1890, p. 249), in which a microbe closely resembling* this variety of bacillus coli was the immediate cause. Now this variety of bacillus coli, viz., Gärtner's bacillus, is somewhat widely distributed. Whenever I have searched for this microbe in sewage, and in the intestinal discharges of diarrhœa cases (not in any way connected with "meat poisoning"), I have been able to isolate it by means of surface phenol gelatine plates. Dr. Mervyn Gordon, in his extensive observations ("Journal of Pathology," June, 1897), has met with this variety of bacillus coli in sewage and in fæcal matter. So too Lorraine Smith has detected it in certain Belfast waters, as well as in the spleen of some typhoid fever cases.

This particular variety of bacillus coli, which is spoken of as Gärtner's bacillus enteritidis, differs from the typical bacillus coli in the following respects: It forms in surface phenol gelatine plates colonies more circumscribed, more opaque also, than those of typical bacillus coli; in gelatine streak it forms a thicker, less crenated, and less spreading band. But in ordinary gelatine shake culture Gärtner's bacillus produces gas bubbles just like bacillus coli; it makes phenol broth at 37°C just as turbid in 24 hours as bacillus coli; it produces acid † (litmus milk, litmus broth) just as readily as bacillus coli; and it produces toxins of the same character and amounts as bacillus coli.‡ It differs, however, from typical bacillus coli in that it is more cylindrical, has more flagella, is more conspicuously motile, does not clot milk, does not give indol in broth, and in that its culture acts more pathogenically (*i.e.*, in smaller doses) on the guinea-pig than does that of typical bacillus coli. It has one further important character, which is this: When the Widal test is applied by adding typhoid blood serum to a broth culture or an emulsion of typical bacillus coli (one part of typhoid blood serum to 20–25 parts of culture), the result is negative as to clumping or agglutinating reaction; whereas the same test applied to Gärtner's variety of bacillus coli gives results distinctly and in an eminent degree positive, as nearly positive, indeed, as with a culture of the typhoid bacillus.

Now the problem to be solved is this: Since Gärtner's variety of bacillus coli is so widely distributed and occurs (or at all events a variety of Gärtner's microbe which in every respect, morphological, cultural, and physiological, agrees with Gärtner's bacillus) in the normal intestine, how and why is it that the intestine does not frequently become subject to its pathogenic action, or that it does not more often infect the system in general? Why, for instance, is this "meat poisoning" not frequently produced by all kinds of meat and other food stuffs that have been, as must frequently occur, exposed to this widely distributed microbe?

Or take as another instance streptococcus infection of the

* Recent comparisons made between it (presence in subculture) and Gärtner's bacillus shows the close resemblance distinctly.

† It has been alleged that Petruschki's bacillus fæcalis and Gärtner's bacillus are identical. But Petruschki expressly rests the differential character of his bacillus on the fact that, unlike bacillus coli and bacillus typhosus, which are distinct acid producers, bacillus fæcalis alkaligenes is an alkali producer. Gärtner's bacillus is, just like bacillus coli, manifestly an acid producer. Sufficient ground has not, I think, been adduced for regarding Petruschki's statement as to the bacillus alkaligenes producing alkali as incorrect.

‡ Sidney Martin, Croonian lectures, Royal College of Physicians, 1898. "British Medical Journal," June and July, 1898.

. B. No. 1.

he Bacillus
eritidis
rogenes;
Dr. Klein.

intestine. Escherich first described fatal cases of diarrhœa in children due to streptococcus accumulation in the small intestine. The intestine in such fatal cases was greatly congested, its contents sanguineous, and in it had accumulated copious multiplication of streptococci. Dr. Andrewes described ("Transactions of the Pathological Society of London," vol. 50, October, 1898) a fatal case of sporadic diarrhœa in which the fluid contents of the small intestine were of a sanguineous character, and the wall of part of the small intestine intensely injected. Here also, in the intestinal contents, abundance of streptococci was noticed. The streptococci in question belonged to the group of streptococcus pyogenes. I have myself described (Report of the Medical Officer of the Local Government Board, 1892–1893) a case of perforation of the bowels in enteric fever, in which the peritoment exudation, as also the blood, contained the streptococcas of erysipelas. There are other cases described, *e.g*, that by Washbourn and Pakes ("British Medical Journal,' June 18th, 1898), in which, in a case of acute enteritis, peritonitis had set in, the peritoneal exudation, as also the intestinal contents and the blood, being crowded with streptococci. Now a careful examination of normal stools, as also of typhoid stools, by means of agar surface plates has in my hand not infrequently yielded colonies of streptococci which, when compared with streptococcus pyogenes, appeared identical therewith Why, then, was there not produced in these cases the severe affection of the small intestine that in others led to congestion hæmorrhage, and even perforation ?

Another and most striking fact in this connexion is th distribution and pathogenicity of the diplococcus pneumoniæ Fraenkel and Weichselbaum (diplococcus capsulatus, diplococcu lanceolatus, streptococcus brevis). It has been abundantl proved that this microbe occurs in the normal saliva and thro (Pasteur, Sternberg), and that it is often present in almost pu culture in the normal nasal cavity and adjoining parts (Besser It has been isolated in a considerable percentage of cases fro catarrhal (bronchial) expectoration (Weichselbaum), also fro normal expectoration (Besser), and from the normal conjuncti (Gasparini). That is to say, it is a microbe very widely distribut on normal mucous membranes. On the other hand, no one doub that it is, on occasion, intimately and causally connected wi typical croupous pneumonia, as well as with a certain percenta of cases of broncho-pneumonia, meningitis, otitis media, per carditis, pleuritis, empyema, and abscess of the neck, leg, & What is it, then, that causes this ubiquitous diplococcus lanceolat in some cases to suddenly rapidly multiply in the lungs, and its life processes there to cause the fibrinous exudation (teemi with diplococci) which fills the alveoli of the lung ? Is it th the lung has become suddenly—say by a chill—transformed fro a previously unfavourable into a highly favourable breedi ground for the microbe ? Or is it that the microbe, owing to so unknown cause, has suddenly developed exceptional virulenc Or is it for both these reasons ? The fact remains that a micro morphologically, culturally, and physiologically identical wi one present in small numbers in many parts in the normal sta and without doing harm, is suddenly found capable of causi severe local infection and disease of one or another organ to whi it has all along had free access (to the lungs from the nose a bronchi, to the middle ear from the throat). The diplococcus pneumoniæ obtained from croupous pneumonia soon loses its virulence—but not altogether its pathogenicity—on rodents when it is cultivated on artificial media, although in many instances a virulence for mice of the normally occurring diplococcus lanceolatus has been demonstrated. It is quite possible that it is necessary that some new condition should be introduced *ab extra* which supplies to the pneumococcus capability to assume a higher virulence, and thus cause disease directly in its host and by transmission cause infection in further persons. The croupous pneumonia referred to this diplococcus is proved by the repeatedly demonstrated transmission of this disease from one individual to another—as, for instance, to a member of the same household, to a person that has been sojourning in the same room with the infected patient, to a nurse or an attendant. But notwithstanding all this, the problem remains still unsolved, *viz.*, what causes this sudden acquisition of exceptional virulence—what causes this sudden invasion of an organ by the rapidly multiplying pneumococcus?

These same difficulties are those that are arising in regard to the spores of bacillus enteritidis sporogenes. What is it that, in one and another instance, causes its sudden assumption of pathogenicity *quâ* human beings; what is the reason that the small intestine becomes suddenly a good ground for the germination and rapid multiplication of the microbe? The wide distribution of its spores in milk used as food (*see* last year's Report as to the occurrence of this bacillus in sewage, horses' dung manure, and most fæcally polluted matters) can in no way preclude it from now and then becoming pathogenic and causing severe enteritis and exudation. Take, for instance, milk. It has been shown, in my Report for 1895–1896, and still more in that for 1897–1898, that in the epidemics of diarrhœa in St. Bartholomew's Hospital, it was clearly the milk which was at fault. In the second epidemic in question, the milk had been taken by the patients between 8 and 11 o'clock in the morning (Sunday, October 6th), the disease manifested itself from about 8 o'clock in the evening till 4 o'clock next morning (12 to 20 hours' incubation.) All the patients that became subjects of the epidemic had partaken of that milk. Of the nurses, one only became affected, and this one nurse was the only nurse that had partaken of that milk; all other nurses having a different milk supply escaped. In the incriminated milk the presence (numerously) of the spores of bacillus enteritidis was proved; in the fluid bowel discharges of the diarrhœa cases their copious presence was demonstrated. I have already in last year's Report, suggested that some condition may have attached to this particular milk which is not ordinarily present in milk, and I have also pointed out that either a rapid multiplication of the bacillus enteritidis, or an increase of its virulence (*quâ homo*), may have been caused by some such new condition in the milk, and that hence the microbe became capable of producing illness. I have pointed out, for instance, that when sterile milk tubes are kept for some time—beyond a week—these become unsuited for the germination of these spores, and for multiplication of this bacillus; the culture is retarded, the bacilli sparse and abnormal looking, and devoid of normal virulence. On the other hand, for obtaining rapidly typical culture of normal virulence, I have insisted on the use of recently sterilized milk. I now find that stale sterilized milk, that is milk which has been kept for a week or longer, can be again rendered a suitable culture medium for bacillus enteritidis by merely re-boiling it. The explanation of this change seems to me very obvious: If the milk be kept sufficiently long after sterilization it becomes again charged with air (oxygen) by absorption; the previous sterilization (heating) had, of course, driven out all air, and the longer after sterilization the milk is exposed to air (is kept, that is) the more oxygen it can absorb, and hence the growth of this strictly anaërobic bacillus enteritidis in such milk is not carried on under favourable conditions. When, however, such stale milk is again re-heated (boiled or kept at a 90° or 95° C. for sufficiently long time) the absorbed air is again driven off, and the milk is thus once again rendered a favourable medium for the growth of this bacillus. It seems to me that this experiment is capable of affording one explanation, at least, why one sample of milk becomes favourable for rapid multiplication and for the rapid production of an actively virulent crop of bacilli enteritidis, whereas another does not. For instance, one sample of milk offers favourable anaërobic conditions, because it is kept well sealed up; another sample because its aërobic microbes find opportunity for multiplication and so consume the oxygen dissolved in the milk. In either case the milk would become suited for the subsequent rapid development of more anaërobic virulent bacilli.

Reading Dr. Andrewes' account of the third epidemic (*The Lancet*, January 7th, 1899) that occurred at St. Bartholomew's Hospital in August, 1898, one cannot escape from his conclusion that "it (the epidemic) was definitely traceable" to the consumption of "rice pudding made with milk, rather than directly to the milk itself." Now Dr. Andrewes shows, by direct test, that in the "making of rice pudding," in the Hospital kitchen, no part of the rice pudding in the various cooking and baking processes becomes heated to a higher temperature than 98° C.; after the baking process, in many parts of the pudding a range of 90°–92° C. was obtained. An article of food that has been exposed to 90°–92° C., and is capable of producing after due incubation definite intestinal infection, must obviously contain a pathogenic microbe capable of multiplying in and causing disease of the intestine. No microbe, however, except in spore form, can pass unscathed a temperature of 90° C., even for a few seconds; spores only can retain vitality under these conditions. The spores of the bacillus enteritidis withstand perfectly unharmed a temperature of 100° C. for several minutes, one of 98° for a considerable time. Thus, Dr. Wild, who worked in my laboratory during 1898, found that in some instances the spores of bacillus enteritidis (from sewage) were capable of withstanding unharmed boiling (100° C.) for as long as fifteen to thirty minutes. Add to this the fact that Dr. Andrewes demonstrated in the same rice pudding that was consumed by the affected persons at St. Bartholomew's Hospital, and in their fluid bowel discharges, the abundant presence of the spores of bacillus enteritidis. The conclusion therefore becomes irresistible that this microbe was the direct cause of the disease,

Experiments as to the "protection" afforded guinea-pigs against bacillus enteritidis sporogenes by injection into them of non-lethal doses of culture of this microbe.

In many of the experiments made with the spores of bacillus enteritidis obtained from typhoid and other intestinal discharges, and from milk, mention was made by me of various degrees of subnormal virulence of the milk cultures. The experimental guinea-pigs, although at first effected with more or less extensive local tumour and made more or less generally ill (quiet and not feeding), gradually recovered, the tumour leading to necrosis and ulceration of the skin, but ultimately quite healing up. Some of these guinea-pigs, which had, therefore, passed through one mild attack of the disease, were then—at varying periods of recovery—subjected to reinjection with milk cultures proved by control experiments to be possessed of the normal degree of virulence, *i.e.*, causing fatal gangrene in the normal dose.

Experiment I.—(1.) Four guinea-pigs, which, having been each injected with the whey (1 cc.) of typical (*a*) milk cultures set going with spores of various origin, had all developed distinct tumour and were generally ill (quiet and off their feed), all completely recovered. After two to three weeks they were reinjected with 0·5 cc. of another typical milk culture (*b*), which had been established 24 hours previously with spores of b. enteritidis that had developed in the anaërobically sealed exudation (*see* last year's report) of a guinea-pig that had died from the typical fatal gangrene.

(2.) Of the above milk culture (*b*) 0·5 cc. and 1 cc. respectively of the whey was also injected subcutaneously into two (No. 1 and 2) control guinea-pigs.

Of these two control guinea-pigs, No. 2, *i.e.*, the one that received 1 cc. of the whey, was found dead next morning, with the typical extensive subcutaneous gangrene; the other (No. 1, injected with 0·5 cc.) though affected generally (quiet and off its feed) and developing distinct tumour, it recovered, the tumour leading to necrosis of the skin. All the four guinea-pigs, however, that had passed through and recovered from a previous injection, and which had been now re-injected with only 0·5 cc. of milk culture (*b*) were dead next morning with the usual characteristic extensive subcutaneous gangrene.

(3.) Four other guinea-pigs that about three to four weeks previously, on injection with typical milk culture of greatly decreased virulence, had only shown slight swelling, and had been barely generally ill, were injected with same milk culture (*b*), each receiving also 0·5 cc. All these four guinea-pigs had tumour, which led to necrosis of the skin, and they all recovered.

(4.) Two further guinea-pigs, which were of the same type as those under heading (3), *i.e.*, had passed through a first injection with only slight swelling and no other general disturbance, five days later received each 0·5 cc. of the same milk culture (*b*). The result was that they showed slight tumour and no other change; one only had ulceration of a small part of the cutis.

Experiment II.—The four guinea pigs of heading (3), after they had quite recovered from the effect of the second injection, were again injected with 0·5 cc. of a milk culture, 0·5 cc. of the whey

PP. B. No. 1.
the Bacillus
teritidis
orogenes;
Dr. Klein.

of which proved on a control guinea-pig of subnormal virulence. All these four guinea-pigs were dead next morning, with typical extensive gangrene.

Of the two guinea-pigs under heading (4) now similarly injected with 0·5 cc. of the same milk culture, one only (that which had had the limited ulceration of the cutis) was dead next morning; the other had big swelling and was ill; but it recovered, the skin ulcerating.

Experiment III.

When this last guinea-pig, heading (4), had quite recovered, it was again re-inoculated with 0·5 cc. of a typical milk culture which on a control guinea-pig proved of subnormal virulence. This guinea-pig also succumbed to this third injection, whereas the control guinea-pig survived.

I think that it is quite clear from these experiments that not only does a previous subnormal infection fail to confer on the guinea-pig immunity, but that, on the contrary, it makes this animal more susceptible. This greater susceptibility, moreover, appears (Experiments I. and III.) to stand in direct proportion to the intensity of the disease produced by the antecedent injection. A number of other experiments, made for a different purpose, have led to the same conclusion, *viz.*, that by antecedent injection of b. enteritidis guinea-pigs are rendered more susceptible to enteritidis infection.

Experiments as to the agglutinating action on culture of the bacillus enteritidis of blood serum of persons or animals aforetime subject to the life process of this microbe.

Experiment (a).—From a patient in the wards of St. Bartholomew's Hospital, one of those affected with rather severe diarrhoea in the epidemic of March, 1898, blood was obtained about three weeks after his complete recovery. The serum of this blood was tested on the subcutaneous exudation of a guinea-pig that had just succumbed (with typical *p.m.* appearances) to injection with the whey of a typical milk culture of bacillus enteritidis of normal virulence. This exudation, which was crowded with the typical short bacilli enteritidis, was diluted with an equal volume of sterile salt solution. By this means, as pointed out in last year's report, the motility of the bacilli becomes very evident. To this diluted exudation the blood serum of the above patient was added in the proportion of one blood serum to [illegible] diluted exudation. Microscopic specimens and naked eye inspection showed that even after an hour and a half no effect was produced; the motility of the bacilli did not suffer and there was no sign [illegible] any agglutination.

Experiment (b).—Blood serum of a guinea-pig that had recovered from a first injection with typical milk culture of bacillus enteritidis, was tested on diluted subcutaneous exudation of [illegible] guinea-pig dead after a typical enteritidis infection, and was tested also on the whey of a typical milk culture of bacillus enteritidis. Proportion 1 in 25 in each instance. Result, after an hour and a half, negative.

Experiment (c).—The blood serum of the same guinea-pig after it had recovered from two successive enteritidis injections, was tested as before on exudation and on the whey of a milk culture. The proportions were again 1 in 25, and the results were negative.

Experiment (d).—Blood serum (heart's blood) of a guinea-pig just dead after enteritidis infection, was tested on subcutaneous exudation from the same animal. The proportions were 1 : 25, and the result negative.

From these experiments it appears then that the blood of an animal that had passed through one or more experimental inoculations of bacillus enteritidis, the blood of a patient who had recovered from a severe attack of epidemic diarrhœa, and the blood of a guinea-pig actually dead from enteritidis infection, failed, all of them, to exert agglutinating action on culture of bacillus enteritidis, whether in the form of the subcutaneous exudation (crowded with the bacilli) or as whey of typical milk cultures containing the enteritidis bacilli in abundance.

A similar series of observations which I have carried out with reference to the variety of bacillus previously referred to as Gärtner's bacillus enteritidis, afford results parallel to the above. Thus :—

William R., age 10, was admitted to St. Bartholomew's Hospital on January 9th. He had been taken ill on January 7th after eating meat pie. On admission his temperature was 102·4 F., but on the 12th January his temperature had become normal, and remained so. He had had no diarrhœa, Widal reaction was negative. On January 16th, the temperature being normal and the patient evidently recovered, I obtained a stool of his, which was semi-solid. With a particle of this stool an emulsion in sterile salt solution was made, and from this a surface phenol gelatine plate was established and incubated at 20·5° C. The plate after three days showed numerous colonies. These were as regards aspect, size, and nature of two kinds : (*a*) a majority of translucent flat crenated filmy colonies of typical bacillus coli—as proved by the various tests in sub-culture; and (*b*), a minority of circumscribed, less transparent, rounded, less flat colonies which on test in sub-culture in every point proved indistinguishable from the bacillus of Gärtner. With the latter the comparison was easily made since Dr. Horton Smith kindly placed at my disposal a culture of this microbe obtained from Dr. Durham.

The blood serum of this boy on January 19th, *i.e.*, 12 days after the onset of his illness, and seven days after his recovery, was tested on a 24 hours old broth culture of Durham's Gärtner bacillus, and on a similar culture of the Gärtner bacillus obtained from the stool of the patient. The proportions used in each instance were one of blood-serum to 20, 30 and 50 of culture of the Gärtner bacillus. The result was that neither bacillus gave with any of the above proportions of serum, microscopically or to the naked eye, any indication of arrest of motility or of clumping, even after five hours' observation: Now this is altogether different from Durham's result. In the case of the Chadderton (Oldham) outbreak of meat poisoning he found that the blood serum of a number of patients that had recovered from the illness had a distinct and conspicuous clumping action on the culture of Gärtner's bacillus ; at any rate "on two of the seven varieties" of this bacillus.

In one other respect the behaviour of the Gärtner bacillus obtained from the stool of this boy was in harmony with the behaviour of the true Gärtner bacillus from Dr. Durham's culture; viz. in that a 24 hours' broth culture, and an emulsion of the bacillus in broth (made with a scraping from a gelatine surface culture), reacted conspicuously and unmistakeably to typhoid blood serum (previously tested with unmistakeably positive result on typhoid culture), 25–50 of the culture or emulsion being mixed with 1 part of the blood serum. The arrest of motility and the clumping of the bacilli was in this experiment, well marked in both cases within 30 minutes to 1 hour; but better with the true Gärtner than with that from the boy's bowels. For comparison, both a broth culture and an emulsion of the typical bacillus coli, which had been isolated from the same phenol plate of the boy's stool, were tested with the same typhoid blood serum, but the result was quite negative even after 5 hours.

As already stated, I have isolated the particular variety of bacillus coli described as Gärtner's bacillus, or at any rate, a bacillus in every point resembling it even down to the positive clumping reaction with typhoid blood, from other stools. The latest sample thus obtained, was from the patient mentioned as *Case* No. 17, (William S.), who suffered from unmistakeable typhoid fever and whose stool when received on the fourteenth day of his illness, was a typical fluid typhoid stool. The subculture of the Gärtner-like bacillus of this stool was also tested, again with negative result, with the blood serum of the above (meat-pie poisoned) boy; whereas, to the typhoid blood serum, it had acted as distinctly positively in clumping as that of Durham's Gärtner culture.

Relation of the Typhoid bacillus to the bacillus enteritidis sporogenes.

Experimental series 1.—In the first place it was attempted to ascertain whether the blood serum of a typhoid patient has any positive effect as to agglutination on culture of bacillus enteritidis sporogenes such as had been shown to be exerted by typhoid blood serum on culture of the Gärtner variety of bacillus coli.

For this purpose, the blood of a typhoid patient (tested at the same time on typhoid culture and producing rapid and unmistakeable clumping), was added to dilute subcutaneous exudation, full of the bacilli enteritidis, of a guinea-pig dead with the typical appearances of extensive enteritidis gangrene; and added also to the whey of a recent typically changed milk culture full of the enteritidis bacilli. The proportions used were: 1 part typhoid blood serum to 20, 30, and 50 parts respectively of exudation or culture. Neither microscopically, nor with the unaided eye, was any change in the direction of arrest of motility and agglutination observed even after 5 hours.

This experiment was repeated on a second occasion on exactly the same lines. The result was the same, viz. completely negative.

Experimental series, 2.—In this series, the blood serum of a guinea-pig dead with the characteristic appearances after injection with the whey of an enteritidis milk culture, was tested on broth culture (24 hours old) of the typhoid bacillus. Also, it was tested on a bouillon emulsion of typhoid bacillus from a recent gelatine surface culture. One part of the blood serum was mixed in each instance with 20 parts of the typhoid culture. Neither microscopically, nor with the unaided eye, was any reaction observed after 1 hour, 5 hours, or even after 24 hours; the motility of the typhoid bacilli remained unimpaired, and there was no sign of agglutination.

In another experiment blood serum was obtained from the ear vein of a guinea-pig that had twice survived the subcutaneous injection with typical enteritidis milk culture of attenuated virulence. This animal after the second injection had a big local swelling which led to ulceration of the skin, but which had healed up. The blood serum of this animal was mixed with typhoid cultures as before in the proportion of 1 : 20. No change either as regards motility or clumping of the bacilli was observable even after 6 hours.

From this it follows that the blood serum of a typhoid patient and the blood serum of an enteritidis guinea-pig have not any agglutinating action—the one on culture of the enteritidis bacillus, the other on the culture of the typhoid bacillus.

Relation of typhoid infection to enteritidis infection.

A last series of experiments was made, which had for its object to ascertain infection of the guinea-pig with bacillus enteritidis, whether simultaneous, or previous, modifies in any way infection of this animal with the typhoid bacillus. For this purpose guinea-pigs, which are susceptible to infection by subcutaneous injection of either microbe, were subjected to experiment.

As is well known, feeding of guinea-pigs with large amounts of typhoid culture (as for instance broth or milk culture) yields no result; the animals remain perfectly normal. But injected subcutaneously with moderate doses of typhoid culture they become distinctly ill. After 24 hours there is soft tumour at the seat of inoculation, which during the second day becomes considerably enlarged; the animal is quiet, huddled up, and refuses food; the tumour subsequently becomes smaller, firmer, and the animal becomes again lively and feeds well. The tumour leads to more or less extensive necrosis and ulceration of the skin, but this in 10 days or a fortnight heals up completely. In a former report these results have been described in detail.

If the dose is large the initial illness leads to a fatal issue in 48 hours or even earlier; and in this case evidence of general septicæmic infection is found in the viscera, viz., congestion of these organs, spleen dark and enlarged; in the heart's blood, as also in the spleen, the typhoid bacillus is easily demonstrated by microscopic specimens and by culture.

The typhoid culture, which I used for these experiments, was obtained about six months previously from the spleen of a person dead of typhoid fever in the second week. It is kept in subcultures in the laboratory, and has on several occasions been tested on guinea-pigs as to its virulence both by intraperitoneal and by subcutaneous injection. The degree of its virulence when obtained, for the experiments now in question, from a surface agar culture, incubated 24–48 hours at 37°C, was as follows:—The whole of the growth on the slanting agar surface (6 centimetres by 2 to 2½ centimetres) having been rubbed down in 5 cc. of sterile bouillon, and an amount of the emulsion, corresponding to $\frac{1}{20}$th to $\frac{1}{10}$th of the whole culture, having been injected *intraperitoneally* into a medium-sized guinea pig, death ensued from intensive peritonitis. Similarly, $\frac{1}{10}$th of the same emulsion injected *subcutaneously* into a guinea-pig not weighing less than 200 and not more than 300 grammes, caused tumour and temporary illness; but the animal completely recovered, was in fact quite lively again on the third day, although the tumour led to ulceration of the skin, which, however, healed up completely.

As a typical enteritidis milk culture, one was used, the virulence of which was tested on control guinea-pigs of about same weight as the above. It was found normal, *i.e.*, 1 cc. of the whey produced fatal result within 20–24 hours with extensive gangrene; 0·5 cc. caused severe illness, but the animal eventually recovered; and 0·25 cc. caused slight transitory tumour, the animal being quiet next day, but was quite lively the day after.

Experimental series 1.—(*a.*) Two guinea-pigs were fed on two successive days with bread, soaked in a mixture of a whole typhoid broth culture (48 hours old) and a whole typical enteritidis milk culture (24 hours old). No disease followed, the animals remained normal.

(*b.*) Two guinea-pigs were fed on two successive days with bread mixed with the contents of a whole agar culture of typhoid bacillus (48 hours old) and with the whole of a typical enteritidis milk culture. Result negative.

(*c.*) As control, two guinea-pigs were fed on bread soaked with a whole broth culture of typhoid bacillus, and two guinea-pigs were similarly fed with bread soaked with a whole enteritidis milk culture. In neither case was any abnormal result produced. In a former report I have adverted to the negative result of *feeding* guinea-pigs either with the sporeless or sporing culture of bacillus enteritidis.

Experimental series 2.—Two guinea-pigs which had recovered after a previous subcutaneous injection with whey of an enteritidis milk culture, were injected subcutaneously with $\frac{1}{10}$th of an agar culture of the above typhoid bacillus. Two other normal guinea-pigs were similarly injected with a $\frac{1}{10}$th of the same agar culture of typhoid bacillus. All 4 guinea-pigs were of about the same size, at any rate as near as they could be selected; they weighed 250, 262, 248, 255 grammes respectively. The result was striking. The two first guinea-pigs became very ill: one was found dead after 36 hours with appearances of general septicæmic infection; the other was very ill with big tumour after 24 hours, and remained ill (quiet, huddled up and refusing food) during the third and fourth day, when it began gradually to recover; subsequently it had extensive ulceration of the skin which it took over three weeks to heal up. The two control guinea-pigs, though a little quiet on the second day, were quite lively and well on the third day, and their inguinal tumours were comparatively small. This tumour led in each instance to a limited ulceration of the skin, which after 10 days had healed up.

Experimental series 3.—Two guinea-pigs had been injected subcutaneously with whey of typical milk culture of bacillus enteritidis, but with subfatal dose; they became affected with tumour and were ill, but recovered again by the end of 6–8 days. They were then injected, at the same time with a control guinea-pig, in the other groin, subcutaneously, with $\frac{1}{10}$th of an agar culture of typhoid bacillus of the same stock as in series 2. The result was interesting, the two experimental guinea-pigs were found dead next day; the control guinea-pig had slight swelling, but was lively. This last named guinea-pig developed a tumour which ulcerated and healed up in the course of ten days or a fortnight.

From these two series of experiments it can, I think, be safely concluded that a previous injection of the guinea-pig with culture of bacillus enteritidis distinctly increases the susceptibility of this animal to subsequent typhoid bacillus infection.

Experimental series 4.—In this series the experiment was reversed. Two guinea-pigs, which had recovered from ulceration after a subcutaneous injection with $\frac{1}{10}$th of an agar typhoid culture, were subjected to a further injection with 0·5 cc. of the whey of a typical milk culture of bacillus enteritidis. At the same time a control guinea-pig was injected with the same dose of whey of milk culture. The result was this: The two experimental guinea-pigs showed no swelling next day, and remained well; the control guinea-pig had tumour and was ill, but it recovered again with ulceration of the tumour.

It appears then that a previous subcutaneous injection with typhoid culture decreases the susceptibility of the guinea-pig to the subsequent infection with bacillus enteritidis.

No. 2.

REPORT on the FATE of PATHOGENIC and other INFECTIVE MICROBES in the DEAD ANIMAL BODY; by DR. KLEIN, F.R.S.

There exist only a few direct and exact experimental observations on the fate of pathogenic bacteria in the dead body.

In the Twelfth Annual Report (1882) of the Medical Officer of the Local Government Board, I described (pp. 209–212) the results of a series of experiments made with bacillus anthracis. Small rodent animals (guineapigs and mice) having been infected with virulent anthrax, died in the typical time. They were then kept unopened above ground, or were buried in earth. Later, at intervals varying from five to fourteen days, the spleens and livers of these dead animals were chopped up along with sterile salt solution and the mixture injected into the subcutaneous tissue of guineapigs. The animals remained alive and well. After a few weeks these experimental guineapigs were tested with virulent anthrax blood and found to have lost none of their susceptibility to anthrax. From these experiments the conclusion (*see* pp. 209, 210) was drawn that in the dead bodies of animals the subjects of anthrax (buried or not buried) the anthrax bacilli, owing to competition of hardier putrefactive microbes, soon degenerate, and that as a consequence the organs and blood of such animals—*i.e.*, animals that had succumbed to virulent anthrax, and whose blood and organs immediately after death, contain, as is well-known, the virulent bacillus anthracis in great number—soon lose all power to infect with anthrax.

Similar results which were obtained by Esmarch (Zeitschrift f. Hygiene, Vol. VII., p. 1) are the only observations in this connexion that hitherto have been quoted by various authors, although they were obtained in 1889, *i.e.*, seven years after the publication of my own results.

My own and Esmarch's observations on rodents go to confirm observations previously made by Feser and recorded by him in Deutsche Zeitschrift f. Thiermedizin, 1877. He experimented with the organs of horses, cattle, goats and sheep dead of anthrax, and which animals had been buried for various periods. As regards one sheep, buried for 14 days in the winter (temperature 6° to 8° C.), he noted that the organs still possessed infective power, *i.e.*, produced anthrax on inoculation into other animals. But as to the rest of the other buried animals (horses, cattle, goats and sheep) observed by him, the organs had long before this (end of 14 days) lost all virulence.

The German Commission (Wolffhügel, Gaffky, Paak, Riedel, Berckholtz, Jäger, Scheurlen), reporter Dr. Petri (Arbeiten aus dem Kais. Gesundheitsamte, Vol. VII., p. 1–32), made similar experiments from 1885 till 1891 with the bodies of guineapigs and mice dead from anthrax, buried in wooden or zinc coffins, and subsequently from time to time tested as regards their infective power. This Commission also found (p. 8) that the infectivity of the buried bodies as a rule passes off in a short time. In exceptional cases it was, however, retained for long periods; in one instance this was the case even after three years and ten months. But, the experiments made in this direction by the German Commission were so contrived that the coffins were placed in a wooden box filled with earth, which box stood in water in a tin basin; and in addition the experimenters poured from time to time water on to the surface of the earth. In the three instances—out of a total of 16—in which the buried bodies proved infective, the possibility, I should say the high probability, that spores had, under the conditions of experiment, been formed would suffice for explanation of the positive results.

Additional experiments which I have recently made with the bodies of a series of six guineapigs dead from typical anthrax, and then enclosed in tin boxes—each body separately in a tin box—and buried under 18–24 inches of moist earth, failed to indicate any survival of living anthrax bacilli in the spleen. The animals were exhumed after 32, 28, 21 and 14 days respectively: when the spleen was in each instance taken out, mashed up in a little sterile salt solution, and the whole emulsion of the spleen in each instance injected into a guineapig. The animal in all instances remained alive and well.

I think therefore that the rule is that the anthrax bacilli in the dead body soon—within at any rate few weeks—degenerate and die, and that the viscera—at first teeming with virulent anthrax bacilli—soon lose all infective power.

The German Commission made also a large number of experiments in similar direction with the cholera and typhoid microbes, and with the tubercle bacillus. The results of these experiments, as also of the numerous experiments that I made in the course of last year with these and other pathogenic microbes, will be presently described in detail. As a preliminary it is necessary to consider the question: Which are the microbes that in the ordinary normal course of things are concerned in the destruction, or gradual reduction and disappearance, of the different organs of the dead body such as is known to occur after burial?

In order to answer this question, a series of experiments was instituted with the bodies of guineapigs that had not been subject to any experiments; that were perfectly normal. These animals were killed by an overdose of chloroform, were then wrapped up in a piece of linen or calico, were placed separately in small wooden or tin boxes, and were buried in ordinary moist earth; earth such as is dug out from a garden (in some instances it was from Wandsworth Common). The experimental burial ground was a large wooden box, and the coffins were buried in the earth contained in it at depths ranging from six inches to two feet. This wooden box was kept at the ordinary temperature of the laboratory. In a second series of experiments the dead bodies of the guineapigs were each wrapped up in a piece of linen or calico and buried, as before in the laboratory, directly in earth or in Thanet sand. The animals' bodies were then at stated periods exhumed and subjected to examination and experiment.

I may state here at once that the observations now to be described as to the nature and distribution of the microbes normally present in dead bodies, and chiefly concerned in the reduction and destruction of the body and its organs, were subsequently confirmed by other observations that I have made (to be

App. B. No. 2.

Report on the Fate of Pathogenic and other Infective Microbes in the dead animal body; by Dr. Klein, F.R.S.

described in detail later on) on the buried bodies of animals (guineapigs) that had died as the result of infection with pathogenic microbes of one and another kind.

MICROBES IN THE CADAVERS OF GUINEAPIGS AFTER BURIAL.

According to current view the putrefactive decomposition of the albuminous substances of the dead body, such as constitute the essential parts of the tissues and organs of man and animals, is effected in a large measure by proteus vulgaris and allied species (proteus mirabilis, proteus Zenkeri), and by bacillus coli and its allies. For both classes of microbes are present in great numbers in the contents of the large intestine (normal and pathological) whence, in articulo mortis and after death, they pass into the abdominal viscera and other organs and decompose the albumen and allied substances constituting such organs. During life even their passage from the intestine and their subsequent growth and multiplication in one and another organ has been observed in cases where the organ (*e.g.*, liver, mesenteric glands, spleen) was in a diseased condition (abscesses, necrotic foci, gangrene, &c.). The literature on this subject is enormous; a considerable number of observers have recorded the presence of bacillus coli or of proteus vulgaris and its allies in diseased or dead tissues of the body during life of an animal, and these persons have rightly concluded that the microbes in question have been conveyed to the tissues from the intestine, either by the lymph stream or by the blood directly. I have also, in a former report, mentioned the occasional passage into, and multiplication of bacillus coli in, the peritoneal cavity of guineapigs, which animals, by the intraperitoneal injection of different microbes (bacillus prodigiosus, staphylococcus aureus, &c.) had been made subjects of severe acute fatal peritonitis.

Now both proteus and bacillus coli grow best under aërobic conditions, *i.e.*, when they have a sufficient supply of free oxygen. Both are, however, capable of multiplying and sustaining life when they are placed under conditions in which no supply of free oxygen is given them. But their power of multiplication is very much lessened under such anaërobic conditions; they, particularly the bacillus coli, are for this reason considered as "facultative anaërobes." Thus, proteus vulgaris, which is considered as the microbe "*par excellence*" of putrid decomposition of albumen, displays under aërobic conditions the power of very rapidly peptonizing and dissolving albumen and albuminoid substances; whereas placed under anaërobic conditions it does this only very tardily and to a very limited degree.

In the bacterioscopic experiments which I carried out with the organs of animals, buried as mentioned above for various periods one of the first and most striking results was that the dissolution and destruction of the viscera of the abdomen and chest, and the muscular walls of the abdomen and chest and of the limbs cannot, to any considerable extent, be ascribed to the action either bacillus coli or proteus vulgaris, or allied forms. number of experiments were made in this direction, viz., for the purpose of ascertaining to what extent the peritoneal cavity a the liver in a buried animal contained proteus vulgaris or bacillus coli, or both these micro-organisms. The materials were prepared in the following manner :—As regards the peritoneum, the animal having been exhumed the wall of the abdomen was cut through, and a few (one to two) cubic centimetres of sterile salt solution poured into the peritoneal cavity. Then, by shaking the body from side to side the peritoneal cavity was thoroughly washed, so that the fluid obtainable therefrom was a strongly turbid emulsion. In the case of the liver a good sized piece of this organ was removed and distributed in one to two cubic centimetres of sterile salt solution, it being found that the liver substance could thus easily be distributed so as to form a fluid, strongly turbid, emulsion. Next, in order to detect in these strongly turbid emulsions the presence of bacillus coli or any coli-like microbe, two kinds of cultures were made—(1) surface phenol-gelatine plates, and (2) phenol-broth cultures. On the former the presence of coli-like colonies could be readily recognised after two to three days' incubation at 20° C., and these were then tested by sub-culture as to their characters: as to gas in shake-gelatine, as to clotting of milk and reddening of litmus milk (acid production), and as to indol formation in broth culture. In phenol-broth cultures, incubated at 37° C., the appearance next day of strong and uniform turbidity would be suggestive of the presence and multiplication of bacillus coli; in which case material from such broth cultures was tested in sub-culture as before. For the detection of the proteus vulgaris, ordinary nutrient gelatine as surface plates, ordinary agar surface plates, and (as in the case of bacillus coli) phenol-gelatine surface plates were used; in the latter, proteus vulgaris grows well. Colonies of proteus vulgaris—as slightly turbid, colourless, liquefied circular areas each with granular central mass—are easily recognised in the gelatine plates after two to three days' incubation at 20° C. From such colonies further sub-cultures were made: (*a*) stab gelatine cultures; (*b*) potato cultures; (*c*) solidified blood-serum cultures; and (*d*) surface agar cultures. In this way the proteus was identified by (*a*) its rapid liquefaction; (*b*) its peculiar filmy, slightly yellow tinted growth; (*c*) its liquefaction of the blood-serum; and (*d*) its smeary, filmy growth on agar. In addition, the microscopic appearances of fresh and stained specimens, the number and distribution of the flagella, and the physiological action of cultures of the microbe on experimental animals were in several instances considered.

App. B. N

Report on Fate of Pa genic and other Infe Microbes i the dead animal bo by Dr. Kle F.R.S

In all instances, both in the search for bacillus coli and for proteus vulgaris, the amount of the turbid emulsion rubbed over the surface of the culture plate was considerable. As a rule five to seven droplets were used for each plate, so that, if either of these microbes were present, the relatively large amount of material used for the plates should yield considerable numbers of colonies.

I proceed now to give the results of some of the experiments made in this direction :—

Experiment 1.—A normal guineapig (No. 1), having been placed, after death, in a small wooden coffin on June 16th, was buried in the earth after manner described. It was exhumed on July 5th, *i.e.*, in 19 days. On the abdominal cavity being opened, a portion of the liver—this organ was greatly reduced in size and had become a brownish, smeary mass—was shaken up in a few cubic centimetres of sterile salt solution, so as to form a strongly turbid emulsion. From this emulsion seven droplets (as they flowed out from a capillary glass pipette) were rubbed over the surface of an ordinary gelatine plate, the same quantity was rubbed over the surface of a phenol-gelatine plate, and a like quantity was added to phenol-broth. The surface of the phenol-gelatine plate on incubation remained free of colonies of bacillus coli and proteus vulgaris, and the phenol-broth remained free of any turbidity. Only the surface of the ordinary gelatine plate showed numerous liquefying colonies. Many of these were those of cocci some few proved to be those of proteus vulgaris. It follows then from this experiment that the liver of this guineapig, buried for 19 days, contained no bacillus coli, and only a limited number of proteus vulgaris.

EXPERIMENT 2.—A normal guineapig (No. 2), had, like guineapig No. 1, been, after death, placed on June 16th in a wooden coffin and buried in earth. This animal was the companion of the former. It was exhumed on July 14th, *i.e.*, 28 days later. The heart muscle, shrunken and thin, was cut out and shaken up in a few cc. of sterile salt solution. It easily broke up into a fine turbid emulsion. With this emulsion (seven drops in each instance), an aërobic surface ordinary gelatine and a phenol-gelatine plate culture were made; also a phenol-broth culture. The result was that no colonies of typical bacillus coli could be identified. In the phenol-gelatine plate there was one single colony of proteus vulgaris.

It follows from this that although the heart muscle was, after 28 days' burial, very much shrunken, it did not contain bacillus coli; and that 14 drops of the emulsion (seven drops for each plate) yielded but one single colony of proteus vulgaris.

I think these two experiments tend to show that so far as the dissolution and decomposition of the viscera are concerned, the rôle of the bacillus coli is nil, and that of proteus vulgaris extremely limited; in fact that practically neither operate in this sense.

A number of guineapigs which had been in experiments to be presently described, injected intraperitoneally with various microbes, and which guineapigs had died within 24 hours from acute peritonitis, were buried as before; some in tin coffins, some in wooden coffins, others directly in the earth or sand. These were, after exhumation at varying periods, tested like the others as to the presence in their viscera of bacillus coli, and of proteus vulgaris: and with like result. I here mention some of these other experiments confirmatory of the above inference.

EXPERIMENT 3.—A guineapig, dead after intraperitoneal injection with emulsion of Koch's cholera vibrio, and a second guineapig, dead after intraperitoneal injection with emulsion of typhoid bacillus, were placed separately in wooden coffins and buried in earth on June 6th. They were exhumed on July 4th, *i.e.*, in 28 days The peritoneal cavity was in each instance opened and was well washed out with a few cc. of sterile salt solution, whereby strongly turbid emulsions were obtained. With these emulsions aërobic surface plates were made on ordinary nutrient gelatine, and phenol-gelatine; as also phenol-broth cultures. Four to five drops of the emulsion were used for each culture. The result was neither bacillus coli nor proteus vulgaris could be obtained from these emulsions.

EXPERIMENT 4.—A guineapig, dead after intraperitoneal injection of emulsion of bacillus prodigiosus, having been placed in a wooden coffin was buried in sand on June 17th. The animal was exhumed on July 13th, *i.e.*, after 26 days. On opening the abdominal cavity it was found that in several places the intestinal wall had become quite decomposed, so that a small amount of fæcal matter had escaped into the peritoneal cavity. The liver was greatly shrunken and reduced to a small, brownish, flabby, soft mass. With a few cc. of sterile salt solution a strongly turbid emulsion of the peritoneal contents was made, and a similar emulsion with a piece of the liver. With seven droplets of each of these emulsions aërobic surface agar plates were inoculated, and incubated at 37° C. The result was that numbers of colonies of a liquefying diplococcus, and some of nondescript bacilli, were obtained both from the peritoneal and liver emulsions. But neither bacillus coli nor proteus vulgaris could be thereby demonstrated by means of these cultures.

EXPERIMENT 5.—A guineapig, dead after intraperitoneal injection of an emulsion of typhoid bacillus, having been placed in a small wooden coffin, was buried in earth on June 18th. The animal was exhumed on July 19th, *i.e.*, 31 days later. On opening the abdominal cavity a great part of the wall of the stomach was found to have become disorganised and dissolved, and partly, but to a lesser degree, the wall of the intestine was in like condition. An emulsion (strongly turbid) of the peritoneal contents; a similar emulsion made by shaking up the heart muscle (much shrunk and decomposed); and a similar turbid emulsion of the much decomposed lung tissue, were employed for making surface plates on ordinary and on phenol-gelatine, four drops of the emulsion being rubbed over the surface of the gelatine in the case of each plate. The result was this: no colonies of bacillus coli; but a small number, not more than four or five, colonies of proteus vulgaris appeared on each of the plates made from the peritoneal emulsion, the heart muscle emulsion, and the lung tissue emulsion. In one of the ordinary gelatine surface plates made from the peritoneal emulsion, there were 11 colonies of a liquefying diplococcus.

As a further experiment with the normal guineapig I mention the following:—

EXPERIMENT 6.—A normal guineapig was, after death, buried on September 16th, directly in sand. It was exhumed on November 10, *i.e.*, 54 days later, or near to the end of eight weeks. Of the liver only a small brownish grey, smeary, shrivelled mass was left. This was taken out as completely as practicable, and the whole shaken up in a few cc. of sterile salt solution, whereby a strongly turbid emulsion was obtained. With seven drops of this emulsion one aërobic surface ordinary gelatine plate and, with a like amount, one aërobic surface ordinary agar plate were made.

The former was incubated at 20°, the latter at 37° C. The result was remarkable, inasmuch as both plates showed no colonies whatever; that is to say, not only were bacillus coli and proteus vulgaris absent, but likewise all other aërobic microbes.

While these further experiments prove that 26 days or more after burying the bodies of guineapigs their abdominal and thoracic viscera contain no bacillus coli, and only comparatively a limited number of proteus vulgaris, other experiments made by me indicate that even at an earlier date than this period—*i.e.*, 16 days, 19 days, 21 days after burial—the number of colonies of typical bacillus coli obtainable by culture in the above fashion from emulsions made of the peritoneal fluid is extremly limited. In similar case proteus vulgaris was somewhat better represented, but by no means to an extent at all suggesting that in the decomposition of these viscera this microbe plays an important or even a trifling part.

The result is, however, different if the skin or the adjacent subcutaneous tissue is subjected to culture test for proteus vulgaris.

The two following series of experiments illustrate these facts:—

Experiment 7.—Three guineapigs, dead after subcutaneous injection with virulent emulsion of agar culture (48 hours old) of diphtheria bacillus, were buried on November 17th; one in a small tin coffin, another, directly in earth, and a third directly in sand. They were exhumed on December 1st, *i.e.*, about 14 days later. The subcutaneous tissue and inguinal lymph glands of the inoculated side were in each instance dark and swollen. With the juice of these lymph glands direct inoculation, by means of a platium loop, of surface serum-agar plates was effected, and these plates, after incubation at 37° C. for 48 hours, were duly scrutinized and subcultured with the following result:—

(*a*) From the guineapig buried directly in earth, the plates showed a general filmy growth of proteus vulgaris; in addition, a few colonies of bacillus diphtheriæ;

(*b*) From the guineapig directly buried in sand, numerous colonies were obtained; some of proteus vulgaris, some of diphtheria bacilli, and some of cocci;

(*c*) From the guineapig in the tin coffin no proteus colonies were obtained; only bacillus diphtheriæ colonies and colonies of cocci.

Bacillus coli was not found in any plate.

Experiment 8.—Three guineapigs, dead after subcutaneous in-injection into the groin of emulsion (from agar surface growth) of virulent plague bacilli, were buried, on November 18th, in the same manner as the guineapigs of Experiment 7. They were exhumed on December 9th, *i.e.* 21 days later. On examining these animals the characteristic hæmorrhagic swelling of the subcutaneous tissue, and enlargement of the inguinal lymph glands, due to plague infection, was still distinct. With juice of the swollen tissue the surface of gelatine and agar plates was directly inoculated by means of a platinum loop. The result was that the plates showed abundance of colonies of proteus vulgaris, but no plague bacilli.

This result, *i.e.*, abundant presence of proteus vulgaris, in parts contiguous to the skin, is, I think, due to these microbes having arrived at the subcutaneous tissue by way of the skin, *i.e.*, from the surface. As a matter of fact, even after being buried for 14 days, and still more distinctly at a later period, the skin is in a putrid condition; it is greenish and discoloured, the hairs and the cuticle of the epidermis easily peel off, and the tissue contains crowds of bacteria, principally, or at any rate to a large extent, proteus vulgaris and cocci.

The above results, then, harmonise with the general view that, under aërobic conditions, or conditions in which to a large extent free oxygen can be obtained by micro-organisms, putrid decomposition of albuminous matters is associated with and caused by the growth and multiplication of the proteus vulgaris. But a like agency cannot be maintained for the decomposition and disappearance of the viscera and deep-seated tissues in the dead body, since here aërobic microbes, such as proteus vulgaris and bacillus coli, do not and cannot maintain and multiply themselves in any degree comensurate with the amount of wasting and decomposition actually going on. From these *a priori* considerations, it would appear likely that the processes of disintegration and decomposition of the deep tissues and viscera are caused by microbes which, under anaërobic conditions, can rapidly grow and multiply. As we shall see presently this view is fully confirmed by direct bacterioscopic analysis.

I make a digression here to note a curious and interesting series of observations which I made in connexion with the proteus vulgaris in the spleen of buried guineapigs.

From three healthy guineapigs which had, after death, been buried for three, four and five weeks respectively, the spleens (much reduced in size so as to appear a mere thin, pale pink, narrow band) were dissected out, finely cut up, and mixed in each instance with several drops of sterile salt solution. The whole mass of each spleen thus dealt with was then injected subcutaneously into the groin of a guineapig. The result was nil, as is usual with the ordinary form of proteus vulgaris; the animals remained well and lively and showed next day only an indication of swelling in the groin, no more than corresponded to the amount of solid matter injected.

But an altogether different result was obtained in the following experiment:— Two guineapigs, dead of plague, were buried on January 9th; one directly in earth, the other directly in sand. They were exhumed on February 6th, *i.e.*, 28 days later. Their spleens—soft, dark and still showing slight enlargement—were dissected out and each separately minced up in a little sterile salt solution. This emulsion was in each instance subcutaneously injected into the groin of a guineapig. One of these guineapigs was found dead in 20 hours, the other was found dead in 40 hours. In both cases the subcutaneous tissue of the groin and the whole of the abdomen was much congested and infiltrated, appearing as a dark red gelatinous mass. On incision, thin sanguineous fluid oozed out freely, and this on microscopic examination and on cultivation proved to contain a pure culture of actively motile proteus vulgaris. The spleen was slightly congested, and not distinctly, if at all enlarged; the bladder and the peritoneum were greatly congested, in the cavity of the latter was sanguineous exudation containing a pure culture of proteus vulgaris. The spleen on microscopic examination and in culture contained a few proteus vulgaris, no other microbe. Figure 1, Plate I., shows a film specimen of the subcutaneous fluid of such an animal.

With the subcutaneous sanguineous fluid from one of the above guineapigs another guineapig was subcutaneously injected, about 0·25 cc. of fluid being used for the injection. The result was in all respects the same as before, viz., extensive hæmorrhagic gelatinous infiltration of the tissues and death of the animal, and the sanguineous thin exudation crowded with proteus vulgaris. A number of culture experiments of the exudation were made: on gelatine plates, on agar plates, in broth, in milk, on blood serum and on potato; and it was found that the microbe in question was the proteus vulgaris of Hauser.

Further and physiological experiments with broth cultures of this microbe were made on guineapigs and mice, and its identity was hereby confirmed. With a broth culture grown at 37° C. for 24 hours, one guineapig and two mice were subcutaneously injected: the guineapig received about 0·5 cc. of the broth culture, and each mouse about 0·25 cc. Both mice were dead in six to eight hours, the guineapig in 20 hours. The guineapig had the extensive hæmorrhagic gelatinous infiltration of the groin and abdomen, the fluid containing a pure culture of proteus vulgaris; the spleen, not visibly enlarged, was congested, containing a few proteus vulgaris; the heart's blood free of proteus vulgaris. That is to say, the post mortem appearances corresponded exactly with those known to be ordinarily produced by the proteus vulgaris of Hauser; they indicated a local though extensive disease, with general intoxication and death due to the metabolic products of the bacteria at the seat of injection. But not so with mice: here no tumour was produced at the seat of inoculation, but the animals died with symptoms of general infection, viz., enlarged dark spleen crammed full of the bacilli, in groups and isolated; the heart's blood crowded with the bacilli, as was shown by cover glass specimens (fresh and after staining), and by culture tests. This different action on mice and guineapigs is well established for the proteus vulgaris of Hauser. It has only to be added that of the proteus vulgaris of Hauser not all cultures—*i.e.*, not all specimens isolated—act with the same degree of virulence. Thus the microbes described by various authors: Babes's bacillus proteus septicus, and bacillus proteus lethalis; Karlinski's bacillus murisepticus pleomorphus; Lanz's bacillus pyogenes foetidus liquefaciens; Roger's bacillus septicus putidus, are no other than proteus vulgaris having somewhat exalted virulence. In this I quite agree with Kruse (Flügge's Mikro-organismen II., p. 280). So too is the proteus vulgaris obtained from the above spleens of the buried plague guineapigs, as Figures 2, 3, 4, 5 and 6, Plates I. and II., show.

As a result of the experiments just related, the important question that presented itself was this: Wherein lies the great difference between the virulence of the spleen emulsion of buried normal and that of buried plague guineapigs? I thought at first that possibly if the spleen of the guineapig becomes, owing to disease, enlarged and congested, it furnishes a better breeding soil for proteus vulgaris growing and immigrating after death into the spleen from the intestine in which it is normally present or from the outside through the skin. I assume it as highly probable that the proteus found in the spleen of the plague-guineapigs came originally from the intestine; this at any rate is the more natural and more feasible explanation, although, as in the case of the subcutaneous tumours of buried guineapigs dead of plague or of diphtheria (*see* antecedent page, experiments 7 and 8) so also in the spleen in these instances, it might have grown from the outside through the skin. But from whichever source it came, the fact remains that it was present after burial abundantly and in virulent form in the set of animals dead of plague, and not in the others. In order to test whether the diseased condition (congestion and enlargement in acute fatal plague) of the spleen has anything to do with this, I infected two guineapigs subcutaneously with culture of virulent anthrax. The animals died in the usual time with typical anthrax, as shown by examining a drop of blood from the ear vein. In such anthrax guineapigs the spleen is always characteristically congested and enlarged. The animals were buried and exhumed after 28 days. The spleens were dissected out; they were dark, soft and still showed the signs of enlargement. Each spleen was minced up in several drops of sterile salt solution, and the whole of the spleen in each instance injected subcutaneously into a guineapig. The result was nil; for next day there was a mere indication of a swelling (referable to the amount of substance injected), and the animals were lively and well. After two days no swelling to speak of could be noticed, the animals being perfectly well; and they remained so. The enlarged and congested condition of the spleen was therefore not the reason why in the three plague guineapigs buried for 28 days that organ became the seat of multiplication of virulent proteus vulgaris.

A point of interest that came out in the above experiments on the guineapigs injected with the proteus cultures was this: As mentioned above the guineapigs died from local disease and general intoxication, whereas the mice died from general septicæmic infection. In the latter cases the heart's blood contained abundance of the proteus injected. But in the guineapig the heart's blood was free of proteus; instead it contained (all the guineapigs yielded the same uniform result) large numbers of motile bacilli (cultures made with a droplet of blood yielded uncountable number of colonies of these bacilli) which on subculture were proved to be a variety of bacillus coli. These subcultures proved virulent when injected into guineapigs, producing even in small doses general acute septicæmic infection and death in 24 to 36 hours. It is clear that since only proteus was injected into the original animals, the numerous bacilli of one and the same variety of coli found in their blood after burial must have got into the blood most probably from the alimentary canal. Probably, owing to the general intoxication of the animal infected with proteus, this coli-bacillus, either in articulo mortis or even earlier, found the altered blood suitable for its multiplication. It is further possible that this condition of the blood (viz., the multiplication in it of virulent bacilli coli) may have brought about the fatal issue. As stated above four guineapigs were examined in this direction. In each case, culture experiment—after subcutaneous injection of pure culture of the proteus of experiment—yielded uniform results; their heart's blood after death was found to contain abundance of a variety of bacillus coli. This variety resembled Gärtner's bacillus enteritidis as regards the opacity and circumscribed form of its colonies, and in its streak culture on gelatine. Also it resembled Gärtner's bacillus in rapidly growing in phenol-broth; in cylindrical shape; in obvious motility; in producing gas in gelatine shake culture; in not curdling milk; in not giving indol reaction; and in being virulent for the guineapig. But it differed from Gärtner's variety in not reddening litmus milk, and in not giving any indication of clumping with typhoid blood serum.

Bacillus Cadaveris Sporogenes.

When examining the intestinal contents, whether of man or of other animals, under the microscope, in fresh specimens (hanging drop) or in stained film preparations, I have often—as must have other observers—noticed here and there a motile, cylindrical bacillus more or less straight and possessed of a terminal spherical or oval enlargement, which is the foremost part in the motile microbe. Such bacilli, both in size, aspect, and shape resemble the tetanus bacillus. I have seen them frequently in the intestinal discharges of persons dead from cholera nostras. Bienstock (Zeitschr; f. Klin. Med. Vol. VIII.) mentions the presence of drumstick-bacilli in fæcal matter. He cultivated some and found them to represent an aërobic species—bacillus putrificus coli. It is, in fact, easily ascertained that there exist in the normal large intestine oval spores and cylindrical bacilli, the latter including terminally similar oval spores or possessing merely a spherical or oval terminal enlargement which gives them the aspect of a drumstick or of something like a spermatozoon. But it is equally easily ascertainable that these bacilli belong to an obligatory anaërobic species. In the dead body they rapidly grow through the intestinal wall on to and into the surrounding tissues and organs—peritoneum, liver, spleen, kidney, and the inner wall of the abdomen. Thence they pass gradually into and through the diaphragm into the viscera of the chest, and further into the subcutaneous muscular tissue of the chest, and into the muscles of the extremities. Not only in the dead body does this take place, but also during life, when owing to one or another cause, either the wall of the intestines, the abdominal wall, or the adjacent subcutaneous tissue is the seat of severe inflammation, necrosis and gangrene. It is an easy matter—the experiment was first made by Gaffky (Mittheilugen des Kais. Gesundheitsamte: Book I., p. 81)—to obtain abundant multiplication of the above microbe, both for making microscopic specimens and for establishing cultures, by placing a dead guineapig (normal or pathological) in the incubator for 12 to 24 hours, and then opening the abdominal cavity and examining the outside surface of the intestine, the peritoneum, liver, &c. Numerous cylindrical and thread-like bacilli are met with, actively motile in the fresh state, some of them possessed of the spherical or oval terminal enlargment. Later (after 48 hours' incubation of the body) the number of such drumstick bacilli is increased, and there are some that have already oval terminal glistening spores (the spore thicker than the bacillus itself); and there are now found also free oval spores. These bacilli, owing to their thread-like form, and owing to their being of about the thinness of the (anaërobic) bacilli of malignant œdema, were considered by Gaffky to be the bacilli of malignant œdema of Koch; and he therefore concluded that the spores of the malignant œdema bacillus are present normally in the abdominal cavity. The above experiment of Gaffky does not, however, warrant this conclusion. The above thread-like bacilli growing out of the intestine under the conditions noted, are not Koch's bacilli of malignant œdema, but belong to an entirely different species, viz., the anaërobic *bacillus cadaveris sporogenes*. As mentioned above they are easily cultivated; and then it is seen —although they, like the bacillus of malignant œdema, produce gas in anaërobic sugar gelatine or sugar agar culture, although they rapidly liquefy sugar gelatine (in which they form spherical colonies of turbid liquefied gelatine), and although they rapidly liquefy solidified blood serum—that they are, nevertheless, a quite different and distinct species. They are distinguished from the malignant œdema bacillus by: (1) forming distinct drumsticks; (2) by forming always terminal spores; and (3) by not acting in any sense pathogenically on the guineapig, even when injected in large doses. Besides this, the liquefied sugar gelatine cultures, the milk cultures, and the liquefied serum cultures of the bacillus cadaveris sporogenes have an offensively putrid odour. Furthermore, bacillus cadaveris sporogenes possesses flagella different in number and arrangement from those described of the bacillus of malignant œdema. The bacillus of malignant œdema resembles the bacillus cadaveris in no more than this; that both are obligatory anaërobes, and that both form threads of the same thinness. The other characters which are common to both microbes, viz., forming spherical liquefying colonies in sugar gelatine, forming copious gas in this and in sugar agar cultures, their motility, and their power to liquefy solidified blood serum, are characters which they share with a considerable number of the known and hitherto described spore-bearing anaërobes. In this sense a distinction and differentiation cannot be made. On the other hand, the formation by bacillus cadaveris sporogenes of drumsticks, its formation of terminal spores, the distribution of its flagella, with the absence of all pathogenic action of it on rodents are sufficiently striking differential characters.

I have made a large number of experiments on the dead bodies of guineapigs—normal or dead from one and another infectious disease—buried in wood or in tin coffins in earth, buried under like conditions in sand, or kept unopened for some days in the laboratory under a bell glass, and I have been able to satisfy myself that bacillus cadaveris sporogenes is the chief microbe which causes the discolouration, decomposition, reduction, decay, disintegration, and disappearance of the organs of the abdomen and chest, of the muscular walls of these cavities, and, further, of the muscles and tissues of the extremities. This microbe it is which in the abdomen and chest causes after death the presence and accumulation of gas and the peculiar cadaveric odour. This microbe starts from the large intestine and gradually (sooner or later according to temperature) pervades the surrounding tissues, multiplies herein, destroying them gradually, and forming as a finish to its life cycle permanent seeds—oval spores. For this reason I have called it *bacillus cadaveris sporogenes*. It is an obligatory anaërobic microbe which, under conditions favouring anaërobiosis, such as obtains in unopened dead bodies and after burial, finds opportunity for its growth and multiplication, and, therefore for effecting the decomposition and destruction of the tissues.

The organs which were examined after exhumation of a number of experimental guineapigs were: The peritoneum, the liver, the spleen, the pancreas, the mesenteric lymph glands, the kidney, the heart, the lungs, the diaphragm, the subcutaneous tissue and the muscles of the thoracic and abdominal walls, the muscles of the extremities, and the subcutaneous tissue of the groin and axilla.

The simplest method, and one which never fails, to obtain the microbe in pure culture is this: The abdominal cavity of a dead guineapig that has been left unopened in the laboratory, or that has been buried in one way or another for two to four weeks, is opened, and a small particle of the liver, or a drop of an emulsion made by putting a little sterile salt solution into the peritoneal cavity, is added to tubes containing sugar gelatine or milk, previously, of course, sterilized. These tubes are then heated to 80° C. for ten to fifteen minutes, by which means everything in the medium contained in the test tubes except spores is killed. After this heating, the tubes are cooled and then incubated anaërobically, the sugar gelatine at 20° C., the milk at 37° C. In a few days the sugar gelatine shows in its deeper parts numerous liquefying spherical colonies all of the same kind; the milk will be found changed and containing crowds of cylindrical bacilli and threads, drumsticks and free spores, which in anaërobic plates can be proved to belong to one and the same species, viz., the bacillus cadaveris sporogenes.

I proceed now to describe the morphological and cultural characters of this microbe.

In examining all the dead tissues mentioned above, as microscopic specimens of the fresh material (in hanging drop) or as stained cover glass films, Figs. 7 and 8, Plate III., the attention is at once arrested by the presence of numerous motile, drumstick-like, long, cylindrical bacilli. In the first week or two after burying motile rods of cylindrical or filamentous shape are numerous; later on these are less numerous, being replaced by drumstick-shaped bodies, many possessing a terminal glistening spore, and in the filaments, corresponding to the cylindrical joints (individual bacilli), are rows of oval glistening spores (one at the end of each joint). Later still free oval glistening spores are copious. The typical bacilli are more or less stiff cylinders, on the average 2–4 μ long, of the same thickness as the malignant œdema bacillus. They are therefore distinctly thinner than the bacillus enteritidis sporogenes, their ends are rounder, and their motility consists in a peculiar wriggling and wobbling locomotion; those that possess a terminal spore or knob move so that this spore end or knobbled end is in front. The bacilli grow rapidly into threads and filaments, which in unstained fresh specimens appear smooth ; on drying and staining they show themselves composed of a series of longer or shorter cylindrical pieces. The spore when fully formed is 1·6 to 1·8 μ long, about 0·8 to 1 μ thick, thicker therefore than the bacillus in which it is formed ; and hence the resemblance to a spore-bearing tetanus bacillus. By the usual methods of staining film specimens for spores, viz., with boiling carbolfuchsin, followed by short counterstaining with methyl-blue, it is seen that the spores of this microbe stain readily with the fuchsin and the bacillary protoplasm with the methyl-blue. In such specimens it becomes also evident that the terminal knob in the drumstick forms is not necessarily a spore, since it is stained blue just like the rest of the bacillus. But all phases of the spore formation *within and at the expense of the knobbed end*—from a minute red dot to a fully-formed big oval red spore—can be found. The free fully-formed spore is surrounded by a thin investment of blue material ; this being the last remnant of the substance of the knob in which the spore had become formed, the rest of the cylindrical bacillus having become granular and disintegrated. There are, however, drumstick forms, of which the terminal knob (stained blue) is two or three times the size of the typical knob, and does not contain any indication of spore formation. These I take to be involution forms. As to the presence of drumsticks, of terminally sporing bacilli, and of free spores, it has to be mentioned that they are present not only in the tissues of the animals buried for two to four weeks, but are easily detected in cultures on solidified blood serum, in the condensation water of agar with slanting surface, and in milk cultures—all, of course, inculated anaërobically, Figs. 9 and 10, Plate III. The process of formation of spores, and also the fully-formed big oval spores, can indeed be observed in a few days, *i.e.*, before the end of a week. The bacillus cadaveris stains readily and well after Gram's method—one minute gentian violet, four minutes iodide iodate of potassium—in its non-sporing cylindrical and filamentous forms, as also as a drumstick and as sporing bacillus, Fig. 11, Plate IV.

The bacilli possess numerous flagella (*see* Fig. 12, Plate IV.) which can be stained after Van Ermengem's method. The flagella are long wavy or spiral filaments surrounding the body and ends of the bacillus, and forming at one end sometimes a bunch of hair-plait, the Haarzopf of Loffler. In cultures in sugar gelatine when liquefaction has become very extensive, one can find these bunches torn off from the bacilli even in fresh and unstained preparations ; they look like beautiful spirals thickened in the middle or more commonly at one end.

When spores are distributed in melted sugar gelatine (heated to 80° C. for 10 to 15 minutes, and then incubated anaërobically at 20 to 21° C.), there are noticed as early as after 24 hours, in the depth of the medium, a plurality of colonies (their number, of course, depending on the amount of material implanted) like granular more or less angular and branched dots. In two to three days the colonies have commenced to liquefy the gelatine, and now appear as smaller or larger turbid spheres which, under a glass in transmitted light, appear brownish, more or less granular, and exhibit by proper illumination a fine dense radial striation in the peripheral portion of the colony, Fig. 18, Plate V. As each colony enlarges into a growing sphere of liquefied gelatine, a distinction between a central granular and a peripheral less granular mass can be made, and from the former numerous threadlike granular prolongations can be traced, Figs. 19, 20, 21, Plate VI., Figs. 29 and 30, Plate VIII. (under a magnifying glass) extending into the peripheral parts, Figs. 13 and 14, Plate IV., Figs. 15, 16, 17, Plate V. The colonies which are situated in the deepest parts of the sugar gelatine culture show sometimes, but not invariably, these threadlike prolongations of the central mass arranged in a radiating manner. Such a colony closely resembles in aspect a colony of the anaërobic non-pathogenic bacillus spinosus of Lüderitz (Zeitschr. f. Hygiene, Vol. V., p. 152); and in respect also of motility, size, and terminal spores these two microbes have much in common. The description given by Lüderitz of the bacillus spinosus is unfortunately rather incomplete, so that an exhaustive comparison between it and the bacillus cadaveris cannot be made; but I am strongly inclined to regard them as, if not identical, at any rate very closely related. In the first place both exhibit in common, in the depth of the anaërobic sugar gelatine culture while liquefaction is progressing, some colonies of radiating character, Fig. 22, Plate VI. With bacillus cadaveris this character is not constant, and according to Kruse (Flügge's Manual of Bacteriology, new Edition, Vol. II.) this "spinosus" character is not constant in Lüderitz's bacillus either; for which reason I do not think for the latter the term bacillus spinosus appropriate. What, however, seems to show that the two microbes are closely related is the fact that Lüderitz obtained his bacillus from the subcutaneous tissue of mice and guineapigs which had died after injection into them of manured earth. In experiments of my own wherein fatal extensive gangrene of the subcutaneous tissue in guineapigs was produced by a normally virulent typically changed milk culture of bacillus enteritidis, the growing out of the bacillus cadaveris from the large intestine into the inflamed peritoneum, and through the gangrenous abdominal wall into the gangrenous subcutaneous tissue, has been an occurrence by no means uncommon. In a certain percentage of cases of such dead guineapigs, particularly if the post-mortem is delayed, there may be noted, here and there amongst the crowds of bacillus enteritidis contained in the sanguinous exudation and in the gangrenous tissues, a very motile bacillus which, by its thinness and length, occasionally by its drumstick shape and terminal spore, can be distinguished from the shorter and thicker bacilli enteritidis. If such exudation is planted in sugar gelatine or in milk, heated to 80° C., and then incubated anaërobically, the result is a growth not of bacillus enteritidis but of bacillus cadaveris. Lüderitz has, as I have said, actually cultivated his bacillus spinosus from similar gangrenous subcutaneous tissue of guineapigs and mice dead as the result of injection of manured earth; that is to say, under conditions in which a growing out and passage through the abdominal wall of bacillus cadaveris from the intestine is highly probable.

Here it will be convenient to point out the cultural and other differences between the bacillus cadaveris and the bacillus enteritidis sporogenes.

App. B. No. 2.

Report on the Fate of Pathogenic and other Infective Microbes in the dead animal body; by Dr. Klein, F.R.S.

Differentiation of Bacillus Cadaveris and Bacillus Enteritidis (Klein).

I have already mentioned that the bacillus cadaveris forms readily drumsticks and terminal thick spores, that it is conspicuously motile, and that it grows out readily into threads and filaments; characters which distinguish it readily from the bacillus enteritidis sporogenes. In addition, the aspect of the colonies in the sugar gelatine in the earlier stages prior to liquefaction, and also when liquefaction has already set in, are sufficiently different with the two microbes. Both, have, however, this in common along with some others, namely, that they are obligatory anaërobic; they cannot therefore grow under aërobic conditions. Both produce much gas in sugar agar without liquefying it; both liquefy sugar gelatine, and liquefy also solidified blood serum. But there are some striking differences to be easily ascertained by making anaërobic cultures of the two microbes on the slanting surface of ordinary nutritive agar or on a surface plate culture (anaërobic) of agar. After two or three days incubation at 37° C., the differences between the two microbes are striking enough to enable a diagnosis to be with certainty made. The difference is this: The bacillus cadaveris forms angular dark granular colonies from which extend numerous radiating dark granular or filamentous longer or shorter processes, which, later on, as incubation proceeds, form a more or less densely reticulated mass of considerable extent; at the same time this dark granular material—centre and peripheral portion—is embedded on a lobulated finely granular translucent filmy basis. In some colonies in the early stages only is this filmy mass noticed, in others there is, in addition, already the central dark granular patch; in others again even in the early stages the dark reticulated expansion of the central mass is well established. But in some cultures the translucent filmy basis is absent altogether or makes its appearance very late. Figs. 23, 24, 25 and 26 of Plates VI. and VII. give accurate representations of the colonies of bacillus cadaveris. The colonies of the bacillus enteritidis, on the other hand, as I have already pointed out in last year's report, are, and remain, circular discs. They are thicker and darker in the centre, translucent and filmy in the peripheral part; and, as incubation proceeds, show this contrast more distinctly owing to the expansion of the peripheral films. There is never the reticulated prolongation and expansion from the centre which is observed with colonies of bacillus cadaveris. With bacillus enteritidis at later stages, after a week and more, the at first sharply defined outline of the peripheral filmy mass becomes a little notched and knobbed (*see* Fig. 28, Plate VIII.). Another striking difference consists in the formation of spores. Whereas the bacillus cadaveris forms readily spores on agar and in milk, the bacillus enteritidis does not do so. I have hitherto altogether failed to obtain spores of bacillus enteritidis either in milk or on agar culture. The changes in milk inoculated with bacillus cadaveris and anaërobically incubated at 37° C. are these: there is no change noticeable for the first two or three days; then there appears a distinct transparency underneath the cream layer, which at this and all later stages remains undestroyed. This transparency extends downwards into the depth, and after ten days or a fortnight the original milk is separated into thin layers: (*a*) the unaltered cream; (*b*) a principal and more or less yellowish watery part containing fine flocculi; and, (*c*) a deeper smaller part containing whitish curds. Later on, this last becomes much reduced in amount, the principal yellowish watery part becoming meanwhile almost clear. In material from the culture motile cylindrical and filamentous bacilli, and extensive spore formation in the single bacilli, as also rows of spores in the threads, can be seen already after several days incubation, certainly within the week. The growth of the microbe in milk and in sugar gelatine is very active also at 20° C., considerably more than that of bacillus enteritidis. The milk culture of the bacillus cadaveris has an offensively putrid odour, its reaction is amphoteric or perhaps a little alkaline: So that the milk culture of the bacillus cadaveris resembles in every detail what in a former report I described as the atypical culture of bacillus enteritidis. Since I became acquainted with the bacillus cadaveris I am inclined to think that what I regarded as *atypical* milk culture of bacillus enteritidis is really the typical milk culture of bacillus cadaveris. That I obtained such cultures last year when working with bacillus enteritidis is explained by the fact that, as stated above, in the gangrenous subcutaneous tissue and exudation of the guineapigs dead after injection of typical milk culture of bacillus enteritidis, a few bacilli cadaveris (derived from the intestine) are, soon after death, occasionally present. If such exudation is collected in capillary glass pipettes (see last year's report) and incubated, the resulting spores will be a mixture of spores of bacillus enteritidis and of bacillus cadaveris; they may be almost entirely those of bacillus cadaveris, if many examples of this microbe had been present already in the gangrenous exudation. I should add that if the post mortem examination of the dead guineapig be made soon after death and the exudation be anaërobically incubated at 37° C., the resulting spores may be and as a matter of fact generally are, those of bacillus enteritidis alone. I have in my possession such preserved subcutaneous exudation in sealed capillary glass pipettes which, after a lapse of over two years, still contains spores only of bacillus enteritidis. When planted in milk, these produce the typical enteritidis change, and the whey of this milk culture acts pathogenic with certainty. It is therefore necessary to be constantly on one's guard, when working with the exudation of the gangrenous subcutaneous tissue of a guineapig dead after injection of however pure a milk culture of enteritidis; for it may happen, and it has happened to me in not a few instances, that the preserved exudation or the sugar gelatine cultures made from such exudation, contain, when examined after some time, only spores of bacillus cadaveris, the bacillus enteritidis having been entirely suppressed by the latter.

The milk cultures, the liquefied gelatine cultures, the agar cultures, and the liquefied serum cultures of the bacillus cadaveris, are without pathogenic effect on the guineapig; as much as 2 and 3 cc. of a liquefied sugar gelatine or of a milk culture can be injected without physiological effect.

. B. No. 2.
rt on the
of Patho-
and
Infective
bes in
and
al body;
. Klein.

It deserves notice that Sternberg described (Manual of Bacteriology p. 492) under the name of bacillus cadaveris an anaërobic bacillus which he found when subjecting to examination pieces of liver, cut out from the body of a yellow fever corpse, wrapped up in a carbolised cover, and incubated at 37° C. But this bacillus of Sternberg cannot be confused with that which I have described; it is much shorter and thicker and does not form spores. Another microbe which Sternberg figures (l.c.) and which he obtained under similar conditions does form spores; but this is also much thicker than that described by me, does not form drumsticks, and above all, is, Sternberg states, very difficult to cultivate anaërobically. The bacillus cadaveris sporogenes that I have isolated grows readily in almost all media (including ordinary alkaline nutrient beef-broth) and without difficulty, so long as anaërobiosis obtains.

The figures in addendum to this report give exact representations of this microbe in its morphological and cultural characters.

Specific Microbes in the Dead and Buried Animal Body.

Having now become acquainted with the bacteriology of the viscera and tissues of the dead body of the experimental animal, I proceeded to investigate the fate of specific microbes in similar dead bodies: bodies in which the microbes had been growing during life and microbes which had been the cause of disease and death of the animal.

Some preliminary experiments were made with bacillus prodigiosus and staphylococcus pyogenes aureus; both being micro-organisms very easily detected by surface agar plate cultures, on which they produce characteristically tinted colonies—those of the former bright pink, those of the latter orange yellow. In each experiment the growth of a recent (48 hours old) culture on the sloping surface of nutrient agar was scraped off, and distributed in sterile bouillon. Of this emulsion a sufficient amount was injected intraperitoneally to produce in each guineapig extensive peritonitis and death within 24 hours. As was described in detail in a former report the peritoneal cavity of such guineapigs contains abundance of viscid to thin fluid exudation crowded with the microbe injected, so that an active multiplication of the microbes had taken place within the peritoneal cavity of the guineapig between the time of its injection and its death. The guineapigs were after death wrapped up each in a piece of linen or calico and buried in earth, some having been first placed in a wooden or tin coffin, others were buried directly in earth [illegible]

After [illegible] the abdominal cavity was opened and a few [illegible] were poured into it, the peritoneal [illegible] emulsion. Of this [illegible] was [illegible] surface of nutrient agar [illegible] dish, the material being carefully and uniformly [illegible] the medium with a sterile [illegible]. The plates were then incubated: those inoculated with the [illegible] of the prodigiosus guineapig at 20·5° C. those inoculated with the peritoneal emulsion of the staphyloccus aureus guineapig at 37° C. The results of these experiments are given in the following Tables 1 and 2 :—

App. B. No. 2.

Report on the Fate of Pathogenic and other Infective Micro-organisms in the dead animal body; by Dr. Klein, F.R.S.

TABLE 1.—(Bacillus Prodigiosus.)

Mode of Infection.	Mode of Burial.	Exhumation.	Result.	Result as to recovery of B. prodigiosus.
1. Intraperitoneally	Earth direct	In 14 days	Positive	Crowds of colonies.
2. "	Sand direct	" 14 "	"	" "
3. "	Wood coffin	" 14 "	"	Limited number of colonies.
4. "	Earth direct	" 21 "	Negative	No colonies.
5. "	Sand direct	" 21 "	Positive	Crowds of colonies.
6. "	Tin coffin	" 21 "	Negative	No colonies.
7. "	Earth direct	" 28 "	Positive	Crowds of colonies.
8. "	Sand direct	" 28 "	"	Limited number of colonies.
9. "	Tin coffin	" 28 "	Negative	No colonies.
10. "	Earth direct	" 6 weeks	"	" "
11. "	Sand direct	" 6 "	"	" "
12. "	Tin coffin	" 6 "	"	" "

From this Table it appears that no prodigiosus colonies were obtained by cultures of five to seven drops of turbid peritoneal emulsion from bodies of guineapigs after six weeks burial in one and another way; and further, that a guineapig buried in a tin coffin yielded no prodigiosus after 21 days, while one buried directly in earth yielded even after 28 days crowds of colonies of bacillus prodigiosus.

TABLE 2.—(Staphylococcus pyogenes aureus.)

Mode of Infection.	Mode of Burial.	Exhumation.	Result.	Result as to recovery of the Staphylococcus.
1. Intraperitoneally	Earth direct	In 14 days	Positive	Crowds of colonies.
2. "	Sand direct	" 14 "	Negative	No colonies.
3. "	Wood coffin	" 14 "	Positive	Crowds of colonies.
4. "	Earth direct	" 22 "	"	" "
5. "	Sand direct	" 22 "	Negative	No colonies.
6. "	Tin coffin	" 22 "	Positive	Crowds of colonies.
7. "	Earth direct	" 28 "	"	" "
8. "	Sand direct	" 28 "	Negative	No colonies.
9. "	Tin coffin	" 28 "	Positive	Numerous colonies.
10. "	Earth direct	" 2 months	Negative	No colonies.
11. "	Sand direct	" 2 "	"	—
12. "	Tin coffin	" 2 "	"	—

From these experiments it appears that the staphylococcus aureus disappeared from the peritoneal cavity of the buried animals in two months, whatever the mode of burial; and that in the case of guineapigs buried directly in sand no colonies appeared after 14 days burial, hereby forming an interesting contrast to the guineapigs buried directly in earth and in coffins.

Now, it has to be remembered that these two microbes are possessed, judging from many experiments which in the course of many years I have had the opportunity of making with various disinfectants, of considerably higher resisting power than other non-sporing microbes, particularly microbes of the pathogenic class. The staphylococcus aureus especially has in my experience a strikingly higher resisting power than any non-sporing pathogenic microbe; and therefore, seeing that the conditions in the dead buried body are so inimical to the vitality of staphylococcus aureus as to cause all trace of it to disappear in two months, it was to be anticipated that the vitality of non-sporing pathogenic microbes would be extinguished in the dead buried body at a much earlier date. This expectation was confirmed by the experiments made with the following microbes; (*a*) the vibrio of cholera; (*b*) the typhoid bacillus; (*c*) the bacillus diphtheriæ; (*d*) the bacillus of bubonic plague; and (*e*) the bacillus of tuberculosis.

(*a*.) *The Vibrio of Asiatic Cholera.*

Guineapigs were injected intraperitoneally with a bouillon emulsion of the cholera vibrio, prepared in the same way as already mentioned in regard of bacillus prodigiosus and staphylococcus pyogenes aureus. The dose injected was always sufficiently large to insure the animal being found dead with intensive peritonitis in 24 hours. As has been mentioned in a former report the (generally) copious peritoneal exudation teems in such case with the vibrio injected. The experimental animals after death were buried in the same manner as in the previous series. After exhumation the peritoneal cavity was washed out with sterile salt solution so as to obtain a strongly turbid fluid. This was then added, in amount to ten or more drops to Dunham peptone-salt solution; the best medium, as has been pointed out in the Report on cholera in England (1893) and in a subsequent report, for obtaining in 24 hours or less a culture of the cholera vibrio.

The peptone salt solution thus inoculated was incubated at 37° C., and next day the superficial layers—always and by predilection containing the young cholera vibrios if there are any at all present—were examined in fresh specimens and in stained cover glass films, Fig. 31, Plate IX. Whether this examination gave positive or negative results, further test of the peptone culture in gelatine plates was in all cases made. The peptone culture, after 48 hours to 3 days incubation, was also tested for cholera red reaction. In the positive cases (when colonies were obtainable from the gelatine plates), subcultures were made on agar; and of growth thereon emulsion in bouillon was made and tested for agglutination with blood serum of a cholera immunised guineapig.

It is to be understood that in the column "Result" in the next Table 3, the term "Positive" refers to *all* the above tests; the term "Negative" to failure to obtain evidence of the presence of the cholera vibrios in the peptone cultures and in the gelatine plates made therefrom.

TABLE 3.—CHOLERA VIBRIO.

Mode of Infection.	Mode of Burial.	Exhumation.	Result as to recovery of Cholera Vibrio.
1. Intraperitoneally ...	Earth direct ...	In 14 days	Positive.
2. " " ...	Sand direct ...	" 14 "	"
3. " " ...	Wood coffin ...	" 14 "	"
4. " " ...	Earth direct ...	" 19 "	Negative.
5. " " ...	Sand direct ...	" 19 "	Positive.
6. " " ...	Tin coffin ...	" 18 "	"
7. " " ...	Earth direct ...	" 28 "	Negative.
8. " " ...	Sand direct ...	" 28 "	"
9. " " ...	Wood coffin ...	" 28 "	"

From these experiments it appears that the cholera vibrios were in 14 days recovered by means of peptone cultures from the peritoneal cavities of the bodies of the buried animals, no matter what the mode of burial had been. Further, that in no case were they recovered from the bodies which had been buried for 28 days; and that as regards bodies buried in the earth direct, they were not recovered after 19 days.

These results fairly coincide with those published by the German Commission already quoted (Arbeiten aus dem Kais-Gesundheitsamte Vol. VII., p. 17), according to which experiments the cholera vibrios were recoverable from the guineapigs buried in wooden coffins after 7 and 19 days, and from the bodies of those buried in zinc coffins after 11 and 12 days.

It would seem then that the cholera vibrios do not retain their vitality in dead and buried bodies for any considerable time, certainly not as long as 28 days; that is to say, in bodies buried in wooden coffins the cholera vibrios are likely to have ceased to live long before any disintegration of the coffins can have taken place.

(b.) *The Typhoid Bacillus.*

The experiments made with cultures of this microbe were on exactly the same lines as those with the cholera vibrio. With the salt washings—very turbid fluid—of the peritoneal cavity of the exhumed and opened guineapig the following cultures were made :—

(*a.*) Phenolated gelatine was allowed to set in a sterile plate dish, and on its surface there were deposited 5–7 drops of the above peritoneal emulsion. These were well and uniformly rubbed over the surface of the set gelatine with a sterile platinum spatula. The plate dish was then incubated at 20·5° C. After two to three days typhoid colonies could, if present, be easily recognised as small, translucent, filmy, more or less roundish or angular patches. These were examined under the microscope in fresh and stained specimens, and in flagella stained specimens; and sub-cultures were made, in streak and shake gelatine, in broth, in litmus milk, and on potato. If these several examinations proved the colony to be possessed of all the morphological and cultural characters of

[illegible]

[illegible] bacillus.* A last test was made by adding blood [illegible] patient known to give positive Widal reaction [illegible] 1 in 25 to 30 of the microbe [illegible]

[illegible] test tubes there were [illegible] emulsion, and the tube[illegible] [illegible] From this broth, [illegible] [illegible] were made, and the broth [illegible] The colonies that appeared o[illegible] [illegible] scrutinised, and those rese[illegible] [illegible] microscopically. If the [illegible] resembled the typhoid bacill[illegible] [illegible] were subjected to sub-c[illegible] [illegible] and in litmus milk, [illegible] the results of the examination of the p[illegible] [illegible]

[illegible] "Positive" in the following T[illegible] [illegible] identification of typhoid bacilli: m[illegible] [illegible] and by Widal's serum test; "Neg[illegible] [illegible] were discoverable that responded [illegible] either morphologically or culturally.

TABLE [illegible]—TYPHOID BACILLUS.

Mode of Infection.	Mode of Burial.	Exhumation.	Result as recovery Typhoid Ba[illegible]
Intraperitoneally	Earth direct ..	In 14 days	Positive
[illegible]	Sand direct ..	„ 14 „	Negative
	Wood coffin ..	„ 14 „	Positive
	Earth direct ..	„ 20 „	Negative
[illegible]	Sand direct ..	„ 20 „	„
[illegible]	Wood coffin ..	„ 20 „	[illegible]
[illegible]	Earth direct ..	„ 28 „	
	Sand direct ..	„ 28 „	
[illegible]	Tin coffin ..	„ 28 „	„

These experiments show that the typhoid bacillus is ev[illegible] resistent than the cholera vibrio. In 20 days after burial, no [illegible] what form this had taken place, typhoid bacilli could [illegible] recovered from any of the bodies: from a guineapig bur[illegible] sand they could not be recovered after 14 days.

The German Commission (l. c. p. 20) exhumed the anim[illegible] earlier than one month after burial, and their examinations [illegible] dead bodies for typhoid bacilli were in all cases negative.

On the other hand, Karlinski (Archiv f. Hygiene Vol. [illegible] 1891) stated that he recovered the typhoid bacillus fro[illegible] spleen of a human corpse exhumed 96 days after burial. [illegible] however, necessary to add that at that time (1891) the diffe[illegible] characters, now so well established, between the typhoid b[illegible] and bacillus coli and coli-like microbes were not sufficient[illegible] distinctly known.

* It is unnecessary to detail here the distinctive characters of the [illegible] bacillus; they have been fully described in a former Report (Oys[illegible] Infectious Diseases).

(c.) The Bacillus Diphtheriæ.

ı these experiments guineapigs were, as is usual, injected cutaneously in the groin with virulent broth or agar ture, 24–48 hours old, of the bacillus diphtheriæ. The animals d between 36 and 48 hours later, with the usual great tumour out the seat of inoculation. Cultural experiments made with e hæmorrhagic and infiltrated tissue of the tumour of a control uineapig proved that there were present in such tissue large umbers of active diphtheriæ bacilli.

The experimental animals were buried and exhumed at lifferent periods. The tissue of the tumour was then exposed y an incision, and material therefrom used for direct cultural experiments, as well as for making a distribution in sterile salt solution. With this mixture fresh guineapigs were subcutaneously njected, and the result watched. If the animals developed umour and died, the tissue of the resulting tumour was further used for cultivation.

As I have pointed out in a former report, the intraperitoneal injection of guineapigs with diphtheria culture is not the surest method for producing acute fatal result, since even with large loses—much larger than those capable of causing acute fatal issue when injected subcutaneously—often only a transitory illness followed by recovery is induced. And I have also pointed out hat in guineapigs dead after subcutaneous injection with culture of he diphtheria bacillus no diphtheria bacilli can as a rule be lemonstrated in the blood or viscera; that herein the disease nduced in this animal resembles diphtheria in the human ubject, being practically a local disease in which the general ymptoms and death are due to the toxins produced by the iphtheria bacilli at the seat of inoculation and afterwards diffused roughout the system.

In the search by culture for the diphtheria bacilli that might be ill present in a living state in the local subcutaneous tumour of e exhumed guineapig, surface plate cultivations and test tube ltivations were in each instance made on a nutritive medium ich was recommended by the late Dr. Kanthack, and ich is now in constant use at St. Bartholomew's Hospital h conspicuous success for the rapid and certain demonstration the true diphtheria bacillus. This medium is a mixture ascites fluid, liquor potassæ, and agar, in proportions as ows:—2 cubic centimetres of a 10 p.c. solution of caustic ash are added to 100 c.c. of the ascitic fluid, and in this ture 2 grams of agar are dissolved and 6 p.c. glycerine dded. This fluid allows of boiling and sterilizing, and then just like ordinary nutrient agar. It is in my experience the medium for isolation of the diphtheria bacillus. The colonies t are of the same characters as on ordinary nutrient agar and the rient material is as transparent as this latter; moreover, other robes do not grow so readily on it as the diphtheria bacillus. iphtheria bacilli are present in any material, a small particle eof rubbed over the surface of the Kanthack's serum, set iously in a plate dish or set with slanting surface in a culture e, will afford, with 24 hours incubation at 37° C., a pure ure of colonies of the diphtheria bacillus.

The colonies that appeared in the cultures made with the material from the exhumed guineapigs were subjected to microscopic examination, with a view to detecting the characteristic nail-shaped and also the club-shaped forms, and the equally characteristic cylindrical or clubbed bacilli with unequally segregated protoplasm. Also they were submitted to sub-culture; the subcultures being tested for their virulence on fresh guineapigs. In the following Table 5, "Positive" comprises all the above tests; "Negative" means the failure to demonstrate either culturally or by animal experiment the presence of diphtheria bacilli. The nature of the colonies, other than those of bacillus diphtheria, that grow on the serum agar plates inoculated from the tumour of the exhumed animals (in the positive cases), have been already mentioned in connection with Experiment 7, and need not therefore be again stated here.

TABLE 5.—DIPHTHERIA BACILLUS.

Mode of Infection.	Mode of Burial.	Exhumation.	Result as to recovery of Diphtheria Bacillus.
1. Subcutaneously ...	Earth direct ...	In 14 days	Positive.
2. " " ...	Sand direct ...	" 14 "	"
3. " " ...	Wood coffin ...	" 14 "	"
4. " " ...	Earth direct ...	" 21 "	Negative.
5. " " ...	Sand direct ...	" 21 "	"
6. " " ...	Wood coffin ...	" 21 "	
7. " " ...	Earth direct ...	" 31 "	
8. " " ...	Sand direct ...	" 31 "	
9. " " ...	Tin coffin ...	" 31 "	"

From these experiments it appears that the diphtheria bacillus has only a short life in the subcutaneous tumour of the dead body of infected animals, since, no matter in what form these were buried, in three weeks they could not any more be recovered by culture, as also the direct animal experiment failed to demonstrate them.

(d.) *The Bacillus of Bubonic Plague.*

In this series guineapigs were inoculated subcutaneously in the groin with virulent culture of the plague bacillus.* As has been pointed out the animals die in between 48 and 72 hours; they show hæmorrhagic tumour about the seat of inoculation; the lymph glands of the groin and pelvis are much swollen; the spleen is enlarged. The tissue of the swollen lymph glands is densely pervaded by the plague bacilli, so is the tissue of the spleen. Descriptions and illustrations having been given of all these points in detail in my previous Report for 1896–97, as also of the cultural characters of the plague bacilli, it is not necessary therefore, here to enter further into this matter.

The dead animals were buried as in the previous experiments, and, after exhumation, the swollen lymph glands of the groin were dissected out, and from them salt emulsion was made; the same was also done with the spleen. From the emulsions surface

* Subcultures of the microbe obtained and reported upon in the Report of the Medical Officer for 1896–97.

plate cultures on gelatine and on agar previously set in plate dishes were established, several drops of emulsion being used for each culture. The emulsion was injected into the subcutaneous tissue of the groin of further guineapigs, and the result watched. If the animal died with inguinal tumour in 48–72 hours, its heart's blood, the inguinal lymph glands, and the spleen were examined microscopically and in culture, to ascertain whether the distribution of the bacilli and their morphological and cultural features were characteristic of true plague.*

TABLE 6.—PLAGUE BACILLUS.

Mode of Infection.	Mode of Burial.	Exhumation.	Result as to recovery of Plague Bacillus.
1. Subcutaneously ...	Earth direct ...	In 14 days	Positive.
2. " " ...	Sand direct ...	" 14 "	"
3. " " ...	Wood coffin ...	" 14 "	
4. " " ...	Earth direct ...	" 17 "	
5. " " ...	Sand direct ...	" 17 "	"
6. " " ...	Wood coffin ...	" 17 "	"
7. " " ...	Earth direct ...	" 21 "	Negative.
8. " " ...	Sand direct ...	" 21 "	"
9. " " ...	Tin coffin ...	" 21 "	
10. " " ...	Earth direct ...	" 28 "	
11. " " ...	Sand direct ...	" 28 "	
12. " " ...	Tin coffin ...	" 28 "	"

From these experiments it appears that the plague bacilli in the dead and buried body cannot be demonstrated after three weeks though up to 17 days they are still present in a living condition. They therefore behave in this respect in very similar fashion to other non sporing pathogenic microbes, inasmuch as they have after three, certainly after four, weeks, burial in a dead body ceased to exist as living infective entities.

(e.) *The Bacillus of Tuberculosis.*

In this series guineapigs were inoculated with purulent tuberculous sputum from the human subject. This material was ascertained by cover glass films to contain abundance of tubercle bacilli; they were stained in boiling carbol fuchsin, decolourised with 33 per cent. nitric acid, and further stained with anilin water methyl blue. The sputum was distributed in sterile salt solution, and of this turbid emulsion half to one cubic centimetre was injected into the peritoneal cavity of each animal. These died in periods varying between three and seven weeks, and were then buried as in the previous experiments.

When guineapigs are thus intraperitoneally infected with tubercular matter, there is invariably found after death the

* I cannot omit here to express my extreme surprise at the mishap that occurred at Vienna with plague cultures in the laboratory there. As in my laboratory, so also in others in England, in almost every laboratory on the Continent, and in India, experiments with plague cultures, as also with cholera, typhoid, tubercle, glanders, tetanus, and a host of pathogenic microbes, have been and are constantly performed. It requires, according to this almost universal experience, but the simplest precautions to avoid mishap.

following characteristic condition of the omentum. Along the large curvature of the stomach, from the porta hepatis to the spleen—and involving the attachment of the omentum majus, the portal glands, the omentum minus and the pancreas—there are observed rows and packets of tubercular deposits which were either still firm and greyish or softened purulent and caseous. Cover glass film specimens made of these tubercular deposits and stained in the customary manner (as above) show enormous numbers of tubercle bacilli.

It was these packets and chains of tubercular deposits along the omentum, which, after exhumation and opening of the abdominal cavity, were dissected out and minced in sterile salt solution, thus yielding a strongly turbid emulsion. Making cover glass film specimens of the emulsion and staining them in the customary manner for tubercle bacilli, very striking specimens are obtained. Two kinds of bacteria only were present: (1) bacillus cadaveris as long bacilli or as threads, as drumsticks, and as free large oval spores—all stained blue by the methyl blue; and (2) tubercle bacilli in small and large groups and also singly, stained bright pink. Fig. 32, Plate IX., gives a good representation of these specimens, but the appearances in the actual specimens owing to the different colouration of the two sets of microbes are far more conspicuous.

With the same emulsion guinea-pigs were injected: one intraperitoneally and one subcutaneously for each buried animal. As the following Table (7) shows, all experiments made in this series were negative; none of the animals became affected with tuberculosis or showed any ill-effects whatever. Unfortunately a statement of Schottelius (Tagblatt der 63 Versammlung deutscher Naturforscher und Aerzte 1889), according to which the tubercular lung of a human body exhumed two years after burial still contained virulent tubercle bacilli, so far misled me that I delayed exhumation of my guinea-pigs and did not start my experiments of testing their tubercular deposits until they had been seven weeks buried. There remains, however, the interesting fact that in no instance could I, after seven weeks, demonstrate any virulent tubercle bacilli in their bodies, although, as was mentioned just now, the tubercular deposits showed in suitably stained cover glass specimens an abundance of seemingly well-preserved tubercle bacilli. In these cases, therefore, the characteristic staining power of the tubercle bacilli was no indication of their being alive and infectious. A curious point about many of these tubercle bacilli in the stained cover glass film specimens was the conspicuous segregation of their protoplasm similar to what is well-known and easily shown as regards the tubercle bacilli of human tubercular sputum. And a further point of interest is that many of the tubercle bacilli in these specimens lost their pink stain, or the stain faded, in the course of a few days. This fading was not due to any fault in the method of staining and washing of the cover films, since other specimens of tubercle bacilli from similar deposits in unburied animals, prepared in exactly the same way and with the dyes from the same bottles, retained their bright pink colour for very long periods; in fact, at the present time, after many months, they look as bright pink as when first made. Whether or not this fading that is the inability of these tubercle bacilli to retain for more than a brief period the fuchsin stain after washing with 33 per cent. nitric acid (retention of the stain being characteristic of active tubercle bacilli), is a sign that tubercle bacilli, while becoming in the dead body deprived of their vitality become deprived also of some essential chemical property, must remain undetermined.

The German Commission, l. c., p. 29, have also failed to obtain infectious tubercle material from rabbits that had died of inoculated tuberculosis and had been buried for periods longer than one month and five days, and three months six days, respectively ; so that this Commission is equally unable to adduce experimental proof supporting Schotelius' above-quoted statement.

TABLE 7.—TUBERCLE BACILLUS.

	Mode of Infection.	Mode of Burial.	Exhumation.	Result as to recovery of Living Tubercle Bacilli.
1.	Intraperitoneally ...	Earth direct ...	In 7 weeks	Negative.
2.	" " ...	Sand direct ...	" 7 "	"
3.	" " ...	Wood coffin ...	" 7 "	"
4.	" " ...	Earth direct ...	" 10 "	"
5.	" " ...	Sand direct ...	" 10 "	"
6.	" " ...	Tin coffin ...	" 10 "	"

In these experiments no living and infectious tubercle bacilli could be demonstrated in the tubercular deposits of the guinea-pigs buried for seven weeks and longer, notwithstanding that the cover glass film specimens showed abundance of stainable tubercle bacilli, and although considerable quantities of the tubercular matter were injected into guinea-pigs.

I may, in conclusion, here mention that from the emulsion used in these experiments aërobic surface plates in gelatine and on agar were made, five to seven drops of the emulsion being used for each plate. But they failed to give any indication of either proteus vulgaris or bacillus coli. Anaërobic cultures on the other hand yielded abundance of colonies of the bacillus cadaveris sporogenes.

(*f*). *The Bacillus Enteritidis Sporogenes.*

In this series it was attempted to ascertain, in the dead and buried bodies of guinea-pigs, the behaviour of a bacillus that is capable of forming spores, and that can be readily detected by culture. I have repeatedly pointed out that the anaërobic bacillus enteritidis fulfils these two conditions : it is capable of forming resisting spores when incubated, under certain conditions, anaërobically, and it is readily detected for the reason that its spores, when planted in milk which is incubated anaërobically at 37° C., rapidly cause a typical and characteristic change of the milk. I have shown that when guinea-pigs are injected subcutaneously with 1 c.c. of the whey of a typically changed and virulent milk culture of the above sort they are found dead within 24 hours with big extensive the dead guinea pigs which were preserved in sealed capillary pipettes) the latter microbe is able to oust the former altogether, so that ultimately the spores remaining present are only those of the bacillus cadaveris.

In addition to the small chance thus afforded the bacillus enteritidis of forming spores in the dead body, the microbe appears to undergo there, as indicated in the table, a distinct and important diminution of its virulence.

SUMMARY.

Direct experiment lends no confirmation to the general and popular belief that the microbes of infectious disease retain their vitality and power of mischief within dead and buried bodies for indefinite periods. On the contrary, these researches show that, as far as the bodies of the guinea pig are concerned, the vitality and infective power of these microbes that have been made the subject of experiment, passes away in a comparatively short time; that in most cases a month is sufficient for this result, that is to say, long before the coffins containing the buried bodies have shown any indication of leakage. A like result is, however, observed in the case of bodies buried directly (wrapped up in a piece of fabric) in earth or in sand. In this case it may be inferred that the vitality and infectiveness of the pathogenic microbes contained in the viscera has passed away long before the outer skin has become permeable by them; and further, the operations of putrefactive microbes effecting their destructive processes from without would be likely to prove a powerful and perhaps efficient barrier for such passage outward of pathogenic microbes.

A question of importance which is directly suggested by the results of this inquiry is this: What is the essential cause or causes of this process of rapid destruction of the vitality and infectiveness of pathogenic microbes in the interior of the dead body? Is it that in the struggle the more favoured bacillus cadaveris soon becomes ubiquitous and ousts the others; or, is it that some chemical metabolic products elaborated by other more favoured microbes in the course of their growth and multiplication, operate as a poison on the pathogenic microbes? These questions can only be answered, or be attempted to be answered, by further and properly devised experiments.

Fig. 1.

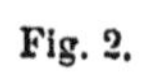

Fig. 2.

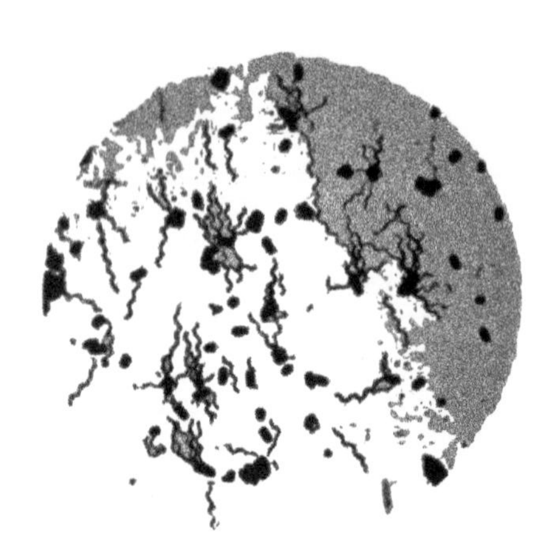

Fig. 3.

BACTERIA AND THE BURIED ANIMAL BODY.

PLATE I.

Fig. 1.

Film specimen of the sanguineous exudation of an experimental guinea-pig that died as the result of inoculation with the spleen-emulsion of a guinea-pig which had been exhumed after burial for 28 days.

Fig. 2.

Film specimen of the spleen-juice of an experimental mouse that died as the result of inoculation with culture of the *proteus vulgaris* which proved fatal to the experimental guinea-pig referred to under Fig. 1.

Fig. 3.

Flagella-stained specimen of the proteus in question.

[Magnifying power, in each instance, 1,000.]

BACTERIA AND THE BURIED ANIMAL

PLATE II.

FIG. 4.

Gelatine stab-culture of the proteus referred to in *two days'* incubation. Liquefaction has commenced of the medium. Near to the middle of the "stal "swarmers" have made their appearance.

[Natural size.]

FIG. 5.

The culture-tube represented in Fig. 4, somewhat magnified.

FIG. 6.

Gelatine stab-culture of the proteus in question at a more advanced stage. Liquefaction actively proceeding, at the surface and throughout the "stab."

[Culture-tube slightly magnified.]

PLATE II.

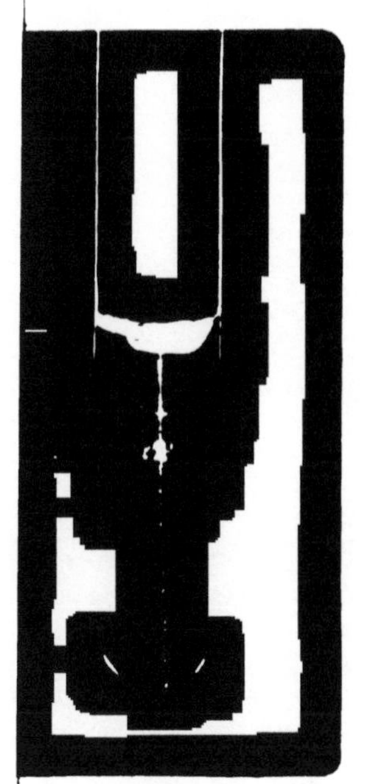

Fig. 4.

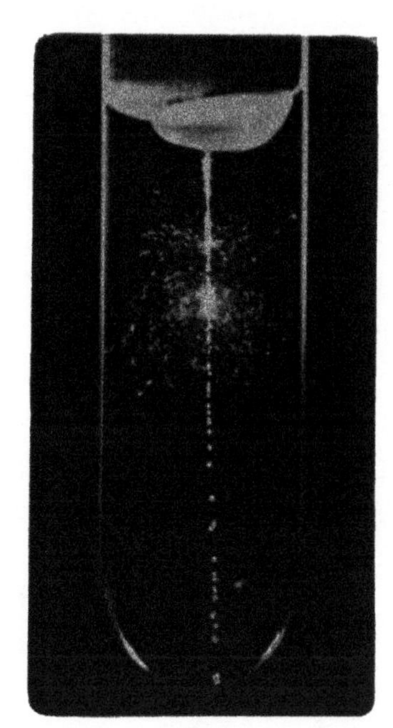

Fig. 5.

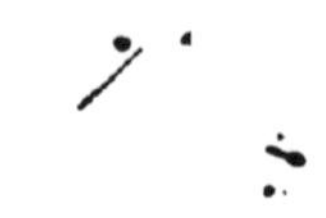

Fig. 7

Fig. 8.

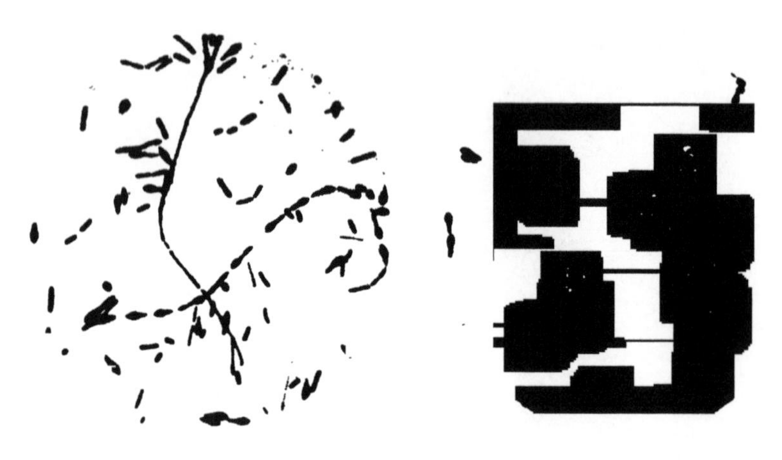

Fig. 9.

Fig. 10.

BACTERIA AND THE BURIED ANIMAL BODY.

PLATE III.

FIG. 7.

Film specimen of the *peritoneal emulsion* of a guinea-pig exhumed *fourteen days* after burial. The specimen had been stained for spores.

There are here seen numerous typical drum-stick forms, some cylindrical forms, and several oval fully-formed free spores of *bacillus cadaveris sporogenes*. The peritoneal emulsion was in fact a "pure culture" of this microbe.

FIG. 8.

Film specimen from the lung of a guinea-pig exhumed *twenty-eight days* after burial. There are exhibited drumsticks, cylindrical forms, and a few free spores of *bacillus cadaveris sporogenes*

FIG. 9.

Film specimen from a pure *milk culture* of *bacillus cadaveris sporogenes*, after incubation, anaërobically, for *seven days* at 37° C. There were seen single bacilli, as also filaments, with and without spores.

FIG. 10.

Film specimen from an *agar culture* of *bacillus cadaveris sporogenes*, after incubation for *four days* at 37° C. Almost all the bacilli contain terminal oval spores.

[Magnifying power, in each instance, 1,000.]

BACTERIA AND THE BURIED ANIMAL BODY.

PLATE IV.

Fig. 11.

Film specimen from a *serum culture* of *bacillus cadaveris sporogenes*, after *six days'* incubation at 37° C. The specimen is stained by the Gram method : Sporing is seen to be proceeding.

[Magnifying power, 1,000.]

Fig. 12.

Flagella-stained specimens of *bacillus cadaveris sporogenes*, from an *agar culture* incubated *two days* at 37° C.

[Magnifying power, 1,000.]

Fig. 13.

Anaërobic *sugar-gelatine culture* of *bacillus cadaveris sporogenes* derived from the heart-emulsion of a guinea-pig exhumed *twenty-eight days* after burial. After inoculation, the culture medium was heated for 15 minutes at 80° C., and then incubated for *three days* at 20° C. There are observed here nine liquefying colonies of the bacillus in question.

[Twice the natural size.]

Fig. 14.

Anaërobic *sugar-gelatine sub-culture* of *bacillus cadaveris sporogenes* derived from a pure milk culture of this micro-organism. After inoculation, the sugar-gelatine was heated for 15 minutes at 80° C.. and then incubated for *two days* at 20° C.

[Natural size.]

PLATE IV.

Fig. 11.

Fig. 12.

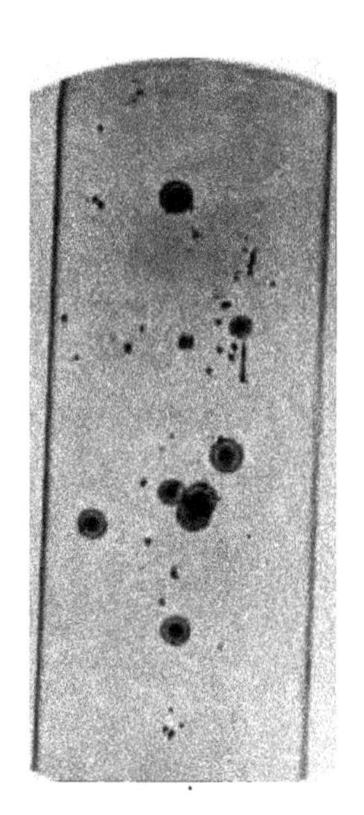

Fig 13.

Fig. 14.

PLATE V.

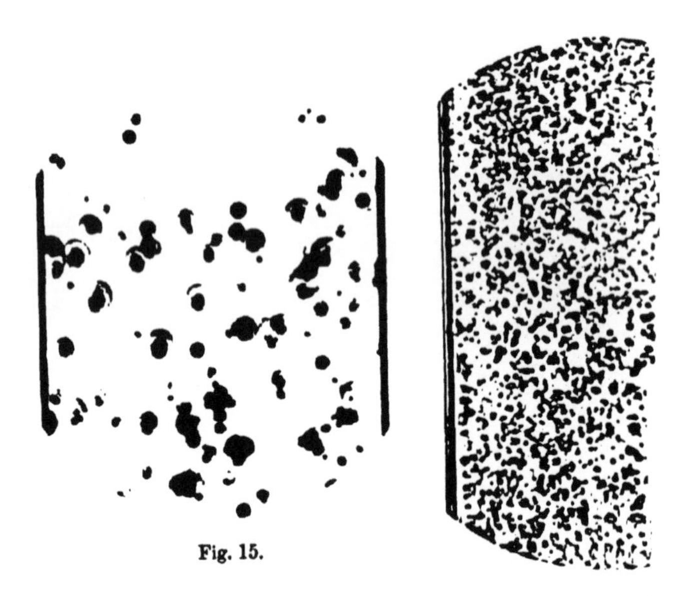

Fig. 15.

Fig. 16.

BACTERIA AND THE BURIED ANIMAL BODY.

PLATE V.

FIG. 15.

The *sugar-gelatine culture* (upper part) of *bacillus cadaveris sporogenes* depicted in fig. 14, further magnified.

[Magnifying power, 4.]

FIG. 16.

Sugar-gelatine culture of *bacillus cadaveris sporogenes*, derived from the liver of a guinea-pig exhumed *two months* after burial: After three days' incubation at 20° C.

[Magnifying power, 3½.]

FIG. 17.

The culture depicted in fig. 16, further magnified.

[Magnifying power, 9.]

FIG. 18.

A single colony of *bacillus cadaveris sporogenes* in *sugar-gelatine*, after 24 *hours'* incubation at 20° C. There is seen the first indication of liquefaction at the marginal part with a fine radial striation.

[Magnifying power, 45.]

BACTERIA AND THE BURIED ANIMAL BODY.

PLATE VI.

Fig. 19.

Three separate colonies of ***bacillus cadaveris sporogenes*** in *sugar-gelatine;* after several days' incubation at 20° C. Liquefaction is set up, and in each colony there is a mass of granules with thready extensions.

[Natural size.]

Fig. 20.

The colonies depicted in fig. 19, further magnified.

[Magnifying power, 3.]

Fig. 21.

The uppermost colony of fig. 20, still further magnified. There is here well seen the granular mass with thready extensions through the turbid liquefied colony.

[Magnifying power, 15.]

Fig. 22.

A *sugar-gelatine culture* of ***bacillus cadaveris sporogenes***, derived from the gangrenous subcutaneous exudation of a guinea-pig which had been kept some time after death. The culture was established by planting in the liquefied medium spores of the bacillus, heating the test tube to 80° C. for 15 minutes, and, after the sugar-gelatine had again set, incubating for ***two to three days*** at 20° C. The nature of the colonies, particularly the "spurious" spinous character of the deepest of them, is well shown.

[Natural size.]

Fig. 20.

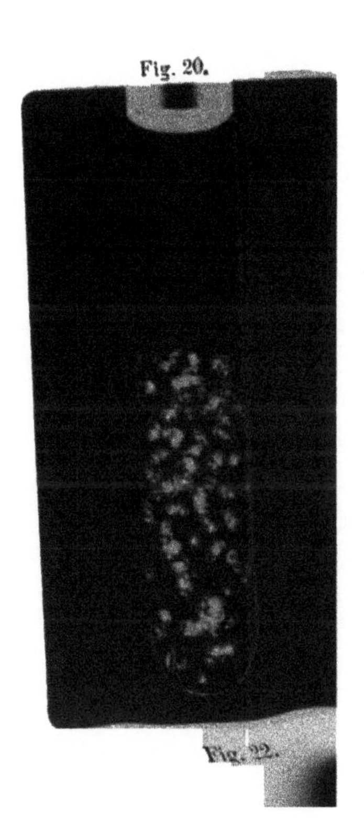

21.

Fig. 22.

PLATE VII.

Fig. 23.

Fig. 25.

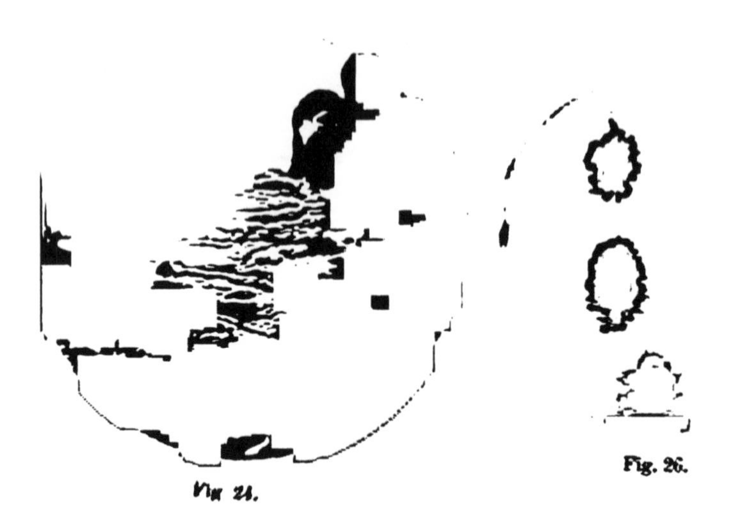

Fig 24.

Fig. 26.

BACTERIA AND THE BURIED ANIMAL BODY.

PLATE VII.

Fig. 23.

Colonies of *bacillus cadaveris sporogenes* on the surface of *agar*, incubated anaërobically at 37°C. The colonies are seen as fine granular patches, with here and there network of projecting fibrils.

Fig. 24.

Specimen of a similar culture to that depicted in Fig. 23.

[Magnifying power, in each instance, 18.]

Fig. 25.

Streak-culture on agar of *bacillus cadaveris sporogenes*, after anaërobic incubation for *several days* at 37° C. The filmy expansion of the growth, its central dark granular mass, and its filamentous extensions are well shown.

[Natural size.]

Fig. 26.

Agar-surface plate of *bacillus cadaveris sporogenes*, incubated anaërobically for *several days* at 37°C. Each of the four colonies represented consisted of a filmy patch with dark granular central mass and filamentous extensions.

[Magnifying power, 10.]

BACTERIA AND THE BURIED ANIMAL BODY.

PLATE VIII.

Fig. 27.

An *agar-surface* plate of *bacillus cadaveris sporogenes*, incubated for *two days* anaërobically at 37°C. The typical character of the colonies of this micro-organism are here well shown; especially the central granular mass, with its network of fibrils extending, in some instances, beyond the outer margin of the filmy

Fig. 28.

An *agar-surface* plate of *bacillus enteritidis sporogenes*, incubated anaërobically for *eight days* at 37°C. This plate resulted from rubbing over the surface of the agar a droplet of diluted subcutaneous exudation of a guinea-pig which died a few hours subsequent to injection into its groin of typical milk culture of the enteritidis bacillus.

The contrast between the chief characters of *bacillus cadaveris sporogenes* and *bacillus enteritidis sporogenes* is well brought out in Figs. 27 and 28.

[Magnifying power, in each instance, 16.]

Fig. 29.

Stab-culture in *sugar-gelatine* of spores of *bacillus cadaveris sporogenes*, incubated anaërobically for *four days* at 20°C.

Fig. 30.

Stab sub-culture, established from the culture depicted in Fig. 29, after incubation anaërobically for *four days* at 20°C.

The rapid liquefaction of the medium effected by this bacillus, the turbid character of the liquefaction, and the granular masses of growth are well exhibited in both stab-cultures.

[Natural size in each instance.]

PLATE VIII.

Fig. 27.

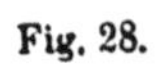

Fig. 28.

Fig. 29.

Fig. 30.

BACTERIA AND THE BURIED ANIMAL BODY.

PLATE IX.

Fig. 31.

Film specimen from peptone salt solution which, after inoculation with the "peritoneal washings" of a guinea-pig, had been incubated for *twenty-four hours* at 37°C.

The guinea-pig in question had died as a result of intra-peritoneal injection with culture of Koch's vibrio, had been buried directly in Thanet sand, and had been exhumed after lapse of *nineteen days*.

The peptone culture contained, it will be observed, the cholera vibrio in "pure culture."

Fig. 32.

Film specimen from a *tuberculous* deposit on the pancreas of a guinea-pig which, dying some weeks subsequent to intra-peritoneal injection of human tubercle, had been buried in earth in a tin box and exhumed after lapse of *ten weeks*.

The specimen shows numerous examples of *bacillus cadaveris sporogenes* (cylindrical bacilli, drumsticks, and spores); and, as well, clumps of *tubercle bacilli*. The latter, on inoculation into guinea-pigs, proved altogether inactive.

No. 2.

REPORT on the GROWTH of the TYPHOID BACILLUS in SOIL; By SIDNEY MARTIN, M.D., F.R.S.

In two previous Reports on this subject* the results were recorded of experiments dealing with the growth of the typhoid bacillus in sterilized soils. The general conclusions arrived at were :—

1. That in sterilized garden soils and in sterilized soils from the [illegible] of houses, [illegible] in [illegible] may be [illegible] organically contaminated soils, the typhoid bacillus lived and multiplied, whether the soil were kept at a uniform temperature of 37° C., at the temperature of the laboratory between 15° and 19° C., or at the temperature of an outside shed between [illegible] and 15° C.

2. That in these soils the bacillus was still alive after [illegible] and that for a short period it remained in vitality after the soil had been so dried that it could be readily reduced to a powder.

3. That in virgin soils, soils which had never been manured or cultivated and which consisted chiefly of sand or p[illegible] the bacillus did not grow nor live: that in the major[illegible] of cases, even on the day following inoculation of the s[illegible] with the bacillus, no evidence of its presence in the s[illegible] could be obtained.

Experiments with sterilized soils were not continued. Th[illegible] that have been recorded in the two previous Reports show qu[illegible] definitely that in certain sterilized soils the typhoid bacillus c[illegible] live and grow; that sterilized garden soils and cultivated s[illegible] can serve as a medium of growth for the bacillus.

In natural conditions, however, soil contains bacteria of vari[illegible] kinds, and it was necessary to ascertain whether the typh[illegible] bacillus could live and grow in soil in the presence of th[illegible] bacteria. The work of the past year was devoted to this subj[illegible] and the results are presented in this Report.

Bacteria in Soil.

The bacteria present in soil may be divided practically i[illegible] three classes :—

1. Aërobic.
2. Anaërobic.
3. Nitrifying.

The last, although both aërobic and anaërobic, deserve a cl[illegible] to themselves, owing to their special action.

---

* Report of Medical Officer of the Local Government Board, 1896–1897, p. [illegible] *Ibid.*, 1897–1898, p. 308.

It was found early in the present investigation that satisfactory result could not be obtained unless the bacteria from the samples of soil used were isolated and in part studied. The complete study of the bacteria of soils would of itself occupy many years of work, and it was found necessary in the present investigation to limit attention to the aërobic forms, which are more readily isolated and the mode of life of which would *a priori* more readily affect the aërobic typhoid bacillus. It was impossible to designate each of the micro-organisms by name: some, such as bacillus mycoides, bacillus subtilis, and bacillus arborescens, were readily recognized; others were forms of proteus, while others could not readily be ascribed to any particular genus. All the bacteria observed were one or other kind of bacillus; no coccus was ever found, and sarcina (probably from the air) only occasionally. In none of the soils was there any bacillus which gave all the cultural characteristics and responded to all the tests of the bacillus coli communis; this point was specially borne in mind. None of the bacteria isolated were like the typhoid bacillus. Although in some cases the colonies on the agar plate superficially resembled the typhoid bacillus, the resemblance as a rule ceased there, and both microscopical examination and cultivation showed marked differences between all these bacteria and the typhoid bacillus. Indeed, in the majority of instances, microscopical examination alone might have been considered sufficient to distinguish the bacteria of the soil from the typhoid bacillus, although it was in practice only one of the tests applied.

The detailed characteristics of the micro-organisms will be given in the account of the experiments.

Some of the bacteria of the soil are spore-bearing; others are both aërobic and anaërobic; others do not appear to form spores. In any case, the number and variety of bacteria present in a particular sample of soil (limiting the statement to garden and cultivated soils) depends on the kind of bacteria present, and on the moisture and the temperature at which the soil is kept. A higher temperature and a certain degree of moisture will encourage the development of the bacteria by aiding the spores to germinate and by stimulating the growth of those that are in the resting stage. This condition is to be taken as directly opposite to that occurring in, say, a dried soil from which indeed bacteria are obtainable in quantities, but not so numerously as when the soil is moist and kept at 37° C. Not only is the number of the bacteria obtainable from a moist soil kept at 37° C. much greater than at lower temperatures, but the kind of bacteria most prevalent is different. This was many times noticed in the course of the investigation; so that it must be concluded that from time to time and according to the degree of moisture present and to the temperature, now one now another particular bacterium would prevail. In the animal body it has been shown that, two or more micro-organisms being present, their action or reaction one on another may differ widely. In some cases they do not affect each other; in others one prevents the development of another; in a third set of instances one helps the action of another. In artificial culture media, no doubt the same effects are to be observed, although the subject is far from worked out. In connexion with the results now to be recorded, this action or reaction of micro-organisms on each other must be borne in mind. It has an important bearing on the growth and the non-growth of the typhoid bacillus in soil containing other bacteria.

Special attention was paid during the investigation to the occurrence of aërobic putrefactive bacteria in the soil, with a view to observing their effect on the typhoid bacillus.

Method of Investigation.

The first object was to test the vitality of the typhoid bacillus in association with other bacteria in a large bulk of soil under natural conditions of temperature, moisture, &c. Before, however, actually proceeding to this, it was deemed advisable to test preliminarily the growth of the bacillus in one and another soil in a flask, the unsterilized soil being moistened and then inoculated with a small portion of a broth culture of the bacillus.

Before inoculation the bacteria already present in the soil were investigated by the method presently to be described; they were isolated and their microscopical and culture characteristics observed.

After inoculation with the broth culture the soil was investigated for the presence of the typhoid bacillus,—in 24 hours, in 2 or 3 days, and later; so that with one sample of soil many experiments were performed at various intervals, the object of frequent investigations being to obviate missing the typhoid bacillus, which might very possibly occur if only one testing were performed.

The typhoid bacillus which was used for inoculation was, in the earlier experiments, obtained 12 months previously from the spleen of a patient dead of typhoid fever; in later experiments, some of them not yet completed, it was isolated from a patient just before it was used. Each culture in broth was tested previous to use in order to make sure that it was uncontaminated. This precaution is a very necessary one. The tests relied on for the diagnosis of the typhoid bacillus are as follows :—

1. Microscopically, rods short and long are seen; chiefly short if the culture be recently obtained from a patient and if it be virulent. In a hanging drop, the bacillus is actively motile, and does not tend to clump. In some broth cultures the bacillus does tend to clump, but these are chiefly from old stock or laboratory cultures, and such were not used in the investigation.

2. (α) The bacillus stains with the ordinary aniline dyes, but not with Gram; the latter characteristic being sometimes useful in differentiation.

   (β) By appropriate staining the bacillus was seen to have from 10–12 longish flagella at the ends and sides.

3. The growth on the surface of agar is copious in 24 hours, and is translucent and fluorescent. The fluorescence is not nearly so well marked as in the case of bacillus coli communis and some putrefactive bacteria.

4. The growth on gelatine does not cause liquefaction, and is not to be distinguished from that of the bacillus coli

communis, but is readily distinguished from that of most of the bacteria in the soil.

5. The growth on potato is almost invisible, and this is especially the case with the virulent bacillus. In some laboratory cultures of the bacillus the growth on potato becomes brownish, although not so brown as that of the B. coli communis.

6. In glucose-gelatine no bubbles of gas are formed as is the case with the B. coli communis and other gas-forming bacteria.

7. Milk is not coagulated by the typhoid bacillus.

8. In broth the bacillus causes a cloudiness and a deposit, but no scum. If the latter occurs it is almost invariably due to a contamination. As regards the many different stocks of the bacillus with which I have from time to time worked I have never yet observed a pure culture form a scum in broth.

   If a micro-organism gives *all* the above characteristics, it is probably the typhoid bacillus. But no one of the individual characteristics is of itself diagnostic; they must be taken altogether. The final test—No. 9 which follows—is, however, the most important, and it was always employed in the investigation when necessary.

9. In an admixture of blood serum from a typhoid patient with a broth culture of the bacillus 24 hours old, the bacilli become motionless and clump. The dilutions used were 1 in 50 and 1 in 100. This test was supplemented by the sedimentation test. the dilution being also 1 in 50 and 1 in 100. These dilutions were considered sufficient in the conditions under which the investigation was carried out.

It is well to insist on all these points, especially as regards the ffect of typhoid blood serum on the suspected bacillus, since by elying merely on microscopical appearances and the cultural haracteristics a mistake may readily be made. The blood serum sed in the testing was always previously tested with a known yphoid bacillus, and its strength estimated.

Experiment I.—Mixed Soils, Unsterilized.

This experiment was a continuation of that described at age 315 of the last Report of Medical Officer of Local Govern-ent Board, 1897–98. As there recorded the soil when erilized, served as an excellent medium for the growth of e typhoid bacillus. A detailed description was not given of e bacteria occurring in the soil; but none of the organisms esent resembled the typhoid bacillus. That which was most umerous in some respects resembled the bacillus coli communis, that it was fluorescent, was motile, and gave a brown growth on otato; but it only slowly coagulated milk. Subsequent investi-ation showed it was not the bacillus coli, and it will be referred again as Chichester I.

The mass of earth, 6 inches deep and 6 inches in diameter, was kept at the temperature of an outside shed, which ranged between 4° and 16° C. After inoculation with a broth culture of the typhoid bacillus, the soil was tested in 3 days and in 14 days, and no typhoid colonies were discovered; but after 50 days (the 3rd testing) a few colonies were found which responded to all the tests of the typhoid bacillus. So much is recorded in the last Report.

Fourth Testing, after 287 days, under the same conditions.—A small portion of the soil (about 1 gramme) was placed in 200cc. of sterile distilled water, and kept at 37° C. for 24 hours. Some of the water was then brushed over the surface of 3 agar plates, which were kept in the incubator 24 hours. The discrete colonies on these plates were subjected to an extensive examination to determine whether any were those of the typhoid bacillus. Some of these colonies were obviously not the typhoid bacillus: they were those of B. subtilis, B. arborescens, and a figure-making bacillus like proteus vulgaris. Others were round or oval, with discrete edges, translucent, and more or less fluorescent. Of these colonies, which were very numerous, 4 groups were selected for testing.

Colony α.—The most numerous in all the plates. This proved to be the same micro-organism (Chichester I.) which was found preponderating in the soil before inoculation with the typhoid bacillus. It was grown in broth, in which it caused a turbidity and a sediment, but no scum. Microscopically it was a large, thick, very motile bacillus, its shape being quite unlike any variety of the typhoid bacillus. On agar it gave a copious fleshy growth in 24 hours, and on potato a dirty yellow fleshy mass consisting of large thick rods containing spores. Milk was not coagulated by it in three days, and it did not liquefy gelatine. The microscopical appearances, as well as the growth on agar and on potato, determined that this bacillus was not the typhoid bacillus, although it somewhat resembled it on the agar plate.

Colonies β.—These colonies rendered the broth slightly turbid; and the microbes resembled the typhoid bacillus in their motility. In the appearances of the growth on agar, by not coagulating milk, and by not liquefying gelatine, the microbes also resembled that bacillus. Microscopically, however, the micro-organism was mainly in the spirillum form.

Colonies γ.—These were pale fluorescent colonies looking like those of the typhoid bacillus; microscopically they consisted of long and short rods, and showed some degree of plasmolysis like the typhoid bacillus.

They were cultivated in broth and then brushed over agar plates. On these, two kinds of colonies developed, one translucent but stippled and radiating, which proved microscopically to be a leptothrix: the others like ground glass and fluorescent,

APP. B. No. 3.

On the Growth of the Typhoid Bacillus in Soil; by Dr. Sidney Martin.

and consisting of long and medium-sized rods with threads, all actively motile. The latter were again cultivated in broth and showed microscopically the appearances just described. The micro-organism, however, was not the typhoid bacillus. It agreed with the typhoid bacillus in giving a filmy transparent growth on potato, in not coagulating milk, and in not liquefying gelatine; but the growth on agar after 24 hours was more copious and more transparent than the typhoid bacillus, the flagella were very few in number, and, most important of all, it did not give the typhoid serum reaction in a dilution of 1 in 100. The serum used was active in much higher dilutions when tested with a known typhoid bacillus.

Colonies δ.—These colonies on the original agar plate looked very much like the last. They were cultivated in broth, and this was brushed over agar plates. The growth was pure, and showed microscopically a motile, short and thick bacillus. There was a copious growth on agar, and a yellowish growth on potato; milk was not coagulated and gelatine not liquefied. The bacillus was readily distinguished from the typhoid bacillus.

Fifth Testing after 322 *days.*—It was thought possible that, in the presence of so many other bacteria, the typhoid bacillus might not in 24 hours have diffused itself in the water used for experiment in the fourth testing. A portion of the soil was therefore now incubated for 24 hours in broth containing 0·05 per cent. phenol. Two drops of this broth from a freshly prepared pipette were then placed in about 10 c.c. of sterile broth in order to dilute the culture, and the mixture was then brushed over the surface of four agar plates, which were incubated for 24 hours. Some of the colonies resembled those of the typhoid bacillus, but microscopically, as well as when grown in broth agar and potato, the microbes proved not to be the typhoid bacillus. They formed spores.

Sixth Testing after 392 *days.*—Another testing was performed to ensure that no mistake had been made. Some of the soil was placed in distilled water as in the fourth testing, and eight agar plates subsequently made. A copious growth of mostly discrete colonies appeared in all the plates. For the most part, however, they were those of the bacillus previously mentioned as preponderating in the natural soil. Four colonies which appeared more like those of the typhoid bacillus were examined; but microscopically they were readily distinguished from it, three of the colonies consisting of thick oval rods, and one of a very minute bacillus.

this experiment, therefore, after inoculation with the typhoid lus of a natural soil containing numerous bacteria, some of h were putrefactive, the bacillus disappeared after 50 days. was a soil in which when sterilized the typhoid bacillus l for over a year (at the last testing). The few colonies of yphoid bacillus which were found after 50 days were definite.

APP. B. No. 3.

On the Growth of the Typhoid Bacillus in Soil; by Dr. Sidney Martin

They responded to all the tests of the bacillus. The serum test for the bacillus is the final and most important test to be employed, and it might be objected that the dilution (1 in 10) which was used last year in this case was not sufficiently high to constitute an accurate test of the bacillus. This is undoubtedly true. But the observation must rest as it stands in the last report, and, considering that the separated bacillus responded to all the other tests of the typhoid bacillus, I must conclude that the two or three colonies which were investigated were those of the typhoid bacillus. Subsequently, however, the bacillus was not obtainable from the soil. It had disappeared entirely.

Experiment II.—Chichester Soil, Unsterilized.

Here the Chichester soil was employed which, when sterilized, had been previously frequently used for the testing of the growth of the typhoid bacillus (see two previous Reports). In this soil, when sterilized, the bacillus will live for a long period, whether at a uniform temperature of 37° C., at the temperature of the laboratory (13°–19° C.), or at the temperature of an outside shed (4°–10° C).

In this experiment, the finely divided unsterilized soil was placed in an Erlenmeyer flask to the depth of about 1 inch, and water was added to make it very moist. A whole broth tube (about 10 cc.) of a recent culture of the typhoid bacillus was added to the flask which was then placed in the incubator. This addition rather flooded the soil, so the india-rubber cap of the flask was removed to allow the excess of liquid to evaporate.

First Testing.—The first testing was some two days after adding the typhoid bacillus culture, and before the excess of liquid in the flask had evaporated.

This testing was done as follows: A small portion of the soil was incubated in a broth tube for 24 hours. At the end of that time the broth was very turbid, and a slight scum had formed on its surface. The culture was shaken up and a small pipetteful removed and put into about 10 cc. of sterile broth. Without incubating, this diluted culture was used for brushing over the surface of four agar plates, the same brush without recharging being used for all the plates. The plates having been incubated for about 48 hours, two of them were found to contain confluent colonies of various bacteria, but in the other two plates the colonies were discrete and readily isolated.

Of these discrete colonies several groups were somewhat fluorescent and translucent, oval or round, with margins not fringed. Four of these groups were further examined. All the other colonies on the plates were quite unlike those of the typhoid bacillus.

Examination of the four groups of colonies.

Colony a.—Actively motile, showing on staining short oval bacilli with plasmolysis. Cultivated in broth 24 hours and, after dilution in sterile broth, brushed over an agar plate. Pure growth obtained of fluorescent colonies, which ...

growing in a gelatine plate gave a pure growth in colonies with a fringed edge, readily distinguished under the microscope from those of the typhoid bacillus.

Colony β.—This consisted also of an actively motile bacillus, as long and short rods, superficially like the typhoid bacillus. But after growing it in broth, and making agar and gelatine plates as with colony *a*, it was readily distinguished from the typhoid bacillus.

Colony γ was proved by the same method of examination to be a mixture of colonies *a* and *β*.

Colony δ gave a pure growth by the same methods, and with all the characteristics of colony *a*.

Colony *a* consisted of the micro-organisms which will be hereafter referred to in this report as Chichester I.

Second Testing, 8 days after adding the typhoid bacillus.—At this time the excess of water had evaporated, leaving no liquid on the surface of the soil, which was just moist throughout. A portion of the soil was placed in 200 cc. sterile distilled water and incubated 24 hours. A sample of this water was then brushed over the surface of four agar plates, one gelatine (stab) plate being also made. Several colonies grew in the gelatine plate, only one of which, for the reason that it resembled that of the typhoid bacillus, was examined: it was found to consist of a short oval bacillus, quite unlike any form of the typhoid bacillus. The agar plates were kept at 37° C. for 24 hours, and at 22° C. for two days longer. There was a copious growth on each plate, chiefly of discrete colonies, most of which were readily recognised as bacteria originally present in the soil. Four colonies were investigated specially, as the only ones which resembled colonies of the typhoid bacillus. Colonies *a*, *β*, *γ* consisted of one and the same micro-organism. Colony δ was very fluorescent and corresponded to the micro-organism of the original soil called Chichester I.

Colonies *a*, *β*, *γ* consisted of rather short stout bacilli, showing plasmolysis: the bacillus was slightly motile, and there were no long rods or threads observed. The three colonies were each grown in broth, samples of which were brushed over agar plates; when they were seen as growing in translucent, slightly fluorescent colonies, consisting of the same short and stout bacillus observed in the original agar plate. Sub-cultures were made as follows :—

Agar: Pure growth : microscopically, short stout bacilli.

Glucose gelatine (shake) : Growth on surface only; unlike the typhoid bacillus : no gas formed.

Potato: No obvious growth in six days.

Milk: Not coagulated in six days.

Broth: Turbid, with a scum: microscopically, short stou bacillus.

The typhoid bacillus was therefore not recovered from the so on this second testing after eight days at 37° C.

Third Testing, after 34 days.—A portion of the soil w incubated in 200 cc. of sterile water for 24 hours: sampl of the water were then brushed over the surface of six ag plates, which were incubated for 24 hours. Numerous coloni developed in each plate, but a careful examination under th microscope revealed only one colony out of the six plat which at all resembled the typhoid bacillus. Microscopicall this colony consisted of short thick rods, readily distinguishe from the typhoid bacillus.

Remarks.—In this experiment (II.) the conditions might hav been considered favourable to the growth of the typhoid bacillu in soil. The soil itself was favourable, inasmuch as the bacillu will grow in it when it is sterilized. The 10 cc. of broth whic had been added to the soil improved the soil as a cultivatin medium. The temperature, 37° C., is one highly favourable t the growth of the bacillus. Yet, even under these conditions, th bacillus could not be obtained from the soil as early as tw days after it had been added. It was this experiment whic directly suggested an investigation of the effect of particula bacteria present in the soil on the growth and vitality of th typhoid bacillus; a subject dealt with later in this Report.

Experiment III.—Chichester Soil, Unsterilized.

This experiment was made with the same soil as the last. Th soil to the depth of about 2½ inches was placed in a sterile beake about 4 inches in diameter, and water added until it was mois throughout. The beaker was then placed in the incubator for si days before the typhoid bacillus was added to the soil containe in it. At this temperature the number of micro-organisms pr sent in the soil greatly increased. Thus, before the moist so was incubated, a small portion had been added to 200 cc. o water and incubated for 24 hours; samples of the water bein then brushed over four agar plates. As a result only a few coloni developed of three or four micro-organisms. After 24 hours' in cubation of the moist soil, however, an enormous number o colonies developed when the soil was tested in the same way. A before stated, after six days' incubation of the moist soil (wat being added from time to time to keep it moist), two broth tub of a pure culture of the typhoid bacillus were added to th soil and well stirred up with a sterile glass rod. One bro tube contained a culture 24 hours old, the other a cultu 10 days old.

First Testing, two days after adding the typhoid bacillus.— A small portion of soil in 200 cc. water, incubated for hours, and samples then brushed over the surface of fo agar plates. These plates were incubated 24 hours. The was a copious growth of micro-organisms in all the plates, mostly of discrete colonies. On examination, however, they consisted chiefly of the bacteria originally contained in the soil, which are named provisionally Chichester I., II., III., IV., V. No colonies like those of the typhoid bacillus were seen. Three of the fluorescent colonies were examined microscopically and by cultivation in broth: they showed the characteristics of Chichester I. and II.

Second Testing, in four days.—A larger quantity of the soil than in the first testing was added to 200 cc. of water, and incubated. Four agar plates made in the way previously described. Copious growth was observed in each plate, the colonies consisting chiefly of Chichester I.; others were Chichester II., IV., and V. None of the colonies were like those of the typhoid bacillus, and those that were examined microscopically turned out to be Chichester I. All the plates had a marked putrefactive odour.

Third Testing, in 13 *days*.—A small portion of the soil was incubated in 0·05 per cent. phenol broth for 24 hours. A copious growth took place. One or two drops of the culture were diluted with about 10 c.c. of sterile broth, and this was brushed over four agar plates. A copious growth occurred on each plate. The colonies consisted chiefly of Chichester I., II., III.; some were of Chichester V. A few colonies were examined, as they were doubtful. One turned out to be Chichester I. Another colony showed microscopically motile rods, rather larger than the typhoid bacillus. Cultivated in broth, this bacillus caused a turbidity and a deposit, but no scum. It was, however, not the typhoid bacillus, as it gave a very copious and fleshy growth on agar, a copious and dirty-looking growth on potato, and it coagulated milk.

Remarks.—The experiment was not continued further. It confirmed Experiments I. and II., in showing that the typhoid bacillus when added to the particular unsterilized soil, even in large quantities, could not be recovered from the soil, even on repeated testings. It had apparently died out.

Experiment IV.—University College Soil, Unsterilized.

The soil used in this experiment had not been previously subjected to investigation. It was a garden soil, obtained from the front of University College, London.

Bacteria present in the soil.—A small portion of the soil was placed in 200 c.c. of water and incubated for 24 hours; four agar plates being then brushed with the water. Six different aërobic bacteria were then isolated and their characteristics studied. None of them were the typhoid bacillus or the bacillus coli communis. These micro-organisms will, for [illegible] reference in the Report, be [illegible] and VI. Their character-[illegible]

	B. Coli, Soil, No. I	[illegible] No. II	[illegible] No. III	[illegible] No. IV	[illegible] No. V	[illegible] No. VI
Agar plate ...	Colonies, small translucent, like ground glass	[illegible] translucent	[illegible]	[illegible]	[illegible]	[illegible]
Microscopical appearance.	Slightly motile, very slender rods and threads	Thick [illegible] motile rods	Slightly motile rods, [illegible]	[illegible]	[illegible] motile	[illegible] motile [illegible]
Broth	Cloudiness, no deposit, no scum	White firm [illegible] deposit	Flaky deposit only	[illegible]	[illegible] deposit	[illegible]
Indol	Absent	Absent	Absent	Absent	Present	Absent
Agar (slope) ...	Gelatinous looking	Dry, white, powdery growth	Fringed, [illegible] colonies	White, opaque fringed	Opaque and fleshy	Porcellaneous growth
Glucose-gelatine (Stab).	Surface growth, no bubbles, no liquefaction.	Diffuse growth, no liquefaction	Surface growth, no liquefaction	Fringed scattered colonies; no liquefaction.	Surface growth, no liquefaction.	Chiefly surface growth; no liquefaction.
Milk	Not coagulated ...	Loosely coagulated	Not coagulated ...	Surface growth, not coagulated.	Not coagulated.	Not coagulated.
Potato	? Transparent growth.	Dry; crinkled growth.	Very crinkled growth.	—	Brownish growth	Dry; smooth, dirty white.

[illegible] bacilli differed therefore markedly in one or other of their characteristics from the typhoid bacillus.

Growth of the typhoid bacillus in University College soil, sterilized.—It was considered advisable to test whether the bacillus would grow in this soil after the soil had been sterilized, before using such soil for experiment in its natural state.

A small amount of soil in an Erlenmeyer flask was moistened with water and sterilized, the completeness of the sterility being afterwards tested. It was then inoculated with about ten drops of a broth culture of the typhoid bacillus four days' old. It was tested in 29 days by placing a small portion of the soil in a sterile broth tube, which was incubated for 24 hours. The broth became turbid, and no scum formed. It was stroked with a needle over the surface of an agar plate and a pure and copious growth of the typhoid bacillus was obtained.

Experiment with the unsterilized University College soil at the temperature of the laboratory (13°–19° C.) :—The sterilized soil having been thus shown to be favourable to the growth of the typhoid bacillus, the experiment with the natural (unsterilized) soil was proceeded with.

The soil was placed in an Erlenmeyer flask to the depth of about one inch and moistened throughout with water. It was inoculated in the centre with five drops of a broth culture of the typhoid bacillus 24 hours old.

First Testing, seven days after adding the typhoid bacillus.—A portion of the soil added to 200 c.c. water, and six agar plates made in the manner previously described. All the bacteria originally found in the soil were present. Some of the colonies were specially examined, as they appeared doubtful; they were round, translucent, and slightly fluorescent.

Colonies α, β, γ were actively motile rods which formed a scum in broth; two gave a reddish growth on potato. Microscopically, they were readily distinguished from the typhoid bacillus.

Colony δ showed large bacilli with transverse markings, quite unlike the typhoid bacillus. Colony ϵ gave a scum in broth.

Colonies ζ, η, θ were those of a very small boat-shaped bacillus.

Eight colonies, therefore, which were considered as slightly like those of the typhoid bacillus when specially examined, were found not to be of that species.

Second Testing, 20 days after inoculation :—The testing was done as before, except that 10 c.c. of sterile broth were added to the 200 c.c. of water to make a better culture medium for the bacteria.

Seven gelatine plates were made, but they were all liquified and useless for investigation in three days.

Eight agar plates were made. Six of these showed no colonies like those of the typhoid bacillus; the colonies appearing were mainly those of the bacteria numbered II., III., IV., V. Only five colonies looked anything like those of the typhoid bacillus, but when tested microscopically they all consisted of large oval bacilli with transverse markings.

Remarks.—This experiment, like those preceding (I. II. III), is negative; in the fact that the typhoid bacillus after addition to the natural soil was not in any case recovered. It seemed to have disappeared.

Experiment V.—Nursery-garden Soil, Unsterilized.

The soil used in this experimentation had not previously been subjected to investigation. It was a garden soil (leafy mould) obtained from a nursery garden. It was difficult to dry and absorbed a large quantity of water. By the methods previously described a large number of bacteria were isolated from this soil: bacillus subtilis, sarcina lutea, and one resembling bacillus Zopfii, as well as others not identified.

Soil to the depth of about one inch was placed in an Erlenmeyer flask, and, after moistening with water, was inoculated in the centre with 2–3 c.c. of a broth culture of the typhoid bacillus 24 hours old. The flask was placed in the incubator at 37°C. for 24 hours, when the first testing was performed; afterwards it was kept at the temperature of the laboratory (13°–19°C.).

First Testing, after 24 hours at 37°C.—A loopful of the soil incubated for 24 hours in 200 c.c. of water, and samples of the water then brushed over six agar plates.

Numerous colonies developed, among which those of the typhoid bacillus were readily recognised and isolated. These colonies were subjected to the following tests:—

1. Microscopically, the bacilli were long and short rods, actively motile, and showing numerous flagella by appropriate staining (van Ermenghen).
2. The growth on agar slope was translucent.
3. Milk was not coagulated.
4. No gas was formed in glucose-gelatine (shake).
5. Gelatine was not liquefied.
6. In broth there was a uniform turbidity, but no scum. No indol was formed.
7. On potato there was a light yellowish brown growth.
8. Tested with typhoid blood serum (which caused marked clumping with known typhoid bacilli in dilutions of 1 : 100) broth cultures showed rapid and marked clumping with dilutions of 1 : 150 and 1 : 100.

The colonies were therefore those of the typhoid bacillus, giving all the characteristic reactions of that bacterium, including the serum test. These colonies of the typhoid bacillus were not so numerous as those of another micro-organism which somewhat closely resembled it on the agar plate. This resemblance was well marked when the colonies were young, but as they grew these other colonies developed a transparent and somewhat scalloped fringe, which the typhoid bacillus colony never does on the agar plate. An investigation of these colonies showed further cultural resemblance to the typhoid bacillus, and indicated how easily a mistake in the diagnosis of the typhoid bacillus may be made unless all the tests are applied. Thus the growth of this particular bacillus on agar, in milk, in glucose-gelatine, on gelatine, in broth and on a potato was exactly like that of the typhoid bacillus; but although motile, the bacillus had as a rule only one terminal flagellum, and did not give the characteristic serum reaction in dilutions of 1 : 50 and 1 : 100, although the preparation was watched for 1½ hours.

Second Testing, 14 *days after adding the typhoid bacillus*, the flask being kept at temperature of laboratory (13°–19° C.). By the same method as before six agar plates were made. Most of the colonies when young were practically indistinguishable from those of the typhoid bacillus; but as they grew they developed a transparent fringe which was distinctive. Six or seven of these colonies were specially investigated, and they all turned out to be the same organism, and not the typhoid bacillus. It was an active, short bacillus, not staining by Gram's method; it gave a translucent growth in agar, a cloudiness in broth, in which no indol formed; it did not cogaglate milk or liquefy gelatine; nor did it form any gas in glucose-gelatine; on potato it formed a light yellowish-brown growth. So far the cultural characteristics were like those of the typhoid bacillus; but it did not give the serum reaction in dilutions of 1 in 50 and 1 in 100, nor was its growth on gelatine plates like that of the typhoid bacillus. It was identical with the bacillus found at the first testing.

Third Testing, after 51 *days at the temperature of the Laboratory.*—Six agar plates being made by the same method as before, numerous colonies developed in each plate, mostly the fringed translucent colonies previously described. Six colonies were examined in detail as doubtful. Of these five were readily distinguished microscopically from the typhoid bacillus: one colony consisting of long oval bacilli with transverse markings; three colonies consisting of short oval bacilli with transverse markings, and in many cases showing bi-polar staining with fuchsin; the fifth colony consisted of short oval bacilli uniformly stained by fuchsin. The sixth colony showed long and short rods with plasmolysis. It agreed with the typhoid bacillus in forming no indol in broth and no gas in glucose-gelatine; but it was not the typhoid bacillus, as it formed a fleshy scum in broth and a gelatinous growth on agar.

Remarks.—In this nursery-garden soil, therefore, the typhoid bacillus was found 24 hours after it had been added, but not at the end of 14 days, nor 51 days afterwards, in spite of most careful searching. It had apparently disappeared.

In these several Experiments, I.–V., the cultures of the typhoid bacillus which were employed were not tested as regards their virulence. They were made from the laboratory stock, and had been obtained about nine months previously from a case of typhoid fever. At that time their virulence had been artificially augmented by passage through guinea-pigs, but it had diminished very greatly by the time the cultures were added to the soil.

Upon the whole the experiments here recorded must be considered rather as indicating the line of investigation that is necessary than as justifying any definite conclusion. It will be important to observe the fate of virulent cultures of the bacillus in unsterilized soil and especially in soil exposed to natural conditions. Experiments in this sense have been commenced, but the results obtained up to the present have been negative. It was thought at first that it would be of great advantage to do these further experiments by adding the typhoid bacillus to the soil in a garden; but it was found impossible to obtain a plot of ground within a reasonable distance for experimental purposes. Accordingly an oblong iron box 7 feet long by 3 feet deep and 2½ feet wide was devised, which had perforated sides and was jacketed to receive the drainage of the soil. Sand, to the depth of about 6 inches, was placed at the bottom of the box and ordinary garden earth to the depth of 2 feet above the sand. After an investigation of the bacteria present in the soil, this was inoculated along a marked line with the typhoid bacillus. As has been said, the results obtained are not sufficiently advanced to be recorded in this report.

The contrast between the results obtained by a study of the growth of the typhoid bacillus in sterilized and in unsterilized soil is marked. In experiments I.-V., newly recorded in this report, only once was the bacillus regained from soil to which it had been added, and that was not later than 24 hours after the addition. For the rest, whether the soil was kept at the temperature of an outside shed, at the temperature of the laboratory, or at a uniform temperature of 37° C., the bacillus had apparently disappeared. This result could not have been due to the methods employed for its detection: the same methods had been successful for its detection in the case of sterilized soil, and in the single instance in which it was obtained from unsterilized soil, its presence was readily recognized. Moreover, in the same soils, when sterilized, the bacillus had been found to live and grow for long periods. It follows then, that the bacteria naturally present in the soil contributed that other element which was inimical to the bacillus.

It was therefore considered advisable to determine what was the effect of each individual bacterium of the soil on the typhoid bacillus when the two micro-organisms were inoculated (1) into a liquid medium in which they could both grow; and (2) into the sterilized soil from which the soil bacterium had been obtained. The results obtained occupy the remainder of the report.

Investigation of the Growth of the Typhoid Bacillus in the Presence of Particular Soil Organisms.

In this investigation the object of research was to ascertain the behaviour and growth of typhoid bacillus in various media and at differing temperatures in presence of one or more organisms found naturally present in soil. For this purpose pure cultivations of a number of soil organisms were prepared. Most of the experiments were performed with bacteria isolated from Chichester soil, a soil which had been used for earlier work; the minority were with organisms isolated from soil removed from the front of University College, London (see Exp. IV.).

The organisms were obtained in pure culture, but no attempt at naming them was made. They will be indicated in the following report as Chich. 1, Chich. 2, U.C.L. 1, etc. The culture media of experiment were a flask containing 200 cc. of sterile distilled water and about 10 cc. of ordinary peptone broth; and sterilized soil,—the soil used in each instance being that from which the particular soil bacillus of experiment had been isolated.

The temperature at which the experiments were performed was either that of the 37° C. incubator, the temperature of an outside shed (3°–14° C.), or, in a few cases, the temperature of the laboratory (13°–19° C.).

The general method of investigation consisted in inoculating the above different media with a definite quantity of a pure broth culture of one of the soil organisms, and at the same time with an equal quantity of pure broth culture of the typhoid bacillus; both cultures being of the same age.

The mixture was grown under the conditions mentioned under each experiment, and later, at stated intervals, the relative proportion of the two sorts of bacteria present was investigated.

This was done in the case of the fluid medium by brushing two or more agar plates with a sterile brush dipped into the nutrient fluid. Surface colonies, generally discrete, were obtained, and from these plates the relative proportion of each organism present could be ascertained.

In the case of the soil medium, a small portion of the soil was added to sterilized water or to sterilized water and broth, and incubated for 24 hours at 37° C. Samples were brushed over agar plates and examined in the same way as for fluid media.

Series A.—Typhoid Bacillus and Chichester I.

In this series of experiments the micro-organisms used consisted of laboratory cultures of the typhoid bacillus and of Chich. 1.

Chich. 1 is a short, almost oval, bacillus, which grows readily on the ordinary media. On agar slope it forms a moist translucent fluorescent or rather opalescent growth. In broth it produces a deposit, but no scum on the surface. In glucose-gelatine-shake culture it produces no gas nor does it liquefy the gelatine even after a week. In this medium the growth is almost all on the surface. The plates on which it is growing have a strong putrefactive odour. It produces no indol in 24 or 48 hours, but gives rise to a slight reaction in indol old broth cultures.

The colonies on agar plates are circular or nearly circular. They are rather opaque. being darker in the centre. In most cases the margin is quite smooth. Their most obvious characteristic is their marked fluorescence.

Experiment No. 1.—Liquid Medium at 37° C.

A flask containing 200 c.c. sterile distilled water and 10 c.c. broth, was inoculated with a loopful of broth culture of Chich. 1 and with a loopful of broth culture of the typhoid bacillus. It was placed in the incubator at 37° C.

First Examination.—*After 24 hours.* Brushed over four plates (agar). Plates incubated at 37°, and examined after two days. Copious growth on all four plates. All have a strong putrefactive smell. The colonies of typhoid bacillus and of Chich. 1 are readily recognised and distinguished from one another.

Microscopic preparations made from each and stained by carbol fuchsin show for the former the typical characters of the typhoid bacillus, and oval or very short bacilli in the case of the latter.

The typhoid bacillus colonies are as numerous as those of Chich. 1.

Second Examination.—*After eight days.* Brushed over three plates. Incubated and examined as before. No colonies of the typhoid bacillus could be seen even after the most careful microscopical examination of the plates under a low power. Four different colonies, the appearances of which were not quite typical of Chich. 1, were sub-cultured on agar; but culturally and microscopically they proved to be that organism. All the colonies present were colonies of Chich. 1.

Experiment No. 2.—Repetition of Experiment I.

This experiment was really a repetition and control of Experiment 1. A flask containing 200 c.c. sterile distilled water and 10 c.c. broth, was inoculated with ¼ c.c. of a 24 hours' broth culture of Chich. 1, and with ¼ c.c. of a 24 hours' broth culture of the typhoid bacillus. The flask was incubated at 37° C.

First Examination.—*After 24 hours.* Brushed over two agar plates. Plates incubated at 37° C. and examined after 24 hours. Copious growth on both plates, of which the great majority are colonies of Chich. 1. Also many typhoid bacillus colonies.

Second Examination.—*After six days.* Needled* over three plates.

Well-marked growth on all three plates after 24 hours' incubation. No colonies of typhoid bacillus to be found after careful search and sub-culture of any doubtful colonies. All Chich. 1 colonies: after six days, only Chich. 1 survived.

Experiment No. 3.—Liquid Medium at 4°-12° C.

The conditions of the experiment were exactly the same as for Experiment 1, except that the flask was kept in an outside shed.

---

* Brushing it was thought might give rise to an excessive number of colonies and that discrete colonies would not be obtained: so instead of a brush a platinum needle bent at a right angle about one-fifth of an inch from the end was used. This was used in the same way as a sterile brush. For such a method of distributing the colonies over the plate the comprehensive word "needled" is used here and elsewhere.

The temperature of the shed during the eight days the experiment lasted was 4°–7° C. for the first 24 hours, 7°–12° C. for the next seven days, as recorded by a maximum and minimum thermometer.

First Examination.—*After 24 hours.* Brushed over two plates. Plates examined after 24 hours incubation at 37° C. Considerable growth on both plates. The great majority of the colonies are Chich. 1. Colonies of the typhoid bacillus present, but not numerous. The colonies of the two organisms were quite readily distinguished, and films were made and examined from each.

Second Examination.—*After three days.* Brushed over two plates. The vast majority of the colonies were now Chich. 1, and comparatively only a few typhoid bacillus colonies present (about 30 altogether on the two plates). Microscopic preparations made to complete the diagnosis.

Third Examination.—*After seven days.* Brushed over two plates. Well marked growth on both plates. All Chich. 1 colonies. No typhoid bacillus colonies could be found, although all colonies at all like typhoid bacillus were subcultured on agar, and subsequently examined. After a week's growth, therefore, only Chich. 1 colonies survived.

Experiment No. 4.—Sterile Soil Medium at 37° C.

A sterile small Erlenmeyer flask was filled to a depth of about half an inch with Chichester soil in a finely divided state. Sufficient distilled water was added to moisten the soil throughout. The whole was then sterilised. To this soil ½ c.c. of 24 hours' broth culture of the typhoid bacillus, and ½ c.c. of a similar broth culture of Chich. 1 were added. The broth was added by means of a sterile pipette to the centre of the soil. The flask was incubated at 37° C.

First Examination.—*After 24 hours.* Two loopfuls of soil were withdrawn from the centre of the flask and added to another flask containing 200 c.c. distilled water and about 10 c.c. broth. This flask was incubated at 37° C. for 24 hours, and then tested by needling over two agar plates, which were incubated for 24 hours and then examined. The great majority of the colonies were Chich. 1 on both plates. On one plate no typhoid bacillus colonies could be found, but on the other about half a dozen typical colonies of this bacillus were present. Films of these confirmed the diagnosis.

Second Examination.—*After six days.* Examined in same way as before, except that three plates were used.
On all three plates there was abundant growth of Chich. 1, but no colonies even remotely resembling those of the typhoid bacillus to be seen on any of the plates.

Third Examination.—*After nine days.* In order to make sure that no typhoid bacilli were present, the soil was again examined in the same way as at the first two testings, except that the plates were brushed with a sterile brush and th six plates were used. There was well-marked growth on a the plates, and on nearly all of them the colonies were di crete. Each plate was most carefully examined for coloni like those of the typhoid bacillus under a low power the microscope. The microscopic differences between tl colonies of the typhoid bacillus and Chich. 1 on agar plat are quite sufficiently marked to render their differentiatio easy. But only one colony could be regarded as a litt doubtful. This was subcultured, and subsequently examine but proved to be Chich 1.

Therefore, in six days and certainly after nine days r typhoid bacilli could be obtained from this soil when in cubated at 37° C.

Experiment No. 5.—Sterile Soil Medium at 8°–12° C.

This experiment was quite similar to No. 4 in its conditior except that the flask containing the soil was kept in an outsid shed. During the two weeks of the experiment the temperatu was from 8°–12° C. The method of examination was the same a for experiment 4, except that the plates were brushed.

First Examination.—*After 24 hours.* Two plates use About 12–14 colonies were those of the typhoid bacillus, th rest of the growth was made up of Chich. 1 colonies. Sul cultures and films were made in confirmation.

Second Examination.—*After six days.* Three plates used copious growth on all three plates. The vast majorit were Chich. 1 colonies; but on all three plates colonies c the typhoid bacillus were easily found. The two colonie were readily distinguished. Films made from the latt showed typical typhoid bacilli.

Here, after six days, there were found more typhoi colonies than in 24 hours after commencement of the ex periment.

Third Examination.—*After two weeks.* Three plates used well marked, but for the most part discrete, growth on a three plates. All colonies of Chich. 1. No colonies lik those of the typhoid bacillus. Three colonies not quit typical of Chich. 1 which were sub-cultured on agar showe on incubation the typical opalescent opaque appearance o Chich. 1 in each case; and in microscopic preparation the short or oval bacilli of this organism were readil identified.

After two weeks only Chich. 1 survived.

Remarks.—These five experiments were all the experiment performed with Chich. 1. In not any of them could the typhoi bacillus be found after two weeks, and in none, except as regard the soil kept at shed temperature, after eight days.

In the same Chichester soil sterilized and moistened with wate the typhoid bacillus was found living after over one year. (Se Report 1897–98.)

With great probability it may be therefore concluded that Chich. 1 is, in its growth, inimical to the growth and life of the typhoid bacillus, and that its presence in a soil tends to destroy any typhoid bacilli which may also be present in that soil.

SERIES B.—TYPHOID BACILLUS AND CHICHESTER 2.

In this series the micro-organisms used consisted of the same culture of the typhoid bacillus and a culture of Chichester 2.

Chich. 2 is a large stout bacillus, single or in short chains, which forms spores freely.

It grows rapidly in the ordinary media. On agar slope it forms an opaque white growth; it forms both a scum and a deposit in broth; abundant pale yellow moist growth on potato. It gives rise to no indol unless cultivated for several weeks, even then the reaction is very faint. In glucose gelatine shake culture it produces no gas. The growth is almost entirely on the surface; the gelatine is slowly liquefied. On agar plates it forms white opaque colonies which are not quite circular. Their margin is very irregular, and in many not well defined. They are dark and opaque in the centre, less opaque near the periphery, and are very granular. They are therefore not in the least like the typhoid bacillus colonies, and are quite readily distinguished from them.

Experiment No. 1.—Liquid Medium at 37° C.

A flask containing 200cc. distilled water and 10cc. broth (the whole sterilized), was inoculated with a loopful of a broth culture of Chich. 2, and with a loopful of broth culture of the typhoid bacillus. Incubated at 37° C.

First Examination.—*After 24 hours.* Needled over two agar plates. These were incubated at 37° C. for 24 hours, and then examined. Considerable growth on both plates. All the colonies appear to be those of the typhoid bacillus. Microscopic preparations show bacilli quite like the typhoid bacillus, and the hanging drop showed actively motile bacilli. Subculture on agar yielded the translucent bluish growth of the typhoid bacillus. Colonies of Chich. 2, if present, would have been quite obvious and prominent among the typhoid colonies.

Second Examination.—*After three days.* Brushed over two plates. Fair amount of growth on the plates when incubated. All colonies of the typhoid bacillus. No Chich. 2 present.

Experiment No. 2.—Repetition of Experiment 1.

This was really a control of Experiment 1. A flask containing 200cc. sterile water and 10cc. broth was inoculated with ¼cc. of young broth culture of Chich. 2, and with ¼cc. of a young broth culture of the typhoid bacillus.

The flask was incubated at 37° C.

First Examination.—*After 24 hours.* Needled over two plates. On the first plate the vast majority of the colonies

. B. No. 2

...e Growth
...e Typhoid
...us in
...; by Dr.
...ey Martin.

are colonies of the typhoid bacillus, but there are also a few colonies of Chich. 2. The two sorts of colony are quite unmistakably distinguishable. On the second plate there is also well marked growth. All the colonies on this plate are those of the typhoid bacillus.

Second Examination.—*After seven days.* Needled over two plates. The colonies of typhoid bacillus form the great majority. There are only a few Chich. 2 colonies.

Third Examination.—*After* 14 *days.* Brushed over two plates. Well marked and for the most part discrete growth on both plates. Colonies of both Chich. 2 and of typhoid bacillus present and easily distinguished. Now, however, the greater number of the colonies are Chich. 2; typhoid bacillus colonies are in a decided minority.

No further examination could be made before the writing of this report.

The method of brushing plates is not accurate in always determining the relative quantities of micro-organisms present; but it may give a valuable indication, and here we seem to have a gradual increase in the amount of the Chich. 2 organism relative to the number of typhoid bacilli present.

Experiment No. 3.—Liquid Medium at 4°–12° C.

Conditions just the same as for experiment No. 1, except that the flask was kept in the outside shed. The temperature of the shed for the first 24 hours was 4°–7° C., and for the rest of the experiment ranged from 7°–12° C.

First Examination.—*After 24 hours.* Brushed over two plates. Both organisms present after 24 hours incubation at 37° C., and easily found and distinguished. Colonies of Chich. 2 slightly more numerous than those of the typhoid bacillus.

Second Examination.—*After three days.* Brushed over two plates. Not much growth on either plate. Both colonies present, but the colonies of Chich. 2 are considerably more numerous than those of the typhoid bacillus.

Third Examination.—*After seven days.* Brushed over two plates. Colonies not numerous on either plate. On the second plate all the colonies were those of Chich. 2. On the first plate two colonies of the typhoid bacillus were seen, the others were Chich. 2 colonies. Films made from these colonies showed the characteristic appearances of the two organisms.

Fourth Examination.—*After two weeks.* Brushed over three plates. On the first plate a fair amount of growth; all were Chich. 2 colonies. On the second plate there were five colonies, all Chich. 2. On the third plate all the three colonies present were also Chich. 2. The Chich. 2 colonies had the characters described above. No typhoid colonies could be found on any of the plates.

Fifth Examination.—*After four weeks.* The flask was removed to the 37° C. incubator 17 days after it was started, so that for the last 11 days the flask contents have been incubating at 37° C.

Two plates brushed. Well marked growth on both plates. All the colonies on the first plate, except one, and the great majority on the other, were typical colonies of the typhoid bacillus. This was confirmed by microscopic specimens, and by examining in hanging drops (very actively motile bacilli). The rest of the colonies were typical Chich. 2 colonies, and in microscopic specimens were indentical with that organism.

Experiment No. 4.—Sterilized Soil Medium at 37° C.

In this case ½cc. of a 24 hours broth culture of Chich. 2, and ›c. of a similar culture of the typhoid bacillus were added to erile Chichester soil in an Erlenmeyer flask in exactly the same ay as Exp. 4, Series A. The soil was incubated at 37° C.

First Examination.—*After 24 hours.* As in all the other experiments in which soil was the medium of growth a small quantity was removed from the centre of the flask, and added to another flask containing 200cc. sterile distilled water and (in this case) 10cc. broth. The mixture was incubated at 37° C. for 24 hours, and then needled over two plates. These in turn were incubated for 24 hours at 37°. Considerable growth on both plates. Nearly all the growth consisted of typhoid bacillus colonies. There are only a comparatively few Chich. 2 colonies. About 38 of these on the first plate, and 9 on the second plate.

Second Examination.—*After six days.* Three plates used; brushed. Copious growth on all three plates, consisting of colonies of Chich. 2 and of typhoid bacillus. The Chich. 2 colonies are the more conspicuous, but the typhoid bacillus colonies are quite as numerous. Films made from the latter colonies show the typical appearances of the typhoid bacillus.

Third Examination.—*After* 14 *days.* Brushed over three plates. The well-marked growth on all three plates consists entirely of colonies of Chich. 2. Though a most careful search was made, no colonies at all like typhoid bacillus colonies, or differing from the typical colonies of Chich. 2, were found.

Here, therefore, at the end of two weeks, only Chichester 2 organism could be found.

Remarks.—In this series no experiments were made with soil ›pt at shed temperatures.

The series is of interest in that it deals with an organism which, its relationship to the typhoid bacillus, seems to stand midway ›tween Chich. 1 and Chich. 3.

In all the experiments conducted at 37° C., *i.e.*, experiments o. 1, No. 3, and No. 4, the typhoid bacillus at first grew most pidly; and indeed, in the case of Experiment No. 1, where only a loopful of each culture was used, killed out its opponent speedily. As growth proceeded, however, the Chich. No. 2 gradually increased more rapidly than the typhoid bacillus, and eventually gained the upper hand; in Experiment 4 killing out its adversary in two weeks. In Experiment 3, in two weeks, it was already in marked excess.

At shed temperatures, Chich. 2, in the only experiment made, maintained the ascendancy at the first, and in two weeks was the only organism that could be found; but later the typhoid bacilli, present, apparently, in too few numbers to be seen on the plates when growing at the shed temperature, gradually increased and in the final experiment conducted at 37° C. assumed the upper hand.

Series C.—Typhoid Bacillus and Chichester 3.

Here the micro-organisms investigated consisted of Chichester 3 and the same culture of the typhoid bacillus that was used for Series A and B. In one or two cases a different culture of the typhoid bacillus was used—this is indicated under the separate experiments.

Chich. 3 is a stout bacillus with rounded ends, and forming long chains. On agar slope it produces a dense white opaque growth. It grows readily in broth, producing no scum but a copious flocculent deposit, the supernatant broth being clear. No indol is produced even after prolonged growth. In glucose-gelatine shake culture it produces no gas, and grows almost entirely on the surface; it liquefies gelatine slowly. On potato it forms an abundant yellow growth, at first moist, later becoming rather dry. On agar plates it forms dense opaque colonies. Hence no confusion could arise between these colonies and those of the typhoid bacillus.

Experiment No. 1.—Liquid Medium at 37° C.

A flask of 200cc. sterile water and 10cc. broth inoculated with a loopful of a broth culture of Chich. 3, and with a loopful of broth culture of the typhoid bacillus. Flask incubated at 37° C.

First Examination.—*After 24 hours.* Needled over two plates. Considerable growth on both plates. All the colonies are typical typhoid bacillus colonies, and no Chich. 3 colonies seen anywhere.

Second Examination.—*After three days.* Brushed over two plates. Fair amount of growth on both plates. All typhoid bacillus colonies. Film made in confirmation shows characters of this bacillus.

Hence, after 24 hours, only the typhoid bacillus survived.

Experiment No. 2.—Liquid Medium at 4°-10° C.

Under exactly the same conditions as Experiment No. 1, except that the flask was kept in the outside shed. Temperature ranged from 7°-10° C. during the experiment, except for the first night, when it was 4°-7° C.

First Examination.—*After 24 hours.* Brushed over two plates. Only a small amount of growth on the two plates.

All the colonies are the same, and all are typhoid bacillus colonies.

Second Examination.—*After three days.* Brushed over two plates. Scarcely any growth on either plate, but the very few colonies present were those of the typhoid bacillus. Film preparations made showed bacilli quite like the typhoid bacillus, and the colonies were quite typical.

Here no Chich. 3 were found, even 24 hours after inoculation of the water and broth.

Experiment No. 3.

In Experiment 1 and Experiment 2 of this series it might possibly be that the culture used of Chich. 3 was dead, or at least not active; or that Chich. 3 would not grow in the medium (*i.e.*, sterile water and broth) provided. To see if this was the case, the following experiments were undertaken. Broth cultures of Chich. 3 and the typhoid bacillus were made, in each case being inoculated from the broth cultures used for Experiments 1 and 2. In both cases well marked growth took place in the 24 hours.

¼cc. of this 24 hours broth culture of Chich. 3, and an exactly similar quantity of the 24 hours broth typhoid bacillus culture, were then added by means of sterile pipettes to a flask containing 200cc. sterile water and 10cc. broth. The flask and contents was incubated at 37° C.

First Examination.—*After 24 hours.* Brushed over two agar plates. Profuse growth on both plates, which consisted of a pure growth of the typhoid bacillus. No Chich. 3 colonies, which, if present, would have been readily distinguished, were to be found after a careful examination.

No further examinations were made, as the object of the experiment was obtained. Here it was found that a recent active culture of Chich. 3 grown with the typhoid bacillus was quite unable to multiply nd propagate itself, if it was not itself killed, within 24 hours.

Experiment No. 4.

Experiment carried out exactly as No. 3, except that the flask was kept in the outside shed. Temperature during the experiment 8°–9° C.

First Examination.—*After 24 hours.* Brushed over two plates. Copious growth on both plates, but almost all the colonies are discrete. All the colonies are the same, and are all typically those of the typhoid bacillus. No Chichester 3 colonies present. The plates were then incubated for a second 24 hours to see if any Chich. 3 would develop, but the growth remained pure typhoid bacillus.

No further examination was made.

Experiment No. 5.

This was a purely control experiment undertaken to see if Chich. 3 would multiply in sterile water and a little broth

App. B. No. 3.

On the Growth of the Typhoid Bacillus in Soil; by Dr. Sidney Martin.

In this experiment ½cc. of a 24 hours' old broth culture of Chich. 3 was added to a flask containing 200cc. sterile water and 10cc. broth. Flask incubated at 37° C.

First Examination.—*After 24 hours.* In the flask there is a copious white flocculent deposit. From this flask needled over one plate. On this plate a well-marked growth of Chich. 3 developed.

Second Examination.—*After 47 days.* A film made from the flask shows fairly numerous stout bacilli. The deposit in the flask had considerably increased.

From this experiment there can be no doubt that Chich. 3 will live and multiply in sterile water and broth, and that the conditions which prevented its multiplication in the above four experiments must be sought for elsewhere than in the medium.

Experiment No. 6.—Sterile Soil Medium at 37° C.

Sterile Chichester soil in Erlenmeyer flask as before, to which was added ½cc. of broth culture of Chich. 3, and a similar quantity of broth culture of the typhoid bacillus, as in the other similar experiments. Incubated at 37° C.

First Examination.—*After four days.* Loopful of soil removed from centre of flask and inoculated into sterile water. From this brushed over three plates. After 24 hours' incubation colonies of Chich. 3 and the typhoid bacillus easily found on the plates, the great majority, however, being Chich. 3 colonies. The two kinds of colonies quite easily distinguished.

Second Examination.—*After seven days.* Four plates used; brushed. On the first plate there is one typical colony of the typhoid bacillus. Film from this shows bacilli similar to typhoid bacilli, and examined in hanging drop actively motile bacilli are seen. All the other colonies on the first plate and all on the other two plates have the characters of Chich. 3.

Third Examination.—*After 15 days.* Three plates used. Colonies not numerous, all are Chich. 3, *i.e.*, dense white opaque colonies. No typhoid bacillus colonies, though a most careful search was made under a low power of the microscope.

Experiment No. 7.—Sterilized Soil Medium at 0°–14° C.

Conditions the same as for Experiment No. 6, except that the flask was kept in the shed, and that the broth culture of the typhoid bacillus used was one isolated quite recently from the spleen of a patient dead of enteric fever. Temperature in shed for first six days was from 10°–14° C., for the next six days it was freezing in the shed, then warmer (8°–12° C.)

First Examination.—*After 48 hours.* Three plates used; plates incubated for 24 hours, but remained sterile.

Second Examination.—*After 13 days.* Three plates used; brushed. Fair amount of growth on all three plates, of which the vast majority consists of typhoid bacillus colonies. Films from these colonies showed bacilli like the typhoid bacillus, except that they were rather short; in this, however, quite in agreement with this brand of typhoid bacillus, which was a very short one. There was also a fair number of other colonies which in microscopic specimens made from them have the characters of Chich. 3, although the characters of the colonies themselves are not quite in accord with those of Chich. 3.

Up to time of writing complete proof was wanting to show whether these were true Chich. 3 colonies or were adventitious colonies due to accidental contamination.

No later examination made up to time of writing.

Remarks.—Considering the experiments of series C. together, it seems to be fairly established that Chich. 3 is an organism which naturally has a difficulty in growing in presence of the typhoid bacillus. Chich. 3 will grow readily enough in a mixture of water and broth when it is the only organism present, but when the typhoid bacillus is also present it is unable to establish itself, and does not grow, or possibly is killed out altogether. This takes place both at low temperature, and at that of 37° C. When, however, this soil micro-organism Chich. 3 is growing in the sterilized soil from which it was obtained, its chances of growth and multiplication are greatly increased, and in one case it so far gained the mastery that no typhoid bacilli could be found after 15 days.

Series D.—Typhoid Bacillus and Chichester 5.

In this series the micro-organisms used consisted of Chichester 5 and of the typhoid bacillus. The first two experiments were performed with the laboratory culture of the typhoid bacillus, used in series A. and B., and most of C.; in the third experiment the culture of typhoid bacillus used was the same as for Series C., Exp. 7.

The main characters of Chich. 5 are as follows:—

It is a thick bacillus with rounded ends, which occurs singly or in short chains. It stains fairly uniformly. On agar slope it produces a white opaque granular growth It grows readily in broth, forming a copious deposit, and a slight surface pellicle, the broth between remaining clear. In glucose gelatine shake culture it produces no gas, but the gelatine is slowly liquefied. It produces no indol. On potato it forms a reddish white growth. Its colonies on agar plates are very characteristic and distinctive. They are dense opaque colonies with a very irregular margin, apparently made up of bands of wavy threads. When free to develop they grow out into fine irregular processes.

Experiment No. 1.—Liquid Medium at 37° C.

A loopful of a young broth culture of Chich. 5 and loopful of young broth culture of the typhoid bacillus, added to 200cc. sterile water and about 10cc. broth. Incubated at 37° C.

First examination.—*After 24 hours.* Needled over two plates. The well marked growth consists almost entirely of Chich. 5 colonies; only one typhoid bacillus colony being found on the first plate, and two colonies on the second plate The two sorts of colony are quite easily distinguished.

Second examination.—*After three days.* Needled over two plates. Only one colony of the typhoid bacillus, all the rest are Chich. 5 colonies. The typhoid bacillus colony was subcultured on agar, and gave a translucent growth which, when examined in hanging drop was actively motile, and microscopic specimens stained by carbol-fuchsine showed bacilli like the typhoid bacillus.

Third examination.—*After seven days.* Needled over two plates. On both copious growth of Chich. 5. The plates were examined most carefully for colonies of the typhoid bacillus; but no colonies even remotely resembling it were seen.

Experiment No. 2.—Liquid Medium at 7°–14° C.

Conditions just the same as for experiment No. 1, except that the flask was kept in the outside shed. The temperature during the time when the flask was in the shed ranged from 7°–14° C.

First examination.—*After 24 hours.* Brushed over two plates. Both sorts of colony on both plates. The typhoid bacillus colonies were, however, the more numerous.

Second examination.—*After three days.* Brushed over two plates. Only one colony present on the two plates; this was a typhoid bacillus colony. No Chich. 5 colonies present.

Third examination.—*After seven days.* Brushed over two plates. No growth on either of the plates.

Fourth examination.—*After two weeks.* Brushed over three plates. No growth on any of the plates.

The flask was then put in the incubator at 37° C., and kept there for 12 days.

Fifth examination.—*After four weeks* (in all), and after 12 days incubation at 37° C.; brushed over two plates: copious growth on both plates of typical colonies of Chich. 5. No typhoid bacillus colonies were observed, although a most careful search was made.

No later examinations made.

This experiment indicates that though no growth takes place on brushing plates, it is not safe to assume that the bacteria not represented are extinct; though in such case it does appear to be reasonable to assume that they are present in only small numbers, and are not actively multiplying.

Experiment No. 3.—Sterilized Soil Medium at 37° C.

Sterile Chichester soil in Erlenmeyer flask inoculated with ½ c.c. of broth culture of Chich. 5, and with equal quantity of broth culture of the typhoid bacillus (the recently isolated growth being used). Incubated at 37° C.

First examination.—*After* 24 *hours*. Brushed over two plates. Marked growth on both plates, which everywhere consists of colonies of Chich. 5; no typhoid bacillus colonies on either plate.

Second examination.—*After three days*. Four plates brushed. Much growth, which everywhere consists of characteristic colonies of Chich. 5. On the second plate, however, there is a single colony which is rather like a typhoid colony. This was subcultured on agar slope and carefully examined. It had the following characters.

Agar slope: Translucent growth.
Milk: No coagulation in two days.
Glucose gelatine shake culture: No liquefaction; no gas developed.
Potato: Invisible growth.
Broth: Cloudy in 24 hours; no pellicle.
Films from agar slope: Short bacilli, which agree with the characters of the typhoid bacillus used.
Hanging drop from 24 hours broth showed an actively motile bacillus.
When mixed with serum from a case of typhoid fever (a serum which clumps and causes cessation of movement in diluted typhoid bacillus cultures) in a 1 : 100 dilution, it causes well-marked clumping after ½ hour though the bacilli are still in part moving at end of 1½ hours.
This, therefore, was a colony of the typhoid bacillus.

Third Examination.—*After* 13 *days*. Brushed over two plates. Well-marked growth on both plates. All typical Chich. 5 colonies. No colonies in the least like those of the typhoid bacillus present.

No further experiments were undertaken with soil kept at shed temperatures.

Remarks.—Chich. 5, like Chich. 1, is an organism which rapidly causes the typhoid bacillus to die out, or at least prevents its increase, when the two organisms are mixed together. In not any of the experiments could the typhoid bacillus be found after three days' incubation along with Chich. 5.

Series E.—Typhoid Bacillus and Chichester 1, 2, 3 and 5.

In this series an attempt was made to ascertain the growth of the typhoid bacillus in presence of all four Chichester organisms together, and also the relative growth rate and persistence of these latter organisms when grown together.

This has only been done partially up to time of preparing this Report.

. B. No. 3.

Growth Typhoid us in by Dr. ey Martin.

Experiment No. 1.—*Liquid Medium at* 37° *C.*

To a flask containing 200cc. sterile water about 10cc. of broth was added. To this flask was then added a loopful from each of broth cultures of Chich. 1, Chich. 2, Chich. 3, Chich. 5. This flask was incubated at 37° C.

First Examination.—*After* 24 *hours.* Brushed over three plates. All the plates have a strong putrefactive odour.

The colonies of Chich. 5 and Chich. 1 are very conspicuous: no colonies to be seen definitely like Chich. 2 or Chich. 3. All doubtful colonies were examined by film preparations or were sub-cultured on agar. These latter were in turn sub-cultured on potato and film preparations made from each. Not any of the colonies were those of Chich. 2 or Chich. 3.

Second Examination.—*After three days.* Three plates brushed. The well-marked growth on all three plates consists mainly of Chich. 5. Chich. 1 easily found, but are much less numerous. No colonies of Chich. 2 or Chich. 3 found. Two colonies not like either Chich. 1 or 5 were sub-cultured, but on further examination appeared to be B. subtilis, with which the flask, or the plates, must have been contaminated.

No further examination made.

Experiment No. 2.

The conditions similar to Experiment No. 1, except that a loop of a culture of the typhoid bacillus was also added to the flask (the recently isolated typhoid bacillus was used). Incubated, as before, at 37° C.

First Examination.—*After* 24 *hours.* Brushed over three plates. Well-marked growth on all three plates. Almost all the colonies were either Chich. 1 or Chich. 5. Only eight colonies in all of the typhoid bacillus were found on the three plates. Confirmed by microscopic preparations and characteristic agar growth when sub-cultured. They were also examined in hanging drop and were actively motile.

Second Examination.—*After three days.* Three plates brushed. Nearly all the growth consisted of Chich. 5 colonies. A few Chich. 1 colonies found. On only one of the plates were any typhoid bacillus colonies seen, about five or six. Sub-cultures from two of these showed on agar well-marked translucent growth, and typical typhoid bacilli were detected in film and hanging drop. No Chich. 2 or Chich. 3 colonies could be found.

No further examination made.

Remarks.—These experiments are very incomplete, but as far as they go they seem to confirm the experiments of Series A, B, C, and D.

Chich. 2 and Chich. 3 are easily inhibited in their growth; or killed, while Chich. 1, and especially Chich. 5, increase.

The typhoid bacillus seems to stand in this respect mid-way between these two groups of organisms.

SERIES F.—TYPHOID BACILLUS AND BACTERIA OF UNIVERSITY COLLEGE SOIL.

In this series are grouped a number of experiments similar to the above which were commenced with micro-organisms isolated from soil obtained from the front of University College. They are very incomplete up to the time of writing, and accordingly they will be summarised.

The experiments were performed in the same way as described under the other experiments. In all these experiments the recently isolated typhoid bacillus sub-cultures were used.

Experiment No. 1.

To 200cc. sterile distilled water and 10cc. broth, one loop of typhoid bacillus broth and one loop of broth culture of U. C. L. 1 were added. Incubated at 37° C.

First Examination.—*After* 24 *hours.* Two plates brushed. A pure growth of the typhoid bacillus, no U. C. L. 1 being present.

Second Examination.—*After three days.* Two plates brushed. Majority of the colonies are those of the typhoid bacillus. There are also present other colonies, apparently U. C. L. 1, but they were not sub-cultured.

No later examinations made.

Experiment No. 2.

Similar to Experiment No. 1, except that a loopful of U. C. L. 3 was added instead of one of U. C. L. 1.

First Examination.—*After* 24 *hours.* Two plates brushed. Almost all typhoid bacillus colonies, a few U. C. L. 3 colonies also present.

Second Examination.—*After three days.* Two plates brushed. About half the growth appears to be typhoid bacillus, the rest is apparently U. C. L. 3.

No later examinations made.

Experiment No. 3.

Sterile Erlenmeyer flask filled to depth of ½ inch with sterile U. C. L. soil; sterilised after addition of sufficient water to keep the soil just moist.

To centre of this soil ½cc. of young broth culture of the typhoid bacillus and ½cc. of similar culture of U. C. L. 1 added. Incubated at the temperature of the laboratory.

First Examination.—*After* 24 *hours.* Brushed over two plates. A few adventitious colonies, neither U. C. L. 1 nor the typhoid bacillus. Otherwise all the colonies are typhoid bacillus colonies.

Second Examination.—*After seven days.* Brushed over three plates. Majority of the colonies are colonies of the typhoid bacillus, but a few other colonies are also present.

No further examinations made.

Experiment No. 4.

Similar to No. 3, only U. C. L. 3 used instead of U. C. L. 1.

First Examination.—*After 24 hours.* Two plates brushed. Well-marked growth which consists of a pure culture of the typhoid bacillus.

Second Examination.—*After one week.* Three plates brushed. Vast majority are colonies of the typhoid bacillus. There are also a few U. C. L. 3 colonies.

No further examination made.

Remarks.—Both in water and broth, and in soil, the typhoid bacillus seems to grow more vigorously than U. C. L. 1 and U. C. L. 3, at any rate at first. The experiments are too incomplete for further inference.

No. 4.

Report on Inoculation of soil with particular Microbes, Pathogenic and other, by A. C. Houston, M.B., D.Sc.

PART I.

Inoculation of Soil in the Laboratory with Bacillus Prodigiosus, Bacillus Diphtheriæ, and Koch's Vibrio.

Series I.—Preliminary Experiments with B. Prodigiosus.

In last year's report (Report of Medical Officer for 1897–98, Appendix B, No. 2, Series I., p. 296–298) a series of preliminary laboratory experiments were recorded respecting the inoculation with B. prodigiosus of sterilised and unsterilised soil placed in Petri's capsules. At the time of writing that report these experiments were still under observation, and they have been continued during the present year. The results of this preliminary investigation are here recapitulated to an extent sufficient only to render intelligible what follows. For full details the report of last year must be consulted :—

Experiment A was started July 30, 1897.—Some soil* was placed in a Petri's capsule to a depth of about ¼-in., and saturated with sterile distilled water containing a loopful of a gelatin prodigiosus culture.

Experiment B.—As above, except that the soil was first sterilised.

Experiment C, started September 22, 1897, corresponded to Experiment A, except that soil V.† was used.

Experiment D was the same as Experiment C, except that the soil (V.) was first sterilised.

The capsules were placed in a dark chamber, and the viability of B. prodigiosus determined from time to time during 1897–98–99 by inoculating "oblique" agar tubes.

---

* Soil II. (town garden soil, not recently manured).

† Virgin sand from the sea shore above high water mark.

Table of Results.

[The sign + signifies a positive result, *i.e.*, the presence of the characteristic r growth of prodigiosus in the agar culture].
[The sign — signifies a negative result, *i.e.*, the absence of the characteristic r growth of prodigiosus in the agar culture].

Date of Testing for the Viability of B. Prodigiosus in the Soils.	A. Garden Soil (unsteril-ised), July 30/97.	B. Garden Soil (sterilised), July 30/97.	C. Sand (unsteril-ised), Sept. 22/97.	D. Sand (sterilised Sept. 22/9'
September 9/97	+	+	..	...
21/97	+	+	...	...
27/97	...	...	—	+
30/97	...	...	—	. .
October 4/97	...	...	—	+
7/97	+	+	...	...
November 8/97	+	+	+	+
January 5/98	+	+	+	+
4/98	+	+	+	—
April 14/98 ... ... (subsequently to testing, soils saturated with sterile distilled water).	—	+	—	+
April 18/98	—	+	—	+ (208 days
June 23/98	—	+	—	—
October 24/98 ... ... (previous to testing, 10 cc. sterile distilled water added to contents of each capsule).	+ (451 days).	—	—	—
October 28/98	...	+ (455 days).	+ (401 days).	—
31/98	...	...	...	—
November 7/98	...	...	...	—
December 22/98	—	—	—	—
March 6/99 ... ... (In each case the whole of the soil was placed in a sterile mortar and bruised with 50 cc. sterile water and "oblique" Agar tubes inoculated).	—	—	—	—

In *Experiment A* (garden soil, unsterilised) the B. prodigiosu remained alive from July 30, 1897, to October 24, 1898, a period o 451 days. But whereas it was readily demonstrated at first, late on it was found only with difficulty. Between the above dates negative result was obtained on three different occasions. O March 6, 1899, it may reasonably be inferred that it was dead, o present in very small number, as *the whole of the soil* was at that dat thoroughly mixed with water, and agar cultures made from th mixture of soil and water.

In *Experiment B* (garden soil, sterilised) the B. prodigiosus re mained alive from July 30, 1897, to October 28, 1898, a period o 455 days. But whereas it was readily demonstrated at first, late on it was found only with difficulty. Between the above date a negative result was obtained on one occasion. On March 6, 1899, there is reason, the whole of the soil having been dealt with as in A, to believe it had lost its vitality, or was present in very small number.

In *Experiment C* (sand, unsterilised) the B. prodigiosus remained alive from September 22, 1897, to October 28, 1898, a period of 401 days. But it was at no time so readily demonstrated as in Experiments A and B. Between the above dates a negative result was obtained on seven different occasions. On March 6, 1899, it was apparently (for reasons of the sort already stated) dead, or present in very small amount.

In *Experiment D* (sand, sterilised) the B. prodigiosus remained alive from September 22, 1897, to April 18, 1898, a period of 208 days. But whereas it was readily demonstrated at first, later on it was found only with difficulty. Between these dates a negative result was obtained on one occasion only. On March 6, 1899, it was in all probability dead.

The conclusions to be drawn from this series of experiments carried out under the above conditions during many months seem to be :—

(1) That the B. prodigiosus can retain its vitality in certain soils under certain conditions for a very considerable time.

(2) That in sterilised soils it is more readily demonstrated than in other soils. [At one time it was thought that the sterilisation of the soil tended to prolong the life of B. prodigiosus, but later results did not entirely confirm this inference.]

(3) That in a pure sandy soil, as well as in a town garden soil, the B. prodigiosus can retain its vitality for long periods. On the whole, however, the sandy soil seemed to be less favourable to it than the other.

Series II.—Further Experiments with B. Prodigiosus.

(1) An earthenware pot (4-in. × 2⅝-in.) was filled with soil taken from a wood; a soil which had not been manured or "treated" in any way. This soil was not sterilised. Distilled water was added to it in excess of saturation, and, after leaving for 24 hours to drip, the soil within the pot was inoculated in the following way :—

A strong platinum needle was charged with B. prodigiosus (from a young agar culture), and the soil inoculated therewith in the same way as in making a gelatine stab culture : that is, the needle was thrust vertically downwards as nearly as possible in the centre of the soil in the pot, and to a depth of about 2-ins.* The pot having been fitted with a loosely adjusted lid, and made to rest in a small shallow dish containing a few ccs. of water, was next placed in a dark chamber. The water in this dish was renewed from time to time to make up for loss by evaporation.

(2) Another experiment similar in every respect to the above, but with a different soil, was also carried out. This soil was a clay garden soil that had been periodically manured.

* This method of inoculation was likewise resorted to in Series III. and IV.

In seeking to determine how long the B. prodigiosus retained its vitality in the soil within the earthenware pots at the site of the inoculation, and as to whether it spread laterally beyond the actual point of inoculation, the following plan † was adopted :—

A sterilised platinum needle was thrust vertically downwards in the centre of the soil in the pot—*i.e.*, along the original line of inoculation—and oblique agar tubes were inoculated by rubbing the needle to and fro over the surface of the medium. Further, the needle, after sterilisation, was again thrust vertically downwards into the soil within the pot, this time at a short distance from its centre, and other agar tubes inoculated. The presence of B. prodigiosus on the agar was readily determined by the characteristic growth which it shows on this medium.

The results of Experiments 1 and 2 are shown in the following table :—

[The sign + signifies a positive result, *i.e.*, a red growth ; and the sign — a negative result, *i.e.*, no red growth.]

Number of days subsequent to inoculation on which soil was tested.	Needle-sample taken—	
	from Centre.	½ inch from Centre.
Experiment 1, 45th day	+	—
" 106th "	—	—
Experiment 2, 34th day	+	—
" 95th "	—	—

In these two experiments there was no evidence of extension of growth in a lateral direction from the centre towards the periphery ; nevertheless in other experiments about to be recorded such extension appeared to have taken place. In Experiment 1 of this series B. prodigiosus was alive after 45 days, and in Experiment 2 after 34 days. In the first experiment, by the 106th day the B. prodigiosus appeared to have lost its vitality; in the second experiment this organism appeared to be dead by the 95th day.

(3.) In another set of experiments soil was inoculated with B. prodigiosus in precisely the same way as above, and in general the procedure was the same. The soil, however, was different (with one exception), and tin vessels were used ; moreover, in two out of the four experiments of this set the soil was sterilised by moist heat before being inoculated. The tin vessels, which were described and figured in last year's report (App. B., No. 2), consisted of tin cylinders having each a *moveable* perforated top and bottom. After inoculation the vessels were placed on a glass disc and covered with a bell jar, so as to prevent the soil from becoming too dry ; and they were kept in the dark at the room temperature. As in the preceding experiments the soil before inoculation was saturated with water.

In Experiments A. and B. virgin soil from a plantation was used. The soil was not sterilised.

† This plan was also used in Series III. and IV.

In Experiment C the same soil was used as in Experiment (2) *ante*, but it was sterilised.

In Experiment D a virgin peaty soil was used. This soil also was sterilised.

The results are shewn in the following table :—

[The sign + signifies a positive result, *i.e.*, a red growth; and the sign — a negative result, *i.e.*, no red growth.]

Experiment.	Number of days subsequent to inoculation on which soil was tested.	Needle-sample taken— from Centre.	½ inch from Centre.	1 inch from Centre.
A. not sterilised	28th day	+	+	No record
B. not sterilised	”	+	—	”
C. sterilised	22nd day	+	+	No record
D. sterilised	”	—	—	”
A. not sterilised	31st day	+	+	—
B. not sterilised	”	No record	No record	No record
C. sterilised	25th day	+	+	+ (one colony)
D. sterilised	”	—	—	No record
A. not sterilised	57th day	—	—	—
B. not sterilised	”	—	—	No record
C. sterilised	51st day	+	+	+
D. sterilised	”	—	No record	No record
A. not sterilised	127th day	—	—	—
B. not sterilised	”	—	—	No record
C. sterilised	121st day	+ (greatest red growth)	+ (less red growth)	+ (least red growth)
D. sterilised		Experiment abandoned.		
A. not sterilised	130th day	—	—	No record
B. not sterilised	”	—	—	”
A. not sterilised		Experiment abandoned.		
B. not sterilised		”	”	
C. sterilised ...	191st day	+ (greatest red growth)	+ (less red growth)	+ (least red growth)
C. sterilised ...	375th day	—	—	—

It will be noted that in Experiment D no growth was at any time obtainable, and this result is probably to be referred to the acidity of the peaty soil.

Experiment C yielded very different results from the results obtained previously with this same soil, in Experiment 2 of this series, in an unsterilised condition. Briefly, B. prodigiosus

APP. B. No. 4.

On Inoculation of Soil with particular Microbes: by Dr. Houston.

survived for 191 days, but not for 375 days. Moreover, it was present although in small amount outside the original line of inoculation, and as late as the 191st day was detected in the cultures made ½-inch and 1 inch from the centre of the soil as well as from the centre itself.

As regards Experiments A and B: in Experiment A a positive result was obtained from the centre of the soil up to the 31st day, but later the results were negative. Further, on the 31st day, B. prodigiosus showed itself in the culture made ½ inch from the eentre of the soil, but not in the culture made 1 inch from the centre.

In Experiment B, on the 28th day, B. prodigiosus was alive at the centre of the soil, but could not be found in the culture made ½ inch from the centre. By the 57th day, B. prodigiosus appeared to be dead, both centrally and ½ inch from the centre of the soil.

[It is difficult to determine whether in Experiments A and C the extension laterally of B. prodigiosus outside the original line of inoculation may not have been due to some mechanical factor, not to actual continuous spread of growth of the microbe through the layers of soil. The physical characters of different soils may have no small concern in determining the possibility of extension mechanically beyond the original area of inoculation. Indeed, it is open to doubt whether in the case of ordinary soils placed under ordinary conditions any *great multiplication* of germs artificially or accidentally introduced really does take place, and I find it difficult to believe in *well-marked and definite spread by actual continuous growth* as compared with mere mechanical extension. From my own observations as to the ratio of spores to total bacteria in sewage, in ordinary surface soil, and in soil grossly polluted with excrement, I am disposed to suspect that in *surface* soil there is ordinarily not only absence of sustained multiplication of imported bacteria but a shorter period of so-called dormancy than seems to be generally accepted, and that, consequently, there is *progressive death of superadded species*, except in the case of sporing forms. Of course I am here assuming that spore formation is *not* the highest stage in the vital activity of a bacterium, but is to be regarded as a means of preserving the species where the conditions are unfavourable.]

Series III.—Experiments with B. Diptheriæ.

The soil in this series of experiments was inoculated in precisely the same way as in the preceding series (Series II.), and was kept under the same conditions. But the micro-organism used in the experiment was the diphtheria bacillus.

(1). The soil was a virgin soil taken from a wood rich in vegetable matter. It was not sterilised. In seeking to ascertain whether or not the diphtheria bacillus had survived in this soil, cultures were made in a variety of media—agar, gelatine, ascitic agar, &c.

On the 5th day after inoculation of the soil a negative result was obtained as regards cultures made with needle-samples ½ inch from the centre of the soil, but as regards the cultures made from the centre (original line of inoculation) colonies appeared on the media closely resembling those of the diphtheria bacillus. These, on subculture and further study, were finally classed as diphtheria, although it is to be noted that when cultivated side by side with the original stock, slight differences showed themselves.

On the 7th day after inoculation further cultures were made with samples from the centre and ½ inch from the centre of the soil. A positive result was obtained in the former case, and a negative one in the latter. Immediately after these cultures had been made the soil was again inoculated centrally from a fresh diphtheria culture.

On the 17th and 61st day after the primary inoculation and the 10th and 54th after the secondary inoculation the testings were repeated. Here a negative result was obtained in both cases, *i.e.*, in the cultures made from the centre and ½ inch from the centre of the soil.

(2). In this experiment the same soil was used as before, but it was sterilised (4 hours at 100° C.) previous to inoculation with the diphtheria bacillus.

On the 4th day after inoculation, cultures were made with samples from the centre and ½ inch from the centre of the soil. No growth appeared in these cultures. The diphtheria bacillus was therefore again inoculated into the soil.

On the 2nd day after this fresh inoculation further cultures were made. On this occasion, however, a growth appeared of sporing bacteria, notwithstanding the prolonged heating of the soil; and the diphtheria bacillus could not be isolated. The soil was therefore again sterilised, this time in the autoclave at 120° C., for 1 hour, and the diphtheria bacillus again introduced into the soil.

On the 7th day after this third inoculation of the soil, cultures were made from the centre, and ½ inch from the centre. A positive result was obtained in the culture made from the centre, and a negative result as regards the culture made ½ inch from the centre.

On the 23rd day after inoculation the diphtheria bacillus could not be isolated by culture from any part of the soil.

These experiments with B. diphtheriæ do not admit of any useful inferences being drawn, even of a provisional nature. Yet they serve to illustrate the extreme difficulty of isolating the diphtheria bacillus in a mixed culture of bacteria, and give rise to the belief that the diphtheria bacillus readily loses its viability in, at all events, some soils.

Series IV.—Experiments with Koch's Vibrio.

The soil in this series was inoculated in precisely the same way as in the preceding series of experiments (Series II. and III.), but the microbe used was Koch's vibrio. The soil was not sterilized.

(1.) The soil was the same as that used in Experiment (1), Series III.

App. B. No. 4.

On Inoculation of Soil with particular Microbes; by Dr. Houston.

In seeking to determine whether or not the vibrio survived the soil, needle-samples were taken from the centre and ½ in from the centre, and inoculated into peptone tubes (1°/o peptor 0·5°/o Na. Cl.). These were incubated for one night at 37° (and the surface layers of liquid then examined in stained micr scopic preparations.

The following table gives the results in a concise form :—

[The sign + signifies a positive result, and the sign — a negative result regards the presence of Koch's vibrio.]

Number of Days subsequent to Inoculation with Koch's Vibrio when the Soil was Tested.	Needle-samples taken—	
	from Centre.	½ inch from Centre
5th day	+	—
17th "	+	
38th "	+	
61st "	—	—

Here Koch's vibrio retained its vitality in the soil for 38 days but not for 61 days.

There was no evidence of any extension of growth having taken place to a distance of ½ inch from the original line of inoculation.

(2.) The soil was the same as that used in Experiment (1) of this series, but it was sterilized before inoculating it with Koch's vibrio.

The results are given in the following table :—

[The sign + signifies a positive, and the sign — a negative result, as regard the presence of Koch's vibrio.]

Number of Days subsequent to Inoculation with Koch's Vibrio when the Soil was Tested.	Needle-samples taken—	
	from Centre.	½ inch from Centre
4th day	+	—
7th "	+	
18th "	—	
20th "	+	
41st "	+	
232nd "	—	—

Here Koch's vibrio retained its vitality for 41 days, but not fo 232 days.

There was no evidence of any extension of growth having taken place beyond the original line of inoculation.

It is unnecessary to make any further comments on the vibrio experiments; they were carried out prior to inoculation of soils in "rure," and are to be regarded as preliminary only.

PART II.

Inoculation of Soil "in rure" with B. Prodigiosus and Koch's Vibrio.

The following is a summary of the several sections of further work, during 1898–99, in the inoculations of soil :—

1. Description of procedure with preliminary remarks.
2. Experimental work.
   - Series A. Inoculation of soil with Koch's Vibrio and B. Prodigiosus.
   - Series B. Second inoculation of soil with Koch's Vibrio.
   - Series C. Third inoculation of soil with Koch's Vibrio.
   - Series D. Fourth inoculation of soil with Koch's Vibrio.
3. Summary of results, and conclusions.
4. Addendum
   - A. Table of Meteorological observations.
   - B. Diagrams 1 and 2 illustrating the inoculated plots.
   - C. Figs. 1 to 12 illustrating the work.

1. Description of Procedure with Preliminary Remarks.

The plot of ground chosen for these experiments, is situated in the country some miles from London, on a gentleman's private estate.

The soil was originally dredged (about 1888), from the bed of a stream and deposited on marshy ground adjoining this stream. It was at one time manured but for the last 4 years no manure or other "dressing" has been applied to it. The plot of ground has a gentle slope of about 1 in 10 in a uniform direction. It lies in the "open," but there are trees and shrubs near which, to some extent and during certain hours of the day, shelter the ground from the direct rays of the sun. The soil may be regarded as of poor quality, and not rich in organic matter. Nevertheless, we have it on the authority of competent observers that not alone soils rich in decaying animal and vegetable organic matter are to be credited with harbouring the specific germs of cholera, but impoverished soils also; soils which so far as can be judged do not widely differ in their general character from the above. The nature of this soil was in some respects unfavourable to the experiments; it was not very open or porous, so that during heavy rains the surface became clogged and the rain was not absorbed sufficiently rapidly to prevent its running off the surface.

An area was marked out measuring 16 × 15 feet, and a trench was dug round it. In addition there was provided a central trench

App. B. No. 4.

On Inoculation of Soil with particular Microbes; by Dr. Houston.

dividing the area into two equal halves. These trenches subsequent to the inoculations of the soil were periodically watered with a strong solution of a powerful germicide. A special plot measuring 5 × 10 feet was marked out with pegs and stout string within each division of the area. The one plot was reserved for experiments with Koch's vibrio and the other for experiments with B. prodigiosus. The dimensions etc. of the plots are shewn in Diagram 1: the left-hand plot (1) was used for inoculation experiments with Koch's vibrio and the right hand one (2) for similar experiments with B. prodigiosus. Diagram 2 is a continuation of Diagram 1 in so far as the vibrio plot (1) is concerned.

The ground was prepared for inoculation by weeding the whole surface and slightly hoeing it to loosen the soil a little, and then raking it as flat as possible.

The experiments with B. prodigiosus, while of considerable interest in themselves, afford in more ways than one a valuable "control" to the experiments with Koch's vibrio. The B. prodigiosus was selected because it is readily isolated and identified in a mixture of bacteria of different sorts, and for the reason that, although a non-pathogenic organism, it, like Koch's vibrio, is a non-sporing microbe and one therefore which might be expected to be influenced by physical agencies to an extent comparable in some measure to Koch's vibrio.

The previously ascertained influence of physical conditions on one or other or on both of these micro-organisms may be described as follows:—

A temperature (moist heat) of about 52° C. for 10 minutes, is commonly fatal to Koch's vibrio, and the thermal death-point of B. prodigiosus is believed to be about 6 degrees higher.

B. prodigiosus does not grow after being frozen for 51 days (Prudden).

A culture of B. prodigiosus may retain its vitality for 18 months if preserved in a hermetically sealed glass tube (Sternberg). Koch's vibrio may be preserved in a moist state for seven months, according to Kitasato.

Dessication is very fatal to non-sporing bacteria and seems to be especially so to Koch's vibrio. Koch found that his "comma bacillus" lost its vitality after being dried upon a cover glass for only three hours.

Lastly, the optimum temperature for the growth of the Koch's vibrio is 37° C. and for the B. prodigiosus about 25° C.

Upon the whole therefore, it was to be anticipated that the B. prodigiosus would retain its vitality for a longer period than Koch's vibrio: That, for instance, the influence of light and particularly of direct sunlight; the effect of extremes of heat and cold; the comparative absence of organic matter and perhaps particularly of decomposing animal organic matter in the soil; the drying effect of wind and heat; and the competing influence of the soil bacteria would together tend to sooner inhibit the growth or destroy the vitality of Koch's microbe than that of the B. prodigiosus. Nevertheless it was believed that the parallel inoculations of the same sort of soil with B. prodigiosus and with Koch's vibrio might be of advantage.

2.—EXPERIMENTAL WORK.

Series A.—Inoculation of Soil with Koch's Vibrio and B. Prodigiosus.

May 15th, 1898.

Koch's Vibrio.*	B. PRODIGIOSUS.
Six oblique Agar tubes were inoculated over the whole surface in each case with a fresh culture of Koch's vibrio, and incubated at 37° C.	Six oblique Agar tubes were inoculated over the whole surface in each case with a fresh culture of B. prodigiosus, and incubated at 20° C.

May 21st.

The growth from these tubes was transferred to 200 cc. sterile water in a sterile bottle, and, on reaching the area to be inoculated, the mixture of bacteria and water was further diluted to 9 pints with rain water and then carefully distributed, by means of a watering can having a fine spray, equally over the whole area reserved for the vibrio experiments, namely, the plot (10 × 5 feet) numbered (1) in Diagram No. 1.	The growth from these tubes was transferred to 200 cc. sterile water in a sterile bottle, and, on reaching the area to be inoculated, the mixture of bacteria and water was further diluted to 9 pints with rain water and then carefully distributed, by means of a watering can having a fine spray, equally over the whole area reserved for the prodigiosus experiments, namely, the plot (10 × 5 feet) numbered (2) in Diagram No. 1.

From certain experiments previously carried out it is believed that in this way at least 100,000 million germs were sown on the plots of soil in each instance.

It cannot be too strongly insisted on that the culture medium on which the growth had taken place was not in either case added to the soil. Therefore, the two sorts of organisms carried with them very little pabulum, and had to depend for their future sustenance to the organic matter and salts in the soil itself.

June 4th.

Samples of soil were collected in sterile test tubes from the spot marked in Diagram No. 1 (1) as 1, 2, 3, 4. A fragment of soil from each of these tubes was added to each of four peptone tubes (peptone 1 %, sodium chloride 0·5 %), and the latter incubated at 37° C.	Samples of soil were collected in sterile test tubes from the spots marked in Diagram No. 1 (2) as 1, 2, 3, 4. Oblique agar tubes were inoculated from portions of these samples, and were incubated at 20° C.

June 6th.

Microscopic preparations, stained with gentian violet, were made from the surface layer of the liquid in the peptone tubes. By careful searching a few comma-like bodies were noted, but neither their appearance nor their numbers allowed of a positive result being recorded.	All showed a copious red growth.

June 10th

Samples of soil were obtained from spots marked 5, 6, 7 on diagram No. 1 (1), and peptone tubes were inoculated from these soils and incubated at 37° C. In some cases the soil was added directly to the peptone; in others the soil was first mixed with water, and the mixture of soil and water added to the peptone.	A sample of soil was collected from the spot marked 5 in Diagram No. 1 (2), and agar tubes inoculated and incubated as before.

* Culture kindly given me by Dr. Klein.

App. B. No. 4.

On Inoculation of Soil with particular Microbes; by Dr. Houston.

June 11th.

A careful microscopic examination of stained cover glass specimens made from the upper layers of the liquid in the peptone tubes yielded negative results.

June 13th.

A copious growth of Prodigiosus was noted.

June 21st.

A stout platinum needle (1¼" long) in a metal holder was, after sterilisation, stabbed downwards into the soil in ten different places in each of the four quarters of the inoculated plot (1) Diagram 1. Between each stab the needle was passed into the liquid contained in a peptone tube, and moved about so as to detach any adhering soil. Four peptone tubes were thus inoculated from each of the four quarters of the plot. They were incubated at 37° C. for one night. On microscopic examination a negative result was obtained.

Another sample of soil was obtained from spot marked 6 on Diagram No. 1 (2), and agar tubes inoculated.

June 23.—The result was negative; so further agar tubes were inoculated from the same sample of soil.

June 29th.—None of the tubes showed any red growth.

July 5th.

Samples of soil were collected from the prodigiosus plot at spots marked 7, 8, 9, 10 on Diagram 1 (2), and agar tubes were inoculated and incubated. The result was negative in each case.

July 12th.

A sample of soil was obtained from spot marked 11 on Diagram No. 1 (2). and on July 15th from spot 12, and agar tubes inoculated. The result was positive. but only a single colony of prodigiosus developed in each case.

July 19th.

A sample of soil was obtained from spot marked 13 on Diagram No. 1 (2). and agar tubes inoculated. The result was positive. but only a single colony of prodigiosus developed.

August 9th.

Another sample was obtained from spot marked 14 on Diagram No. 1 (2), and agar tubes inoculated. The result was negative, although the experiment was repeated.

October 26th.

The plot was by now overgrown with weeds to such an extent that it was necessary to have it weeded. Scrapings of surface soil from the spots marked × on Diagram No. 1 (2) were mixed with an equal amount of sterile water, namely, 300 grammes of soil and 300 cc. of water, and agar tubes inoculated from the mixture. A single colony of B. prodigiosus developed.

The B. prodigiosus had then survived 158 days.

December 30th.

Surface scrapings (341 grammes) were taken from the whole of the upper right quarter (marked with horizontal lines and circles on Diagram No. 1 (2)) of the prodigiosus area. The soil thus obtained was mixed with an equal amount of sterile water, and agar tubes were inoculated from the mixture. The result was negative.

January 14th, 1899.

Surface scrapings (530 grammes) were taken from the whole of the left lower quarter (marked with oblique lines running from right to left from above downwards on Diagram No. 1 (2)) of the prodigiosus area. The soil thus obtained was mixed with an equal bulk of sterile water, and a number of "oblique" agar tubes were inoculated from the mixture. The result was again negative.

February 23rd.

Surface scrapings of soil (210 grammes) were taken from the whole of the upper left quarter (marked with oblique lines running from left to right from above downwards on Diagram No. 1 (2)) of the prodigiosus area. The soil thus obtained was thoroughly mixed with 500 cc. sterile water, and a number of "oblique" agar tubes were inoculated from the mixture of soil and water. The result was negative.

March 2nd.

Surface scrapings of soil (495 grammes) were taken from the whole of the right lower quarter (marked with vertical lines and black dots on Diagram No. 1 (2)) of the prodigiosus area. The soil thus obtained was thoroughly mixed with 500 cc. sterile water, and a number of "oblique" agar tubes were inoculated from the mixture of soil and water. The result was negative.

Series B.—Second inoculation of Soil with Koch's Vibrio.*

June 25th.

Six oblique agar tubes were inoculated over the whole surface in each case with a fresh culture of Koch's vibrio and incubated at 37° C. for one night and afterwards at 20° C.

July 5th.

The growth, was as before, transferred to sterile water, and on reaching the vibrio plot, further diluted to nine pints. But on this occasion, instead of inoculating the whole plot, only the upper left quarter (marked with oblique lines running from right to left from above downwards on Diagram No. 1 (1)) was seeded with Koch's microbe. Immediately after inoculation a sample of soil was taken from spot marked 8 on Diagram No. 1 (1) and in the evening of the same day three peptone tubes were inoculated and incubated at 37° C.

July 6th.—A positive result was obtained on microscopic examination without any difficulty. This would seem to indicate that the method was not at fault, and that when Koch's vibrio was alive and present in any number it could be *readily identified in a mixture of bacteria of different sort.*

July 7th.

Another portion of soil was obtained from the spot marked 9 on Diagram No. 1 (1), and a peptone tube was inoculated therewith and placed in the incubator at 37° C.

July 8th.—On microscopic examination a positive result was obtained without any difficulty.

July 12th.

Another sample of soil was taken from the spot marked 10 on Diagram No 1 (1), and two peptone tubes inoculated and incubated as before.

July 13th.—Here a negative result was obtained on microscopic examination.

On July 14th and 19th

fresh samples were taken from spots marked 11 and 12 on Diagram No. 1 (1), and peptone tubes inoculated as in the previous experiments. But in both cases the result was negative.

August 9th.

The whole of the area last inoculated with Koch's vibrio, namely, the area marked with oblique lines running from right to left from above downwards on Diagram No. 1 (1), was scraped with a sterile tin scoop into a sterile tin vessel. 250 grammes of this soil were then mixed with 250 cc sterile water, and 1 cc of the mixture of soil and water was added to each of five peptone tubes. The tubes were incubated as before at 37° C.

---

* Antecedent to each further inoculation (2nd, 3rd and 4th) the soil was prepared by digging, hoeing, and raking. The weeds were burnt.

APP. B. No. 4.

On Inoculation of Soil with particular Microbes; by Dr. Houston.

August 10th.—The result was practically negative, although a few organisms were noted in the microscopic stained preparations which might conceivably be the progeny of Koch's microbe somewhat altered in its morphological character by sojourn in soil.

On August 25th the remainder of the soil (380 grammes) was mixed with 380 cc. of water, and seven peptone tubes were inoculated with the mixture of soil and water. After incubation for one night at 37° C., microscopic stained preparations were made from the tube, and very carefully examined. The result, however, was quite negative.

The stoppered bottle containing the above mixture of soil and water was kept until October 20th at room temperature, when fresh peptone tubes were examined. The result was negative.

Series C.—Third inoculation of soil with Koch's Vibrio.*

October 23rd,

Six oblique agar tubes were inoculated as previously with a fresh culture of Koch's vibrio.

October 26th.—The growth, as before, was transferred to a small bulk of sterile water and subsequently diluted to 9 pints. The mixture of bacteria and water was then watered over the right upper eighth part of the vibrio plot—marked with black cross hatching on diagram No. 1 (1). Immediately after inoculation, a sample of soil was taken from the spot marked A on the diagram and a peptone tube inoculated and incubated at 37° C.

October 27th.—On microscopic examination of stained cover glass specimens made from the surface layer of liquid in the peptone tube a positive result was obtained without any difficulty.

October 27th.

Another sample of soil was taken from spot marked B on diagram No. 1 (1), and a peptone tube inoculated as before. The result was disappointing as on microscopic examination of the liquid on the following day a positive diagnosis could not be arrived at, although there were present a few organisms, which were not unlike Koch's microbe. Another fragment of this B soil (so-called) was thereupon placed in another peptone tube and after incubation for one night at 37° C., examined for the presence of the vibrios. The result was quite negative.

On October 28th, 29th, and 31st,

fresh samples of soil were obtained from spots marked C, D, E, on diagram No. 1 (1), and three peptone tubes were inoculated and then incubated at 37°. After one night at 37° C., the contents of the tubes were examined microscopically. As regards C and D, a practically negative result was obtained, but in the case of E, the result was beyond doubt positive.

November 8th.

Portions of soil were taken from spots marked F, G, H, I, on diagram No. 1 (1), and four peptone tubes were inoculated. These, after one night's incubation at 37° C., were examined microscopically, but with negative result. On repeating experiments a similar negative result was obtained.

November 15th.

The surface layers of soil from the whole of the area last inoculated, namely, from the right upper eighth part of the plot (marked with black cross hatching, on Diagram No. 1 (1)), were, by means of a sterile tin scoop, transferred to a sterile tin vessel. Of this soil, 500 grammes were mixed with 500 cc. of sterile water and shaken-up in a wide mouthed sterilised glass bottle. Three peptone tubes were then inoculated with the mixture, and after being incubated at 37° C. for 15 hours, were examined microscopically. The result in each case was negative. So 1 cc. of the supernatant fluid in the bottle (the gross particles having subsided to the bottom of the bottle) was inoculated into a peptone tube. After incubating the tube for one night at 37° C., stained cover glass specimens were made from the surface layers of the liquid. Here, a positive result was obtained. See Fig. 1, Plate X. . The experiment was repeated on the following day, namely, November 18th, 1 cc. of the supernatant fluid being again used. The result, however, was practically negative.

* Antecedent to each further inoculation (2nd, 3rd, and 4th) the soil was prepared by weeding, hoeing, and raking. The weeds were burnt.

Series D.—Fourth inoculation of soil with Koch's Vibrio.

January 14th, 1899.

The whole of the left upper quarter (marked with black shading on Diagram No. 1 (1), and with a rectangular black line on Diagram No. 2) of the vibrio plot was watered with eight quarts of diluted liquid horse manure, namely, liquid horse manure one part, water nine parts. Afterwards this area was watered with nine pints of water containing a large number of Koch's vibrio: the bacterial growth having been obtained, as before, by the inoculation of six oblique agar tubes with this microbe. The growth on the surface of these agar cultures, after incubation for three days at 37° C., was transferred to 200 cc. distilled water, and, on reaching the site of experiment, further diluted with about nine pints of rain water. The "dressing" of the soil with manure was repeated daily until January 28th, afterwards the manure was applied twice weekly until February 23rd.

A sample of soil was taken from the spot marked I. on Diagram No. 2, immediately after the inoculation of the soil had been made. A peptone tube was inoculated therewith in the usual way, and, after one night's incubation at 37° C., examined microscopically. Koch's vibrios were present in enormous numbers. See Fig. 2, Plate X.

January 15th.

Another sample of soil was obtained from spot marked II, and on January 16 a sample from place marked III, on Diagram No. 2. Peptone tubes were inoculated as before. The result was somewhat disappointing in the case of sample II, as comparatively few "commas" could be found; but, in sample III, they were present in great numbers. See Fig. 3, Plate X.

January 17th.

A sample of soil was taken from the spot marked IV, on Diagram No. 2, a peptone tube inoculated therewith, and after incubation at 37° C. for one night, examined as in the previous experiments. The result, however, was negative; and a repetition of the experiment yielded no better result.

January 18th.

A sample was obtained from spot marked V on Diagram No. 2, and in this instance a positive result was obtained by the peptone method. See Figs. 4 and 5, Plates X. and XI.

On January 19th, 20th, 21st, 23rd and 24th,

samples were obtained from spots marked VI., VII., VIII., IX., X. on diagram No. 2, and peptone tubes inoculated therewith. In the case of samples VIII. and X., the soil was ground in a mortar with a small quantity of sterile water, and a small amount of the mixture of soil and water added to peptone tubes. The other samples were either added directly in small amount to the peptone tubes, or else first shaken up with a little sterile water and several platinum loopfuls of the mixture added to the peptone.

The result was negative as regards VI., VII., VIII. and IX. As regards soil X., a positive diagnosis was arrived at, but only with great difficulty. See Fig. 6, Plate XI. , showing a long spiral, possibly (perhaps probably) a spirillum of Koch's microbe.

January 25th.

A sample from spot marked XI. on Diagram No. 2, yielded a positive result. See Fig. 7. Plate XI. . In this instance, the soil was shaken with 100 cc. sterile water and filtered, and to the filtrate there was added an amount of peptone and sodium chloride corresponding to 1 per cent. of the former and 0·5 per cent. of the latter.

January 26th.

Surface scrapings were taken from the whole of the right lower quarter (marked with black shading on Diagram No. 2) of the area last inoculated. Of this soil 190 grammes were shaken with 500 cc. of sterile water and filtered. To the filtrate there was added peptone (1·0 per cent.), and sodium chloride (0·5 per cent.) After incubation at 37° C. for one night, the surface layers of the liquid were examined microscopically. No great difficulty was experienced in arriving at a positive diagnosis; there were present "commas," S-shaped forms, and spirilla. See Figs. 8, 9, 10, Plates XI. and XII.

3, No. 4.

ation of
th
ilar
es; by
uston.

February 2nd.

Surface scrapings of soil (355 grammes) were taken from the left lower quarter (marked with black cross hatching on Diagram No. 2) of the area last inoculated, and treated as in the previous experiment. Here, however, the result was negative.

February 16th.

Surface scrapings of soil (400 grammes) were taken from the upper left quarter (marked with wavy lines on Diagram No. 2) of the area last inoculated, and on February 17th, this soil was thoroughly mixed with 600 cc. of sterile water, in a tall sterilised glass cylinder. After standing for 4 hours, 100 cc. of the water were siphoned off into a sterile vessel from near the surface, and 1 per cent. of peptone and 0·5 per cent. of sodium chloride were added.

February 18th.—After 15 hours incubation at 37° C., stained cover glass preparations were made from the surface layer of the liquid and examined microscopically. No "commas" or S-shaped forms, or spirilla could be found even after prolonged searching.

February 23rd.

Surface scrapings of soil (270 grammes) were taken from the upper right quarter (marked with black dots on diagram No. 2) of the area last inoculated, and on February 29th the soil was thoroughly mixed with 730 cc. sterile water in a tall sterilised glass cylinder. After standing for about 4 hours, 100 cc. of the water were siphoned off from near the surface into a sterile vessel, and 1 per cent. of peptone and 0·5 per cent of sodium chloride added. After incubation overnight at blood-heat, the upper layers of liquid were examined microscopically in stained cover glass preparations. Although at first a negative result was obtained, after prolonged searching some organisms were seen, which might conceivably be those of Koch's microbe somewhat altered by sojourn in soil, and a very few others which were sufficiently like this bacillus as to suggest possible identity. On the whole the diagnosis was only doubtfully positive, because in order to arrive at a satisfactory conclusion it is necessary to find a number of "commas" or S-shaped forms or spirilla, unless indeed those present are altogether typical of Koch's vibrio. Nevertheless, a few forms were noted which were very like the comma bacillus, and had these been present in any number the diagnosis would have been positive, and that without any slight element of doubt. See Figs. 11 and 12, Plate XII.

With reference to the micro-photographs in addendum to this report, it is particularly to be noted that they represent in each case only one "field" or portion of a field, whereas in the actual examination of the specimens a very large number of different "fields" were subjected to strict scrutiny under the microscope. So that it by no means necessarily follows that the diagnosis as stated in the text was arrived at by the appearance shown in a single field and represented in a particular micro-photograph; on the contrary, it was usually arrived at by the prolonged observation of a plurality fields under the microscope.

Before giving a summary of the results it is useful to contrast the two methods employed in judging of the presence of the Koch's vibrio and B. prodigiosus respectively.

There can be little doubt that good as the peptone method is for the demonstration of Koch's vibrio in a mixture of bacteria of different sorts, the characteristic bright red growth produced by B. prodigiosus in agar culture, would enable the observer to discover the latter when present in such mixture in smaller proportion than the former. In the case of B. prodigiosus a single germ implanted on a solid medium, would develop in a few days into a visible and characteristic colony of prodigiosus. Whereas a single Koch's vibrio inoculated into a liquid together with a host of other bacteria of various sorts might readily lose its vitality, or at all events multiply so slightly as to be lost amid the numerous other micro-organisms. It might have been possible in the above experiments to have isolated Koch's vibrio from the soil samples by using solid media, either directly, or indirectly after preliminary growth in peptone. But the number of germs in soil (*rarely less than 1 million per gramme*), leads one to doubt if this end could readily be achieved *except in those cases where the comma bacillus was present in such numbers as to enable one to arrive at a positive diagnosis by the peptone method without any difficulty*. It is perhaps, worthy of note, that the object of these experiments was not so much to strain at the demonstration of a single vibrio after so many days, but rather to determine broadly whether, in the case of a soil inoculated on the surface with many millions of Koch's vibrio, these same vibrios or their progeny retained their vitality for any considerable period of time.

Whether the peptone method is altogether satisfactory or not, may be open to question, but it seems quite clear that if Koch's vibrio is present in any quantity in a mixture of bacteria of various sorts it can be demonstrated with ease and certainty by this method. See Figs. 2, 3, and 4, Plate X.

Before leaving this subject it is important to note the fact that many microbes when placed under a variety of conditions, but chiefly when placed under unfavourable conditions, tend to alter in their morphological and biological characters; and that such lateration, howsoever it may show itself, may persist even when the microbe is restored to a suitable medium, sometimes even when subcultures covering a considerable period of time and allowing of the development of numerous progeny are resorted to. Dr. Klein has shown this as regards vibrios, and it is known that the B. prodigiosus may be so grown as to show white instead of red colonies. It is quite possible therefore that in the above experiments, when Koch's vibrio could no longer be demonstrated, the microbe had during its sojourn in soil become so altered, morphologically, as not to retain any close resemblance to its progenitors. Even however if this had happened, it would not have been safe to infer that there would likewise have occurred diminution in its virulence, although, perhaps, loss rather than exaltation of pathogenic power is likely to be associated with a departure from the normal, so far as may be judged from experience of cholera stools and subculture therefrom. Certainly in the above experiments—and this is shown to a small extent in some of the photographs—it almost seemed as if each successive sample of soil displayed microbes having less and less morphological resemblance to the original stock while yet showing sufficient similarity to merit a positive result being recorded. As it seemed to me the length and thickness of the spirilla increased and the "commas" became thicker and plumper as time went on.

As regards B. prodigiosus, it is, although very unlikely, conceivable that the failure to detect its presence after a considerable period of time was due not to loss of vitality, but to the micro-organism having lost its chromogenic property. Surmises, however, such as the above, must needs be advanced tentatively only; they are not to be entertained unless supported by the results of a long series of investigations.

It is convenient at this stage to draw attention to the table of Meteorological observations in addendum to this report.

These observations were taken at a spot within ½ mile of the plot of land used for the inoculation experiments, and the data were kindly supplied by the gentleman who is in charge of the meteorological station. They have an important bearing on the results of the inoculation experiments; for they indicate that the physical conditions at or near the time when the inoculations were made, and afterward during the time that samples of soil were being collected for examination, were presumedly often very unfavourable. To make these references easily understood, the plan has been adopted of marking on the table the dates of inoculation and subsequent collection of samples of soil in thicker type.

3.—SUMMARY OF RESULTS.

This Summary is of the briefest; reference should to be made to the details stated in the text in order to form a correct estimate of the value of the observations.

KOCH'S VIBRIO.

A.—First Inoculation.

14th day after inoculation—negative.
20th „ „ ditto.
31st „ „ ditto.

B.—Second Inoculation.

Same day. Positive; diagnosis made without any difficulty.
2nd day after inoculation, ditto.
7th day—negative.
9th day „
14th day „
35th day—negative, although a few organisms were noted in the microscopic stained preparations which may have conceivably represented the progeny of Koch's microbe somewhat altered morphologically by its sojourn in soil.

C.—Third Inoculation.

Same day—positive; diagnosis made without any difficulty.
1st day after inoculation—practically negative.
2nd day—practically negative.
3rd day—ditto.
5th day—positive, beyond doubt.
13th day—negative.
20th day—positive, but result arrived at with great difficulty.

B. PRODIGIOSUS.

A.—First and only Inoculation.

14th day after inoculation—positive; copious red growth.
20th day—positive; copious red growth.
25th day—negative.
39th day „
46th day—positive, but only one colony.
49th day „ „ „
53rd day „ „ „
74th day—negative.
158th day—positive, but only one colony
223rd day—negative.
238th day „
278th day „
285th day „

D.—Fourth Inoculation

[In this case the soil was periodically dressed with liquid horse manure.]
Same day—positive, diagnosis made without any difficulty.
1st day after inoculation—positive, but diagnosis made with some difficulty.
2nd day—positive, diagnosis very readily made.
3rd day—negative.
4th day—positive.
5th day—negative.
6th day „
7th day „
9th day „
10th day—positive, diagnosis made with great difficulty.
11th day—positive.
12th day „
19th day—negative.
33rd day „
40th day—positive (?)

It will be seen from the above summary that B. prodigiosus survived and was present in sufficient number in the soil to be capable of demonstration during 158 days, but that there was distinct evidence of a rapid diminution in the number of these germs. Assuming for the moment that no multiplication of the bacillus took place it is evident that the rain did not, as a matter of fact, wash off the surface, or deeper down into the soil, during that period all the prodigiosus germs originally sown on the soil. Wherefore, it cannot well be maintained that the difficulties experienced in demonstrating the cholera vibrio soon after the inoculations were made was due simply to a mechanical washing away of the microbes. Nor, since B. prodigiosus did not soon die out, can it readily be held that the physical conditions were so eminently unfavourable that no microbes other than those present in spore form, or those "peculiar" to the soil, could possibly survive. Of course it may have happened that the number of B. prodigiosus was kept up to some extent by repeated multiplication of the microbe, and that this did not take place in the case of the vibrio. Probably the loss from unfavourable physical conditions such as direct sunlight, extreme cold, lack of humidity, &c., was greater with Koch's vibrio than with B. prodigiosus, and possibly also the final extinction of the latter was delayed for a longer period by multiplication later on of the surviving germs when the conditions became after a time less unfavourable. It must also be remembered that the peptone method used for demonstrating the presence of the vibrio is probably less delicate as a test than mere recognition of the characteristic red colonies in oblique agar culture in the case of B. prodigiosus.

Turning now to the experiments with Koch's vibrio,* it is to be

* In judging these results it is important to bear in mind that the experiments were not always strictly parallel. Thus, sometimes only a fragment of soil was used, and at other times a large quantity of soil was mixed with water and a portion of this mixture of soil and water added to the peptone broth. Again, sometimes the surface scrapings of soil from the whole or a large portion of the inoculated area were mixed with water, and the comparatively clear liquid (after most of the soil had sunk to the foot of the vessel) used to breed the bacteria in, by adding peptone and salt to it instead of adding a small quantity of the soil mixture to previously prepared peptone and salt solution.

APP. B. No. 4.

On Inoculation of Soil with particular Microbes: by Dr. Houston.

noted that as early as the 14th day after the *first inoculation* (and subsequently on the 20th and 31st day) the vibrio could not be demonstrated. Yet in the prodigiosus series on the 14th day that bacillus was still present in great numbers, and was isolated as late as the 158th day. A reference to the table of Meteorological Observations reveals nothing of note about the time of this sowing as regards the temperature, but that as regards the rainfall, very heavy rain fell two days after the germs were sown.

Immediately after the *second inoculation* of the vibrio plot had been made a sample of the soil was taken, chiefly with the object of proving that in a mixture of soil bacteria of varied sort Koch's vibrio could be readily demonstrated if present in any number. This sample and another sample yielded distinctly positive results. Nevertheless, on the 7th, 9th, and 14th day after inoculation the result was quite negative. On the 35th day the result of examination was not wholly negative; that is, a positive diagnosis was permissible only by inferring morphological changes in the vibrio as a result of its sojourn in soil.

A reference to the meteorological table does not, as to conditions obtaining during this series of experiments, throw much light on the subject, unless it be considered that the rainfall was deficient in amount in relation to the somewhat warm weather prevailing at the time.

In the case of the *third inoculation* of the vibrio plot, although there was no difficulty in obtaining a positive result from the examination of a sample of soil collected immediately after the act of inoculation, yet on the very next day a further sample yielded a practically negative result. On the 5th day the result was beyond doubt positive, and on the 20th day after inoculation a like result was obtained, but with great difficulty. As regards the meteorological conditions at this time it may be worthy of note that 0·85 inches of rain fell on the 3rd day after inoculation, and that on several occasions the temperature fell to about the freezing point. At the date of this, the third vibrio inoculation experiment, the B. prodigiosus was still alive, 158 days after the primary and only inoculation of the soil with this microbe.

As regards the *fourth inoculation* of the vibrio plot, the soil preliminary to inoculation and afterwards (first daily and later bi-weekly) was watered with a diluted liquid horse manure. The climatic conditions were unfavourable. The soil at the time of inoculation was saturated with water, and the day after the inoculation 0·36 inch of rain fell. On the two days following the rainfall was 0·14 and 0·10 inch. Moreover, the conditions of temperature were likewise unfavourable, the minimum thermometer registering 32° F. or under on a number of occasions. A positive result was however recorded on the 1st, 2nd, 4th, 10th, 11th and 12th days after inoculation. The result was negative on the 19th and 23rd day, but on the 40th day after inoculation a few micro-organisms were found in the microscopic stained preparations which resembled Koch's vibrio somewhat closely.

It is difficult to arrive at a correct estimate of the value of these results, and it would be unwise to indulge in inferences which a more prolonged study of the subject might prove to be premature.

It is clear indeed that the prodigiosus bacillus can survive in the *surface layers** of some soils for at least 158 days; and, making every possible allowance for mere mechanical washing away of the germs originally sown on the soil, it seems equally certain that if multiplication of the microbe takes place at all it does so at an unequal rate as compared with the rate of decay and death of the species.

Judging by the results obtained with B. prodigiosus it would be rash to infer that the failure to demonstrate the presence of Koch's vibrios in the soil a comparatively short time after inoculation was due to the mere mechanical washing away of the implanted germs; and it has to be borne in mind that the presence of B. Prodigiosus in soil is more easily demonstrated and with greater certainty than the presence of Koch's microbe.†

Making due allowance for the difference of method of test being probably most favourable to the demonstration of B. prodigiosus, it seems likely that if the original germs of the two species of microbes (vibrio and prodigiosus) multiplied at all in the soil, multiplication was greatest in the case of B. prodigiosus; and further, it appears all but certain that decay and death affected the prodigiosus germ least, and that unfavourable physical conditions operated most powerfully against the pathogenic vibrio of Koch.

A provisional interpretation of these results leads to the inference that both the Koch's vibrio and bacillus prodigiosus rapidly decrease in number in the *surface layers* of soil; that the B. prodigiosus may retain its vitality there for 158 days; but that Koch's vibrio dies quickly, or becomes so reduced in number as to be no longer capable of being demonstrated. This occurs to the vibrio in a few days in "undressed" soils; in about 12 days (possibly, however, 40 days—*see* notes) in soils periodically watered with liquid manure.

Any such inference must, however, be regarded as applying only to the particular soil chosen for these experiments; it does not necessarily apply to other soils, other seasons of the year, other climates, and other conditions of experiment.

With regard to some, at all events, of the micro-photographs, it may be contended that other vibrios in soil besides Koch's microbe might possibly have given rise to the appearance presented in the photographs; that a positive identification of this vibrio could only be arrived at by isolation of the microbe in pure culture, the examination of microscopic stained preparations being open to fallacy. All this is true. Yet it cannot be without significance that on so many occasions an entirely negative result was obtained, and that on others the appearances were so far characteristic of Koch's microbe as to leave no reasonable ground for doubt. Moreover, a positive result was not recorded, or was recorded only with a query attached to it, whenever only a few suspicious germs were noticed in the preparations.

---

* It must be insisted on that these experiments deal only with the vitality of B. prodigiosus and Koch's vibrio in the surface layers of soil.

† Nevertheless, it is beyond dispute that the peptone method will reveal the presence of Koch's vibrios when these are present in any number. See, however Figs. 2, 3, 4, Plate X.

T. B. No. 4.

...oculation of ... with ...ticular ...robes; by Houston.

TABLE of METEOROLOGICAL OBSERVATIONS taken within half a mile of the place where the KOCH'S VIBRIO and B. PRODIGIOSUS EXPERIMENTS were carried out.

Height of Station above O.D. 140 feet. Height of Rain Gauge above ground 4 feet 1 inch.

	May 1898.			June 1898.			July 1898.			August 1898.			September 1898.		
	Rain.	Max.	Min.	Rain.	Max.	Min.	Rain.	Max.	Min.	Rain.	Max.	Min.	Rain.	Max.	Min.
	Ins.	F. °	F. °	Ins.	F. °	F. °	Ins.	F. °	F. °	Ins.	F. °	F. °	Ins.	F. °	F. °
1	·21	54·0	41·2	·15	55·9	40·2	·15	71·0	53·0	..	76·3	48·2	..	67·8	40·2
2	..	66·7	45·3	·04	59·4	44·4	..	73·0	57·3	..	74·3	52·4	..	73·7	44·8
3	·13	63·9	47·5	..	61·9	42·3	..	69·0	49·4	·03	72·2	55·8	..	82·0	55·2
4	·01	63·8	48·3	..	**61·9**	**47·4**	..	64·0	49·4	..	70·9	49·3	..	83·1	53·4
5	·25	55·0	44·6	..	70·6	47·0	..	**70·8**	**47·1**	·01	71·1	54·4	..	77·7	56·1
6	..	62·8	45·4	·10	67·8	51·1	..	75·0	57·2	·01	68·0	52·3	..	80·5	61·4
7	·01	70·4	40·2	..	68·9	51·4	..	**75·9**	**55·1**	·56	54·6	51·2	..	87·6	58·4
8	·05	56·8	47·4	..	64·9	46·6	..	70·0	52·3	..	58·3	44·5	..	88·9	63·2
9	·03	61·8	49·1	·11	66·9	55·3	..	62·0	48·5	..	**67·2**	**56·9**	..	86·1	66·6
10	·13	60·9	48·5	**·03**	**62·8**	**54·1**	..	58·0	51·1	..	66·2	50·9	..	74·3	56·5
11	..	56·8	48·5	..	70·9	48·0	..	70·0	42·1	..	76·1	57·3	·12	75·6	54·2
12	·10	53·9	44·3	..	55·0	49·2	..	**76·9**	**48·2**	..	81·1	55·2	..	69·4	56·5
13	·19	54·2	35·6	..	55·9	46·6	·03	68·0	57·2	..	82·6	63·8	·01	68·2	45·4
14	·07	61·2	44·2	..	54·0	47·5	..	**76·2**	**42·5**	..	83·6	59·7	..	77·2	58·4
15	·09	54·4	41·3	..	59·2	46·4	..	**82·3**	**51·9**	·13	83·6	63·0	..	80·2	51·1
16	..	55·3	42·5	..	61·8	45·5	..	80·6	58·4	..	79·5	60·5	..	82·8	51·2
17	..	57·4	35·2	..	67·8	45·3	..	76·1	54·1	..	75·1	58·4	..	86·8	59·1
18	·02	55·6	37·1	..	72·3	50·6	·04	78·7	57·3	·02	76·2	55·3	·11	66·9	61·3
19	·00	49·8	39·0	..	69·7	59·4	..	**71·4**	**50·4**	..	82·5	58.7	·01	64·1	39·2
20	·34	53·6	41·5	..	68·9	52.3	..	65·8	53·4	..	79·7	57·3	..	73·2	55·9
21	..	**61·3**	**48·5**	..	**67·2**	**59·5**	..	71·3	42·6	..	80·2	56·1	..	72·9	48·0
22	..	68·7	45·2	..	65·9	55·3	·37	70·9	59·0	·02	86·7	60·9	..	67·7	45·3
23	·41	72·3	44·5	·04	64·7	46·2	..	71·2	57·3	..	74·1	59·1	..	64·5	40·5
24	..	63·9	53·0	·10	59·3	52·5	..	74·3	52·4	..	70·1	56·4	..	58·1	34·5
25	..	58·3	46·5	·27	61·3	51·5	..	71·7	52·3	..	69·8	50·0	..	61·5	40·4
26	..	54·8	44·0	·38	60·2	47·1	..	71·1	60·9	..	69·9	47·5	..	63·2	34·2
27	·03	61·2	43·0	·18	59·9	46·5	·19	75·9	58·3	·09	66·2	60·8	..	64·0	36·2
28	..	58·8	46·1	..	64·9	49·7	·24	71·8	58·1	·11	66·9	51·1	..	64·0	43·0
29	·11	61·8	45·1	·02	66·4	51·6	..	58·0	51·3	·06	65·9	48·3	·24	63·0	35·7
30	·03	59·8	48·3	·01	64·9	56·3	..	64·5	49·5	·06	75·7	56·2	..	57·0	52·2
31	·25	59·8	51·4	..	..	..	..	76·4	44·8	..	69·2	53·9	..	..	..

NOTE.—The thermometers are certified at Kew and are exposed in a Stephenson's screen. The rain gauge is certified by Mr. Symons. The station is also regularly inspected.

[The above data were kindly supplied by a friend of the gentleman on whose estate the inoculation experiments were carried out.]

The dates of inoculation and subsequent collection of samples of soil are shown in thicker type.

October 1898.		November 1898.			December 1898.			January 1899.			February 1899.		
Max.	Min.	Rain.	Max.	Min.	Rain.	Max.	Min.	Rain.	Max.	Min.	Rain.	Max.	Min.
F. °	F. °	Ins.	F. °	F. °	Ins.	F. °	F. °	Ins.	F. °	F. °	Ins.	F. °	F. °
63·2	35·7	..	**55·8**	**34·3**	..	51·3	38·5	·08	42·7	13·4	..	40·9	28·7
65·0	38·5	·20	59·4	44·0	·06	54·7	49·0	·09	41·9	35·0	..	**38·2**	26·7
67·2	41·7	·06	69·8	53·1	..	55·2	48·0	..	49·1	34·4	..	40·8	26·9
61·9	54·9	..	55·1	42·4	·02	56·9	49·0	..	52·7	39·4	·35	39·6	24·7
59·3	57·1	..	57·7	46·4	..	54·2	52·2	..	47·8	33·5	·03	39·1	34·2
60·7	57·3	..	57·3	35·7	1·02	55·2	52·6	..	45·4	30·3	·31	46·1	35·3
58·9	53·3	..	52·7	42·7	·43	45·4	45·0	·03	50·4	41·2	·17	55·2	35·6
56·9	52·3	..	**60·1**	**37·8**	·07	50·9	40·2	..	53·9	44·2	·59	53·6	46·1
60·9	39·5	..	57·2	43·2	·04	52·1	39·2	·03	52·8	45·2	·11	58·0	47·9
55·7	47·2	·01	56·7	48·8	..	53·7	42·6	·14	49·4	46·4	..	63·1	50·1
51·9	41·3	·01	56·6	41·1	·05	54·6	51·3	·26	50·8	36·5	·15	54·4	49·4
57·0	39·3	·01	55·1	39·4	..	53·7	49·5	·60	54·2	37·3	·23	50·6	42·4
51·7	37·2	·01	56·8	45·8	..	46·2	40·5	·41	53·8	41·2	·09	52·7	43·3
56·8	40·3	..	55·1	31·8	**·03**	**50·9**	**35·1**	..	**47·7**	**38·2**	·01	50·9	44·3
50·7	48·1	**·02**	**52·8**	**32·3**	..	49·3	42·1	**·36**	**54·2**	**37·1**	·35	50·1	42·1
58·4	47·9	..	56·8	48·6	..	51·4	36·3	**·14**	**52·0**	**42·6**	..	**51·2**	**38·1**
61·9	49·2	..	53·9	48·5	..	58·7	48·2	**·10**	**50·1**	**32·5**	..	55·9	[illegible]
60·2	53·4	..	52·3	47·1	·01	54·9	49·4	**·01**	**52·3**	**33·2**	..	55·1	34·2
56·8	49·4	..	47·4	42·2	..	47·2	48·4	**·05**	**52·8**	**48·4**	·03	53·2	34·9
60·6	46·5	·11	48·6	41·5	·08	40·7	31·2	**·20**	**53·9**	**45·2**	..	45·7	42·3
64·7	52·3	·64	46·7	43·6	..	41·4	31·0	**·20**	**54·3**	**46·9**	..	46·4	39·2
63·2	57·9	·02	38·8	35·9	..	46·3	27·2	·05	52·1	49·3	..	49·7	31·8
64·2	49·3	·61	40·4	27·2	..	42·2	27·2	**·04**	**41·9**	**41·4**	..	**53·1**	**27·7**
60·2	49·4	·40	46·3	37·1	..	48·2	30·2	..	**40·1**	**34·2**	..	53·4	26·8
58·2	46·2	·49	50·3	40·4	..	49·8	25·2	..	**38·5**	**25·2**	..	43·3	26·2
61·2	**51·5**	·07	45·6	40·5	..	54·1	42·3	..	**40·8**	**25·5**	..	43·8	23·5
59·9	**53·1**	..	45·9	40·1	·56	54·2	46·5	..	40·9	33·7	[illegible]	43·8	24·5
60·6	**50·2**	·38	40·9	32·0	·05	46·3	39·7	..	39·3	27·4	..	49·8	22·2
59·8	**50·9**	·01	40·3	32·5	·16	50·8	36·6	·05	41·0	32·4	..	..	..
55·4	47·4	·05	49·0	28·3	..	**46·3**	**37·5**	..	42·0	37·1	..	..	..
56·3	**48·2**	..	..	..	·30	44·4	26·5	..	36·7	28·8	..	..	..

March 1899.	
Max.	Min.
F. °	F. °
56·1	26·3
51·8	**38·2**
52·0	28·3

NOTE.—The thermometers are certified at Kew and are exposed in a Stephenson's screen. The rain gauge is certified by Mr. Symons. The station is also regularly inspected.

[The above data were kindly supplied by a friend of the gentleman on whose estate the inoculation experiments were carried out.]

The dates of inoculation and subsequent collection of samples of soil are shown in thicker type.

APP. B N

On Inoculatio Soil with particular Microbes; Dr. Housto

SOIL INOCULATION.

PLATE X.

Koch's Vibrio sown on soil *in rure.*

FIG. 1.

Stained film specimen from a peptone* culture of a sample of the soil collected on the *twentieth day* after the third sowing.

FIG. 2.

Stained film specimen from a peptone culture of a sample o the soil collected *on the day* of fourth sowing.

FIG. 3.

Stained film specimen from a peptone culture of a sample o the soil collected on the *second day* after the fourth sowing.

FIG. 4.

Stained film specimen from a peptone culture of a sample of the soil collected on the *fourth day* after the fourth sowing.

[Magnifying power, in each instance, 1,000.]

---

* Peptone, 1 per cent.; sodium chloride, 0·5 per cent.

PLATE X.

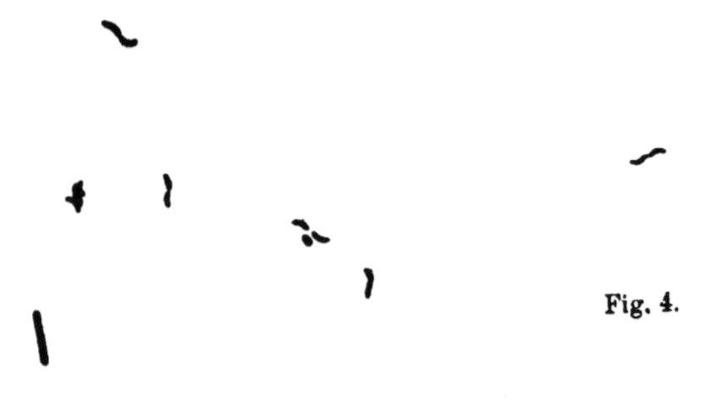

Fig. 1.

Fig. 2.

Fig. 3.

Fig. 4.

PLATE XI.

Fig. 5.

Fig. 6.

Fig. 7

Fig. 8.

SOIL INOCULATION.

PLATE XI.

Koch's Vibrio sown in soil *in rure.*

FIG. 5.

Stained film specimen from a peptone* culture of a sample of the soil collected on the *fourth day* after the fourth sowing.

FIG. 6.

Stained film specimen from a peptone culture of a sample of the soil collected on the *tenth day* after the fourth sowing.

FIG. 7.

Stained film specimen from a peptone culture of a sample of the soil collected on the *eleventh day* after the fourth sowing.

FIG. 8.

Stained film specimen from a peptone culture of a sample of the soil collected on the *twelfth day* after the fourth sowing.

[Magnifying power, in each instance, 1,000.]

---

* Peptone, 1 per cent. ; sodium chloride, 0·5 per cent

SOIL INOCULATION.

PLATE XII.

Koch's Vibrio sown on soil *in rure*.

FIG. 9.

Stained film specimen from a peptone* culture of a sample of the soil collected on the *twelfth day* after the fourth sowing.

FIG. 10.

Stained film specimen from a peptone culture of a sample of the soil collected on the *twelfth day* after the fourth sowing.

FIG. 11.

Stained film specimen from a peptone culture of a sample of the soil collected on the *fortieth day* after the fourth sowing.

FIG. 12.

Stained film specimen from a peptone culture of a sample of the soil collected on the *fortieth day* after the fourth sowing.

[Magnifying power, in each instance, 1,000.]

---

* Peptone, 1 per cent.; sodium chloride, 0·5 per cent.

PLATE XII.

Fig. 9.

Fig. 10.

Fig. 11.

Fig. 12.

No. 5.

REPORT on the CHEMICAL and BACTERIOLOGICAL EXAMINATION of the "WASHINGS" of SOILS with reference to the AMOUNT and NATURE of the ORGANIC MATTER and the Number and Character of the BACTERIA contained in them; by A. C. HOUSTON, M.B., D.Sc.

In the concluding remarks of my last year's report (App. B. No. 2) the following statement was made :—

"The general tendency of this soil enquiry is to encourage research having for its object the parallel examination, chemical and bacteriological, of the 'washings' from surface soils, and more particularly those soils which are to be thought of as dangerous adjuncts to water supplies. Judging by the results which have been obtained in the enquiry, such a research seems to promise definite information as to how far chemistry and how far bacteriology can be relied on to detect dangerous contamination of water through the medium of soil. There would seem to be more than an indication that though chemistry might fail in certain cases, dangerous pollution could in most instances readily be demonstrated by modern bacteriological methods."

The practical utility, apart altogether from any intrinsic scientific interest which may attach to it, of a research of the above sort lies chiefly in the fact that epidemics of certain diseases are sometimes clearly to be traced to the flood water from polluted soil gaining access to waterworks. The history of so-called "water epidemics" shows that chemistry has failed in many cases, even when the water supply was being periodically examined, to detect evidence of any pollution, much less of a dangerous and specific pollution, where contaminating material had beyond doubt been gaining access to the water through the medium of soil. It cannot be said that bacteriology has likewise signally failed under circumstances of the above sort because the records of periodic bacteriological examination *properly carried out* are so few that it is impossible to speak with certainty as to the value or otherwise of the results which have been obtained.

Last year's work consisted in the parallel examination, chemical and bacteriological, of samples of soil obtained from a number of diverse sources, and the general outcome of the work went to show that cultivated as compared with virgin soils; for instance, soils rich in animal organic matter as compared with soils rich in vegetable organic matter, and soils fed with sewage as compared with soils of similar general character but not polluted with excrement; do not always differ markedly the one from the other on chemical examination. At the same time these several soils were found to differ *inter se* in the most striking fashion when examined bacteriologically. So much was this found to be the case that in speaking of the chemical results it was stated that "it would seem from these experiments that it is the amount of vegetable organic matter relative to the amount of inorganic matter originally present in the soil which determines its chemical status, and not so much the addition of animal organic matter in the form of manure or sewage. Thus a pure sandy soil grossly polluted with sewage would, in a short time, probably contain less organic nitrogen, less even of ammonia, and certainly absorb

APP. B. No. 5.

On the Chemical and Bacteriological Examination of "Washings" of Soils; by Dr. Houston.

less oxygen from permanganate, than a pure virgin soil rich in peat but containing very little inorganic matter."

Again, in dealing with the bacteriological results it was said: "It has been shown that in soils XIII. and XIX. (virgin peaty soils) the B. enteritidis was apparently not present. All moorland waters draw their supply largely from the surface drainage of peat, and such waters, therefore, which often yield *high* results chemically, and which only escape condemnation from a knowledge of the source of the organic matter,* are likely to be free from the spores of B. enteritidis. On the other hand, a water supply contaminated with 'surface' water from cultivated soils would in all probability contain the B. enteritidis, and yet such a water might not, on chemical analysis, reveal evidence of recent pollution or of the presence of an objectionable amount of organic matter. Ten milligrams, for instance, either of soil XVI.† or of soil XIX.‡ in 1 litre of pure water would not affect the chemical composition of the water sufficiently to allow of its being considered impure; yet in the first case the water would contain 100 and in the second case no spores of B. enteritidis. The detection of the B. enteritidis in such a water as XVI. would be easy; namely, by filtering 1 litre of the water through a sterile Pasteur's filter, brushing the surface of the filter with a sterile brush into 5cc. sterile water, and adding of the filter brushing suspension 1 or more cc. to a sterile milk tube, heating the latter to 80° C. for 10 minutes, and cultivating its contents anaerobically, by Buchners' method, at a temperature of 37° C."

In short, last year's work seemed to presage precise information from a research having for its object the discovery of how far chemistry and how far bacteriology can be relied on to detect evidence of contamination of water in those cases where the source of the pollution is from the "washings" of soils; and appeared even to indicate that though chemistry might fail in certain cases, pollution, and possibly dangerous pollution, could in most instances readily be demonstrated by appropriate bacteriological methods.

As regards the work now in hand, it needs to be insisted on that the object of this research has not been merely to show that a bacteriological difference exists between the "washings" from one soil as compared with another soil, nor to demonstrate the presence of micro-organisms characteristic of objectionable pollution in the case of contaminated soils, while determining the absence of these same germs in virgin soils. Its purpose has been also to ascertain whether the bacteria present in the "washings" of soils in general differ from the micro-organisms usually found in our water supplies. Much confusion exists on this subject, it being very generally assumed that the bacterial flora of water must necessarily, from the intimate connection existing between soil

* Doubtless a water yielding a large amount of albuminoid ammonia (slowly given off) absorbing much oxygen from permanganate, and yielding but little free ammonia and chlorine *suggests* vegetable contamination. But this is a *suggestion* or *hint* merely, and gives no precise and definite information as to the source and nature and significance of the organic matter present.

† Pasture soil fed with sewage.

‡ Virgin peaty soil.

and water supply, resemble very closely the bacterial flora of soil, and it is even considered by some that the species of germs in air, water, and soil are very much alike. If we consult the lists of microbes isolated respectively from water, soil, and the atmosphere, we do indeed find that the same names repeat themselves in the separate tables to a bewildering extent, and such repetition is apt to lead to a misconception of the true facts of the case. But while it is true that the bacterial flora of soil, water, and air are in some respects comparable, it also is true that the *relative abundance* of the different species of germs found in the atmosphere, in our water supplies, and in the earth is widely different. Thus a particular micro-organism may be found in soil, in water, and in air, and yet be characteristic of soil alone, since it is sparsely distributed elsewhere in nature. Another microbe may be discovered in water, in soil, and in air, and still be almost peculiar to water, occurring only rarely in the soil and in the air. Yet a third germ may be ever present in the air and occur only in soil and in water as a chance and occasional visitor. It has been the failure in the past to recognize the importance of determining not only the *mere presence* but also the *relative abundance* of particular germs in this that or the other substance, which has hampered the progress of bacteriology, particularly in the bacterioscopic analysis of waters. Of course there is great difficulty in obtaining such records, but even approximate data are, in this connexion, of considerable value.

In last year's report some stress was laid on this point, and the work carried out showed conclusively that certain organisms are present in soils in numbers hitherto unsuspected.

For example, in one soil there were present 99,000 germs of B. mycoides per gramme, and in nearly all the soils examined it was present in great numbers (on an average about 32,000).

Again, cladothrix was likewise present in great numbers; in nine soils there were on an average 34,000 germs present per gramme of soil. There are different forms of cladothrix, but the one referred to in this report is the common variety, namely, the cladothrix which stains the gelatine in the neighbourhood of its colonies a bismarck brown colour, and which resembles very closely, and is probably identical with, cladothrix dichotoma. The published descriptions, however, of this latter micro-organism are somewhat incomplete.

Another point to which great attention was directed was the number of bacteria present in the spore form, and their nature; and also the *ratio between the number of spores and the total number of bacteria inclusive of spores.**

Thus in one soil there were present, per gramme, 4,000, in another 5,000, and in another 40,000 spores of B. mycoides.

* In the case of a water supply recently examined it was found that there were present, per cc., 2,630 germs, and yet there were *no* spores in 1cc. of the water. This is no doubt an extreme case, but it serves to illustrate the great difference in the ratio of spores of bacteria to the total number of bacteria in water as compared with soil. In raw sewage 15 samples gave on an average a ratio of 1 to 10,000.

APP. B. No. 8.

On the Chemical and Bacteriological Examination of "Washings" of Soils; by Dr. Houston.

The number of spores relative to the total number of bacteria in 13 different soils was approximately 1 to 2, 1 to 10, 1 to 8, 1 to 2, 1 to 3, 1 to 4, 1 to 24, 1 to 2, 1 to 3, 1 to 1,000, 1 to 2, 1 to 43, and 1 to 5.

In this connexion it is not out of place to show the importance of the subject by an illustration of what might take place in actual practice.

Let it be supposed that the periodic bacteriological examination is being carried out of the filtered water from a large river supplying an important town. It might be that week after week there were present in the cultures either no colonies (or very few) of B. mycoides and cladothrix, and also that the number of spores of bacteria was very small as compared with the total number of germs inclusive of spores. Then suddenly a change is noticed in the cultures, namely, the presence of B. mycoides and cladothrix in considerable numbers, and an increased number of spores relative to the total number of bacteria. To the mind of the bacteriologist accustomed to habitually examine samples of soil and water this would at once suggest "flood water," overtaxing of the filters, with resulting inefficient filtration; and such a condition of things would probably be found to be associated with heavy rainfall. The bacteriologist in question would be led to suspect danger, and to resort to other and special methods of detecting bacteria of more objectionable sort,* and the above state of affairs might perhaps be the first indication leading him eventually to pronounce unfavourably on the water supply. The B. mycoides, be it noted, is itself a harmless microbe, and "flood water" may be free from specific germs. But since "flood water" is especially liable to contain the germs of disease, any indication of the presence of "flood water" passing into a public supply is of some value; the fact, indeed, may turn out one of considerable importance.

How little chemistry can be relied on in cases of the above sort, the history of "Water Epidemics" already shows. Meanwhile the records of bacteriological examination periodically carried out, and by competent observers, have been too few to allow of their value being accurately gauged.

The following is a summary of the several sections of this Report:—

A. Description of the samples of soil.

B. Description of the method used in obtaining "washings" of the various soils.

C. Description of the chief chemical and bacteriological methods adopted in the investigation, under—

(A.) Chemical methods.

(1) Free ammonia.

(2) Albuminoid ammonia.

(3) Oxygen absorbed from permanganate.

---

* For example, phenol gelatine plate cultures for B. coli, and anaerobic cultures for B. enteritidis sporogenes (Klein); although as a matter of fact making of such cultures would be a matter of routine in the case of responsible bacteriologists.

(B). Bacteriological methods.
(1) Total number of bacteria.
(2) Spores of bacteria.
(3) B. coli. communis.
(4) B. enteritidis sporogenes (Klein).
(5) Streptococci.

D. Results of chemical examination of "soil waters."
(*a*) Free ammonia.
(*b*) Albuminoid ammonia.
(*c*) Sum of the ammonias (free and albuminoid).
(*d*) Oxygen absorbed from permanganate.

E. Results of bacteriological examination of "soil waters."
(*a*) General results—
(1) Total number of bacteria.
(2) Spores of bacteria.
(3) Sorts of bacteria.
(*b*) Results as regards the presence of B. coli.
(*c*) " " " Spores of B. enteritidis sporogenes.
(*d*) Results as regards the presence of Streptococci.

A. Description of the Samples of Soil* used in the Investigation.

I. Sewage field; from a pasture soil fed with sewage at Settle, Yorkshire. The sample was collected on the bank of the river Ribble, about 10 yards above the waters' edge.

II. Garden soil, from a garden at Hackbridge, Surrey. The soil was originally re-claimed from marshy land, but now consists of a mixture of different kinds of soil. It was last manured in November, 1897, with farmyard manure (chiefly cow dung), which was dug into the surface layers of the soil. The plot of ground from which the sample was collected is within a few yards of an artificial lake.

III. Pasture soil, from a field adjacent to the sewage field (No. I.). Manured three years previously, with farmyard manure, and seven years previously with slaughter-house manure. The sample was collected from the bank of the river Ribble, a few yards above the waters' edge.

IV. Meadow land, draining into a stream which eventually joins the river Ribble. Yearly manured with farmyard manure. The sample was collected from a spot adjacent to a small brook, which runs through the meadow and receives its drainage.

V. Pasture soil, from a field at Giggleswick, Yorkshire. This field is not manured. The sample was collected from a spot adjacent to a big drain, which leads into the meadow (No. IV.).

* It will be noted that all the samples of soil bear some relation to water, although not necessarily to water supply. In some cases the impurities in the soil were much greater than the impurities in the adjacent water, and in these cases it may be inferred that a flood would wash some of the polluting materials out of the soil into the water, and thereby contaminate it to a greater or less extent. But in other cases, notably, as regards soils, VII. and VIII., the foul character of the soil may properly be ascribed to the rise and fall of grossly polluted rivers depositing their own filth on their banks.

'P. B. No. 5. — the Chemi- and Bacte- logical Ex- ination of 'ashings" Soils; by Houston.

VI. Mud, from Thames bank at Sunbury, above the "intake" of the East London Water Company. The sample was collected below high water mark, and not far from the edge of the water.

VII. Mud, from bank of river Brent, about 100 yards below the point at which the Hendon sewage effluent enters the river. The sample collected below flood water mark, and not far from the waters' edge.

VIII. Mud, from Thames bank near Blackfriars, taken at low tide.

IX. Peaty soil, from peat bog land bordering the river Ribble at Swarthmoor, near Austwick, Yorkshire. The sample was collected from a spot close to a small stream leading into and about 40 yards distant from the Ribble.

X. Mud, from bank of river Colne, at London Colney, 17½ miles from London. Below flood water mark, and not far from the waters' edge.

B. DESCRIPTION OF THE METHOD USED IN OBTAINING "WASHINGS" OF THE VARIOUS SOILS.

The washings of the soils, termed in the report "*soil water*," I., II., III., &c., were, in each case, obtained in the following way :—

10 grammes of soil, weighed in a sterile watch glass, were transferred to a sterile mortar, bruised, and carefully mixed with 100cc. of pure sterile distilled water. After leaving for a quarter of an hour, so as to allow the grosser particles of various sorts to settle, 50cc. of the mixture of soil and water were decanted into a sterile glass cylinder (9 x 1"). After a quarter of an hour 10cc. of the mixture were drawn off from near the surface with a sterilized 10cc. pipette, and placed in a sterile vessel. 990cc. of pure sterile distilled water were next added, so as to bring the total bulk of soil and water to 1000cc. This 1000cc. was called "*soil water*," I., II., III., &c., as the case might be, and it was subjected to the chemical and bacteriological tests.

C. DESCRIPTION OF THE CHIEF CHEMICAL AND BACTERIOLOGICAL METHODS ADOPTED IN INVESTIGATING THE VARIOUS "SOIL WATERS."

(A.) *Chemical Methods.*

(1.) *Free Ammonia.*—A small quantity of distilled water is poured into a flask, and the flask connected with a condenser. After boiling for a few minutes so as to get rid of any traces of ammonia from the apparatus, the flask is disconnected and its contents thrown away. 300cc. of "soil water" are then added and the flask is connected again with the condenser and distillation proceeded with. 100cc. are distilled over and an aliquot part tested for ammonia with Nessler's reagent and a standard solution of ammonium chloride.

As to reagents in this connection :—(1.) Nessler's reagent: potassium iodide 35 grammes, mercuric chloride 13 grammes dissolved in 1 litre distilled water. A saturated solution of mercuric chloride is then added until the precipitate is permanent

160 grms. caustic potash are added and dissolved, and mercuric chloride again added to render the solution sensitive. (2.) Ammonium chloride solution: dissolve 3·15 grms. ammonium chloride in 1 litre of distilled water, and dilute this concentrated solution with 100 times its bulk of pure distilled water. 1cc. equals 0·01 milligram of ammonia.

(2.) *Albuminoid Ammonia.*—To the liquid remaining in the flask after the determination of the free ammonia, 50cc. of alkaline permanganate solution (previously boiled "free" of ammonia) are added, and 100cc. distilled over; an aliquot part is tested for ammonia with Nessler's reagent and a standard solution of ammonium chloride.

As to reagents in this connection:—(1.) Nessler's reagent, as above. (2.) Ammonium chloride solution, as above. (3.) Alkaline permanganate solution: dissolve 8 grms. potassium permanganate, and 200 grms. caustic potash in 1 litre of distilled water. Immediately before using take 50cc. of above solution, and 100cc. of distilled water, and boil in an open vessel until the bulk of the liquid is reduced two-thirds. The remaining 50cc. (now free from ammonia) is added to the contents of the flask.

(3.) *Oxygen absorbed from permanganate.*—300cc. of "soil water" are poured into a flask and 10cc. in each instance of standard potassium permanganate solution, and dilute sulphuric acid added. After heating to 100° C. for one hour and subsequently cooling, 2 cc. potassium iodide solution are added, and the flask contents titrated with standard sodium hyposulphite solution. When the brown colour has all but disappeared, 2 cc. starch solution are added, and titration proceeded with until the blue colour vanishes. A blank experiment with the same amount of distilled water, permanganate, and acid is carried out in the same way as the above. The difference between the amount of hyposulphite used in the blank experiment and the amount in the case of the soil mixture is multiplied by the amount of available oxygen contained in the permanganate added, and the product, divided by the amount of hyposulphite used in the blank experiment, equals to the amount of oxygen absorbed by 300 cc. of "soil water."

As to reagents in this connection:—(1.) Standard solution of potassium permanganate: dissolve 1·975 grms. in 1 litre of distilled water; 10 cc. equal 5 mgrms. available oxygen. (2.) Dilute sulphuric acid, 1 in 10. (3.) Solution of potassium iodide, 10 per cent. (4.) Solution of sodium hyposulphite: dissolve 3·875 grms. in 1 litre of distilled water. (5.) Starch solution, about 1 grm. in 200 cc. water, boiled and filtered.

There is an objection to this process, namely, that nitrites, ferrous salts, and chlorides reduce the permanganate. The amount of organic matter, however, in the "soil waters" is so great in relation to the amount of these reducing agents likely also to be present that the value of the process is not seriously affected.

(B.) *Bacteriological Methods.*

(1.) ***Total number of bacteria.***—Gelatine plate cultures containing 0·1 and 1·0 cc. "soil water" were used in estimating the total number of micro-organisms.

App. B. No. 5.

On the Chemical and Bacteriological Examination of "Washings" of Soils; by Dr. Houston.

(2.) *Spores of bacteria.*—1·0 cc. of "soil water" was added to 10 cc. sterile gelatine in a test tube. After heating to 80° C. for 10 minutes, the mixture of gelatine and "soil water" was poured into a Petri's capsule.

(3). B. *Coli Communis.*—0·1 cc. of phenol (5 per cent.) was added to 10 cc. sterile gelatine in a test tube. After liquefying the gelatine the mixture of phenol and gelatine was poured into a Petri's capsule. After the gelatine had become quite solid 0·1 cc. of "soil water" was added, and a "surface" plate culture made. All colonies resembling B. Coli were sub-cultured in broth (for diffuse cloudiness and indol reaction); in litmus milk (for acidity and clotting); and in gelatine "shake" culture (for gas formation).

(4.) B. *Enteritidis Sporogenes* (Klein).—2·0, 1·0, and 0·1 cc. of "soil water" were added severally to each of three milk tubes. The tubes were heated to 80° C. for 10 minutes, and then cultivated anaërobically by Buchner's method at a temperature of 37° C. Usually, in less than 24 hours, and practically always in 36 to 48 hours, the B. Enteritidis, when present, produces certain characteristic changes in the milk culture; namely, the casein is precipitated and torn into masses, which (the masses) are frequently forced up against the cotton wool plug, owing to the copious development of gas that occurs; the whey (which on microscopic examination shows the presence of numerous bacilli) remaining transparent, or nearly so.

(5.) *Streptococci.**—A test tube, containing 10 cc. of agar, was heated until the medium was liquefied. The melted agar was then poured into a Petri's capsule. After it had become quite solid, 0·1 cc. of "soil water" was added, and a "surface" plate culture made. The culture was incubated at 37° C. for 24 hours (sometimes 36 to 48 hours), and the plate carefully examined under a low power of the microscope (Leitz, 2 eyepiece, 3 objective). All colonies in any way resembling streptococci were sub-cultured, primarily in broth, and, subsequently, if the result was satisfactory, in other media. The reason for this procedure is this: it is often quite impossible, even for the most skilled observer, to say whether a particular colony is a streptococcus or not; and so minute are the colonies that it is always difficult, sometimes impossible, to obtain a sufficient amount of growth to inoculate a broth tube and at the same time leave enough for the making of a satisfactory cover-glass preparation for microscopic examination. Hence it is advisable to make as certain as possible of perpetuating the growth, and, if the examination of the broth culture is satisfactory, to proceed to further cultivations in other media. In searching for streptococci many difficulties are met with. Frequently there is no growth in the broth tubes, the streptococci having (presumably) lost their vitality. Sometimes the growth is greatly delayed, and on resorting to further sub-culture a

* The result of these investigations on the streptococci occurring in soil were of a sufficiently encouraging nature to lead me to turn my attention to their relations with water. The results so far obtained in this direction are given in a separate report, Appendix B, No. 6.

egative result is obtained, the organism having been obtained in state of feeble vitality. Again, it not uncommonly happens ıat a growth occurs, but the micro-organism turns out on xamination not to be a streptococcus.

). RESULTS OF THE CHEMICAL EXAMINATION OF TEN SOILS AS REGARDS THE AMOUNT OF ORGANIC MATTER IN THE "WASHINGS" (TERMED "SOIL WATERS") FROM EACH SAMPLE.

It is convenient to give the results in the form of a table Table I.), and to make certain comments thereon.

TABLE I.

HOWING the results of Chemical Examination of Ten Soils, obtained from various sources, as regards the amount of Free and Albuminoid Ammonia, and Oxygen absorbed from Permanganate in the "washings" (termed "soil waters") from each sample. Result stated as part per 100,000 parts.

Experiment.	Description of the sample of soil.	Free Ammonia.	Albuminoid Ammonia.	Sum of the Ammonias, Free and Albuminoid.	Oxygen absorbed from Permanganate (1 hour at 100° C.).
	1	2	3	4	5
1	Soil I., Sewage field ...	0·04	0·0147	0·0547	0·305
2	Soil II., Garden soil (last manured about 14 months previously).	0·04	0·0134	0·0534	0·219
3	Soil III., Pasture soil (last manured three years previously with farm yard manure. Seven years previously with slaughter house manure).	0·022	0·0169	0·0389	0·34
4	Soil IV., Meadow (manured yearly with farm yard manure).	0·027	0·0227	0·0497	0·59
5	Soil V., Pasture soil (not manured).	0·027	0·0227	0·0497	0·372
6	Soil VI., Mud from Thames bank above "intake" for water supply.	0·0227	0·024	0·0467	0·298
7	Soil VII., Mud from bank of river Brent about 100 yards below the point at which sewage effluent enters the river.	0·0134	0·020	0·0334	0·278
8	Soil VIII., Mud from Thames bank near Blackfriars.	0·0187	0·0134	0·0321	0·174
9	Soil IX., Peaty soil from Moorland.	0·02	0·025	0·045	0·927
10	Soil X., Mud from bank of river Colne at London Colney.	0·03	0·01134	0·04134	0·089

waters" VIII. and X. would be regarded as of "great purity," and "soil waters" I., II., III., V., VI., VII. of "medium purity."

Moreover, soil I (from sewage field) was to the oxygen test *apparently* purer than soil V. (pasture soil, *not* manured).

Again the "washings" from soil VIII. (mud from the bank at Blackfriars of the grossly polluted Thames) simulated, as regards oxygen absorbed, a water of "great purity."

The chemical results are shown diagrammatically in diagram 1. (at end of this report).

E. RESULT OF THE BACTERIOLOGICAL EXAMINATION OF TEN SOILS AS REGARDS THE NUMBER AND CHARACTER OF THE BACTERIA IN THE "WASHINGS" (TERMED "SOIL WATERS") FROM EACH SAMPLE.

It is proposed, in this connection, to give in the first place a table (Table II.) showing the general results obtained, and in the second place to make certain comments thereon. Lastly, a number of tables dealing with special bacteriological experiments will be given, and the results contained in them will be briefly considered.

TABLE II.

SHOWING the results of Bacteriological Examination of Ten Soils, obtained from various sources, as regards the total number of bacteria and the number of spores of bacteria in the "washings" (termed "soil waters") from each sample.

Experiment.	Description of the sample of soil. 1	Total number of aërobic bacteria (inclusive of spores) in 1 cc. of "soil water." 2	Number of spores of bacteria in 1 cc. of "soil water." 3	Ratio of spores (col. 3) to bacteria (col. 2). 4
1	Soil I., Sewage field	The number of colonies in 0·1 cc. "soil water" culture was so great as to make it impossible to count them with any approach to accuracy.	50 [B. Mycoides 5 colonies; B.Subtilis, or closely allied form, 2 colonies.]	...
2	Soil II., Garden soil (last manured about 14 months previously).	650 [4 colonies of B. mycoides, 1 of B. Fluorescens non-liquefaciens and 1 of Cladothrix in 0·1 cc. "soil water" culture.]	121 [At least 15 of these B. Mycoides; B. Mesentericus and B. Subtilis also present.]	1 : 5

No. 8.

Experiment.	Description of the sample of soil 1	Total number of aërobic bacteria (inclusive of spores) in 1 cc. of "soil water." 2	Number of spores of bacteria in 1 cc. of "soil water." 3	Ratio of spores (col. 3) to bacteria (col. 2). 4
3	Soil III, Pasture soil (last manured three years previously with farm yard manure, seven years previously with slaughter-house manure).	2,280 [As many as 60 colonies of B. Mycoides in 0·1 cc. "soil water" culture. In culture containing 1·0 cc. "soil water" too many colonies to allow of accurate counting; here there were 4 colonies of B. Fluorescens liquefaciens.]	127 [Notwithstanding the large number of colonies of B. Mycoides in the ordinary cultures none were present in the plate for spores. 1 colony of B. Mesentericus (variety L.) and a number of (variety E).]	1 : 18
4	Soil IV., Meadow (manured yearly with farm yard manure).	800 [1 colony of B. Fluorescens non-liquefaciens in 0·1 cc. "soil water" culture. In culture containing 1·0 cc. "soil water" 5 colonies of B. Mycoides and 3 of B. Fluorescens liquefaciens.]	200 [3 colonies of B. Mycoides and several of B. Mesentericus.]	1 : 4
5	Soil V., Pasture soil (not manured).	440 [The majority of the colonies were those of B. Mycoides.]	216 [The majority of the colonies were those of B. Mycoides.]	1 :
6	Soil VI., Mud from Thames bank above "intake" for water supply.	540 [1 colony of B. Mycoides in 0·1 cc. "soil water" culture. In culture containing 1·0 cc. "soil water" 2 colonies of B. Fluorescens liquefaciens.]	432 [5 colonies of B. Mycoides.]	1 :
7	Soil VII., Mud from bank of river Brent, about 100 yards below point at which sewage effluent enters the river.	3,470 [In culture containing 1·0 cc. "soil water" the colonies were too numerous to allow of accurate counting, here there were 2 moulds, 3 colonies of B. Fluorescens liquefaciens and 7 colonies of B, Mycoides.]	140 [2 colonies of B. Mycoides.]	1 :

APP. B. No. 5.

On the Chemical and Bacteriological Examination of "Washings" of Soils; by Dr. Houston.

ri- t.	Description of the sample of soil. 1	Total number of aërobic bacteria (inclusive of spores) in 1 cc. of "soil water." 2	Number of spores of bacteria in 1 cc. of "soil water." 3	Ratio of spores (col. 3) to bacteria (col. 2). 4
	Soil VIII., Mud from Thames bank near Blackfriars.	2,200 [B. Mycoides present but in small proportion.]	50 [2 colonies of B. Mesentericus (variety I.)]	1 : 44
	Soil IX., Peaty soil from Moorland.	211 [18 colonies of B. Mycoides in the 1·0 cc. "soil water" culture. No colonies of the Fluorescent bacteria or cladothrix noted on 3rd day.]	117 [3 colonies of B. Mycoides and 2 of B. Mesentricus (variety I.)]	1 : 1·8
	Soil X., Mud from bank of river Colne at London Colney.	800 [12 colonies of cladothrix in the 0·1 cc. "soil water" culture. In the 1·0 cc. culture 5 colonies of B. Fluorescens liquefaciens, 6 of B. Fluorescens non-liquefaciens, 3 of B. Mycoides, and 3 of cladothrix.]	324 [1 colony of B. Mycoides; none apparently of B. Mesentericus. Several of granular bacillus of soil (2b).]	1 : 2·4

(a.) *General bacteriological results.*

ı considering the above general bacteriological results, the ›wing points are to be noted .—

.) *Total number of bacteria* (col. 2, Table II.).—The number acteria in soil I.* (sewage field) was so enormous that in a tine plate culture containing 0·1 cc. "soil water," the number d not possibly be counted. It is safe to assert that there ; present at least several thousand microbes per cc. of "soil 'r."

her soils conspicuous as regards number of contained bacteria ': soil III. (pasture soil) containing 2,280 germs per cc. of water"; soil VII. (mud from bank of river Brent below ; at which sewage effluent enters) containing 3,470 microbes;

ere, as elsewhere in report, it is convenient to speak as if the soils them- had been examined, although in reality it was the "washings" from the hat were submitted to examination.

and soil VIII. (mud from Thames bank near Blackfriars) yielding 2,200 micro-organisms on cultivation. The smallest number was found in soil IX. (virgin peaty soil), namely, 211 per cc. The manured pasture soil (III.) contained more than five times as many microbes as the pasture soil which had never been manured (V.) the figures being respectively 2,280 and 440. The manured garden (II.) and meadow (IV.) soils contained respectively 650 and 800 germs per cc. of soil water. The mud from the Thames bank above the "intake" of a water supply (soil VI.) contained less than a quarter of the number found lower down the river in the mud near Blackfriars. Lastly, soil X. (mud from the bank of Colne) contained the same number of micro-organisms as the manured meadow soil (IV.).

(2.) *Number of Spores of Bacteria* (col. 3, Table II.).—It is especially to be noted that in soil I. (sewage field), the number of spores* was only 50 per cc. of "soil water," notwithstanding the large total number of micro-organisms. It is probable that in the moist character of the soil and its richness in animal organic matter, lies the explanation of this observed fact. The largest number of spores was found in soil VI. Next in order came soil X., soil V., soil IV., soil VII., soil III., soil II., soil IX. Soils I. and VIII. contained an equal number. It will be noted that the soils most impure as regards the total number of bacteria contained relatively (and in some cases actually) very few spores. Thus in soil I. the number of germs were too great to count, yet the number of spores was only 50. In soil III. the total number was 2,280 and the number of spores 127 (1 : 18). In soil VII. the total number was 3,470 and the spores 140 (1 : 24). In soil VIII. the total number of microbes was 2,200 and the number of spores 50 (1 : 44). The ratios as regards the other soils, namely, soils II., IV., V., VI., IX., X., were 1 : 5; 1 : 4; 1 : 2; 1 : 1·2; 1 : 1·8; 1 : 2·4, respectively. As I have said, the amount of moisture present in the various soils, as well as the amount of pabulum, is in this connection a factor of some importance. Within certain limits it is believed that the greater the amount of moisture in a soil, the larger is likely to be the total number of bacteria, and the smaller the number of spores, relatively, if not actually, to the total number of germs.

In all the soils, the number of spores, both actually and relatively, to the total number of bacteria, was far in excess of the number found in potable waters.

Figs. 15 and 16, Plate XIV.†, show two gelatine plate cultures, the one containing 1·0 cc. "soil water" V. ("total number"), and the other 1·0 cc. (heated 80° C. for 10 minutes) "soil water" V. ("spores"). Yet the number of colonies is only about twice as great in the first case, *i.e.*, in the culture which was not heated to 80° C. for ten minutes, as in the second, where all germs not present as spores were destroyed by such heating.

---

* A considerable number of experiments as regards the total number of bacteria and the number of spores of bacteria, carried out with Barking and Crossness raw sewage, showed that the ratios of spores to bacteria was 1 to 11,744 and 1 to 9,662, respectively, in Barking and Crossness crude sewage.

† The Figures (13 to 17) illustrating this report will be found at the end of Appendix B. No. 6.

(3.) *As to the Sorts of Bacteria* found in the Cultivations (cols. 2 and 3, Table II.).

B. *Mycoides.*—This microbe was commonly found in the cultures of the various "soil waters," and usually it was present in great abundance. Moreover, its spores were very numerous in the cultivations previously heated to 80° C. for 10 minutes. In soil I. (spore plate, col. 3, Table II.) one-tenth of the colonies were those developing from spores of B. mycoides. In soil II. the numbers per cc. were 40 in the ordinary plate culture, and at least 15 in the "spore" cultivation. In soil III. the numbers per cc. were 600 in the ordinary plate, but in this soil there were no spores of this microbe detected. Possibly in this soil the B. mycoides was, at the time of the collection of the sample, in an active phase of existence.* In soil IV. the numbers were 5 in the ordinary plate and 3 in the spore plate. In soil V. as regards both plates the majority of the colonies were those of B. mycoides. This is beautifully shown in Figs. 15 and 16, Plate XIV. In soil VI. the numbers were 10 and 2, respectively. In soil VII. there were 7 colonies in the ordinary plate and 2 in the spore plate. In soil VIII. none were observed in the spore plate, and in the ordinary plate B. mycoides was present, but only in small number. In soil IX. the numbers were 18 and 3, respectively. This is in excess of the number usually found in pure peaty soils. In soil X. the numbers were 3 and 1 respectively.

At the risk of repetition it must be insisted on that pure potable waters, whether filtered or unfiltered, do not contain this organism (either as bacilli or as spores of bacilli) or *contain it only in small numbers.* Judging by the above results, therefore, the presence of B. mycoides in a potable water implies the presence of surface "washings" of soil, not improbably the "washings" of cultivated soil; and in the case of a filtered water presence of this microbe points strongly to the over-taxing of the filters with flood water. As was insisted in last year's report (App. B., No. 2), its presence, at all events in any abundance, may be, in the case of filtered water, of signal importance as showing inefficiency of filtration, and always is significant in the case of all potable waters as giving some indication of the possible presence of other and perhaps more objectionable "soil bacteria." It is too much to assert that, because it is non-pathogenic its presence is of little moment. So far as we know organic matter *per se* is harmless, yet it is on the amount of organic matter that chemists judge of the "fitness" of a water for domestic use.

From last year's work (App. B., No. 2), it appears that B. mycoides is commonly present in great numbers in soils, but is especially abundant in garden and pasture soils, and may be absent or present only in small numbers in pure peaty and sandy soils. Nevertheless soil IX. (peaty soil), as has been already

* It is difficult, however, to speak on this subject with certainty, because some bacteriologists hold that bacteria proceed to sporulation, chiefly or solely when the cells have attained their highest development. Others assert that bacteria change from a transitory "growth form" to a "persistent spore-form," in order to avoid extinction when the conditions for their growth are unfavourable.

noted, contained a rather large number of this micro-organism. B. mycoides is found in raw sewage, but only in very small number relatively to the total number of bacteria.

Cladothrix.—It is probable that in all the soils cladothrix was present, though its presence was not always detected. The reason for this probably lies in the fact that in the plate used for the estimation of the "total number of aërobic bacteria" the amount of "soil water" added to the gelatine was such as to give rise to the growth of a considerable number of colonies, and that these, by their growth, and by liquefying the gelatine masked the presence of the slowly developing colonies of cladothrix. In soil X. cladothrix was present in the proportion of 120 per cc. of "soil water." In soil II. there were 10 per cc. The cladothrix is a prominent soil micro-organism, and is absent, or present only in small number, in *potable* waters. Its presence then in a given water may point to the presence of washings from surface soil or "flood water," but, judging from last year's work (App. B., No. 2), gives no indication of the nature of the soil from which it is derived, *e.g.*, whether polluted or not. Cladothrix is occasionally found in crude sewage, but in number very small relative to the total number of bacteria of all sorts.

Other bacteria.—Other bacteria commonly found in the "soil waters" were B. mesentericus (two varieties); B. fluorescens liquefaciens (more than one variety); B. fluorescens non-liquefaciens (chiefly two varieties described in last year's report as 4*d* and 5*e*.); proteus-like microbes, B. subtilis (and allied forms); granular bacillus (*see* last year's report, App. B., No. 2); a bacillus somewhat closely allied to the granular bacillus, and noted in laboratory note-book as bacillus from soil (10*j*); and a considerable number of other microbes which could not be identified, and some of which have doubtless not been described.

In soil III. there were 4; in soil IV., 3; in soil VI., 2; in soil VII., 3; and in soil X., 5 colonies of B. fluorescens liquefaciens per cc. "soil water." "Soil waters" II. and IV. contained 10 colonies of B. fluorescens non-liquefaciens per cc., and "soil water" X., 6 colonies. B. mesentericus was frequently present in the "spore cultures," from the various "soil waters." *

In sequence to these general bacteriological results it is of advantage to consider the special experiments as to the presence of B. coli, B. enteritidis sporogenes (Klein), and streptococci in minimal quantities of the various "soil waters."

---

* *See* Fig. 17, Plate XV.

(*b*.) *Results as regards the presence of B. coli.*

The results as regards B. coli. are given in the following table :—

TABLE III.

SHOWING the results of Bacteriological Examination of Ten Soils, obtained from various sources, as regards the presence of B. coli in a minimal quantity of the "washings" (termed "soil waters") from each sample.

Experiment.	Description of the sample of soil.	Colonies resembling (more or less closely) B. Coli in surface phenol (0·05%) gelatine plate cultures containing 0·1 cc. "soil water."	Sub-culture of Colonies resembling B. Coli in—		
			Broth; for (*a*) Diffuse cloudiness in 24 hours at 37° C., and (*b*) indol test 5th day at 37° C.	Gelatine "shake" cultures; for gas formation, 20° C.	Litmus milk; for (*a*) Acidity, and (*b*) Clotting at 37° C.
1	Soil I. Sewage field.	A large number of colonies resembling B. Coli.	(*a*) + (*b*) −	+ in 24 hours.	(*a*) + in 24 hours. (*b*) No clot until 5th day.
2	Soil II. Garden soil (last manured about 14 months previously).	No colonies resembling B. Coli in 0·1 cc. "soil water."	...	...	...
3	Soil III. Pasture soil (last manured 3 years previously with farmyard manure, 7 years previously with slaughter-house manure).	A few colonies bearing some resemblance to B. Coli, but not sufficiently "film-like" in the character of their growth.	Practically no growth at 37° C.	+ in 24 hours.	Practically no growth at 37° C.
4	Soil IV. Meadow (yearly manured with farm yard manure).	No colonies resembling B. Coli in 0·1 cc. "soil water."	...	...	...
5	Soil V. Pasture soil (not manured).	No colonies resembling B. Coli in 0·1 cc. "soil water."	...	...	...
6	Soil VI. Mud from Thames bank above "intake" for water supply.	No colonies resembling B. Coli in 0·1 cc. "soil water."	...	...	...

App. B. No. 5.

On the Chemical and Bacteriological Examination of "Washings" of Soils; by Dr. Houston.

Experiment.	Description of the sample of soil.	Colonies resembling (more or less closely) B. Coli in surface phenol (0·05%) gelatine plate cultures containing 0·1 cc. "soil water."	Sub-culture of Colonies resembling B. Coli in—		
			Broth; for (a) Diffuse cloudiness in 24 hours at 37° C., and (b) indol test 5th day at 37° C.	Gelatine "shake" cultures; for gas formation, 20° C.	Litmus milk; for (a) Acidity, and (b) Clotting at 37° C.
7	Soil VII. Mud from bank of River Brent, about 100 yards below point at which sewage effluent enters the river.	At least eight colonies indistinguishable from B. Coli in 0·1 cc. "soil water."	(a) + (b) −	+ in 24 hours.	(a) + in 24 hours. (b) No clot 5th day.
8	Soil VIII. Mud from Thames bank, near Blackfriars.	A few colonies bearing some resemblance to B. Coli, but not sufficiently "film-like" in the character of their growth.	(a) + (b) −	—	(a) + in 24 hours. (b) No clot 5th day.
9	Soil IX. Peaty soil from Moorland.	No colonies resembling B. Coli in 0·1 cc. "soil water."	...	...	...
10	Soil X. Mud from bank of River Colne, at London Colney.	One colony bearing a remote resemblance to B. Coli.	(a) + (b) −	+ but not until 2nd day, and not well marked. Moreover, liquefication set in quite early.	(a) No change 24 hours; later, trace of acid. (b) No clot 8th day, and only trace of acid.

It is to be noted that only a small quantity, namely, 0·1 cc. of the "soil water" was used in each case, and, no doubt, if a larger quantity had been employed, either by direct culture or by the "filter brushing" method, B. coli might have been demonstrated in some cases where it was not detected when using the above small amount. Still, in relation to the total number of germs of all sorts found in the various "soil waters," 0·1 cc. was considered a sufficient quantity. In this connexion, it may be worthy of note that in crude sewage the total number of germs commonly exceed one million and the number of B. coli (or closely allied forms) 100,000 per cc.

Last year's work (App. B. No. 2) showed conclusively that B. coli is not abundant everywhere as is generally stated; and that, as a matter of fact, it cannot be readily isolated from soil (which, if B. coli is ubiquitous, might be supposed to harbour it in great numbers); unless, indeed, we consider as B. coli organisms bearing only a superficial resemblance to the B. coli of the intestinal tract. That B. coli may become altered by sojourn in soil is very likely, but it would be unwise to class as such, organisms bearing only a remote resemblance to it, on the supposition that, at some antecedent time, their morphological and biological characters were altogether "typical." If it were possible to measure, in any way, modifications in this sense, they might be of advantage as affording some evidence as to whether the source of the original polluting material was remote or recent. If it is permissible to hazard a conjecture, I am disposed to suspect that B. Coli outside the animal body and under natural conditions as opposed to artificial laboratory conditions, tends to lose in the first place the power of forming indol, of producing strong acid, and of rapidly clotting milk; or to retain these properties in diminished degree. Afterwards other qualities may also disappear, as, for example, ability to grow luxuriantly at blood heat.

This is no mere digression, because it is commonly stated, as an argument against the significance of the presence of B. coli in potable waters, that the intimate connexion existing between soil and water supply is quite sufficient to explain its appearance in a water. No doubt where soil has been recently polluted with fœcal matter, B. coli is likely to be abundant in washings therefrom; but last year's work, and this year's work also, go to show that soil, other than recently and grossly polluted soil, is not likely to yield to water B. coli in any large number.

Reviewing the results stated in the table, it is to be noted that there were no colonies resembling B. coli in 0·1 cc. "soil water" in the case of soils II., IV., V., VI., and IX. In soil I. there were a large number of colonies resembling B. coli. One of these colonies, was, on subculture, found to be typical in all respects, except that broth cultures gave no indol on the 5th day and that the clotting of the milk was delayed until the 5th day. In soil III. there were a few colonies bearing a remote resemblance to B. coli in the phenol plate culture. One of these on subculture, although giving gas in a gelatin shake culture in 24 hrs. at 20°C., failed, or nearly so, to grow at blood heat.* In soil VII. there were at least eight colonies indistinguishable from B. coli in the phenol plate culture. One of these on subculture proved to be "typical" in all respects, except that no indol was formed in broth culture, and there was no clotting of milk. Possibly, as I have said, indol formation and acid clotting of milk are properties easily lost by B. coli when it is separated from the animal body. In soil VIII. there were a few colonies bearing some resemblance to B. coli. One of these gave, on subculture, no gas in gelatin shake culture, no clot in milk, and no indol in broth; but it gave rise to diffuse cloudiness in broth in 24 hrs. at 37°C, and also to acidity in milk culture. Lastly, soil X. showed one colony bearing a remote resemblance to B. coli in 0·1 cc. "soil water." This however liquified* the gelatin at an early date, and so need not be considered further.

Thus, out of these ten samples of soil, five (II., IV., V., VI., IX.), contained, in 0·1 cc. "soil water," no colonies at all resembling

* When a microbe simulating B. coli does not grow luxuriantly at blood heat, or liquifies, however slightly, the gelatin within five days, it is doubtful whether it ought to be regarded as being even *remotely* akin to B. coli.

App. B. No. 5.

On the Chemical and Bacteriological Examination of "Washings" of Soils; by Dr. Houston.

B. coli. In the remaining five soils (I., III., VII., VIII., X.), colonies showed themselves, in the phenol plate cultures, sufficiently like B. coli to make it advisable to resort to subculture. Of these bacilli suspected of kinship to coli, III. was discarded as growing only with great difficulty at blood heat, and X., because it liquified the gelatin at an early date. In VIII. there was a considerable element of doubt; but in I. and VII. the result was positive, except that in I. there was no indol formation and there was delayed clotting of milk, and that in VII. there was no indol formation and no clot in milk (5th day). Soil I. was from a sewage field and soil VII. from the bank of the grossly-polluted river Brent.

(c.) *Results as regards the presence of B. Enteritidis Sporogenes, (Klein).*

The results as regards the presence of spores of B. Enteritidis Sporogenes (Klein) in the various soil waters are given in the following table:—

TABLE IV.

SHOWING the results of bacteriological examination of ten soils, obtained from various sources, as regards the presence of spores of B. enteritidis sporogenes (Klein) in small quantities of the "washings" (termed "soil waters"), from each sample.

The sign + signifies positive result, *i.e.*, typical changes in the milk.
The sign — signifies negative result, *i.e.*, no changes (or not characteristic changes) in the milk.

Experiment.	Description of the Sample of Soil.	Amount of the "Soil water" added to the milk tubes, with result.		
		2 cc.	1 cc.	0·1 cc.
1	Soil I. Sewage field	No culture.	+	+
2	Soil II. Garden soil (last manured about 14 months previously).	,,	+	—
3	Soil III. Pasture soil (last manured three years previously with farm yard manure, seven years previously with slaughter-house manure).	,,	+	—
4	Soil IV. Meadow (manured yearly with farm-yard manure).	,,	—	—
5	Soil V. Pasture soil (not manured) ...	,,	—	—
6	Soil VI. Mud from Thames bank, above "intake" for water supply.	+	—	—
7	Soil VII. Mud from bank of river Brent, 100 yards below where effluent from sewage works enters.	+	+	—
8	Soil VIII. Mud from Thames bank, near Blackfriars.	+	+	—
9	Soil IX. Peaty soil from moorland... ...	—	—	—
10	Soil X. Mud from bank of river Colne, at London Colney.	+	+	—

In the first five experiments no "2 cc. milk cultures" were made. As regards the first three experiments, culture of the sample in this amount was unnecessary since B. Enteritidis was present in a smaller quantity of "soil water." But as regards experiments 4 and 5 although B. Enteritidis was not present in 1 cc. it may possibly have been present in 2 cc. "soil water."

Soil IX. (peaty soil) contained no Enteritidis, even in 2 cc. "soil water." Soils IV. (meadow) and V. (pasture not manured) contained, as has been said, no spores of Enteritidis in 1 cc. "soil water." Soil VI. contained no spores in 0·1 or in 1·0 cc., but a positive result was obtained as regards the 2 cc. culture. Soils II., III., VII., VIII., and X. yielded positive results with 1 cc.; but soil I. (sewage field) alone gave a positive result with 0·1 cc. "soil water." Of the two pasture soils the manured soil (III.) contained B. Enteritidis in 1 cc., whereas a similar culture in the case of the unmanured soil (V.) yielded a negative result. Again soil VI. (Thames mud) contained B. Enteritidis in 2 cc., but not in 1 cc; whereas soil VIII. (Thames mud lower down the river) contained the spores of this anaërobe in 1 cc. "soil water." Soil I. (sewage field) contained about ten times as many of these spores as any of the other soils and 20 times more than soil VI.; and necessarily more than 20 times as many as soil IX., this latter soil giving a negative result with 2 cc. "soil water."

It is evident from the above results that the spores of B Enteritidis are present in the "washings" of *impure* soils and absent (or present in fewer numbers) in the "washings" of *pure* soils. It cannot, therefore, well be said that the surface drainage from *pure* soils equally with *polluted* soils would contaminate a water supply with the spores of this pathogenic anaërobe.

Last year's report (App. B., No. 2) was equally definite in this respect since, generally speaking, it was found that *impure* soils contained B. Enteritidis in great numbers and *pure* soils in very small amount. Thus some polluted soils contained 10,000 spores per gramme of soil and some virgin soils none in 10 mgrms of soil.

That B. Enteritidis Sporogenes may be present in considerable numbers in soil is doubtless to be traced to the fact of its being a "sporing anaërobe."

It may be worthy of note that in crude sewage the number of spores of B. Enteritidis Sporogenes varies usually from 100 to 1,000 per cc. This statement is made as the result of a very large number of experiments.

Lastly, *pure* waters do not contain B. Enteritidis, even in 10cc., and in many pure waters the spores of this microbe cannot be demonstrated in as much as 500 cc. (obtained by the "filter brushing" method) of water.

(*d*.) *Results as regards the presence of Streptococci.*

The results as regard the presence of streptococci in the various "soil waters" are given in Table V., and in Table VI., in addendum to this report, there are set forth the chief morphological and biological characters of these streptococci.

TABLE V.

B. No. 5.

e Chemi-
d Bacte-
ical Ex-
ation of
hings"
ls; by
ouston.

SHOWING the results of Bacteriological Examination of Ten Soils, obtained from various sources, as regards the presence of streptococci in the "washings" (termed "soil waters") from each sample.

Experiment.	Description of sample of soil.	Results, as regards the presence of streptococci, in 1·0 cc. "soil water." Surface Agar plate culture, incubated at 37° c.
1	Soil I. Sewage field	The Agar plate was somewhat crowded and thus a negative result was obtained. But in a surface gelatine culture a large number of streptococci were found. *See* Table VI.
2	Soil II. Garden soil (last manured about 14 months previously).	Negative result.
3	Soil III. Pasture soil (last manured 3 years previously with farm yard manure, 7 years previously with slaughter-house manure).	" "
4	Soil IV. Meadow (manured yearly with farm yard manure).	" "
5	Soil V. Pasture soil (not manured).	" "
6	Soil VI. Mud from Thames Bank above "intake" for water supply.	" "
7	Soil VII Mud from bank of River Brent about 100 yards below the point at which sewage affluent enters the river.	A large number of streptococci were found. Streptococci II. to V. (both inclusive). *See* Table VI.
8	Soil VIII. Mud from Thames bank, near Blackfriars.	A large number of streptococci were found. Streptococci VI. and VII. *See* Table VI.
9	Soil IX. Peaty soil from moorland.	Negative result.
10	Soil X. Mud from bank of river Colne at London Colney.	" "

It will be seen that a negative result is recorded in seven out of ten experiments, namely, in soils II., III., IV., V., VI., IX., and X. In the remaining three soils a positive result was obtained, namely, in soils I. (sewage field); VII. (Brent mud); and VIII. (Thames mud). Speaking in general terms it may be said that streptococci could not be found in the pure or moderately pure soils, but that in the grossly polluted soils they were found to be present in large numbers.

These results must be considered not unimportant if Streptococci (*as a class*) are derived from the intestinal tract, and tend, outside the animal body, not only not to multiply, but to rapidly lose their vitality and to die. Their presence would seem to indicate *recent* and *objectionable* pollution. Of course there may be exceptions to the rule; *e.g.*, some streptococci may be able to multiply in nature outside the animal body, while others may be peculiarly resistant, and so avoid extinction when the conditions are unfavourable. Again, there may be streptococci in nature which do not owe their origin to excremental matter. However this may be, detection of streptococci in polluted soils, and failure to detect them in non-polluted soils, is of interest in view of the circumstance that, as a result of examination of a large number of samples of crude sewage and effluents from biological filters, I have found streptococci present in great numbers (1,000 or more per cc.) in such material; and I have been encouraged accordingly to go on to test water-supplies as to the presence or absence in them of these micro-organisms. The results obtained in this connexion are recorded in Appendix B. No. 6.

Table VI. in addendum to this report gives an account of the chief morphological and biological characters of the streptococci isolated from the "soil waters." It shows that streptococcus I. was isolated from "soil water" I. (*see* figs. 13 and 14, Plate XIII.). Streptococci II., III., IV., and V. from "soil water" VII. Streptococci VI. and VII. from "soil water" VIII.

These streptococci were all capable of being differentiated from each other even on the basis of a few characters, as is shown by the subjoined "tree," although, no doubt, some of them were so closely similar as almost to suggest identity of species.

The results, as regards pathogenicity, were, on the whole, negative. Thus, mice inoculated subcutaneously with 1 cc. broth culture were apparently unaffected in the case of streptococci I., II., VI., and VII.; and in the case of streptococci III. and V. they died as late as the thirteenth and twelfth day respectively, and were, moreover, affected with other parasitic disease. Streptococcus IV. produced (apparently) death on the sixth day, but no streptococci could be isolated from the splenic juice. Of course, it does not follow from these results that the streptococci were necessarily non-virulent to all animals, indeed, we know that mice are insusceptible to certain streptococci which are virulent as regards human beings, or that at a stage prior to their isolation they were non-pathogenic. It may have been on the contrary, that they all were pathogenic at one time, but had lost their virulence through sojourn in soil and outside the animal body. Still there is no gainsaying the fact that, if these streptococci had been distinctly pathogenic in character, their importance would have been greatly increased.

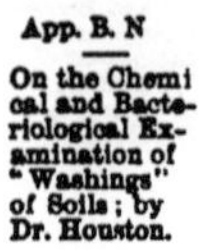
App. B. N

On the Chemical and Bacteriological Examination of "Washings" of Soils; by Dr. Houston.

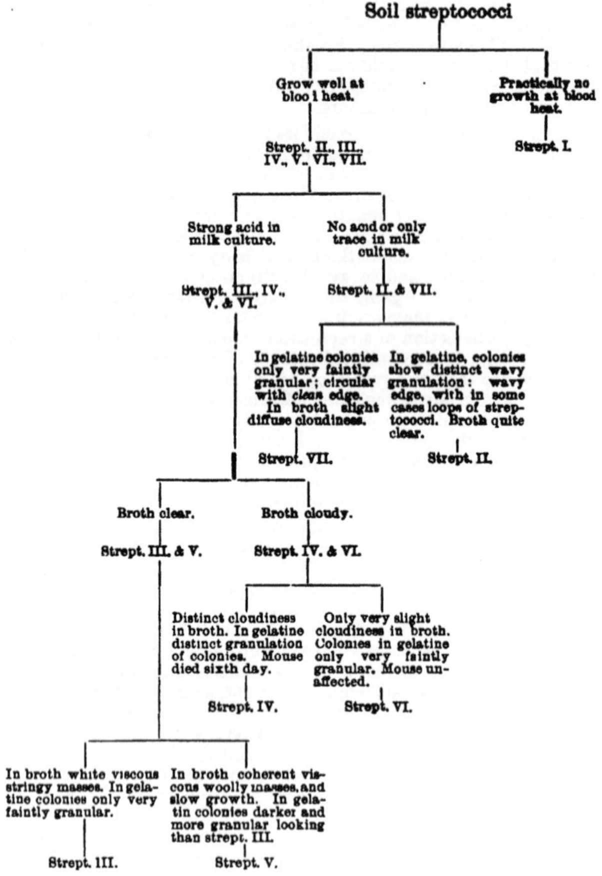

The bacteriological results are shewn diagrammatically in diagram 2 (at end of this report).

Concluding Remarks.—The results obtained in this enquiry clearly show that chemistry can not always be relied on to detect in a water supply the presence of "flood water," much less of "flood water" of objectionable or dangerous sort. It is true that some of the "soil waters" in question would have been condemned on chemical analysis; but it is also evident that the "washings" from some soils of unobjectionable character yielded higher results (chemically as regards organic matter) than those known to be grossly and, perhaps, dangerously polluted.

The bacteriological results, if not entirely satisfactory, yielded at all events, results of an encouraging nature.

It is quite clear that the access of "flood water" from surface soils to a water supply would be likely to be associated with an increased ratio of spores to total number of bacteria, along with the presence in the water of bacteria almost peculiar to soil, *e.g.*, B. mycoides and cladothrix. If in addition B. enteritidis sporogenes (spores of), and possibly B. coli and streptococci be detected, there are obtained, apart from the total number of bacteria, valuable data in drawing conclusions as to the fitness of such a water for domestic use. That the pollution of a water supply *directly with sewage* might give rise, at all events, as regards B. coli, B. enteritidis, and streptococci to somewhat similar results, is of little moment, since such contamination would be more dangerous than the *indirect* pollution of water through the medium of impure soils.

At the risk of repetition, attention may be again drawn to certain of the chemical and bacteriological facts obtained.

Soil I., taken from a sewage field, yielded, as regards its "washings," the following results. Chemically examined the results were :—0·04, 0·0147, 0·305 pts. per 100,000 respectively of free ammonia, albuminoid ammonia, and oxygen absorbed from permanganate (1 hr. at 100° C.).

Bacteriologically examined the results were :—Bacteria in 0·1 cc. so numerous that they could not be counted. A large number of B. coli (or closely allied forms) present in 0·1 cc. Spores of B. enteritidis sporogenes present in 0·1 cc. Streptococci present in 0·1 cc.

Contrast these results with those obtained in the examination of soil IX., a virgin peaty soil.

Chemically, the figures in this case were :—0·02, 0·025, 0·927 respectively of free ammonia, albuminoid ammonia, and oxygen absorbed from permanganate.

Bacteriologically the results were :—211 bacteria per cc. 117 spores of bacteria per cc. No colonies of B. coli in 0·1 cc. No spores of B. enteritidis in 2 cc. No streptococci in 0·1 cc.

On the other hand soils VII., and VIII., taken from the banks of two grossly polluted rivers, are roughly in agreement. Chemically and as regards soil VII., the results were :— 0·0227, 0·024, 0·298 respectively of free ammonia, albuminoid ammonia and oxygen absorbed from permanganate. The corresponding figures in the case of soil VIII. were 0·0187, 0·0134, and 0·174. Bacteriologically the results were :—3,470 and 2,200 bacteria per cc. of "soil waters" VII. and VIII. Colonies of a bacillus nearly akin to B. coli, if not of B. coli itself, in, 0·1 cc. "soil water" VII. As regards 0·1 cc. "soil water" VIII., the colonies were less nearly related to B. coli. Spores of B. enteritidis present in 1 cc. of both "soil waters." Streptococci present in 0·1 cc. of both "soil waters."

Lastly, soil V., a pasture soil which had never been manured, gave the following figures :—

Chemically, 0·027, 0·0227, 0·372 respectively of free ammonia, albuminoid ammonia, and oxygen absorbed from permanganate.

. B. No. 5.

e Chemi-
d Bacte-
cal Ex-
tion of
ings"
ls; by
ouston.

Bacteriologically, 440 bacteria and 216 spores of bacteria per cc.: the majority of the colonies developing both in the ordinary plate and in the plate for spores being those of B. mycoides. No coli in 0·1 cc. No spores of B. enteritidis in 1·0 cc. No streptococci in 0·1 cc.

It is evident that notwithstanding the amount of work that has been carried out in this connexion, further information is desirable. In particular, the results seem to encourage a research, having for its object the parallel examination, chemical and bacteriological, of the "washings" of soil in *more dilute* form than was here practised. In short, there would seem, judging from the results so far obtained, to be an indication that if water was contaminated with an amount of soil, or the "washings" of soil, so minute as altogether to defy detection by chemical means, that such pollution could still be demonstrated bacteriologically if a sufficiently large bulk (*e.g.* 1000 cc.) of the water were dealt with. Such amount should be filtered through a sterilised Pasteurs' filter, the surface of the filter brushed with a sterile brush into 5 to 10 cc. sterile water, and the "filter brushing suspension" of bacteria and water submitted to detailed bacteriological examination.

At the end of this report will be found Table VI. showing the chief morphological and biological characters of certain streptococci isolated from the "washings" (termed "soil waters") of soils.

Diagrams 1 and 2 showing the results of the chemical and bacteriological examination of the various "soil waters."

Figures 13 to 17, Plates XIII., XIV., and XV. (in illustration of Appendix B. No. 5) will be found at the end of Appendix B. No. 6.

DIAGRAM I. Showing as regards free Ammonia, Albuminoid Ammonia, Sum of the Ammonias (free & Albuminoid), & Oxygen absorbed from permanganate, the results of the chemical examination of the various "Soil Waters"

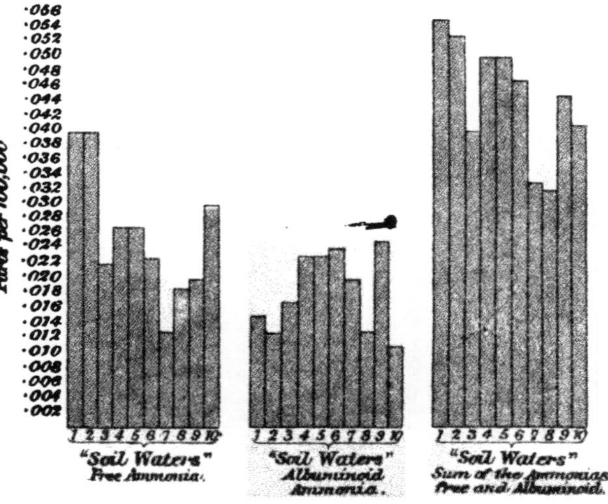

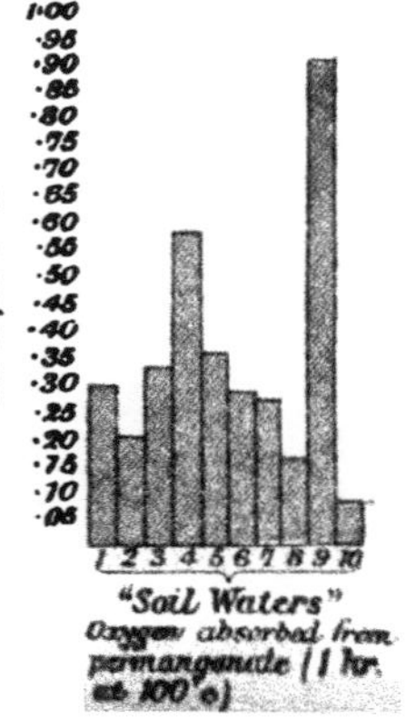

B. Coli, Spores of B. Enteritidis Sporogenes (Klein), & Streptococci, the results of the Bacteriological examination of the various "Soil Waters."

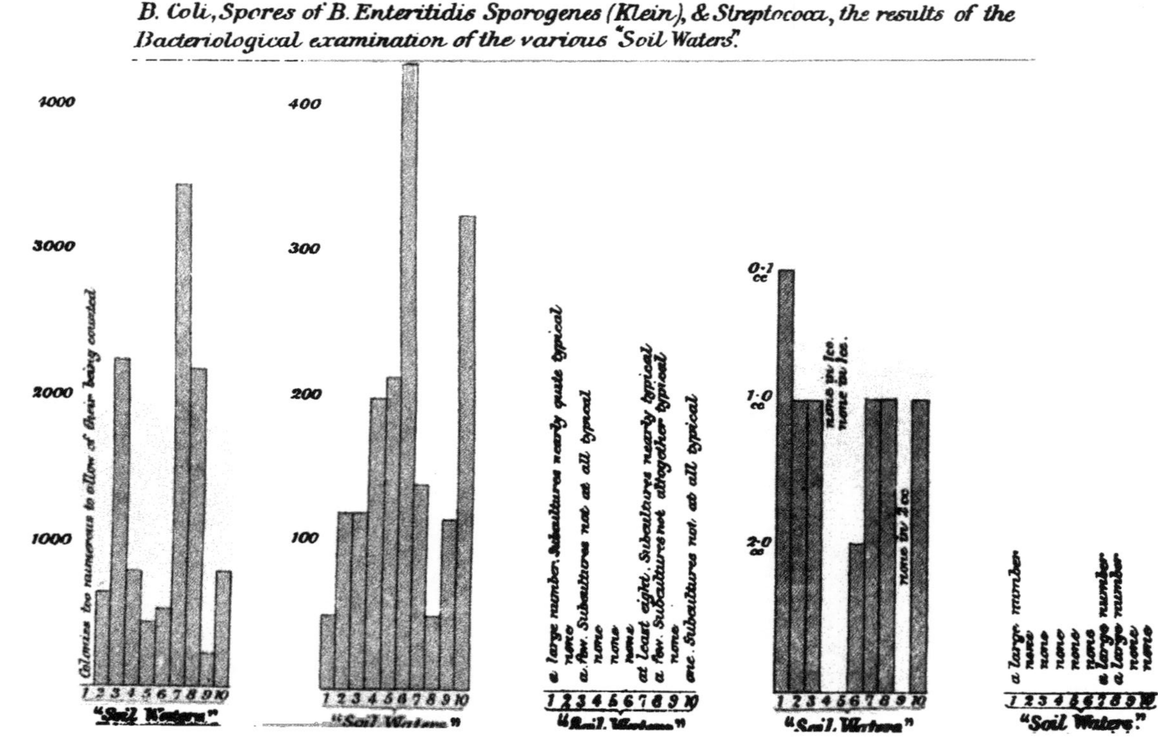

ue agar tures.	Broth cultures.		Remarks.
ansparent s.greyish- in colour, aving a ar appear- n old cul- me of the s showed spot cen-	Appearance very characte The broth remained per transparent (even in cu repeatedly shaken, and for some months) and s a conglomerate, white, w mass was to be seen.	h .)	The gelatine cultures were incubated at 20° C., and the agar broth and litmus milk cultivations at 37° C. There was, however, a single exception, namely streptococcus I. This organism grew either not at all or only in an imperfect manner at blood heat.
semi- rent col- greyish- n colour. granu- l usually lar in	The broth remained quite c and at the foot of the tu few white, woolly masses c be seen, but the mass of growth was joined togeth form a larger viscous, st white mass.	h .)	The streptococci were isolated from "surface" agar plates inoculated with 0·1 cc. "soil water," and incubated at 37° C. There was, however, one exception, namely, streptococcus I. This organism was isolated from a "surface" gelatine plate, incubated at 20° C. and inoculated with 0·1 cc. "soil water."
greyish- colonies, ansparent, granu- l of cir- ape.	Clear broth, but at foot along lower sloping sid tube, stringy, viscous m and threads. On ge shaking the tubes these flo throughout the medium white, thread-like growth	h) n d r d o	
es closely III.	Distinct but not abun diffuse cloudiness ; at foo tube, viscous white depos	h) s 's n n s,	
is strepto.	Absolutely clear broth, wit foot of tube, a small, coher viscous, white, woolly mas Grows slowly.	h .) s d ve re	
greyish- ranspar- nies.	Broth only very sligh clouded. At foot of tube on lower sloping side wh streak-like growth. On gen shaking the tube these come diffused throughout medium as stringy masses.	h .)	
s strep- VI.	Slight diffuse cloudiness. V like growth on lower slop side of tube. On gently sh ing the tube, a white, visc growth rose from foot of t in spiral fashion.		

e fig. 13, plate XIII.

NOTE by DR. HOUSTON on BACTERIOSCOPIC EXAMINATION of DRINKING WATER, with PARTICULAR REFERENCE to the RELATIONS of STREPTOCOCCI and STAPHYLOCOCCI with Water of this Class.

Not many years ago, the bacteriological examination of a water was practically limited to the enumeration of the total number of germs of all sorts present in one and another sample.

Later, bacteriologists began to insist that not the mere number, but the *character* of the microbes present in the sample was of essential importance in determining the wholesomeness or otherwise of a water. And, in support of this view, it was pointed out that of two different waters, one might contain a host of germs, none of them harmful, whereas the other, presenting only a few microbes, might contain hardly any but micro-organisms of dangerous kind. It is a curious fact that this almost self-evident proposition, although strenuously advocated, was for long unaccompanied by definite indication as to the kinds of microbes that were adventitious, and to be regarded therefore with suspicion, and those to be looked upon as peculiar, proper as it were, to pure water. It is true that putrefactive bacteria were placed in the former category, and that certain other bacteria were considered by one or another worker to belong to the latter class; nevertheless, as ability to identify pathogenic bacteria increased, it was too easily assumed that the failure to discover the germs of disease in a water implied "purity and safety." But soon it came to be recognised that all bacteria are in a sense putrefactive, and that certain well known putrefactive microbes may be present in water, apparently above suspicion of pollution; and, further, that in a given water pathogenic micro-organisms are not discoverable, or can be isolated only on rare occasions, even when such water is known to be grossly, and not long since specifically, contaminated.

At the earlier period referred to, Bacillus coli was unknown, or, at all events, had not assumed the great importance attached to it at a later date. But when B. coli came to be recognised as a microbe characteristic of, and especially abundant in, intestinal discharges, further interest and impetus was given to the bacteriological examination of water. Unfortunately, however, the *mere presence* of B. coli in a water, not its *relative abundance there*, was the point at first insisted on; so that when other authorities asserted that B. coli was not only abundant everywhere, but present also in water that was above reproach, a strong reaction set in, and this micro-organism as an indicator of danger fell into comparative disfavour.

The chief arguments advanced against acceptance of the presence of B. coli in water, as indicative of dangerous fouling of such water, may be briefly stated as follows:—

B. coli (1) is abundant everywhere; (2) it multiplies readily outside the animal body; (3) it occurs in the excreta of mammals and birds, as well as in the intestine of man.

APP. B., No. 6.

On Bacterioscopic Examination of Drinking Waters; by Dr. Houston.

It is not proposed in this brief note to deny the above propositions as regards B. coli. All that I am proposing, in reference to this microbe, is to adduce certain considerations tending to the view that demonstration of its presence in a water may be of no small value as indicating the antecedents of that water.

In the first place, if B. coli is abundant everywhere, it might be inferred that it would be specially abundant in soils, since soil is regarded as the great resting place for micro-organic life. But this does not appear to be the case, for, out of 21 samples of surface soil* obtained from different sources (for example, soil from orchards, gardens, moorland, pasture, sand pits, sea shore, sewage, fields, &c.; in brief, pure soils, and soils grossly polluted either recently or at no very distant date with manure, urine, and fæces), only four yielded unequivocal evidence that they contained B. coli. It is true that 13 (two doubtful) out of the 21 samples of soil yielded colonies in phenol gelatine plate culture (made from a phenol broth culture, inoculated with $\frac{1}{3}$ of a gramme of soil, and incubated at 37° C. for 24 hours), sufficiently like B. coli to suggest identity with that microbe; but of these 13 doubtful micro-organisms six failed to respond in 24 hours to the gelatine "shake" culture test, and the other three failed in one or more respects to identify themselves with that bacillus. And, of course, it might be argued that B. coli was really present in more than half of these 21 soils, but had become so altered by a long sojourn in the soil as to be no longer identifiable. However this may be as regards soil, it is certainly not customary to condemn a water because it contains germs bearing a *remote* resemblance to B. coli, and which, therefore, might conceivably be regarded as the descendants of a strain of B. coli typical in generations gone by of this micro-organism.

In the second place, if B. coli, under favourable conditions, multiplies outside the animal body, there is every reason to believe that when the physical conditions are unfavourable its growth is not only inhibited but that it loses its vitality and dies.

In the third place, in the absence of conclusive proof to the contrary, it is unsafe to assume that the excreta of healthy (much less of diseased) mammals and birds are harmless to man.

Lastly, it is not the mere presence of B. coli that should tend to condemn a water, but its relative abundance therein. It is an incontestable fact that B. coli is absent, or present in small amount only, in waters above suspicion of recent pollution, whereas in sewage†, which is the most usual, and presumably the most dangerous, source of contamination of water supply, it is present in enormous numbers. My own records, which extend over a year, and deal with a large number of samples, show that in raw London sewage the numbers usually exceed 100,000 per cubic

---

* Report of the Medical Officer, Local Government Board, 1897–8, App. B., No. 2.

† Sewage is not, of course, the only source of pollution of water; but no substance (which is not *equally objectionable* in character) at any time likely to contaminate a water supply contains this micro-organism in numbers at all comparable to sewage.

APP. B., No. 6.

On Bacterioscopic Examination of Drinking Waters, by Dr. Houston.

centimetre. And yet the same sewage, in the proportion of 0·1 per cent. would contaminate a pure water, from the chemical point of view, to an extent hardly if at all appreciable. Such a contaminated water would nevertheless yield on culture 100 coli per cubic centimetre, a number altogether beyond that which the least hostile of responsible bacteriologists would consider permissible in a drinking water.

Passing now to other criteria as to dangerous contamination of water.

Quite recently a fresh impetus has been given to the bacterioscopic analysis of water, by the discovery by Dr. Klein of an anaërobic pathogenic spore-forming micro-organism, termed by him B. enteritidis sporogenes.* It is safe to anticipate that this discovery is destined to largely enhance the value of bacteriological test of potable waters. For the spores of this anaërobe cannot be demonstrated in pure water,† whereas in impure water it can readily be shown to be present, and in raw sewage it is specially abundant. My own records, which, as I have said, extend over a year, and deal with a very large number of samples of sewage, and in particular of raw London sewage, show that the spores of this microbe are present in numbers varying usually from 100 to 1,000 or more per cubic centimetre. It cannot fairly be said of this anaërobe that it is likely to multiply outside the animal body, and the only possible objection to accepting its presence as an index of dangerous contamination of water is that the fact that its being a sporing anaërobe weakens somewhat its usefulness as evidence of recent,‡ and therefore presumably specially dangerous, pollution. Nevertheless, and so far as can be seen at present, it not only equals, but far excels any known test, chemical or bacteriological, for inferring the wholesomeness or otherwise of a drinking water.

In studying the bacteriology of polluted soils, of crude sewage and sewage effluents, and of impure waters, I have endeavoured to find some organism, or class of micro-organisms, which might be of value, if present in water, as indicating *recent and objectionable pollution.*

In polluted soils, in crude sewage, in sewage effluents, and in impure waters, I have found streptococci and staphylococci to be present, often in great numbers.

Although not discarding the class of germs known as staphylococci,§ I lay less stress upon them for the reason that they comprise hardy germs capable of persisting under conditions the reverse of favourable. Streptococci on the other hand may, as a class, be thought of as germs especially liable to discouragement

* "Centralblatt für Bakteriologie und Parasitienkunde," Band xviii., No. 241. Band xxii., No. 5, 20, 21. Reports of Medical Officer. Local Government Board. 1895-6, and 1897-8.

† By this it is not only meant that B. enteritidis sporogenes is absent in 1cc. of a pure water, but that it is likewise absent in the bacterial contents (obtained by the Pasteur filter brushing method) of it may be as much as 500cc. or more of such water.

‡ Dr. Klein's later results lead one to believe that an observed variability of this bacillus as regards virulence may possibly suffice to indicate a close or less close relation to recent pollution.

§ *See*, however, table of results.

[illegible]

[illegible] follows. In [illegible] large quantities of the [illegible]

[illegible] streptococcus, showing [illegible] Found in Freiburg water by Tils.

[illegible] streptococcus, showing [illegible] Found in Freiburg water by Tils.

[illegible] table [illegible] page [illegible].

water through a sterile Pasteur filter, and to brush, with a sterile brush, the material left on the surface of the filter into a few cubic centimetres (5 to 10cc.) of sterile water. Of this suspension of bacteria and water 0·1 to 0·005cc. is added to an agar plate, and a surface culture made. In the case of impure water it was often sufficient to make a surface agar plate direct from the water, and, conversely, in the case of foul water it was occasionally necessary first to dilute the sample with a definite quantity of sterile water. The agar plate is incubated at 37° C., because, although there are certain streptococci and staphylococci which will not grow at this temperature, it is, for the majority, the most favourable temperature, and it is precisely these latter micro-organisms that it is most important to investigate. The plate was examined after 24 hours, and all the minute colonies examined under a low power of the microscope (Leitz No. 2 eyepiece, No. 3 objective). Sometimes it was possible to find in this way colonies which could with tolerable certainty be diagnosed as streptococci, but in most cases the diagnosis was doubtful, and, as matter of fact, in all cases sub-cultures were made. The colonies are so minute that it is almost impossible to obtain an amount of growth sufficient to make satisfactory microscopic cover glass preparations. This being so, it was found best to inoculate a broth tube from the colony to be tested, using a fine platinum needle for the purpose; to incubate it at 37° C.; and afterwards to resort to further sub-cultures from this tube if a seemingly positive result was obtained on microscopic examination of its contents. Frequently a negative result was obtained; there being either no growth* (the streptococcus having already lost its vitality), or the growth manifested proving that of some other microbe, the solation of which was not aimed at nor desired. But, commonly, if a number of tubes are inoculated from separate doubtful colonies, and the water under examination be impure, streptococci are usually found in one or more of the sub-cultures.

Results of Experiments.

A brief record of the chief results obtained is given in the following table :—

* I have noted this so frequently that the absence of growth in the culture media employed cannot well be referred to some error of manipulation. Dr. Gordon, too, informed me that he has more than once cut out the fragment of agar on which a streptococcus colony rested, and transferred it bodily into a broth tube, and yet no growth occurred therein.

TABLE showing the RESULT of BACTERIOLOGICAL EXAMINAT of STREPTOCOC

Experiments.	Name and Source.	Morphology.	Surface Gelatine Plates at 20° C. under low power (Leitz, 2 eyepieces, 3 obj.).	Surface Agar Plat under low p (Leitz, 2 eyepiec
1 (a)	Staphylococcus A_1... River (Z) above intake of water supply. Present in 10 c.c. of water.	Stains with Gram. Cocci matted together; also short chains.	See remarks	See Remarks ..
1 (b)	Staphylococcus B_2... From the same culture as A_1; possibly the same organism, and certainly a closely allied form.	Stains with Gram. Cocci matted together; also short chains. Matting less distinct than in A_1.	Ditto	Ditto .. ..
2 (a)	Streptococcus C_3. .. River (Y) above intake of water supply. Present in 10 c.c. of water.	*Stains with Gram. Chains of cocci of medium length, not matted in masses.	Circular, transparent, minute, faintly granular, yellowish-white colonies. No loops of cocci visible at edge of colonies; no liquefaction.	More or less circu colonies of yellov colour, faintly with unbroken e
2 (b)	Streptococcus D_4. .. From the same culture as C_3.	Stains with Gram. Chains of cocci of short length; also little masses of cocci.	See under oblique gelatine	No record ..
3 (a)	Staphylococcus E_5... River (Z) above intake of water supply. Sample collected at a later date than in previous experiment. Present in 10 c.c. of water.	Stains with Gram. Cocci matted together; also short chains.	See remarks	See remarks ..
4 (a)	Staphylococcus F_6... River (Y) above intake of water supply. Sample collected at a later date than in previous experiment. Present in 10 c.c. of water.	Stains with Gram. Cocci of somewhat irregular shape and more or less clumped together, also short chains.	Ditto	Ditto .. ..
5 (a)	Staphylococcus G_7... River (Y) above intake of water supply. Sample collected from the same source as the previous sample, but at a later date. Present in 0.1 c.c.	Stains with Gram. Rather large cocci in masses; ? also short chains.	Ditto	Ditto . ..
6 (a)	Staphylococcus H_8... River (Z) above intake of water supply. Sample collected from the same source as in Experiment 3 (a), but at a later date. Present in 0.1 c.c.	Stains with Gram. Cocci aggregated together, but ? also short chains.	Ditto	Ditto .. ..

* See Figure 18, Plate .

…es of Water, obtained from Various Sources, for the Presence …ntally) for Staphylococci.

… at 37° C.	Litmus Milk at 37° C.	Oblique Gelatine at 20° C.	Animal Experiments.	Remarks.
…loudiness in … At foot of …ite and very …eposit.	Strongly acid, but no clot even after 15 days' incubation.	An opaque, white, more or less confluent growth (staphylococcus-like). Very slow liquefaction.	Mouse inoculated subcutaneously with 1 c.c. of broth culture (24 hours at 37° C.); died third day. Cocci in splenic juice, and agar plates therefrom showed colonies of apparently the same organism.	Although at first in some doubt whether to consider this organism a staphylococcus or a streptococcus showing (morphologically) staphylococcus formation, it was finally classed as a staphylococcus, and its further biological study abandoned. Possibly remotely akin to staphylococcus albus.
…itto.	Ditto.	Ditto.	Mouse inoculated subcutaneously with 1 c.c. 24 hours' broth culture; died sixth day.	Ditto.
…ght diffuse …s: bacterial …t foot of tube …y abundant, very viscous. …r sloping side … streaky, …ng white	Very feeble acidity; no clot (15th day).	Minute greyish-white circular colonies, nearly transparent. No liquefaction.	Mouse apparently unaffected by subcutaneous inoculation of 1 c.c. 24 hours' broth culture.	Unquestionably a streptococcus.
… diffuse …s. At foot of …hite viscous	? trace of acidity; no clot (15th day).	*No growth occurred.*	Ditto.	Unquestionably a streptococcus. (Apparently an organism of *feeble vitality.*)
… diffuse …ss in 24 hours. …f tube white, …scous bacte…sit.	No visible change up to eighth day. Eventually alkali production well marked.	Grows very slowly. Thin transparent growth which eventually assumes a faint citron colour. By 13th day slight liquefaction.	Ditto	Same remarks as in Experiment 1 (*a*). (Possibly remotely akin to staphylococcus citreus.)
…loudiness in … At foot of …ite bacterial not very vis…	No clot; at first only a slight alkaline change, but later alkalinity extremely well marked.	Opaque, white confluent, staphylococcus-like growth. No liquefaction.	Mouse inoculated subcutaneously with 1 c.c. broth culture (24 hours at 37° C.); died sixth day.	Same remarks as in Experiment 1 (*a*).
…oudiness in 24 … At foot of …ery viscous …l deposit.	Strong acidity, but no clot (15th day).	Golden-yellow growth, and rather rapid liquefaction.	Mouse inoculated subcutaneously with 1 c.c. broth culture (three days at 37° C.) remained well, and was apparently unaffected.	? Staphylococcus pyogenes aureus, or closely allied form.
Ditto	No visible change 15th day.	Citron-coloured growth, accompanied by rapid liquefaction.	Animal not inoculated.	? Staphylococcus pyogenes citreus, or closely allied form.

Experiments.	Name and Source.	Morphology.	Surface Gelatine Plates at 20° C. under low power (Leitz, 2 eyepiece, 3 obj.).	Surface Agar Plates under low pow (Leitz, 2 eyepieces,
7 (a)	Streptococcus L[illegible] .. River (Z) above intake of water supply. Sample collected from the same source as in Experiment 6 (c), but at a later date. Present in 0·5 c.c.	Stains with Gram. Long chains of cocci; also masses made up of chains.	†Yellowish-grey, transparent colonies showing wavy granulation. Edge sinuous, and made up of loops of streptococci. Sometimes the chains of cocci run in parallel rows, and sometimes as single loops. Resembles streptococcus longus. No liquefaction.	Yellowish-brown, transparent colo irregular shape a size. The chains give the colonies granulated ap The loops of c readily seen at t phery. Resemble coccus longus.
8 (a)	Streptococcus J_{10} .. River (X), a foul water, not used for domestic purposes. Sample collected 100 yards below the point of entrance of a sewage effluent. Present in 0·001 c.c.	Stains with Gram. Cocci chiefly in adherent masses, but also separate chains.	Colonies minute, circular, transparent, yellowish-grey in colour, showing very faint granulation. Each colony usually shows a dark spot centrally placed. The edge is *clean*; there are no loops visible. The minute size and transparency of the colonies alone suggests that they are streptococci. No liquefaction.	Here the colonies a less transparent, granulation, whic in character, is m marked. The col usually more or l lar in shape. The loops visible at t the colonies. Th usually *clean*, b times has a frayed ance.
8 (b)	Streptococcus (?) K_{11}. From the same culture as J_{10}.	Stains with Gram. Cocci chiefly in masses, many of them of large size and of irregular shape; ? also chains.	Minute, transparent, yellowish-grey, nucleated, faintly granular colonies, with *clean* edge. No liquefaction.	Here the colonies a and less transpa show a wavy a granulation. Som colonies are quite with *clean* or c slightly fraye Other colonies (u smaller ones) are in shape, with b edge and a sus loops but these lo open out as in I whole colony is n pact.
8 (c)	Streptococcus (?) L_{12}. From the same culture as J_{10}.	Stains with Gram. Cocci chiefly in masses, many of them of large size and of irregular shape; ? also chains.	Colonies very similar to K_{11}, but, although eventually attaining the same size, they grow more slowly at the start.	Colonies very simi but although e attaining the s they grow more the start.

† *See* Figures 19 and 20, Plate XV.

...at 37° C.	Litmus Milk at 37° C.	Oblique Gelatine at 20° C.	Animal Experiments.	Remarks.
...h. At foot of ...olly masses ...ad to cohere, ... somewhat ...o platinum	Practically no change even after six days. ? however, very feeble acidity.	Transparent, coarsely-granular colonies, which tend to remain separate, and not to form a continuous growth. No liquefaction.	Mouse inoculated subcutaneously with 1 c.c. broth culture remained apparently unaffected.	Unquestionably a streptococcus, and one bearing a close resemblance in many respects to streptococcus longus. It did not, however, produce distinct acidity in milk.
...it not very ... cloudiness ...in 24 hours. ...of tube a ...l somewhat ...posit.	In three days, distinct acidity but no clot; no clot sixth day.	Minute, circular, transparent colonies. No liquefaction.	Mouse apparently unaffected by subcutaneous inoculation of 1 c.c. broth culture (24 hours at 37° C.).	Unquestionably a streptococcus.
...ains quite ...n sides of ...at the foot, ...hite masses ...seen, which ... somewhat ...ture.	Practically no change visible even after several days (eight). ? however, trace of acid.	Ditto.	Mouse apparently unaffected by subcutaneous inoculation of 1 c.c. broth culture (96 hours at 37° C.).	Difficult, if not impossible, to say whether a streptococcus showing (morphologically) staphylococcus formation, or a staphylococcus simulating (biologically) a streptococcus. The slow rate of growth, clear broth, minute transparent colonies in gelatine and agar, and absence of visible change in milk, is in favour of K_{11} being a streptococcus. Morphologically, K_{11} is more like a staphylococcus.
...ains quite ... the original ...e growth at ...of the tube ... the size of a ...ad on the ...y, and con... a small, ...uffy viscous	No visible change even after several days (eight).	Same as K_{11}, but the colonies take longer to show themselves. No liquefaction.	Mouse apparently unaffected by subcutaneous inoculation of 1 c.c. broth culture (three days at 37° C.).	Practically the same as K_{11}.

Before dealing with the results recorded in the above table, all of which refer to unfiltered waters, it is important to refer to certain other experiments which yielded a negative result.

Experiment A.

River T., after filtration: the same water as in experiment [illegible] of the Table, but after being filtered. No streptococci found in 10cc. of water.

Experiment B.

River T., a pure upland water. No streptococci found in 10cc. of water.

Experiment C.

River T., above intake for water supply: the same water as in experiment [illegible] of the Table. No streptococci found in 10cc. of water.

Summarising the results given in the Table:—

(1.) Eight samples of different waters, or of the same water obtained on different occasions, yielded positive results either as regards the presence of streptococci or staphylococci, or of both.

(2.) Three separate river waters contained streptococci. Of these:—

(a) Two were impure waters, to be used (after filtration) for domestic purposes.

(b) One was from a polluted river, quite unfit for use for drinking purposes. Here, streptococci were found in ·001cc.; but it may reasonably be affirmed that the contamination was not so gross as to be 1,000 times greater in amount, or 1,000 times more recent in character, than what pertains in the case of some at all events of our water supplies.

(3.) So far none of the streptococci isolated from water have proved pathogenic under the conditions of experiment.

(4.) Certain of the staphylococci killed mice.

In conclusion, I would urge that the results obtained encourage belief that further work in this direction is called for; that prolonged research of the above sort is likely to demonstrate, in a much more conclusive manner than this brief note can hope to do, that the subject is one which, notwithstanding its manifest importance, has not yet been accorded the attention it deserves.

Figs. 18, 19, and 20, Plate XV., are submitted in illustration of streptococci C_2 and I_9. As regards Fig. 20, it is open to the critic to suggest that it pourtrays a bacillus rather than a streptococcus. To this the answer is that *undoubted* streptococci under certain conditions and in some media normally assume a bacillary form, and that the streptococcus longus is a case in point.

PLATE XIII.

Fig. 13.

Fig. 14.

BACTERIA IN SOIL WASHINGS.

PLATE XIII.

FIG. 13.

Surface colony of soil streptococcus I. : from an agar culture.

[Magnifying power, 60.]

FIG. 14.

'Impression' preparation of soil streptococcus I.: from a gelatine culture. Stained by Gram's method.

[Magnifying power, 1,000.]

BACTERIA IN SOIL WASHINGS.

PLATE XIV.

Fig. 15.

Gelatine plate culture, for *total number* of bacteria, containing 1 cc. "soil water," V. Second day at 20°C.

[Natural size.]

Fig. 16.

Gelatine plate culture, for *spores* of bacteria, containing 1 cc. "soil water," V. Previously heated to 80°C. for 10 minutes. Second day at 20°C.

[Natural size.]

It is to be noted, as regards Figs. 15 and 16, that nearly all the colonies are those of B. Mycoides, and that the number of these microbes in the "spore" plate (fig. 16.) were nearly as numerous as the number in the "ordinary" plate (fig. 15.)

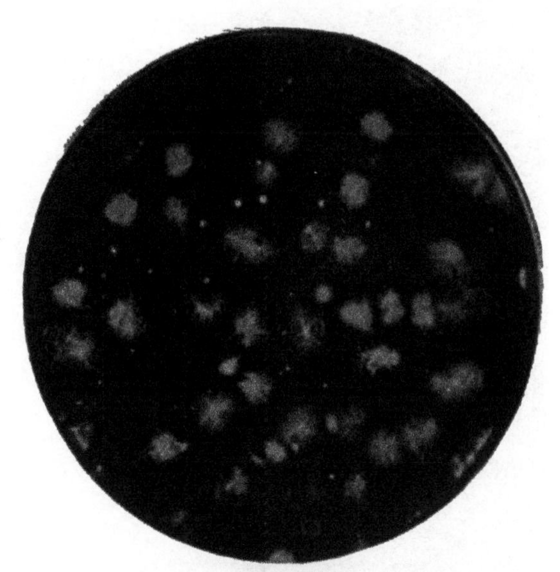

Fig. 15.

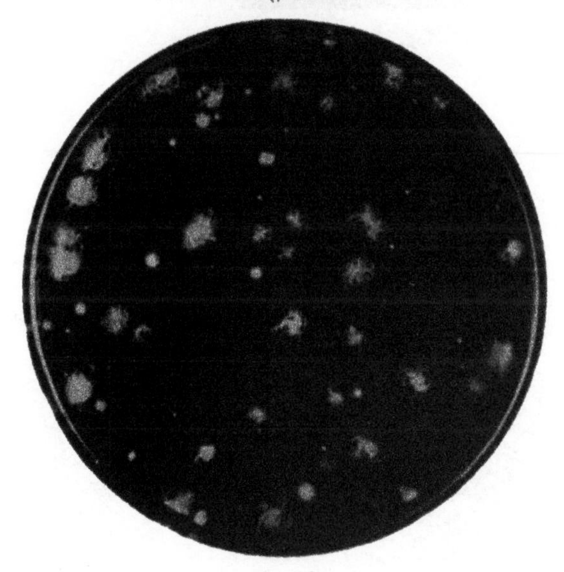

Fig. 16.

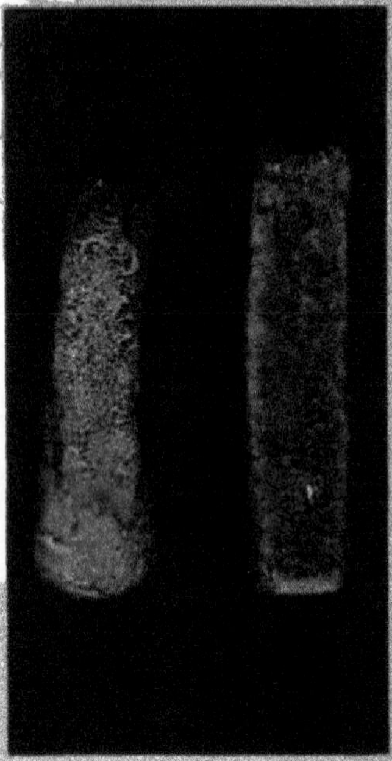

PLATE XV.

Fig. 17.

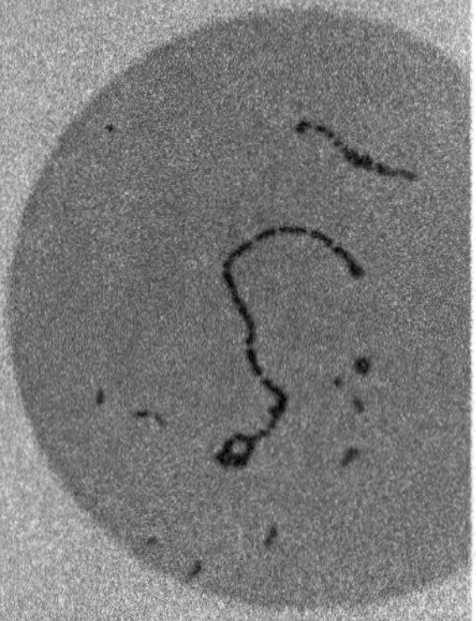

Fig. 18.

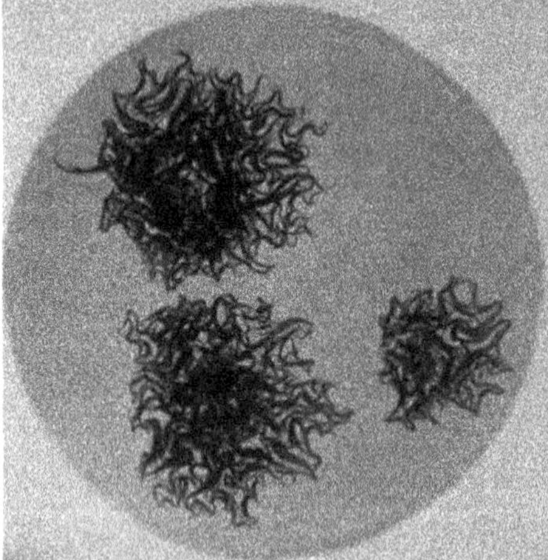

Fig 19.

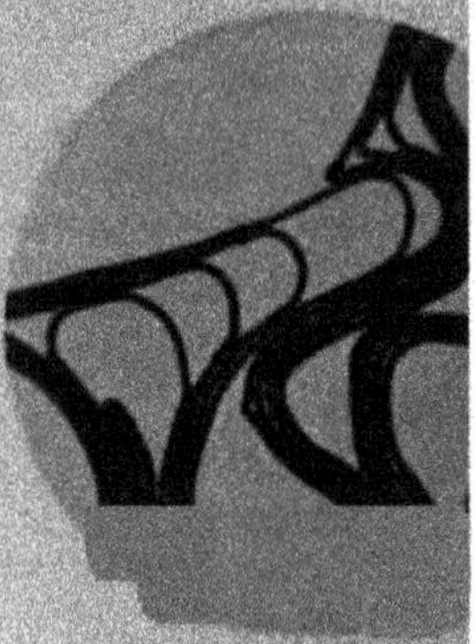

Fig. 20.

BACTERIA IN SOIL WASHINGS.

PLATE XV.

FIG. 17.

Potato culture of B. mesentericus isolated from soil III. Left side, variety I. Right side, variety E. 24 hours at 37°C.

[Natural size.]

STREPTOCOCCI FROM WATER.

FIG. 18.

Microscopic preparation from a broth culture (30 hours at 37° C.) of water streptococcus, C_3. Stained by Gram's method.

[Magnifying power, 1,000.]

FIG. 19.

'Impression' preparation from a surface gelatine plate culture of water streptococcus, l_8. Stained by Gram's method. 2 days growth at 20·C.

[Magnifying power, 50.]

FIG. 20.

The same preparation as Fig. 19, but more highly magnified.

[Magnifying power, 1,000.]

hand. I have managed to obtain some little information on the subject for which it was undertaken.

The Morphological and Cultural Identification of Streptococcus Scarlatinæ.

In few groups of micro-organisms have the individuals more features in common than streptococci; and yet, by careful attention to morphological and cultural details, considerable differences may be detected between representatives of this class. Streptococci, in fact, are not merely cocci growing in chains; chain-formation, though an important and obvious characteristic, especially in fluid media, is by no means the only form of their growth. For example, on solid media the ordinary streptococcus of pus forms a considerable number of colonies that display no chain-formation at all, and, on the other hand, streptococcus scarlatinæ shows at the edges of many of its younger agar colonies a chain-formation that is in marked contrast to the growth of the same organism on gelatine or even, in many cases, in broth. Again, a streptococcus obtained by me from a normal throat, though it formed fair-sized chains in broth, and grew on gelatine in colonies entirely composed of an open lattice-work of chains, developed on agar colonies that were proved by impression-preparations to show no tendency whatever to form chains.

Chain-formation, therefore, may be only an incident in the life-history of a streptococcus. Another feature demanding comment is the occasional production of large spherical elements which may exceed two or even three times the average size of those with which they are associated. These forms, which retain the stain more deeply than the others, as may be well seen in preparations from old cultures stained by Gram's method, have been mentioned before. It may be noted, however, that the opposite morphological phase is also sometimes present, namely, elements, or chains of them, several times *less* than the average size. A further important character of the streptococci which, so far as I am aware, has hitherto received little if any attention, is the tendency which many of them show here and there to form spindle and even rod-shaped elements. At times this may be so remarkable that it would hardly be surprising if the organism were actually to be mistaken for a true bacillus by observers insufficiently acquainted with the cultural morphology of streptococci. The resemblance of some of these occasional forms to diphtheria bacilli is especially striking. Moreover this tendency of streptococci to bacillary phases cannot be dismissed by calling them involution forms; for in some of the most prominent instances the culture was of only 17 hours' growth. This eccentricity is not limited to one kind of streptococcus: it may be seen in specimens of widely different origin and character. It is mentioned here chiefly to draw attention to a feature which has been lost sight of, though it is of primary import in understanding the true bacteriological position of the streptococci.

It is not my present purpose, however, to describe features which streptococci possess in common, but rather those in which they differ. At the same time it may be truly said that there is no morphological characteristic possessed by these organisms in

App. B. No. 7.
On the Bacteriology of Scarlatina: by Dr. Gordon.

common, in which they do not also show considerable differences in regard to the degree in which they individually possess it. The classification of the streptococcus group by Lingelsheim was based on their growth in broth. According to him streptococcus *Brevis* made broth turbid, grew in that medium in short chains, and was not pathogenic to either mice or rabbits; whereas streptococcus *Longus* left the broth clear, grew in long chains, and was always pathogenic to rabbits and sometimes to mice also. In the category of *Longus* came the streptococcus of pus, and also that of erysipelas.

This division of the streptococci into two classes is of great convenience, as a beginning; but, on further acquaintance with the group, its value is found to be somewhat limited. In the first place, diplococcus pneumoniæ grows so similarly to *Brevis* in broth that the diagnosis between them rests chiefly on the result of injecting a mouse. It is, however, the description of *Longus* that is the more inadequate of the two; for instance, it is not uncommon to find the broth turbid and to detect in it microscopically long chains. But in spite of many exceptions, Lingelsheim's bisection of the group might be provisionally adopted but for the fact that the streptococci of pus, erysipelas, and sepsis, that is to say, perhaps the best known streptococci, are, as regards the length of their chains, intermediate between the longest and shortest types.

It has been necessary, therefore, both to modify and to amplify the classification of Lingelsheim. Taking the length of the chains in broth as a primary criterion, the growth is also investigated in other fluids, such as litmus-milk, and gelatine at 37° C. The colonies on agar and also on gelatine plates are next observed, and their structure determined by means of impression-preparations. Proceeding thus, three main divisions of streptococci are distinguished, *Longus*, *Medius*, and *Brevis*. In this sense *Brevis* includes Lingelsheim's *Brevis* and also diplococcus pneumoniæ; the two being distinguished from other streptococci by the fact that they both make gelatine at 37° C. uniformly turbid in one night, while they differ from each other (1) in the pneumococcus chains being the straighter of the two, and (2) in the pneumococcus colonies on gelatine being much more slowly developed. In the class of *Medius* come the majority of streptococci, particularly those of pus, sepsis, and erysipelas; so that Lingelsheim's *Longus* class is included in this category. On the other hand, the term *Longus* is restricted to an organism that from the length and profusion of its chains is at the very head of the streptococcus race. Sub-divisions of both *Longus* and *Medius* make broth turbid. They do not, however, render gelatine at 37° C. turbid, as examples of *Brevis* do. (Plate XVI, Fig. 1, Tube 4.)

I have not included streptococcus scarlatinæ in the above triple division of the group, because, strictly speaking, this organism belongs to another class of the streptococci altogether. Indeed, owing to the marked coherency and grouped arrangement of a large proportion of its cocci in broth culture, streptococcus scarlatinæ or conglomeratus might, in some instances, be mistaken for a staphylococcus. The agar colonies, however, particularly if impressed, speedily dispel this error. Suffice it for the present to note that while streptococcus scarlatinæ shows a tendency in some of its characters to emulate streptococcus *Longus*, in others it approximates to *Medius*. As examples of both the latter classes, and, indeed, of *Brevis* also, exist in normal and in scarlatina throats, it is advisable, in order to avoid confusion, to describe the main differences between Longus, Medius, and Scarlatinæ in the various culture media in which I have been in the habit of growing them.

Streptococcus Longus.

The particular example now to be described I have termed streptococcus longus buccalis, from the fact that it is present in the normal mouth and fauces, and sometimes is the most numerous micro-organism in plate cultures obtained from thence.

Broth, 24 hours at 37° C.—The fluid remains absolutely clear. At the bottom of the tube is a white weedy mass from which streamers, not unlike strings of mucus, shoot up into the body of the fluid. On slightly rotating the tube, the growth is seen to have no coherency, a fact that is confirmed by distributing a portion of it on a cover-glass, over which it can be spread out with great facility. After a few days' incubation the growth may show rather more coherency. Microscopically, as in the case of other streptococci, the best preparations are obtained by staining with Gram's method. The growth consists entirely of cables, which are remarkable both for their length and for their comparative straightness. There is practically no group arrangement, which explains in some degree the lack of coherency. (Plate XVII, Fig. 2.)

Gelatine, 37° C. 24 hours.—The growth is similar to that in broth, except that it is more profuse. The macroscopic characters observed in broth-culture are consequently exaggerated, especially with regard to streamers. This is a useful medium as a control on the broth, and also because specimens may be fixed and kept for reference. (Plate XVI, Fig. 1, Tube 1.)

Litmus-milk, 37° C., 24 hours.—Again the character of the growth in broth is reproduced. At the bottom of the tube, and clearly outlined against the already faintly pink tint of the medium, is a white mass which may be proved to be entirely composed of long streptococcus chains. The slightly acid reaction of the milk after one night's incubation, is increased during the next few days. There is, as a rule, no clotting, not even in a week. Microscopically the growth is similar to that in broth.

The agar colonies grow rapidly, and in one night, at 37° C., often reach full size. They are of two kinds: (1.) Those constituting the majority are flat, feathery-edged, grey, and granular to the hand-lens. Under a low power they are seen to be composed of streptococci, which twist interminably to form a network of convoluted tendrils. (Plate XVII, Fig. 3) The chains are sharply defined, and are seen particularly well at the edges of the colonies which they frill with their tangles and returning loops. (2.) The colonies in the minority are more heaped up and limited. Occasionally no cables are visible at their edges, but a whole colony forms a dark coarsely granular mass which may exhibit nodulation. It is not always easy to distinguish these colonies from those of streptococcus scarlatinæ, except by a broth subculture.

Gelatine colonies, 20° C.—By the second day a plentiful crop is obtained of grey colonies having characteristic shapes and frilled edges. Impression specimens show that chain formation is also the main characteristic of the growth at this lower temperature. (Plate XVII, Fig. 4.)

In all these subsidiary media, therefore, the type of growth is consistent with that in the original broth-culture. It is curious to note that streptococcus longus closely resembles the anthrax bacillus in respect to the *shape* of many of its gelatine and agar colonies.

Streptococcus Medius.

The particular micro-organism now to be described as representative of this class came from pus.

Broth, 24 hours at 37° C.—The fluid is clear, and at the bottom of the tube is a collection of flakes, flocculi, and powder. By slightly rotating the tube it is seen that these particles show no special tendency to cohere. On spreading out some of the mass on a cover-glass the material is easily distributed. Microscopically, the specimen consists mostly of medium-sized curling chains, which are seen to be more disjointed and more sinuous than those of longus. Occasionally, where the chains happen to be closely set, some grouped arrangement occurs. It is, however, in many cases due to chains twisting on themselves, or crossing each other. (Plate XVIII., Fig. 5.)

Gelatine, 24 hours at 37° C.—The fluid is clear, even in the case of medius (B) where the broth is turbid. Small flakes and flocculi are scattered throughout the medium and collect especially at the bottom of the tube, where they may form a skin in some cases. (Plate XVI., Fig. 1, Tube 3.)

Litmus-milk, 24 hours at 37° C.—There is no white mass at the bottom as in the case of longus. The acid-production is faint and as a rule very slow, at the end of the week not equalling the amount which S. scarlatinæ produces in a night. Some streptococci of this class seem to produce no acid at all in litmus-milk. There is no clot even in a month in the majority of cases. Microscopically, the appearance of the growth is similar to that in broth.

The agar colonies are of two kinds. The majority are round, firm-edged, grey, and of either smooth or finely granular consistency. Sometimes the edge is slightly ragged from the presence of a few medium-sized chains. The minority of the colonies may show rather more of the network type. Impression specimens, however, prove that the chains are disjointed and of only medium length. It is not uncommon to see no network at all in agar plates of medius. (Plate XVIII., Fig. 6.)

Gelatine colonies.—By the second day, round or oval, greyish blue spots appear on the plate. These colonies are composed almost entirely of cocci arranged in group or staphylococcus formation. It is not common to see a chain at the edge of a colony. After some days, however, these may be more plentiful. At no time is anything approaching the longus type of gelatine colony produced. (Plate XVIII, Fig. 7.)

The medius class embraces the streptococcus of pus, pyæmia, septicæmia, ulcerative endocarditis, and puerperal septicæmia.

The examples of the streptococcus of erysipelas that I have at present seen come under this class. Medius is also represented in the normal and in the scarlatina throat.

Streptococcus Scarlatinæ.

The samples submitted to culture were in each instance obtained from the tonsil of persons suffering from Scarlatina.

Broth, 24 hours, at 37° C.—As noticed in previous reports, the fluid is quite clear. The growth occurs at the bottom of the tube as one large or several smaller white masses which when the tube is slightly shaken show a marked tendency to maintain their coherency. In the case of the single mass, it floats through the fluid as a flattened bun-like body. When spreading a piece of this mass on a cover-glass difficulty is often experienced in separating the material. It may even be necessary to squeeze the mass between two cover-glasses in order to obtain a layer thin and separated enough for microscopical examination. Microscopically, the mass formation is very marked, a point emphasised by Kurth. (Plate XIX., Fig. 8.) As a rule chain formation is not so obvious as in the case of medius, and may in some cases be difficult to make out at all distinctly.

Gelatine, 24 hours at 37° C.—The same appearance as in the broth but often exaggerated. In some instances of streptococcus scarlatinæ obtained from the throat, the growth approaches closely to the longus type in the same medium. In Plate XVI., Fig. 1, Tube 2 is seen a coherent mass which has been shaken off the bottom of the tube and become inverted. This culture of S. scarlatinæ was several days old.

Litmus-milk, 24 hours at 37° C.—As a rule there is a firm solid clot. In one night the acid-production is also very strong, and, in the common case where there is a clot as well, the lower half of the tube is yellowish white, the top layer being pink. This decolorisation of the lower half of the litmus-milk is due to a reducing action of the streptococcus; for, on breaking up the clot and allowing the air to get at the decolorised portion, it assumes the pink colour of the higher layer. Microscopically, there is more chain-formation than is usually observed in broth, but massed cocci are also evident.

The agar colonies, after 24 hours at 37° C., are of three types:—(1) the grey, granular, irregularly outlined tuberculated colony, excellent photographs of which appeared in last year's report: (2) colonies of similar kind and having a coherent appearance, but without tubercles: (3) younger and smaller colonies which have a fine frilling of chains around a more compact coherent centre. This chain-work may at times be so marked and continuous in detail that a near approach to the longus type is made. Perhaps, the most useful point to remember in searching for S. scarlatinæ colonies on agar plates is the granular, glossy, coherent centre. The tubercles again are a great aid. Taken as a whole, conglomeratus colonies on agar-plates from scarlatina tonsils are smaller, more opaque, and more irregular than those of the other streptococci present. An impression of an agar colony intermediate between types (2) and (3) is figured in Plate XIX., Fig. 9.

Gelatine colonies at 20° C.—Generally in two to three days, sometimes later, grey dots may be seen appearing on the surface of the gelatine.* These colonies are either circular or oblong, have as a rule a firm edge, and consist of a closely set coherent mass of cocci (Plate XIX., Fig. 10). In this case, as also in the case of other streptococci, the individual cocci appear to grow to a larger size at this lower temperature. Chain-formation is at times more marked than in the case of medius. Older colonies, as Dr. Klein originally noted, develop marked nodulation.

The coherency of the growth in broth, microscopically the marked mass-formation, the rapid clotting and the marked acid-production in litmus-milk are strong distinguishing features of S. scarlatinæ. The microscopical appearance of the growth in milk is also of importance. Add to these the growth in gelatine at 37° C. where it approaches longus; the gelatine colonies at 20° C. where it comes nearer to medius except that the colonies are slower to develop; and it seems that streptococcus scarlatinæ is between the two. The agar colonies confirm this view. In all media the coherency and excessive group-formation are particularly striking, and distinguish S. scarlatinæ completely from either of these two allied classes of streptococci.

Scarlatina Cases.

Much of the scarlatina material now to be reported on was obtained from the London Fever Hospital, by the kind co-operation of Dr. Hopwood. To Dr. Washbourn also I am indebted for allowing me to examine his patients there. Similarly I am beholden to Dr. MacCombie, the Medical Superintendent of the Brooke Hospital, and to the Assistant Medical Officer, Dr. Willcox, for the facilities they have afforded me for examining cases. I may add that all the cases herein reported on were undoubted examples of scarlatina in the opinion of the fever hospital experts.

Ear Discharges.

(1) R. T., male, aged 5½. Second to third week since the attack of scarlatina began, first day of the ear discharge. Cultures showed a few colonies of staphylococcus albus liquescens and a great majority of streptococcus medius.

(2) W., male, 3. Third to fourth week since the attack, second day of the ear discharge. Cultures showed staphylococcus albus and albus liquescens, and a small bacillus culturally resembling the xerosis bacillus.

(3) J. B., male, 19. Third to fourth week since the attack, second day of the ear discharge. Cultures showed only staphylococcus aureus and albus.

(4) G., male, 19. Fourth to fifth week since the attack, first day of the ear discharge. The only organism obtained on culture was staphylococcus albus liquescens.

(5) R., male, 2. Fifth to sixth week since the attack, fourth

* In some instances of S. scarlatinæ obtained from the tonsil in scarlatina, it is only with considerable difficulty that a gelatine culture of 20° C. is established. Once started, however, at 20° C. it maintains its vitality for a longer time than at 37° C.

week of the ear discharge. Staphylococcus aureus and albus liquescens, and a large number of colonies of streptococcus medius were obtained on culture.

(6) W. S., male, 6¾. Fifth week since the attack, third week of the ear discharge. Cultures showed staphylococcus aureus and albus liquescens, and many colonies of streptococcus medius.

(7) J. H., male, 5½. Fifth week since the attack, third week of the ear discharge. Cultures showed staphylococcus aureus and albus liquescens. There were a considerable number of colonies of an organism resembling the xerosis bacillus and also many colonies of a virulent example of Friedländer's bacillus.

(7*b*) The same case exactly one month later. The xerosis bacillus and Friedländer's bacillus alone were obtained.

(8) E. A., female, 4. Sixth week since the attack, sixth day of the ear discharge. Cultures showed several colonies that proved to be streptococcus medius, a few colonies of a bacillus resembling the xerosis bacillus, and many colonies of the true diphtheria bacillus.

(9) L. A., female, 3. Fifth week since the attack, fourth week of the ear discharge. Cultures showed a majority of the xerosis bacillus and a minority of staphylococcus aureus.

(10) E. F. B., female, 9 months. Ninth week since the attack, fourth week of the ear discharge. Cultures showed staphylococcus aureus, an organism resembling the xerosis bacillus, and streptococcus medius.

(11) G. K., female, 11 months. Tenth week since the attack, seventh week of the ear discharge. Cultures showed a majority of staphylococcus aureus and a minority of streptococcus brevis.

(12) S. S., female, 6. Tenth week since the attack, eighth week of the ear discharge. Cultures showed a few colonies of staphylococcus albus liquescens.

In six of these twelve cases therefore streptococci were obtained. In five instances the streptococci were culturally and morphologically of the medius class, and in one the organism was identical with brevis. In no case was a streptococcus which I could identify with streptococcus scarlatinæ recovered.

Examples of both medius and brevis thus obtained were injected into mice and in two instances a rabbit's ear was also inoculated.

Experiments with streptococcus medius.—Case (1). R. T. First day of the discharge. The mouse injected with the S. medius obtained from this case was dead in 48 hours. Post-mortem, the picture was one of acute general septicæmia. There was congestion of the lungs and viscera and also around the point of inoculation. Streptococci were demonstrated in situ and recovered on culture.

Case (5). R. Fourth week of the ear discharge. The injected mouse died in 48 hours and had the same post-mortem appearance as in case (1). A rabbit's ear inoculated showed slight localised inflammation next day and later on an abscess the size of a large pea.

Case (6). W. S. Third week of the ear discharge. The mouse died in 48 hours with the same post-mortem appearance as in cases (1) and (5). A rabbit's ear inoculated showed next day considerable œdema, which by the second day had spread over all the ear; this was now pendulous and dark but without the bright blush of erysipelas. The rabbit died on the third day of general septicæmia, and its organs showed the same congested appearance as in the case of the mice. There was also pericarditis with effusion. Pure cultures of the streptococcus were obtained from the pericardial fluid, heart's blood, kidney, spleen, and from the exudation of the affected ear.

Experiments with streptococcus brevis.—Case (11). G. K. Seventh week of the ear discharge. A mouse injected showed no symptoms, but was alive and well three months later.

Experiments with bacilli from ear discharges.—In case (7) the organism resembling the xerosis bacillus was injected into a guinea-pig, but without further effect than a transient œdema next day at the point of inoculation. On the other hand the diphtheria bacillus obtained in case (8) killed a guinea-pig in 48 hours, and the micro-organism was recovered in pure culture. All the specimens of the xerosis bacillus obtained from ear discharges gave negative results with Neisser's diphtheria stain. The diphtheria culture of case (8) however gave a faintly positive result with this test. The Friedländer's bacillus of case (7) killed a mouse in 48 hours with general septicæmia, and in the blood well-marked capsules were demonstrated around the bacilli. From this fact, and also from its cultural and staining characters, I conclude that this organism was identical with that generally known as Friedländer's bacillus.

Nasal Discharges.

(1) W., male, aged 3. Second week of the nasal discharge and third week since the attack of scarlatina began. Cultures showed staphylococcus aureus, several colonies of streptococcus medius, and several of *streptococcus scarlatinæ.*

(2) M. K., female, 2. Second week of the nasal discharge and third week since the scarlatina attack. Staphylococcus citreus and albus liquescens, streptococcus brevis, and bacillus mesentericus were obtained in culture.

(3) J. M., male, 5. Sixth day of the nasal discharge and third week since the attack. Cultures showed staphylococcus aureus and albus, and streptococcus medius. The latter was in the majority.

(4) R. S., female, 5. Tenth day of the nasal discharge and fifth week since the attack. Cultures showed staphylococcus aureus and albus only. A culture from the throat this week also failed to give streptococcus scarlatinæ, which, however, was recovered the following week. *Vide* throats (4) and (4*B*).

(5) A. B., male, 7. Second day of the nasal discharge and fifth week since the attack. Cultures show staphylococcus albus liquescens and citreus liquescens, and streptococcus brevis.

(6) E. B., female, 4. Third week of the nasal discharge and seventh week since the attack. Cultures showed staphylococcus aureus and albus liquescens, many colonies of a small bacillus morphologically and culturally resembling the xerosis bacillus, and a majority of streptococcus medius.

(7) A. F., male, 2. Eighth week of both attack and nasal discharge. Cultures showed staphylococcus aureus and albus, and several colonies of streptococcus medius.

(8) H. W., male, 4. Seventh week of the nasal discharge and eighth week since the scarlatina attack. Cultures showed staphylococcus albus, albus liquescens, and sarcinæ.

(9) D. P., female, 1½. Second week of the nasal discharge and ninth week since the scarlatina attack. The patient has had adenitis of the glands in the neck, and was admitted with a submental abscess from which streptococcus scarlatinæ and staphylococcus albus liquescens were recovered at that date. Dr. Hopwood was good enough to furnish me with the broth culture which he had made from the pus when the abscess was opened. Cultures from the nasal discharge now yield staphylococcus albus, albus liquescens, citreus, citreus liquescens, and aureus. Several colonies of a small bacillus resembling Hofmann's bacillus, and two tuberculated colonies of *streptococcus scarlatinæ.*

(10) R. P., female, 3. Tenth day of the nasal discharge and ninth week since the attack. Cultures yield staphylococcus albus liquescens and aureus. Several colonies of the small xerosis bacillus, and a majority of streptococcus medius.

(11) W. B., male, 1½. Eleventh week of the nasal discharge and thirteenth week since the attack. Cultures show two colonies of sarcina lutea, and several colonies of a small bacillus that appeared to be of the xerosis type but which refused to grow in subculture.

(12) R. T., male, 3. Fourth week of the nasal discharge and sixteenth week since the attack. Cultures showed only staphylococcus albus and aureus.

Thus streptococci were obtained in eight of twelve cases of nasal discharge. In two they were of the brevis class alone, in four of the medius class alone, and in one case S. scarlatinæ was the only streptococcus present. In one case streptococcus medius and S. scarlatinæ were both present.

Experiments with streptococci from nasal discharges.—Streptococcus medius. Case (3) J. M. Sixth day of the nasal discharge. A mouse hypodermically injected died on the fourth day, and had the same post-mortem appearance as in the case of mice injected with medius from the ear-discharge. A rabbit's ear showed next day some localised œdema at the seat of inoculation, and later on a pea-like abscess.

Case (7) A. F. Eighth week of both attack and nasal discharge. A mouse injected died in 24 hours with the usual post-mortem appearance of mice dead from streptococcus medius.

Case (10) R. P. Tenth day of the nasal discharge, and ninth week since attack. A mouse injected died in three days with the same post-mortem appearance as the above mice.

Streptococcus brevis.—Case (2) M. K. Second week of the nasal discharge, and third of scarlatina attack. A mouse injected presented no symptoms, but was alive and well two months afterwards. A rabbit's ear showed next day slight œdema and redness at the point of inoculation. This cleared up later without forming an abscess.

Streptococcus scarlatinæ.—Case (9) D. P. Third week of the nasal discharge, and ninth of scarlatina attack. The mouse injected died on the seventh day. Post-mortem, both lungs were congested at their bases, the liver was blackened, especially at its edges, and much congested elsewhere. The spleen was enlarged and blackened at one end, and the cortex of the kidneys also showed a congested appearance. Streptococci were demonstrated *in situ* in all these organs, arranged as diplococci, in groups, and in short chains. The micro-organism was also recovered in culture.

In this case (9) the Hofmann pseudo-diphtheria bacillus was tested on a guinea-pig with the result that the animal appeared to be quite unaffected. This micro-organism, and also those of the xerosis type obtained from the nasal discharges, all gave a negative result with Neisser's diphtheria stain.

Throats.

(1) E. P., female, aged 18. Patient had an irregular rash, some fever, but no sore throat. She showed desquamation later. Her brother had typical scarlatina at the same time. Cultures from the tonsil during the third week since attack gave staphylococcus albus and citreus, several colonies of streptococcus brevis and medius, and two colonies of *streptococcus scarlatinæ*. (1*B*). A week later other cultures showed staphylococcus albus liquescens, several colonies of streptococcus brevis and medius, but no S. scarlatinæ.

(2) C., female, 18. Patient had both diphtheria and scarlatina at first, and diphtheria bacilli were then recovered. There is no doubt as to her having scarlatina as well as diphtheria, for the rash and desquamation were typical. Cultures taken during the third week showed staphylococcus albus, aureus, and citreus; and streptococcus brevis, medius, and *scarlatinæ*. The latter was plentiful. No diphtheria bacilli were obtained.

(3) P., female, 8, has had two attacks. It is now over three weeks since the second attack, which proved to be a mild one, began. Cultures from the tonsil show staphylococcus albus liquescens, and citreus; also streptococcus medius and *streptococcus scarlatinæ*.

(4) R. S., female of 5. Now in the fifth week since the attack began. A nasal discharge also began 10 days since (*vide* Nasal Discharge, No. (4). Cultures from the tonsil show streptococcus brevis and medius, but no S. scarlatinæ.

(4*b*) Cultures from the tonsil in this case made a week later showed staphylococcus albus liquescens, streptococcus medius, and *streptococcus scarlatinæ*. The nasal discharge had by now cleared up.

(5) E. O., female, 4. Now in the fifth week since the attack. Cultures from the tonsil show staphylococcus albus liquescens and citreus; streptococcus brevis, medius, and many colonies of *S. scarlatinæ*.

(6) E. R., female, aged 3. Now in the fifth week since the attack. Cultures from the tonsil show staphylococcus albus and albus liquescens. Streptococcus medius several colonies, and one colony of *streptococcus scarlatinæ*.

(7) W., female of 5. Now in the fifth week since the attack. Cultures from the tonsil showed staphylococcus albus and streptococcus brevis. No streptococcus scarlatinæ.

(7*B*) Cultures from the tonsil in this case made a week later showed staphylococcus albus and albus liquescens, a few colonies of streptococcus medius and many colonies of streptococcus longus, but no S. scarlatinæ.

(8) J. B., female, 18. Now in the sixth week since the attack. Cultures from the tonsil show staphylococcus albus and albus

liquescens, a few colonies of streptococcus brevis, several of medius, and several of *S. scarlatinæ.*

(9) R., female of 3. Now in the sixth week since the attack. Cultures from the tonsil show staphylococcus albus, albus liquescens, citreus, and several colonies of *streptococcus scarlatinæ.* B. buccalis minutis (Vignal) was also present.

(10) C. P., female of 8. It is over six weeks since the patient was attacked with scarlatina and she is about to be discharged. Cultures from the tonsil yield staphylococcus albus, and citreus, several colonies of streptococcus brevis, and of streptococcus medius. Colonies of *streptococcus scarlatinæ* were also present.

(11) —— female, aged 6. Now in the seventh week since the attack. Cultures from the tonsil show staphylococcus albus and citreus, streptococcus brevis, medius, and *S. scarlatinæ.*

(12) L. P., female, aged 3. Eighth week since admission. Cultures from the tonsil show staphylococcus albus liquescens and citreus; *streptococcus scarlatinæ*, however, was in the great majority.

(13) L., female, 5. Ninth week since admission. Cultures from the tonsil show staphylococcus albus, streptococcus medius and *streptococcus scarlatinæ.*

(14) M. C. D., female, 2. Ninth week since admission, patient about to go out. She had adenitis as a complication. Cultures from the tonsil show staphylococcus albus, albus liquescens, citreus, and aureus; streptococcus medius, brevis, and *S. scarlatinæ.*

(15) —— female, aged 5. Ninth week since admission. She has had a nasal discharge which has ceased. Cultures from the tonsil show staphylococcus albus, streptococcus brevis, medius, and longus, but no scarlatinæ.

The next three cases were the subject of several examinations.

(16) M., male, 10. Sixth day since the attack began. The rash has just faded away. Cultures from the tonsil show staphylococcus albus and aureus, streptococcus medius, and *streptococcus scarlatinæ.*

(16*B*) The same case was examined on the 20th day and cultures from the tonsil showed staphylococcus albus, albus liquescens, and aureus; streptococcus longus, and *S. scarlatinæ.*

(16*c*) A third examination of this case made on the 34th day showed staphylococcus albus and aureus; streptococcus longus and *S. scarlatinæ.*

(17) M. B., female, 2. Fourth day since attack began, and the rash is now fading. Cultures from the tonsil show staphylococcus albus, streptococcus longus, and a majority of *streptococcus scarlatinæ.*

(17*b*) The same case on the 18th day showed staphylococcus albus liquescens and aureus; streptococcus longus, but no S. scarlatinæ.

(17*c*) A third examination made on the 32nd day showed staphylococcus albus liquescens and aureus; streptococcus longus, but again no S. scarlatinæ.

(18) N. M., female, 12. 19th day since the attack began, and the patient is now desquamating. Cultures from the tonsil showed staphylococcus albus; streptococcus longus, medius, and *S. scarlatinæ.*

B. No. 7.
—
iology of
tina;
Gordon.

(18*b*) The same case showed in cultures from the tonsil made on the 32nd day, staphylococcus albus liquescens, and aureus; streptococcus longus, medius, and *scarlatinæ.*

(18*c*) A third examination, made on the 47th day, showed staphylococcus albus and citreus; streptococcus longus, medius, and *S. scarlatinæ.*

Thus in 27 examinations of the tonsils of persons in all stages, but especially in the later, stages of scarlatina, streptococcus scarlatinæ was detected 20 times. Of the seven cases in which this organism was not obtained, two were in the third and fourth weeks, and five were after the beginning of the fifth week from the date of attack. On the other hand, streptococcus scarlatinæ was present in 13 out of 18 cases examined after the beginning of the fifth week. In two of the latter cases (Nos. (10) and (14)) the patients were about to be discharged.

The following table shows in a condensed form the bacteriological results of the throat examinations.

Summary of Scarlatina Throats.

No.	Stage.	Staphylococci.				Streptococci.				Remarks.
		Albus.	Alb. Liq.	Aureus.	Citreus.	Brevis.	Medius.	Longus.	Scarlatinæ.	
17	4th day	+	...	...	..	...	...	+	+	
16	6th day	+	...	+	...	...	+	...	+	
17*b*	18th day	...	+	+	...	...	...	+	...	
18	19th day	+	...	...	...	...	+	+	+	
16*b*	20th day	+	+	+	...	...	...	+	+	
1	3rd week	+	...	...	+	+	+	...	+	No sore throat.
2	"	+	...	+	+	+	+	...	+	Diphtheria also at fi[illegible]
1*b*	4th week	...	+	...	...	+	+	...	...	
3	"	...	+	...	+	...	+	...	+	Second attack.
17*c*	32nd day	...	+	+	...	...	...	+	...	
18*b*	"	...	+	+	...	...	+	+	+	
16*c*	34th day	+	...	+	...	...	...	+	+	
4	5th week	...	...	...	...	+	+	...	...	Rhinorrhœa presen[illegible]
5	"	...	+	...	+	+	+	...	+	
6	"	+	+	...	...	...	+	...	+	
7	"	+	...	...	...	+	...	...	...	
4*b*	6th week	...	+	...	...	...	+	...	+	S. scarlatinæ not[illegible] tained previous we[illegible] q. v.
7*b*	"	+	+	...	...	...	+	+	...	
8	"	+	+	...	...	+	+	...	+	
9	"	+	+	...	+	...	...	...	+	
10	7th week	+	...	...	+	+	+	...	+	Just going out.
11	"	+	...	...	+	+	+	...	+	
18*c*	47th day	+	...	...	+	...	+	+	+	
12	8th week	...	+	...	+	...	...	...	+	
13	9th week	+	...	...	...	...	+	...	+	
14	"	+	+	+	+	+	+	...	+	Had adenitis. Jus[illegible] going out.
15	"	+	...	...	...	+	+	+	...	Had rhinorrhœa.
	Total 27	18	14	8	10	11	19	10	20	

Experiments on mice with streptococcus scarlatinæ obtained from the throat.—Case (18) above. Eighteenth day since the onset of scarlatina. A mouse hypodermically injected from a 48 hours' broth culture died on the 37th day. Post-mortem, the lungs, especially at their bases, the liver, and the kidneys, were congested and dark. The spleen was much atrophied. Streptococcus scarlatinæ was demonstrated *in situ* chiefly in groups, or as diplococci, or rarely in short chains, in the liver, spleen and kidneys. Cultures from the latter three sources, and also from a lymphatic gland in the axilla, all yielded streptococcus scarlatinæ.

Case (18*b*) The same case, 32nd day. The culture of streptococcus scarlatinæ obtained at this stage killed a mouse on the fifth day after injection. Post-mortem, the lungs showed patches of congestion. The liver and kidneys were also congested. The spleen was considerably enlarged and blackened. The streptococcus was recovered from all these sources as well as from the heart's blood.

Case (2) Third week since onset and the case complicated at first with diphtheria. The injected mouse, though it got thin and was less lively for some days, eventually recovered and was alive and well ten weeks later.

Case (5) Fifth week from onset. The injected mouse died on the 31st day with the same appearance as seen in the mouse of case (18). The spleen in this case however was not definitely atrophied. The streptococcus scarlatinæ was recovered in cultures from the liver and spleen. It was also present in a culture from one kidney though not from the other.

Case (10) Seventh week from onset and patient about to be discharged. The injected mouse died on the tenth day with the usual congestion of its organs. S. scarlatinæ was recovered therefrom, though with some difficulty owing to putrefaction having commenced.

Case (11) Seventh week from onset. The injected mouse recovered.

Case (12) Eighth week from onset. The injected mouse died on the tenth day with the usual congested appearance of its organs; and the streptococcus scarlatinæ was recovered from its tissues.

Case (14) Ninth week from onset, and patient about to be discharged. The injected mouse died in four to five days. Post-mortem, the organs were found to be congested, and streptococcus scarlatinæ was demonstrated *in situ* in large numbers, in the case of the kidneys especially. It was also recovered from these organs on culture.

Thus it appears that streptococcus scarlatinæ from the throats of the above patients when hypodermically injected into mice brought about a fatal result on eight out of ten occasions, the interval between the date of injection and the death of the animal being from four to 37 days. Both the mice that survived showed signs of illness for a time. These physiological results, which form a further distinguishing character of streptococcus scarlatinæ, are in accord with those obtained by Dr. Klein when experimenting on mice with the original micro-organism of the now well-known Hendon Outbreak.

STREPTOCOCCUS TYPES.

PLATE XVI.

FIG. 1.

Showing mode of growth in gelatine, at 37° C., of four types of Streptococcus.

Tube 1.—*Streptococcus longus*, after *twenty-four hours'* incubation.

Tube 2.—*Streptococcus scarlatinæ*, after *several days'* incubation.
The coherent mass of growth having been shaken from the bottom of the tube, has become inverted.

Tube 3.—*Streptococcus medius* from pus, after *twenty-four hours'* incubation.

Tube 4.—*Streptococcus brevis*, after *twenty-four hours'* incubation.
In this case alone has the gelatine become turbid.

PLATE XVI.

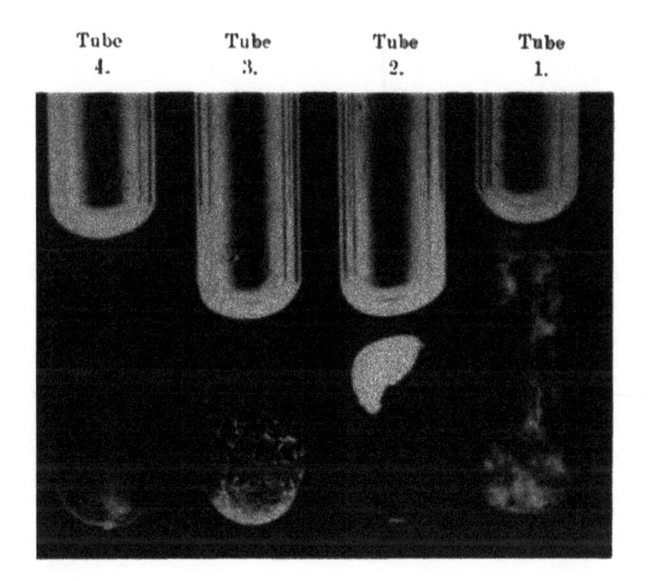

Fig. 1.

STREPTOCOCCUS MEDIUS.

PLATE XVIII.

FIG. 5.

Film specimen from a broth culture of *streptococcus medius*, after 24 hours' incubation at 37° C. The chains are shorter, more disjointed, and more sinuous than in the case of S. longus. There is, too, a slight effect of "grouping" where a chain curls sharply on itself.

[Multiplying power, 1000.]

FIG. 6.

Impression specimen of colonies of *streptococcus medius* on agar, after 24 hours' incubation at 37° C. Showing group-arrangement.

[Multiplying power, 60.]

FIG. 7.

Impression specimen of a small colony of *streptococcus medius* on gelatine, after 48 hours' incubation at 20° C. Showing group-arrangement.

[Multiplying power, 1000.]

Plate XVIII.

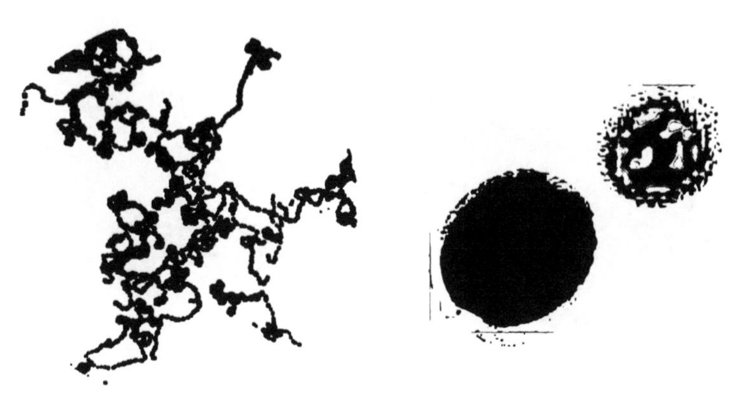

Fig. 5.

Fig. 6.

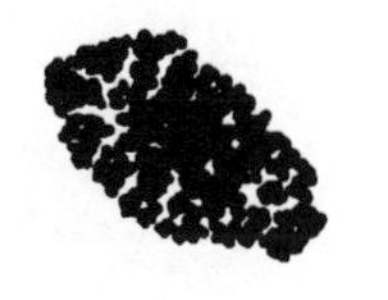

Fig. 7.

PLATE XIX.

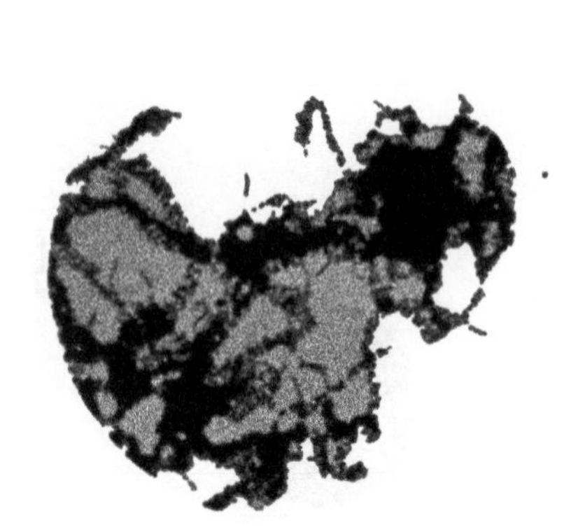

Fig 8.

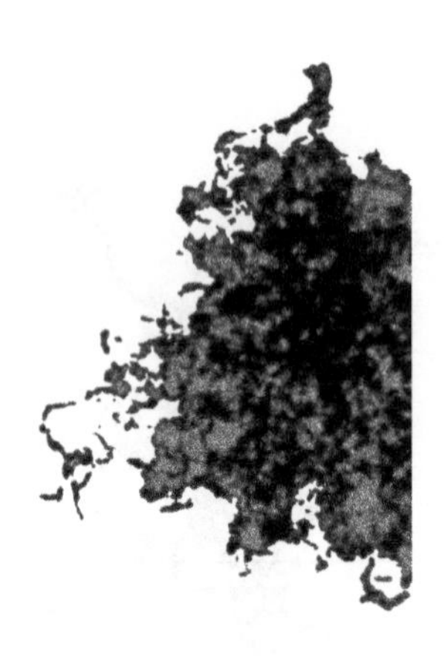

Fig. 9.

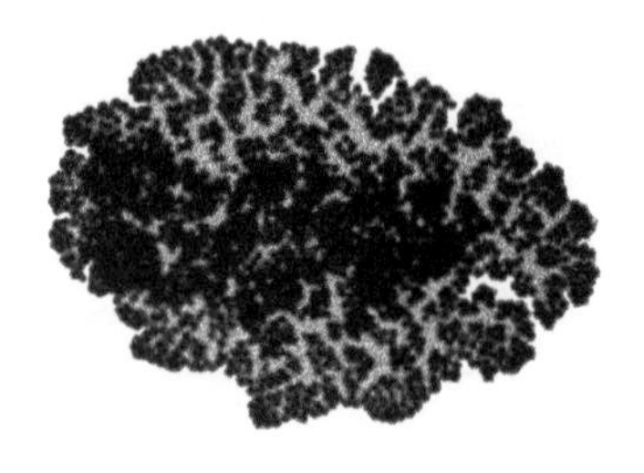

Fig. 10.

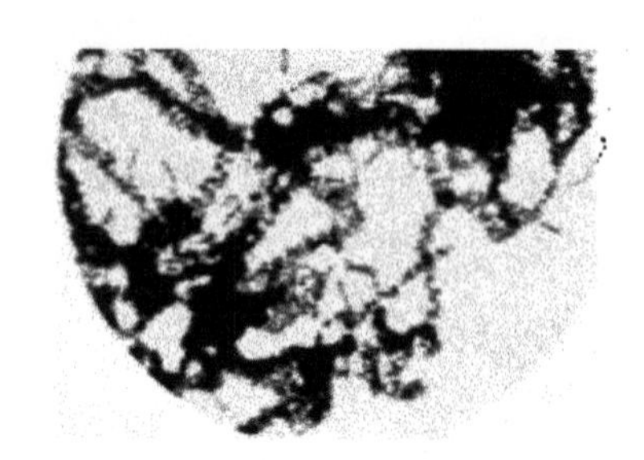

Fig 8.

Fig. 9.

STREPTOCOCCUS SCARLATINÆ.

PLATE XIX.

FIG. 8.

Film specimen from a broth culture of *streptococcus scarlatinæ* or *conglomeratus*, after 24 hours' incubation at 37° C. Showing marked coherency and mass-formation.

[Multiplying power, 1000.]

FIG. 9.

Impression specimen of a colony of *streptococcus scarlatinæ* on agar, after 24 hours' incubation at 37° C. Centrally the mass-formation is marked; peripherally, the chain-formation is well seen.

The Colony is intermediate between the second and third types described in the text.

[Multiplying power, 500.]

FIG. 10.

Impression specimen of a small colony of *streptococcus scarlatinæ* on gelatine, after 48 hours' incubation at 20° C. Showing, generally, the group-arrangement; and, towards the centre the coherency referred to in the text.

[Multiplying power, 1000.]

No. 8.

F. B. No. 8.

Bacteriological Evidence of Recent Sewage Pollution of Potable Waters; by Drs. Klein and Houston.

FURTHER REPORT on BACTERIOLOGICAL EVIDENCE of RECENT, and therefore DANGEROUS, SEWAGE POLLUTION of elsewise POTABLE WATERS: by Drs. KLEIN and HOUSTON.

Holding B. coli and B. enteritidis sporogenes as microbes not, it is true, *peculiar* to sewage but much more abundant* in it than in any other substance not equally objectionable in character, we have, in last year's report,† shewn that, by means of bacterioscopic methods, a demonstration of the presence of sewage in water can be carried to an incomparably higher degree than by means of the hitherto employed chemical processes. Thus, we have shewn that sewage may be added to water in the proportion of 1 part of sewage to 1,000 parts of water, and be scarcely detectable by delicate chemical tests, or, at all events, without affecting its composition sufficiently to allow of such a mixture being condemned on chemical grounds alone. Further, we experienced no difficulty in detecting the presence of B. coli (by *direct* culture) even when the dilution was 0·005 per cent. or *1 part of sewage in 20,000 parts of water.*‡

Finally, we succeeded in demonstrating the presence of the spores of the virulent B. enteritidis when only 0·002 cc. of sewage was added to 1,000 cc. of sterile water, i.e., *a dilution of 1 in 500,000.*§

These results are the more remarkable since, in our experience, B. coli is either not present or present in very small numbers in *pure* waters, while the spores of B. enteritidis are often absent in as much as 500 cc.‖ of a water free from any likelihood of pollution.

Obviously then, though it be admitted that the *mere presence* of B. coli and B. enteritidis in a water is not absolute proof of objectionable pollution, and that their *relative abundance* therein is the all important question, it follows from what has been stated above, that sewage present in a water in amount inappreciable by the usual chemical tests would yet yield to bacterioscopic tests a number of these microbes altogether beyond that which the least responsible of bacteriologists might consider permissible in a potable water.

It appeared, therefore, desirable to obtain further information on the following points:—

1. Whether samples of sewage diluted in each case with varying amounts of pure sterile water yielded to chemical and bacteriological analysis figures in harmony with the degree of such dilutions.

2. Whether sewage obtained from a plurality of diverse sources yielded on chemical and bacterioscopic examination comparable results.

---

* Our records show that in raw sewage there are usually present more than 100,000 B. coli and from 10 to 1,000 spores of B. enteritidis per cc.

† Report of Medical Officer of Local Government Board, 1897–98, App. B, No. 4, p. 318–325.

‡ Expt. 3. Series I.

§ Expt. 4. Series II.

‖ Ascertained by the "filter brushing method."

. The extent to which dilutions of sewage could be carried rt of the point at which bacterioscopic testing failed to d a positive result as to the presence of the microbes of eriment.

7ith these ends in view samples of sewage were obtained from it different sources. Some of the samples were from manuuring towns and contained trade effluents of varied sort. As rds other samples, the sewage was wholly domestic in character. sample was the sewage coming from a large hospital.

1 Series I. to IV., samples of sewage (A. B. C. D.) from four erent sources were in each case diluted 100, 1,000, 10,000, and 000 times with pure sterile water.

1 Series V., four samples of sewage (E. F. G. H.) from four erent sources were in each case diluted 200 times prior to nical, and 10,000, and 100,000 times antecedent to bacterioic tests.

A.—*Chemical Analysis.*

1 analysing the experimentally polluted water, two processes e chosen which are in general use among chemists at the ent day.

mmonia process (Wanklyn). 500 ccs. of the polluted water e examined in the ordinary way for free and albuminoid nonia.

xygen permanganate process (Tidy and others). 500 ccs. of polluted water were dealt with and the experiments were ied out at the room temperature and for a period of four rs.

hese processes are too well known to require any description.

he chemical results are shewn in Table I.

1 each case 1,000 ccs. of distilled water were polluted with a nite amount of crude sewage. Of the contaminated water ccs. were used for the ammonia process and 500 ccs. for the gen permanganate process.

s has been said, all the samples of sewage, namely, A. B. C. E. F. G. H. were obtained from different sources. Generally aking, they all differed from each other in one or other respect were chosen for this very reason, so that the results might e as wide a range of usefulness as possible.

1 Series I., crude sewage (A.) the dilutions were 1·0, 0·1, 0·01, 01 per cent.; experiments 1, 2, 3, 4 respectively.

1 Series II., III., IV., the dilutions were the same as before, the samples of sewage (B. C. D.) were obtained each from a erent source.

n Series V., the dilution was 0·5 per cent. in each instance, but h sample of sewage (E. F. G. H.) was obtained from a different rce; experiments 1, 2, 3, 4 respectively.

'hus, as regards Series I. to IV., four experiments were carried in each series making a total of sixteen experiments with four ples of sewage. And, as regards Series V., four experiments re carried out each with a different sewage. So that altogether, nty experiments were made with eight samples of sewage.

TABLE I.

[illegible]

[illegible]	[illegible]	Free Ammonia.	Albuminoid Ammonia.	Sum of the Am- [illegible]
[illegible]	[illegible]	[illegible]	[illegible]	[illegible]
[illegible]	[illegible]	[illegible]	[illegible]	[illegible]
[illegible]	[illegible]	[illegible]	[illegible]	[illegible]
[illegible]	[illegible]	[illegible]	[illegible]	[illegible]
[illegible]	[illegible]	[illegible]	[illegible]	[illegible]
[illegible]	[illegible]	[illegible]	[illegible]	[illegible]
[illegible]	[illegible]	[illegible]	[illegible]	[illegible]
[illegible]	[illegible]	[illegible]	[illegible]	[illegible]
[illegible]	[illegible]	[illegible]	[illegible]	[illegible]
[illegible]	[illegible]	[illegible]	[illegible]	[illegible]
[illegible]	[illegible]	[illegible]	[illegible]	[illegible]
[illegible]	[illegible]	[illegible]	[illegible]	[illegible]
[illegible]	[illegible]	[illegible]	[illegible]	[illegible]
[illegible]	[illegible]	[illegible]	[illegible]	[illegible]
[illegible]	[illegible]	[illegible]	[illegible]	[illegible]
[illegible]	[illegible]	[illegible]	[illegible]	[illegible]
[illegible]	[illegible]	[illegible]	[illegible]	[illegible]
[illegible]	[illegible]	[illegible]	[illegible]	[illegible]
[illegible]	[illegible]	[illegible]	[illegible]	[illegible]
[illegible]	[illegible]	[illegible]	[illegible]	[illegible]

[illegible] A. [illegible]
B. mixed.
C. [illegible]
D. mixed with large amount of trade liquid.
E. [illegible]
F. domestic, with large amount of trade liquid.
G. wholly domestic.
H. mixed.

In considering from the chemical view point the results recorded in Table I., it is to be noted that—

Judged by the oxygen permanganate process (four hours at the room temperature), Experiment 1, Series I., and Experiment 1, Series IV., denote waters of medium organic purity. The rest of the experiments denote waters of great organic purity.

Judged by the ammonia process (amount of albuminoid ammonia), nearly all the experiments denote waters of great organic purity. The exceptions are, Experiment 1, Series II., and Experiment 4, Series V., which denote waters organically safe; and Experiment 1, Series I.; Experiment 1, Series III.; and Experiment 1, Series IV.; which denote dirty waters.

The albuminoid ammonia in a water may be held to indicate *actual* evil, and the free ammonia *potential* evil. Nevertheless, even if the sum of the ammonia (free and albuminoid) be taken as all representing objectionable pollution, the figures go to show that the testimony of chemical analysis is not to be trusted in cases where the contaminating matter consists of minute quantities of raw sewage. How different are the indications from the bacteriological view point will presently be shown. It must be remembered, however, that most natural waters contain more than mere traces of organic matter. Indeed, many of them contain an amount which is not far short of placing them in the class of waters of doubtful purity. Hence additional pollution in the way above practised might bring many a natural water from the class of waters of great purity into the class of waters which are of medium purity, or from the class of waters of medium purity into that which includes waters to be regarded with suspicion or even to be condemned. Still, it is obvious that if this is true from the chemical view point it is likely also to be true bacteriologically considered.

B.—*Bacterioscopic Analysis.*

In the bacteriological examination of the artificially polluted waters, the same samples of sewage were, as regards Series I., II., III., IV., used, and the same dilutions made as in the chemical experiments. As regards Series V., the amount of sewage added to the water in each case, for the chemical test was 0·5 per cent.; an amount which, while yielding results appreciable chemically, yet did not contaminate the water sufficiently to allow of its condemnation. In this Series (V.), and for the bacteriological experiments, the dilutions were the same as in the preceding Series (I. to IV.), except that the 1 per cent. and 0·1 per cent. dilutions were omitted.

The dilutions were as follows:—

(1)	1	per cent.	1 : 100
(2)	0·1	"	1 : 1,000
(3)	0·01	"	1 : 10,000
(4)	0·001	"	1 : 100,000

Of the above dilutions, in the case of (1) and (2) the presence of B. coli and B. enteritidis sporogenes was usually demonstrated *directly* without resorting to the "filter brushing method." B. coli was directly demonstrated by making *surface* phenol (0·05 per cent.) gelatine plates and phenol (0·05 per cent.) broth cultures, and afterwards resorting to further sub-cultures and the application of certain tests, to all of which B. coli gives a positive result. And B. enteritidis was directly demonstrated by inoculating milk tubes, which, after having been heated to 80°C. for ten minutes, were cultivated anaerobically at blood heat.*

As regards dilutions (3) and (4), the amount of sewage added to the water was so small that the *indirect* method was employed in detecting the microbes in question. It was found necessary to resort to filtration of the polluted water through a sterile

* For further details, reference may be made to Report of Medical Officer of Local Government Board, 1897–98, Appendix B, No. 4.

Pasteur's filter. After filtration, the surface of the filter was brushed, with a sterile brush, into a definite quantity of sterile water (5 to 10cc.), and of the filter brushing suspension (each cc. of which represented the bacterial contents of a large quantity of the original liquid) definite amounts were used to inoculate, as regards B. coli, phenol broth and phenol gelatine, and, as regards B. enteritidis, milk tubes.

In addition, and as regards each different sample of sewage, the total number of aerobic bacteria was estimated. But the total number of bacteria in each of the four dilutions of the eight different samples of sewage was not ascertained. Nor did this appear to be in any way necessary.

On comparing the results as to the number of B. coli and spores of B. enteritidis obtained by the *direct* and *indirect* methods, it will be seen that the figures, corresponding to the number of each of these microbes per cc. of the raw sewage, are higher in the first case (*i.e.*, direct method) than in the second (*i.e.*, indirect method). Nevertheless, the two sets of figures obtained by the direct method, and when working with the two dilutions (1·0 and 0·1 per cent.), harmonised in each case fairly well. And the same may be said as regards the comparison between the two sets of figures obtained by the indirect method and when working with the two dilutions of 0·01 and 0·001 per cent.

That the indirect method should yield lower results is, of course, only what might be anticipated. Take, for example, the experiments where 0·1 cc. sewage was added to 1,000 cc. of water. Here the polluted water, which contained about two drops of sewage, was filtered through a Pasteur filter, and the surface of the filter, which would arrest all, or nearly all, the particulate matter, was brushed repeatedly with a sterile brush, and the "brushings" mixed with 5 cc. of sterile water contained in a sterile dish. This 5 cc. might be expected to contain the major portion, but not quite all, of the particulate matter contained in the 1,000 cc., and, therefore, in the two drops of sewage. As a matter of fact, the records contained in the bacterioscopic table show that the loss by this process might be as great as about 50 per cent. and as low as 15 per cent.

The table of bacterioscopic results indicates that the number of B. coli and spores of B. enteritidis, as judged by the direct method (Experiment 1 in each of the Series I. to IV.), was on an average, and in round numbers respectively, 500,000 and 1,000 per cc.

It is apparent, then, that in dilutions 3 and 4 (1 : 10,000 and 1 : 100,000), it was not absolutely necessary to resort to a special filtration process in order to demonstrate the presence and estimate the number of B. coli in the polluted water, because each cc. of the litre of water contaminated in the above proportions was likely to contain in the first case 50, and in the second case 5, germs of B. coli.

But as regards B. enteritidis the case was different, since the corresponding figures would be 1 spore per 10 cc. in the first case and 1 spore per 100 cc. in the second : so that the addition directly of 100, or even 10 cc., to an ordinary milk tube would be either impossible or would dilute the medium to a very serious extent. Here, then, the indirect method was necessary, and by its use positive results were usually readily obtained. For, and still dealing ures, 1 cc. of the 5 cc. "filter brushing susntain 20 spores in the 0·01 per cent. and 2 spores nt. dilution.

what has been said, that positive results, as oscopic tests, might have been obtained in cases n of the sewage with water was even greater nt. With a dilution of 0·001 per cent. there een 100 B. coli in 1 cc. of the 5 cc. filter brush· the polluted litre of water, and there would re of B. enteritidis in the whole of the filter latter case, however, it would of course be more than one litre of water polluted in the n order to hope to demonstrate the presence of

n it must be remembered that, although some contain more than 500,000 B. coli and 1,000 c., many contain much less.

cal results are given in the following table :—

TABLE II.

g Results of Bacterioscopic Analysis.

Dilution.	Number of Microbes per 1 cc. of Crude Sewage.	Number of Bacillus Coli per 1 cc. of Crude Sewage.	Number of Spores of Virulent Bacillus Enteritidis per 1 cc. of Crude Sewage.
er cent., or 1 in 100	14,240,000	260,000	2,000
„ „ 1 in 1,000	—	230,000	2,000
„ „ 1 in 10,000	—	180,000	1,560
„ „ 1 in 100,000	—	174,000	1,500
„ „ 1 in 100	7,800,000	180,000	200
„ „ 1 in 1,000	—	180,000	200
„ „ 1 in 10,000	—	120,000	170
„ „ 1 in 100,000	—	110,000	150
„ „ 1 in 100	4,800,000	500,000	2,000
„ „ 1 in 1,000	—	420,000	2,000
„ „ 1 in 10,000	—	280,000	1,600
„ „ 1 in 100,000	—	240,000	1,520
„ „ 1 in 100	36,000,000	1,100,000	400
„ „ 1 in 1,000	—	—	330
„ „ 1 in 10,000	—	680,000	260
„ „ 1 in 100,000	—	—	240
„ „ 1 in 10,000	2,800,000	200,000	30
„ „ 1 in 100,000	—	150,000	—
„ „ 1 in 10,000	4,100,000	500,000	56
„ „ 1 in 100,000	—	—	—
„ „ 1 in 10,000	28,000,000	2,000,000	50
„ „ 1 in 100,000	—	1,500,000	—
„ „ 1 in 10,000	21,000,000	1,000,000	35
„ „ 1 in 100,000	—	1,600,000	—

APP. B. No. 8.

On Bacteriological Evidence of Recent Sewage Pollution of Potable Waters; by Drs. Klein and Houston.

It will be noted that the total number of microbes, the numb of B. coli, and the number of spores of B. enteritidis varied a considerable extent in the different samples of sewage.

Thus, Sample D contained 36,000,000 micro-organisms an Sample E 2,800,000, per cc.

Sample G contained 2,000,000 B. coli (dil. 0·01 per cent.) an Sample B (dil. 0·001 per cent.) 110,000, per cc.

Samples A (dils. 1 and 0·1 per cent.) and C (dils. 1 and 0·1 pe cent.) contained 2,000 spores of B. enteritidis, and Sample only 30, per cc.

There appears then to be no definite parallelism between th total number of bacteria and the number either of B. coli or o spores of B. enteritidis.

And similarly no parallelism can be traced between the numbe of B. coli and the number of spores of B. enteritidis.

Since a number of the samples were obtained from manufacturing towns and contained therefore trade effluent of varied sort, the facts above recorded tend to suggest that the addition of trade waste to sewage does not *necessarily* greatly alter the biological qualities of the liquid. Indeed, in some cases the number of bacteria, of B. coli, and of B. enteritidis was greater in "trade" sewage than in sewage of purely "domestic" character As however, the length of time between the collection and subsequent examination of the samples varied to some extent, and as also the records are not very numerous, it is well to hesitate in drawing inferences.

In four samples, E, F, G, and H, the presence of B. enteritidis could not be detected in the 1 : 100,000 dilutions. Possibly, and indeed probably, a positive result might have been obtained if a much larger volume of the water (polluted in the same proportion) had been dealt with by the Pasteur filter. Indeed, in two cases (B and D) this was done, and with positive result, 3,000 cc. being filtered in the first case and 2,000 cc. in the second.

With reference to the more general question, namely, the relative value of chemistry and bacteriology in detecting water pollution, the figures speak for themselves. They show that, at any rate under the conditions of our experiments, chemistry, as usually applied, is powerless to detect a degree of pollution of water with sewage which, from the view point of the bacteriologist, would be considered gross in amount. And further they show that not only is bacteriology capable of detecting, in a water in such circumstances, microbes characteristic of sewage, but is capable also of detecting these bacteria when the degree of sewage pollution of the water is *from ten to one hundred times less than* that in which the organic matter contributed by the sewage to the water has failed to get recognition by the methods commonly in use by the chemist.

APPENDIX C.

THE HISTOLOGY OF VACCINIA; by S. MONCKTON COPEMAN, M.A., M.D., F.R.C.P., and GUSTAV MANN, M.D., B.Sc.

PART I.—INTRODUCTION; by DR. COPEMAN.

On the Histology of Vaccinia; by Drs. Copeman and Mann.

It is a somewhat remarkable fact that although, for a century past, practically every medical man has had more or less opportunity of becoming acquainted with the clinical appearances resulting on the inoculation of the human subject with cow-pox, or vaccinia, comparatively little attention has been paid to the minute anatomy of the various stages of the eruption typical of this affection. I have long been desirous that this omission should be repaired, being convinced that an accurate knowledge of the histological changes involved could hardly fail to throw some further light on the pathology of vaccinia, concerning which, even at the present time, so much divergence of opinion exists. But the time at my disposal has proved insufficient to enable me to carry through the necessary investigations unassisted. Several years ago I suggested to Mr. Stanley Kent that he should undertake, with me, a detailed investigation of the histology of the vaccine vesicle, and for this purpose I provided him with a complete series of the necessary specimens of skin from a previously vaccinated calf. But, for various reasons, he has been unable to proceed further with the research than sufficed to enable him to read at the Bristol meeting (1894) of the British Medical Association a short preliminary paper on the subject, an abstract of which was subsequently published in the Journal of the Association.* Under these circumstances I asked Dr. Mann, who I found was also interested in the subject, to undertake the histological examination of material which I collected for the purpose. The scope of the work and the manner of carrying it out, together with the probable interpretation of the results obtained, have been matters of frequent discussion between us. But the actual histological work (the account of which forms the chief bulk of the paper, and nearly all that is interesting in it) was performed by Dr. Mann alone.

We desired to embrace, in our investigation, microscopic examination of the skin throughout the sequence of events following on vaccination, from the time of implantation of the virus beneath the surface of the epidermis up to that at which the vesicle attains maturity and onwards through the phase of desiccation of the pustule and the subsequent healing of the skin beneath the "crust."

This being so, it was obviously impossible to obtain the necessary material from the human subject, although Dr. Cory, by vaccinating, as occasion served, the supernumerary digits of such children as came under his observation with this abnormality and subsequently removing the vaccinated digit, had been enabled, some years ago, to investigate to some extent the changes occurring as the result of vaccination in human skin.

* British Medical Journal, Vol. II., 1894, page 633.

For the purpose, therefore, of making sure of obtaining specimens of skin at the precise periods required, the material, on which the results of the present research are based, was obtained from the calf, small pieces of skin being excised from a previously vaccinated area at such subsequent intervals of time as had been determined on. For this purpose deep anæsthesia was found to be unnecessary, the operation being of momentary duration only. In order to avoid, as far as might be, fallacy arising from the slightly differing rate at which the process of vaccination reaches its height in different calves and under varying circumstances, the influence of which are at present but imperfectly understood, several sets of specimens were procured.

In all instances the vaccination was carried out by means of linear incisions on the abdomen after the manner now ordinarily adopted at the Government Animal Vaccination Establishment. Each piece of skin, excised so as to include little else than the tissue immediately bordering on the line of incision, was, on removal, at once divided transversely into two portions, one of which was placed in absolute alcohol, the other in a watery solution of picric acid and corrosive sublimate. These small specimens of skin were then wrapped in absorbent cotton wool, placed each separately in a tiny wide-mouthed stoppered bottle, carefully packed, and despatched to Dr. Mann, at the Physiological Laboratory, Oxford.

In order to assist the reader in following the minuter histological details set out in Part II. of this paper it may be well here to give a brief account of the macroscopic and coarser microscopic appearances which are to be observed in the skin during the local development of the vaccine vesicle. It must, however, be borne in mind that in the calf the whole process of vaccination runs a distinctly shorter course than in the human being, a fact that is probably in some measure dependent on the normally higher body temperature of the calf as compared with that of man. Thus in the human subject the vaccine vesicle ordinarily attains maturity on the *8th day* [(24 hours × 7) = 168 hours], while in the calf the vesicle will have attained a similar stage of development on the *6th day* [(24 hours × 5) = 120 hours].

Within an hour after vaccination, more particularly in the calf, the skin immediately bordering on the inoculation wound not infrequently becomes somewhat raised owing to a transient local urticaria. This, however, rapidly passes off, and, in the human subject, practically no further change becomes obvious before the third day after the insertion of vaccine lymph, by which time a small inflamed spot or "papule" may usually be observed at the point where the vaccination was performed. Next day this spot appears more florid, and on passing the point of the finger over it a certain degree of hardness and swelling is perceptible. By the fifth day the papule develops into a small pale vesicle. This vesicle has a milky white colour, it is depressed in the centre, and its edges are distinctly elevated above the level of the surrounding skin. As yet the vesicle has no inflammatory zone around it.

For the next two days the vesicle increases in size; assuming, if the vaccination was performed by the method of puncture, a circular form; if done by an incision, an oval shape. But in both cases the margin is regular and well defined. About the eighth day an inflammatory zone of a bright red colour, termed the "areola," begins to appear around the base of the vesicle; this increases in extent for two or perhaps three days more, by which time it may extend for about a couple of inches from the vesicle. The latter still retains its concave appearance, but a crust of a brownish colour will have commenced to form in the centre. By about the eleventh day the vesicle has attained its greatest magnitude, and the surrounding inflammation begins to abate. The fluid contained in the vesicle, or "pustule" as it is now called, which before was thin and transparent, becomes more viscid and somewhat turbid. After this period the whole becomes quickly converted into a smooth shining dry crust of a dark brownish colour. This crust, unless forcibly removed, will adhere for a week or more and then fall off, leaving the skin beneath apparently sound, but livid for a time, and afterwards more or less permanently scarred.

During the evolution of the local changes which result from the insertion of vaccine lymph beneath the surface of the skin it is possible, as previously mentioned, to recognise three more or less definite stages of papule, vesicle, and pustule.

The same statement holds good with reference to the eruption of small-pox, whether this be local, *i.e.*, due to intentional inoculation of the virus, or general, as the result of casual infection.

In each instance the appearance of the first or *papular* stage is brought about by inflammatory reaction, causing an increase of intercellular fluid, together with concomitant increase in volume and number of epithelial cells, of the rete Malpighii more particularly. The papule gradually becomes enlarged by a circumferential extension of the same process, and owing to further changes in the cells first affected, vacuoles arise in the central portion of the papule, by the extension of which this ultimately becomes a vesicle.

The *vesicle* is a multi-locular structure, the dissepiments, by means of which its interior is divided up, being formed from the thinned and extended remains of the original epithelial cells. Owing to the fact that the process of vacuolation increases, for a time, more extensively at the advancing edge of the vesicle, the central portion remains somewhat less elevated, thus giving rise to the appearance termed umbilication.

At a somewhat early stage of the process an outflow of leucocytes takes place towards the point of injury. In time each blood vessel becomes the centre of an aggregation of leucocytes, which by the rapid increase in their numbers eventually transform the originally clear inflammatory exudation into a purulent fluid. The vesicle is said now to have become converted into a *pustule*.

By the thinning and ultimate rupture of its trabeculæ the pustule finally becomes unilocular. The turbid fluid contained in it now gradually dries up, and, together with the necrosed remains of epidermal cells, takes part in the formation of the *crust*, which, under the microscope, appears as a homogeneous mass very deeply coloured by the ordinary stains.

Meanwhile a regeneration goes on underneath the crust, the new epidermis being formed by an ingrowth from the surrounding *stratum lucidum*. The extent to which the cutis vera has been involved determines the depth of the resulting scar

PART II.—HISTOLOGY; Plates XX.–XXXIII.;

By DR. GUSTAV MANN

(From the Physiological Laboratory, Oxford).

Methods Employed in the Present Research.

The pieces of skin, as soon as excised, were placed for 24 hours into either absolute alcohol or picro-corrosive solution, this latter consisting of one part of saturated $Hg\,Cl_2$ in $\frac{3}{4}$ per cent. $Na\,Cl$ (= 10 per cent. solution of $Hg\,Cl_2$) mixed with three parts of a saturated watery solution of picric acid (solubility of picric acid = 0·6 per cent.).

The specimens were passed in the usual way through chloroform and paraffin. Sections were cut varying in thickness from 2·5-25μ, and were fixed to slides by a special albumen method;* and also, without any fixative, by Gulland's method. This latter precaution was taken to meet the possible objection that the albumen on the slide might have given rise to appearances which have nothing to do with vaccination. I am, however, able to state that absolutely no difference can be seen between sections fixed by one or the other method, while by the use of albumen one can make sure of sections adhering to the slide during the prolonged staining with alkaline or acid solutions involved in certain of the methods adopted.

After the removal of the paraffin from the sections with xylol and absolute alcohol, they were first examined unstained, in the following media, which are arranged according to their refractive indices: water = 1·333, absolute alcohol = 1·361; glycerine and water equal parts = 1·397, chloroform = 1·449, glycerine = 1·456, Bergamot oil = 1·464, xylol = 1·497, Canada balsam = 1·535, metacinnamene with an equal weight of phenylthiocarbimide = 1·639. Phenylthiocarbimide ($C_6\,H_5\,CNS$) = 1·654, methylenediiodide ($CH_2\,I_2$) = 1·743°D. The three substances last mentioned were kindly suggested and given me by H. G. Madden.

Of these substances it is well always to use water, Bergamot oil, balsam and the metacinnamene-phenylthiocarbimide mixture; The last of these forms also a valuable addition to our list o mounting media, for which purpose it should be employed i conjunction with Bergamot oil, balsam and xylol balsam.

In addition to the investigation of tissue elements by examina- tion in media with a higher or a lower refractive index than th possess themselves, recourse was had to staining reagents, a again, sections were first examined in water to ascertain wh elements give up their colour on being treated with alcohol.

It is difficult to arrange the dyes which were used, systemati ally, but the following list may serve as an indication of the li of research adopted :—

For histological, apart from bacteriological, purposes the chi methods employed were :—

A. Substantive ones (Bancroft), *i.e.*, without previous mor danting.

---

* Anat. Anz. Bd. VIII., 1893, p. 442.

(1) Acid dyes :—

(a) Mann's biacid mixture of methylblue and eosin. Methylblue is the triphenylpara-rosanilintrisulfoacid ($C_{37}H_{26}N_3S_3O_9Na_3$).

1 per cent. methylblue in distilled water ... 35 c.c.
1 per cent. eosin in distilled water ... ... 45 c.c.
Distilled water ... ... 100 c.c.

Sections are left in this mixture for five to ten minutes, washed in water, dehydrated and mounted in balsam. This constitutes what may be termed the "short method." Or they are dealt with by the "long method," *i.e.*, are left in the stain 12–24 hours, then washed in distilled water, thoroughly dehydrated, and placed in a vessel containing absolute alcohol 30 c.c. to which, previously, five drops of a 1 per cent. solution of KOH in absolute alcohol have been added. When the sections have turned a reddish tint, the slide is washed with absolute alcohol to remove the alkaline alcohol, then rinsed in distilled water till differentiated. If the sections are not blue enough, a drop of acetic acid added to the water in which they are being rinsed will restore the colour.

(b) Picro-nigrosin, in a watery solution.
(c) Ehrlich's triacid mixture.
(d) Ehrlich's acid haematoxylin.

(2) Neutral dyes :—

(a) Ehrlich-Biondi mixture.
(b) Jenner's methyleneblue-eosin precipitate dissolved in pure methylic-alcohol (Merck). Stain for one hour. Rapidly wash in absolute alcohol and clear.
(c) Delafield's haematoxylin made with haematein. Two drops of this solution in 30 c.c. of distilled water. Stain for three to seven days.

(3) Alkaline dyes :—

Toluidinblue, thionin, methyleneblue (the polychrome methyleneblue of Unna is a mixture of methyleneblue and dimethylamin hydrochlorate or methylene violet and methylene red).

Methylviolet 6B was used in Kromayer's modification of Weigert's fibrin stain for the demonstration of the fibrils in epithelial cells.

B. Adjective methods used were :—

(1) M. Heidenhain's iron-alum haematoxylin.
(2) Rawitz' tannin and tartar emetic, followed by fuchsin, safranin, methylviolet, gentian-violet and smaragd-green.
(3) Wasielewski's mixture; fuchsin one part, potash alum three parts, water 100 parts, followed by $\frac{1}{2}$ per cent. $K_2Cr_2O_7$ in 70 per cent. alcohol.

Special bacteriological methods used included the following :— Gram's original method, and its modifications according to Nicolle and Claudius; Löffler's methyleneblue (methyleneblue 0·5 parts, absolute alcohol 30 parts, and KOH (1 : 10,000 of water) 100 parts); Löffler's formula, but with 1 : 1,000 of KOH, *i.e.*, containing ten times the ordinary amount of alkali. Further, carbol-thionin and carbol methyleneblue were employed, also saturated solutions of toluindinblue, thionin and methyleneblue in 5 per cent. formaldehyde solution (calculated from formol, which is supposed to contain 40 per cent. formaldehyde, and

[illegible]

[illegible] XXI. and XXII. is the [illegible] if we take the [illegible] from which the affection is spreading, then we find. probably because of the direction of the lymph flow, that one side of the skin undergoes its characteristic changes more rapidly and to a greater extent than the other.

During the first 48 hours the hypodermis is apparently normal, but in the course of the third day (Fig. 5, Plate XXI., representing changes after 72 hours) it begins to swell underneath the inoculated area. On the fourth and fifth days the œdema loses its local character, and is for this reason less evident in Figs. 6 and 7, Plates XXI. and XXII.

Whatever the nature of the virus may be, it does not give rise to much leucocyte infiltration during the first two days. Comparing Figs. 3 and 4, Plate XX. (48 hours), with Fig. 5, Plate XXI. (72 hours), there will be noted in the latter considerable leucocyte infiltration in the dermal layer, which during the fourth (Fig. 6, Plate XXI.) and fifth days (Fig. 7, Plate XXII.) becomes very marked, and on the sixth day (Fig. 8, Plate XXII.) results in the dermis having swollen to such an extent as to bulge into the hypoderm, and now the latter also becomes invaded, the thick collagenous bundles composing it being pressed apart.

The umbilicated appearance during the fourth and fifth days (Figs. 6 and 7, Plates XXI. and XXII.) appears, as Unna has suggested, to be mostly due to the original injury caused by the process of vaccination.

During the sixth day (Fig. 8, Plate XXII.) both the epidermis and the dermis are seen to undergo considerable necrotic changes. The size of the affected area can be ascertained without any difficulty, as all the photographs (Figs. 1–8, Plates XX.–XXII.) are taken to the same scale, viz., 15 diameters, and they are therefore directly comparable.

Minute Structure of the Epidermis.

To understand the changes which the epithelium undergoes as the result of inoculation with the virus of cowpox, it is necessary to first refer to the normal appearances.

In preparations fixed in alcohol and stained by Kromayer's method, or fixed in picro-corrosive solution and stained by Wasielewski's method, Ehrlich's tri-acid, or Jenner's fluid, the stratum corneum forms one-fifth of the total thickness of the epidermis, the stratum granulosum being represented by only one layer of cells, which latter are so few in number as not to be usually in contact with one another. The rete Malpighii is from three to four cells deep (*vide* later, Fig. 9, Plate XXIII.). The lowermost cells of this layer, *i.e.*, those in direct contact with the dermis, are columnar, with long, finger-like processes which enter the basal membrane (*vide* later). What is quite characteristic of all the cells in this layer is the presence of fibrils which in the lowest columnar cells run vertically and at the apices curve round into the neighbouring cells. The various cells are, in this way, firmly united to one another, the weakest spot being the area between the adjacent basal halves of the columnar cells where the bridging fibrils are very few in number.

On examining the fibrils lying between neighbouring cells (*i.e.*, the so-called "prickles") each is seen to have on its centre a minute granule, which was first noticed by Reinke. The inter-fibrillar matter in the intercellular region is formed by lymph derived from the dermis; while the interfibrillar substance in the cell is a viscous cytoplasm, which in the rete Malpighii is fairly abundant, in the stratum granulosum undergoes a transformation into the basophil granules, and in the stratum corneum for the greater part, along with the fibrils, gives an eosinophilous reaction. That there exists a basophil substance seems to be demonstrated by specimens stained for one hour in Jenner's fluid, then rapidly washed in absolute alcohol and cleared in xylol, as by this method the stratum corneum is stained pale blue.

The nuclei in the rete Malpighii, when stained by either the short or the long methylblue-eosin method, stain a bright blue (80 per cent.) or brilliant red (20 per cent.). The blue nuclei usually show feebly developed nucleolar material.

Perinuclear spaces are very rare in normal skin.

Changes in the Epidermis due to Vaccination.

In all the pieces of vaccinated skin examined, the epidermis was for the most part completely divided along the whole track of inoculation. In any case the vaccine matter is brought directly into contact with both epidermis and dermis.

It is customary to distinguish in vaccinia three distinct histological appearances in the epidermis, viz., (1) Weigert's primary coagulation necrosis, due to the great virulence of the poison at the seat of lesion; (2) Unna's reticulating fibrinoid change, affecting chiefly the upper layers of the epidermis; (3) Unna's "ballooning colliquation," which is most evident in the lower strata of the rete malpighii.

1. *The primary coagulation necrosis* one would naturally expect to find close to the seat of infection; in the region where, during the early stages of infection, because of the temporary resistance offered by the healthy cells, and because of the want of circulation, the poison must accumulate. When the cells, therefore, once commence to fall victims to the irritant, they quickly succumb. This condition is best seen at the end of the second and commencement of the third day, close to the line of inoculation, but only one or two cells in each section are to be found which can be said to have degenerated without having passed through the typical changes immediately to be described.

It may be doubted, however, whether the explanation suggested is the whole explanation; for it is quite possible that the cells which have succumbed have done so in consequence of direct mechanical irritation, the result of the process of vaccination. The cells referred to do not show the typical fibrillar arrangement, but are smaller than normal, the cytoplasm is granular, amphophil, either collected centrally round the nucleus or occupying with the latter an excentric position, the remainder of the cell being empty and not traversed by fibrin threads.

2. *The reticulating fibrinoid change*, according to Unna, takes place thus:—Near the nucleus, or near the periphery of the cell, small vacuoles filled with fluid appear. These, becoming confluent, form a net which joins the nucleus to the cell wall. This net then undergoes a fibrinoid change, and has precipitated on it fibrin-granules from the intra-cellular fluid, which granules Renant supposed to be micrococci. The long preservation of the

nucleus and the cell wall with its prickles is stated by him to be characteristic of this change.

3. *The "ballooning colliquation"* of epidermal cells, according to Unna, is characterised (1) by the cytoplasm, as a whole, undergoing degenerative changes, there being thus no demarcation into a peripheral and central zone; (2) by the loss of the prickles and rounding off of the cells in the rete Malpighii; (3) by the amitotic division of the cell-nucleus giving rise to numerous nuclei; (4) by an increased plasticity of the cell which, adapting itself to its surroundings, will assume a globular, flattened, or pointed shape; (5) by an ultimate fibrinoid change affecting both the cytoplasm, or its remains, and the nuclei. In vaccinia complete degeneration usually results, while in varicella regeneration of these cells is possible.

Examination of the specimens shows that the earliest change is an increase in the diameter of the intercellular channels (*vide* Fig. 9, Plate XXIII.), but how this is brought about is doubtful. We certainly find evidence of œdema in the subjacent dermis, and might therefore suppose that the intercellular spaces become distended with lymph, which, being unable to escape on the surface, forces the cells apart, and thus renders the "prickles" much more evident. On the other hand a possible power of contraction inherent in the cells, and called forth by the action of the virus, must be taken into account. That during later stages this tendency does show itself will become evident. On studying Fig. 9, Plate XXIII., it will be noticed that the prickles are in reality the intercellular portions of fibrils coursing from cell to cell and uniting them to one another. Unna in his textbook has paid no attention to Ranvier's discovery of these fibrils, and speaks of the "homogeneous protoplasm of the younger cells," when describing the ballooning colliquation of varicella. In addition to these fibrils there is found in normal cells the interfibrillar cytoplasm which has a mesh-like arrangement, as is seen in Fig. 12a, Plate XXIV. In describing the epidermal changes, it is necessary to consider separately the fibrillar and the non-fibrillar elements, and also to realise that the latter can retract from the nucleus, although under normal conditions they are in contact with it. By the retraction of the non-fibrillar elements a perinuclear space is formed.

In tracing the changes which the cells in the upper layers of the rete Malpighii undergo under normal conditions, and those which they show under the action of the virus, a sharp distinction must be drawn between bodies formed inside the perinuclear space and those found outside it in the cytoplasm. In Fig. 9, Plate XXIII., the cytoplasm of the cells, No. 1, is shown in direct contact with the nucleus, while in No. 2 a perinuclear space has been formed. This latter appearance is normally found in the upper layer of the skin, and becomes especially well marked in those changes which accompany the formation of nails and hoofs, but although I have looked through the large collection of epidermal sections kindly lent me by Prof. Arthur Thomson, I failed to find any structures lying within the perinuclear sac. In the cell, No. 4, Fig. 9, a pointed body lies against the periphery of the nucleus, apparently not within a space, and in the cytoplasm is found an oval body looking like a diminutive nucleus. The significance of these bodies is not known, but that element lying close to the nucleus may be a very early stage of Guarnieri's supposed parasite.

The various papers dealing with epithelial changes said to be due to the action of parasites may thus be summarised:—

In 1886–7, Van der Loeff, in three papers, described in vaccine lymph and in variola vera, "proteid" (protozoa) or amœba-like bodies.

L. Pfeiffer, 1887–91, arrives at the conclusion that the contagium is a parasite belonging to the class of sporozoa (Leuckart), and calls it *Monocystis epithelialis*. There cannot be any doubt, if one takes into consideration the figures accompanying the above papers, that both authors mistook epithelial cells for sporozoa, viz., those cells which Unna describes as having undergone a ballooning colliquation. This explanation of Unna's is fully confirmed by the results of the present research.

Entirely different structures were subsequently described in 1892 by Guarnieri as parasites. During the prepustular stages small granules the size of cocci are seen at the periphery of the lesion while more centrally they may reach in size one-half the diameter of cell nuclei. Usually they are placed at a certain distance from the nucleus but when lying close to it they are in most cases invaginated by it, owing to their plastic condition. On very rare occasions they may indent the nucleus. These later stages are supposed to represent a rhizopod, capable of amœboid movement, with an evident nucleus and capable of dividing "without any doubt" by fission and probably also by endogenous gymnospore formation. Guarnieri applied to his supposed parasite the name of *Cytoryctes vaccinæ* or *variolæ* for the reason that it hollows out the cytoplasm of the epidermal cells.

Ferroni and Massari in 1893 reinvestigated the subject in Grassi's laboratory, and all three agreed that appearances as described by Guarnieri could be obtained by the application of such substances as croton oil, osmic acid, iodine, or Indian ink, and therefore held that Guarnieri's bodies were partly nuclear derivatives and partly leucocytes.

Guarnieri repeated the experiments of the two authors just mentioned with negative results (1894) and was confirmed in this by L. Pfeiffer, who states that no purely chemical reagents will produce such uniform results as vaccine, leaving the nuclei at first unaffected. He then states that Guarnieri's bodies are young or early stages of the bodies previously (1887, *see* above) described by him under the name of *Monocystis epithelialis*.

Monti (1894) described bodies similar to those of Guarnieri in the rete Malpighii, but classes them under the group of Lobosi rather than true Protozoa. According to this observer the elements measure 2–3 μ in diameter.

Piana and Galli-Velerio (1894) and J. Jackson Clarke (1893–95) all accept Guarnieri's view as did also Ruffer and Plimmer in 1894 at the International Medical Congress in Rome, with this restriction, that they were not convinced of distinct spore-formation.

In 1895 Sicherer confirmed Guarnieri and L. Pfeiffer, and E. Pfeiffer confirmed J. Clarke's statement as to the peripheral processes of his coccidia, which were stated by him to penetrate into the clear zone surrounding them, and even to pass through it,

He does not agree with the conclusions of Guarnieri and L. Pfeiffer, nor do his researches with glycerine, croton oil, osmic acid, or silver nitrate, confirm Ferroni and Massari.

APPENDIX C.

On the Histology of Vaccinia; by Drs. Copeman and Mann.

The three most recent papers on this subject are by Jackson Clarke (1895), Wasielewski (1897) and Häckel (1898). Clark states that the variola parasites "find their most complete homology in those found in cancer and sarcoma." He distinguishes (a) hyaline globules 2–4 μ, (b) spherical bodies 5–7 μ, with a hyaline nucleus, finely granular protoplasm and one or more highly refractile granules, (c) ellipsoid bodies 7·5 μ, with two small nuclei, (d) amoeboid bodies 7·5 μ in their largest diameter, but with no distinct nucleus, (e) spheres about 8 μ, with pseudopodia, which can be fixed by a temperature of 20°C, and containing one or more large hyaline nuclei, (f) ovoid bodies 3–5 μ enclosed in a capsule possessing a fine opening at the pointed end. a, b, and c, usually occur free, occasionally in lymph cells or in the deep cells of the rete malpighii, d-bodies are always free, while "e" are partly free and partly found between epidermal cells; f-elements are found most frequently in the middle layers of the epidermis. His Fig. 7 shows an enormous "parasite" nearly filling the whole of the epithelial cell, with peripheral granules, and a central fragmented nucleus, and his Fig. 12 free globular "parasites" with peripheral granules.

Wasielewski, a pupil of L. Pfeiffer, worked on the rabbit's cornea, fixed preferably in picro-corrosive solution, stained in alum-fuchsin, and differentiated by bichromate-alcohol. He also is an upholder of the protozoan theory and figures the smallest granular "parasite" at the periphery of the lesion and the largest ones near the point of inoculation. He especially points out that the bodies in question always lie in the perinuclear space. The nucleus at first spherical, may become indented by 1–3 bodies, which may either by a degenerative change give rise to ringlike structures encircling the apparently normal nucleus, or they may multiply and form spherical bodies with red granules at their periphery which lie in a concavity of the nucleus.

The most thorough and apparently unbiassed work which has been undertaken on this subject is that of Armand Häckel.

According to this observer the results obtained by inoculating the cornea of rabbits, show that Guarnieri's "parasites" give the following micro-chemical reactions:—

1. Distilled water has no action.

2. Potassium iodide stains them in the same manner as other cell constituents.

3. With saturated NaCl, the bodies and nuclei disappear.

4. Very dilute KOH causes both nuclei and bodies to disappear, and subsequent application of acetic acid does not restore them.

5. A one per cent. sodium carbonate solution destroys all detail in the cells, the nucleoli are no longer visible, and the bodies with their envelopes disappear. Subsequent treatment with acetic acid leads to the reappearance of the nuclei and bodies.

6. Five per cent. acetic acid renders the previously invisible small spherical bodies very evident, while the granules and droplets round the larger bodies disappear.

7. Flemming's solution renders the small bodies and the nuclear chromatin very distinct, while the protoplasmic granules round the larger cells disappear. The bodies are surrounded by a cleft.

8. One-half per cent. Na Cl saturated with Hg Cl, shows spaces round the bodies and nuclei in a few cells only. The same solution with the addition of acetic acid renders the bodies much more distinct and many pericorpuscular and perinuclear spaces become visible.

9. One per cent. osmic acid occasionally shows nucleoli in the epithelial cells, whether the bodies are present or not.

10. None of the reagents mentioned above will cause contraction in those bodies which have an amœboid shape.

11. With Biondi stain, Guarnieri's bodies stain blue and the leucocyte nuclei green, while with iodine-green-fuchsin mixture, the former stain red while the latter stain blue, as do the nuclei of epithelial cells.

Häckel divides the appearances met with in vaccinated areas into several groups :—

A. Homogeneous bodies varying in size from those which are just visible, to others having a diameter of 3 μ. They are usually found in the perinuclear cleft but occasionally lie in the cellplasm, and, if there, they are usually surrounded by a clear zone. The older the cell and the more cramped in position, the deeper the nuclear pocket in which the bodies lie.

B. Elements similar in position to those already described showing a distinct differentiation into a central blue and a peripheral red zone, the line of demarcation between the two being more or less sharp. The erythrophil external portion may undergo a granular or threadlike degeneration.

C. Similar to B, but the erythrophil granules are usually very regularly arranged and reach occasionally the size of granules A. These forms resemble the "daisies" of Guarnieri.

D. Spheroidal bodies surrounded by a broad clear zone, with threads extending through the latter into the cytoplasm (Clarke's and E. Pfeiffer's peripheral processes of coccidia). Fine red granules may be seen along the course of the fibrils, especially in cells lying towards the centre of the lesion.

E. Demilunes, spindles, &c., which appear later than the spheroidal forms, and always lie close to the nucleus.

F. Triangular or pyramidal forms which are very rare.

G. Lobular forms looking like blue amœbæ with red granules.

All the forms above mentioned are derived from the bodies first described under "A". The original small cyanophil body by imbibition of fluid begins to swell, and simultaneously the blue colour becomes violet. In addition to this, infiltration changes occur, affecting the periphery of the mass, and leading to the formation of erythrophil granules. Inasmuch as many granules protrude from the margin, the latter appears corroded. Instead of only the periphery being affected by the imbibition one finds occasionally the vaccine body has undergone this hyaline change in toto. What is significant is that while the body shows the signs of hyaline degeneration, neither the cytoplasm nor the nucleus appear to suffer directly by these changes. Apart from the hyaline degeneration one may find a vacuolation of the vaccine corpuscles which leads to the formation of, at first, coarse, then fine trabeculæ and ultimately fine granules. The early stages in this transformation induced Guarnieri to speak of filamentous forms of protozoa.

In addition to degenerative changes in the vaccine bodies, cells as a whole may undergo chromatoplasmolysis and form spheroidal bodies which Clarke mistook for the formation of sporogoniæ (with and without chromatin) and spores.

Häckel therefore considers Guarnieri's bodies not to be parasites, but to result from a peculiar transformation of a portion of the cellplasm due to the specific stimulus of the vaccine virus.

Three other suggestions which have been made are these: Babes (1894 International Medical Congress, Rome) referred to a possible nucleolar origin; Leoni (ibid) supposed Guarnieri's bodies to represent necrotic parts of the nucleus; and Salmon, working in Metschnikoff's laboratory, holds that they represent the hyperchromatic residue of nuclein derived from the nucleus of leucocytes.

Examination of our own specimens shows that after the preliminary dilatation of the intercellular lymph channels (Fig. 9, Plate XXIII.) the vaccine virus produces definite changes in the different layers of the epidermis, which will be described from below upwards. The general effect after 48–60 hours is seen in Fig. 20, Plate XXVII. Those cells which lie in direct contact with the dermis and belong to either the skin proper or to the root-sheaths of the hairs, are affected later than those lying in the middle layers of the rete Malpighii. This difference is due probably to the lowermost cells being more vigorous, as they contain a greater quantity of nonfibrillar cytoplasm, and also to the fact that pressure being exerted on them from all sides, will tend to prevent their increase in size. Already at an early period, e.g., 48 hours after vaccination in animals which have reacted well, the basal cells have increased from 13–15μ in length to 30–35μ; their nuclei showing a corresponding increase from 10μ to 13·5μ. The subsequent changes are not the same for all cells, for a certain number cease to react further and soon become atrophic owing to the pressure exerted on them by their neighbours. The remaining cells either undergo a reticular change which will be described later in connexion with the middle layer of cells or they give rise to what Unna has termed "balloons." The latter are formed (*vide* Fig. 14, Plate XXV., representing a section 72 hours after vaccination) by the epidermis first breaking up into a number of strands of epithelial cells, which appear as pillars surrounded by lymph, many of which, later on, because of a still greater accumulation of lymph, are torn across. There are thus formed *bullæ*, separating a superficial set of epidermal cells, from a deep set, the latter being still in contact with the dermis. Owing to this strain on the epidermal cells the intercellular bridges are torn across, and cells are liberated in groups of from one to ten. Amongst the larger groups there are some cells which are more vigorous and which for a time appear to thrive, while others are undergoing atrophic and reticular alterations.

The epithelial cells in addition to being affected as a whole, show also characteristic changes in their interior. At the early period, when the epidermal cells have commenced to increase in size, one or two small bodies appear at the poles of the cell (Fig. 12 k, Plate XXIV.). This occurrence bears out Häckel's observation on the cornea. The perinuclear space is not well marked, but there cannot be any doubt of the fact that these small bodies lie close to the nucleus within a clear space. Although hundreds of serial sections have been examined with the hope of tracing the origin of the granules, this has not been satisfactorily accomplished. When they first appear they measure about 0·5 μ in diameter, but soon they rapidly increase in size to 2μ (Fig. 12 l, Plate XXIV.); if there are two bodies present, the larger one is always found on the side nearest the surface.* The structure in question is one of the early stages of Guarnieri's supposed parasite.

The lowermost epidermal cells have a marked tendency towards nuclear proliferation, which expresses itself in one of the following three types:—

A. If the cell has become a "balloon," *i.e.*, if it is isolated, it may continue to grow till it reaches five times the normal diameter (Fig. 11, Plate XXIII.), the measurements for the cell No. 1 being 43μ × 75μ. What is characteristic of these cells is the feeble development of a perinuclear space, and the power of occasionally breaking up into a number of daughter cells (*vide* Fig. 13, Plate XXIV.) corresponding to the number of nuclei. This phenomenon has several analogues both in the vegetable and animal world.

The usual fate of these big cells is a reticular transformation, commencing round the nucleus and then spreading outwards. Fig. 11 represents the same cell stained by three different methods. The section in which it was first noticed was originally stained by Löffler's method for twelve hours, differentiated in alcohol, and mounted in balsam. This stain brought out the nuclei, and a number of basophil granules scattered throughout the cytoplasm. A reticular appearance could be seen on using a very small aperture of the diaphragm, but to make sure whether all the "prickles" had disappeared, the section was restrained in acid fuchsin and glacial acetic acid, and differentiated in bichromate alcohol, by which method the reticulum was revealed and a faintly stained granular appearance. To bring out the latter more fully, the section was treated by Möller's method for staining spores. The appearances which were seen by each of these methods were carefully recorded by the camera lucida, and ultimately combined into one picture. In the same Fig. 11 will be seen an epithelial cell which atrophied from the first, with a long collapsed nucleus (No. 2), and further, a binucleated cell (No. 3) with very regular reticulate transformation. The coarse fibrils extending upwards from the basal membrane and fixing the cell *in situ* are fibrin threads.

Belonging also to this type A, one finds, during the later stages of vaccination (120–144 hours), the appearances figured on Plate XXIV. (Figs. 12, o, q, and lower part of p, (p[1])). It is by no means certain

* Fig. 12 l is printed upside down.

as to what may be the correct interpretation of what is to be seen, but the following views may be suggested :—We are either dealing with leucocytes, which having entered the giant cells, are rendered inert by them through the deposit of a capsule, or the leucocyte, to protect itself, becomes surrounded by an envelope. This leucocyte hypothesis, however, does not seem so probable as the following theory :—As already pointed out above, the formation of daughter cells undoubtedly does occur, and the question arises, is it possible for that nucleus, which is the most vigorous, to surround itself by a capsule and thus to endeavour to safeguard itself. Appearances, such as are represented by these figures would seem to support the latter hypothesis, for we see an apparently normal nucleus surrounded by granular protoplasm, which communicates with the exterior by a definite narrow channel, while the greater part of the cell is surrounded by a thick capsule, showing a distinctly radial arrangement. In studying epithelial changes we should, as far as possible, endeavour to give a physiological explanation of the appearances met with, and not draw hasty conclusions as to their representing sporozoa. J. Clarke rightly points out that many cells appear similar to those found in cancer, but surely this is the best reason for not pinning one's faith to a parasite theory. On rare occasions I have succeeded in demonstrating, to my own satisfaction at any rate, that the capsule surrounding the cells gives the keratin reaction, and in one case the typical reaction of the granules met with in the stratum granulosum was present. We have, therefore, some grounds for the view that one or more of the daughter cells derived from a giant cell will subsequently pass through the same set of changes, which, under normal conditions, an epidermal cell undergoes in the course of its existence.

To refer once more to the giant cell in Fig. 11, Plate XXIII. :—On the right side of the perinuclear space will be seen a dark mass which represents the remains of a Guarnieri's body. The changes which this element undergoes will be detailed fully later on.

The ultimate fate of the cells belonging to the type A is shortly this :—The vacuolation and consequent reticulation commencing around the nucleus, spreads towards the periphery, the cytoplasm lying inside the vacuoles becomes gradually diminished, the cell loses its definite outline, leucocytes enter the body of the cell, and the nuclei derived from the original epidermal nucleus shrivel and lose their affinity for basic dyes, they become eosinophilous, and eventually give the same staining reactions as the reticulated cytoplasm. Unna calls the change a fibrinoid one, but the trabeculæ which are left never stain as deeply as true fibrin, and it is doubtful as to whether fibrin granules are really deposited on the surface of the cell network. Photographic views of these giant cells will be seen in Fig. 14, Plate XXV., and Fig. 16, Plate XXVI.

B. The second type of giant cells is characterised by the cell body undergoing condensation, due to the absorption of its fluid contents. This type approaches in its staining reactions much more closely to those of true fibrin. Here again one or more of the nuclei tend to surround themselves with cytoplasm and thus form daughter cells within the confines of the mother cell. Fig. 12, Plate XXIV., represents an elongated cell with two atrophied nuclei at the apex (p^3), with four nuclei arranged in series in the middle (p^2) and a daughter cell enclosed in its capsule at the base (p^1). The four central nuclei are stained a brilliant red by Wasielewski's method, and possess a strong resemblance to Russell's "fuchsin bodies." They represent a typical change which nuclei in the uppermost layers of the rete Malpighii are apt to undergo on the fifth and sixth day after vaccination. Instead of passing from a vesicular into the usual contracted state, the nuclei become condensed into spherical masses, which, during the early periods, enclose a distinct nucleolus, but ultimately become homogeneous. The lowest cell in Fig. 12p., Plate XXIV., has become binucleated, and shows a smaller body towards the right, resembling in its staining reactions Guarnieri's bodies.

While the types A and B are formed, chiefly during the second, third, and fourth day, and always mark the height of the vaccine reaction, the next type is formed amongst apparently, as yet, quite healthy epithelium, principally during the fourth and fifth day.

C. The third type of giant cell differs from the two previous ones in being usually represented by a large number of vesicular nuclei free in a cavity among the surrounding cells. This type does not as a rule show any Guarnieri's bodies. What has become of the cell body of the original epidermal cell it is diffcult to say. The nuclei seem to have the power of division after the cytoplasm has completely disappeared. Should the epithelial cells which surround the giant cell undergo the typical change due to the vaccine virus which have been described above, the nuclei of type C are set free and soon undergo degeneration.

Unna has stated that the formation of the many nuclei in the epidermal cells takes place amitotically, but while this certainly seems to hold good for the majority of cases, there is no doubt that true mitosis also occurs.

It has already been pointed out that the epidermal cells in the upper part of the rete mucosum react differently to the vaccine virus from the lowest cells. The account of the changes to be given is based on the examination of sections fixed in picro-corrosive solution and stained at 30°C for 12 hours in Löffler's blue containing KOH in the proportion of 1 : 1,000. The solution employed was a fortnight old and contained therefore in addition to methyleneblue some methyleneviolet, due to the decomposition of the methyleneblue (Unna). Compare Fig. 12,A–E, Plate XXIV., seventy-two hours after vaccination. The sections were dehydrated with alcohol and mounted in xylol balsam.

Normal epithelium shows the nuclei stained deep blue, while the nonfibrillar cytoplasm is violet. The fibrillæ running through the cell remain unstained for which reason the individual cells stand out very sharply. The cytoplasm (Fig. 12A, Plate XXIV.) appears as a distinct sponge-work, the apertures or holes in which serve for the reception of the epithelial fibrillæ. In those cells close to the stratum granulosum where a contraction of the nucleus has taken place a delicate fibril seems to stretch from either of the two poles of the nucleus and to extend into the cytoplasm. The nucleus at mid-focus seems to be drawn out into a fine point connected with these fibrils (Fig. 12B, Plate XXIV.)

The first change due to vaccination is an alteration in the regular arrangement of the cytoplasmic network close to the nucleus, some of the meshes becoming larger, and thus the appearance of vacuolation is produced. During this stage delicate strands of cytoplasm still connect the nucleus with the cell body proper (Fig. 12C, Plate XXIV.) Soon afterwards, during the stage of enlargement of the cell, the violet-coloured plasm assumes the appearance of a coarse meshwork, each corner of the meshes being markedly thickened (Fig. 12D[1] and D[2], Plate XXIV.) Still later the delicate strands joining the thickened nodal points tear across and thus give rise to little globules and rod-like structures. At this period the cells have a certain resemblance to the well-known picture of Nissl's granules in nerve cells (Fig. 12E, Plate XXIV.)

By the staining method employed the early stages in the formation of Guarnieri's bodies are not readily studied, but the later stages are well seen, as is also a peculiarity regarding the relative position of these bodies to the site of inoculation. It must be remembered that the calf is vaccinated by longitudinal incisions into the skin and that our sections were taken at right angles to this incision. Guarnieri's bodies are invariably situated in that half of the cell farthest away from the line of incision, thus in Fig. 14, Plate XXV., the cells all appear darker towards the right side of the photograph, the site of inoculation being on the left. On the parasitic theory this is very difficult to explain, but what proves definitely that the supposed parasites are not the cause of the change in the cells is the fact that the latter have undergone a complete vacuolar change in that half which is looking towards the point of inoculation, while that portion which contains Guarnieri's bodies is by far the more normal. In Figs. 12 F-I, Plate XXIV., five cells are delineated which exhibit this remarkable change. One half of each of these cells has lost all affinity for methylene-violet, while the other half stains deeply. Close to the site of inoculation the whole cell is unstained, or rather has a pale greenish tint resembling that of the fibrin threads. A little farther out the cell nuclei are still deeply coloured and in cells not quite as much affected, the degenerative change is spreading to that side of the cell farthest away from the inoculated area, and the supposed parasite is found to have undergone degeneration before the cytoplasm on which it is supposed to be feeding has completely disappeared.

It has been mentioned above that the early stages in the formation of Guarnieri's bodies connot be traced by this method, but that the later stages may be. It is well, however, here to point out that by the staining method employed, Guarnieri's bodies, as described by Häckel, stain exactly in the same way as the nonfibrillar cytoplasm.

If one studies the early formation of Guarnieri's bodies by entirely different staining methods, such as those of Wasielewski

APPENDIX C.

On the Histology of Vaccinia; by Drs. Copeman and Mann.

or Ehrlich, or by means of Unna's polychromemethyleneblue-tannin method, or by Mann's biacid mixture, small bodies (0·5-1 μ) will be noticed, usually at one or both poles of the cell, sometimes indenting the nucleus, but, as a rule, lying some distance from it. These gradually increase in size till they reach 3-4 μ in diameter. The general appearance may be gathered from the photograph No. 13 (Plate XXV). It will be noticed that the body always lies in the perinuclear space. On two occasions appearances were found suggesting that the body had just left the nucleus, the latter being distinctly drawn out and destitute of nucleolar matter. (Fig. 12M, Plate XXIV.)

The CHANGES IN THE EPIDERMAL CELLS, apart from those affecting the cytoplasm (Fig. 12, Plate XXIV.) are best studied in sections of tissue removed during the early part of the third day of vaccination and stained according to the Möller or Wasielewski methods. A typical group of cells is seen in Fig 10, Plate XXIII; they appear two to three times larger in diameter than in normal skin, and adhere firmly together by means of the fibrils which bridge the somewhat dilated intercellular lymph channels. The perinuclear cleft, originally small (No. 1), increases considerably in extent (Nos. 4, 5 and 3), till ultimately the walls of the sac almost touch the periphery of the cell. There is thus brought about an accumulation of the fibrils at the periphery of each cell. The cells derived from a common ancestor are united together more firmly by those fibrils which run through them from base to apex. By the end of the third and fourth day these cells, owing to the pressure exerted by the accumulation of lymph, will have become much stretched. The lymph cannot escape either downwards or laterally, and hence, being pent up in the normal intra-cellular channels, exerts pressure on the resisting stratum corneum, owing to which the latter becomes elevated above its normal level. Inasmuch as the only elements which have resisted the disintegration changes, apart from the contents of the nuclear sac, are the fibrils above mentioned, they will act as ties which tend to bind down the stratum corneum. The latter being, however, pushed upwards by the accumulation of lymph, the bundles of fibrils are rendered tense and will compress any cells which may lie between them and thus lead to their atrophy. In this manner the greater number of the giant cells formed in the upper layers of the epidermis eventually become destroyed.

It is difficult to offer a satisfactory explanation of the enormous enlargement of the perinuclear sac which is found to take place. If it arise in the first instance by colliquation, *i.e.*, the flowing together of a number of small vacuoles, this process cannot explain the subsequent increase. On careful focussing, the cytoplasm is seen to be so sharply marked off from the space as to suggest the presence of a special membrane. The theory which at present appears the most feasible is, that by the rarefaction of the cytoplasm, a number of substances are set free which are capable of giving rise to a process of endosmosis, which latter brings about a dilatation of the sac.

In the stratum granulosum the cells increase considerably in their vertical diameter and the granules become very evident when stained by polychrome-methyleneblue (Fig. 15, Plate XXVI.) If it were not for slight differences in the size of these granules and their occasionally irregular outline, they might readily be mistaken for diplococci, as they frequently are arranged in pairs. The same figure (15, Plate XXVI.) also shows to what an enormous extent the intercellular channels may become dilated, thereby leading to subsequent atrophy of the cells.

Fig. 16, Plate XXVI., is taken from the same preparation but somewhat nearer to the site of inoculation. Many of the cells in the upper layers of the epidermis show a premature conversion into those distinctive of the stratum granulosum. The granules in the latter are derived from the remains of the cytoplasm and not from the cell-fibrils as has been stated.

The cells in immediate contact with the fully keratinised cells of the stratum corneum become considerably distended and, owing to the rich lymph supply, the nuclei, which under normal conditions would only stain feebly or not at all, now stain deeply, owing to the formation of nuclein. This rejuvenescence is, however, but short-lived, as necrosis soon sets in and the cells take part in the formation of the "crust." At the margin of the affected area, many nuclei, as already pointed out, contract into apparently homogeneous spheres which possess a strong affinity for fuchsin.

During the fourth and fifth day of vaccination, necrosis of the lowermost cells in contact with the dermis sets in, and by the sixth day extensive retrogressive changes become noticeable in the cells lying at a somewhat higher level. What is the cause of these necrotic changes will become apparent later on.

To return to the consideration of Guarnieri's bodies. Are all the different structures which have been described under this heading derived from one and the same original element? Have the various forms described by Häckel a common origin? It would appear to be necessary to distinguish a distinct eosinophilous element which is of nucleolar origin (Fig. 12 L and M, Plate XXIV.), and usually placed at the poles of the nucleus, and secondly, a basophilous structure, cytoplasmic in origin, which is usually seen best in those cells in which the nuclei stain deeply in polychrome-methyleneblue-tannin preparations (Fig. 12K, Plate XXIV. and possibly the body marked G.B. in No. 4, Fig. 10, Plate XXIII). There is also a third body which is neither of nucleolar nor cytoplasmic origin, but which consists of matter secreted by the nucleus. It may happen that this secretion is precipitated round the bodies of nucleolar origin, or both may be found separately. Thus the cell No. 2, in Fig. 10, Plate XXIII., had two eosinophilous nucleolar bodies lying in the next serial section, while the large crescentic-shaped body, which is depicted, represents the nuclear secretion.

As the staining reactions of these crescents are exactly the same as that of the cytoplasm (*vide* description of Fig. 12, Plate XXIV.) the suggestion may be hazarded that these demilunes, crescents, or ringlike bodies represent material which under normal conditions would pass directly into the cell-plasm, but which is prevented from so doing owing to the formation of the perinuclear space. The secretion in adapting itself to

c
ogy of
ia;
s.
ian and

the surface of the nucleus produces the peculiar appearances met with.

The degenerative changes which these crescentic masses eventually undergo have been well described by Häckel.

That, very rarely, leucocytes have the power of entering epithelial cells, but only to undergo a peculiar degeneration, is shown by the figure 12R, Plate XXIV., the spherical masses having the same appearance as those already described by M. Heidenhain for salamander leucocytes.

As to the nuclear division induced by the vaccine virus, the cell No. 3, in Fig. 10, Plate XXIII., contains four nuclei, two of which lie in the next serial section; No. 2 possesses two nuclei only, while Nos. 4 and 5, are mononucleated. It is thus evident that the cells close to the stratum corneum may increase considerably in size without undergoing nuclear division. During the fifth and sixth day of vaccination one commonly finds at the margin of the vaccine vesicle giant cells containing as many as a dozen nuclei, which occasionally are arranged in a definitely crescentic form.

If then Guarnieri's bodies do not constitute the organism specific to vaccinia, is it possible to demonstrate anything else that may more probably do so? With all reserve, attention may be drawn to certain appearances found during the second and third day after vaccination in the cell-plasm of the epithelial cells. In Fig. 10, Plate XXIII., 48 hours, and in Fig. 11, Plate XXVI., 120 hours after vaccination, stained by Möller's method for spores and well differentiated, a number of exceedingly small granules, varying from 0·2–0·25μ in diameter, can be seen. These elements are most distinct close to the perinuclear sac, because there the thinness of the cell is more marked. The granules are usually arranged in pairs, they lie between the epithelial fibrils, and are very numerous. Only a few of the most conspicuous granules have been drawn, so as to simplify the figure.

Are these bodies micro-cocci or are they merely a granular precipitate? The question cannot be definitely settled until a dependable method of cultivating the vaccine virus in artificial media outside the animal body has been devised.

Minute Structure of the Dermis.

In the dermis three distinct layers may be distinguished, the basal membrane, the dermis proper, and the hypoderm. The first of these is in direct contact with the epidermis and consists of a very dense feltwork of collagenous fibres giving the characteristic methylblue reaction with Mann's bi-acid mixture.

These fibres fit into the finger-like epithelial processes, and occasionally fine fibrils may be seen to extend upwards within the inter-epithelial lymph spaces, to one-half the length of the lowermost layer of columnar cells.

Underneath this layer is the dermis proper which may conveniently be divided into a superficial vascular and a deep elastic layer. The former consists of collageneous bundles measuring from one-fifth to thirty μ in diameter, between which are the hair follicles, blood vessels and lymphatics. The blood vessels, even down to the finest subepithelial capillaries possess a very distinct reticular envelope of white fibrous tissue, which stains a deep blue colour in marked contrast to the bright red of the endothelium, fixed connective-tissue cells, sebaceous glands and erector pili muscles.

The deeper strata of the dermis are particularly rich in coarse elastic fibres (*vide* Fig. 21, Plate XXVIII.) which run parallel to the surface, giving off finer twigs which pass more or less vertically upwards to terminate under the basal membrane, occasionally even running for a short distance in the latter.

The dermis, as compared with the hypoderm, is richly supplied with blood vessels and lymphatics, and has in consequence a much more open texture. The hypoderm consists of coarse bundles of white fibrous tissue (20–100μ in diameter) which are arranged so as to form thick strands. The arrangement of the connective tissue cells is similar to that in tendon bundles. Between these strands of white fibrous tissue are found a few elastic fibres.

One hour after vaccination (Fig. 19, Plate XXVII.) the epidermis is seen divided, the dermis is laid bare, and at the bottom of the wound is a fibrin clot with red and white corpuscles entangled in it. The blood vessels in the neighbourhood are engorged, and in some cases completely blocked by leucocytes. Between the epidermal edges is seen a clot representing escaped serum which appears to possess a granular structure due to the action of the fixing solution employed. Stained by Gram's method the granules are so resistant to the process of decolorisation and so regular as to closely simulate micro-cocci.

This peculiarity led Kent to speak of large numbers of micro-organisms between the lips of the wound. By a special modification of Gram's method, which he has not as yet published, he states that he has succeeded in staining the specific diplo-bacillus of vaccinia, which is said to be always contained in special cells in the deeper layers of the dermis. I have no doubt that he has been misled by appearances, such as shown in the lower portion of Fig. 24, Plate XXIX., which represents a collection of plasma cells stained by Claudius' modification of Gram's method. By the same process some of the nuclei in the deeper layers of the epidermis (Fig. 24) and dermis (Fig. 24) are very resistant to the decolorisation by Gram, and the gentian violet is apt to be precipitated in forms which resemble bacilli. That these appearances are not pathogenic is shown by their occurrence in perfectly healthy tissues.

Changes in the Dermis due to Vaccination.

The first change to be noticed in the fixed connective tissue elements is a necrosis of the basal membrane, 48 hours after vaccination (*vide* Fig. 9, Plate XXIII.). This swells up and if stained by the short bi-acid method, exhibits a great affinity for

APPENDIX C.

On the Histology of Vaccinia; by Drs. Copeman and Mann.

the eosin rather than the methylblue. What is remarkable is that this change should take place beneath the vaccinated area, even at a point distant from that at which the epidermis is divided, and at a time when the changes in the epithelium are as yet but slight.

Commencing with the second day and reaching its height on the fifth day, an œdema of the dermis occurs accompanied by leucocyte infiltration. Primarily in the centre, and later, at the margin of the affected area, some special chemotaxis seems to call forth an emigration of the eosinophilous and neutrophilous leucocytes. Thus we find in Fig. 22, Plate XXVIII., a small blood vessel from which a considerable number of cells have escaped. Before discussing the probable fate of these cells it will be best to shortly summarise some papers dealing with the question.

In 1892, P. Doehle described as to be found in the blood of patients suffering from smallpox, measles, scarlet fever and syphilis, small spheres (0·5–1μ) which were either homogeneous or contained a highly refractile nucleus surrounded by a clear zone. They showed movements, which in some instances could be observed to be due to the action of a flagellum four to five times the length of the body. Occasionally two spheres were enclosed in a common capsule. He also noticed granular and amœboid elements (2·5μ), and ill-defined rodlike protoplasmic bodies possessing small flagella. The structures referred to were found from about the third to the tenth day after the eruption. In the clear lymph removed shortly after the formation of the vesicle he found, in addition to white and a few red corpuscles, bodies resembling those present in the blood. Many of the larger ones showed flagella 0·5–1μ long. During this period he also occasionally observed bodies (2–2·5μ) which constantly changed their shape, being sometimes rounded and sometimes bean-shaped. He expressed the opinion that the smaller elements decrease in number, while the larger ones increase during the later stages of development of the pustule. They, further, never lie included in epithelial cells, during the early stages, but may do so later on.

In 1893 Buttersack described what were apparently the same elements, 50 hours after vaccination, although his view of vaccine fibrils breaking up into globules (sporulation) was due to his method of examining films of lymph dried by exposure to the air.

In 1894 Kanthack and Hardy studied the behaviour of the leucocytes of the frog when brought into contact with anthrax bacilli, and arrived at the conclusion that the eosinophilous cells discharge their granules over the bacilli, and that the work of destruction is completed by the hyaline cells.

In 1896 Weber made further investigation of the blood of patients suffering from variola, and enumerates as occurring therein :—

(1) Spherical highly refractile granules, 1·8μ in diameter, which show Brownian and occasionally progressive movement; (2) granules measuring 5·6–7·2μ, found in the interior of non-mobile,

homogeneous and spherical elements, which he supposed to represent a phase in the life-history of a protozoon, and so termed *Sirenenkörperchen*; (3) large granules occasionally containing smaller ones, a transition towards the formation of the previously mentioned stage; (4) granules joined by threads; (5) stages in the formation of larger bodies from the growth of small granules, and of the division of fully formed bodies; (6) forms measuring 15μ, with one or two opaque and one clear nucleus.

Weber states that his bodies are very like eosinophilous cells, and that they only differ from leucocytes in possessing actively moving granules. He obtained cultures in alkaline semi-solid agar solution, which were invisible macroscopically. As the bodies described by him were also found in measles and scarlatina, it appeared doubtful as to whether they could in any way be regarded as specific.

Walter Reed also found in the blood of monkeys and calves, during the active stages of vaccinia, granular amœboid bodies, having one-third the diameter of the red corpuscles. He later discovered them in normal blood, and therefore believed that no causative relationship existed between these bodies and vaccinia.

H. F. Müller, too, discovered, independently of Reed, in normal blood, granules which differed essentially from blood platelets.

In 1897 Stokes and Wegefarth arrived at the conclusion that the bactericidal power of the leucocytes and of the blood serum of man and many animals is due to the presence of specific granules, especially those of the eosinophil and neutrophil type. When called upon to resist the action of invading bacteria the granular leucocytes can give up their granules to the surrounding fluids and tissues.

The rapidity with which, after vaccination, changes occur in the epidermis and dermis varies enormously in different calves. In this respect it may be pointed out that Figs. 2 (Plate XX.) and 22 (Plate XXVIII.) represent the appearance of the dermis in a calf which had been vaccinated 48 hours previously, and which "had not taken well." The corresponding epidermis is shown in Fig. 9 (Plate XXIII.), and it will become apparent that in this region the vaccine lymph was not brought into direct contact with the dermis. Notwithstanding this fact the basal membrane is necrosed beneath the site of inoculation, and, moreover, marked changes extend for nearly 3mm. to either side of this point.

Sections stained in methylblue-eosin mixture show large numbers of eosinophil and neutrophil leucocytes. In most of these (Fig. 23 No. 1, Plate XXIX.), the granules are arranged round one side of the nucleus, some cytoplasm occasionally lying outside the granular zone.

No. 3, in which the nucleus lies to the right, while the cytoplasm with its reticular arrangement is on the left, probably marks the initial step towards a shedding of the granules. Whether No. 4 represents a further step in the transformation of the leucocyte or whether it precedes the condition seen in No. 3 is doubtful, but ultimately the nucleus leaves its excentric position and becomes central, a change accompanied by a rarefaction of the nuclear contents (No. 5). These leucocytes cannot be seen to multiply by either direct or indirect division. It may be well to state that precautions were observed to exclude the possibility of appearances such as those seen in Nos. 3 and 4 being due merely to the nonstaining of basophil granules.

In Fig. 23, Plate XXIX. in addition to the leucocytes, numerous granules are seen in the lymph spaces between the collageneous bundles. These bodies vary in diameter from 0·75 to 1·5μ; a few are spherical but the greater number are irregular in outline. They are not improbably identical with those which the various authors above-mentioned have found in the blood and also in the lymph of vaccinia and variola during the febrile stages and the further suggestion may be hazarded that they are in reality leucocyte granules which have been discharged from the cells. As it is, however, as yet premature to make dogmatic assertion on this point, these elements may for the present be conveniently termed "Z granules."

Whatever the origin of these Z granules may be, they seem to possess the powers of amœboid movement and growth, as Figs. 18 (Plate XXVI.). 25, 26 (Plate XXX.), and 28 (Plate XXXI.) appear to show. These figures represent the appearances observed 72 hours after vaccination. The photograph (Fig. 28) shows how numerous these elements may be, while Fig. 25 demonstrates the great differences in their diameter; the larger granules measuring fully 3μ. The preparation here figured was treated as follows: (1) Ten minutes in 1 per cent. acetic acid, (2) three days at 30° C. in Löffler's methyleneblue containing ten times the normal amount of KOH, (3) 2½ hours in 1 per cent. acetic acid, (4) absolute alcohol and xylol. The section has a uniform bluish-black tint, quite different from that ordinarily obtained with methyleneblue.

That the granules are amœboid seems probable because of the various shapes which they exhibit (Fig. 18, Plate XXVI.), because they may be found at considerable distances away from leucocytes, and because they are met with also in the narrow septa which separate the bundles of white fibrous tissue (Fig. 26, Plate XXX.)

Sections stained for 10 minutes in undiluted polychrome-methyleneblue, washed for 20 seconds in distilled water, and subsequently differentiated in a 33 per cent. solution of tannin (Unna) for two to five minutes, show the appearances depicted in Fig. 26. The Z granules, with high focus, stand out as highly refractile elements, either not stained at all or of a faint green colour. In mid-focus they exhibit a minute spherical or elongated granule of a deep blue colour. Whether this appearance can be explained on purely physical grounds as being only due to a partial decolourisation of the Z granules is not certain, for in many instances the central globule will appear of the same size although the Z granules, as a whole, differ greatly in their diameter.

Unna's staining method is specially useful for tracing the changes in the fixed and wandering connective tissue cells, both nuclei and the cell bodies with their winglike processes being well stained.

By it the leucocyte, Fig. 15 (Plate XXIV.), was discovered stretching throughout the thickness of the epidermis, and measuring 126·7μ in length.

By yet another method, namely, by staining for 12 hours in Löffler's methyleneblue solution containing 1 : 1000 KOH, then differentiating in 2 per cent. H_2SO_4 it is possible to demonstrate in sections of skin, removed 72 hours after vaccination, cells apparently identical with various forms of Waldeyer's plasma cells. These are represented on Plate XXIV., Fig. 14 a–f. In the sections they appear as cherry-red elements, the remainder of the section being stained a bright blue. What is characteristic of the plasma cells is the presence in them of granules which are smaller than those of the eosinophilous leucocytes. The cells may appear rounded or provided with processes, their average length is 14–15μ and their nucleus under normal circumstances measures about 5 × 7·5μ. Therefore they are much smaller than the winged connective-tissue cells, some of which (Fig. 25, Plate XXX.) extend with their processes over 50μ.

The granules of these plasma cells stain of a bright cherry-red colour, as long as the nucleus stains blue, but in most cells the nucleus has lost its blue tint and is apparently undergoing degenerative changes. The healthy appearance of the cell is seen in Fig. 14a, Plate XXIV., in which the nucleus has distinct chromosomes, and the perinuclear area contains but few granules which, however, stain deeply; in Fig. 14b, Plate XXIV., the nucleus still stains with methyleneblue, but has a vacuolated appearance, and the granules are less deeply stained than in a; in b, d, e and f the nucleus stains with methylene violet of the same tint as the granules (some of which are feebly stained or quite invisible) and has a more or less vacuolated appearance.

The reason for mentioning the plasma cells here is to point out a possible relationship between the appearances obtained by the method above detailed and that seen in methyl-blue-eosin sections (Figs. 22 and 23, Plates XXVIII. and XXIX). In the former figure a number of cells are surrounded by eosinophilous granules, which are somewhat smaller than those of leucocytes, and which show a greater affinity for eosin. They correspond to the granules figured in the upper fourth of Fig. 23, Plate XXIX. Can it be that plasma cells set free their granules?

Commencing with the second day of vaccination and reaching its height at about 120 hours, the œdema of the dermis leads to the marked changes represented in Figs. A, B and C, Plate XXXIII. These three photographs are taken from the same section (120 hours after vaccination). Fig. A, Plate XXXIII., is farthest away from, and Fig. C, Plate XXXIII., nearest to the point of inoculation. The section was stained in a mixture of gentian violet one part, glacial acetic acid two parts, alcohol five parts, made up to 100 parts with water. The epidermis stained red while the collagenous bundles stained a deep blue. This method is specially adapted for studying the relationship of the dermis to the epidermal cells, the connective tissue processes springing from the basal layer, standing out sharply. In Fig. A., Plate XXXIII, the normal connective tissue appears very darkly stained as the section was 10μ thick, but nevertheless indications of lymphatics and blood vessels can be seen. The epidermal cells are normal in appearance. In Fig. B, Plate XXXIII., the connective tissue elements have been pushed asunder by the accumulation of lymph, the basal membrane being, however, as yet not perforated. The epidermal cells are markedly enlarged and stain less readily. In Fig. C, Plate XXXIII., the collagenous bundles appear as very delicate trabeculæ, the basal membrane is giving way in various places, and the epidermal elements have been completely destroyed.

In comparing the three figures with one another it is well to bear in mind that the delicate reticular arrangement seen in Fig. C, Plate XXXIII., is not due to an actual diminution in the total amount of white fibrous tissue present, but owing to the œdema the tissue elements have become spread out over a larger area. Actual destruction of the connective tissue is not one of the primary effects of the vaccine virus, and when it eventually occurs is due partly to the direct action of extraneous micro-organisms which first become noticeable about the third day, and partly to the simultaneous accumulation of leucocytes which, by the pressure they exert tend to bring about atrophy of the white fibrous tissue.

As Unna points out, during the vesicular stage of vaccinia the inflammatory changes are small and leucocyte infiltration is at first remarkably scanty.

To demonstrate the presence of extraneous micro-organisms subsequent to the third or fourth day, either the method detailed above for Z-granules, by the use of the modified Löffler stain may be adopted, leaving the sections at 30° C. or the sections may be stained for 12 hours at 45° C. without previous treatment with acetic acid, the stain being rapidly washed off before the slide has had time to cool and the section then cleared in xylol and mounted in balsam. By this second method the appearance shown in Fig. 17, Plate XXXVI., was obtained 96 hours after vaccination. Diplococci 0·75–1 μ

ɪr are seen surrounded by a large number of To the left of the preparation is an epithelial arly stage of necrosis.

Plate XXXI., stained by the acetic acid modified ethod was taken from a section 72 hours after . In it big bullae are seen which have resulted aration of the lowermost epithelial cells A from the er B. Free in the lymph minute diplo-bacilli will l.

description of the reticular change in epidermal previously, certain granules, best seen in Fig. 10, III., were described which resemble diplococci. the same staining method, namely that of Möller, appearance can be demonstrated during the second, fourth day in the œdematous connective tissue, seen in Fig. 18, Plate XXXVI, taken from a hours after vaccination. Numerous small granules, arranged in pairs, are found between the bundles ve tissue.

ute these granules are may be gathered by comparing such undoubtedly extraneous micro-organisms as are . 17, Plate XXVI.

usion reference must be made to the Figs. 29–31 ., which show that up to the 72nd hour after vac- ucocyte infiltration is limited to the dermis proper, extraneous organisms which are the exciting cause coagulum filling the mouth of the wound and also e dermis more readily permits of such emigration the hypoderm. In the latter but few blood vessels as its structure is virtually that of tendon. Fig. 30, II., shows that for the first few days after vaccina- lastic fibres are quite unaffected and this is so till 96th or 120th hour, when some of the fibres lying e line of incision can be seen to break up into granules, subsequent to which they fail to give staining reactions with Tänzer-Unna's orcein, but with Weigert's resorcin, fuchsin, and ferric-chloride

APPENDIX C

On the Histology of Vaccinia: by Drs. Copeman and Mann.

—:

Chronological order of events :—

fter vaccination : Effects of injuries due to inoculation, viz.: Vasodilation; emigration of a few leucocytes from vessels directly injured. Stasis of leucocytes in blood and lymph vessels close to site of inoculation.

rs after : Marked emigration of eosinophilous leucocytes. Shedding of the Z-granules. Necrotic swelling of the basal membrane.

x C of and

48–72 hours after	: Swelling of the epithelial structures and hence bulging of the epidermis into the dermis. Accumulation of lymph in the epidermis, causing a breaking up of its cells into strands. Formation of bullae Some cells atrophy while others hypertrophy. Proliferation of nuclei, especially in the deeper layers. Various elements differing in their origin, but generally described as Guarnieri's bodies, make their appearance. These are not parasites. The Z-granules show amœboid movement and growth. Both in the epidermal cells and in the lymph spaces of the dermis very minute cocci-like bodies are found in large numbers, and in the dermis the Z-granules are found close to these small bodies. The dermis shows localised swelling. Leucocytes with several nuclei are rare.
72–120 hours after	: Invasion of inoculated area by micro-organisms. Few eosinophilous, large numbers of hyaline leucocytes wandering towards the place of inoculation, but as yet they are restricted to the dermis proper. The epithelial "balloons" reach their height of development. Many encapsuled cells resembling the structures described by some authors as cancer parasites are present.
120–144 hours after	: Marked necrotic changes in the epidermis and dermis. Invasion of the hypoderm by leucocytes. Necrosis of the leucocytes in the dermis.

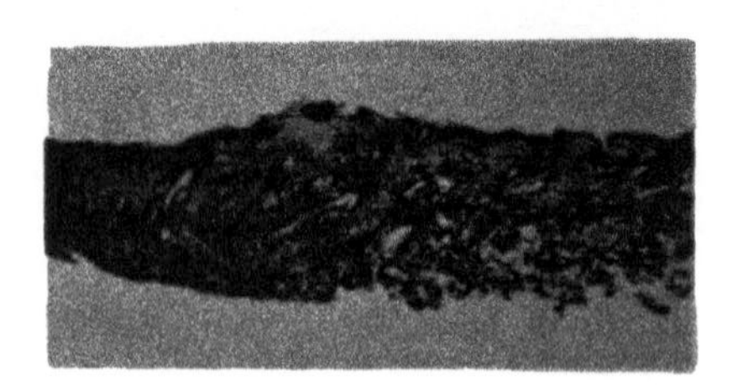

PLATE XX.

Fig. 1.

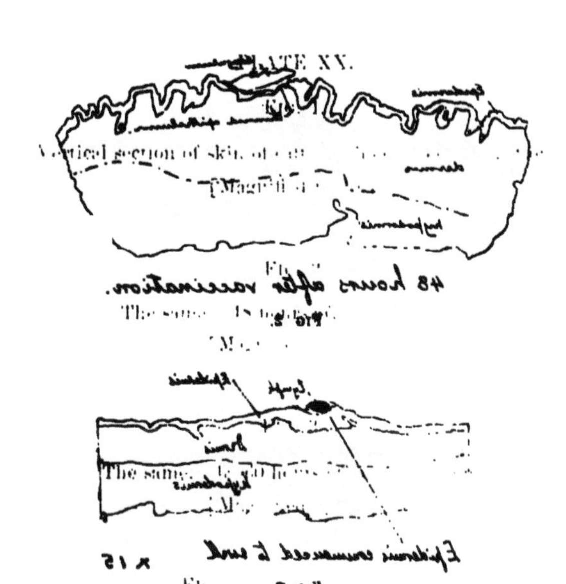

Fig. 2.

Fig. 3.

x 15.

Fig. 4.

PLATE XX.

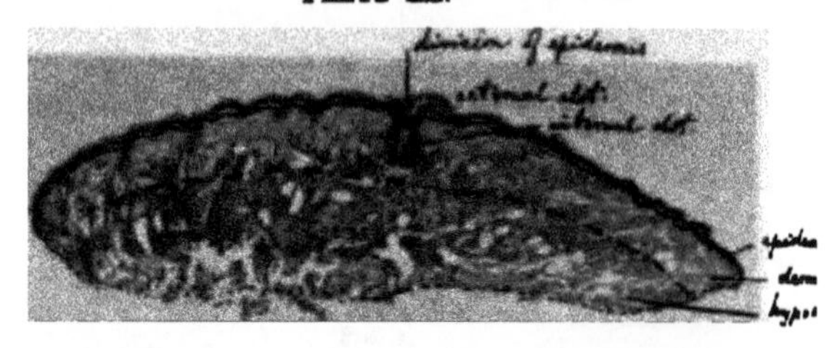

FIG. 1.

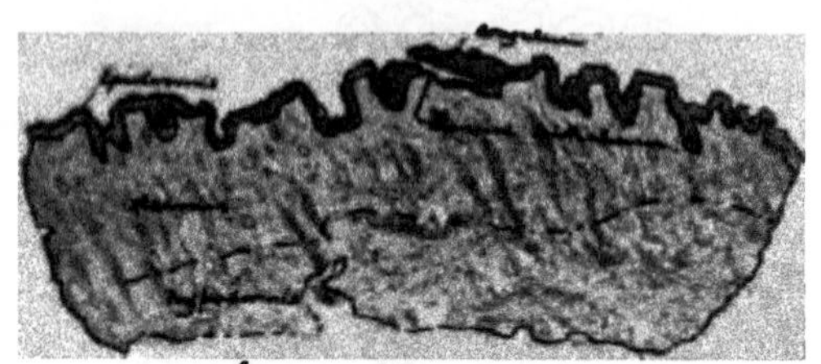

48 hours after vaccination.

FIG 2.

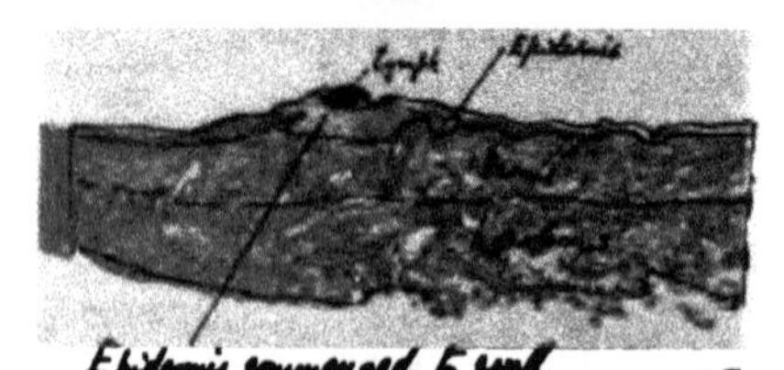

F.G. 3.

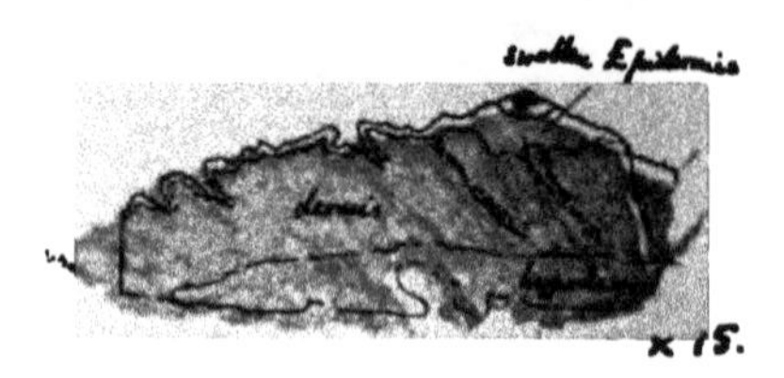

FIG. 4.

PLATE XX.

Fig. 1.

Vertical section of skin of calf one hour after vaccination.
[Magnified by 15.]

Fig. 2.

The same. 48 hours after vaccination.
[Magnified by 15.]

Fig. 3.

The same. 48–60 hours after vaccination.
[Magnified by 15.]

Fig. 4.—Same as Fig. 3.

PLATE XXI.

FIG. 5.

Vertical section of skin of calf 72 hours after vaccination.
[Magnified by 15.]

FIG. 6.

The same. 96 hours after vaccination.

PLATE XXI.

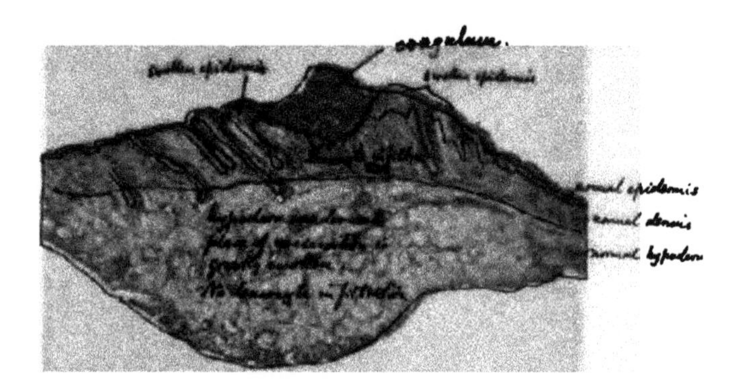

FIG. 5.

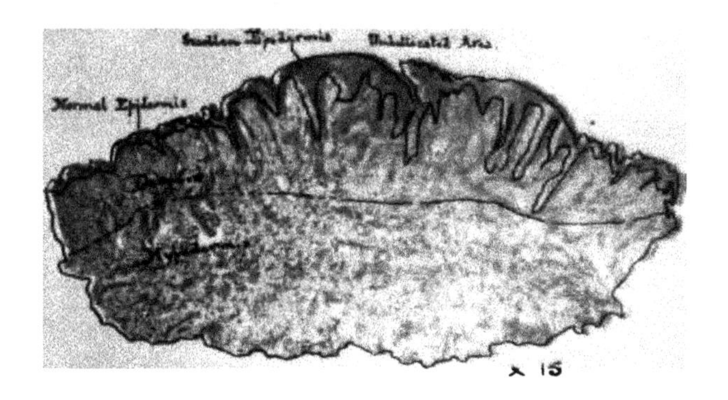

FIG 6.

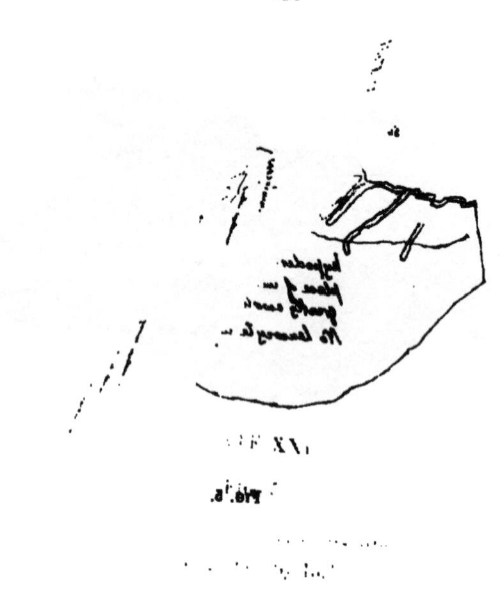

Fig. 5.

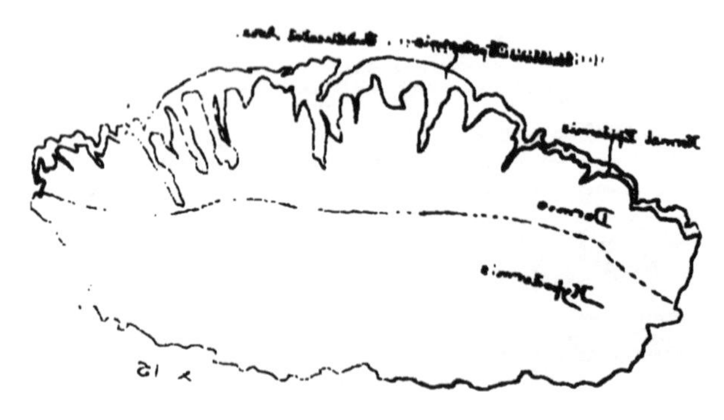

Fig. 6.

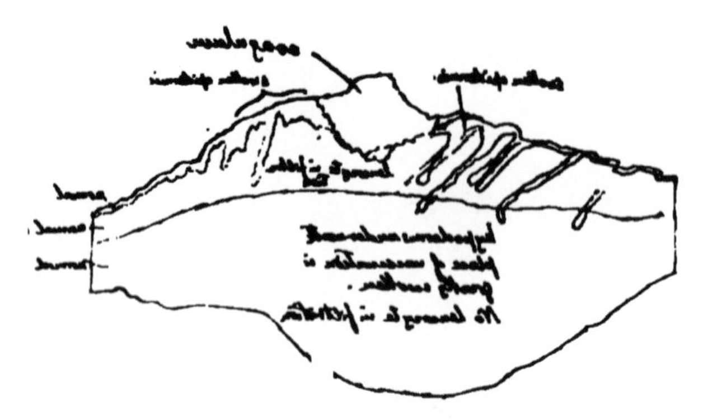

PLATE XXI.

FIG. 5.

[illegible] section of skin of [illegible] 72 hours after [illegible]

[Magnified by 15.]

FIG. 6.

[illegible] ion.

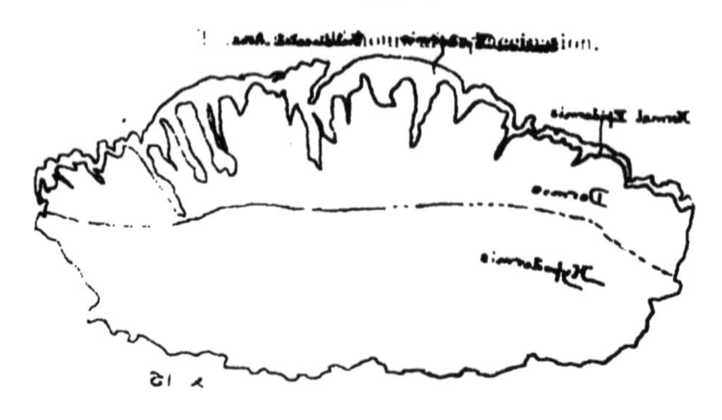

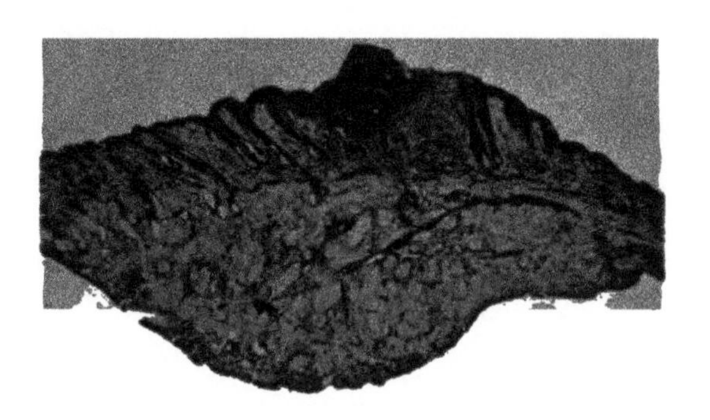

PLATE XXII.

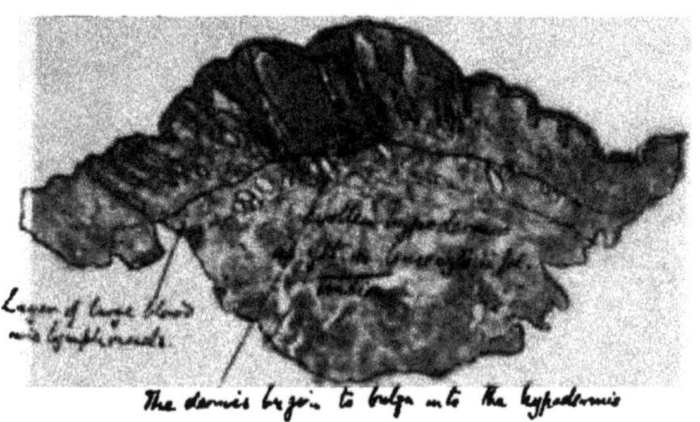

FIG. 7

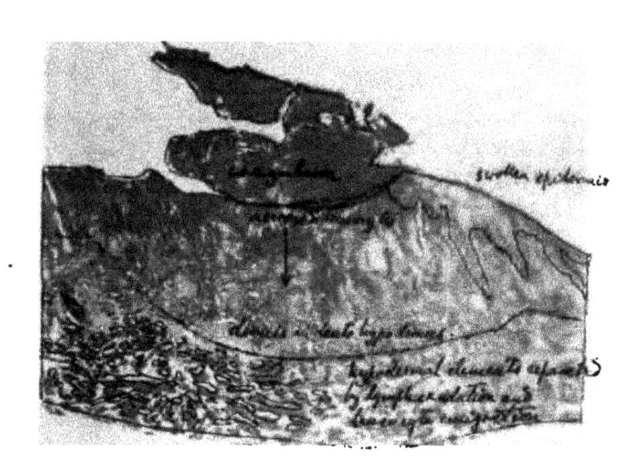

FIG 8.

Fig. 7.

Fig. 8.

PLATE XXII.

FIG. 7.

Vertical section of skin of calf 120 hours after vaccination.

FIG. 8.

The same. 144 hours after vaccination.

PLATE XXIII.

Fig. 9.

Vertical section of skin of calf 48 hours after vaccination.

Epidermal cells with fibrils. Intercellular lymph channels increased in diameter. Necrosis of basal membrane.

Fig. 10.

The same. 72 hours after vaccination. Unna's "reticulating colloquation."

Fig. 11.

The same. 120 hours after vaccination. Unna's "ballooning colloquation."

PLATE XXIII.

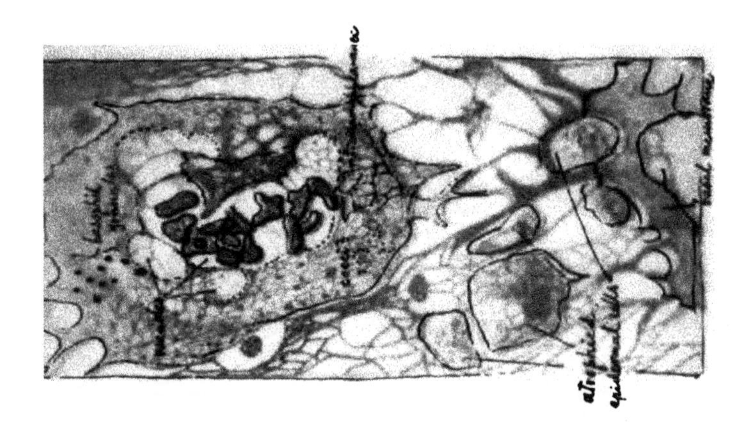

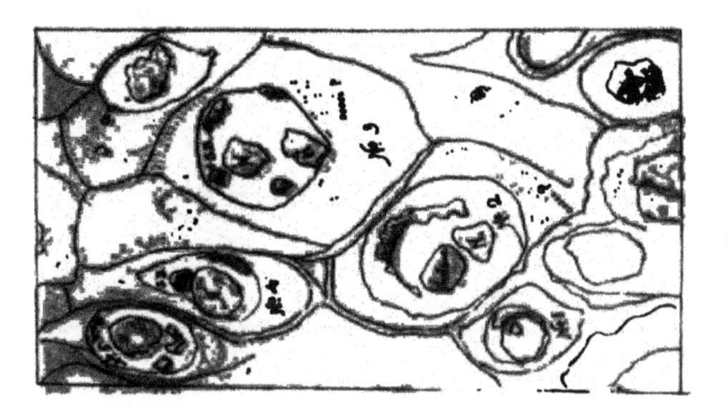

Fig. 10.

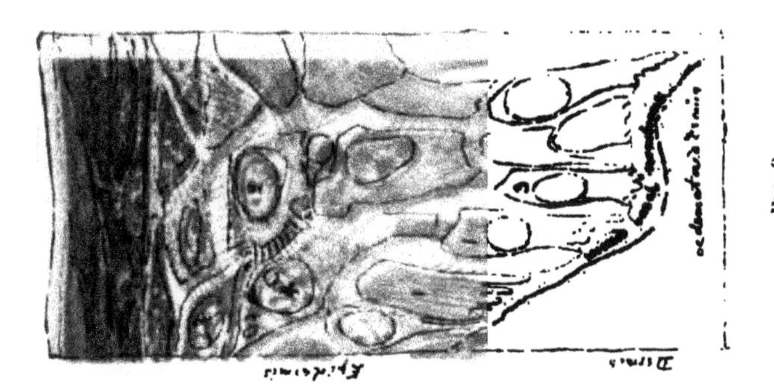

Fig. 9.

PLATE XXIII.

Fig. 11.

Fig. 10.

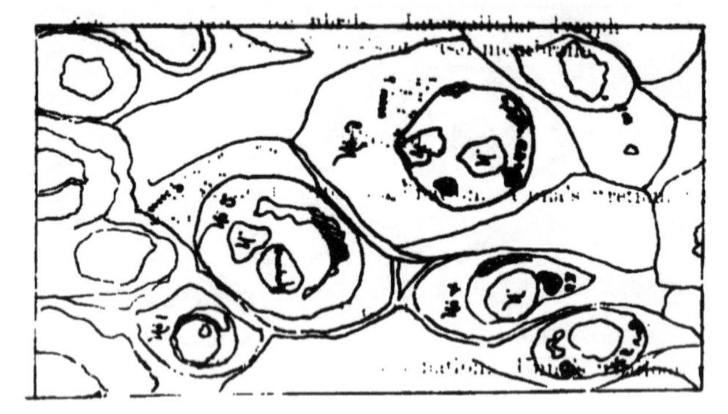

Fig. 9.

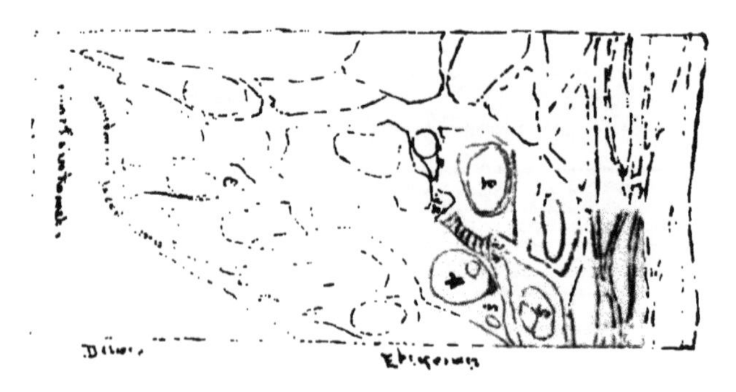

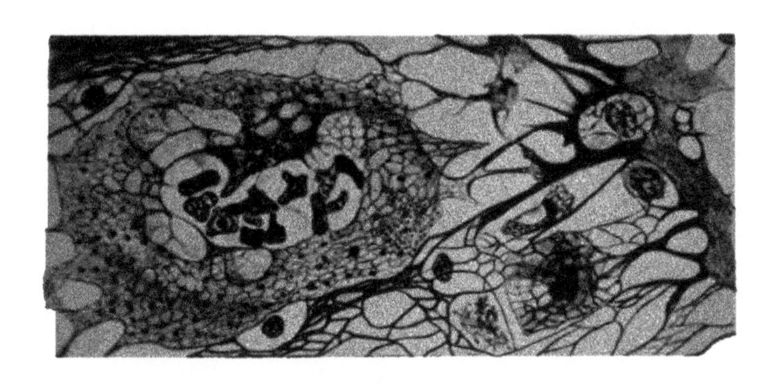
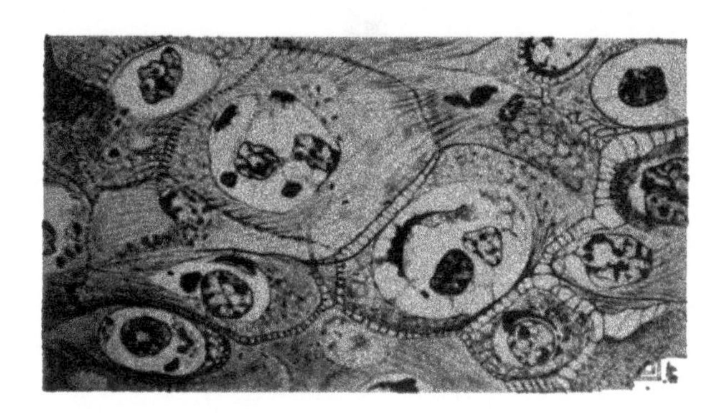
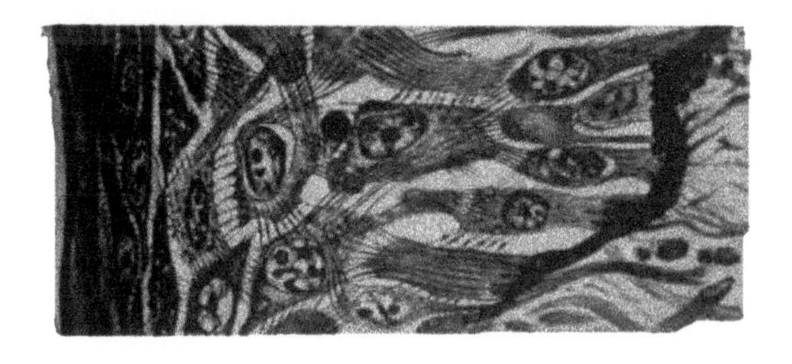

PLATE XXIV.

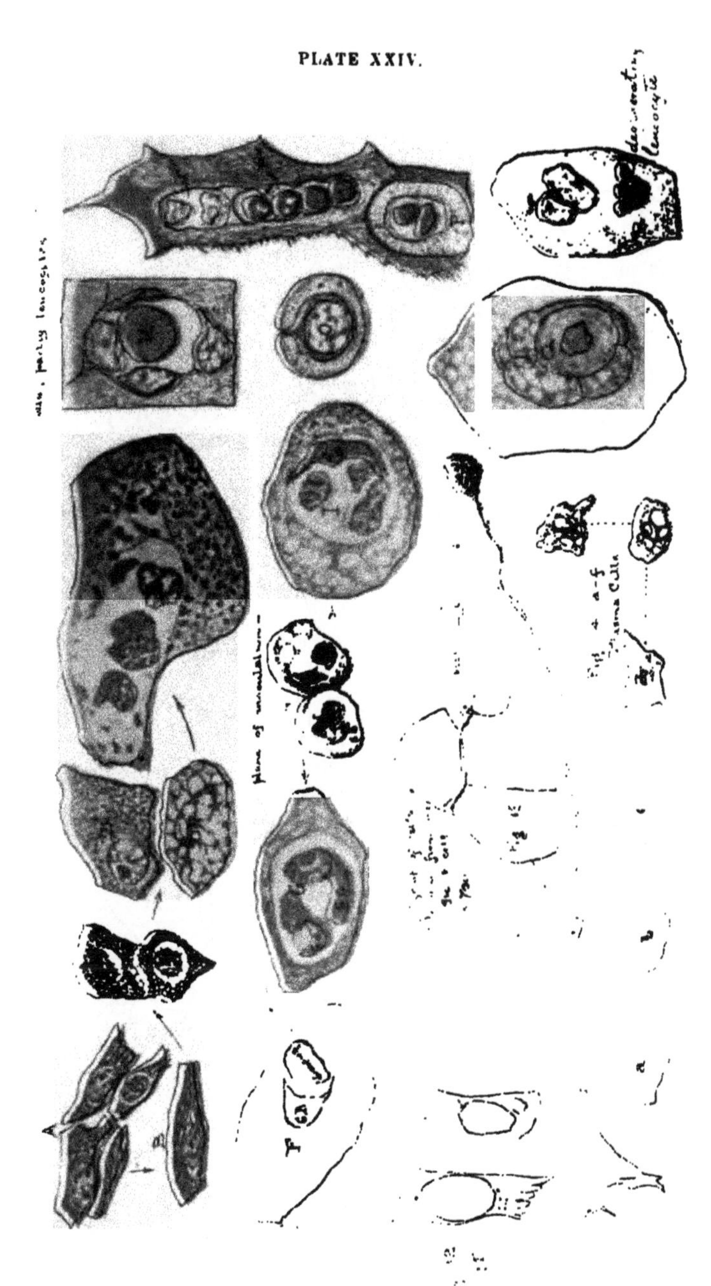

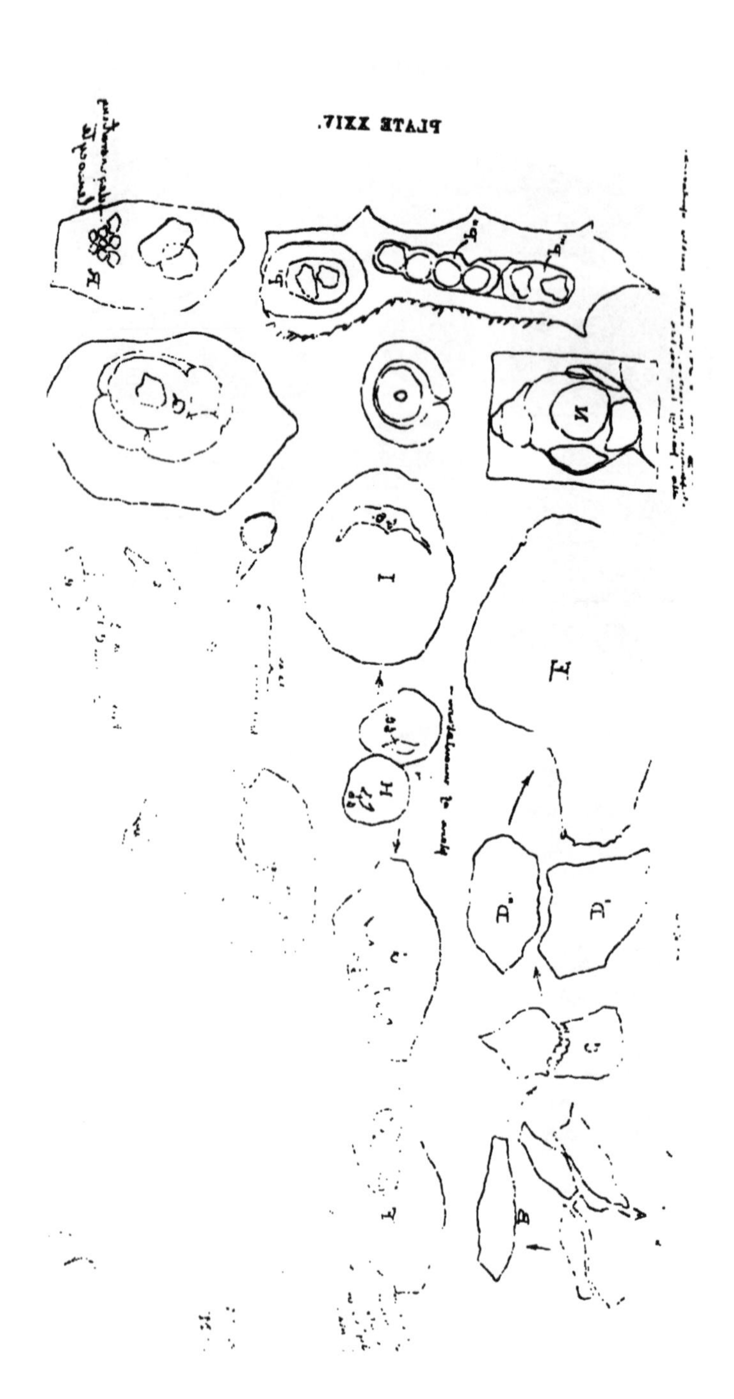

PLATE XXIV.

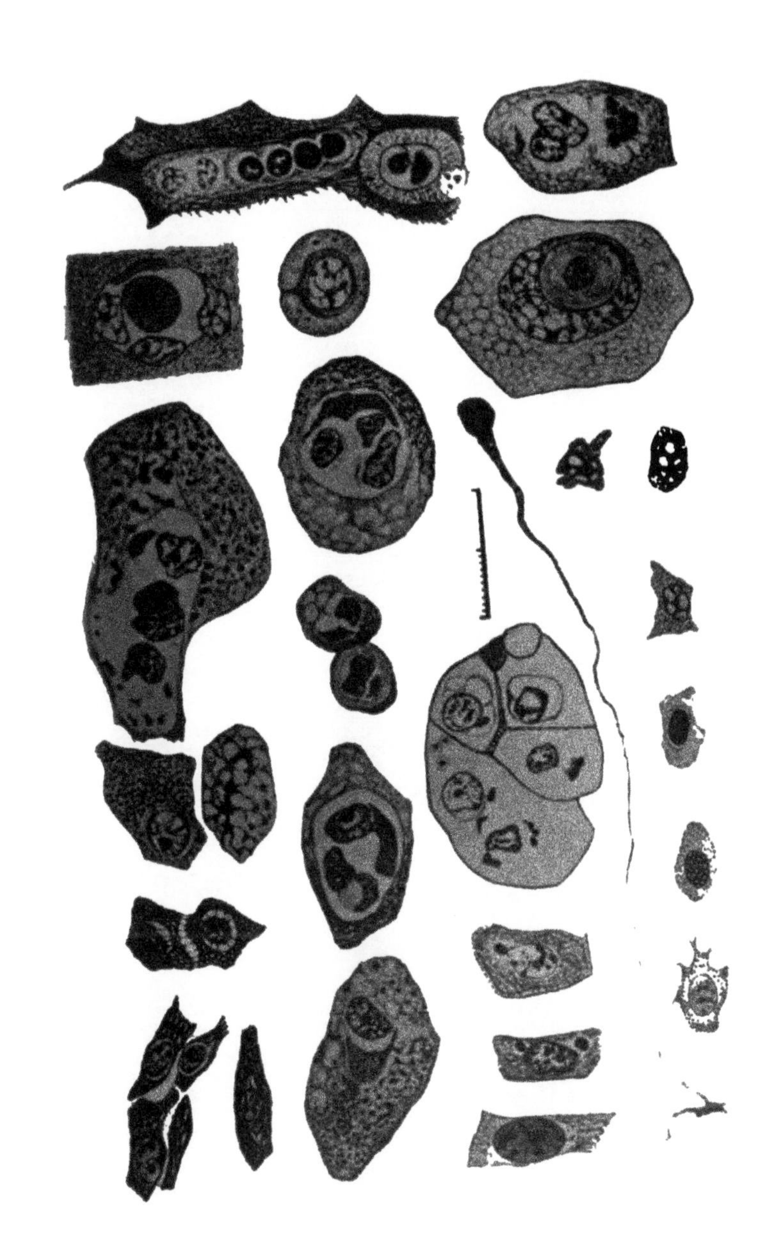

PLATE XXIV.

FIG. 12, A—E.

Changes in the cytoplasm of epidermal cells 72 hours after vaccination.

FIG 12, F—I.

Nuclear secretions forming the crascentic variety of Guarnieri's bodies.

FIG. 12, K.

A basophil Guarnieri's body of cytoplasmic origin.

FIG. 12, L & M.

Nucleolar origin of Guarnieri's bodies

Fig. 12, N—Q.

Epithelial cells simulating parasites.

FIG. 12, R.

Degenerating leucocyte in epidermal cell.

FIG. 13.

Nest of daughter cells derived from a single giant cell.

FIG. 14, *a—f*.

Plasma cells.

FIG. 15.

Leucocyte 126·7 μ in length.

All figures magnified by 1,000, except Fig. 13, which is only magnified by 700.

PLATE XXV.

Fig. 13.

Epidermis, 72 hours after vaccination, showing Guarnieri's bodies.

[Magnified by 300.]

Fig. 14.

Epidermis, 72 hours after vaccination: Formation of bullæ, epithelial strands, and "balloons."

[Magnified by 100.]

PLATE XXV.

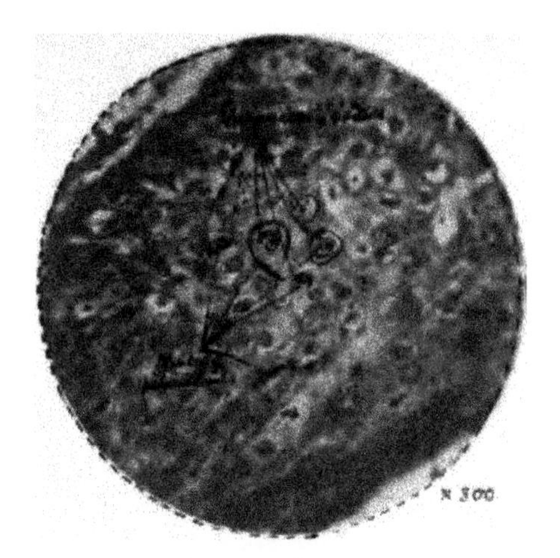

FIG. 13

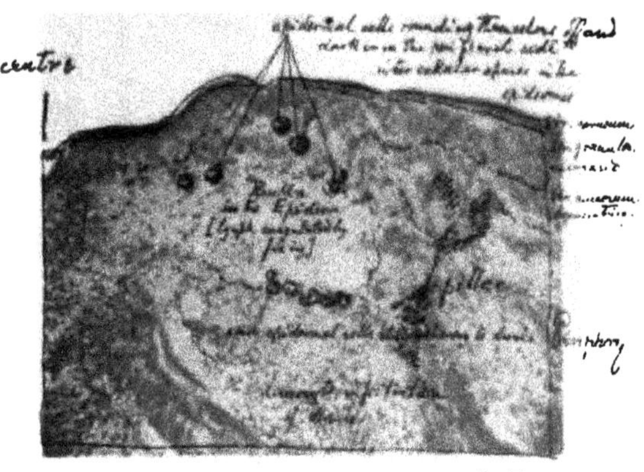

FIG. 14

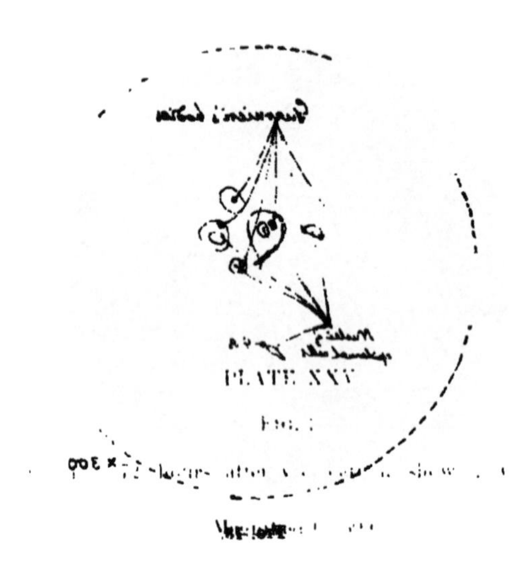

PLATE XXV

FIG. 1

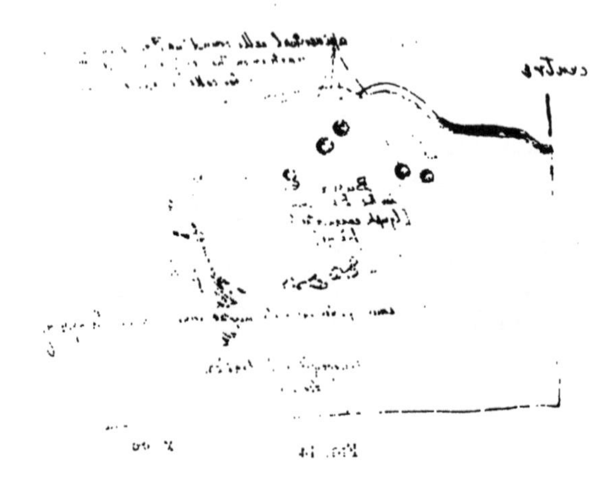

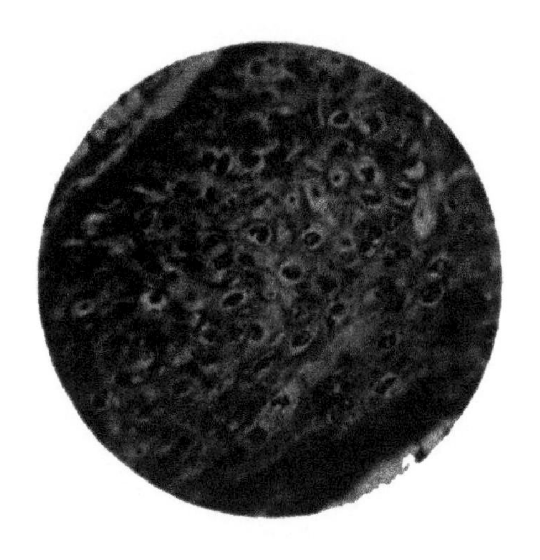

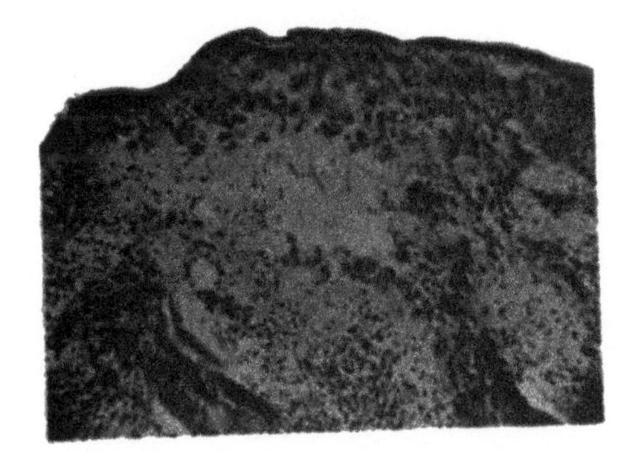

PLATE XXVI.

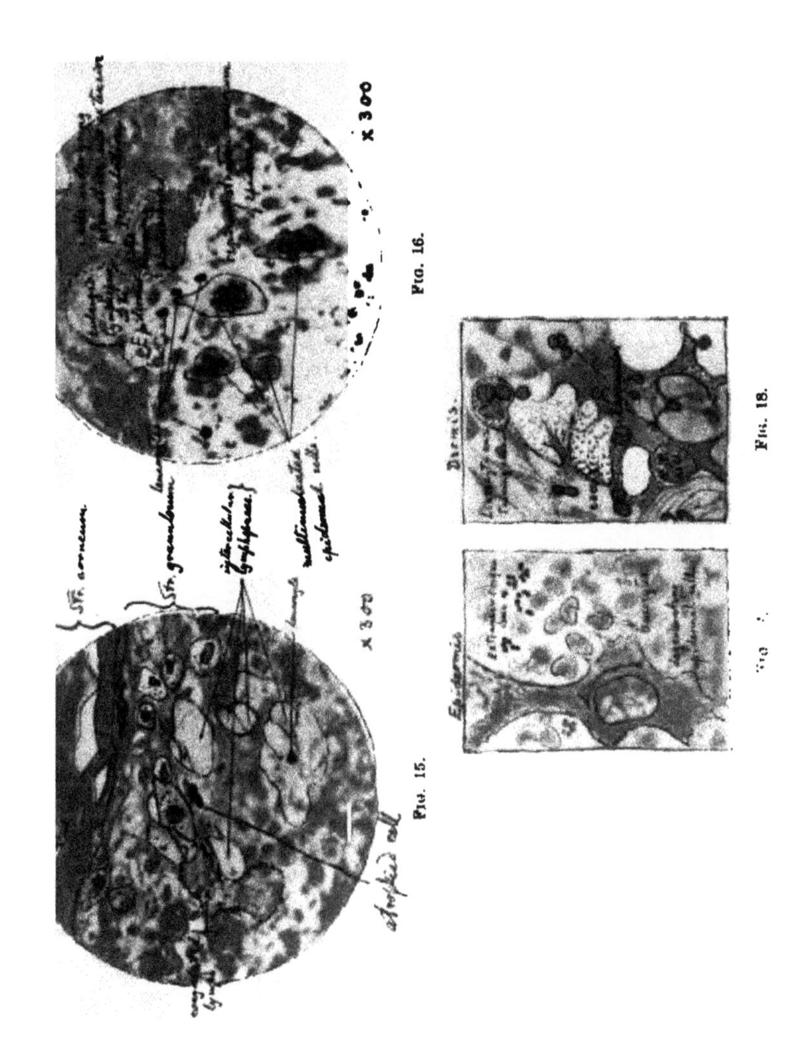

Fig. 15.

Fig. 16.

Fig. 18.

PLATE XXVI.

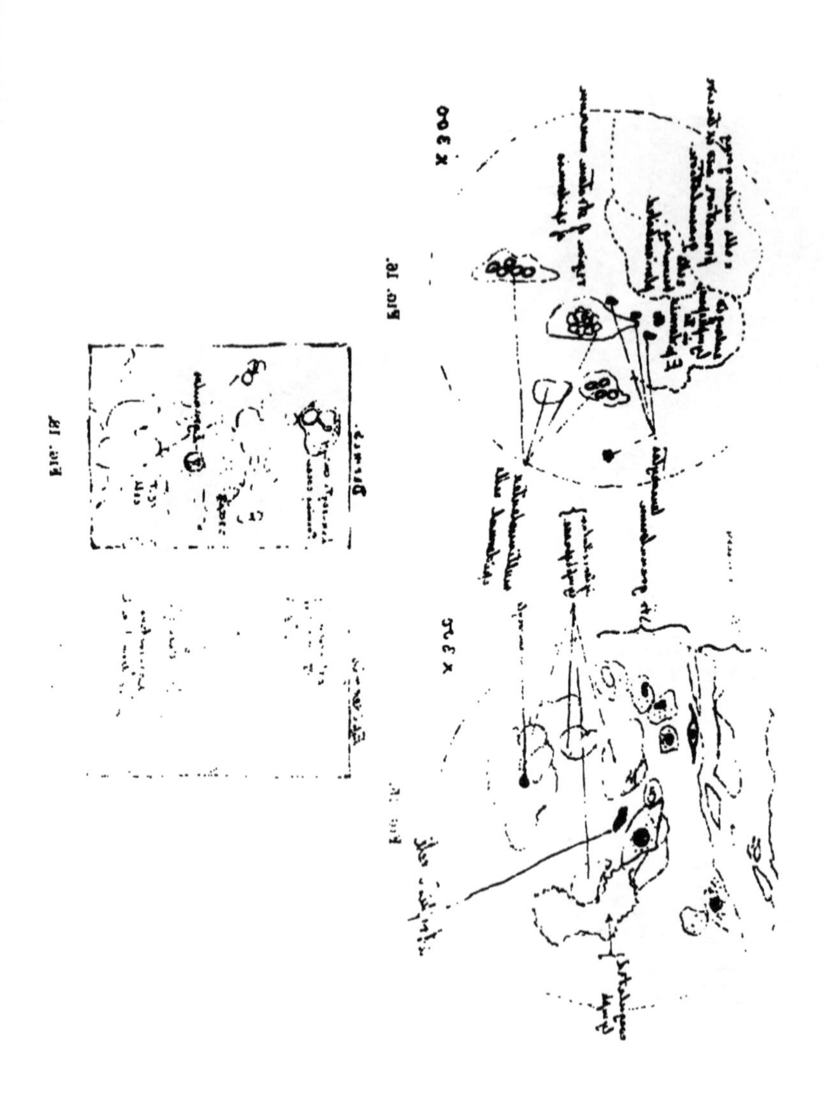

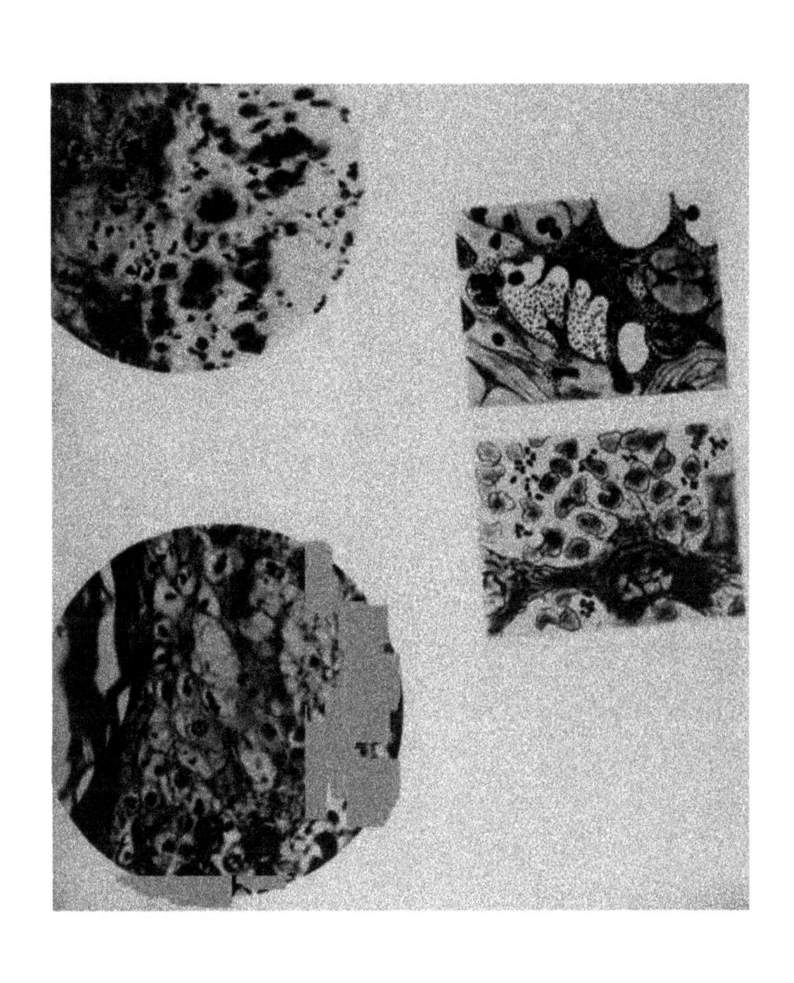

PLATE XXVI.

Figs. 15 & 16.

Upper part of *stratum mucosum* in the epidermis, 72 hours after vaccination. Fig. 15 is nearer the site of inoculation, and in both figures the right-hand portion is that furthest from the plane of inoculation.

[Magnified by 300.]

Fig. 17.

Micro-organisms surrounded by leucocytes 96 hours after vaccination.

Fig. 18.

Minute granules surrounded by the Z-granules, 72 hours after vaccination.

PLATE XXVII.

Fig. 19.

Vertical section of skin of calf one hour after vaccination.
[Magnified by 50.]

Fig 20.

The same. 48 hours after vaccination.
[Magnified by 50.]

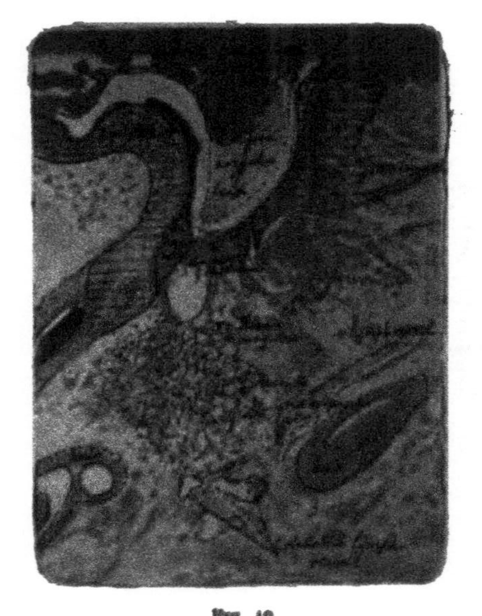

FIG. 19.

FIG. 20.

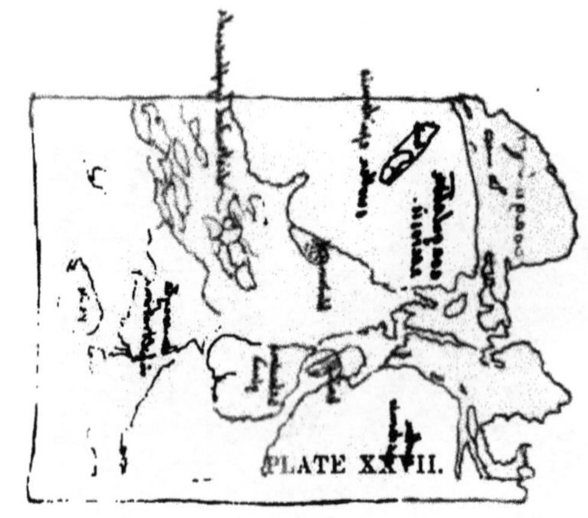

Fig. 19.

Vertical section of skin of calf one hour after vaccination.
[Magnified by 50.]

Fig. 20.

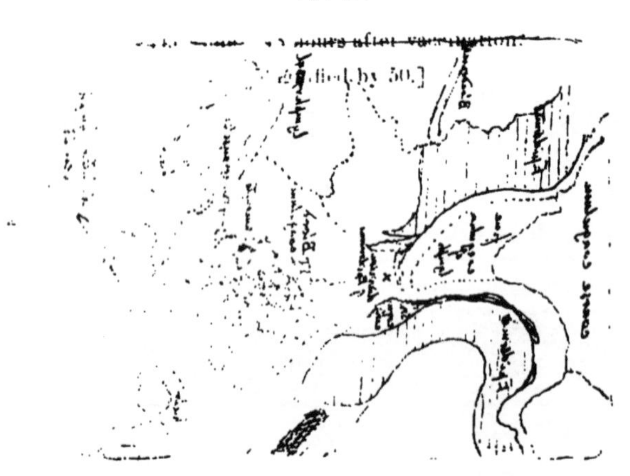

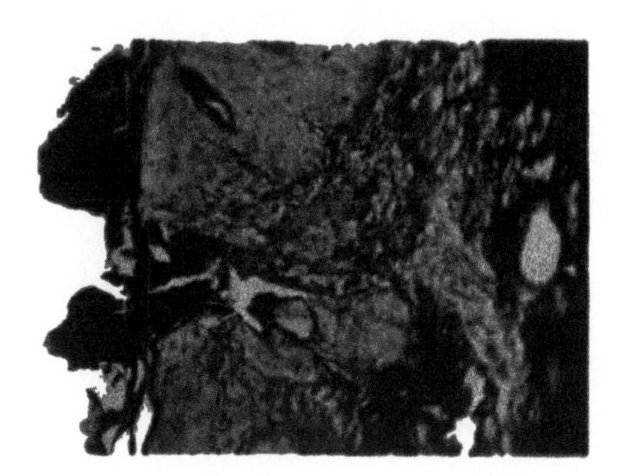

PLATE XXVIII

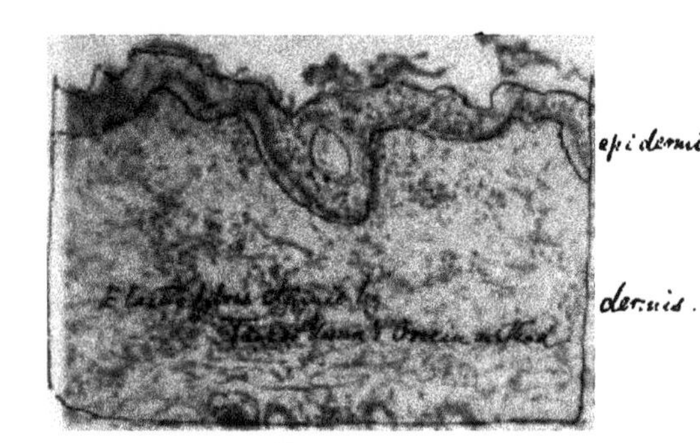

FIG. 21

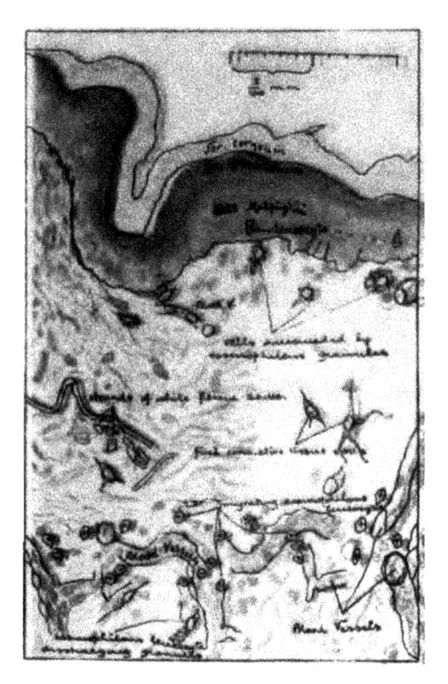

FIG. 22

PLATE XXVIII.

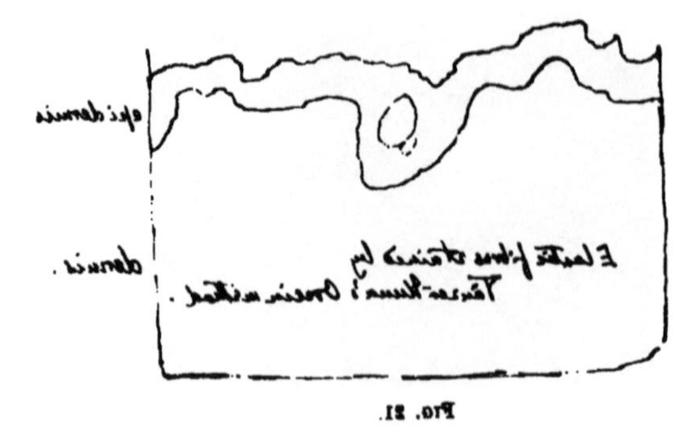

FIG. 21.

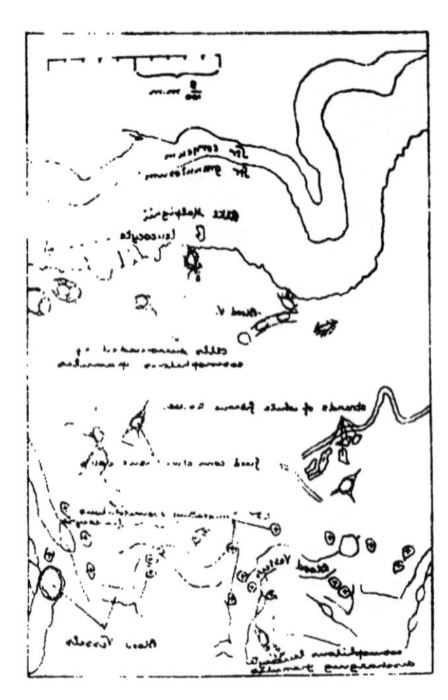

FIG. 22

PLATE XXVIII.

Fig. 21.

Periphery of skin-section 72 hours after vaccination. Commencing leucocyte infiltration. The distribution of the elastic fibres is also shown.

Fig. 22.

Periphery of skin-section 48 hours after vaccination.

PLATE XXIX.

FIG. 23.

Z-granules.

[Magnified by 1,000.]

FIG. 24.

Pseudo-bacilli and cocci produced in the eosinophilous epiderm cells, and in plasma cells by Gram's method of staining.

[Magnified by 1,000.]

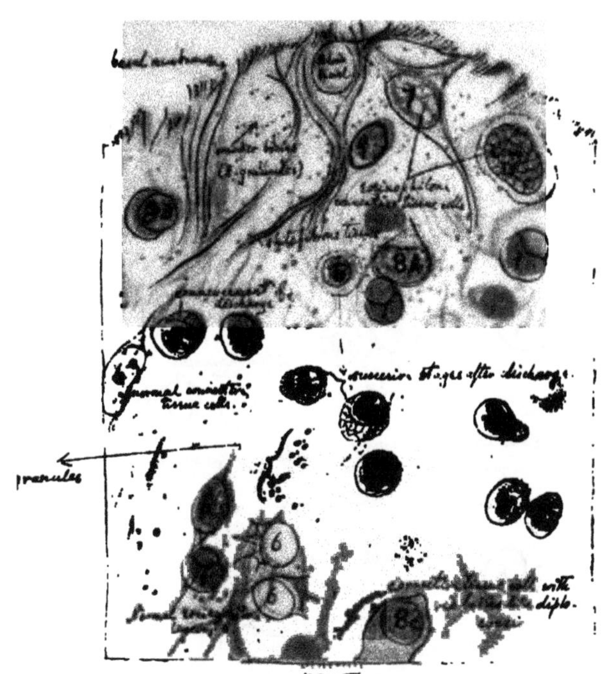

FIG. 2a

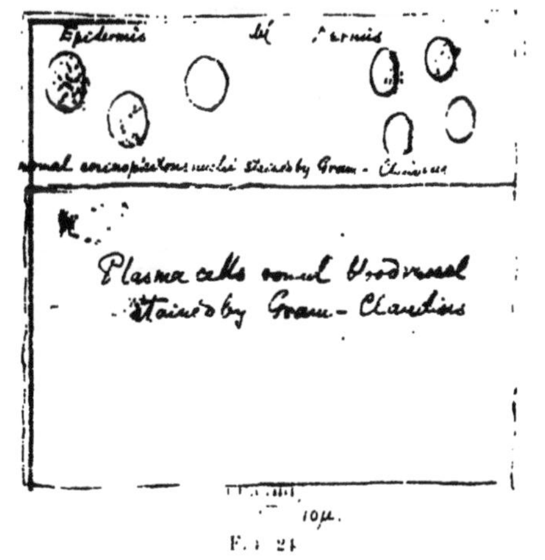

10μ.

FIG. 2b

PLATE XXIX.

542

FIG. 23.

10μ

10μ

FIG. 24

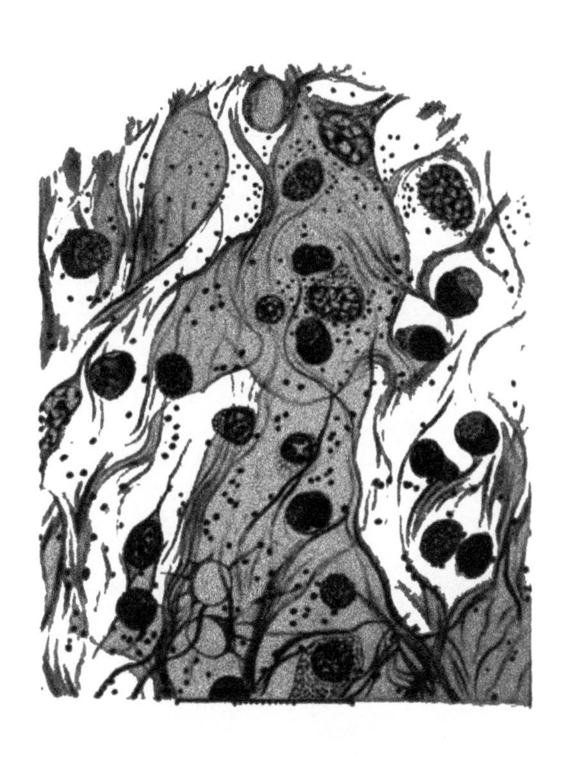

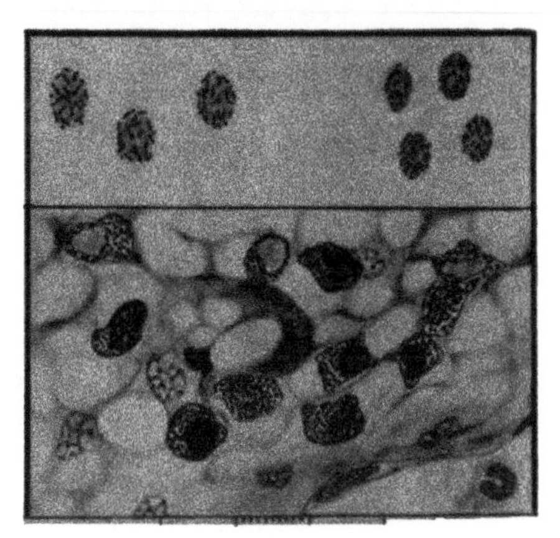

PLATE XXX

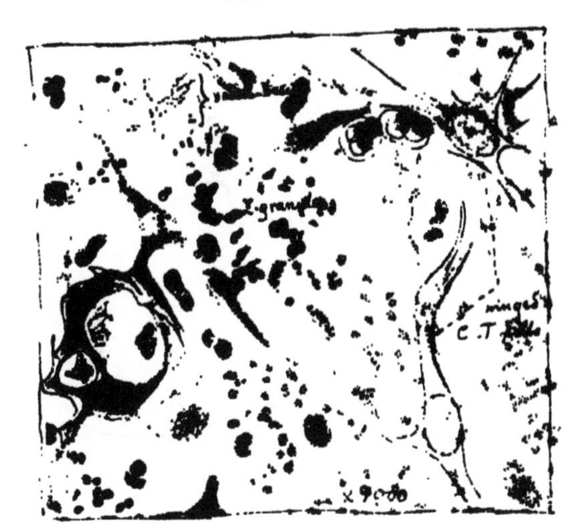

Fig. 25.

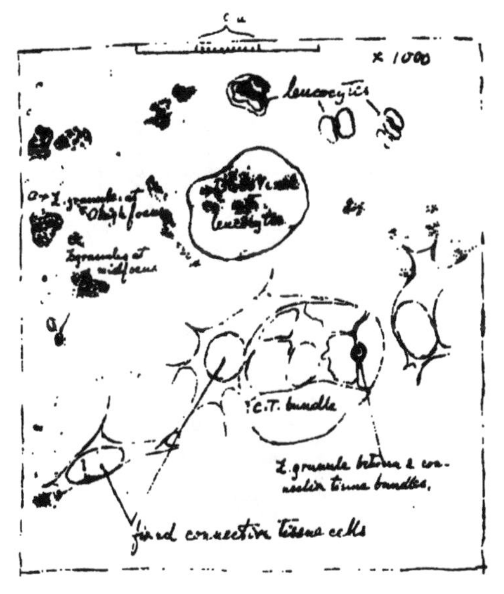

Fig. 26.

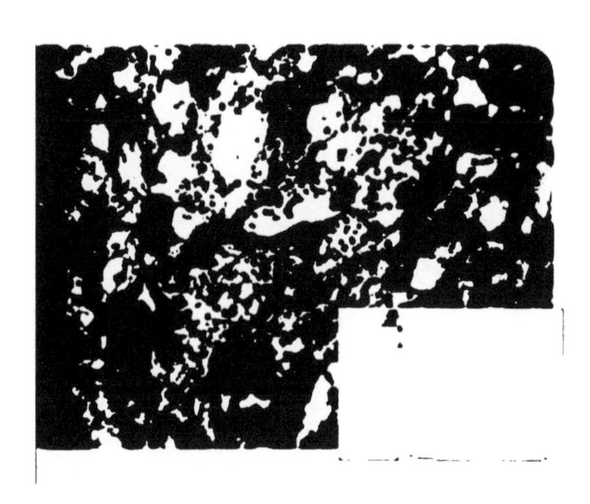

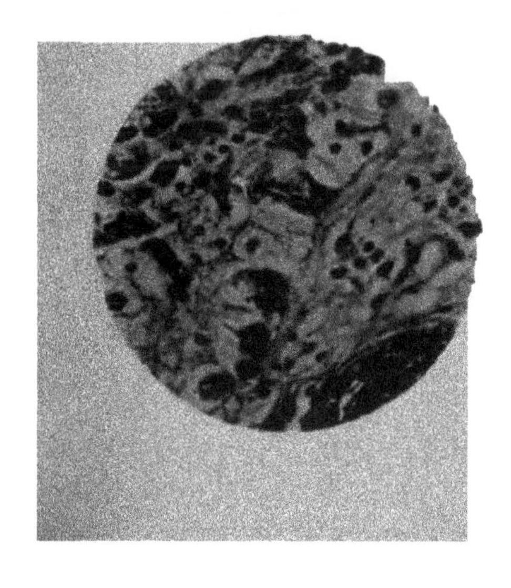

PLATE XXX.

Fig. 25.

Section, stained by Löffler's modified method, showing increase in size of Z-granules 72 hours after vaccination.

Fig. 26.

Similar section stained by Unna's polychrome methylene-blue tannin method.

PLATE XXX.

FIG. 25.

Section, stained by Löffler's modified method, showing increase in size of Z-granules 72 hours after vaccination.

FIG. 26.

Similar section stained by Unna's polychrome methylene-blue tannin method.

PLATE XXXI.

FIG. 27.

Junction of epidermis and dermis 72 hours after vaccination.
[Magnified by 400.]

FIG. 28.

Microphotograph of Z-granules 72 hours after vaccination.
[Magnified by 500.]

PLATE XXXI

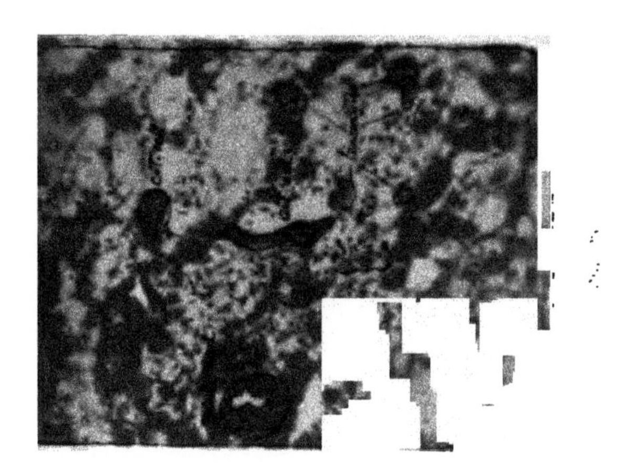

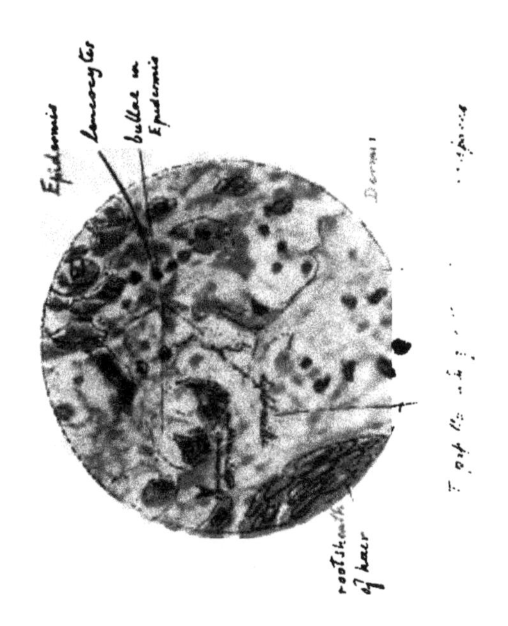
Epidermis
Leucocytes
bullae in Epidermis
Dermis
rootsheath of hair

Ten of the domains
are not included altogether

PLATE XXXII.

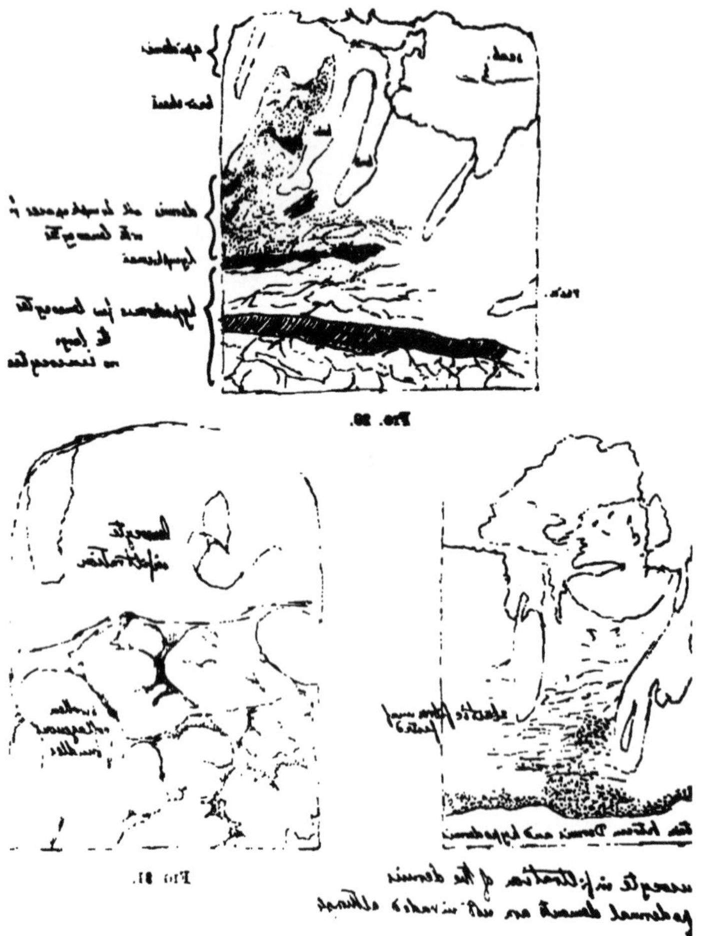

FIG. 29.

FIG. 31.

FIG. 30.

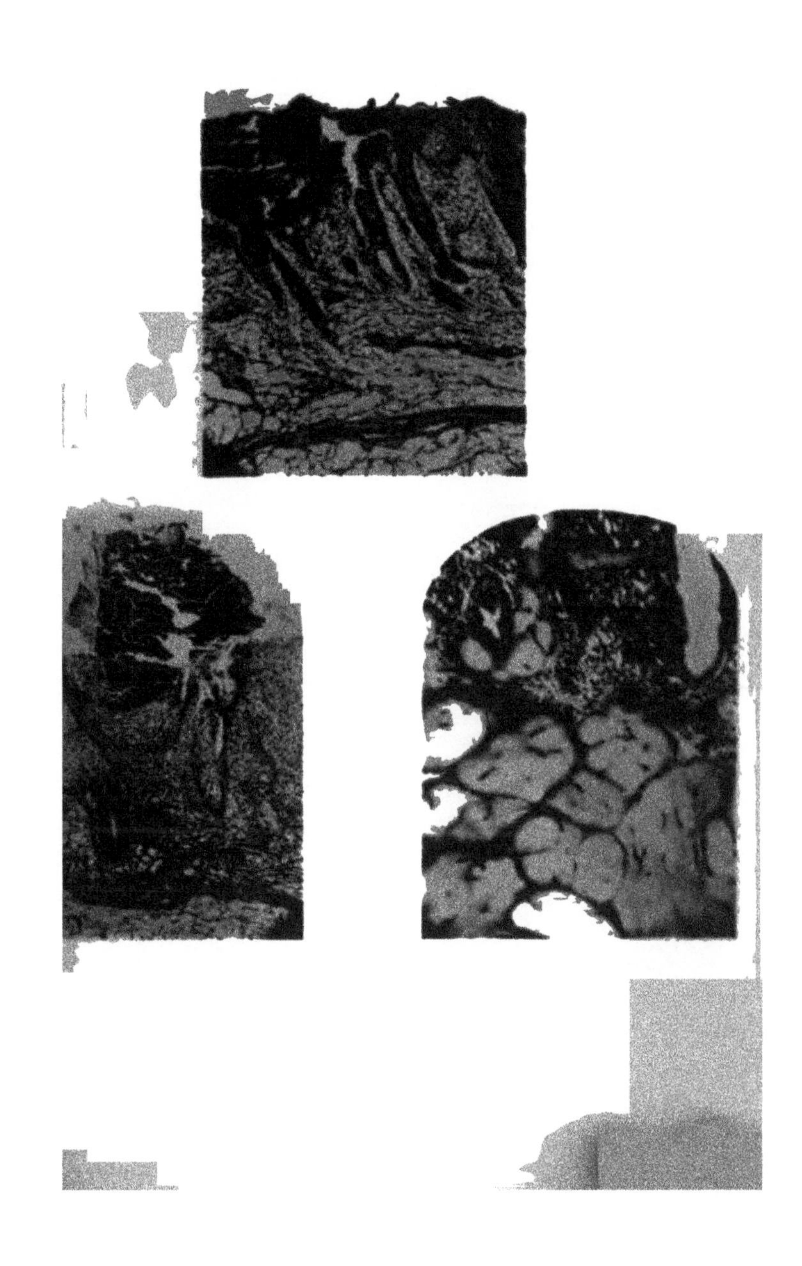

PLATE XXXII.

Fig. 29–31.

Leucocyte infiltration of the dermis 72 hours after vaccination.

Fig. 30 shows the elastic fibres as yet unaffected.
Figs. 29 and 30 are magnified by 50.
Fig. 31 is magnified by 300.

PLATE XXXIII.

Figs. A.–C.

Vertical section of skin of calf 120 hours
All three photographs are taken from the same
is taken at the periphery of the section, and Fig. C. close to
line of inoculation.

[Magnified by 300.]

PLATE XXXIII.

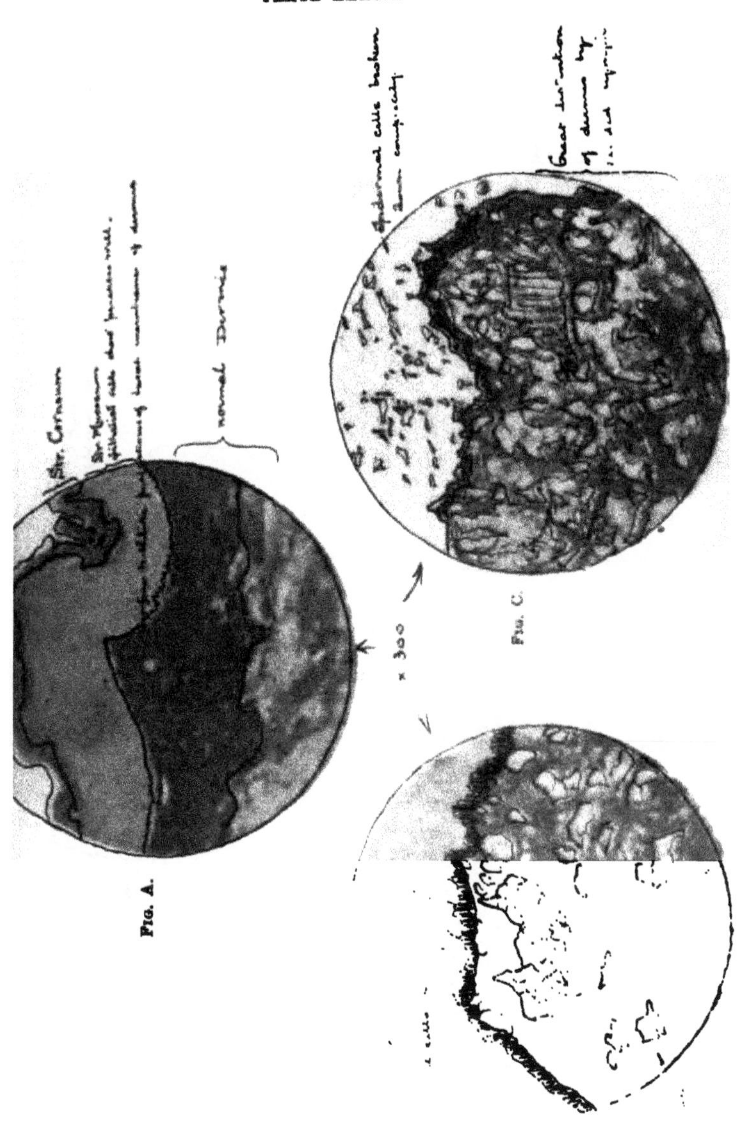

Fig. A.

Fig. C.

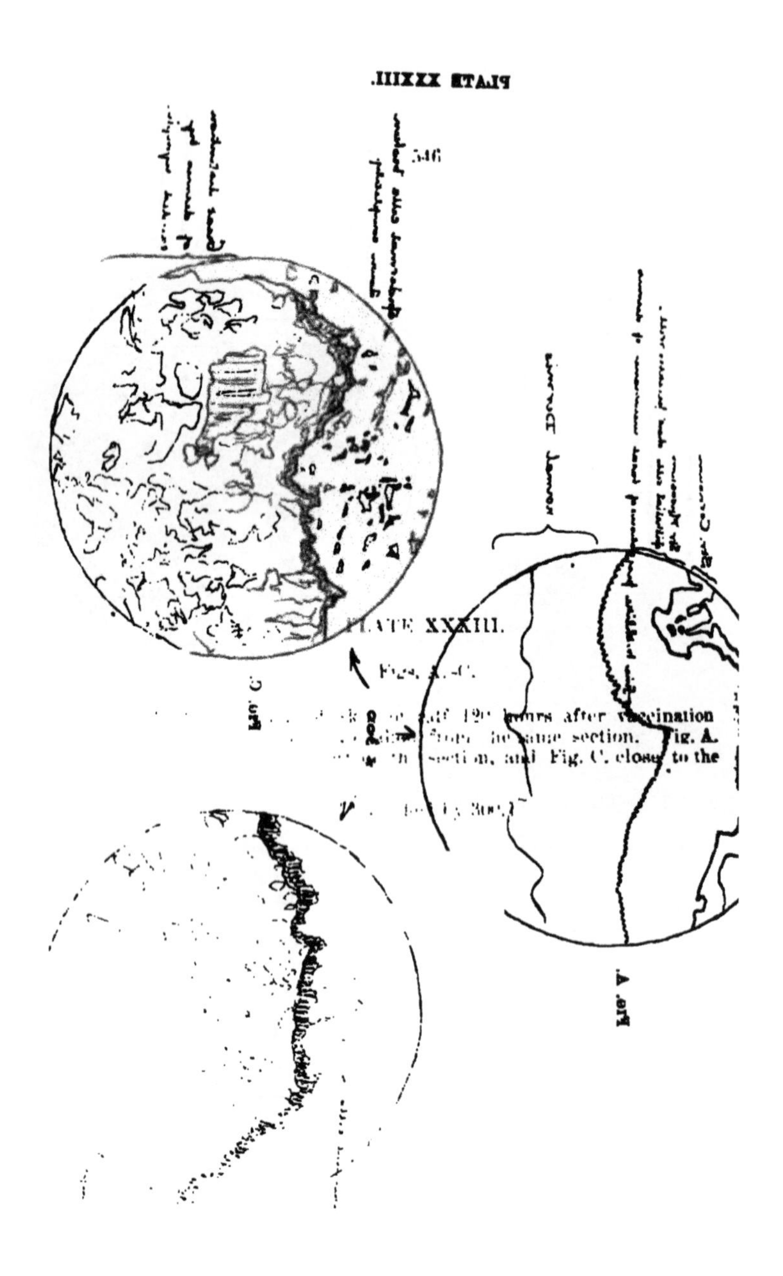

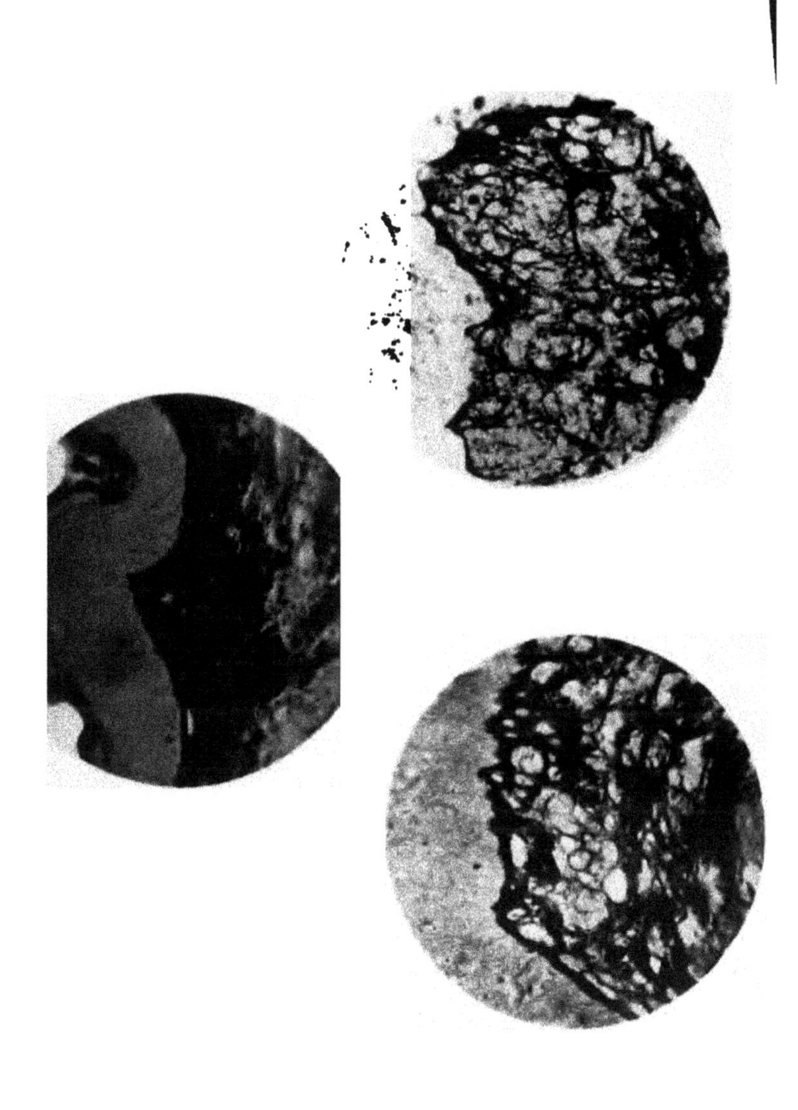

APPENDIX D.

APPX

Index to Visitatio Medical Inspecto 1871-98.

Index to the LOCAL VISITATIONS which have been made by MEDICAL INSPECTORS under the direction of the LOCAL GOVERNMENT BOARD, from the date of the ESTABLISHMENT of the BOARD to the end of 1898, with regard to the INCIDENCE of DISEASE on particular places, and to questions concerning LOCAL SANITARY ADMINISTRATION.

The names of localities inspected are arranged alphabetically. The reports printed in *italics* have been reproduced in the medical officer's annual volumes; those marked* have been separately placed on sale; and those marked† have not been printed.

[*See* further index on page .]

The inspections made in connexion with the Cholera Survey of 1885–86 *and the Inland Sanitary Survey of* 1893–95 *are tabulated at the end of the above Index, and will also be found tabulated in the Report of* 1885 *on Cholera* [*C.* 4873], *in the Annual Volume of* 1886 *of the Medical Officer* [*C.* 5171], *and in the Report of* 1897 *on the Inland Sanitary Survey of* 1893–95 [*C.* 8215].

No tabulation has been made of the Port and Riparian Districts inspected in 1893–94, *in connexion with the Port and Riparian Sanitary Survey of that period. The inspection of such districts forms the subject of a special Report, issued in* 1895 [*C.* 7812].

No.	Locality Inspected.	Medical Inspector.	Year.
1	**Abergele and Pensarn U.**	**Dr. Parsons** ...	1881
2	**Abingdon U.**	**„ Thorne...** ...	1872
3	**Alcester Town†**	**„ Ballard** ...	1875
4	**Aldbrough***	**„ Reece**	1897
4a	**Aldeburgh-on-Sea U.**	**„ Sweeting** ...	1898
5	**Aldershot U.†**	**„ Turner...** ...	1886
6	**Alnwick and Canongate U.***	**Mr. Spear**	1884
6a	*Alnwick U.**	**Dr. Buchanan** ...	1898
7	**Alton Registration District**	**„ Sweeting** ...	1893
8	**Alvaston and Boulton U.†**	**„ Parsons** ...	1890
9	**Amble U.†**	**Mr. Power...** ...	1883
10	**Amlwch**	**„ Evans**	1893
11	**Ampthill R.†**	**Dr. Thorne...** ...	1882
12	**Andover R.†**	**„ Blaxall** ...	1882
13	**Andover U.**	**„ Thorne...** ...	1872
13a	**Appleby (not issued)**	**„ Corfield** ...	1871
14	**Appleby**	**„ Ballard** ...	1873
15	**"Arethusa" Training Ship†**...	**Mr. Power...** ...	1880
16	*Arlford (near Chester)*	**Dr. Ballard** ...	1881
17	*Armley*	**Do.** ...	1872
18	*Arnesby (printed only in annual volume)* ...	**Mr. Spear**	1887
19	*Ascot*	**Dr. Ballard** ...	1877
20	**Ash†**	**Mr. Power...** ...	1876
21	**Ash†**	**Dr. Turner** ...	1886
22	*Ashbourne U.**	**„ Bruce Low** ...	1888
23	**Ashby de la Zouch R.**	**„ Parsons** ...	1892
24	**Ashby Woulds U.**	**Do.** ...	1892
25	**Ashton-in-Makerfield U.**	**Mr. Radcliffe** ...	1872
26	Ashton-in-Makerfield U.	Dr. Wheaton ...	1893
27	Ashton-under-Lyne U.	„ Ballard ...	1882
28	Atherstone and Polesworth	„ Airy	1878
29	Atherstone R.*	„ Wheaton ...	1893
30	Atherton Registration Sub-District ...	Mr. Power... ...	1877
31	Atherton U.	Dr. Parsons ...	1886
32	Auckland Registration District	„ Thorne ...	1874
33	Aveley†	Mr. Spear	1880
34	*Avon Llwyd Valley**	„ T. W. Thompson	1895
35	Axminster†	Dr. Blaxall ...	1874
36	Aylesbury R.	„ Thomson ...	1891
37	*Aylesbury U.*	„ Gresswell ...	1888
38	*Aylesbury U.**	Mr. Spear	1888
39	Balby-cum-Hexthorpe	Dr. Thorne... ...	1873
40	Baldock U.	Do.	1874
41	Bangor U.†	Dr. Barry	1884
42	*Bangor U. and R. and Bethesda U.*	Do.	1882
43	Barking	Dr. Harries ...	1873
44	Barking (Lodge Farm)	„ Buchanan ...	1873
45	Barkingside	„ Thorne ...	1880
46	Barnham Broom*	Mr. T. W. Thompson	1894
47	Barnstaple, Bideford, and Ilfracombe ...	Dr. Reece	1894
48	Barnstaple R.*	„ Bruce Low ...	1890
49	*Barrowby*	„ Parsons ...	1890
50	Barrow in-Furness U.	Mr. Radcliffe ...	1872
51	Barrow-on-Soar Registration District ...	Dr. Barry	1883
52	Barrow-on Soar R.†	Do.	1883
53	Barrow-on-Soar R.†	Dr. Parsons ...	1886
54	Barton-on-Irwell Registration District ...	„ Stevens ...	1874
55	Basingstoke U....	„ Ballard ...	1871
56	Bath R.†	„ Thorne ...	1880

No.	Locality Inspected.	Medical Inspector.	Year
57	Battle Registration District (part of) ...	Dr. Airy	1878
58	Bedale	Mr. Power	1877
59	Beddington Royal Female Orphan Asylum*	Dr. Gresswell ...	1886
60	Bedlingtonshire U.	" Parsons ...	1879
61	Bedlingtonshire U.*	Do.	1889
62	Bedwellty Registration District	Mr. Spear	1882
63	Bedwellty Registration District	Do.	1884
64	*Berkhampstead R.**	Dr. Blaxall ...	1890
65	Berwick-upon-Tweed U.*	" Page	1888
65a	Bettws-y-Coed R.*	" Wheaton ...	1898
66	*Beverley U.**	" Page	1884
67	*Bewdley U.*	" Thomson ...	1897
68	*Bicester U.**	Do.	1896
69	Biddenden U.	Dr. Thorne ...	1879
70	Bierley Lane	Do.	1874
71	Bilston Registration Sub-District	Dr. Ballard ...	1874
72	Bingham†	" Thorne ...	1872
73	*Birmingham and Aston*	" Buchanan ...	1874
74	Bishop's Lydiard, &c.	" Blaxall ...	1882
75	Bishop's Stortford R.	Mr. Sweeting ...	1885
76	Bishop's Stortford U.	Dr. Thorne... ...	1873
77	Blaby R....	" Blaxall ...	1880
78	Blackburn R.	" Parsons ...	1889
79	*Blackburn U. and R.*	" Airy	1881
80	Blackwater†	" Stevens ...	1876
81	Blackwater River†	" Turner ..	1886
82	Bleadon	Mr. Power	1879
83	Blyth Registration Sub-District	Dr. Airy ...	1872
84	Bodmin Registration District	" Parsons ...	1881
85	Bodmin R.†	Do.	1885
86	Bodmin U.†	Do.	1885
87	Bognor U.	Dr. Stevens ...	1874
88	Bolton Registration District	" Ballard ...	1871
89	Bourne Bridge and Stapleford Abbotts ...	" Airy	1881
90	Bourton-on-the-Water (*not issued*)	" Ballard ...	1874
91	Box†	" Blaxall ...	1880
92	*Bradford, &c., &c., &c. (Woolsorters' disease)*	Mr. Spear	1880
93	Bradford U. (Wilts)	Dr. Thorne ...	1877
94	Bradford U. (Yorks)	Mr. Radcliffe ..	1871
95	Brailes	" Power ...	1876
96	Brecknock R.	Dr. Fletcher ...	1895
97	Brecknock U.	" Harries ...	1878
98	Breedy Butts Farm	" Wilson... ...	1893
99	Brent River	" Copeman ...	1893
100	Brick Garth†	" Fletcher ...	1895
101	*Bridgend Registration District*	Mr. Spear	1888
102	Bridgwater U.	Dr. Blaxall ...	1874
103	Bridlington U.	" Parsons ...	1881
104	Brierley Hill U.	Do.	1882
105	*Brightlingsea U.**	Dr. Buchanan ...	1897
106	Brinkworth	" Bruce Low ...	1890
107	Brixworth R.	" Thorne ...	1874
108	Brixworth R.	" Parsons ...	1885
109	Broadstairs U.*	" Bruce Low ...	1894
110	*Bromley and Beckenham Joint Hospital District.*	Mr. T. W. Thompson.	1894
111	*Brownhills U.*	Do.	1892
112	Brownhills U.	Do.	1893
113	Brynmawr U.†	Do.	1895
114	Buckingham R.	Dr. Gresswell ...	1889
115	Buckingham U.	" Parsons ...	1888

No.	Locality Inspected.	Medical Inspector.	Year.
116	Bulwell	Dr. Harries ...	1871
117	Burnham U.	" Blaxall ...	1883
118	Burnley U.	" Beard	1873
119	Burnley U.	" Airy	1884
120	Burton Latimer	" Thorne ...	1872
121	Burton-on-Trent Registration District (part of).	" Airy	1878
121a	*Burton-on-Trent U.**	" Thomson ...	1898
122	Bury U.*	" Copeman ...	1894
123	*Bury U.**	" Bruce Low ...	1897
124	Bushey Parish	Mr. Royle	1897
125	*Bush Hill Park (printed only in annual volume).*	Dr. Copeman ...	1891
126	Buxted and Maresfield*	" Airy	1887
127	Bywell Registration Sub-District†	Do.	1874
128	*Caius College, Cambridge*	Dr. Buchanan ...	1874
129	Calne U.	" Blaxall ...	1874
130	Calne U.	Do. ...	1884
131	Calstock	Do. ...	1876
132	Calvert†	Dr. Bruce Low ...	1890
133	Camberley and York Town	" Parsons ...	1889
133a	*Camborne U. and neighbourhood**	" Bruce Low ...	1898
134	Cameley	" Sweeting ...	1892
135	Camelford R†	" Home	1872
136	*Camelford R.*	" Ballard ...	1888
137	Campden	" Harries ...	1873
138	Cannock U. and R. (Registration Sub-District).†	" Ballard ...	1874
139	Carlisle U.	Mr. Power	1874
140	Carlton	Dr. Harries ...	1871
141	Carmarthen R....	Mr. Power	1878
141a	*Carnarvon U.**	Dr. Wheaton ...	1898
142	Carnarvonshire Combined Sanitary Districts	" Bruce Low ...	1895
143	Castleford Registration Sub-District ...	" Parsons ...	1880
144	Catshill	" Ballard ...	1873
[145	*Cerebro Spinal Meningitis*]* *Eastern Counties*	" Bruce Low ...	1890
146	Chalton	" Buchanan ...	1896
147	Chalvey	Mr. Power	1876
148	Charles Registration Sub-District	Dr. Home	1872
149	Chatteris U.	Mr. Radcliffe ...	1875
150	Chelmsford U*	Dr. Reece	1896
151	*Chepping Wycombe U.*	" Wheaton ...	1895
152	Chesham	Mr. Power	1871
153	Cheshunt U.†	Dr. Thorne ...	1875
154	Cheshunt U.*	Mr. Sweeting ...	1885
155	Chester U*	Dr. Ballard ...	1889
156	Chesterfield Registration District	" Thorne ...	1874
157	*Chester-le-Street, R.**	" Wilson... ...	1893
158	Chichester U.	" Airy	1879
159	*Chichester U.**	" Bulstrode ...	1896
159a	*Chichester U.**	" Thomson (with Col. Marsh.)	1898
160	Chippenham R. (part of)	" Blaxall ...	1884
161	Chippenham U. and R. (Registration District).	Do. ...	1874
162	Chipping Norton R.	Dr. Fletcher ...	1897
163	Chipping Sodbury R.	" Ballard ...	1872
164	Chittlehampton	" Home	1872
165	Chorley (Lancs.) U.	" Parsons ...	1881

Locality Inspected.	Medical Inspector.	Year.
Christchurch (Hants.) U.*	Dr. Mivart... ...	1898
Clapham	" Parsons ...	1882
Clayton, West, U.	" Thorne ...	1874
Cleator Moor U.†	" Page	1888
Clitheroe R.	" Airy	1880
Coggeshall*	Do.	1882
Colne and Marsden U.†	Do.	1884
Combrooke	Dr. Ballard ...	1873
Congleton U.†	" Page	1888
"Cornwall" Training Ship	Mr. Radcliffe ...	1877
"Cornwall" Training Ship	" Power	1879
"Cornwall" Training Ship †	Do.	1881
Corwen R.	Dr. Parsons ...	1881
Coslany Registration Sub-District * ...	" Airy	1888
Cowbridge U.	" Bruce Low ...	1895
Cowpen U.	" Parsons ...	1889
*Cradley**	" Gresswell ...	1889
Cranbrook R.*	" Airy	1887
Cranfield	" Cory	1879
Cricklade and Wootton Bassett R.†... ...	" Thorne ...	1880
Crofton	" Copeman ...	1894
Crompton U.	" Swete	1879
Croyde	" Home	1872
Croydon R.†	" Page	1883
Croydon U.	" Buchanan ...	1875
Cumberworth U.	" Barry	1891
Dalham†	Dr. Turner... ...	1886
Dartford Registration Sub-District† ...	Mr Spear	1884
Dartford R.	Dr. Thorne ...	1879
Dartford U. and R.	Mr. Spear	1882
Darton Registration Sub-District†	Dr. Bruce Low	1889
Dee Watershed*	Do. ...	1895
Denbigh	Dr. Parsons ...	1881
Denbigh U.	" Thorne... ...	1877
Derry Hill*	" Horne	1893
Desborough U.*	" Fletcher ...	1894
*Devonport**	" Parsons ...	1883
Dewsbury†	" Stevens ...	1874
Dewsbury Registration District	" Thorne... ...	1878
Dingestow Registration Sub-District* ...	Mr. Spear	1888
Dolgelley U.	Dr. Parsons ...	1888
Doncaster U.	" Thorne... ...	1873
Donington and Moulton	" Page	1883
Donington and Moulton †	Do.	1884
Dore R.	Dr. Fletcher ...	1898
Draycott†	" Beard	1872
Droylesden U.†	" Fletcher ...	1897
Dudley U.	" Ballard ...	1874
Durham County	Mr. T. W. Thompson	1894
Durham Registration District	" Spear	1881
Eagley	Mr. Power... ...	1876
*Ealing U.**	Do. ...	1887
Easington, R.	Do. ...	1879
Easingwold, R....	Dr. Barry ...	1890
East Haddon	" Bruce Low ...	1889
Eastry R.	Mr. Spear	1887
Eaton Bray R.	Dr. Wheaton ...	1897

No.	Locality Inspected.	Medical Inspector.	Year.
221	*Ebbw Fach Valley**	Mr. T. W. Thompson	1895
222	*Ebbw Fawr Valley**	Do. ...	1895
223	*Ebbw Main Valley**	Do. ...	1895
224	Ecton	Dr. Buchanan ...	1872
225	Edmondsley	" Harries ...	1873
226	Edmonton U.	" Parsons ...	1880
227	Ely R.†	Do. ...	1885
228	Enfield U.	Do. ...	1880
229	*Enfield U**	Dr. Bruce Low ...	1888
230	*Enfield Workhouse**	" Copeman ...	1893
231	Erpingham R.	" Gresswell ...	1885
231a	Eton R.*	" Johnstone ...	1898
232	Eversholt*	" Parsons ...	1884
233	Exeter U.	" Fletcher ...	1895
234	Exmouth U	" Parsons ...	1888
235	Faldingworth and Barlings	Dr. Gresswell ...	1885
236	*Fallowfield*	" Airy	1879
237	*Fareham Registration District**	Mr. Spear	1888
238	*Faringdon R.*	Do. ...	1889
239	Farnborough†	Dr. Turner ..	1886
240	Farnham Registration District	Mr. Sweeting ...	1885
241	Farnham U†	Dr. Turner ...	1886
242	Faversham R.	Mr. Power ...	1880
243	Faversham U.	Do. ...	1880
244	Faversham U. and R. (Registration Sub-District).	Do. ...	1875
245	Felstead	Dr. Airy	1880
246	Festiniog Registration Sub-District (R) ...	" Blaxall ...	1875
247	Festiniog U.*	Mr. Evans	1894
248	Fleetwood-on-Wyre U.	Dr. Harries ...	1873
249	Flint U.†	Mr. Spear	1888
250	*Flint U.**	Dr. Reece	1895
251	Foleshill R.†	" Ballard ...	1874
252	Foleshill R.	" Parsons ...	1891
253	Folkestone	Do. ...	1882
254	Forest Row and East Hoathley	Dr. Airy	1880
255	Frimley†	" Turner... ...	1886
256	Fulbeck	" Wheaton ...	1896
257	Gainsborough Union	Mr. Radcliffe ...	1876
257a	Gainsborough U.*	Dr. Mair	1898
258	Galgate	" Barry	1882
259	Gateshead U.*	Do.	1883
260	*Gillingham U.**	Dr. Sweeting ...	1896
261	Glanaber	" Airy	1880
262	Glanford Brigg R.	" Gresswell ...	1885
262a	Glyncorwg U.	Mr. Royle	1898
263	Godalming and Farncombe	" Power	1874
264	Godmanchester U.	Dr. Parsons ...	1884
265	Goole U..	" Home	1871
266	Grampound	" Corfield ...	1871
267	Gravesend U	Mr. Radcliffe ...	1877
268	Gravesend U.*	" S. F. Murphy...	1885
269	Grays Registration District	Dr. Airy	1889
270	Great Baddow Registration Sub-District (part of).	Do.	1873
271	Great Coggeshall	Dr. Thorne... ...	1876
272	Great Dunmow*	" Airy	1883

No.	Locality Inspected.	Medical Inspector.	Year.
273	Great Grimsby U.	Dr. Home	1871
274	Great Grimsby U.	" Parsons ...	1881
275	Great Massingham	" Thorne ...	1877
276	Great Milton	Do.	1872
277	Great Ormond Street Hospital	Mr. Power	1880
278	Great Ouseburn R.†	Dr. Wheaton ...	1897
279	Great Tey†	" Copeman ...	1893
280	Great Yarmouth	" Airy	1875
281	Guildford R.†	Mr. Power	1882
282	Guisbrough Registration District (part of)	Dr. Thorne ...	1875
283	Guisbrough U.	" Harries ...	1873
284	Gunnislake	" Blaxall ...	1881
285	*Halifax*	Dr. Ballard ...	1881
286	*Halstead Registration District**	" Bruce Low ...	1889
287	Hambledon, &c *	" Parsons ...	1884
288	Hambledon R.*	" Airy	1887
289	Handsworth U.	" Ballard ...	1875
290	Hanley U.*	Mr. Spear	1889
291	Hanwell	Dr. Thorne ...	1876
292	Hastings (St. Mary-in-the-Castle Sub-District)†	Mr. Spear	1890
293	*Hastings U. and R.**	Dr. Bruce Low ...	1894
294	Hatfield R.	Do.	1889
295	Hatley Cockayne	Dr. Buchanan ...	1896
296	Haverfordwest R. (*not issued*)†	" Swete	1879
297	Haverfordwest	" Parsons ...	1880
298	Hayfield R.	Do. ...	1886
299	Hayle U.*	Dr. Mivart ...	1897
300	Heage U.	" Parsons ...	1883
301	Heath Town U.	Mr. Spear	1884
302	Hebden Bridge Registration Sub-District ...	Dr. Gresswell ...	1885
303	Helmsley R.*	" Bruce Low ...	1895
304	Helston, Falmouth, and Redruth U. and R.†	" Ballard ...	1882
305	*Helston R.**	" Parsons ...	1887
306	*Helston U.**	Do. ...	1887
307	Hemel Hempstead R.	Mr. Sweeting ...	1885
308	Hendon R.	Dr. Bruce Low ...	1892
309	*Hendon U.**	Mr. Power	1883
310	Henley and Barham	Dr. Airy	1880
311	Hepworth U.	" Barry	1891
312	Hersham†	" Thorne ...	1872
313	Hetton-le-Hole Registration Sub-District ...	Mr. Power	1874
314	Hetton-le-Hole†	Dr. Fletcher ...	1893
315	Heywood U.	Mr. T. W. Thompson.	1892
316	Higham Ferrers	Dr. Home	1871
317	Hinckley U.†	" Airy	1881
318	*Hinckley U. and R.**	" Wheaton ...	1893
319	Hinckley U. and R.*	Do.	1894
320	Hindley U.†	Mr. Spear	1882
321	Hindley U.	Dr. Parsons ...	1886
322	Hitchin R.	Mr. Sweeting ...	1885
323	*Hitchin U.**	" Power	1883
324	Holbeach R.	Dr. Parsons ...	1882
325	Holbeach U.*	Mr. Evans	1895
326	Hollingbourn R.	Dr. Parsons ...	1886
327	Holme Cultram U.	Do.	1888
328	Holyhead R.†	Mr. Harvey ...	1886

No.	Locality Inspected.	Medical Inspector.	Year.
329	Holyhead U. and R. (Registration Sub-District).	Dr. Ogle	1879
330	Holywell Parish	„ Mivart... ...	1896
331	Holywell Registration District	„ Blaxall ...	1875
332	Holywell R.	„ Parsons ...	1881
333	Holywell U.	Do	1881
334	Hoo R.	Dr. Airy	1881
335	Hoo R.	Mr. Spear	1889
336	Horningsham†	„ Power	1876
337	Horsforth U.*	Dr. Bruce Low ...	1897
338	Horwich Registration Sub-District ...	Mr. Spear	1889
339	*Houghton-le-Spring R.**	Dr. Page	1889
340	Howden R.	Mr. Spear	1881
341	Hucknall Torkard U....	Dr. Harries ...	1872
342	Hucknall Torkard U.*	„ Horne	1894
343	Hucknall Torkard U.*	„ Buchanan ...	1896
344	Hucknall-under-Huthwaite U.†	„ Harries ..	1873
345	Huddersfield U.	„ Buchanan ...	1872
346	Hull U.	„ Airy	1882
347	Huntingdon Registration District	„ Parsons ...	1880
348	Huntingdon U....	Do.	1884
349	Huntingfield	Dr. Mivart... ...	1896
350	Hythe and Bramshaw*	„ Bulstrode ...	1894
351	Ilkeston U.	Dr. Blaxall ...	1881
352	Ilminster	Do.	1871
353	Ince-in-Makerfield U....	Dr. Parsons ...	1879
354	*Ivybridge U. (Reprinted in Volume for 1886)* (Small-pox among Rag-sorters).*	Do.	1887
355	Keighley and Oakworth†	Dr. Stevens ...	1875
356	Kempston*	Mr. Sweeting ...	1885
357	Kessingland*	Dr. Bruce Low ...	1896
358	Keynsham R.	„ Blaxall ...	1888
359	Kidderminster U.*	„ Parsons ...	1884
360	Kidderminster U.	Do.	1885
361	Kilburn and St. John's Wood	Mr. Power... ...	1878
362	*Kilkhampton*	Dr. Parsons ...	1888
363	Killingworth	„ Airy	1872
364	Kingsbridge R.†	„ Ballard ...	1882
365	Kingsclere and East Woodhay R.	„ Blaxall ...	1884
366	King's Lynn	„ Airy	1882
367	*King's Lynn and Gaywood*	„ Bruce Low ...	1892
368	*King's Lynn and Gaywood**	„ Mivart... ...	1897
369	*Kirkheaton U.*	„ Barry	1891
370	Knighton Registration District	„ Airy	1878
371	Lakenheath	Dr. Copeman ...	1892
372	Lambeth†	„ Buchanan ...	1873
373	*Lancaster U.**	„ Thomson ...	1897
374	*Laxfield**	„ Bruce Low ...	1894
375	Launceston R.†...	„ Parsons ...	1884
376	Leek R.	Do.	1889
377	Leigh (Lancashire)	Mr. Power... ...	1872
378	Lenton U. (now part of Nottingham)† ...	Dr. Thorne ...	1875
379	Lepton U.	„ Barry	1891
380	Lewes Registration District	„ Thorne ...	1874
381	Leyton†	„ Sweeting ...	1893

No.	Locality Inspected.	Medical Inspector.	Year.
382	Limsfield	Dr. Copeman ...	1892
383	Linslade	" Bulstrode ...	1894
384	Littleport	" Thorne ...	1873
385	Llandewyrcwm	" Airy	1880
386	Llandissilio Registration Sub-District ...	Mr. Spear	1888
387	Llanelly	Dr. Harries ...	1872
388	Llanelly†	" Blaxall ...	1876
389	Llanelly U.	" Parsons ...	1880
390	Llanfrechfa, Upper, U.†	Mr. Spear	1883
391	Llanfyllin R.	Dr. Parsons ...	1881
392	*Llanfynydd*	" Sweeting ...	1895
393	Llangollen U.	" Parsons ...	1881
394	Llanllyfni and Llanwnda	" Airy	1880
395	Llanrhaiadr	" Thorne ...	1877
396	Llanwddyn	" Parsons ...	1888
397	*Loddon**	" Copeman ...	1895
398	London, Southern Districts†...	Mr. Radcliffe ...	1872
399	*London, Part of*†	Dr. Buchanan ...	1896
400	Long Benton	" Sweeting ...	1894
401	Long Buckby*	" Bruce Low ...	1896
402	Long Eaton U.†	" Turner ...	1886
403	Longton U.*	Mr. Spear	1889
403a	Longton U. and Fenton U.*	Dr. Fletcher ...	1898
404	*Loughton*	Mr. Power	1878
405	Loughton†	" Spear	1880
406	*Lower Brixham U.*	Dr. Blaxall ...	1888
407	Lower Sheringham†	" Airy	1885
408	Lowestoft U.*	" Copeman ...	1896
409	Ludgvan U.*	" Sweeting ...	1896
410	Ludlow U. and R. (Registration District)...	" Airy	1876
410a	Lunesdale R.†	" Fletcher ...	1898
411	Lurgashall	" Airy	1880
412	Lymington U.	Do.	1878
413	Lympsham	Mr. T. W. Thompson	1895
414	Macclesfield Registration District (part of)	Dr. Thorne ...	1873
415	Machynlleth Registration District	" Airy	1876
416	Maesteg U.*	Mr. Spear	1889
417	Manchester Royal Infirmary...	" Radcliffe ...	1876
418	Manorbier	Dr. Airy	1880
419	*Mansfield District**	" Buchanan ...	1896
420	Mansfield Registration District	" Gresswell ...	1885
421	Margate U. (*not issued*)	" Harries ...	1873
422	Margate U.*	" Page	1887
423	Marham†	" Parsons ...	1887
424	Market Weighton†	Mr. Royle ...	1885
424a	Marston, Hingham, and Long Bennington...	Dr. Wheaton ...	1898
425	*Marylebone, St.*	Messrs. Radcliffe and Power.	1873
426	Maryport U.	Mr. Spear	1882
427	Mathry and Llanrian	Dr. Parsons ...	1880
428	Melton Mowbray R.†	" Blaxall ...	1881
429	*Melton Mowbray U.*	Do.	1880
430	Mendham	Dr. Airy	1873
431	Merrow	" Horne	1893
432	Messingham†	" Bruce Low ...	1897
433	Mexborough U.	" Thorne ...	1873
434	*Middlesbrough U.**	" Ballard ...	1888
435	Middlesbrough U.*	" Bruce Low ...	1896
436	*Midsomer Norton U.**	" Blaxall ...	1888

No.	Locality Inspected.	Medical Inspector.	Year.
221	*Ebbw Fach Valley**	Mr. T. W. Thompson	1895
222	*Ebbw Fawr Valley**	Do. ...	1895
223	*Ebbw Main Valley**	Do. ...	1895
224	Ecton	Dr. Buchanan ...	1872
225	Edmondsley	" Harries ...	1873
226	Edmonton U.	" Parsons ...	1880
227	Ely R.†	Do. ...	1885
228	Enfield U.	Do. ...	1880
229	*Enfield U.**	Dr. Bruce Low ...	1888
230	*Enfield Workhouse**	" Copeman ...	1893
231	Erpingham R.	" Gresswell ...	1883
231a	Eton R.*	" Johnstone ...	1898
232	Eversholt*	" Parsons ...	1884
233	Exeter U.	" Fletcher ...	1895
234	Exmouth U	" Parsons ...	1888
235	Faldingworth and Barlings	Dr. Gresswell ...	1885
236	*Fallowfield*	" Airy	1879
237	*Fareham Registration District**	Mr. Spear	1888
238	*Faringdon R.*	Do. ...	1889
239	Farnborough†	Dr. Turner ..	1886
240	Farnham Registration District	Mr. Sweeting ...	1885
241	Farnham U†	Dr. Turner ...	1886
242	Faversham R.	Mr. Power ...	1880
243	Faversham U.	Do. ...	1880
244	Faversham U. and R. (Registration Sub-District).	Do. ...	1875
245	Felstead	Dr. Airy	1880
246	Festiniog Registration Sub-District (R) ...	" Blaxall ...	1875
247	Festiniog U.*	Mr. Evans	1894
248	Fleetwood-on-Wyre U.	Dr. Harries ...	1873
249	Flint U.†	Mr. Spear	1888
250	*Flint U.**	Dr. Reece	1895
251	Foleshill R.†	" Ballard ...	1874
252	Foleshill R.	" Parsons ...	1891
253	Folkestone	Do. ...	1882
254	Forest Row and East Hoathley	Dr. Airy	1880
255	Frimley†	" Turner... ...	1886
256	Fulbeck	" Wheaton ...	1896
257	Gainsborough Union	Mr. Radcliffe ...	1876
257a	Gainsborough U.*	Dr. Mair	1898
258	Galgate	" Barry	1882
259	Gateshead U.*	Do.	1883
260	*Gillingham U.**	Dr. Sweeting ...	1896
261	Glanaber	" Airy	1880
262	Glanford Brigg R.	" Gresswell ...	1885
262a	Glyncorwg U.	Mr. Royle	1898
263	Godalming and Farncombe	" Power	1874
264	Godmanchester U.	Dr. Parsons ...	1884
265	Goole U..	" Home	1871
266	Grampound	" Corfield ...	1871
267	Gravesend U.	Mr. Radcliffe ...	1877
268	Gravesend U.*	" S. F. Murphy...	1885
269	Grays Registration District	Dr. Airy	1889
270	Great Baddow Registration Sub-District (part of).	Do.	1873
271	Great Coggeshall	Dr. Thorne... ...	1876
272	Great Dunmow*	" Airy	1883

Locality Inspected.	Medical Inspector.	Year.
Great Grimsby U.	Dr. Home	1871
Great Grimsby U.	" Parsons ..	1881
Great Massingham	" Thorne ...	1877
Great Milton	Do.	1872
Great Ormond Street Hospital	Mr. Power	1880
Great Ouseburn R.†	Dr. Wheaton ...	1897
Great Tey†	" Copeman ...	1893
Great Yarmouth	" Airy	1875
Guildford R.†	Mr. Power	1882
Guisbrough Registration District (part of)	Dr. Thorne ...	1875
Guisbrough U.	" Harries ...	1873
Gunnislake	" Blaxall ...	1881
Halifax	Dr. Ballard ...	1881
*Halstead Registration District**	" Bruce Low ...	1889
Hambledon, &c.*	" Parsons ...	1884
Hambledon R.*	" Airy	1887
Handsworth U.	" Ballard ...	1875
Hanley U.*	Mr. Spear	1889
Hanwell	Dr. Thorne ...	1876
Hastings (St. Mary-in-the-Castle Sub-District)†	Mr. Spear	1890
*Hastings U. and R.**	Dr. Bruce Low ...	1894
Hatfield R.	Do.	1889
Hatley Cockayne	Dr. Buchanan ...	1896
Haverfordwest R. (*not issued*)†	" Swete	1879
Haverfordwest	" Parsons ...	1880
Hayfield R.	Do. ...	1886
Hayle U.*	Dr. Mivart ...	1897
Heage U.	" Parsons ...	1883
Heath Town U.	Mr. Spear	1884
Hebden Bridge Registration Sub-District ...	Dr. Gresswell ...	1885
Helmsley R.*	" Bruce Low ...	1895
Helston, Falmouth, and Redruth U. and R.†	" Ballard ...	1882
*Helston R.**	" Parsons ...	1887
*Helston U.**	Do. ...	1887
Hemel Hempstead R.	Mr. Sweeting ...	1885
Hendon R.	Dr. Bruce Low ...	1892
*Hendon U.**	Mr. Power	1883
Henley and Barham	Dr. Airy	1880
Hepworth U.	" Barry	1891
Hersham†	" Thorne ...	1872
Hetton-le-Hole Registration Sub-District ...	Mr. Power	1874
Hetton-le-Hole†	Dr. Fletcher ...	1895
Heywood U.	Mr. T. W. Thompson.	1892
Higham Ferrers	Dr. Home	1871
Hinckley U.†	" Airy	1881
*Hinckley U. and R.**	" Wheaton ...	1893
Hinckley U. and R.*	Do.	1894
Hindley U.†	Mr. Spear	1882
Hindley U.	Dr. Parsons ...	1886
Hitchin R.	Mr. Sweeting ...	1885
*Hitchin U.**	" Power	1883
Holbeach R.	Dr. Parsons ...	1882
Holbeach U.*	Mr. Evans	1895
Hollingbourn R.	Dr. Parsons ..	1886
Holme Cultram U.	Do.	1888
Holyhead R.†	Mr. Harvey ...	1886

No.	Locality Inspected.	Medical Inspector.	Year.
329	Holyhead U. and R. (Registration Sub-District).	Dr. Ogle	1879
330	Holywell Parish	" Mivart... ...	1896
331	Holywell Registration District	" Blaxall ...	1875
332	Holywell R.	" Parsons ...	1881
333	Holywell U.	Do	1881
334	Hoo R.	Dr. Airy	1881
335	Hoo R.	Mr. Spear	1889
336	Horningsham†	" Power	1876
337	Horsforth U.*	Dr. Bruce Low ...	1897
338	Horwich Registration Sub-District ...	Mr. Spear	1889
339	*Houghton-le-Spring R.**	Dr. Page	1889
340	Howden R.	Mr. Spear	1881
341	Hucknall Torkard U.	Dr. Harries ...	1872
342	Hucknall Torkard U.*	" Horne	1894
343	Hucknall Torkard U.*	" Buchanan ...	1896
344	Hucknall-under-Huthwaite U.†	" Harries ..	1873
345	Huddersfield U	" Buchanan ...	1872
346	Hull U.	" Airy	1882
347	Huntingdon Registration District	" Parsons ...	1880
348	Huntingdon U....	Do.	1884
349	Huntingfield	Dr. Mivart... ...	1896
350	Hythe and Bramshaw*	" Bulstrode ...	1894
351	Ilkeston U.	Dr. Blaxall ...	1881
352	Ilminster	Do.	1871
353	Ince-in-Makerfield U....	Dr. Parsons ...	1879
354	*Ivybridge U. (Reprinted in Volume for 1886)* (Small-pox among Rag-sorters).*	Do.	1887
355	Keighley and Oakworth†	Dr. Stevens ...	1875
356	Kempston*	Mr. Sweeting ...	1885
357	Kessingland*	Dr. Bruce Low ...	1896
358	Keynsham R.	" Blaxall ...	1888
359	Kidderminster U.*	" Parsons ...	1884
360	Kidderminste U.	Do.	1885
361	Kilburn and St. John's Wood	Mr. Power	1878
362	*Kilkhampton*	Dr. Parsons ...	1888
363	Killingworth	" Airy	1872
364	Kingsbridge R.†	" Ballard ...	1882
365	Kingsclere and East Woodhay R.	" Blaxall ...	1884
366	King's Lynn	" Airy	1882
367	*King's Lynn and Gaywood*	" Bruce Low ...	1892
368	*King's Lynn and Gaywood**	" Mivart... ...	1897
369	*Kirkheaton U.*	" Barry	1891
370	Knighton Registration District	" Airy	1878
371	Lakenheath	Dr. Copeman ...	1892
372	Lambeth†	" Buchanan ...	1873
373	*Lancaster U.**	" Thomson ...	1897
374	*Laxfield**	" Bruce Low ...	1894
375	Launceston R.†...	" Parsons ...	1884
376	Leek R.	Do.	1889
377	Leigh (Lancashire)	Mr. Power	1872
378	Lenton U. (now part of Nottingham)† ...	Dr. Thorne ...	1875
379	Lepton U.	" Barry	1891
380	Lewes Registration District	" Thorne ...	1874
381	Leyton†	" Sweeting ...	1893

No.	Locality Inspected.	Medical Inspector.	Year.
2	Limsfield	Dr. Copeman ...	1892
3	Linslade	" Bulstrode ...	1894
4	Littleport	" Thorne ...	1873
5	Llandowyrcwm	" Airy	1880
6	Llandissilio Registration Sub-District ...	Mr. Spear	1888
7	Llanelly	Dr. Harries ...	1872
8	Llanelly†	" Blaxall ...	1876
9	Llanelly U.	" Parsons ...	1880
0	Llanfrechfa, Upper, U.†	Mr. Spear	1883
1	Llanfyllin R.	Dr. Parsons ...	1881
2	*Llanfynydd*	" Sweeting ...	1895
3	Llangollen U.	" Parsons ...	1881
4	Llanllyfni and Llanwnda	" Airy	1880
5	Llanrhaiadr	" Thorne ...	1877
6	Llanwddyn	" Parsons ...	1888
7	*Loddon**	" Copeman ...	1895
8	London, Southern Districts†...	Mr. Radcliffe ...	1872
9	*London, Port of*†	Dr. Buchanan ...	1896
0	Long Benton	" Sweeting ...	1894
1	Long Buckby*	" Bruce Low ...	1896
2	Long Eaton U.†	" Turner ...	1886
3	Longton U.*	Mr. Spear	1889
3a	Longton U. and Fenton U.*	Dr. Fletcher ...	1898
4	*Loughton*	Mr. Power	1878
5	Loughton†	" Spear	1880
6	*Lower Brixham U.*	Dr. Blaxall ...	1888
7	Lower Sheringham†	" Airy	1885
8	Lowestoft U.*	" Copeman ...	1896
9	Ludgvan U.*	" Sweeting ...	1896
0	Ludlow U. and R. (Registration District)...	" Airy	1876
0a	Lunesdale R.†	" Fletcher ...	1898
1	Lurgashall	" Airy	1880
2	Lymington U.	Do.	1878
3	Lympsham	Mr. T. W. Thompson	1895
4	Macclesfield Registration District (part of)	Dr. Thorne ...	1873
5	Machynlleth Registration District	" Airy	1876
6	Maesteg U.*	Mr. Spear	1889
7	Manchester Royal Infirmary...	" Radcliffe ...	1876
8	Manorbier	Dr. Airy	1880
9	*Mansfield District**	" Buchanan ...	1896
0	Mansfield Registration District	" Gresswell ...	1885
1	Margate U. (*not issued*)	" Harries ...	1873
2	Margate U.*	" Page	1887
3	Marham†	" Parsons ...	1887
4	Market Weighton†	Mr. Royle ...	1885
4a	Marston, Hingham, and Long Bennington...	Dr. Wheaton ...	1898
5	*Marylebone, St.*	Messrs. Radcliffe and Power.	1873
6	Maryport U.	Mr. Spear	1882
7	Mathry and Llanrian	Dr. Parsons ...	1880
8	Melton Mowbray R.†	" Blaxall ...	1881
9	*Melton Mowbray U.*	Do.	1880
0	Mendham	Dr. Airy	1873
1	Merrow	" Horne	1893
2	Messingham†	" Bruce Low ...	1897
3	Mexborough U.	" Thorne ...	1873
4	*Middlesbrough U.**	" Ballard ...	1888
5	Middlesbrough U.*	" Bruce Low ...	1896
6	*Midsomer Norton U.**...	" Blaxall ...	1888

No.	Locality Inspected.	Medical Inspector.	Year.
437	Millbrook	Dr. Ballard ...	1880
438	Millbrook†	Do.	1882
439	Moggerhanger	Dr. Buchanan ...	1896
440	Mold U.*	" Wheaton ...	1894
441	Monmouth Registration District	" Fletcher ...	1892
442	Monmouth U.	" Pirrie	1872
443	*Moseley and Balsall Heath*	" Ballard ...	1873
444	Motcombe	" Simpson ...	1885
445	Moulton†	" Turner... ...	1886
446	*Mountain Ash U.**	Mr. Spear	1887
447	Mytholmroyd (in four Sanitary Districts)*	Dr. Page	1888
448	*"Nazareth House," Hammersmith*	Mr. Spear	1883
449	Neath Registration District	Dr. Airy	1877
450	*Neath R. (part of)*	Mr. Spear	1890
451	Necton†	Dr. Airy	1876
452	Newark U.	" Parsons ...	1885
453	*New Brighton**	Mr. Spear	1888
454	*New Clee**	Dr. Page	1888
455	New Delaval*	" Sweeting ...	1894
456	New Hinckey	" Thorne ...	1873
457	Newlyn†	" Blaxall ...	1877
458	Newlyn East	" Ballard ...	1880
459	New Malden U.†	Do. ...	1871
460	*Newport U., &c.**	Dr. Thomson ...	1894
461	Newport Pagnell R. (part of)	" Thorne ...	1872
462	Newport Pagnell R.	" Parsons ...	1884
463	New Quay (Cardigan) U.†	" Fletcher ...	1893
464	New Quay (Cornwall) U.	" Ballard ...	1879
465	New Shoreham U.†	" Thorne ...	1882
466	Newton Heath U.†	Do. ...	1880
467	Newtown and Llanllwchaiarn U.	Dr. Blaxall ...	1877
468	Newtown and Llanllwchaiarn U.†	Do. ...	1880
469	Nocton Rise Farmhouse†	Dr. Gresswell ...	1885
470	Normandy†	" Horne	1893
471	Northallerton U.†	" Page	1887
472	Northampton Lunatic Asylum	" Buchanan ...	1875
473	North and South Tawton	" Blaxall ...	1881
474	*North London (3 Districts)**	Mr. Power	1885
475	*Northop Hall*	" Spear	1890
476	Norwich : Norfolk and Norwich Hospital...	" Radcliffe ...	1875
477	*Norwood (Uxbridge R.)*	" Power	1882
478	Nottingham	Dr. Ballard ...	1881
479	Nottingham U....	" Thorne ...	1871
480	Nunney	" Ballard ...	1872
481	*Oakley*	Dr. Downes ...	1883
482	*Okehampton R.*...	" Blaxall ...	1887
483	Okehampton U.	Do. ...	1879
484	Oldbury U.	Dr. Ballard ...	1875
485	Oldham Registration District (part of) ...	" Stevens ...	1875
486	Oldham U.	" Beard	1872
486a	Ormskirk U.*	" Copeman ...	1898
487	Ormskirk R	" Parsons ...	1883
488	Orsett R.	Mr. Spear	1883
489	Oswaldtwistle U.	Do.	1887
490	Ovenden U.	Dr. Thorne ...	1873
491	Over Darwen U.	" Stevens ...	1874

No.	Locality Inspected.	Medical Inspector.	Year.
492	Padstow U.	Dr. Blaxall ...	1877
493	Pemberton U.	„ Airy	1880
494	Pemberton U.	Mr. Spear	1890
495	*Pembroke R.*	Do.	1890
496	Pembroke U.	Dr. Airy	1878
497	Pembroke U.	„ Reece	1895
498	Penistone Registration Sub-District * ...	Mr. Spear	1889
499	Penistone U.	Dr. Thorne ...	1879
500	Penkridge	„ Airy	1876
501	Penrhynside	„ Bruce Low ...	1896
502	Penrith U.*	Do. ...	1894
503	Perry Street, Kent	Dr. Thorne ...	1871
504	Peterchurch	Do.	1877
505	Petersfield R.†	Dr. Turner ...	1886
506	Phillack *	„ Mivart... ...	1897
507	Phillack U.†	„ Reece	1895
508	*Pirbright*	Mr. Power	1882
509	Pirbright †	Dr. Horne	1893
510	Plymouth U. (*Two Reports*)	„ Blaxall ...	1878
511	Plymouth	Do.	1882
512	Plympton St. Mary R.†	Dr. Ballard ...	1882
513	Pocklington	Mr. Evans	1893
514	Pontardawe Registration District (R.) ...	Dr. Parsons ...	1880
515	Pontefract Registration Sub-District (part of) † (*not issued*)	„ Beard	1875
516	Pontypool Registration District (part of) ...	„ Ogle	1879
517	Pontypridd Registration District	„ Airy	1876
518	Pontypridd Registration Sub-District * ...	Mr. Spear	1889
519	Poole *	Dr. Bulstrode ...	1893
520	Porchester	„ Stevens ...	1872
521	Portland U.†	„ Blaxall ...	1886
522	Portsmouth	„ Thomson ...	1896
523	Portsmouth U....	„ Thorne ...	1876
524	Potterne *	„ Copeman ...	1894
525	Potterspury	„ Bulstrode ...	1895
526	Potterspury R.†	„ Thorne ...	1875
527	Potton *	„ Parsons ...	1882
528	Putney †...	„ Blaxall ...	1883
529	Pwllheli R.	„ Parsons ...	1887
530	Pwllheli U.*	Do.	1887
531	Quarry Bank U.	Dr. Wheaton ...	1895
532	Queen Camel *	„ Parsons ...	1888
533	Quickmere U.	Mr. Spear	1879
534	Radford	Dr. Thorne ...	1872
535	Radwinter	Mr. Power	1877
536	Rainham*	„ Evans	1894
537	Ramsey U.	Dr. Airy	1875
538	Raunds	Do.	1880
539	Raunds*...	Dr. Bruce Low ...	1895
540	Raunds and Heyford	Do.	1891
541	*Rawdon U.*	Dr. Barry	1890
542	Rawdon U.†	Do.	1891
543	Redditch U.†	Dr. Ballard ...	1873
544	*Redhill and Caterham*	„ Thorne ...	1879
545	Redruth Registration District	„ Blaxall ...	1876
546	*Retford, East, R.* (*printed only in annual volume*).	Mr. Spear	1887

No.	Locality Inspected.	Medical Inspector.	Year.
547	Rhayader R.†	Mr. Spear	1889
548	Rhymney U.†	„ Power	1880
549	Risca	Dr. Fletcher ...	1893
550	Rochdale	„ Thorne ...	1882
551	*Rochester U.*	Mr. Spear	1889
552	Romford R.	Dr. Parsons ...	1884
553	*Rotherham, Rawmarsh, and Greasborough (1892 vol.).*	„ Thomson ...	1891
554	Rotherham U.	„ Ballard ...	1873
555	Rotherhithe	„ Thomson ...	1893
556	Rothwell U.†	„ Parsons ...	1882
557	Rowley Regis U.	„ Ballard ...	1874
558	Royston R.	„ Thorne ...	1876
559	*Rumney Valley**	Mr. T. W. Thompson	1895
560	Runcorn U.	Dr. Airy	1874
561	Runcorn U.*	Mr. Spear	1890
562	Ruthin R.	Dr. Parsons ...	1881
563	Ruthin U.	„ Home	1871
564	Ruthin U.	„ Parsons ...	1881
565	Rye R.*	Mr. Spear	1889
566	*Rycdale**	Dr. Bruce Low ...	1893
567	*St. Albans U.**	Mr. S. F. Murphy...	1884
568	St. Anne's Chapel	Dr. Blaxall ...	1881
569	St. Asaph R.	„ Parsons ...	1881
570	*St. Austell R.*	Do.	1888
571	St. George, Hanover Square†	Dr. Barry	1889
572	St. George, Hanover Square	„ Sweeting ...	1891
573	St. Helen's (Lancs.)	Mr. Spear	1882
574	St. Ives (Hunts.) R.†	Dr. Airy	1881
575	St. John's Wood†	Mr. Radcliffe ...	1872
576	*St. Joseph's Industrial School, Manchester**	Dr. Page	1888
577	*St. Mary Cray, &c. (Small-pox among Rag Sorters).*	„ Parsons ...	1881
578	*St. Pancras (printed only in annual volume)*	Mr. Power	1882
579	Sale U.	Dr. Parsons ...	1882
580	Samford R.	„ Airy	1889
581	Sandal Registration Sub-District	Mr. Sweeting ...	1885
582	*Scotter and Scotton**	Dr. Parsons ...	1890
583	Sculcoates R.	„ Blaxall ...	1877
584	Sculcoates R.*	„ Barry	1890
585	*Sedgley Registration Sub-District*	„ Ballard ...	1874
586	Seend†	Mr. Sweeting ...	1885
587	Seisdon R.	Dr. Ballard ...	1874
588	Selborne...	„ Blaxall ...	1879
589	Sevenoaks U.	„ Bruce Low ...	1896
590	Shadwell	Mr. Spear	1880
591	Shaftesbury U.	Dr. Simpson ...	1885
592	Sheerness U.†	Mr. Spear	1884
593	*Sheffield U. (Separate Volume, Small-pox Epidemic*, 1887–88).	Dr. Barry	1887
594	Shepton Mallet R.†	„ Gresswell ...	1885
595	*Sherborne U.*	„ Blaxall ...	1873
596	*Sherborne U.*	Do. ...	1882
597	Shifnal R.	Dr. Wheaton ...	1897
598	Shildon and East Thickley*	„ Bruce Low ...	1893
599	Sidmouth U.	Mr. Radcliffe ...	1877
600	*Sirhowy Valley**	„ T. W. Thompson	1895
601	Skegby and Fulwood†	Dr. Harries ...	1873
602	Skelmanthorpe U.†	„ Bruce Low ...	1889

Locality Inspected.	Medical Inspector.	Year.
Smalley†	Dr. Page	1887
Southborough U.	Mr. Spear	1882
Southend U.	Dr. Thorne... ...	1880
Southend U.*	„ Bruce Low ...	1896
South Stoneham R.	„ Blaxall ...	1879
Sowerby Bridge U.	„ Parsons ...	1889
Spennymoor U.	Do. ...	1884
Spennymoor U.*	Dr. Page	1888
Spilsby R.	Mr. Spear	1884
*Staines R.**	Dr. Blaxall ...	1890
Stalybridge U.	Mr. Spear	1890
Standish with Langtree U.†	Dr. Page	1888
Stapleford	„ Thorne... ...	1877
Stevenage U.	Do.	1873
Stoke (extra urban)†	Dr. Horne	1893
Stoke-on-Trent Registration District ...	„ Ballard ...	1872
Stone R....	Do. ...	1874
Stourbridge Registration District	Do. ...	1873
Stourbridge Registration District	Dr. Parsons ...	1879
Stourbridge R (Staffs. Division) *	Do. ...	1888
Stourbridge R. (Worcester Division)† ...	Do. ...	1888
Stow-on-the-Wold R.	Dr. Sweeting ...	1898
Sunderland U.	Mr. Radcliffe ...	1871
Sunderland U.†	„ T. W. Thompson	1895
Sutton Bridge	Dr. Parsons ...	1879
Sutton-in-Ashfield U.	„ Harries ...	1873
Sutton-in-Ashfield U.	„ Parsons ...	1882
Sutton-in-Ashfield U.	Do. ...	1883
Sutton Veny	Dr. Airy	1872
Swanage U.	„ Home	1872
Swanage U*	Mr. Harvey ...	1886
Swanage U.*	Dr. Wheaton ...	1897
Swansea R. (part of)	Mr. Radcliffe ...	1875
Swimbridge	Dr. Ogle	1879
Swindon Registration Sub-District (part of)	„ Blaxall ...	1879
Swinton U.	„ Ballard ...	1872
*Swinton and Pendlebury U.**	„ Thomson ...	1898
Taunton	Dr. Blaxall ...	1882
Tavistock R.†	Do. ...	1881
Tees Valley	Dr. Barry	1890
Tees Valley	Do. ...	1891
Tempsford†	Dr. Parsons ...	1882
Tenbury R.	„ Airy	1893
Tendring R.†	Mr. Spear	1890
Thirsk R†	Dr. Parsons ...	1889
Thorne Registration District	Do. ...	1883
Thorpe Saint Andrew	Dr. Copeman ...	1893
Thurmaston	„ Fletcher ...	1895
*Tichmarsh**	Mr. Power	1883
Tideswell	Dr. Thorne ...	1876
Tipton U.	„ Ballard ...	1874
Todmorden	„ Buchanan ...	1872
Torpoint†	„ Ballard ...	1882
Totnes U. and R.	„ Parsons ...	1880
Tottenham U.	Mr. Radcliffe ...	1873
Tottenham U.	„ Sweeting ...	1885
Trawden U.†	„ T. W. Thompson	1891
Tredegar U.	„ Radcliffe ...	1878
Tredegar U.	„ Spear	1889

No.	Locality Inspected.	Medical Inspector.	Year.
660	*Trent River**	Dr. Bruce Low ...	1893
661	Trimdon Colliery and Trimdon Grange† ...	„ Page	1884
662	Trotterscliffe (*not issued*)	„ Ballard ...	1879
663	Truro Registration District	„ Blaxall ...	1874
664	Truro R.†	„ Parsons ...	1880
665	Tudhoe, Elvet, &c., &c.	Mr. T. W. Thompson	1891
665*a*	Tunbridge Wells U.*	Dr. Buchanan ..	1898
666	Tynemouth Registration District*	„ Barry	1883
667	Uckfield U.	Mr. Power	1881
668	Ulverston U.	„ Spear	1882
669	*Upton**	Dr. Parsons ...	1889
670	*Ure River*	Mr. Sweeting ...	1885
671	Usk	Dr. Horne	1893
672	Uxbridge R.*	Mr. Evans	1893
673	Uxbridge U.†	Dr. Bruce Low ...	1889
674	Vron†	Mr. Spear	1887
675	Wadebridge*	Dr. Buchanan ...	1897
675*a*	Wakefield	Mr. Radcliffe ...	1871
676	Wakefield: West Riding House of Correction.	Dr. Ballard ...	1875
677	Walker U.†	„ Airy	1872
678	Walsall R.†	„ Thorne ...	1881
679	Walsingham R.	„ Gresswell ...	1885
680	Walthamstow U.	„ Airy	1888
681	Walton-on-the-Naze U.	„ Parsons ...	1881
682	Walworth†	„ Copeman ...	1895
683	Ware R.	„ Parsons ...	1884
684	Ware U.†	„ Thorne ...	1874
685	Wareham U.	„ Home	1872
686	Warrington U.	„ Ballard ...	1871
687	Watford U.	Mr. Sweeting ...	1885
688	Watford U. and R.	„ Royle	1897
689	Wath U.†	Dr. Ballard ...	1872
690	Wednesbury U. (Registration Sub-District)	Do.	1875
691	*Welbeck*	Do.	1880
692	Wellingborough R.†	Mr. Spear	1889
693	Wellingborough U.	Dr. Thorne ...	1874
694	Wellington (Somerset) R.	„ Simpson ...	1885
695	Wellington (Somerset) U.	„ Blaxall ...	1872
696	Wellington (Somerset) U.	„ Simpson ...	1885
697	Wells (Norfolk) Registration Sub-District (part of) †	Mr. Power	1875
698	*Wells (Norfolk) U.*	„ Spear	1890
699	Wells (Somerset) R.	„ Langdon ...	1879
700	West Auckland	Dr. Harries ...	1872
701	West Bromwich U.*	„ Buchanan ...	1895
701*a*	*West Bromwich U.**	Do.	1898
702	West Bromwich U. and R.	Dr. Ballard ...	1875
703	West Cowes U.	Mr. Spear	1886
704	Weston-super-Mare U.	Dr. Blaxall ...	1882
705	Wetherby	„ Thorne ...	1877
706	Whitby U. and R.	„ Wilson ...	1893
707	Whitchurch (Hants.)	„ Thorne ...	1872
708	Whitford Registration Sub-District ...	Mr. Power	1871
709	Whitford Registration Sub-District ...	Dr. Parsons ...	1888

No.	Locality Inspected.	Medical Inspector.	Year.
710	Whitstable	Mr. Power	1883
711	Whixley Registration Sub-District* ...	" Sweeting ...	1885
712	Wickford	Dr. Thorne ...	1879
713	Widnes U.*	" Bruce Low ...	1894
714	Wigan and Ince-in-Makerfield U.	" Copeman ...	1892
715	Wigan U.	Mr. Radcliffe ...	1873
716	Wight, Isle of	Dr. Ballard ...	1880
717	Willenhall Registration Sub-District ...	Do.	1874
718	Williton R.†	Do.	1881
719	*Wimbledon U.*	Mr. Power	1886
720	Wincanton R	Dr. Home	1872
721	Wincanton R.†	" Airy	1884
722	Wincanton R.	" Parsons ...	1886
723	*Winchester U.*	Mr. Power	1877
724	Winchester U.	Dr. Parsons ...	1887
725	Windsor U.	" Airy and Mr. A. Taylor.	1886
726	Windsor U.	Do.	1889
727	Wing†	Dr. Blaxall ...	1879
728	Wisbech R.	" Page	1888
729	Witney U.	" Fletcher ...	1897
730	Wiveliscombe U.	" Blaxall ...	1875
731	Wolstanton and Burslem Registration District.	" Ballard ...	1872
732	Wolverhampton U.	Do. ...	1874
733	Wombwell U.	Dr. Parsons ...	1886
734	Woolfardisworthy	" Airy	1880
735	Woolwich U.	Mr. Spear	1884
736	Worcester U.†	Dr. Parsons ...	1889
737	Worcester U.	" Fletcher ...	1897
738	Workington U.†	" Page	1888
739	*Worthing**	" Thomson ...	1893
739a	Wortley R.†	" Reece	1898
740	Wrexham U.*	" Wheaton ...	1897
741	Wrexham U. and R. (Registration District —part of).	" Airy	1878
742	Wribbenhall	" Thomson ...	1897
743	*Wycombe Marsh**	" Buchanan ...	1895
744	Yateley†	Dr. Turner... ...	1886
745	Yeadon U.	Mr. Spear	1879
746	*York U.**	Dr. Airy	1884
747	*York Town and Camberley**	Mr. Power	1886
748	York Town R.	" T. W. Thompson	1892
749	Ystradgunlais	Dr. Harries ...	1873
750	Ystradyfodwg U.† (now Rhondda) ...	" Airy	1876

The following districts have been inspected, wholly or in part, in connexion with the inquiries in the foregoing Table, the number of which is placed against their names:—

Reference No.	District Visited.	Medical Inspector.	Year.
449	Aberavon U.	Dr. Airy	1877
516	Abersychan U.	„ Ogle	1879
62	Abertillery U.	Mr. Spear	1882
63	Abertillery U.	Do. ...	1884
3	Alcester R. (part of)	Dr. Ballard ...	1875
240	Aldershot U.	Mr. Sweeting ...	1885
81	Aldershot U.	Dr. Turner ...	1886
581	Altofts U.	Mr. Sweeting ...	1885
588	Alton R.	Dr. Blaxall ...	1879
7	Alton U. and R.	Mr. Sweeting ...	1893
152	Amersham R. (part of)	„ Power	1871
183	Ampthill R. (part of)	Dr. Cory	1879
10	Anglesey R.	Mr. Evans	1893
88	Astley Bridge U.	Dr. Ballard ...	1871
73	Aston	„ Buchanan ...	1874
28	Atherstone R. (part of)	„ Airy	1878
30	Atherton U.	Mr. Power	1877
32	Auckland R.	Dr. Thorne ...	1874
610	Auckland R.	„ Page	1888
82	Axbridge R. (part of)	Mr. Power	1879
413	Axbridge R.	„ T. W. Thompson	1895
35	Axminster R. (part of)	Dr. Blaxall ...	1874
670	Aysgarth R.	Mr. Sweeting ...	1885
650	Bakewell R. (part of)	Dr. Thorne ...	1876
132	Bakewell R.	„ Bruce Low ...	1890
196	Bala R.	Do. ...	1895
196	Bala U.	Do. ...	1895
443	Balsall Heath U.	Dr. Ballard ...	1873
142	Bangor and Beaumaris R.	„ Bruce Low ...	1895
142	Bangor U.	Do. ...	1895
235	Barlings	Dr. Gresswell ...	1885
195	Barnsley R.	„ Bruce Low ...	1889
187	Barnstaple R. (part of)	„ Home	1872
635	Barnstaple R.	„ Ogle	1879
47	Barnstaple U. and R.	„ Reece	1894
54	Barton-on-Irwell R.	„ Stevens ...	1874
54	Barton-on-Irwell U.	Do. ...	1874
140	Basford R. (part of)	Dr. Harries ...	1871
203	Batley U.	„ Thorne ...	1878
57	Battle R.	„ Airy	1878
58	Bedale R. (part of)	Mr. Power	1877
670	Bedale R.	„ Sweeting ...	1885
261	Beddgelert	Dr. Airy	1880
356	Bedford R.	Mr. Sweeting ...	1885
62	Bedwellty R.	„ Spear	1882
63	Bedwellty R.	Do. ...	1884
51	Belgrave U.	Dr. Barry	1883
603	Belper R.	„ Page	1887
142	Bethesda U.	„ Bruce Low ...	1895
734	Bideford R.	„ Airy	1880
47	Bideford U.	„ Reece	1894
527	Biggleswade R.	„ Parsons ...	1882
642	Biggleswade R.	Do.	1882
146	Biggleswade R.	Dr. Buchanan ...	1896
295	Biggleswade R.	Do. ...	1896

Reference No.	District Visited.	Medical Inspector.	Year.
489	Biggleswade R.	Dr. Buchanan ...	1896
712	Billericay R.	„ Thorne ...	1879
71	Bilston U.	„ Ballard ...	1874
203	Birkenshaw U....	„ Thorne ...	1878
203	Birstall U.	Do. ...	1878
32	Bishop Auckland U.	Do. ...	1874
710	Blean R....	Mr. Power ...	1883
647	Blofield R.	Dr. Copeman ...	1893
349	Blything R.	„ Mivart ...	1896
84	Bodmin U. and R.	„ Parsons ...	1881
414	Bollington U.	„ Thorne ...	1873
88	Bolton R.	„ Ballard ...	1871
338	Bolton R.	Mr. Spear	1889
88	Bolton U.	Dr. Ballard ...	1871
310	Bosmere and Claydon R.	„ Airy	1880
271	Braintree R. (part of)	„ Thorne ...	1876
170	Braintree R.	„ Airy	1882
156	Brampton and Walton U.	„ Thorne ...	1874
213	Brandon and Byshottles U.	Mr. Spear	1881
291	Brentford R. (part of)	Dr. Thorne ...	1876
101	Bridgend R.	Mr. Spear	1888
101	Bridgend U.	Do. ...	1888
620	Brierley Hill U.	Dr. Ballard ...	1873
449	Briton Ferry U.	„ Airy	1877
218	Brixworth R.	„ Bruce Low ...	1889
577	Bromley R.	„ Parsons ...	1881
185	Bromley R.	„ Copeman ...	1894
144	Bromsgrove U. (part of)	„ Ballard ...	1873
720	Bruton	„ Home	1872
387	Brynmawr U. (part of)	„ Harries ...	1872
385	Builth R.	„ Airy	1880
731	Burslem U.	„ Ballard ...	1872
121	Burton-on-Trent R. (part of)	„ Airy	1878
192	Calne R.	Mr. Spear	1884
747	Camberley	„ Power	1886
578	Camberwell	Do.	1882
304	Camborne U.	Dr. Ballard ...	1882
500	Cannock R. (part of)	„ Airy	1876
394	Carnarvon R.	Do.	1880
142	Carnarvon R.	Dr. Bruce Low ...	1895
142	Carnarvon U.	Do. ...	1895
720	Castle Cary	Dr. Home	1872
143	Castleford U.	„ Parsons ...	1880
544	Caterham	„ Thorne ..	1879
352	Chard R. (part of)	„ Blaxall ...	1871
270	Chelmsford R. (part of)	„ Airy	1873
571	Chelsea Parish	„ Barry	1889
312	Chertsey R. (part of)	„ Thorne ...	1872
196	Chester R.	„ Bruce Low ...	1895
196	Chester U.	Do. ...	1895
156	Chesterfield R....	Dr. Thorne ..	1874
156	Chesterfield U....	Do. ...	1874
225	Chester-le-Street R. (part of)	Dr. Harries ...	1873
91	Chippenham R.	„ Blaxall ...	1880
199	Chippenham R.	„ Horne	1893
169	Chipping	„ Airy	1880
196	Chirk R.	„ Bruce Low ...	1895
156	Clay Lane U.	„ Thorne ...	1874
256	Claypole R.	„ Wheaton ...	1896

Reference No.	District Visited.	Medical Inspector.	Year.
424a	Claypole R.	Dr. Wheaton ...	1898
454	Clee-with-Weelsby U.	" Page	1888
380	Cliffe U.	" Thorne ...	1874
134	Clutton R.	Mr. Sweeting ...	1892
441	Coleford U.	Dr. Fletcher ...	1892
142	Colwyn Bay U.	" Bruce Low ...	1895
142	Conway R.	Do. ...	1895
501	Conway R.	Do. ...	1896
142	Conway U.	Do. ...	1895
585	Coseley U.	Dr. Ballard ...	1874
101	Cowbridge U.	Mr. Spear	1888
83	Cowpen U.	Dr. Airy	1872
666	Cowpen U.	" Barry	1883
666	Cramlington U.	Do.	1883
142	Criccieth U.	Dr. Bruce Low ...	1895
387	Crickhowell R. (part of)	" Harries ...	1872
388	Crickhowell R. (part of)	" Blaxall ...	1876
646	Crowle U.	" Parsons ...	1883
59	Croydon R.	" Gresswell ...	1886
719	Croydon R.	Mr. Power ...	1886
20	Dartford R.	Mr. Power	1876
192	Dartford R.	" Spear	1884
192	Dartford U.	Do.	1884
195	Darton U.	Dr. Bruce Low ...	1889
401	Daventry R.	Do.	1896
524	Devizes R.	Dr. Copeman ...	1894
203	Dewsbury U.	" Thorne ...	1878
39	Doncaster R. (part of)	Do.	1873
504	Dore R. (part of)	Do.	1877
281	Dorking R.	Mr. Power	1882
423	Downham R.	Dr. Parsons ...	1887
156	Dronfield U.	" Thorne ...	1874
287	Droxford R.	" Parsons ...	1884
245	Dunmow R.	" Airy	1880
272	Dunmow R.	Do.	1883
610	Durham R.	Dr. Page	1888
665	Durham R.	Mr. T. W. Thompson	1891
213	Durham U. and R.	" Spear	1881
716	East Cowes U.	Dr. Ballard ...	1880
254	East Grinstead	" Airy	1880
81	East Hampstead	" Turner ...	1886
14	East Ward R. (part of)	" Ballard ...	1873
62	Ebbw Vale U.	Mr. Spear	1882
63	Ebbw Vale U.	Do.	1884
196	Edeyrnion R.	Dr. Bruce Low ...	1895
229	Edmonton U.	Do. ...	1888
125	Edmonton U.	Dr. Copeman ...	1891
230	Edmonton U.	Do. ...	1895
384	Ely R. (part of)	Dr. Thorne ...	1873
230	Enfield U.	" Copeman ...	1895
404	Epping R. (part of)	Mr. Power	1878
405	Epping R.	" Spear	1880
407	Erpingham R.	Dr. Airy	1885
434	Eston U....	" Ballard ...	1888
147	Eton R. (part of)	Mr. Power	1876

Reference No.	District Visited.	Medical Inspector.	Year.
304	Falmouth R.	Dr. Ballard ...	1882
304	Falmouth U. (town and parish)	Do. ...	1882
520	Fareham R. (part of)	Dr. Stevens ...	1872
237	Fareham R.	Mr. Spear	1888
237	Fareham U.	Do. ...	1888
263	Farncombe	Mr. Power	1874
578	Farnham R.	Do.	1882
240	Farnham R.	Mr. Sweeting ...	1885
81	Farnham R.	Dr. Turner ...	1886
81	Farnham R.	Do. ...	1886
255	Farnham R.	Do. ...	1886
747	Farnham R.	Mr. Power	1886
133	Farnham R.	Dr. Parsons ...	1889
240	Farnham U.	Mr. Sweeting ...	1885
88	Farnworth U.	Dr. Ballard ...	1871
618	Fenton U.	Do. ...	1872
403a	Fenton U.	Dr. Fletcher ...	1898
261	Festiniog R.	" Airy	1880
142	Festiniog R.	" Bruce Low ...	1895
142	Festiniog U.	Do. ...	1895
331	Flint U....	Dr. Blaxall ...	1875
46	Forehoe R.	Mr. T. W. Thompson	1894
275	Freebridge Lynn R. (part of)	Dr. Stevens ...	1877
367	Freebridge Lynn R.	" Bruce Low ...	1892
480	Frome R. (part of)	" Ballard ...	1872
448	Fulham Board of Works District (one institution).	Mr. Spear	1883
96	Fylde R....	Dr. Wilson... ...	1898
582	Gainsborough R.	Dr. Parsons ...	1890
660	Gainsborough R.	" Bruce Low ...	1893
324	Gedney Drove End	" Parsons ...	1882
582	Glanford Brigg R.	Do. ...	1890
432	Glanford Brigg R.	Dr. Bruce Low ...	1897
263	Godalming U.	Mr. Power ...	1874
347	Godmanchester U.	Dr. Parsons ...	1880
544	Godstone R.	" Thorne ...	1879
382	Godstone R.	" Copeman ...	1892
203	Gomersal U.	" Thorne ...	1878
49	Grantham R.	" Parsons ...	1890
269	Grays Thurrock U.	" Airy	1889
553	Greasborough U.	" Thomson ...	1891
270	Great Baddow	" Airy	1873
454	Great Grimsby U.	" Page	1888
670	Great Ouseburn R.	Mr. Sweeting ...	1885
711	Great Ouseburn R.	Do. ...	1885
263	Guildford R. (part of)	Mr. Power	1874
282	Guisbrough R.	Dr. Thorne ...	1875
496	Gunthwaite	Mr. Spear	1889
92	Halifax	Mr. Spear	1880
286	Halstead R.	Dr. Bruce Low ...	1889
286	Halstead U.	Do. ...	1889
281	Hambledon R.	Mr. Power	1882
618	Hanley U.	Dr. Ballard ...	1872
80	Hartley Wintney R. (part of)	" Stevens ...	1876
81	Hartley Wintney R.	" Turner... ...	1886
239	Hartley Wintney R.	Do. ...	1886
744	Hartley Wintney R.	Do. ...	1886

Reference No.	District Visited.	Medical Inspector.	Year.
288	Haslemere	Dr. Airy	1887
292	Hastings R.	Mr. Spear	1890
292	Hastings U.	Do. ...	1890
522	Havant R.	Dr. Thomson ...	1896
522	Havant U.	Do. ...	1896
427	Haverfordwest R.	„ Parsons ...	1880
196	Hawarden R.	„ Bruce Low ...	1895
545	Hayle U.	„ Blaxall ...	1876
304	Hayle U.	„ Ballard ...	1882
717	Heath Town U.	Do. ...	1874
302	Hebden Bridge U.	Dr. Gresswell ...	1885
447	Hebden Bridge U.	„ Page	1888
203	Heckmondwike U.	„ Thorne ...	1878
183a	Helston R.	„ Bruce Low ...	1898
474	Hendon	Mr. Power	1885
242	Herne Hill (Faversham R.)	Do. ...	1880
127	Hexham R. (part of)	Dr. Airy	1874
708	Holywell R.	Mr. Power	1871
331	Holywell R.	Dr. Blaxall ...	1875
709	Holywell R.	„ Parsons ...	1888
475	Holywell R.	Mr. Spear	1890
196	Holywell R.	Dr. Bruce Low ...	1895
330	Holywell R.	„ Mivart ...	1896
331	Holywell U.	„ Blaxall ...	1875
330	Holywell U.	„ Mivart... ...	1896
196	Hoole U.	„ Bruce Low ...	1895
88	Horwich U.	„ Ballard ...	1871
338	Horwich U.	Mr. Spear	1889
313	Houghton-le-Spring R. (part of)	„ Power	1874
100	Houghton-le-Spring R.	Dr. Fletcher ...	1895
314	Houghton-le-Spring R.	Do. ...	1895
666	Howden U.	Dr. Barry	1883
430	Hoxne R. (part of)	„ Airy	1873
374	Hoxne R.	„ Bruce Low ...	1894
498	Hoylandswaine U.	Mr. Spear	1889
420	Hucknall-under-Huthwaite U.	Dr. Gresswell ...	1885
346	Hull, Port of	„ Airy	1882
347	Huntingdon U. and R.	„ Parsons ...	1880
47	Ilfracombe U.	Dr. Reece	1894
714	Ince-in-Makerfield U.	„ Copeman ...	1892
88	Kearsley U.	Dr. Ballard ...	1871
355	Keighley R.	„ Stevens ...	1875
355	Keighley U.	Do. ...	1875
92	Keighley U.	Mr. Spear	1880
571	Kensington Parish	Dr. Barry	1889
120	Kettering R. (part of)	„ Thorne ...	1872
742	Kidderminster R.	„ Thomson ...	1897
731	Kidsgrove U.	„ Ballard ...	1872
443	King's Norton R. (part of)	Do. ...	1873
370	Knighton R.	Dr. Airy	1878
370	Knighton U.	Do. ...	1878
258	Lancaster R.	Dr. Barry	1882
17	Leeds U. (part of)	„ Ballard ...	1872
377	Leigh U. (part of)	Mr. Power	1872
727	Leighton Buzzard R.	Dr. Blaxall ...	1879

Reference No.	District Visited.	Medical Inspector.	Year.
383	Leighton Buzzard R.	Dr. Bulstrode ...	1894
380	Lewes U. and R.	„ Thorne ...	1874
279	Lexden and Winstree R.	„ Copeman ...	1893
670	Leyburn R.	Mr. Sweeting ...	1885
235	Lincoln R.	Dr. Gresswell ...	1885
469	Lincoln R. (one house)	Do. ...	1885
88	Little Hulton U.	Dr. Ballard ...	1871
88	Little Lever U....	Do. ...	1871
203	Liversedge U.	Dr. Thorne ...	1878
329	Llandausaint Registration Sub-District ...	„ Ogle	1879
392	Llandilofawr R.	„ Sweeting ...	1895
142	Llandudno R.	„ Bruce Low ...	1895
142	Llanfairfechan U.	Do. ...	1895
396	Llanfyllin R.	Dr. Parsons ...	1888
196	Llangollen R.	„ Bruce Low ...	1895
196	Llangollen U.	Do. ...	1895
142	Llanrwst R.	Do. ...	1895
400	Loddon and Clavering R.	Dr. Sweeting ...	1894
15	London, Port of	Mr. Power	1880
176	London, Port of	Do. ...	1881
618	Longton U.	Dr. Ballard ...	1872
18	Lutterworth R.	Mr. Spear	1887
414	Macclesfield R. (part of)	Dr. Thorne... ...	1873
669	Macclesfield R....	„ Parsons ...	1889
669	Macclesfield U....	Do. ...	1889
415	Machynlleth R.	Dr. Airy	1876
101	Maesteg U.	Mr. Spear	1888
662	Malling R.	Dr. Ballard ...	1879
481	Malmesbury R.	„ Downes ...	1883
106	Malmesbury R....	„ Bruce Low ...	1890
196	Malpas R.	Do. ...	1895
576	Manchester U. (one institution)	Dr. Page	1888
601	Mansfield R. (part of)	„ Harries ...	1873
420	Mansfield R.	„ Gresswell ...	1885
691	Mansfield U.	„ Ballard ...	1880
126	Maresfield (Uckfield R.)	„ Airy	1887
575	Marylebone, St. (part of)	Mr. Radcliffe ...	1872
670	Masham U.	„ Sweeting ...	1885
586	Melksham R.	Do. ...	1885
142	Menai Bridge U.	Dr. Bruce Low ...	1895
719	Merton	Mr. Power	1886
582	Messingham	Dr. Parsons ...	1890
143	Methley U.	Do. ...	1880
447	Midgley U.	Dr. Page	1888
411	Midhurst R.	„ Airy	1880
720	Milborne Port	„ Home	1872
371	Mildenhall R.	„ Copeman ...	1892
203	Mirfield U.	„ Thorne ...	1878
331	Mold U....	„ Blaxall ..	1875
196	Mold U....	„ Bruce Low ...	1895
204	Monmouth R.	Mr. Spear	1888
441	Monmouth R.	Dr. Fletcher ...	1892
441	Monmouth U.	Do. ...	1892
203	Morley U.	Dr. Thorne ...	1878
517	Mountain Ash U.	„ Airy	1876
518	Mountain Ash U.	Mr. Spear	1889
357	Mutford and Lothingland R.	Dr. Bruce Low ...	1896

Reference No.	District Visited.	Medical Inspector.	Year.
364	Narberth R.	Mr. Spear	1888
449	Neath U. and R.	Dr. Airy	1877
660	Newark U.	„ Bruce Low ...	1893
156	Newbold-cum-Dunston U.	„ Thorne ...	1874
618	Newcastle-under-Lyme U. (part of) ...	„ Ballard ...	1872
350	New Forest R.	„ Bulstrode ...	1894
682	Newington U.	„ Copeman ...	1895
191	Newmarket R.	„ Turner... ...	1886
445	Newmarket R.	Do. ...	1886
716	Newport (I. W.)	Dr. Ballard ...	1890
282	Normanby U.	„ Thorne... ...	1875
434	Normanby U.	„ Ballard ...	1888
581	Normanton U.	Mr. Sweeting ...	1885
540	Northampton R.	Dr. Bruce Low ...	1891
70	North Bierley U. (part of)	„ Thorne... ...	1874
508	Northfleet U. (part of)	Do. ...	1871
178	Norwich U.	Dr. Airy	1888
116	Nottingham U. (part of)	„ Harries ...	1871
534	Nottingham U....	„ Thorne... ...	1872
378	Nottingham U.	Do. ...	1875
145	Oakley, Brome, Scole, &c.	Dr. Bruce Low ...	1890
355	Oakworth U.	„ Stevens ...	1875
101	Ogmore and Garw U....	Mr. Spear	1888
473	Okehampton R.	Dr. Blaxall ...	1881
485	Oldham R.	„ Stevens ...	1875
89	Ongar U.	„ Airy	1881
282	Ormesby U.	„ Thorne ...	1875
434	Ormesby U.	„ Ballard ...	1888
493	Orrell	„ Airy	1880
33	Orsett R.	Mr. Spear	1880
269	Orsett R.	Dr. Airy	1889
203	Osset-cum-Gawthorpe U.	„ Thorne ...	1878
196	Oswestry R.	„ Bruce Low ...	1895
196	Overton R.	Do.	1895
456	Oxford U. (part of)	Dr. Thorne ...	1872
516	Panteg U.	Dr. Ogle	1879
418	Pembroke R.	„ Airy	1880
498	Penistone R.	Mr. Spear	1889
498	Penistone U.	Do.	1889
142	Penmaenmawr U.	Dr. Bruce Low ...	1895
304	Penryn U.	„ Ballard ...	1882
545	Phillack U.	„ Blaxall ...	1876
304	Phillack U.	„ Ballard ...	1882
148	Plymouth U. (part of)	„ Home	1872
424	Pocklington R.	Mr. Royle	1885
749	Pontardawe R. (part of)	Dr. Harries ...	1873
515	Pontefract R. (part of)	„ Beard	1875
516	Pontypool U.	„ Ogle	1879
517	Pontypridd R.	„ Airy	1876
518	Pontypridd R.	Mr. Spear	1889
517	Pontypridd U.	Dr. Airy	1876
518	Pontypridd U.	Mr. Spear	1889
618	Pool's Dam	Dr. Ballard ...	1872
525	Potterspury R....	„ Bulstrode ...	1895
142	Pwllheli R.	„ Bruce Low ...	1895
142	Pwllheli U.	Do.	1895

Reference No.	District Visited.	Medical Inspector.	Year.
620	Quarry Bank U.	Dr. Ballard ...	1873
51	Quorndon U.	" Barry	1883
203	Ravensthorpe U.	Dr. Thorne ...	1878
553	Rawmarsh U.	" Thomson ...	1891
282	Redcar U.	" Thorne ...	1875
304	Redruth R.	" Ballard ...	1882
133a	Redruth R.	" Bruce Low ...	1898
304	Redruth U.	" Ballard ...	1882
545	Redruth U. and R.	" Blaxall ...	1876
544	Reigate U.	" Thorne ...	1879
750	Rhondda	" Bruce Low ...	1893
670	Ripon R.	Mr. Sweeting ...	1885
670	Ripon U.	Do. ...	1885
43	Romford R. (part of)	Dr. Harries ...	1873
44	Romford R. (part of)	" Buchanan ...	1873
45	Romford R.	" Thorne ...	1880
536	Romford R.	Mr. Evans	1894
553	Rotherham U.	Dr. Thomson ...	1891
485	Royton U.	" Stevens ...	1875
236	Rusholme U.	" Airy	1879
395	Ruthin R. (part of)	" Thorne ...	1877
196	Ruthin R.	" Bruce Low ...	1895
716	Ryde (I.W.)	" Ballard ...	1880
535	Saffron Walden R. (part of)	Mr. Power	1877
567	St. Albans R.	" S. F. Murphy...	1884
266	St. Austell R. (part of)	Dr. Corfield ...	1871
457	St. Columb Major R. (part of)	" Blaxall ...	1877
458	St. Columb Major R.	" Ballard ...	1880
675	St. Columb Major R.	" Buchanan ...	1897
178	St. Faiths R.	" Airy	1888
578	St. George, Hanover Square	Mr. Power	1882
437	St. Germans R....	Dr. Ballard ...	1880
438	St. Germans R....	Do. ...	1882
653	St. Germans R....	Do. ...	1882
578	St. Giles, Bloomsbury...	Mr. Power	1882
716	St. Helens (I.W.)	Dr. Ballard ...	1880
361	St. John's Wood	Mr Power	1878
578	St. Martin-in-the-Fields	Do.	1882
581	Sandal Magna U.	Mr. Sweeting ..	1885
716	Sandown U.	Dr. Ballard ...	1880
346	Sculcoates R.	" Airy	1882
661	Sedgefield R.	" Page	1884
585	Sedgley, Upper, U.	" Ballard ...	1874
666	Seghill U.	" Barry	1883
444	Shaftesbury R....	" Simpson ...	1885
716	Shanklin U.	" Ballard ...	1880
209	Shardlow R. (part of)	" Beard	1872
615	Shardlow R. (part of)	" Thorne ...	1877
137	Shipston-on-Stour R. (part of)	" Harries ...	1873
95	Shipston-on-Stour R. (part of)	Mr. Power	1876
717	Short Heath U.	Dr. Ballard ...	1874
327	Silloth (Holme Cultram)	" Parsons ...	1888
282	Skelton U.	" Thorne ...	1875
203	Soothill, Nether, U.	Do. ...	1878
203	Soothill, Upper, U.	Do. ...	1878
83	South Blyth U.	Dr. Airy	1872
666	South Blyth and Newsham	" Barry	1883

Reference No.	District Visited.	Medical Inspector.	Year.
164	South Molton R. (part of)	Dr. Home	1872
447	Sowerby U.	„ Page	1888
207	Spalding R.	Do.	1883
208	Spalding R.	Do.	1884
32	Spennymoor U.	Dr. Thorne	1874
270	Springfield	„ Airy	1873
683	Stanstead Abbots	„ Parsons	1884
498	Stocksbridge U.	Mr. Spear	1889
618	Stoke-on-Trent R.	Dr. Ballard	1872
618	Stoke-on-Trent U.	Do.	1872
620	Stourbridge R.	Do.	1873
181	Stourbridge R.	Dr. Gresswell	1889
620	Stourbridge U.	„ Ballard	1873
90	Stow-on-the-Wold R.	Do.	1874
172	Stratford-on-Avon R. (part of)	Do.	1873
362	Stratton R.	Dr. Parsons	1888
54	Stretford U.	„ Stevens	1874
420	Sutton-in-Ashfield U.	„ Gresswell	1885
121	Swadlincote U.	„ Airy	1878
451	Swaffham R. (part of)	Do.	1876
636	Swindon, New, U.	Dr. Blaxall	1879
636	Swindon, Old, U.	Do.	1879
54	Swinton and Pendlebury U.	Dr. Stevens	1874
196	Tarvin R.	Dr. Bruce Low	1895
74	Taunton R.	„ Blaxall	1882
131	Tavistock R. (part of)	Do.	1876
284	Tavistock R	Do.	1881
568	Tavistock R.	Do.	1881
69	Tenterden R.	Dr. Thorne	1879
276	Thame R. (part of)	Do.	1872
646	Thorne R.	Dr. Parsons	1883
660	Thorne R.	„ Bruce Low	1893
203	Thornhill U.	„ Thorne	1878
538	Thrapston R.	„ Airy	1880
649	Thrapston R.	Mr. Power	1883
540	Thrapston R.	Dr. Bruce Low	1891
539	Thrapston R.	Do.	1895
498	Thurlstone R.	Mr. Spear	1889
51	Thurmaston U.	Dr. Barry	1883
302	Todmorden R.	„ Gresswell	1885
447	Todmorden R.	„ Page	1888
406	Totnes R.	„ Blaxall	1888
415	Towyn U.	„ Airy	1876
663	Truro U. and R.	„ Blaxall	1874
731	Tunstall U.	„ Ballard	1872
88	Turton U.	Do.	1871
214	Turton U. (part of)	Mr. Power	1876
30	Tyldesley-cum-Shakerley U.	Do.	1877
83	Tynemouth R. (part of)	Dr. Airy	1872
363	Tynemouth R. (part of)	Do.	1872
400	Tynemouth R.	Dr. Sweeting	1894
455	Tynemouth R.	Do.	1894
666	Tynemouth U. and R.	Dr Barry	1883
126	Uckfield R.	Dr. Airy	1887
196	Uwchaled R.	„ Bruce Low	1895
477	Uxbridge R.	Mr. Power	1882

Reference No.	District Visited.	Medical Inspector.	Year.
716	Ventnor U.	Dr. Ballard ...	1880
581	Wakefield R.	Mr. Sweeting ...	1885
666	Walker U.	Dr. Barry	1883
453	Wallasey U.	Mr. Spear	1888
666	Wallsend U.	Dr. Barry	1883
697	Walsingham R. (part of)	Mr. Power	1875
166	Wandsworth Board of Works District ...	Dr. Parsons ...	1882
528	Wandsworth Board of Works District ...	" Blaxall ...	1883
631	Warminster R. (part of)	" Airy	1872
336	Warminster R. (part of)	Mr. Power ...	1876
717	Wednesfield U.	Dr. Ballard ...	1874
316	Wellingborough R. (part of)	" Home ...	1871
224	Wellingborough R. (part of)	" Buchanan ...	1872
716	West Cowes U.	" Ballard ...	1880
88	West Houghton U.	Do. ...	1871
338	West Houghton U.	Mr. Spear ...	1889
705	Wetherby R. (part of)	Dr. Thorne... ...	1877
590	Wetherby U.	Mr. Spear	1880
324	Whaplode Drove	Dr. Parsons ...	1882
707	Whitchurch (Hants) R. (part of)	" Thorne ...	1872
196	Whitchurch (Salop) U.	" Bruce Low ...	1895
666	Whitley and Monkseaton, U.	" Barry	1883
156	Whittington U.	" Thorne ...	1874
143	Whitwood, U.	" Parsons ...	1880
714	Wigan U.	" Copeman ...	1892
716	Wight, Isle of, R.	" Ballard ...	1880
460	Wight, Isle of, R.	" Thomson ...	1894
717	Willenhall U.	" Ballard ...	1874
666	Willington Quay U.	" Barry	1883
720	Wincanton	" Home	1872
532	Wincanton R.	" Parsons ...	1888
19	Windsor R. (part of)	" Ballard ...	1877
271	Witham R. (part of)	" Thorne ...	1876
232	Woburn R.	" Parsons ...	1884
731	Wolstanton and Burslem R.	" Ballard ...	1872
691	Worksop U.	Do. ...	1880
674	Wrexham R.	Mr. Spear	1887
196	Wrexham R.	Dr. Bruce Low ...	1895
196	Wrexham U.	Do. ...	1895
743	Wycombe R.	Dr. Buchanan ...	1895
280	Yarmouth U.	Dr. Airy	1875
576	York U.	" Page	1888
133	York Town	" Parsons ...	1889
515	Ystradyfodwg U.	" Airy	1876
514	Ystradyfodwg U.	" Parsons ...	1880

Name of District.	Urban, Rural or Port.	Date of Report.	Name of Inspector.
Chesterfield	R.	June 1886 ...	Mr. Spear.
Chester le Street	U.	January, 1886...	Dr. Page.
Chesterton	U.	June, 1885 ...	Mr. Spear.
Chesterton	R.	May, 1885 ...	Do.
Chatham	U.	March, 1885 ...	Dr. de Chaumont.
Chichester	U.	June, 1885 ...	Mr. Spear.
Chailey...	R.	December, 1885	Do.
Chilvers Coton	U.	July, 1886 ...	Do.
Chippenham	U.	August, 1886 ...	Do.
Chippenham	R.	Do. ...	Do.
Chorley (Cheshire)	U.	May, 1885 ...	Dr. Airy.
Chorley...	U.	November, 1886	" Page.
Clay Lane	U.	June, 1886 ...	Mr. Spear.
Clayton West	U.	February, 1887..	Dr. Gresswell.
Cleethorpes with Thrumscoe	U.	May, 1886 ...	" Airy.
Cockermouth	U.	August, 1885 ...	" Page.
Colchester	U.	January, 1886...	" Airy.
Colchester	P.	October, 1885 ...	Do.
Colchester (Maldon)	P.	Do. ...	Do.
Colne Valley Combination ..	...	January, 1887...	Dr. Gresswell.
Consett	U.	October, 1886 ...	" Page.
Conway	U.	December, 1885..	" Davies.
Coseley	U.	July, 1886 ...	" Gresswell.
Cowes	P.	March, 1886 ...	" Davies.
Cowpen	U.	June, 1885 ...	" Page.
Crediton	U.	April, 1885 ...	Mr. Spear.
Crediton	R.	Do. ...	Do.
Crickhowell	R.	March, 1886 ...	Do.
Crowle	U.	January, 1886 ...	Do.
Cumberworth	U.	January, 1887 ..	Dr. Gresswell.
Darlaston	U.	August, 1886 ...	Dr. Barry.
Darlington	U.	March, 1885 ...	" Page.
Darlington	R.	Do. ...	Do.
Dartford	U.	June, 1885 ...	Mr. Spear.
Dartford	R.	Do. ..	Do.
Dartmouth	U.	March, 1885 ...	Dr. Davies.
Darton	U.	December, 1886..	Mr. Spear.
Deal	P.	August, 1884 ...	Dr. de Chaumont.
Deal	U.	January, 1885 ...	" Davies.
Denby Dale	U.	February, 1887...	" Gresswell.
Deptford	U.	Do ...	" de Chaumont.
Derby	U.	August, 1886 ...	Mr. Spear.
Devonport	U.	June, 1885 ...	Dr. Blaxall.
Dewsbury	U.	December, 1886..	" Barry.
Dodworth	U.	Do. ...	Mr. Spear.
Dorchester	R.	March, 1885 ...	Dr. Davies.
Dover	P.	August, 1884 ...	" de Chaumont.
Dover	P.	January, 1885 ...	" Davies.
Dover	P.	Do. ...	Do.
Driffield	R.	December, 1886..	Dr. Airy.
Dronfield	U.	June, 1886 ...	Mr. Spear.
Dudley	U.	March, 1885 ...	Dr. Barry.
Durham	U.	January, 1886 ..	" Page.
Durham	R.	Do. ...	Do.
Easington	R.	February, 1885 ..	Dr. Page.
East Cowes	U.	May, 1885 ...	" Blaxall.

Name of District.	Urban, Rural or Port.	Date of Report.	Name of Inspector.
Blackrod	U.	November, 1886	Dr. Parsons.
Blean	R.	March, 1885 ...	„ de Chaumont.
Bodmin	R.	May, 1885 ...	„ Davies.
Bognor	U.	July, 1885 ...	Mr. Spear.
Bolton	R.	May, 1886 ...	Dr. Page.
Bootle	U.	May, 1885 ...	„ Blaxall.
Boston	U.	November, 1886	Dr. Airy.
Boston	P.	August, 1885 ...	Do.
Brampton and Walton ...	U.	June, 1886 ...	Mr. Spear.
Branden and Byshottles ...	U.	December, 1885	Dr. Page.
Bridgend	U.	October, 1885 ...	Dr. Davies.
Bridgend	R.	Do. ...	Do.
Bridgend	R.	April, 1886 ...	Do.
Bridgwater	U.	January, 1885 ..	Dr. Blaxall.
Bridgwater	P.	Do. ...	Do.
Bridlington	U.	August, 1886 ...	Dr. Airy.
Bridlington	R.	January, 1887 ...	Dr. Davies.
Bridport	U.	February, 1885	Do.
Bridport	R.	January, 1887 ...	Do.
Brierley Hill	U.	May, 1886 ...	Dr. Gresswell.
Bristol	U.	January, 1885...	Dr. Blaxall.
Bristol	P.	Do. ...	Do.
Briton Ferry	U.	August, 1885 ...	Dr. Davies.
Broadstairs	U.	January, 1885 ...	Do.
Bromborough	U.	July, 1886 ...	Mr. Spear.
Brynmawe	U.	March, 1886 ...	Do.
Budleigh Salterton	U.	August, 1885 ...	Do.
Bulkington	U.	September, 1886	Do.
Burnley	U.	October, 1886 ...	Dr. Page.
Burslem	U.	September, 1886	Dr. Barry.
Bury	U.	Do. ...	Dr. Page.
Calne	U.	August, 1886 ...	Mr. Spear.
Calne	R.	Do. ...	Do.
Cambridge	U.	December, 1885	Dr. Airy.
Camelford	R.	July, 1885 ...	„ Davies.
Cardiff	U.	Do. ...	„ Blaxall.
Cardiff	P.	Do. ...	Do.
Cardigan	U.	January, 1886 ...	Dr. Davies.
Cardigan	P.	Do. ...	Do.
Carlisle	U.	August, 1885 ...	Dr. Blaxall.
Castleford	U.	October, 1886 ...	Mr. Spear.
Carmarthen	U.	October 1885 ...	Dr. Davies.
Carmarthen	R.	November, 1885	Do.
Carnarvonshire	Combd.	December, 1885	Do.
Carnarvon	U.	Do. ...	Do.
Carnarvon	P.	Do. ...	Do.
Carnarvon	R.	August, 1886 ...	Mr. Spear.
Castle Ward	R.	October, 1885 ...	Dr. Page.
Chard	U.	May, 1885 ...	Mr. Spear.
Chard	R.	Do. ...	Do.
Chepstow	U.	February, 1885	Dr. Blaxall.
Chepstow	P.	Do. ...	Do.
Chester	U.	October, 1886 ...	Dr. Page.
Chester	P.	January, 1886 ...	Do.
Chesterfield	U.	June, 1886 ...	Mr. Spear.

Name of District.	Urban, Rural or Port.	Date of Report.	Name of Inspector.
Harwich	P.	October, 1885 ...	Dr. Airy.
Haslingden	U.	September, 1886.	" Page.
Havant	U.	September, 1885.	Mr. Spear.
Havant	R.	September, 1885.	Do.
Haverfordwest	U.	September, 1885.	Dr. Davies.
Haverfordwest	R.	Do. ...	Do.
Hayle	U.	July, 1885 ...	Do.
Heath Town	U.	July, 1886 ...	Dr. Barry.
Heckmondwike	U.	November, 1886.	Do.
Helston	R.	July, 1885 ...	Dr. Davies.
Hemsworth	R.	April, 1885 ...	" Airy.
Hepworth	U.	January, 1887 ...	" Gresswell.
Herne Bay	U.	January, 1885 ...	" Davies.
Heywood	U.	October, 1886 ...	" Page.
Higher Bebington	U.	July, 1886 ...	Mr. Spear.
Hinderwell	U.	December, 1885.	Do.
Holme Cultram	U.	August, 1885 ...	Dr. Blaxall.
Holme Valley Combination ...	...	January, 1887 ...	" Gresswell.
Holmfirth	U.	Do. ...	Do.
Holme	U.	Do. ...	Do.
Holyhead	U.	December, 1885.	Dr. Davies.
Holywell	U.	June, 1885 ...	Do.
Holywell	R.	Do. ...	Do.
Honley	U.	January, 1887...	Dr. Gresswell.
Horwich	U.	December, 1886.	" Parsons.
Hoo	R.	March, 1885 ...	" de Chaumont.
Houghton le Spring	U.	Do. ...	" Page.
Houghton le Spring	R.	Do. ...	Do.
Howden	R.	June, 1885 ...	Dr. Airy.
Hoylandswaine	U.	February, 1887..	" Gresswell.
Huddersfield Combinations ...	...	January, 1887...	Do.
Hull	U.	March, 1886 ...	" Page.
Hull and Goole	P.	August, 1885 ...	" Blaxall.
Huntingdon	U.	May, 1885 ...	Mr. Spear.
Huntingdon	R.	Do. ...	Do.
Hythe	U.	January, 1885...	Dr. Davies.
Ilfracombe	U.	July, 1885 ...	Dr. Davies.
Ince in Makerfield	U.	October, 1886 ...	" Parsons.
Ipswich	U.	June, 1886 ...	" Airy.
Ipswich	P.	Do. ...	Do.
Jarrow	U.	March, 1885 ...	Dr. de Chaumont.
Keighley	U.	April, 1886 ...	Dr. Barry.
Kettering	U.	July, 1886 ...	Mr. Spear.
Kettering	R.	Do. ...	Do.
Kidsgrove	U.	September, 1886	Dr. Barry.
Kingsbridge	R.	March, 1885 ...	" Davies.
Kings Lynn	P.	August, 1885 ...	" Airy.
Kirk Burton	U.	February, 1887..	" Gresswell.
Kirkheaton	U.	Do. ...	Do.
Kirkleatham	U.	December, 1885.	Mr. Spear.

Name of District.	Urban, Rural or Port.	Date of Report.	Name of Inspector.
Lancaster	U.	August, 1885 ...	Dr. Blaxall.
Lancaster Port	P.	Do. ...	Do.
Lanchester	R.	November, 1885	Dr. Page.
Lathom	U.	April, 1886 ...	Do.
Leadgate	U.	October, 1886 ...	Do.
Lepton	U.	February, 1887..	Dr. Gresswell.
Lewes	U.	November, 1885	Mr. Spear.
Lexden	R.	May, 1886 ...	Dr. Airy.
Limehouse	U.	February, 1885..	„ de Chaumont.
Lincoln...	U.	October, 1886 ...	„ Airy.
Linthwaite	U.	January, 1887...	„ Gresswell.
Liskeard	R.	April, 1885 ...	„ Davies.
Littlehampton	U.	July, 1885 ...	Mr. Spear.
Liverpool	U.	May, 1885 ...	Dr. Blaxall.
Liverpool	P.	April, 1885 ...	Do.
Liversedge	U.	November, 1886	Dr. Barry.
Llanelly	U.	August, 1885 ...	„ Davies.
Llanelly	R.	Do. ...	Do.
Loftus	U.	December, 1885.	Mr. Spear.
London...	P.	February, 1885..	Dr. de Chaumont.
Long Eaton	U.	June, 1886 ...	Mr. Spear.
Longton	U.	September, 1886	Dr. Barry.
Longwood	U.	January, 1887...	„ Gresswell.
Lower Bebington	U.	July, 1885 ...	Mr. Spear.
Lower Brixham	U.	March, 1885 ...	Dr. Davies.
Lowestoft	P.	November, 1885	„ Airy.
Lyme Regis	U.	March, 1886 ...	Dr. Davies.
Lytham	U.	August, 1885 ...	„ Blaxall.
Lytham	P.	Do. ...	Do.
Macclesfield	U.	August, 1886 ...	Dr. Page.
Maesteg	U.	October, 1885 ...	„ Davies.
Maidstone	U.	March, 1885 ...	Mr. Spear.
Maidstone	R.	Do.	Do.
Maldon	U.	January, 1887...	Dr. Airy.
Maldon	R.	Do. ...	Do.
Malling	R.	March, 1885 ...	Mr. Spear.
Malton	U.	December, 1885	Do.
Malton	R.	Do. ...	Do.
Margam	U.	September, 1885	Dr. Davies
Margate	U.	January, 1885 ...	Do.
Marsden	U.	January, 1887 ...	Dr. Gresswell.
Maryport	U.	August, 1885 ...	„ Blaxall.
Meltham	U.	January, 1887 ...	„ Gresswell.
Merthyr Tydfil	U.	September, 1885	„ Ballard.
Methley	U.	November, 1886	Mr. Spear.
Mexborough	U.	December, 1886	Do.
Middlesbrough	U.	June, 1885 ...	Do.
Middlewich	U.	August, 1886 ...	Do.
Milford	U.	August, 1885 ...	Dr. Davies.
Milford Port	P.	October, 1885 ...	Do.
Millom	U.	August, 1885 ...	Dr. Blaxall.
Milton	U. & R.	Do. ...	„ de Chaumont.
Mirfield	U.	November, 1886	„ Barry.
Mold	U.	June, 1885 ...	„ Davies.
Monk Bretton...	U.	December, 1886	Mr. Spear.
Morecambe	U.	December, 1885	Dr. Page.

Name of District.	Urban, Rural or Port.	Date of Report.	Name of Inspector.
Morley	U.	November, 1886	Dr. Barry.
Morpeth	U.	August, 1885...	" Page.
Morpeth	R.	June, 1885 ...	Do.
Mossley	U.	October, 1886 ...	Dr. Gresswell.
Moss-side	U.	March, 1886 ...	" Page.
Mountain Ash	U.	May, 1885 ...	" Davies.
Narberth	R.	November, 1885	Dr. Davies.
Neath	U.	August, 1885 ...	Do.
Neath	R.	October, 1885 ...	Do.
Neath	R.	April, 1886 ...	Do.
Neston and Parkgate ...	U.	July, 1886 ..	Mr. Spear.
Netherthong	U	January, 1887...	Dr. Gresswell.
Newbiggin	U.	June. 1885 ...	" Page.
Newbold and Dunston ...	U.	" 1886 ...	" Spear.
Newcastle-upon-Tyne ...	P.	March, 1885 ...	" de Chaumont.
Newcastle-upon-Tyne ...	U.	Do. ...	Do.
Newcastle-under-Lyne ...	U.	September, 1886	Dr. Barry.
Newhaven	U.	June, 1885 ...	Mr. Spear.
Newhaven	P.	Do. ...	Dr. Davies.
Newhaven	R.	November, 1885	Mr. Spear.
Newport (Isle of Wight) ...	U.	September, 1885	Do.
Newport (Mon.)	U.	July, 1885 ...	Dr. Blaxall.
Newport (Mon.)	P.	Do. ...	Do.
New Shoreham	U.	Do. ...	Mr. Spear.
New Shoreham	P.	June, 1885 ...	Do.
Newquay	U.	Do. ...	Dr. Davies.
Newquay	U.	January, 1886 ...	Do.
Newton Heath	U.	April, 1886 ...	Dr. Page.
Normanton	U.	September, 1886	Mr. Spear.
Northam	U.	July, 1885 ...	Dr. Davies.
Northam	P.	Do. ...	Do.
Northampton	U.	December, 1886	Mr. Spear.
Northwich	U.	July, 1886 ...	Do.
Northwich	R.	September, 1886	Do.
Norwich	U.	January, 1886 ...	Dr. Airy.
Nuneaton	U.	July, 1886 ...	Mr. Spear.
Oldbury	U.	May, 1886 ...	Dr. Gresswell.
Openshaw	U.	April, 1886 ...	" Page.
Ormskirk	U.	October, 1886 ...	" Parsons.
Ormskirk	R.	Do. ...	Do.
Orrell	U.	Do. ...	Do.
Orsett	R.	June, 1885 ...	Mr. Spear.
Ossett-cum-Gawthorpe ...	U.	November, 1886	Dr. Barry.
Ovenden	U.	Do. ...	" Parsons.
Over Darwen	U.	Do. ...	" Page.
Oxford	U.	October, 1886 ...	" Airy.
Oystermouth	U.	August, 1885 ...	" Davies.
Padstow	U.	June, 1885 ...	Dr. Davies.
Padstow	P.	Do. ...	Do.
Pemberton	U.	December, 1886 ...	Dr. Parsons.
Pembroke	U.	September, 1885	" Davies.

Name of District.	Urban, Rural or Port.	Date of Report.	Name of Inspector.
Pembroke	R.	October, 1885 ...	Dr. Davies.
Penistone	U.	February, 1887	" Gresswell.
Penistone	R.	Do. ...	Do.
Penryn	U.	July, 1885 ...	Dr. Davies.
Penzance	U.	April, 1885 ...	Do.
Penzance	P.	Do. ...	Do.
Penzance	R.	Do. ...	Do.
Peterborough	U.	July, 1886 ...	Mr. Spear.
Peterborough	R.	August, 1886 ...	Do.
Pickering	U.	December, 1885	Do.
Pickering	R.	Do. ...	Do.
Plomesgate	R.	July, 1885 ...	Dr. Airy.
Plymouth	U.	June, 1885 ...	" Blaxall.
Plymouth	P.	June, 1885 ...	Do.
Plymouth	P.	March, 1886 ...	Dr. Davies.
Pontardawe	R.	November, 1885	Do.
Pontefract	U.	October, 1886 ...	Mr. Spear.
Pontefract	R.	November, 1886	Do.
Poole	U.	February, 1885	Dr. Davies.
Poole	P.	Do. ...	Do.
Poplar, North	U.	Do. ...	Dr. de Chaumont.
Poplar, South	U.	Do. ...	Do.
Pontypridd	U.	May, 1885 ...	Dr. Davies.
Pontypridd	R.	Do. ...	Do.
Portsmouth	U.	January, 1885 ...	Dr. Blaxall.
Portsmouth	P.	Do. ...	Do.
Portsmouth	P.	February, 1886	Dr. Davies.
Prescot	U.	June, 1886 ...	" Page.
Prescot	R.	Do. ...	Do.
Preston	U.	July, 1886 ...	Do.
Prestwich	R.	May, 1886 ...	Do.
Pwllheli	U.	December, 1885	Dr. Davies.
Quarry Bank	U.	July, 1886 ...	Dr. Gresswell.
Quickmere	U.	October, 1886 ...	Do.
Radcliffe	U.	October, 1886 ...	Dr. Page.
Ramsey	U.	May, 1885 ...	Mr. Spear.
Ramsgate	U.	January, 1885 ...	Dr. Davies.
Ravensthorpe	U.	October, 1886 ...	" Barry.
Rawmarsh	U.	December, 1886	Mr. Spear.
Redcar	U.	December, 1885	Do.
Rhymney	U.	November, 1885	Dr. Davies.
Rochester	U. & P.	March, 1885 ...	" de Chaumont.
Rochford	R.	December, 1885	Mr. Spear.
Romney Marsh	R.	January, 1885 ...	Dr. Davies.
Rotherham	U.	April, 1886 ...	Do.
Rotherhithe	U.	February, 1885	Dr. de Chaumont.
Rowley Regis	U.	June, 1886 ...	" Gresswell.
Runcorn	U.	June, 1885 ...	" Airy.
Runcorn	R.	Do. ...	Do.
Ryde	U.	September, 1885	Mr. Spear.
Saddleworth	R.	October, 1886 ...	Dr. Gresswell.
St. Austell	R.	April, 1885 ...	" Davies.
St. Columb Major	R.	January, 1885...	Do.
St. George's in the East ..	U.	February, 1885..	Dr. de Chaumont.

Name of District.	Urban, Rural or Port.	Date of Report.	Name of Inspector.
St. Germans	R.	April, 1885 ...	Dr. Davies.
St. Helens	U.	October, 1885 ...	Mr. Spear.
St. Helens (Lancs.)	U.	May, 1886 ...	Dr. Page.
St. Ives...	U.	September, 1885	„ Davies.
St. Neots	U.	May, 1885 ...	Mr. Spear.
St. Thomas	U.	August, 1885 ...	Do.
St. Thomas	R.	Do. ...	Do.
Salcombe	U.	March, 1885 ...	Dr. Davies.
Salford...	U.	August, 1886 ...	„ Page.
Saltburn	U.	December, 1885..	Mr. Spear.
Samford	R.	January, 1887 ...	Dr. Gresswell.
Sandgate	U.	January, 1886 ...	„ Davies.
Sandown	U.	October, 1885 ...	Mr. Spear.
Sandwich	U.	January, 1885 ...	Dr. Davies.
Scammonden	U.	January, 1887...	„ Gresswell.
Scarborough	U.	July, 1885 ...	„ Blaxall.
Scholes...	U.	January, 1887...	„ Gresswell.
Seaham Harbour	U.	June, 1885 ...	„ Page.
Sedgefield	R.	February, 1885..	Do.
Sidmouth	U.	September, 1885	Mr. Spear.
Shanklin	U.	Do. ...	Do.
Shardlow	R.	May, 1886 ...	Do.
Sheerness	U.	March, 1885 ...	Dr. de Chaumont.
Sheffield	U.	May, 1886 ...	„ Davies.
Shepley	U.	January, 1887...	„ Gresswell.
Shelley...	U.	Do. ...	Do.
Sheppey	R.	March, 1885 ...	Dr. de Chaumont.
Sherborne	U.	April, 1885 ...	Mr. Spear.
Sherborne	R.	May, 1885 ...	Do.
Shildon and East Thickley ...	U.	June, 1885 ...	Dr. Page.
Shrewsbury	U.	June, 1886 ...	Mr. Spear.
Sittingbourne...	U.	March, 1885 ...	Dr. de Chaumont.
Skelmanthorpe	U.	January, 1887...	„ Gresswell.
Skelmersdale	U.	October, 1886 ...	„ Parsons.
Skelton and Brotton... ...	U.	December, 1885..	Mr. Spear.
Slaithwaite	U.	January, 1887...	Dr. Gresswell.
Smethwick	U.	May, 1886 ...	Do.
Soothill, Nether	U.	November, 1886	Dr. Barry.
Soothill, Upper	U.	Do.	Do.
Southampton	U.	August, 1886 ...	Dr. Blaxall.
Southampton	P.	February, 1886..	„ Davies.
South Blyth	U.	June, 1885 ...	„ Page.
Southborough...	U.	March, 1885 ...	Mr. Spear.
South Cave and Wallingfen..	U.	May, 1885 ...	Dr. Airy.
South Crosland	U.	January, 1887...	„ Gresswell.
Southend	U.	November, 1885	Mr. Spear.
South Gosforth	U.	October, 1885 ...	Dr. Page.
Southport	U.	October, 1886 ...	„ Parsons.
South Shields...	U.	March, 1885 ...	„ de Chaumont.
South Stockton	U.	Do. ...	„ Page.
Sowerby Bridge	U.	December, 1886..	„ Parsons.
Spalding	U.	October, 1886 ...	„ Airy.
Spennymoor	U.	May, 1885 ...	„ Page.
Standish	U.	December, 1886.	„ Parsons.
Steyning	R.	July, 1885 ...	Mr. Spear.
Stockport	U.	July, 1886 ...	Dr. Page.
Stocksbridge	U.	February, 1887.	„ Gresswell.
Stockton-on-Tees	U.	March, 1885 ...	„ Page.
Stockton	R.	March, 1885 ...	Do.
Stanhope	U.	April, 1885 ...	Do.

Name of District.	Urban, Rural or Port.	Date of Report.	Name of Inspector.
Stoke-upon-Trent	U.	September, 1886	Dr. Barry.
Stoke-upon-Trent	R.	Do. ...	Do.
Stourbridge Improvement Act	U.	May, 1886 ...	Dr. Gresswell.
Stourbridge	R.	June, 1886 ...	Do.
Stratton	R.	July, 1885 ...	Dr. Davies.
Sunderland	U.	March, 1885 ...	„ de Chaumont.
Sunderland	U.	July, 1885 ...	„ Blaxall.
Sunderland	P.	Do. ...	Do.
Sunderland	P.	March, 1885 ...	Dr. de Chaumont.
Swansea	P.	August, 1885 ...	„ Ballard.
Swansea	U.	Do. ...	Do.
Swansea	R.	November, 1885	Dr. Davies.
Swinton	U.	November, 1886	Mr. Spear.
Tavistock	R.	August, 1885 ...	Mr. Spear.
Teesdale	R.	May, 1885 ...	Dr. Page.
Teignmouth	U.	March, 1885 ...	„ Davies.
Tenby	U.	September, 1885	„ Ballard.
Tendring	R.	January, 1887 ...	„ Airy.
Thorne	R.	January, 1886 ...	Mr. Spear.
Thornhill	U.	November, 1886	Dr. Barry.
Thrapstone	R.	October, 1885 ...	Mr. Spear.
Thurlstone	...	February, 1887	Dr. Gresswell.
Thurstonland	...	Do. ...	Do.
Thanet, Isle of	R.	January, 1885...	Dr. Davies.
Tipton	U.	August, 1886 ...	„ Gresswell.
Todmorden	U.	November, 1886	Dr. Barry.
Tonbridge	U.	March, 1885 ...	Mr. Spear.
Tonbridge	R.	Do. ...	Do.
Torquay	U.	Do. ...	Dr. Davies.
Totnes	R.	Do. ...	Do.
Tow Law	U.	April, 1885 ...	Dr. Page.
Towyn	U.	January, 1886 ...	„ Davies.
Toxteth Park	U.	May, 1885 ...	„ Blaxall.
Tredegar	U.	November, 1885	„ Davies.
Truro	U.	April, 1885 ...	Do.
Truro	R.	Do. ...	Do.
Tunstall	U.	September, 1886	Dr. Barry.
Tynemouth	U.	September, 1885	„ Page.
Tynemouth	R.	Do. ...	Do.
Upholland	U.	December, 1886	Dr. Parsons.
Upper Sedgley	U.	July, 1885 ...	„ Gresswell.
Ventnor	U.	September, 1885	Mr. Spear.
Wakefield	U.	August, 1886 ...	Mr. Spear.
Wakefield	R.	October, 1886 ...	Do.
Walker	U.	November, 1885	Dr. Page.
Walsall...	U.	July, 1886 ...	„ Barry.
Walton-on-the-Hill	U.	May, 1885 ...	„ Blaxall.
Wath	U.	December, 1886.	Mr. Spear.
Wavertree	U.	May, 1885 ...	Dr. Blaxall.

Name of District.	Urban, Rural or Port.	Date of Report.	Name of Inspector
Weardale	R.	April, 1885 ...	Dr. Page.
Wednesbury	U.	July, 1886 ...	„ Gresswell.
Wells	P.	„ 1885 ...	„ Airy.
Westbourne	R.	September, 1885	Mr. Spear.
West Cowes	U.	May, 1885 ...	Dr. Blaxall.
West Derby	U.	Do. ...	Do.
West Firle	R.	November, 1885.	Mr. Spear.
West Ham	U.	February, 1885.	Dr. de Chaumont
Westhampnett	R.	July, 1885 ...	Mr. Spear.
West Hartlepool	U.	February, 1885.	Dr. Blaxall.
West Worthing	U.	November, 1885.	Mr. Spear.
Weymouth	U.	February, 1885.	Dr. Davies.
Weymouth	P.	Do. ...	Do.
Worthing	U.	November, 1885.	Mr. Spear.
Whitby...	U.	July, 1885 ...	Dr. Blaxall.
Whitby...	R.	December, 1885.	Mr. Spear.
Whitechapel	U.	February, 1885.	Dr. de Chaumont.
Whitehaven	U.	August, 1885 .	„ Blaxall.
Whitley, Upper	U.	February, 1887	„ Gresswell.
Whittington	U.	June, 1886 ...	Mr. Spear.
Whitwood	U.	November, 1886	Do.
Widnes	U.	September, 1886	Dr. Page.
Wigan	U.	November, 1886	Do.
Wigan	R.	October, 1886 ...	Dr. Parsons.
Wight, Isle of...	R.	Do.	Mr. Spear.
Willenhall	U.	July, 1886 ...	Dr. Barry.
Willington	U.	November, 1885	„ Page.
Williton	R.	December, 1885	„ Davies
Winsford	U.	September, 1886	Mr. Spear.
Wirral	R.	Do.	Do.
Wisbech	P.	August, 1885 ...	Dr. Airy.
Withington	U.	March, 1886 ...	„ Page.
Woodbridge	R.	December, 1885	Mr. Spear.
Workington	U	April, 1885 ...	Dr. Page.
Workington	U.	August, 1885 ...	„ Blaxall.
Workington	P.	Do.	Do.
Worksop	U.	May, 1885 ...	Mr. Spear.
Worksop	R.	„ 1886	Do.
Worsborough	U.	December, 1886	Do.
Wortley	R.	February, 1887	Dr. Gresswell.
Yarmouth, Gt.	P.	November, 1885	Dr. Airy.
Ynyscynhaiarn	U.	December, 1885	„ Davies.
Ystradyfodwg	U.	May, 1885 ...	Do.

Inland Sanitary Survey, 1893–95.

Rural Districts of England and Wales inspected during 1893–95 as regards their Sanitary Circumstances and Administration.

Name of District.	Population, Census 1891.	County.	Inspector.
1. Atherstone... ...	15,441	Warwickshire ...	Dr. Wheaton.
2. Axbridge	24,965	Somersetshire ...	Do.
3. Barnstaple... ...	19,071	Devonshire ...	Dr. Reece.
4. Barton-upon-Irwell	26,832	Lancashire... ...	„ Bruce Low.
3. Beverley	10,519	Yorkshire	„ Wheaton.
4. Bideford	7,322	Devonshire ...	„ Reece.
5. Bodmin	11,644	Cornwall	„ Wheaton.
6. Calne	6,014	Wiltshire	„ Wilson.
7. Carmarthen ...	23,873	Carmarthenshire ...	„ Reece.
8. Cheadle	22,302	Staffordshire ...	„ Fletcher.
9. Chesterfield ...	59,192	Derbyshire... ...	„ Wilson.
10. Cosford	12,369	Suffolk	„ Fletcher.
11. Crickhowell ...	7,464	Brecknockshire ...	„ Wilson.
12. Daventry	13,709	Northamptonshire..	„ Wheaton.
13. Docking	16,030	Norfolk	Do.
14. Doncaster	28,364	Yorkshire	Do.
15. Downham	15,840	Norfolk	Do.
16. Driffield	13,140	Yorkshire	Do.
17. Durham	32,686	Durham	Dr. Wilson.
18. Eastington... ...	36,782	Do.	Do.
19. Hastings	11,243	Sussex	Dr. Bruce Low.
20. Hemsworth ...	14,631	Yorkshire	„ Wilson.
21. Holbeach	8,599	Lincolnshire ...	Do.
22. Holywell	29,843	Flintshire	Mr. Evans.
23. Houghton-le-Spring	31,445	Durham	Dr. Wilson.
24. Kendal	21,606	Westmoreland ...	„ Fletcher.
25. Ludlow	13,196	Shropshire... ...	„ Bruce Low.
26. Mansfield	21,500	Derbyshire... ...	„ Wilson.
27. Narberth	18,190	Pembroke and Carmarthen.	„ Reece.
28. Newcastle - under - Lyme.	6,174	Staffordshire ...	„ Fletcher.
29. Pembroke	11,763	Pembrokeshire ...	Do.
30. Rochford	17,938	Essex	Dr. Wheaton.
31. Rothbury	6,083	Northumberland ...	„ Fletcher.
32. Runcorn	22,467	Cheshire	„ Bruce Low.
33. Sculcoates	8,786	Yorkshire	„ Wheaton.
34. Spalding	12,719	Lincolnshire ...	„ Wilson.
35. Stone	13,872	Staffordshire ...	„ Fletcher.
36. Sunderland ...	17,552	Durham	Mr. T. W. Thompson.
37. Thanet, Isle of ...	9,466	Kent	Dr. Bruce Low.
38. Warrington ...	12,783	Lancashire... ...	Do.
39. Wharfedale ...	7,551	Yorkshire	Dr. Horne.
40. Williton	15,470	Somersetshire ...	Mr. T. W. Thompson.
41. Wolstanton and Burslem.	32,387	Staffordshire ...	Dr. Fletcher.
42. Woodbridge ...	19,960	Suffolk	„ Copeman.

INDEX TO COUNTY BOROUGHS, MUNICIPAL BOROUGHS, and URBAN DISTRICTS.

[C.B. signifies County Borough. M.B., Municipal Borough. Imp. Act Dist., Improvement Act District. U.D., Urban District.]

Name.	Date of Report.	Name of Inspector.
Aberystwith M. B.	July, 1893	Dr. Reece.
Aldeburgh M. B.	April, 1893 ...	" Copeman.
Alfreton U. D.	May, 1894	" Wilson.
Alvaston and Boulton U. D.	June, 1894	Do.
Amble U. D.	February, 1893 ...	Mr. T. W. Thompson.
Ashbourne U. D.	June, 1893	Dr. Wilson.
Ashby-de-la-Zouch U. D. . .	December, 1893 ...	" Fletcher.
Ashton-in-Makerfield U. D.	September, 1893 ...	" Wheaton.
Audley U. D....	April, 1894... ...	" Fletcher.
Bacup M. B.	July, 1893	Dr. Wheaton.
Banbury M. B.	" 1894	Do.
Barmouth U. D.	" 1893	
Barnstaple M. B.	May, 1894	Do.
Baslow and Bubnell U. D....	June, 1894	Dr. Wilson.
Beaconsfield U. D.	N[illegible]ember, 1893 ...	" Fletcher.
Bedlingtonshire U. D. ...	Ju[illegible], 1893	" Wilson.
Bedwellty U. D.	September, 1893 ...	Mr. Evans.
Belper U. D.	May, 1894	Dr. Wilson.
Benfieldside U. D.	November, 1893 ...	Do.
Berwick-upon-Tweed M. B.	June, 1893 ...	Mr. T. W. Thompson.
Beverley M. B.	May, 1893	Dr. Wheaton.
Bideford M. B.	June, 1894 ...	" Reece.
Bingley Imp. Act Dist. ...	February, 1894 ...	" Wheaton.
Bingley Outer U. D. ...	January, 1894 ...	Do.
Bishop Auckland U. D. ...	May, 1893	Dr. Sweeting.
Blaydon U. D.	July, 1893	" Wilson.
Boston M. B.	September, 1893 ...	" Bulstrode.
Brandon and Byshottles U.D.	Do. ...	" Wilson.
Bridlington U. D.	April, 1894 ...	" Wheaton.
Bridport M. B.	Do. ...	" Reece.
Broadstairs and St. Peter's U. D.	August, 1894 . .	" Bruce Low.
Budleigh Salterton U. D. ...	May, 1894	" Reece
Burslem M. B.	April, 1894 ...	" Fletcher.
Burton-upon-Trent M. B. ...	August, 1893 ...	" Wilson.
Calne M. B.	January, 1895 ...	Dr. Wilson.
Calverley U. D.	July, 1894	Do.
Camborne U. D.	Do. ...	Dr. Reece.
Carlisle City and M. B. ...	April, 1894 ...	" Wilson.
Carmarthen M. B.	" 1893 ...	" Reece.
Chepstow U. D.	March, 1894 ...	Do.
Chesterfield M. B.	June, 1894	Dr. Wilson.
Clay Cross U. D.	May, 1894	Do.
Clayton U. D.	September, 1894 ...	Do.
Cleckheaton U. D.	August, 1894 ...	Do.
Cleethorpes-with-Thrunscoe U. D.	May, 1893	Mr. T. W. Thompson.
Cockermouth U. D. ...	April, 1894... ...	Dr. Wilson.

Name.	Date of Report.	Name of Inspector.
Consett U. D.	November, 1893 ...	Dr. Wilson.
Coseley U. D.	„ 1894 ...	„ Fletcher.
Cottingham U. D.	June, 1893	„ Wheaton.
Cowpen U. D.	„ 1893	„ Wilson.
Criccieth U. D.	August, 1893 ...	„ Reece.
Cromer U. D.	March, 1893 ...	„ Copeman.
Crowle U. D.	June, 1893 ...	„ Wheaton.
Darlaston U. D.	July, 1894	Dr. Fletcher.
Dawlish U. D.	May, 1894	„ Reece.
Deal M. B.	June, 1893	„ Thomson.
Doncaster M. B.	March, 1894 ...	„ Wheaton.
Drighlington U. D. ...	August, 1894 ...	„ Wilson.
Dronfield U. D.	May, 1894	Do.
Dudley C. B.	August, 1894 ...	Dr. Fletcher.
Durham City and M. B. ...	June, 1894	„ Wheaton.
East Retford M. B.	February, 1894 ...	Dr. Wheaton.
Eccles M. B.	November, 1894 ...	„ Bruce Low.
Eccleshill U. D.	July, 1894	„ Wilson.
Ely City and U. D.	Do. ...	„ Wheaton.
Exeter City and C. B. ...	July, 1893	„ Bulstrode.
Faversham M. B.	April, 1894... ...	Dr. Thomson.
Felling U. D.	September, 1893 ...	„ Wilson.
Fenton U. D.	January, 1894 ...	„ Sweeting.
Filey U. D.	August, 1893 ...	Mr. T. W. Thompson.
Gainsborough U. D. ...	December, 1893 ...	Dr. Fletcher.
Gildersome U. D.	June, 1894	„ Wheaton.
Godmanchester M. B. ...	„ 1893	„ Wilson.
Gravesend M. B.	October, 1893 ...	„ Thomson.
Grays Thurrock U. D. ...	June, 1893	„ Horne.
Great Clacton U. D. ...	July, 1893	„ Reece.
Great Driffield U. D. ...	June, 1893	„ Wheaton.
Great Yarmouth C. B. ...	November, 1893 ...	„ Copeman.
Hanley C. B.	February, 1894 ...	Dr. Sweeting.
Harwich M. B.	June, 1894	„ Wheaton.
Hastings C. B.	March, 1894 ...	„ Bruce Low.
Havant U. D.	September, 1894 ...	„ Thomson.
Haverfordwest M. B. ...	June, 1893	„ Reece.
Heage U. D.	August, 1894 ...	„ Wilson.
Heanor U. D.	October, 1894 ...	Do.
Heath Town U. D.	July, 1894	Dr. Fletcher.
Hedon M. B.	June, 1893	„ Wheaton.
Hereford City and M. B. ...	December, 1893 ...	„ Bruce Low.
Herne Bay U. D.	September, 1894 ...	Do.
Holyhead U. D.	June, 1893	Dr. Reece.
Hoole U. D.	December, 1893 ...	„ Fletcher.
Hornsea U. D.	April, 1894 ...	„ Wheaton.
Houghton-le-Spring U. D.	October, 1893 ...	„ Wilson.
Howdon-on-Tyne U. D. ...	„ 1894 ...	„ Fletcher.
Hoyland Nether U. D. ...	August, 1894 ...	„ Wheaton.
Hunstanton U. D.	March, 1893 ...	„ Copeman.
Hunsworth U. D.	May, 1894	„ Wheaton.
Huntingdon M. B.	September, 1894 ...	Do.
Hythe M. B.	... Do. ...	Dr. Bruce Low.

Name.	Date of Report.	Name of Inspector.
Idle U. D.	July, 1894	Dr. Wilson.
Ilfracombe U. D.	May, 1894	" Reece.
Ilkeston M. B.	September, 1893] ...	" Wheaton.
Ipswich C. B.	July, 1894	Do.
Keighley M. B.	February, 1894 ...	Dr. Wheaton.
Kendal M. B.	March, 1894 ...	" Fletcher.
Kidwelly M. B.	April, 1893 ...	" Reece.
Knaresborough Imp.Act.Dis.	April, 1894 ...	" Wheaton.
Knottingly U. D.	January, 1894 ...	" Wilson.
Leadgate U. D.	November, 1893 ...	Dr. Wilson.
Levenshulme U. D.... ...	July, 1894	" Bruce Low.
Lincoln City and C. B. ...	May, 1894	" Wheaton.
Llandudno Imp. Act Dist. ...	February, 1894 ...	" Reece.
Llanelly U. D.	April, 1893... ...	Do.
Long Sutton U. D.	September, 1894 ...	Dr. Wheaton.
Longton M. B.	January, 1894 ...	" Sweeting.
Loughborough M. B. ...	November, 1893 ...	" Fletcher.
Louth M. B.	September, 1893 ...	" Bulstrode.
Lower Brixham U. D. ...	October, 1894 ...	" Wheaton.
Ludlow M.B....	June, 1894... ...	" Bruce Low.
Lydd M. B.	July, 1894	Do.
Lyme Regis M. B.	April, 1894... ...	Dr. Reece.
Maesteg U. D.	September, 1893 ...	Mr. Evans.
Margate M. B.	August, 1894 ...	Dr. Bruce Low.
Maryport Imp. Act Dist. ...	May, 1894	" Wilson.
Menai Bridge U. D.... ...	February, 1894 ...	Mr. Evans.
Milford Imp. Act Dist. ...	June, 1893... ...	Dr. Reece.
Millom U. D....	April, 1894... ...	" Wilson.
Milton-next-Sittingbourne Imp. Act Dist.	Do. ...	" Thomson.
Minehead U. D.	June, 1893 ...	Mr. T. W. Thompson.
Mold U. D.	December, 1893 ...	" Evans.
Morley M. B....	May, 1894	Dr. Wheaton.
Morpeth M. B.	July, 1894	" Wilson.
Newark-upon-Trent M. B. ...	July, 1893	Dr. Bruce Low.
Newbold and Dunston U. D.	June, 1894	" Wilson.
Newburn U. D.	March, 1894 ...	Do.
Newcastle-under-Lyme M. B.	February, 1894 ...	Dr. Sweeting.
Newquay (Cornwall) U. D.	July, 1894	" Reece.
New Romney M. B.... ...	Do. ...	" Bruce Low.
Northam U. D.	May, 1894	" Reece.
North Bierley U. D.... ...	March, 1894 ...	" Wheaton.
Norwich City and C. B. ...	May, 1893	" Copeman.
Oldbury U. D.	August, 1894 ...	Dr. Fletcher.
Ormskirk U. D.	March, 1894 ...	" Bruce Low.
Paignton U. D.	October, 1894 ...	Dr. Wheaton.
Pemberton U. D.	February, 1894 ...	" Bruce Low
Pembroke M. B.	August, 1893 ...	" Reece.
Penrith U. D.	January, 1894 ...	" Bruce Low.
Penryn M. B.	June, 1893	" Bulstrode.
Poole M. B.	June–July, 1893 ...	.. Do.

Name.	Date of Report.	Name of Inspector.
Pudsey U. D.	July, 1893	Dr. Horne.
Pwllheli M. B.	August, 1893 ...	„ Reece.
Ramsey U. D.	September, 1894 ...	Dr. Wheaton.
Ramsgate M. B.	August, 1894 ...	„ Bruce Low.
Ripley U. D.	May, 1894	„ Wilson.
Ross Imp. Act. Dist. ...	December, 1893 ...	„ Bruce Low.
Runcorn Imp. Act. Dist. ...	December, 1894 ...	Do.
Rye M. B.	September, 1894 ...	Do.
Ryton U. D.	July, 1893	Dr. Wilson.
St. Ives (Cornwall) M. B. ...	October, 1894 ...	Dr. Wheaton.
St. Neots U. D.	September, 1894 ...	Do.
Salford C. B.	November, 1894 ...	Dr. Bruce Low.
Sandgate U. D.	September, 1894 ...	Do.
Sandwich M. B.	May, 1893	Dr. Thomson.
Seaham Harbour U. D. ...	Do.	Mr. T. W. Thompson.
Seaton U. D.	May, 1894	Dr. Reece.
Sedgley U. D.	May, 1893	„ Wilson.
Shildon and East Thickley U. D.	September, 1893 ...	Do.
Shipley U. D.	July, 1894	Do.
Shrewsbury M. B.	June, 1894	Dr. Bruce Low.
Sidmouth U. D.	May, 1894	„ Reece.
Sittingbourne U. D. ...	April, 1894 ...	„ Thomson.
Skegness U. D.	April, 1893 ...	„ Copeman.
Smallthorne U. D.	April, 1894 ...	„ Fletcher.
South Blyth U. D.	June, 1893	„ Wilson.
South Cave and Wallingfen U. D.	May, 1893	„ Wheaton.
Southend-on-Sea M. B.	February, 1894 ...	„ Thomson.
South Shields C. B. ...	May, 1894	Mr. T. W. Thompson.
Southwick U. D.	October, 1893 ...	Dr. Wilson.
Southwold M. B.	April, 1893 ...	„ Copeman.
Spalding Imp. Act Dist. ...	Do.	„ Wheaton.
Spennymoor U. D.	September, 1893 ...	„ Wilson.
Stalybridge M. B.	June, 1893	Do.
Stanhope U. D.	October, 1893 ...	Do.
Stanley U. D.	November, 1893 ...	Do.
Stockport C. B.	Do.	Dr. Bruce Low.
Stoke-upon-Trent M. B. ...	February, 1894 ...	„ Sweeting.
Stretford U. D.	December, 1894 ...	„ Bruce Low.
Sunderland C. B.	April, 1894 ...	Mr. T. W. Thompson.
Swadlincote U. D.	June, 1894	Dr. Wilson.
Swanage U. D.	June, 1893	„ Bulstrode.
Teignmouth U. D.	August, 1893 ...	Dr. Bulstrode.
Tenby M. B.	June, 1893... ...	„ Reece.
Thornton U. D.	September, 1894 ...	„ Wilson.
Tipton U. D.	August, 1894 ...	„ Fletcher.
Tong U. D.	Do.	„ Wilson.
Tow Law U. D.	September, 1893 ..	Do.
Towyn U. D.	July, 1893	Dr. Reece.
Truro City and M. B. ..	January, 1894 ...	„ Fletcher.
Tunstall U. D.	May, 1894	Do.
Tynemouth M. B.	June, 1894 ...	Dr. Wheaton.
Walker U. D.	September, 1894 ...	Dr. Fletcher.
Wallsend U. D.	Do.	Do.

Name.	Date of Report.	Name of Inspector.
Walsall C. B....	August. 1894 ...	Dr. Fletcher.
Walsoken U. D.	October, 1894 ...	" Wilson.
Walton on-the-Naze Imp. Act Dist.	August, 1893 ...	" Reece.
Warblington U. D.	September, 1894 ...	" Thomson.
Warrington M. B.	November, 1894 ...	" Bruce Low.
Wednesbury M. B.	September, 1894 ...	" Fletcher.
Whickham U. D.	July, 1893	" Wilson.
Whitley and Monkseaton U. D.	October. 1894 ...	" Fletcher.
Widnes M. B.	November, 1894 ...	" Bruce Low.
Wigan C. B.	February, 1894 ...	Do.
Willenhall U. D.	November, 1893 ..	Dr. Wheaton.
Willington U. D.	Do.	" Wilson.
Willington Quay U. D. ...	September, 1894 ...	" Fletcher.
Wilsden U. D.	August, 1894 ...	" Wilson.
Workington M. B.	March, 1894 ...	Do.
Yeadon U. D.	July, 1893	Dr. Horne.
Ynyscynhaiarn U. D. ...	August, 1893 ...	" Reece.

Lightning Source UK Ltd.
Milton Keynes UK
UKHW021608051118
331792UK00010B/2182/P